FLORA OF MADEIRA

THE NATURAL HISTORY MUSEUM

FLORA OF MADEIRA

edited by
J.R. Press & M.J. Short

assisted by
N.J. Turland

Intercept

© The Natural History Museum, London, 1994

© Illustrations, Fundacao Berardo, Madeira, Portugal, 1994

All rights reserved. No part of this publication may be reproduced, stored in a retrieval system, or transmitted, in any form or by any means, electronic, mechanical, photocopying, recording, or otherwise, without the prior permission of the publisher.

The right of the Authors to be identified as the Authors of this work has been asserted by the Authors in accordance with the Copyright Designs and Patents Act 1988.

A catalogue record for this book is available from the British Library.

ISBN 1 898298 80 7

Printed by St Edmundsbury Press Ltd, Bury St Edmunds, Suffolk, England

Cover designed by Michael Morey

Cover photograph by M.J. Short, *Helichrysum melaleucum*.

Reprinted 2001 for

Intercept Limited
PO Box 716
Andover
Hampshire SP10 1YG
United Kingdom

Tel: +44 (0)1264 334748
Fax: +44 (0)1264 334058
Email: intercept@andover.co.uk
Website: http://www.intercept.co.uk

CONTENTS

List of Contributors vi

Preface vii

Acknowledgements viii

Maps of the Madeiran and Salvage Islands ix

Introduction 1

Organization of the Flora 8

Flora 13

Illustrations 471

Index to Scientific Names 529

Index to Portuguese Names 569

LIST OF CONTRIBUTORS

S. Andrews, Royal Botanic Gardens, Kew

M.J. Cannon, c/o The Natural History Museum, London

T.A. Cope, Royal Botanic Gardens, Kew

R. Khan, c/o The Natural History Museum, London

M. Gibby, The Natural History Museum, London

D.A. Goyder, Royal Botanic Gardens, Kew

D. McClintock, Platt, Sevenoaks, Kent

J.M. Mullin, c/o The Natural History Museum, London

J. Ormonde, Instituto Botanico, Coimbra

A.M. Paul, The Natural History Museum, London

J.R. Press, The Natural History Museum, London

M.J. Short, The Natural History Museum, London

B.R. Tebbs, Lambeth College, London

M.C. Tebbs, c/o Royal Botanic Gardens, Kew

N.J. Turland, The Natural History Museum, London

C. Whitefoord, c/o The Natural History Museum, London

A.R. Vickery, The Natural History Museum, London

PREFACE

The Macaronesian region comprises the archipelagos of the Azores, Madeira, the Salvages, the Canaries and the Cape Verdes and has attracted the attention of botanists because of the uniqueness and diversity of the flora. Madeira is especially interesting since it contains some of the most extensive and best preserved natural vegetation remaining in Macaronesia. It also has a long history of species introductions from all over the world and is increasingly popular with visitors who are attracted by the wealth of plants to be seen there.

Extensive botanical collecting in the Madeiran archipelago began in the mid-eighteenth century and the flora is reasonably well-known although some inaccessible areas would bear further investigation. Rock-ledges in the high central peaks, the large valley of the Ribeira de Janela in the north-west of the island and some of the coastal cliffs and gorges have all recently yielded new records or confirmations of old ones known from only one or very few collections. Many populations of endemic species are very small, making every new additional record valuable.

It is 70 years since any comprehensive work devoted to the flora of the Madeiran islands has been produced and the excellent, recent checklists for Macaronesia by Hansen & Sunding (1985, 1991) have emphasized the need for a new *Flora* incorporating the considerable advances in our knowledge of botany in this area. The Natural History Museum has a long tradition of interest in the Macaronesian Islands and the compilation of the *Flora of Madeira* was undertaken there by the European Plant Information Centre (EPIC). Many individuals, both inside and outside the Museum, have contributed to the *Flora* and we are grateful for all their efforts.

This book aims to provide a concise yet comprehensive guide encompassing all of the wild (i.e. native and naturalized) vascular flora known to occur in the Madeiran and Salvage Islands and will be of use to anyone with an interest in the botany of the region. English was chosen as the most appropriate language in part for the convenience of the authors but principally to make the *Flora* accessible to the widest possible audience.

The *Flora of Madeira* draws heavily on the works of previous Madeiran botanists, in whose debt we remain. We hope our work will prove to a be a valuable and informative addition to Madeiran and Macaronesian literature.

JRP
MJS

ACKNOWLEDGEMENTS

We would like to thank all of the authors who provided accounts in the *Flora*, and Margaret Tebbs for producing the illustrations. Many people, as well as those named below, have contributed in other ways and we gratefully acknowledge their help.

In Madeira, the staff of the Museu Municipal do Funchal, the Jardim Botanico and the Serviços Florestais generously provided facilities for both field and herbarium work; we are particularly indebted to Dr J.M. Biscoito, Senhor G. Maul and Senhor M. Nobrega for much help and encouragement. The following people provided information and advice; Peter Boyce (Kew), Eng. Henrique Costa Neves (Funchal), Prof. Alfred Hansen (Copenhagen), Dr. Charlie Jarvis (The Natural History Museum, London), Graham Quinn (Funchal), Dr Tim Rich (Surrey), Dr Ken Robertson (Champaign, Illinois) and Eng. Rui Viera (Funchal). Nicoletta Cellinese provided editorial support.

We are grateful to the Fundacão Berardo (Funchal) for financing the production of the fifty-seven plates of original illustrations, and to the Park Fund (The Natural History Museum) which provided support for N.J. Turland.

Finally, we acknowledge the efforts of Nick Turland; without his considerable assistance this work would have been harder and much longer.

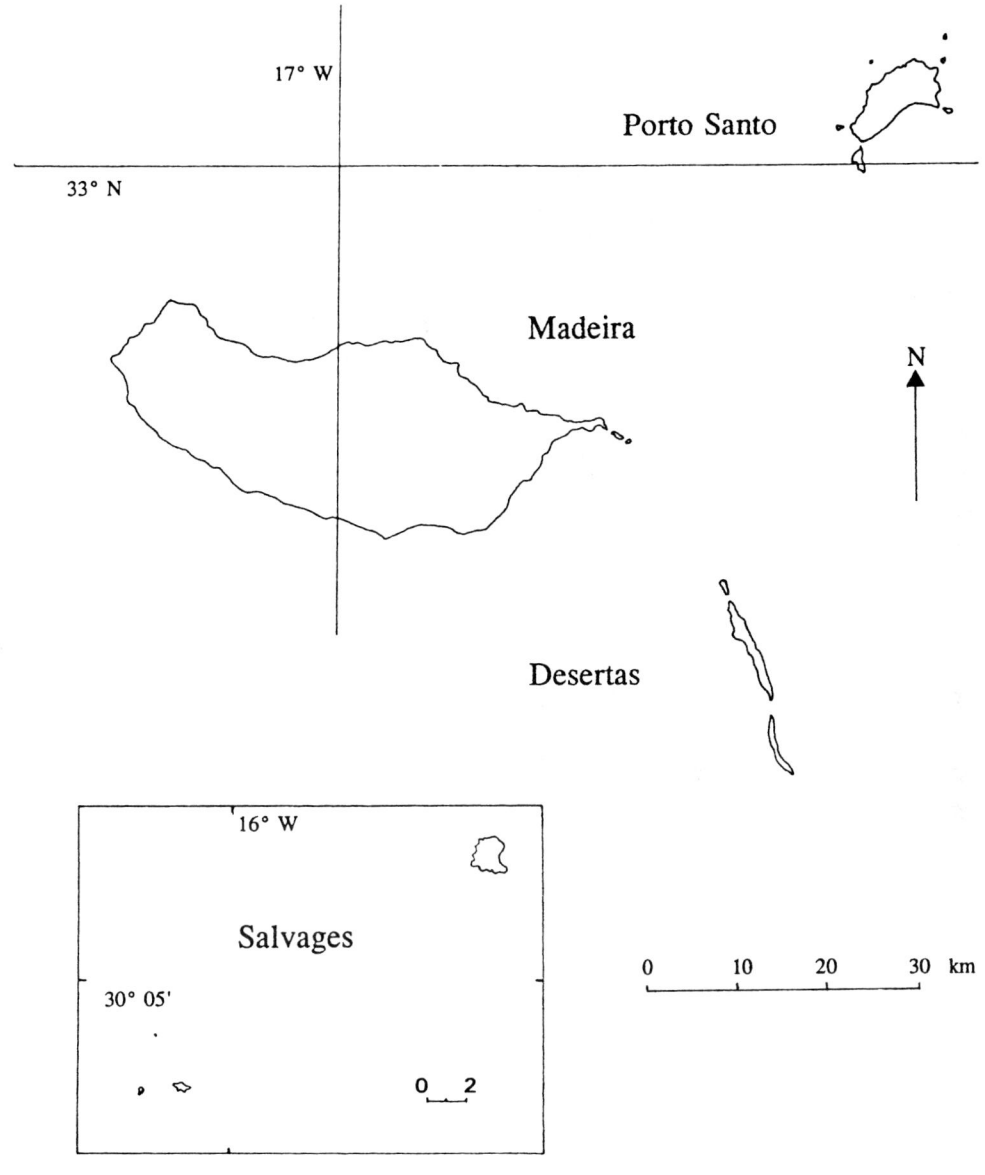

Fig. 1 The Madeiran and Salvage island archipelagos

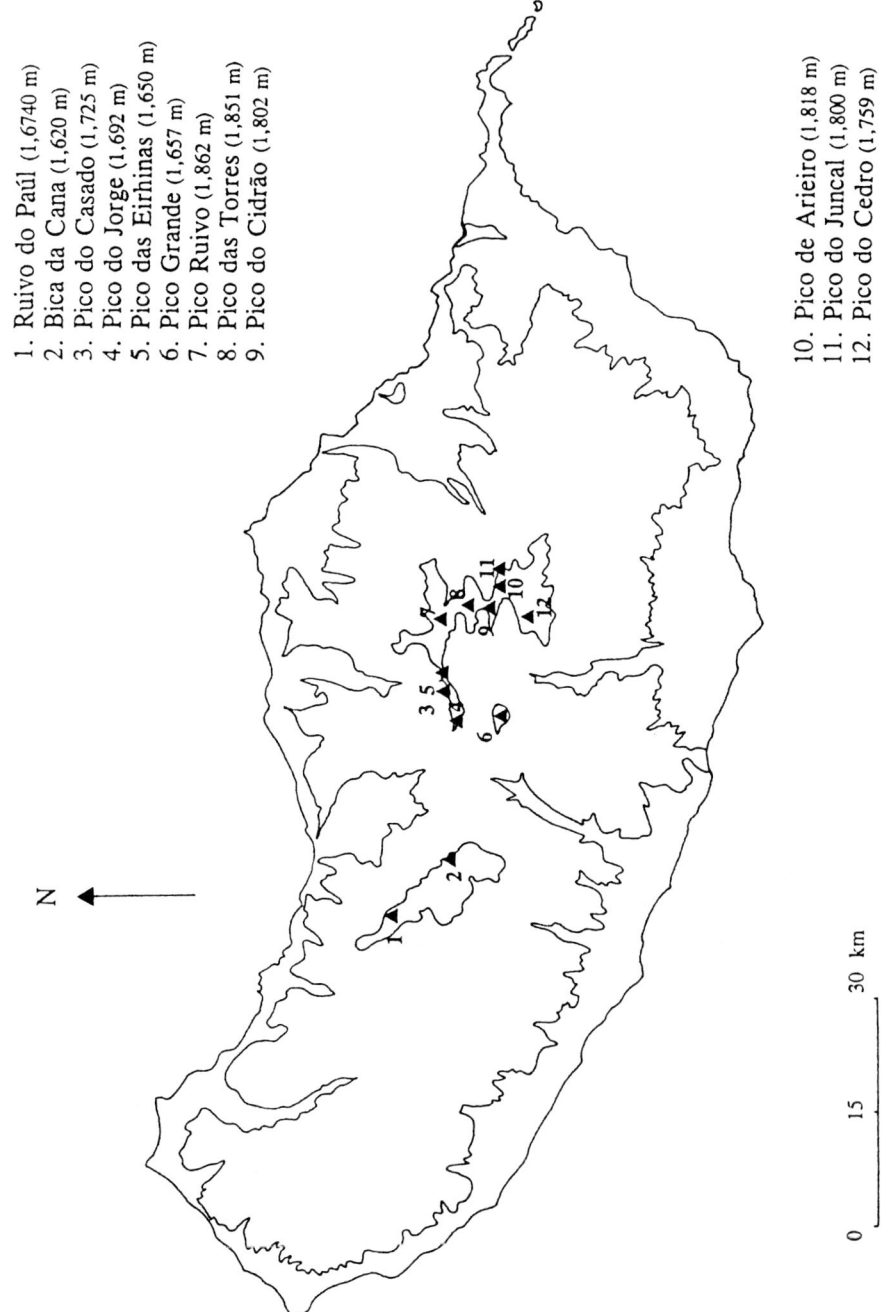

Fig. 2 Madeira: topography
1,000 m and 1,500 m contours and spot heights over 1,600 m

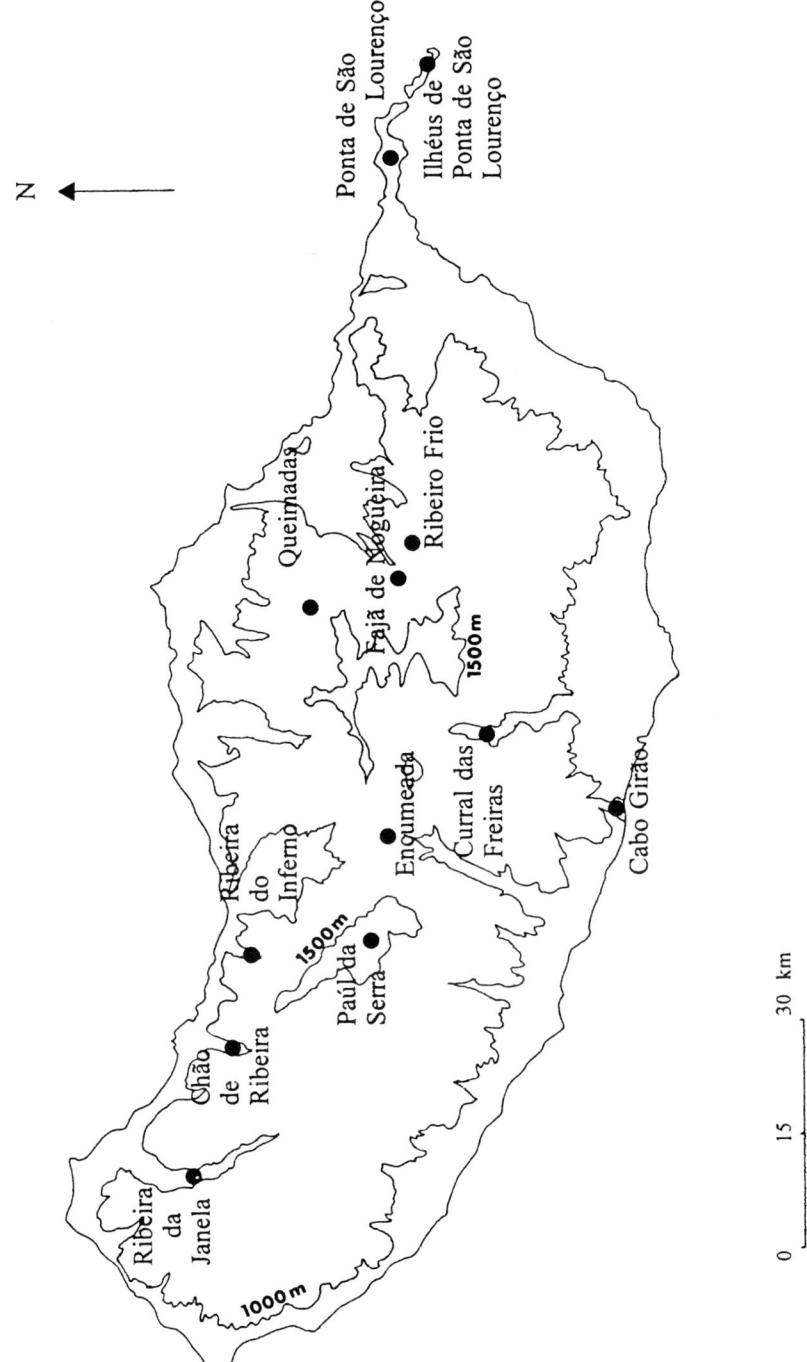

Fig. 3   Additional localities in Madeira

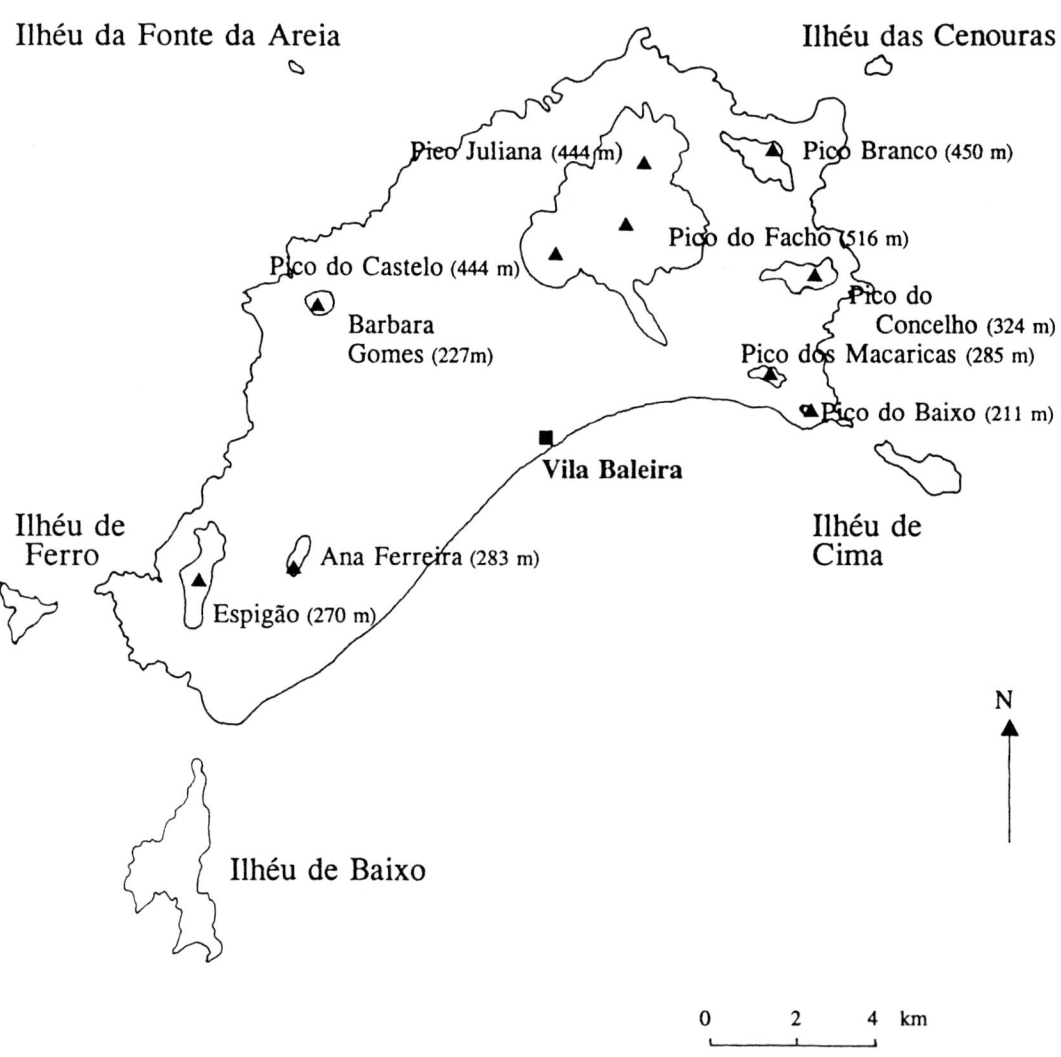

Fig. 4 Porto Santo and associated islets.
200 m contour and spot heights shown

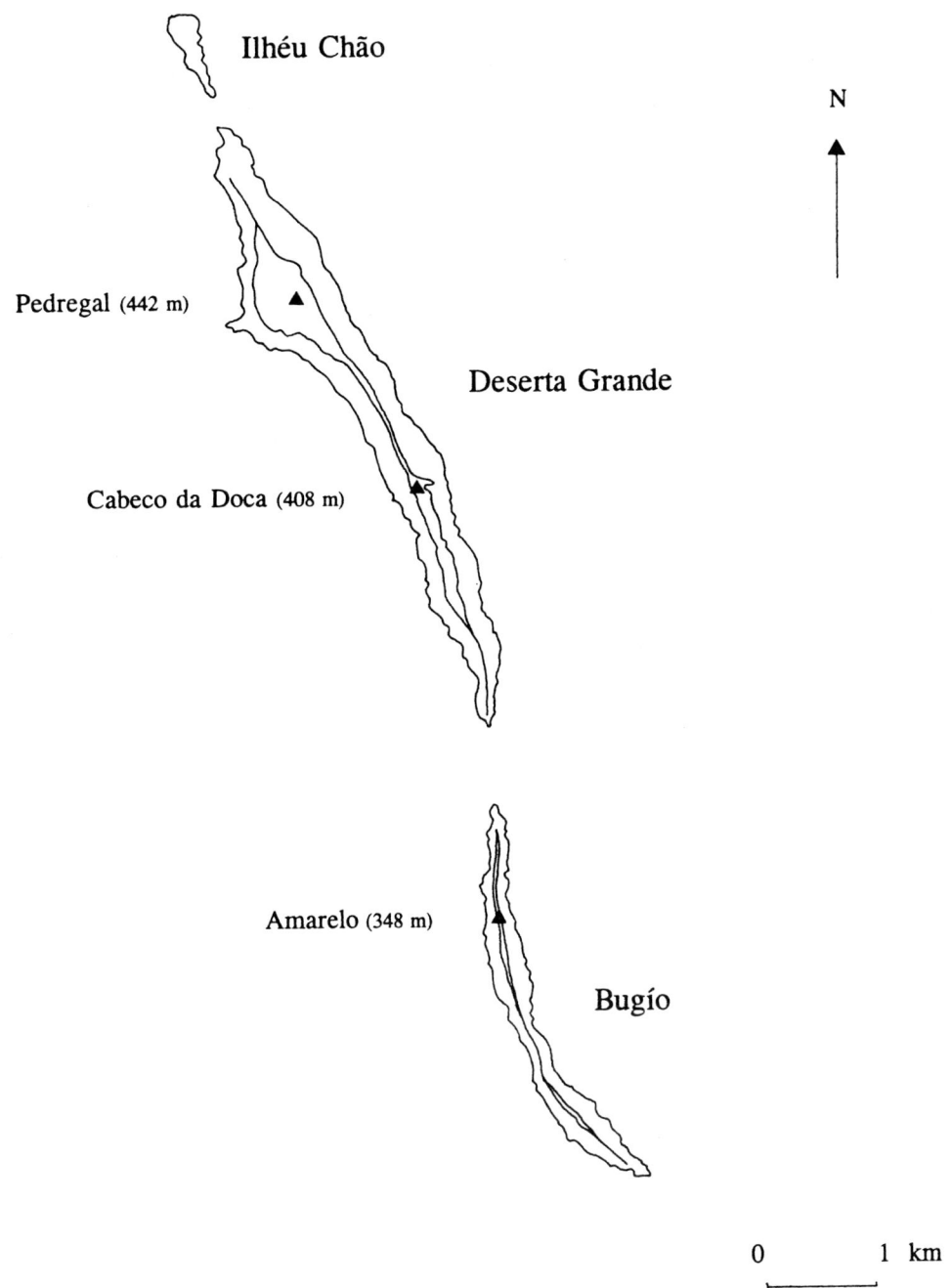

Fig. 5 The Desertas
400 m contour (Deserta Grande), 300 m contour (Bugío) and spot heights

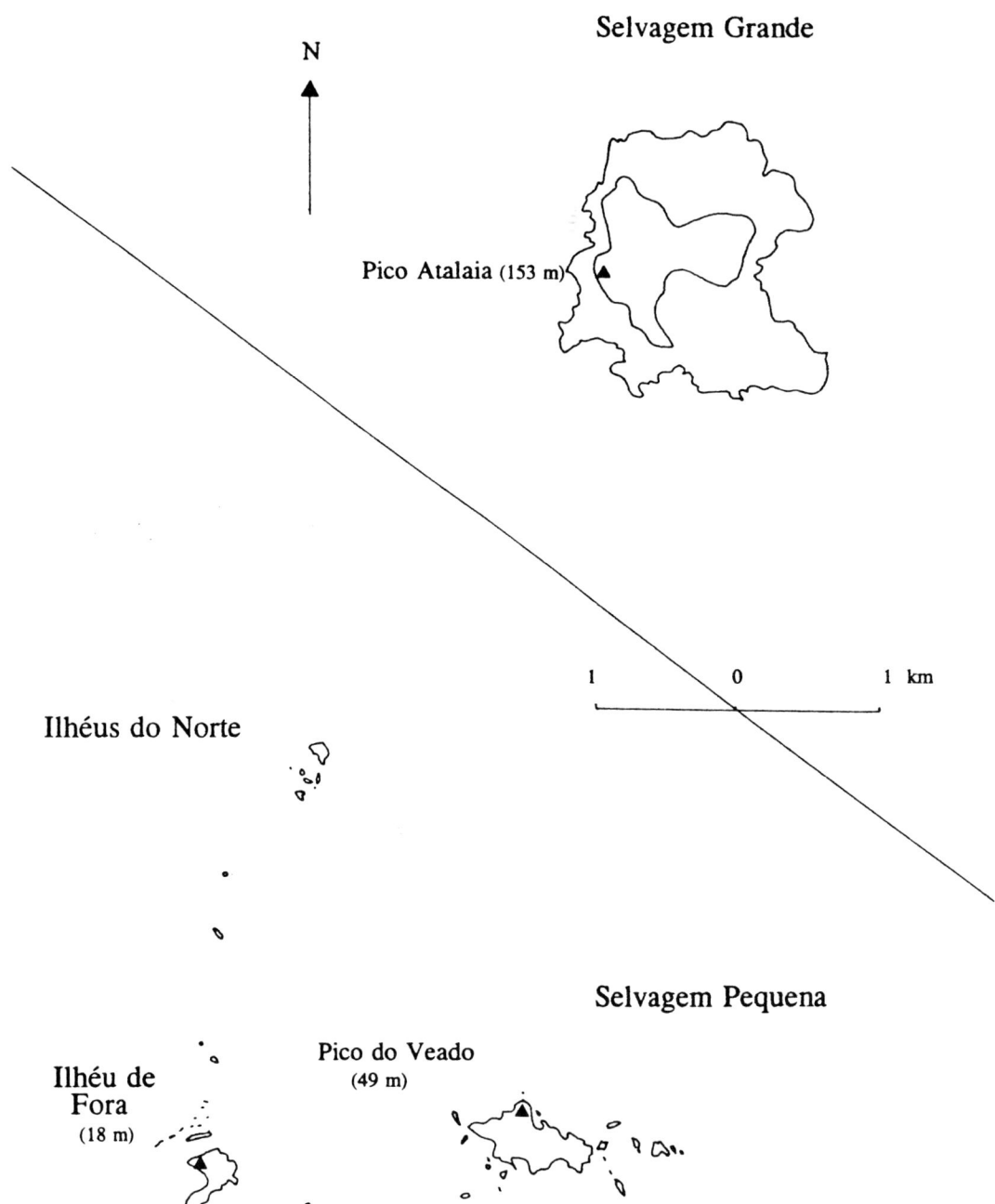

Fig. 6 The Salvages
100 m contour and spot heights

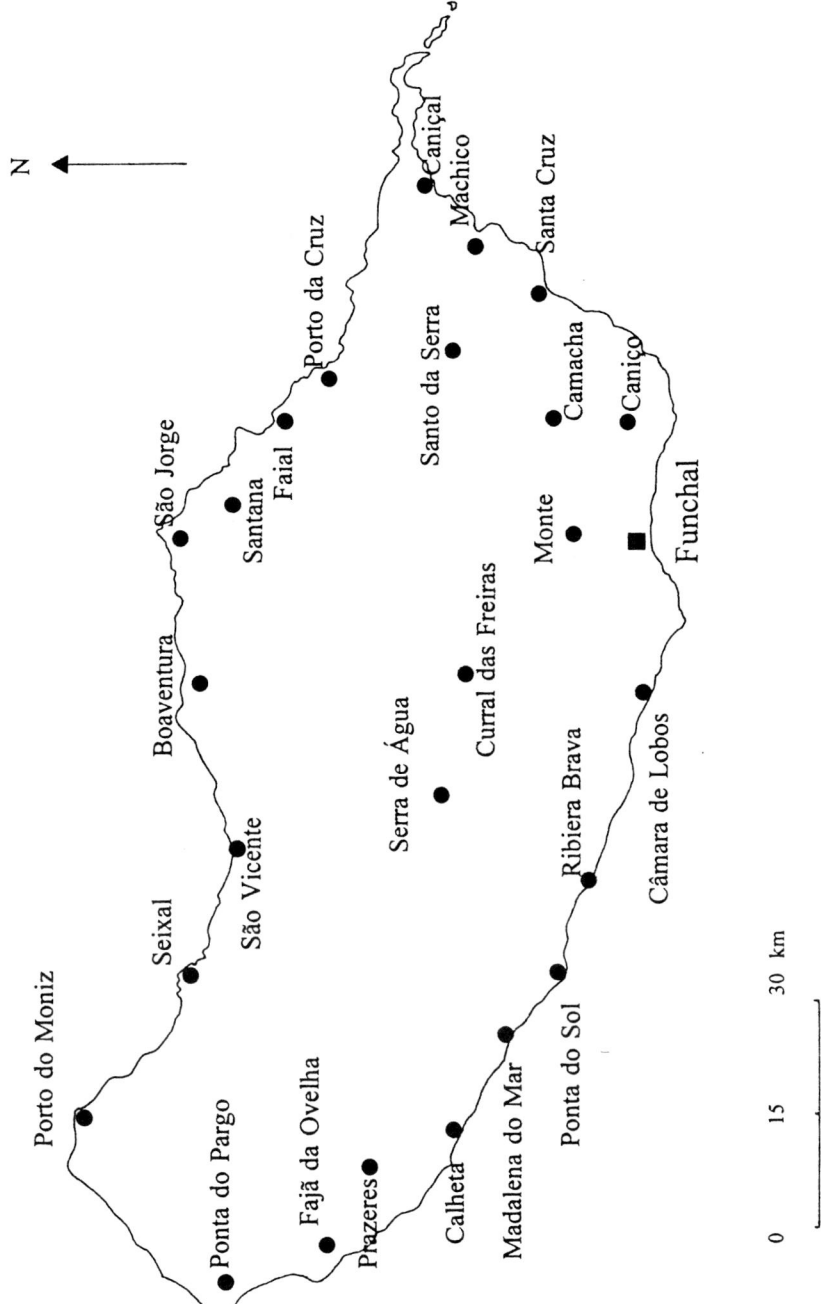

Fig. 7   Population centres in Madeira

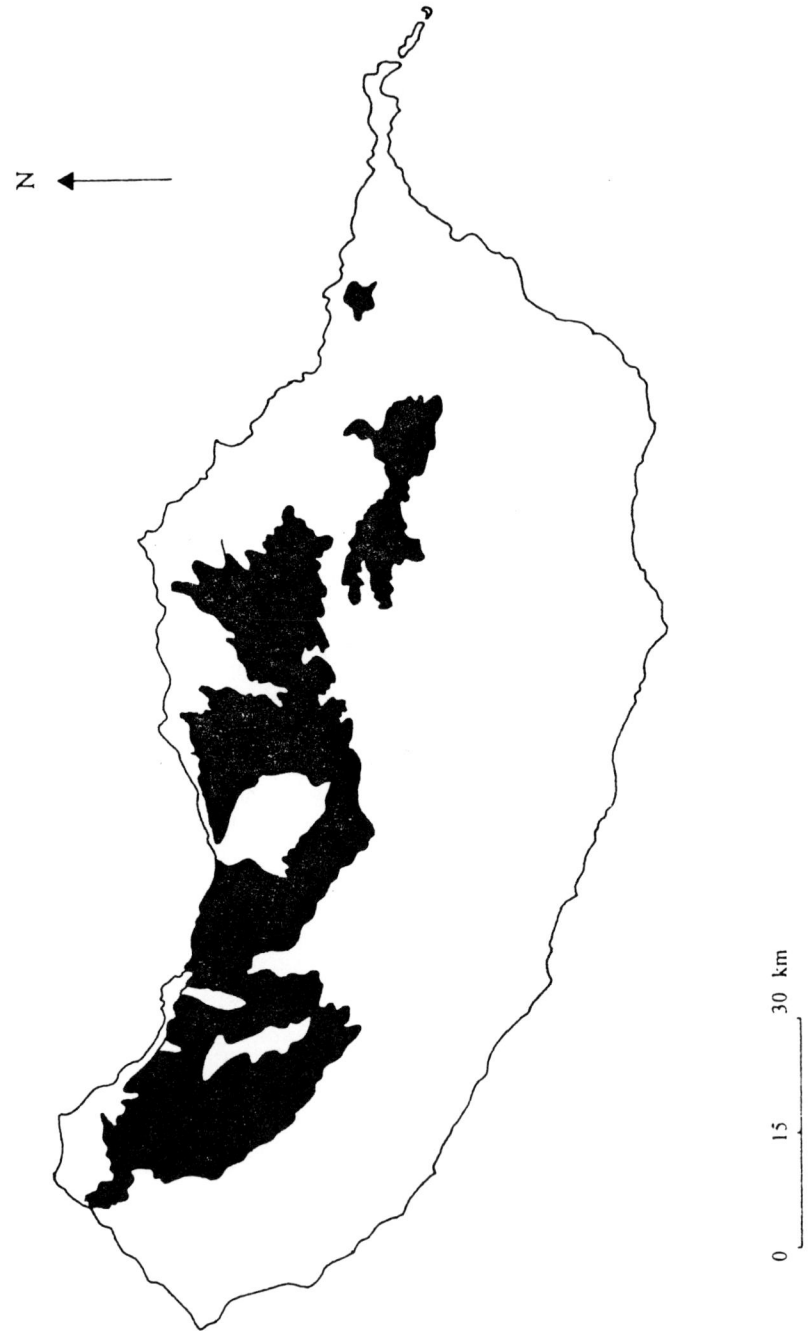

Fig. 8　Current distribution of evergreen forest (laurisilva) in Madeira

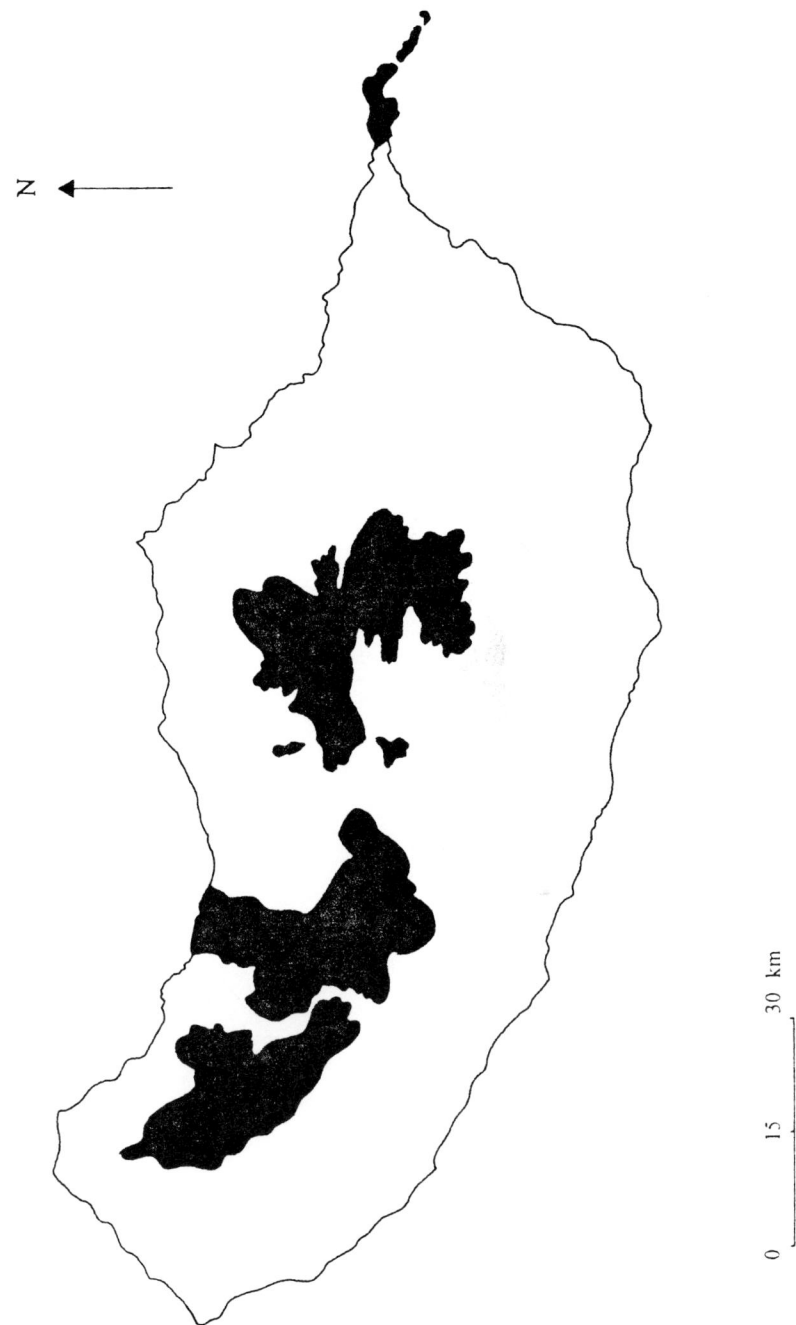

Fig. 9 Areas wholly or partially protected in Madeira as natural or upland reserves

INTRODUCTION

GEOGRAPHY AND GEOLOGY

The islands of the Madeiran archipelago lie in the eastern Atlantic about 978 km from Lisbon and 630 km off the west coast of Morocco, between latitudes 33° 10'-32° 20' N and longitudes 16° 10'-17° 20' W (Fig. 1). All of the islands are of Tertiary volcanic origin with an estimated age of between 60 and 70 million years and are mainly of basaltic rock. Unlike the Azores and Canaries there has been no recent volcanic activity. The soils are mostly of the pedalfer type, poor in calcium and potassium but acidic and rich in humus.

Madeira itself is very much the largest and highest of the islands, with an area of 728 sq. km and a highest point (Pico Ruivo) 1861 m above sea-level. The terrain is generally rugged and dissected by deep ravines and gorges. A mountainous spine runs east to west across the island and is split almost at its central point by a high pass at Encumeada. The eastern half of the spine contains most of the highest peaks of Madeira, centred around Pico Ruivo, Pico de Arieiro and Pico das Torres (Fig. 2). This area is very steep, with numerous inaccessible rock ledges. The western half of the spine is generally somewhat lower and includes the high plateau of Paúl da Serra from which descend a number of steep-sided gorges. There are also many high and steep sea cliffs, especially on the north coast, while on the south coast the 633 m high Cabo Girão is one of the highest sea cliffs in the world. The Ponta de São Lourenço forms a long 'tail' at the extreme eastern end of the island. This is an area of gentler, rolling slopes terminating in two small islets, Ilhéu de Santo Agostinho and Ilhéu de Fora which together are known as the Ilhéus de Ponta de São Lourenço (Fig. 3).

Porto Santo lies some 57 km to the north-east of Madeira and is the second largest island, covering *c.* 50 sq. km It is much lower than Madeira, with a smoother landscape, lacking sharp ridges and deep gorges. There are seven small, conical peaks in the eastern part of the island, clustered around Pico do Facho, at 517 m the highest point of the island (Fig. 4). The west of the island has only a few, rounded heights and these two areas of higher ground are separated by a sandy plain. The southern coast of Porto Santo boasts the only extensive (9 km) sandy beach in the Madeiran archipelago. It is a major feature of the island but is narrow along its whole length and only the most meagre of dunes develop. There are a number of small, offshore islets, all relatively flat. The largest of these, Ilhéu de Baixo and Ilhéu de Cima, lie off the south-west and south-east corners respectively of Porto Santo.

The Madeiran archipelago is completed by three small, rocky islands 24 km off the south-east coast of Madeira which are collectively known as the Desertas (Fig. 5). These islands lie along a north-south axis and are separated from each other by narrow channels. The northenmost and smallest is Ilhéu Chão, an almost flat, table-like island about 100 m above sea-level and with an area of approximately 0.5 sq. km The central, and largest, island is Deserta Grande with an area of about 10 sq. km while the southern island of Bugío has about 3 sq. km Both are very rugged, each consisting of a single, knife-like ridge rising to 442 m and 348 m respectively.

The Salvages (Fig. 1) form a separate archipelago of very small islands 300 km south of Madeira but with strong biogeographic links with the Canary Islands which lie a further 180 km to the south. The three largest islands are Selvagem Grande, Selvagem Pequena and Ilhéu de Fora; the remainder are little more than sea rocks (Fig. 6). Lying *c.* 300 km off Africa, between latitudes 30° 00'-30° 10' N and longitudes 15° 50'-16° 05' W, they cover a land area of less than 3 sq. km and reach a height of only 153 m (Pico Atalaia on Selvagem Grande).

CLIMATE AND HYDROLOGY

The climate of the Madeiran archipelago is typically Mediterranean (Fig. 10), the differences between precipitation and summer drought being the result of variation in altitude and wind

2 INTRODUCTION

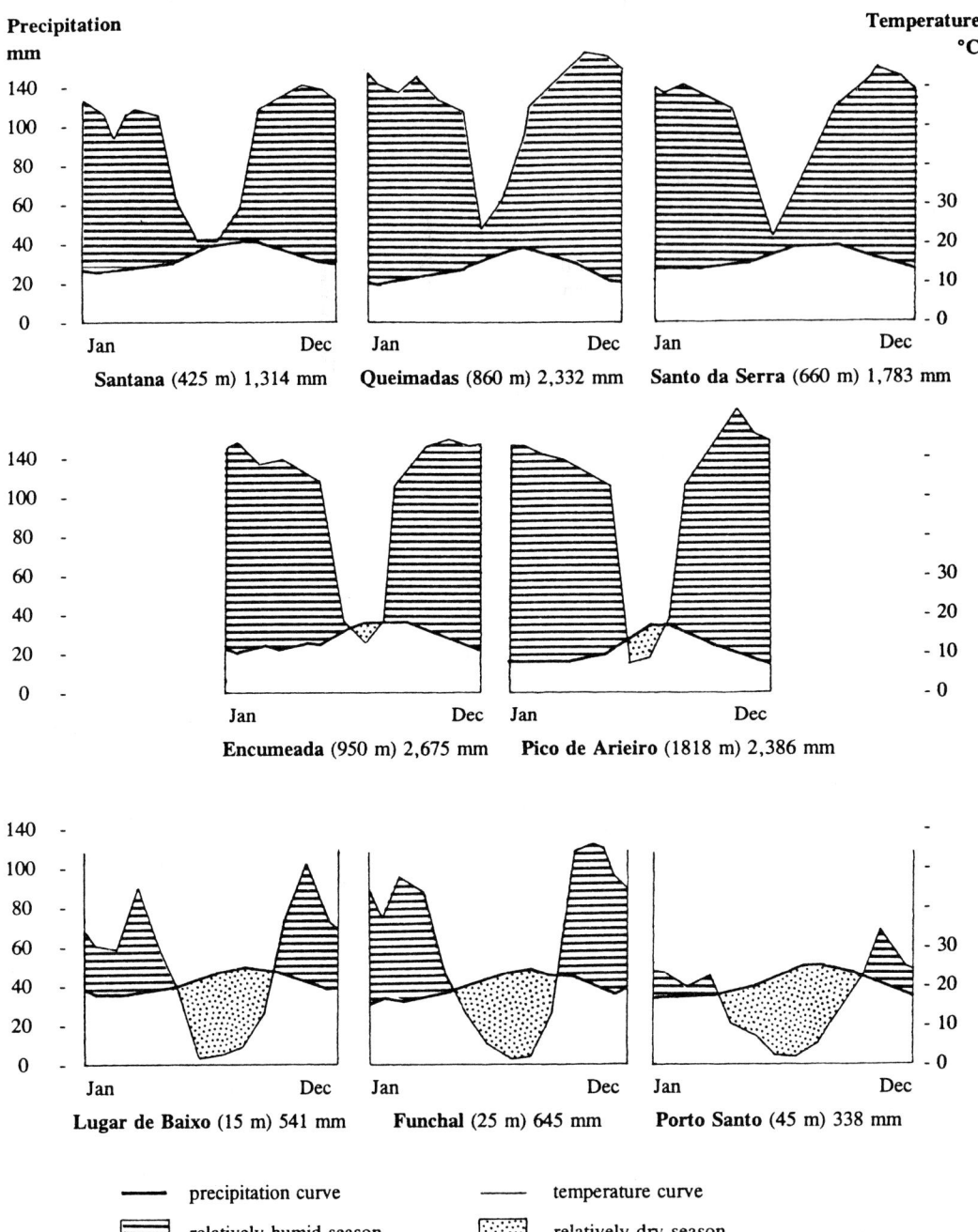

Figures following station names are height above sea-level, mean annual temperature and mean annual rainfall.

Fig. 10 Diagrams showing mean monthly rainfall and mean monthly temperatures for stations in Madeira and Porto Santo (based on data in Walter, Harnickell & Mueller Dombois, *Climate-Diagram Maps* (1975)).

exposure. It is strongly influenced by the Azores anticyclone and the trade winds which, in Madeira, are principally from the north and north-east. Unusually among the Macaronesian islands, these winds affect southern slopes in Madeira as well as northern slopes. The winds are the most important factor for the formation of coastal fogs which are a major influence on the formation of dominant and climax communities of native species, principally the evergreen forest. The forest plays a major role in the hydrology of the island, particularly in precipitating out water droplets from the dense fogs which often wreath the upper regions and which contribute considerably to the ground water supply. At even higher altitudes *Erica arborea* has a similar effect. The result is an abundance of water, particularly in northern and central parts and the higher regions of the southern half of the island. It supplies the enormous number of small rivers (*ribeiras*), streams and springs encountered everywhere. The effects of erosion are easily seen when the appearance of streams in the northern and southern halves of the island are compared. Those in the north flow permanently clear, while many of those in the south are turbid with silt after heavy rains but may become almost dry in the summer. In addition to to the natural water courses, a vast system of man-made aquifers (*levadas*) extends across all but the highest parts of the island. These levadas may be many kilometres long and range in size from narrow, trough-like structures to a few the size of small canals. The driest parts are sheltered areas in the south and east of the island, especially the Funchal region and the low peninsula of Ponto de São Lourenço at the eastern end of the island.

Mean monthly temperatures are equable, with few extremes during the year. Typically they range from around 15-20°C in the sheltered bay of Funchal area to around 6-10°C at Pico de Arieiro. Snow occasionally falls in the high peaks but frost is virtually unknown.

Porto Santo has a similar range of temperatures to the warmer parts of Madeira but it is considerably drier (Fig. 10). It is too low for the frequent formation of coastal fogs and the few ribeiras are seasonally dry. Other than man-made reservoirs and irrigation systems, the only permanent water comes from five natural springs dispersed about the island.

Like Porto Santo, the Desertas gain little water from coastal fogs in the way that Madeira does. Thereise a considerable number of ribeiras on the two larger and higher islands but most are temporary or seasonally dry. The steepness of the terrain and the lack of soil to retain water means that most is quickly lost as run-off. The much lower Ilhéu Chão is a very dry island. There is no permanent water on any of the Salvage Islands, which although they are frequently swept by rain, can be prone to long periods of drought.

POPULATION

The population of Madeira fluctuates during the year due to the large numbers of vistors but the permanent population has remained remarkably stable in recent years, rising by only 1.3% between 1981 and 1991. In 1991 (the last year for which figures are available) it had a population of 257,692, with 126,000 of the people living in the island's only large town, the capital Funchal. Most of the other, very much smaller, population centres are found around the coast. They include Calheta, Câmara de Lobos, Machico, Ponta do Sol, Ribeira Brava and Santa Cruz in the south and south-east and Faial, Porto da Cruz, Porto do Moniz, Santana and São Vicente in the north (Fig.7).

The population of Porto Santo is only 5,000, the greatest proportion living in the main town of Vila Baleira. The Desertas and Salvages are uninhabited except for the Natural Park wardens stationed on Deserta Grande and Selvagem Grande.

VEGETATION

Madeira was discovered by the Portuguese in 1475 at which time the islands were considered to have been in a pristine state. Since then, the vegetation of nearly all the islands has undergone rapid and profound changes. In Madeira, indigenous vegetation types are closely linked to altitude, with many characteristic species confined to specific altitudinal bands because of narrow ecological tolerances. There is also a marked difference in the respective altitudinal limits for the same vegetation in the northern and southern halves of the island due to climatic effects. The pattern is somewhat complicated by local variations in these limits, for instance in deep ravines where the relative humidity is higher and the temperature lower than at comparable altitudes on exposed slopes. A number of plant communities have been recognized in Madeira (Tab. 1) and these can be grouped into three principal vegetation types: coastal vegetation, evergreen forest and upland vegetation.

The herb and shrub community (*Aeonio-Lytanthion* alliance) which occurs primarily within the coastal regions of Madeira is rarely fully developed at altitudes above 300 m in the south and 100 m in the north. It is broken up by areas of urban development and intensive cultivation. The shrub layer is not always present and in some areas dominant native species such as *Globularia salicina*, *Euphorbia piscatoria* and *Echium nervosum* have been replaced by introduced species such as *Opuntia tuna*, *Ulex europaeus* and *Cytisus scoparius*. The vegetation of the steepest cliffs and coastal ravines is worthy of special mention. The vertical faces of those along most of the northern coast and parts of the southern coast have a striking flora which is particularly rich in chasmophytic and endemic taxa.

The vegetation characteristic of Macaronesia as a whole is the evergreen forest or laurisilva. This forest once covered some 60% of Madeira but is now reduced to about 16% (10,000 ha) of the main island. The four dominant tree species, *Apollonias barbujana*, *Laurus azorica*, *Ocotea foetans* and *Persea indica*, are all members of the family Lauraceae and, with the exception of *L. azorica*, are endemic to Macaronesia.

The evergreen forest represents a relic of the subtropical evergreen forests which in the Tertiary era surrounded the Tethys Sea in what is now southern Europe and north-western Africa. It is thought that the oceanic position of the islands has had a moderating influence on climatic changes there, with the result that the forests have escaped significant alteration for millions of years. Furthermore, their isolation has protected them from aggressive, climate-induced plant migrations. Although the Macaronesian forests are much poorer than those earlier forests, at least in tree species, the shrub and herb layers have numerous endemics and bryophyte and lichen communities are highly developed and diverse, especially in epiphytes. The forest develops in conditions of high precipitation, high humidity and mild temperatures and is especially dependent on the formation of coastal fogs. The most crucial factor determining the distribution of the forest in Madeira is temperature, unlike the Canaries where it is rainfall, and the Azores where it is altitude.

Two types of indigenous evergreen forest formerly occurred in Madeira. Dry evergreen forest is nowadays exclusive to the Canaries. It develops in areas with a combination of high solar radiation and high mean temperatures, low annual precipitation (350-500 mm) and sporadic sea fogs. Important arboreal species are *Apollonias barbujana*, *Visnea mocanera* and *Picconia excelsa* but there are few ferns and lianes and an absence of hygrophilous species such as *Ocotea foetans* and *Persea indica*. Little is known about this forest type in Madeira but it has been suggested that it occupied much of the southern slopes between 300-700 m and scattered, drier parts of the north between 100-400 m. These areas were earliest and most extensively changed by exploitation and agriculture, and are now occupied by a transitional vegetation, cultivated land and exotic forest of *Pinus* or *Eucalyptus* and *Acacia* species

The remaining evergreen forest in Madeira is humid evergreen forest (*Clethro-Laurion* alliance). It occurs in areas receiving reduced solar radiation levels and mean temperatures, high (500-1200 mm) annual precipitation and where coastal fogs are frequent. The dominant trees are *Laurus azorica*, *Ocotea foetans* and *Persea indica* and it is characterised by the presence of another endemic tree species, *Clethra arborea*. The forest covers large areas between 300-1300 m on the northern slopes and 700-1200 m on the southern slopes, and reaches its richest development in deep, extensive ravines such as Ribiero Frio and Ribeira da Janela (Fig. 8). Although the altitudinal limits may have contracted due to the activities of man, the natural limits are probably not more than 100 m above or below the present ones.

Upland vegetation occurs above the altitude limits for the evergreen forest. The main communities here are widespread *Erica arborea* and *Vaccinium padifolium* scrub, grazed grassland and, in the high peaks, *Erica cinerea* scrub.

Table 1. Plant communities in Madeira.

Alliance **Aeonio-Lytanthion**
Association Hyparrhenietum hirtae (coastal grassland throughout island)
Association Euphorbietum piscatoriae (sea-cliffs throughout island)
Association Biserrulae-Scorpiurietum (coastal grassland on Ponta da São Lourenço)
Alliance **Clethro-Laurion**
Association Deschampsietum argenteae (exposed ravines habitats with permanent water supply)
Association Vaccinio-Sibthorpietum (widespread)
Association Campylopo-Airetum (grazed areas above the cloud zone)
Association Ericetum cinereae (high peaks)

The original vegetation of Porto Santo appears to have had *Juniperus phoenicea* as a dominant species, together with *Dracaena draco*; a thermophilous type of evergreen forest with *Apollonias barbujana* was also present. None of this vegetation remains and indeed few trees now grow on the island except for some small-scale plantations of exotic conifers, managed by the Servicos Florestais (Forestry Service), and a few introduced species such as *Tamarix gallica*. For the most part the island supports a dry type of grassland of Mediterranean appearance. The introduction of rabbits by early settlers is claimed to have been the cause of the initial degradation of the native vegetation and a continuing regime of grazing by both rabbits and stock, coupled with the lack of water, maintains the situation.

The vegetation of the Desertas and the Salvages has been similarly degraded by rabbits and goats, except on Selvagem Pequena and Ilhéu de Fora where grazing animals have never been introduced. The Desertas are unlikely ever to have supported a true laurisilva type of evergreen forest as they are below the altitudinal limits required for its full development. However, it seems likely that that there was once at least some tree cover and reports of the appearance of seedlings of species such as *Apollonias barbujana* following the removal of grazers lends support to this hypothesis. The Salvages and various islets have never supported forest of any kind; even on the untouched islands the vegetation is low, open and somewhat monotonous, dominated by a few halophytes such as *Frankenia laevis*, *Limonium papillatum*, *Mesembryanthemum crystallinum*, *M. nodiflorum*, *Senecio incrassatus* and *Suaeda vera*, but also with a surprizingly high number of endemic taxa for such a small area.

The numerous ornamental species which have been introduced to the archipelagos and become naturalized over the centuries occur mainly in Madeira and Porto Santo but are not confined to these islands. These species are most common in the lowland regions where the success of aggressive invaders such as *Acacia* species, *Papaver somniferum* and *Pittosporum undulatum* is clearly evident. A few, such as *Ageratina adenophora* and *Erigeron karwinskianus* have even penetrated the higher regions where, doubtless, more will follow.

ENDEMISM

Like many oceanic islands, both the Maderian and Salvage Island archipelagos exhibit a high level of endemism (Tab. 2). The respective floras of the archipelagos, and to some extent the floras of the individual islands, have evolved in isolation, both from each other and from the continental floras, for millions of years. Genera such as *Aeonium*, *Argyranthemum* and *Lotus* have groups of very closely related, vicariant taxa with discrete and often narrow distributions among the islands, and sometimes with further representatives in the closest parts of continental Europe and Africa.

Two levels are usually recognized, one for taxa endemic to individual islands or island groups (archipelagic endemics) and another for taxa endemic to the Macaronesian region as a whole (regional endemics). This latter category is sometimes taken to include taxa which have outliers in the so-called 'Macaronesian enclave' around the Ifni region of Morocco.

Species endemism in the Maderian archipelago is about 10%. For the Salvages the figure is only 4% but rises to 10% when species shared with the Madeiran archipelago are included. There are a further 69 (6%) regional endemic species represented in the Madeiran archipelago and 11 (10%) in the Salvages.

Important groups with a high proportion of archipelagic and regional endemics combined include the Compositae (132 species, 24 endemic), Cruciferae (47 species, 11 endemic), Crassulaceae (229 species, 11 endemic), Labiatae (39 species, 9 endemic), Liliaceae (24 species, 7 endemic) and Gramineae (139 species, 8 endemic). The pteridophyte flora (75 species, 14 endemic) is particularly rich in the evergreen forest. The only endemic genera are *Chamaemeles* (Rosaceae, 1 species), *Musschia* (Campanulaceae, 2 species), *Parafestuca* (Gramineae, 1 species) and possibly *Monizia* (Umbelliferae, 1 species).

Table 2. Endemism in the Madeiran and Salvage archipelagos.

	Number of species	Island endemics (%)	Regional endemics (%)
Madeira	1,163	113 (10%)	60 (5%)
Porto Santo	448	29 (6%)	18 (4%)
Desertas	199	30 (15%)	19 (10%)
Salvages	105	11 (10%)	11 (10%)
All islands	1,226	123 (10%)	69 (6%)

CONSERVATION

Fortunately for the flora of the region conservation and management strategies are being actively pursued in all of the islands. The Parque Natural da Madeira is the authority responsible for advising on and overseeing conservation measures in the islands. Its primary aims are *1)* conserving and regenerating the evergreen laurisilva forest, and *2)* protecting and managing sensitive, high altitude zones. These aims are being achieved by a combination of legal instruments to restrict exploitation and damaging agricultural practices and a government policy of acquiring land for reserves in the more sensitive areas (Table 3). As well as covering a variety of vegetation types and landscapes (Table 4), the Natural Park of Madeira contains several categories of reserve which receive different levels of protection. The most important are the six Wholly Protected Reserves, covering a total area of 2,322 ha and the ten Partially Natural Reserves which cover 6,400 ha (Fig. 9). The former are completely protected areas with limited access. The latter have a lower level of protection and a greater degree of access allowed. They include those areas of evergreen forest which are still well-preserved, despite human interference.

Table 3. Reserves in the Madeiran Islands.

Reserve	Date legally established	Area in hectares
Salvage Islands Natural Reserve	29.10.1971	9,455 (9,172 sea area)
Natural Park of Madeira	12.11.1982	56,700
Desertas Natural Reserve	23.05.1990	9,672 (8,249 sea area)

Table 4. Vegetation and other protected zones within the Natural Park of Madeira

Vegetation/ Zonal type	Area in hectares
Evergreen forest (laurisilva)	12,000
Geological reserve and natural alpine vegetation	3,912
Exotic forest	18,000
Transitional vegetation, pastures, recreation zones and protected landscapes	22,788
Total area	56,700

ORGANIZATION OF THE FLORA

SCOPE

The *Flora of Madeira* covers the Madeiran Islands, i.e. Madeira itself, Porto Santo and its neighbouring islets, and the three islands of the Desertas (Ilhéu Chão, Deserta Grande and Bugío). The Salvages are also included in the *Flora*. An important nature reserve, the Salvages fall within the political and administrative boundaries of Madeira and are thus treated as a justifiable extension of the Madeiran area.

As far as possible, all native and naturalized vascular plants known to occur in the area are included in the *Flora* but it has proved difficult to develop a consistent treatment for other categories of plants. Casuals and aliens which are recorded only occasionally and which do not persist are omitted for the most part, as are the numerous garden ornamentals which are grown in the islands. Crop plants generally are omitted but those which persist for some time on field and terrace margins after the crop has been cleared, species used extensively in roadside plantings and the more widespread timber, street and amenity trees are mentioned, usually as un-numbered entries in the text.

LITERATURE

A large body of literature has been consulted during the compilation of this *Flora*, ranging from monographs and floras to checklists and floristic notes. Only publications which have been used extensively in preparing descriptions are cited in the text.

A few works of special importance in Madeiran botany form the core literature for the *Flora* and these have been consulted throughout. The publications of Lowe (1859-72), Menezes (1914) and Hansen & Sunding (1989, 1993) are especially relevant. To allow users easily to trace the taxa described in the core works, all accepted names of species, subspecies and varieties appearing in them are included in this *Flora*, either as currently accepted names or as synonyms. The full list of core works is given below.

A. Hansen & P. Sunding, Checklist of the vascular plants of Macaronesia. 4. Revised edition, in *Sommerfeltia* **17**: 1-295 (1993).

R.T. Lowe. *Manual Flora of Madeira*.I & II. (1859-72).
- *Florulae Salvagicae Tentamen* (1869).

C.A. Menezes. *Flora do archipelago da Madeira.* (1914).

T. Monod, Conspectus Florae Salvagicae, in *Bolm Mus. munic. Funchal* Sppl. 1: 1-79 (1990).

C. Romariz. Flora da ilha da Madeira - Peridófitos, in *Revta Fac. Sci. Univ. Lisboa* IIC, **3**: 53-112 (1953).

TAXONOMIC CATEGORIES AND SEQUENCE OF TAXA.

The taxonomic categories used in the *Flora* are: Family, Genus, Species, Subspecies and Variety. Subfamilies, Tribes and other supra-generic categories are also used where they are felt to be helpful. In a few instances, where several species are particularly difficult to distinguish from each other, the concept of a 'species group' is used. In these cases, a general description of the group is given, followed by shortened descriptions of the constituent species. Taxa below the level of variety have not been used anywhere in the *Flora*.

Families are numbered with upper case roman numerals, genera and species with arabic numerals and varieties with lower case roman numerals; subspecies are lettered. The sequence of families closely follows that of Melchior in Engler, *Pflanzenfamilien*, ed. 12 (1964), the same as that used in Tutin et al. (eds) *Flora Europaea* (1964-93). The sequences of the remaining taxonomic categories are those deemed most appropriate by the individual authors of the accounts. In all cases the intention is to facilitate use of the account in the identification of taxa and how they relate to each other.

In addition to the numbered and lettered taxa the *Flora* contains unmarked entries in all categories. These are usually reduced descriptions for plants whose taxonomic status is in doubt, those which have not yet been shown categorically to belong to the established wild flora of the islands, or are otherwise felt to be worthy of mention. They are placed at the appropriate points in the taxonomic sequence but are omitted from keys.

CRITICAL GROUPS

Fortunately only two critical groups of plants are represented in the Madeiran islands, *Rubus* and *Taraxacum*. The relatively few difficult *Rubus* taxa covered in the *Flora* are dealt with as a single 'species group'. More problematic is *Taraxacum*, which is notorious as a taxonomically difficult genus throughout its range. Ten species have been claimed as growing in Madeira but material for all of them is scanty and we believe it is not yet possible to provide an adequate account for this genus. As a compromise, the account describes the sections to which the species are currently assigned by *Taraxacum* taxonomists; the species themselves and their purported distributions are simply listed within the sections, together with the names of the *Taraxacum* specialists (where known) who have determined the material.

AUTHORS

The name of the author of an account follows the appropriate family or genus name. Where the majority of a family account has been written by one author, his or her name appears under the family only; any additional authors' names appear under the relevant genus names. A list of contributors is also provided at the begining of the *Flora* on p. vi.

KEYS

With the exception of un-numbered taxa mentioned above, keys are provided for all families, genera and species, and for some infraspecific taxa. The keys are dichotomous and non-indented. The characters used in the keys for genera and species are repeated in the descriptions but this is not always true for characters used in the family key.

That part of the family key dealing with pteridophytes, together with the key to pteridophyte genera, are also special cases. In the family key the pteridophytes are separated as a single group and not as individual families. This in turn leads to a single key for all pteridophyte genera (see Descriptions below). However, to aid the user, the genera have been placed under their normally accepted family names in the text and numbered accordingly. The numerical reference given for each genus in the key is therefore given in the form:

IV.2. Botrychium

where *Botrychium* is genus **2** in family II. Ophioglossaceae in the text.

NAMES

Family names are given without authority or bibliographic reference. Genus names are given with the authority but without bibliographic reference. Accepted names of species, subspecies and varieties are given with the authority and for taxa which are found in the wild the bibliographic reference is also given.

No attempt has been made to provide full synonomy. Synonyms are only cited when they have been used as accepted names in the core literature or when their omission would cause confusion. They are given in alphabetical order, together with their authority but without bibliographic reference. Synonyms at specific level are separated by semi-colons. Where infraspecific taxa belong to the same higher rank the name of that rank is not repeated and the synonyms are separated by commas. Thus, for **Parietaria judaica** L., the synonyms *P. officinalis* auct., non L., *P. officinalis* subsp. *judaica* (L.) Beguinot, *P. officinalis* var. *diffusa* (Mert. & Koch) Weddell and *P. ramiflora* Moench are rendered in the form *P. officinalis* auct., non L., subsp. *judaica* (L.) Beguinot, var. *diffusa* (Mert. & Koch) Weddell; *P. ramiflora* Moench. Because of the many recent changes in pteridophyte taxonomy and nomenclature, basionyms are indicated for some species in these accounts.

Vernacular names are also given where known but no attempt has been made to provide a complete list. Where a name is used frequently or exclusively in Madeira but not in Porto Santo, or vice versa, this is indicated by the acronym M (Madeira) or P (Porto Santo). The case does not apply to the remaining islands, none of which are permanently inhabited.

AUTHORITIES, CITATIONS AND ABBREVIATIONS

Authors' names are abbreviated according to R.K. Brummit & C.E. Powell, *Authors of plant names*, Royal Botanic Gardens, Kew (1992). Book titles are cited according to A. Stafleu & R.S. Cowan, *Taxonomic literature*, ed. 2 (1976-1988); Supplement A. Stafleu & E.A. Mennega (1992-1993); journal titles follow *Serial publications in the British Museum (Natural History) library* ed. 3 (1980). Only standard English abbreviations are used in the text.

DESCRIPTIONS

The terminology used in the descriptions follows that found in modern English language *Floras* but an attempt has been made to avoid using the most obscure terms, replacing them with simpler phrases. Unqualified measurements refer to length. Where measurements are given in the form 5 × 2 cm, they refer to length versus width. Figures given in parentheses indicate exceptional values outside the normal range.

Descriptions are brief and diagnostic. At genus level and below they apply only to taxa as they occur in the Madeiran and Salvage Islands so the full range of variation of a taxon worldwide may not be covered. In family descriptions this rule has been relaxed somewhat since some large families are poorly represented in our area. In these cases additional characters referring to the family in general may be given to avoid creating a distorted impression but it should not be assumed that such descriptions apply to the whole family as it occurs outside Madeira. For pteridophytes, no family descriptions are given as the current state of pteridophyte taxonomy does not allow diagnoses suitable for easy identification at the family level to be produced.

To avoid superfluous repetition, characters given in the family descriptions are not necessarily repeated in the genus descriptions; the same is true for genus characters in species descriptions. Where one species is very similar to another a comparative description may be given in which only the differences between the species and that with which it is being compared are provided. Comparative descriptions always begin in the same way, i.e

Like **3** but ...

Here the description of species **3** applies except for the differences stated. A comparative description does not necessarily imply that the two plants have the same overall appearance.

For infraspecific taxa only diagnostic characters are given.

CHROMOSOME NUMBERS

These are given only when the count has been made from material originating from the *Flora* area. They are expressed in the form $2n=8$.

ILLUSTRATIONS

The fifty-seven plates of original line drawings at the end of the *Flora* section of the book illustrate x of the y endemic taxa. Only those taxa for which suitable material was unavailable or whose diagnostic characters, e.g. difference in colour, cannot be depicted in this way have been omitted. Many of these plants have never before been illustrated. The drawings mostly show the habit of the plants but details of diagnostic characters are also shown where these are especially helpful. The cross-reference to an illustration is given at the end of the relevant description.

FLOWERING PERIOD

This immediately follows the species descripion and is given in the following form:

Fl. II - VI.

The roman numerals indicate the months covering the main flowering period only, i.e. when the majority of plants flower. In many species isolated flowering individuals may occur outside this period. Where a species flowers throughout the year this is stated. When the main flowering period is unknown this is indicated by an interrogative (?).

STATUS, HABITAT AND DISTRIBUTION

Information on these three categories is given in a single, composite statement following the flowering period. Because of the difficulty in differentiating beyond doubt between native and historically introduced plants, no status is given for these taxa. The status of endemic taxa or those recently introduced to the islands is given. Where the former are regional (Macaronesian) endemics, the other islands groups in which they occur are shown in square brackets at the end of the statement. For the latter, an indication of their native range is also provided. Information on habitat and ecology has been kept to a minimum. The distribution of taxa is variously described but whenever possible the islands on which it occurs are named individually; otherwise the island group is named, e.g. the Desertas. For a number of the rarest and most endangered plants precise localities are deliberately withheld and a more generalized distribution is given. The names of the island groups are given in English. All other place names are given their Portuguese spellings.

The statement is completed by a series of symbols and acronyms summarising endemic status (if applicable) and overall distribution among the islands. The symbols and acronyms are explained on page 12.

ADDITIONAL INFORMATION

When additional data such as commercial, local or historical uses are given they are included in the general statement covering status, habitat and distribution. Any further taxonomic information and explanatory notes on the treatment of a taxon are given in an appended note following the entry for that taxon. In an attempt to provide the user with as much information as possible, many such notes have been included, particularly for taxa whose treatment in this *Flora* is controversial or in doubt.

SYMBOLS USED IN THE FLORA

M   Madeira.

D   Desertas (Ilhéu Chão, Deserta Grande, Bugío).

P   Porto Santo and associated islets.

S   Salvages (Selvagem Grande, Selvagem Pequena, Ilhéu de Fora).

●   archipelagic endemic (i.e. Madeiran and Salvage islands).

■   regional endemic (i.e. Macaronesia); the island groups outside the Madeiran and Salvage Islands in which it the taxon occurs are shown after the distribution acronyms e.g. ■ M [Azores].

?   preceeding an island symbol indicates that the presence of the taxon on that island is doubful.

ARTIFICIAL KEY TO FAMILIES

This key is for all the families covered in the *Flora*. In an attempt to make the key as useful as possible, allowance has been made for characters which may be interpreted as fitting either statement in a couplet, or those where the parts described are fleeting and, therefore, may be absent from the specimen. Plants in these categories may be keyed out by following either of the statements in the relevant couplets. Where a family is represented at a point in the key by an atypical genus, the genus is indicated.

1. Plants reproducing by spores . . . . **I-XVIII. PTERIDOPHYTA**
- Plants reproducing by means of seeds . . . . (SPERMATOPHYTA) 2

2. Plants with naked seeds borne in cones; always trees or shrubs with needle- or scale-like, evergreen leaves . . . . (XIX-XXII. GYMNOSPERMAE) 3
- Plants with ovules contained in a structure; herbs, shrubs or trees, rarely trees with needle- or scale-like, evergreen leaves . . . . (XXIII - CXXIII. ANGIOSPERMAE) 6

3. Leaves opposite, whorled or absent 4
- Leaves spirally arranged or in pairs 5

4. Leaves free, scale- or needle-like; stems not articulated . . **XX. CUPRESSACEAE**
- Leaves scale-like, fused to form a sheath around each node; stems articulated . . . . **XXII. EPHEDRACEAE**

5. Leaves all spirally arranged but often twisted to lie in 2 spreading ranks . . . . **XX1. TAXACEAE**
- Some leaves in pairs, each pair with a basal sheath of brown, scale-like leaves . . . . **XIX. PINACEAE**

6. Plants aquatic, leaves all submerged or floating on surface of water 7
- Plants terrestrial, sometimes growing in water margins but then at least some leaves emergent above surface of water 10

7. Plants free-floating, stems and leaves reduced to a small, plate-like, green thallus . . . . **CXIX. LEMNACEAE**
- Plants rooted in mud, stems and leaves clearly differentiated 8

8. Flowers held above water surface; perianth 4-lobed, greenish . . . . **CVIII. POTAMOGETONACEAE**
- Flowers minute, completely submerged; perianth absent 9

9. Flowers 1-2 in leaf axils; floating leaves in rosettes (plants of freshwater) . . . . **XCIII. CALLITRICHACEAE**
- Flowers in pedunculate spikes; floating leaves absent (plants of brackish water) . . . . **CIX. RUPPIACEAE**

10. Inflorescence consisting of many tiny flowers packed into a cylindrical spike, the whole structure enveloped by a large, bract-like and often coloured structure (*spathe*) . . . . **CXVIII. ARACEAE**
- Inflorescence various but not as above 11

11. Perianth absent, flowers in a greenish *cyathium* consisting of a stalked ovary and several stamens surrounded by a 4- to 5-lobed, calyx-like involucre, the lobes alternating with conspicuous glands . . . . **LIV. EUPHORBIACEAE** (*Euphorbia*)
- Perianth present or absent but flowers never in a cyathium 12

12. Flowers small, often reduced to small, tubular florets, with or without a strap-shaped extension to the tube, all crowded into a single head backed by a calyx-like ruff of bracts (*involucre*), the whole structure (*capitulum*) resembling a single flower 13
- Not this combination of characters 18

13. Flowers with an obvious calyx, at least the calyx-teeth green 14
- Flowers lacking an obvious calyx though sometimes with a whorl of hairs or scales or a small rim at base of corolla 17

14. Leaves in whorls of 4 or more . . . . **LXXXIX. RUBIACEAE** (*Sherardia*)
- Leaves not in whorls 15

15. Shrubs . . . . **XCVII. GLOBULARIACEAE**
- Herbs, sometimes with a woody stock 16

16. Tubular, membranous sheath enclosing top of flowering stem beneath capitulum. . . . . **LXXXIII. PLUMBAGINACEAE** (*Armeria*)
- Flowering stem lacking tubular sheath . . . . . **CV. CAMPANULACEAE** (*Jasione*)

17. Stamens clearly exserted beyond corolla, anthers free . . . . **CIV. DIPSACACEAE**
- Stamens not clearly exserted, anther cohering to form a tube around style . . . . **CVI. COMPOSITAE**

18. Perianth of 1 whorl, or of 2 or more similar whorls, or absent 19
- Perianth of 2 dissimilar whorls 68

19. Perianth absent 20
- Perianth present 26

20. Flowers enclosed by chaffy bracts; leaves narrow and grass-like 21
- Flowers not enclosed by chaffy bracts; leaves broader 22

21. Flowers with a bract above and below; leaves ± jointed at junction with sheath, often with a projecting ligule; sheaths usually open; stems oten circular in cross-sectyion . . . . **CXII. GRAMINEAE**
- Flowers with a bracxt below only; leaves not jointed at junction with sheath, ligule (if present not projecting; sheaths usually closed; stems often triangular in cross-section . . . . **CXX. CYPERACEAE**

22 At least the uppermost leaves in whorls . . . . **XCIII. CALLITRICHACEAE**
- Leaves all alternate 23

23. Leaves palmately lobed; flowers completely enclosed in an urn-shaped receptacle . . . . **XXV. MORACEAE**
- Leaves entire; flowers not completely enclosed 24

24.	Flowers hermaphrodite, solitary or in clusters	**LXXV. MYRTACEAE** (*Eucalyptus*)
-	Flowers in catkins	25
25.	Plants with brownish glands, especially on young twigs and leaves; fruit a fleshy drupe	**XXIV. MYRICACEAE**
-	Plants lacking glands; fruit a dry capsule	**XXIII. SALICACEAE**
26.	Flowers in simple or compound umbels, the flower stalks all arising from the same point	27
-	Flowers variously arranged but not in umbels	30
27.	Woody climbers	**LXXVII. ARALIACEAE**
-	Herbs or small shrubs	28
28.	Perianth-segments 6; leaves entire	**CX. LILIACEAE**
-	Perianth-segments 4-5; leaves usually divided	29
29.	Petals 4	**XL. PAPAVERACEAE** (*Chelidonium*)
-	Petals 5	**LXXVIII. UMBELLIFERAE**
30.	Silvery or brownish, thinly membranous sheaths (*ochreae*) present at nodes of stems	**XXVIII. POLYGONACEAE**
-	Membranous sheaths absent	31
31.	Leaves compound	32
-	Leaves simple, though sometimes very deeply divided	33
32.	Leaves pinnate	**XLVII. ROSACEAE**
-	Leaves 2-pinnate	**XLVIII. LEGUMINOSAE (MIMOSOIDEAE)**
33.	Evergreen trees with large, leathery leaves	34
-	Herbs or shrubs, rarely trees and then deciduous	35
34.	Leaves in rosettes at tips of otherwise naked branches	**CIX. AGAVACEAE** (*Dracaena*)
-	Leaves arranged along length of branches	**XXXIX. LAURACEAE**
35.	Flowers borne in centre or along edges of leaf-like cladodes	**CX. LILIACAEAE** (*Ruscus, Semele*)
-	Flowers not borne on leaf-like cladodes	36
36.	Perianth sepaloid	37
-	Perianth petaloid	50
37.	Leaves opposite	38
-	Leaves alternate or basal	41
38.	Perianth-segments in 1 whorl of 3	**LIV. EUPHORBIACEAE**
-	Perianth-segments in 1 or more whorls of 4 or 5	39
39.	Trees; leaves palmately lobed	**LVIII. ACERACEAE**
-	Herbs; leaves not lobed	40

40. Plants with stinging hairs; perianth-segments 4; leaves ovate-lanceolate or broader . . . . **XXIV. URTICACEAE**
- Plants without stinging hairs; perianth-segments 5, or if 4 than leaves very narrow . . . . **XXXVI. CARYOPHYLLACEAE**

41. Plants climbing . . . . **CXIII. DIOSCORIACEAE**
- Plants not climbing 42

42. Leaves narrow, up to 5(-10) mm wide, cylindrical or flat 43
- Leaves broader, generally more than 5 mm wide, flat 44

43. Perianth-segments 6 . . . . **CXV. JUNCACEAE**
- Perianth-segments 5 . . . . **XXIX. CHENOPODIACEAE**

44. Epicalyx present . . . . **47. ROSACEAE**
- Epicalyx absent 45

45. Stipules present . . . . **LIV. EUPHORBIACEAE**
- Stipules absent 46

46. Perianth dry, membranous . . . . **XXX. AMARANTHACEAE**
- Perianth not dry, membranous 47

47. Flowers in racemes; plants not farinose . . . . **XXXII. PHYTOLLACACEAE**
- Flowers solitary, or in cymes, panicles or rarely spikes; plants often farinose . . . 48

48. Perianth-segments 4; stigma 1 . . . . **XXVI. URTICACEAE**
- Perianth-segments usually 5; stigmas 2 or more 49

49. Perianth-segments green on outside, yellowish within; fruit with 4 large tubercles or horns on top . . . . **XXXIII. AIZOACEAE** (*Tetragonium*)
- Perianth-segments concolorous; fruit an achene lacking large tubercles or horns . . . . **XXVIII. CHENOPODIACEAE**

50. Corolla strongly zygomorphic 51
- Corolla actinomorphic 54

51. Corolla with a swollen utricle at the base but lacking a spur; leaves entire . . . . **XXVII. ARISTOLOCHIACEAE**
- Corolla with a spur or conical projection at base: leaves deeply lobed or divided . 52

52. Leaves entire or slightly toothed . . . . **CIII. VALERIANCEAE**
- Leaves deeply lobed or finely divided 53

53. Leaves finely pinnately divided; flowers with a blunt, rounded spur . . . . **XL. PAPAVERACEAE** (*Fumaria*)
- Leaves palmately divided; flowers with a tapering spur . . . . **XXXVII. RANUNCULACEAE** (*Consolida, Delphinium*)

54. Plants climbing 55
- Plants not climbing 56

55.	Plants climbing by means of tendrils	**LXIX. PASSIFLORACEAE**
-	Plants climbing by means of twining stems	**XXXV. BASELLACEAE**

56.	Plants succulent	57
-	Plants not succulent	59

57. Stems ± leafless, consisting of flattened pads with cushion-like structures (*areoles*) usually bearing spines . . . . **LXXIII. CACTACEAE**
- Stems leafy, leaves sometimes with spinose margins but not as above 58

58. Leaves small (less than 10 cm); plants papillose or with crystalline hairs . . . .
. . . . **XXXIII. AIZOACEAE**
- Leaves large (up to 2 m); plants not papillose . . . . **CXI. AGAVACEAE**

59.	Shrubs	60
-	Herbs	63

60.	Plants spiny	**XXXVIII. BERBERIDACEAE**
-	Plants unarmed	61

61. Leaves large, broad; at least the outer flowers more than 5 mm across, sterile . . . . .
. . . . **XLV. HYDRANGACEAE**
- Leaves small, needle-like; all flowers less than 5 mm across, fertile 62

62.	Flowers in clusters, yellow	**LXV. THYMELEACEAE**
-	Flowers in racemes, pink	**LXXX. ERICACEAE**

63.	Leaves deeply divided	64
-	Leaves entire	65

64.	Perianth-segments 4	**XL. PAPAVERACEAE**
-	Perianth-segments 5 or more	**XXXVII. RANUNCULACEAE**

65.	Perianth-segments 5	**XXXI. NYCTAGINACEAE**
-	Perianth-segments 6	66

66.	Stamens 3	**CXIV. IRIDACEAE**
-	Stamens 6	67

67.	Ovary superior	**CX. LILIACEAE**
-	Ovary inferior	**CXII. AMARYLLIDACEAE**

68. Flowers papillionaceous, strongly zygomorphic with the adaxial petal (*standard*) outermost, the 2 lateral petals (*wings*) free and the 2 lower petals (*keel*) innermost and partly adhering to each other, concealing the stamens and ovary; leaves usually pinnately divided . . .
. . . . **XLVIII. LEGUMINOSAE (LOTOIDEAE)**
- Flowers not papillionaceous; leaves entire or divided 69

69.	Petals all joined above the base to form at least a short tube	70
-	Petals free or united only at the base	108

70.	Corolla at least slightly zygomorphic, often 1- or 2-lipped	71
-	Corolla actinomorphic	87

71.	Plants lacking chlorophyll; leaves reduced to scales	C. OROBANCHACEAE
-	Plants with chlorophyll; leaves not scale-like	72

72.	Plants climbing	73
-	Plants not climbing	74

73.	Plants climbing by means of twining stems; leaves simple	CII. CAPRIFOLIACEAE
-	Plants climbing by means of tendrils; leaves pinnate	XCVIII. BIGNONIACEAE

74.	Flowers with a backward or downward projecting spur	75
-	Flowers lacking a spur	77

75.	Leaves finely pinnately divided; flowers laterally compressed .... XL. PAPAVERACEAE (*Fumaria*)	
-	Leaves entire; flowers not laterally compressed	76

76.	Sepals 3; spur formed by the lowermost, petaloid sepal	LX. BALSAMINACEAE
-	Sepals 5; spur formed by the lowermost petal	XCVI. SCROPHULARIACEAE

77.	Flowers 3-merous	78
-	Flowers 4- or 5-merous	79

78.	Sepals free	CXXII. CANNACEAE
-	Sepals connate into a tube	CXXI. ZINGIBERACEAE

79.	Sepals 2, large, petaloid, ± concealing corolla	XXVIII. POLYGONACEAE
-	Sepals 5, sepaloid, not concealing corolla	80

80.	Ovary deeply 4-lobed; fruit consisting of 4 small nutlets	81
-	Ovary not deeply lobed; fruit a capsule or berry-like	83

81.	Leaves alternate	XCI. BORAGINACEAE
-	Leaves opposite	82

82.	Style arising from base of ovary, between the 4 lobes	XCIV. LABIATAE
-	Style arising from top of ovary	XCII. VERBENACEAE

83.	Shrubs, stems often with conspicuous prickles	XCIV. VERBENACEAE (*Lantana*)
-	Herbs, stems unarmed	84

84.	Bracts spinose-dentate; ovules and seeds 4	XCIX. ACANTHACEAE
-	Bracts unarmed; ovules and seeds numerous	85

85.	Ovary inferior	CV. CAMPANULACEAE (*Lobelia*)
-	Ovary superior	86

86.	Calyx campanulate, enlarging in fruit	XCV. SOLANACEAE
-	Calyx usually not campanulate, not enlarging in fruit .... XCVI. SCROPHULARIACEAE	

KEY TO FAMILIES   19

87. Leaves 2-pinnate; flowers reduced to small florets borne in globose heads . . . .
    . . . . **XLVIII. LEGUMINOSAE (MIMOSOIDEAE)**
-   Leaves simple or , at most, 1-pinnate; flowers not as above 88

88. Corolla 4-lobed 89
-   Corolla 5-lobed 93

89. Leaves compound, with 3 or more leaflets . . . . **LXXXV. OLEACEAE**
-   Leaves simple 90

90. Evergreen trees . . . . **LXI. AQUIFOLIACEAE**
-   Herbs or small shrubs 91

91. Calyx an inconspicuous, toothed rim; stamens 3 . . . . . **CIII. VALERIANACEAE**
-   Calyx larger, obvious; stamens 4-many 92

92. Flowers large, more than 2 cm long, pendent in lax cymes . . . .
    . . . . **XLIII. CRASSULACEAE**
-   Flowers small, less than 1 cm long, in long or short, capitate spikes . . . .
    . . . . **CI. PLANTAGINACEAE**

93. Shrubs or small trees 94
-   Herbs 97

94. Flowers with a corona of 5 free segments . . . . **LXXXVIII. ASCLEPIADACEAE**
-   Corona absent 95

95. Stamens alternating with petaloid staminodes . . . . **LXXXIV. SAPOTACEAE**
-   Staminodes absent 96

96. Stamens equal in number to corolla-lobes . . . . **XCV. SOLANACEAE**
-   Stamens twice as many as corolla-lobes . . . . **XXXVIII. ERICACEAE**

97. Leaves 3-foliolate 98
-   Leaves simple, though sometimes deeply divided 99

98. Leaves 3-foliolate, leaflets radiating from tip of petiole . **XLIX. OXALIDACEAE**
-   Leaflets 3 or more, pinnately arranged . . . . **XCV. SOLANACEAE**

99. Leaves opposite 100
-   Leaves alternate 103

100. Leaves peltate, plants succulent . . . . **XLIII. CRASSULACEAE** (*Umbilicus*)
-    Leaves not peltate, plants not succulent 101

101. Evergreen perennials . . . . **LXXXVII. APOCYNACEAE**
-    Annuals or deciduous perennials 102

102. Stamens opposite corolla-lobes . . . . **LXXXII. PRIMULACEAE**
-    Stamens alternating with corolla-lobes . . . . **LXXXVI. GENTIANACEAE**

103. Ovary deeply 4-lobed; fruit consisting of 4 small nutlets . **XCI. BORAGINACEAE**
- Ovary not deeply lobed; fruit a capsule or berry 104

104. Stems creeping or climbing . . . . **XC. CONVOLVULACEAE**
- Stems ascending or erect, not creeping or climbing 105

105. Fertile stamens alternating with sterile staminodes . . . **LXXXII. PRIMULACEAE**
- All stamens fertile 106

106. Styles 5; papery, erect in 1-sided spikes . . . . . . **LXXXIII. PLUMBAGINACEAE**
- Styles 1-2; flowers not papery, not in 1-sided spikes 107

107. Ovary superior . . . . **XCV. SOLANACEAE**
- Ovary inferior . . . . **CV. CAMPANULACEAE**

108. Petals 3, 4, 6 or more 109
- Petals 5 136

109. Petals 3, 6 or more 110
- Petals 4 122

110. Flowers zygomorphic 111
- Flowers actinomorphic 113

111. Flowers in cymes subtended by a bract-like spathe . . **CXVI. COMMELINACEAE**
- Flowers in racemes or spikes 112

112. Upper and lateral petals divided into 3-numerous lobes . . . **XLII. RESEDACEAE**
- Upper and lateral petals not divided . . . . **CXXIII. ORCHIDACEAE**

113. Spiny shrubs . . . . **XXXVIII. BERBERIDACEAE**
- Unarmed herbs 114

114. Flowers more than 5 cm across; styles petaloid . . . . . **CXIV. IRIDACEAE** (*Iris*)
- Flowers less than 5 cm across; styles not petaloid 115

115. Petals 6 116
- Petals 3 119

116. Plants succulent 117
- Plants not or scarcely succulent 118

117. Epicalyx present . . . . **LXXIV. LYTHRACEAE**
- Epicalyx absent . . . . **XLII. RESEDACEAE**

118. Sepals 2 . . . . **XXXIV. PORTULACACEAE**
- Sepals 4-6 or more . . . . **XLIII. CRASSULACEAE**

119. Leaves opposite 120
- Leaves alternate or basal 121

120.	Petals equal	**XLIII. CRASSULACEAE**
-	Petals unequal	**LLXXXII. PRIMULACEAE** (*Pelletiera*)

121.	Flowers solitary or in few-flowered cymes	**CXXII. COMMELINACEAE**
-	Flowers in much-branched racemes with whorled branches	**CVII. ALISMATACEAE**

122. Flower with a slender hypanthial tube or inferior ovary (0.5-)1 cm or more long and which is often coloured . . . . **LXXVI. ONAGRACEAE**
- Flowers not as above 123

123. Leaves opposite or whorled 124
- Leaves alternate or basal 129

124. Shrubs or trees 125
- Herbs 126

125. Flowers white . . . . **LXXV. MYRTACEAE**
- Flowers pink, reddish or purplish . . . . **LXXVI. ONAGRACEAE**

126. Leaves in whorls of 3 or more . . . . **LXXXIX. RUBIACEAE**
- Leaves opposite 127

127. Sepals fused to form a tubular calyx . . . . **LXXII. FRANKENIACEAE**
- Sepals free 128

128. Stamens twice as many as petals; plants glabrous . . . . **LIII. LINACEAE**
- Stamens equalling or fewer than petals, or if as many then plants hairy . . . .
. . . . **XXXVI. CARYOPHYLLACEAE**

129. Leaves with tendrils . . . . **LIX. SAPINDACEAE**
- Leaves lacking tendrils 130

130. Shrubs 131
- Herbs 133

131. Leaves less than 15 mm; calyx longer than corolla . . . . **LXV. THYMELIACEAE**
- Leaves more than 15 mm; calyx shorter than corolla 132

132. Plants glandular-punctate; petals hooded . . . . **LV. RUTACEAE**
- Plants not glandular-punctate; petals ± flat . . . .
. . . . **XLI. CRUCIFERAE** (*Crambe, Sinapidendron*)

133. Sepals 2 134
- Sepals 4 135

134. Leaves entire . . . . **XXXIV. PORTULACACEAE**
- Leaves lobed or divided . . . . **XL. PAPAVERACEAE**

135. Petals entire or emarginate; stamens 6 . . . . **XLI. CRUCIFERAE**
- Petals divided into numerous lobes; stamens 7-nmumerous . **XLII. RESEDACEAE**

136. Flowers in simple or compound umbels, the flower stalks all arising from the same point; leaves usually pinnately divided 137
- Flowers not in umbels; leaves entire or divided 138

137. Ovary inferior .... LXXVIII. UMBELLIFERAE
- Ovary superior .... L. GERANIACEAE

138. Flowers with a spur 139
- Flowers lacking a spur 141

139. Sepals prolonged backwards into appendages .... LXVIII. VIOLACEAE
- Sepals lacking appendages 140

140. Leaves peltate .... LI. TROPAEOLACEAE
- Leaves not peltate .... L. GERANIACEAE

141. Plants succulent 142
- Plants not or scarcely succulent 143

142. Sepals 2 .... XXXIV. PORTULACACEAE
- Sepals 5 .... XLIII. CRASSULACEAE (*Sedum*)

143. Leaves in whorls of 3 or more 144
- Leaves not whorled 145

144. Fruit a capsule .... XXXVI. CARYOPHYLLACEAE
- Fruit of 2 appressed mericarps or berry-like .... LXXXIX. RUBIACEAE

145. Leaves compound 146
- Leaves simple though sometimes deeply lobed 150

146. Leaves opposite 147
- Leaves alternate 148

147. Leaflerts 3 or more, inlforescence amny-flowered .... CII. CAPRIFOLIACEAE
- Leaflets 2; flowers solitary .... LII. ZYGOPHYLLACEAE

148. Stamens 5 .... LVII. ANACARDIACEAE
- Stamens 9 or more 149

149. Flowers slightly zygomorphic; stamens 9-10 ....
.... XLVIII. LEGUMINOSAE (CAESALPINIOIDEAE)
- Flowers actinomorphic; stamens (10-)numerous .... LXVII. ROSACEAE

150. Leaves opposite 151
- Leaves alternate 157

151. Trees or shrubs 152
- Herbs 155

152. Stamens grouped into 3-5 discrete bundles .... LXVII. GUTTIFERAE
- Stamens 5-many, not grouped into discrete bundles 153

153.	Leaves with at least some stellate hairs	**LXX. CISTACEAE**
-	Stellate hairs absent	154

154.	Stipules present; stamens 5	**LXIII. RHAMNACEAE**
-	Stipules absent; stamens numerous	**LXXV. MYRTACEAE**

155. Petals unequal, each divided into numerous, narrow lobes; carpels free, stellately spreading in fruit . . . . **XLII. RESEDACEAE**
- Petals equal, entire or divided into only 2 lobes; fruit a capsule 156

156.	Stamens 6, in 2 whorls; style 1	**LXXII. FRANKENIACEAE**
-	Stamens 8-10 in 1 whorl; styles 2-5	**XXXVI. CARYOPHYLLACEAE**

157.	Trees or shrubs	158
-	Herbs	165

158.	Leaves needle- or scale-like	159
-	Leaves not needle- or scale-like	160

159. Leaves needle-like; stamens numerous; capsule woody, dehiscing by 5 valves . . . . **LXXV. MYRTACEAE**
- Leaves scale-like; stamens 5; capsule dehiscing by 2 valves . . . . **LXXI. TAMARICACEAE**

160.	Fruit a dehiscent capsule	161
-	Fruit a ± fleshy pome or drupe, or indehiscent	162

161.	Flowers in 1-sided racemes; stamens 10	**LXXIX. CLETHRACEAE**
-	Flowers solitary or few in clusters; stamens 5	**XLVI. PITTOSPORACEAE**

162.	Stamens 5	163
-	Stamens numerous	164

163. Flowers usually in axillary fascicles; stamens attached at base of petals . . . . **LXXXI. MYRSINACEAE**
- Flowers usually solitary; stamens alternating with petals . **LXII. CELASTRACEAE**

164.	Stipules absent; seeds *c.* 6	**LXVI. THEACEAE**
-	Stipules present (though sometimes falling early); seed 1(-2) .	**XLVII. ROSACEAE**

165.	Stamens 2-10	166
-	Stamens numerous	169

166.	Flowers white	167
-	Flowers yellow, blue, pink or purplish	168

167. Leaves lobed, or orbicular with sinuate margins; fruit a capsule . . . . **XLIV. SAXIFRAGACEAE**
- Leaves linear to oblanceolate, entire; fruit a 1-seeded achene enclosed in the persistent calyx . . . . **XXXVI. CARYOPHYLLACEAE**

168. Leaves palmately lobed; flowers pinkish or purple .... **L. GERANIACEAE**
- Leaves entire; flowers yellow or blue .... **LIII. LINACEAE**

169. Stamens united into a tube surrounding the style; epicalyx usually present; fruit a schizocarp .... **LXIV. MALVACEAE**
- Stamens free; epicalyx absent; fruit a head of achenes or an inflated capsule .... **XXXVII. RANUNCULACEAE**

PTERIDOPHYTA

M. Gibby & A.M. Paul

Literature: I. Manton, J.D. Lovis & M. Gibby, Cytology of the fern flora of Madeira, in *Bull. Br. Mus. nat. Hist.* (Bot.) **15**(2): 123-161 (1986).

1. Stems jointed; leaves united in whorls to form a sheath at nodes . . **III.1. Equisetum**
- Not as above 2

2. Plants with aerial stems and branches bearing numerous leaves less than 1 cm, without stipe 3
- Plants with erect or creeping rhizomes bearing larger leaves with distinct stipe and lamina 6

3. Stems slender; leaves thinly herbaceous; heterosporous . . . . **II.1. Selaginella**
- Stems robust; leaves coriaceous; homosporous 4

4. Stems erect, branching dichotomously; fertile leaves in zones along stem . . . . **I.3. Huperzia**
- Stems creeping, with short lateral branches; fertile leaves in terminal strobili 5

5. Leaves appearing opposite and decussate; strobili usually stalked, often borne in pairs . . . . **I.2. Diphasiastrum**
- Leaves spirally arranged; strobili sessile, borne singly . . . . **I.1. Lycopodiella**

6. Fronds not spirally coiled when young, fleshy; sporangia large and thick-walled, borne on a special structure arising from lamina base 7
- Fronds spirally coiled when young, membranous, herbaceous or coriaceous; sporangia with walls only 1 cell thick 8

7. Sterile lamina and fertile spike both simple . . . . **IV.1. Ophioglossum**
- Sterile lamina pinnate, fertile spike compound . . . . **IV.2. Botrychium**

8. Terminal pinnae bearing dense clusters of large, globular sporangia replacing the lamina; fronds large, 2-pinnate with linear-oblong pinnae and pinnules . . . . . **V.1. Osmunda**
- Sporangia borne on underside or margins of lamina; fronds various 9

9. Rhizome long-creeping, wiry; lamina membranous, translucent; sporangia in marginal pockets 10
- Rhizome creeping or erect; lamina herbaceous to coriaceous; sporangia on underside of lamina or marginal 11

10. Fronds small, less than 10 cm; lamina oblong to lanceolate, pinnate; sporangia borne in orbicular, 2-valved pockets . . . . **VIII.1. Hymenophyllum**
- Fronds larger, up to 40 cm; lamina triangular to ovate, 2- to 3-pinnate; receptacle bearing sporangia persistent and eventually protruding, bristle-like, from flask-shaped pocket . . . . **VIII.2. Trichomanes**

11. Fronds simple and entire 12
- Fronds divided 15

12. Lamina reniform . . . . **VI.5. Adiantum**
- Lamina linear or lanceolate 13

13. Lamina linear; fronds up to 9 × 0.25 cm . . . . **XIII.1. Asplenium**
- Lamina lanceolate; fronds up to 45 cm 14

14. Upper surface of lamina glossy, mid-green and glabrous; fronds not dimorphic; sporangia in paired, linear sori . . . . **XIII.3. Phyllitis**
- Both surfaces of lamina matt and densely scaly; fronds dimorphic; sporangia covering entire lower surface of lamina . . . . **XVI.1. Elaphoglossum**

15. Fronds palmately lobed or irregularly forked . . . . **XIII.1. Asplenium**
- Fronds pinnatisect or more divided 16

16. Fronds pinnatisect to pinnate, pinnae simple 17
- Fronds more divided 24

17. Lamina densely scaly beneath . . . . **XIII.2. Ceterach**
- Lamina glabrous or sparsely scaly 18

18. Plants usually epiphytic; rhizomes long-creeping; stipes articulated to rhizome; sori ± round, without indusium . . . . **IX.1. Polypodium**
- Plants usually terrestrial; rhizomes erect or short-creeping; stipes not articulated to rhizome; sori various, with indusium 19

19. Fronds oblanceolate, widest towards apex, with a very long apical pinna; margins of fertile segments deflexed and modified to form an indusium . . . . **VII.1. Pteris**
- Not this combination of characters 20

20. Fronds dimorphic; sori oblong to linear, parallel to and opening towards midrib . . 21
- Fronds not dimorphic; sori ± round, or if oblong then not parallel to midrib 22

21. Pinnae entire or crenate; sori linear, continuous, indusium attached at pinna margin . . . . **XVIII.1. Blechnum**
- Pinnae finely serrate; sori oblong, not continuous, not marginal . . . **XVIII.2. Doodia**

22. Indusium elliptical to linear; scales of lattice appearance (*clathrate*); stipe and at least lower part of rachis shiny dark brown . . . . **XIII.1. Asplenium**
- Indusium peltate; scales not clathrate; stipe and rachis green, pale or mid brown . . 23

23. Pinnae narrow, auricled; sori in 1 row on each side of midrib . . **XV.1. Polystichum**
- Pinnae ovate-acuminate, not auricled; sori scattered over lower surface of lamina . . . . **XV.2. Cyrtomium**

24. Ultimate segments stalked, ± wedge-shaped; margin of fertile segments deflexed and modified to form an indusium . . . . **VI.5. Adiantum**
- Not as above; if margin of fertile segments deflexed and modified to form an indusium then ultimate segments usually sessile and not wedge-shaped 25

25. Lamina thinly herbaceous, 2-pinnate-pinnatisect, juvenile fronds less divided, ultimate segments cuneate; rhizome rudimentary; sori spread along veins, lacking indusium .... **VI.4. Anogramma**
- Not this combination of characters 26

26. Lamina obscured beneath by dense covering of light brown to colourless scales .... **VI.2. Notholaena**
- Not as above 27

27. Fronds covered with a woolly indumentum of white to brown, multicellular hairs .... **VI.3. Cosentinia**
- Not as above 28

28. Stipe and at least lower rachis dark red-brown to black, usually shining 29
- Not as above 30

29. Scales of lattice appearance (*clathrate*); sori linear, borne on veins **XIII.1. Asplenium**
- Scales not clathrate; sori marginal, indusium comprising deflexed, modified margin of lamina .... **VI.1. Cheilanthes**

30. Fronds 1-pinnate, pinnae lobed to pinnatisect; abundant unicellular hairs on both surfaces of lamina or mainly on lower surface of costae and then yellow glands abundant .. 31
- Fronds variously divided; unicellular hairs and yellow glands absent 33

31. Lamina ovate-lanceolate; most pinnae adnate to rachis and deeply lobed, ± entire towards frond apex .... **XII.2. Stegnogramma**
- Lamina lanceolate; pinnae lobed at least half-way to costa, sessile but not adnate . 32

32. Fronds gradually narrowed to base, with many spherical yellow glands beneath, smelling of lemon, unicellular hairs mainly on underside of costae; sori at sides of pinna segments; indusium inconspicuous .... **XII.1. Oreopteris**
- Fronds abruptly narrowed to base with the lowest 1-2 pairs of pinnae very reduced; lamina with abundant unicellular hairs on both surfaces, lacking yellow glands; sori mid-way between mid-vein and margin of pinna segments; indusium reniform, hairy . **XII.3. Christella**

33. Sori pouch-shaped; rhizome long-creeping, densely scaly; lamina lacking scales and hairs, 3- to 4-pinnate .... **XVII.1. Davallia**
- Not this combination of characters 34

34. Fronds large (more than 1 m); sori marginal, ± spherical with 2-valved indusium; stipe ± lacking scales and hairs but bases densely clothed with long brown hairs 35
- Not this combination of characters 36

35. Arborescent, erect stem up to 1 m high; sori 1-2 mm .... **X.2. Dicksonia**
- Rhizome prostrate; sori 2-4 mm .... **X.1. Culcita**

36. Fronds large, up to 3 m; veins anastomosing; proliferous buds often present towards frond apex; sori oblong, in 2 rows, 1 on each side of and adjacent to mid-vein .... **XVIII.3. Woodwardia**
- Not this combination of characters 37

37. Lamina 2-pinnate to 3-pinnate-pinnatifid; scales lacking; lamina, especially below, with many slender, colourless and scattered reddish hairs; sori when present marginal and continuous, protected by inrolled modified lamina margin . . . . **XI.1. Pteridium**
- Not this combination of characters; lacking hairs 38

38. Ultimate segments long, tapering, at least twice as long as broad; sori marginal, oblong to linear, protected by deflexed modified lamina margin . . . . **VII.1. Pteris**
- Not as above 39

39. Fronds thinly herbaceous; stipe brittle, ± glabrous even at base; sori orbicular with hood-like indusium . . . . **XIV.4. Cystopteris**
- Fronds herbaceous to coriaceous; stipe scaly, at least at base; sori various, but if orbicular then indusium reniform, peltate or lacking 40

40. Lamina 3- to 5-pinnate; lowest pinnae asymmetrical with longer and more developed basiscopic pinnules; sori orbicular 41
- Lamina pinnate-pinnatifid to 2-pinnate-pinnatisect; lowest pinnae ± symmetrical; sori orbicular, oblong or J-shaped 42

41. Rhizome erect; lamina herbaceous, narrowly triangular to triangular-ovate; indusium less than 1 mm diameter . . . . **XV.4. Dryopteris**
- Rhizome creeping; lamina subcoriaceous to coriaceous, broadly deltoid to pentagonal; indusium more than 1 mm diameter . . . . **XV.3. Arachniodes**

42. Rhizome erect; sori orbicular; stipe with 5-7 vascular strands; rachis scaly 43
- Rhizome creeping, upright or ascending; sori oblong or J-shaped; 2 vascular strands at base of stipe uniting to form 1 above; rachis glabrous or bearing occasional scales 44

43. Indusium reniform . . . . **XV.4. Dryopteris**
- Indusium peltate or lacking . . . . **XV.1. Polystichum**

44. Rhizome upright or ascending, fronds borne in a shuttlecock; lamina lanceolate; sori J-shaped . . . . **XIV.1. Athyrium**
- Rhizome creeping, fronds borne singly; lamina triangular-ovate; sori oblong . . . . **XIV.2. Diplazium**

I. LYCOPODIACEAE

1. Lycopodiella Holub

Stems creeping, the main stems subterranean or not, rooting at intervals, the variously branched shoots horizontal, erect, or both. Leaves simple, overlapping, subulate, curved upwards and arranged spirally. Fertile leaves (*sporophylls*) arranged in terminal, erect to pendent, sessile strobili, bearing sessile sporangia at bases; homosporous, spores subtetrahedral.

1. L. cernua (L.) Pic. Serm. in *Webbia* **23**: 166 (1968).
Lycopodium cernuum L. [basion.]; *Lycopodiella veigae* (Vasc.) A. Hansen & Sunding; *Palhinhaea cernua* (L.) Franco & Vasc.; *P. veigae* Vasc.
Main stem horizontal, woody, with erect, much-branched, tree-like shoots up to 35 cm. Leaves up to 3×0.4 mm, entire, ± circular in section. Strobili numerous, solitary, terminating ultimate branchlets, to 14 mm long, ultimately pendulous, with appressed sporophylls; sporophylls rhomboid,

acuminate, with lacerate-fimbriate margins.
Recorded only once in Madeira, locality unknown. **M**

2. Diphasiastrum Holub

Stems long, creeping, rooting at intervals, with ascending or erect, much-branched lateral branches, often in fan formation (*flabellately-branched*). Leaves simple, thick and scale-like, spirally arranged but appearing opposite and decussate on main branches, in 4 ranks, ± dimorphic, lateral leaves keeled. Strobili terminal, usually stalked, borne singly or in pairs; sporophylls not resembling vegetative leaves, bearing sporangia at bases; homosporous, spores subtetrahedral.

1. D. madeirense (J.H. Wilce) Holub in *Preslia* **47**: 108 (1975).
Lycopodium madeirense J.H. Wilce [basion.]; *Diphasium madeirense* (J.H. Wilce) Rothm.; *Lycopodium complanatum* sensu auct. mad., non L. (1753)
Erect shoots 3-5 × branched and up to 20(-40) cm high. Sterile branches distinctly flattened, up to 4(-5) mm wide (including leaves). Leaves narrowly lanceolate to subulate with acute to rounded apices and decurrent bases. Fertile branches ± terete, each bearing 2-5 strobili, on slender, often 1- to 3-forked peduncles up to 6 cm long with sparse leaves; sporophylls with decurrent bases, broadly ovate, narrowing abruptly to a long, gradually acuminate apex, margins ± scarious; sporangia reniform, spores trilete. **Plate 1**
Extremely rare on heathland over 1000 m. Known from a single area in central Madeira. ■ **M** [*Azores*]

Records of *D. tristachyum* (Pursh) Holub have not been verified.

3. Huperzia Bernh. in Schrad.

Stems erect, rooting at base, branching dichotomously. Leaves simple, spirally arranged but appearing in many rows on the stem, overlapping, spreading, linear to ovate-lanceolate. Fertile leaves (*sporophylls*) similar to sterile leaves, not in terminal strobili but in zones along the branches. Sporangia shortly stalked, in axils of sporophylls, reniform; homosporous, spores subtetrahedral.

1. Leaf margin entire or, rarely, with few inconspicuous teeth; gemmae usually present in leaf axils . . . . **1. suberecta**
- Leaf margin distinctly toothed; gemmae absent . . . . **2. dentata**

1. H. suberecta (Lowe) Tardieu in *Adansonia* II, **10**: 20 (1970).
Lycopodium suberectum Lowe [basion.]; *Huperzia selago* sensu auct. mad., non (L.) Bernh. ex Schrank & Mart.(1829); *H. selago* (L.) Bernh. ex Schrank & Mart. subsp. *suberecta* (Lowe) Franco & Vasc. *comb. illegit.*; *Lycopodium selago* sensu auct. mad., non L. (1753)
Stems stout, up to 41 cm. Leaves up to 9 × 2 mm, linear- to ovate-lanceolate, acute, entire, rarely with few inconspicuous teeth. Gemmae usually present in axils of some leaves. **Plate 1**
Rather common in upland regions of the northern half of Madeira over c. 500 m in damp, shady, rocky places in wooded slopes, gullies and banks above levadas. **M**

2. H. dentata (Herter) Holub in *Folia geobot. phytotax.* **20**: 72 (1985).
Lycopodium dentatum Herter [basion.]; *Huperzia selago* subsp. *dentata* (Herter) Valentine; *Lycopodium selago* subsp. *suberecta* sensu Romariz
Like **1** but leaves up to 1.2 mm wide, linear-lanceolate, margins with numerous long fine teeth; gemmae absent. **Plate 1**
In similar habitats to 1, and often growing with it, but possibly less common. ■ **M** [*Azores*]

The name *Huperzia suberecta* has often been misapplied to the taxon *H. dentata*. The holotype of *H. suberecta*, (Lowe, s.n., Madeira) does not, however, have dentate leaves.

II. SELAGINELLACEAE

1. Selaginella P. Beauv.

Stems creeping, irregularly branched, producing roots at branching points. Leaves thinly herbaceous, simple, minutely toothed, with a ligule on the adaxial surface, spirally arranged but those on branches appearing in four rows, dorsiventral and dimorphic. Sporangia solitary on bases of sporophylls arranged in sessile terminal strobili; heterosporous, megasporangia usually at base of strobilus, microsporangia towards apex.

1. Roots arising on same side of stem as axillary leaves; lateral leaves up to 2.5 mm long, less than twice as long as broad; sporophylls dimorphic, microspores orange . . . . **1. denticulata**
- Roots arising on opposite side of stem to axillary leaves; lateral leaves up to 3(-4) mm long, 2-3 × as long as broad; sporophylls not dimorphic, microspores pale yellow . . . . **2. kraussiana**

1. S. denticulata (L.) Spring in *Flora, Jena* **21**: 149 (1838).
Lycopodium denticulatum L. [basion.]
Stems up to 25 cm, no swelling at branching points, roots arising on the same side of stem as axillary leaves, green, often tinged with red especially in drier situations. Leaves usually ± contiguous even at base of main stem. Lateral leaves up to 2.5(-3) mm, less than twice as long as broad, slightly asymmetrical, broadly elliptical to almost round, apiculate. Strobili occasionally branched, not clearly differentiated from sterile part of stem; megasporangia and microsporangia in ± equal proportions, both borne on both types of leaves; microspores orange.
Very common throughout Madeira, especially in the mountains of the interior, on banks, levada walls and damp rock faces; also in Porto Santo and on Deserta Grande. **MDP**

2. S. kraussiana (Kunze) A. Braun, *Ind. Sem. Horti. Berol.* **1859** (App.): 22 (1860).
Lycopodium kraussianum Kunze [basion.]
Stems commonly up to 32 cm, occasionally much longer, slight swelling at branching points, roots arising on opposite side of stem to axillary leaves, green. Leaves often widely spaced on main stems especially towards base. Lateral leaves spreading, up to 3(-4) mm, 2-3 × as long as broad, asymmetrical, acute. Strobili clearly distinguished from sterile part of stem, with small, narrow sporophylls, lowest 1 or 2 somewhat larger and bearing megasporangia, those above bearing microsporangia; microspores pale yellow.
Widespread and locally abundant throughout Madeira in damp woods, on banks beside levadas and damp roadsides. Possibly native, perhaps originating as an escape from cultivation. **M**

III. EQUISETACEAE

1. Equisetum L.

Rhizome subterranean, branched, creeping, dark brown. Aerial stems jointed and grooved. Branches usually in whorls, arising at the nodes just below the sheaths. Small, simple leaves united in whorls to form a sheath around stem. Sporangia borne on peltate sporangiophores arranged in a terminal cone; homosporous, spores green, spherical with 4 strap-like elaters.

1. E. telmateia Ehrh. in *Hannover. Mag.* **21**: 287 (1783).   [*Cavalinho, Pinheirinho*]
E. maximum sensu auct., non Lam. (1799)
Sterile and fertile stems present, ± ivory-coloured, jointed, with many inconspicuous grooves.

Sterile stems up to 75 × (1-)3-5(-7) mm, with whorls of numerous green, jointed branches at nodes. Branches up to c. 20 cm × less than 1 mm, stiff at first, becoming less rigid, usually with 4 longitudinal, bi-angular ridges bearing tiny saw-like teeth. Stem sheaths greenish white with 12-18 long teeth, dark brown with pale margins. Fertile stems up to 26 cm × 5-8 mm, unbranched, usually appearing before the sterile stems. Stem sheaths large and loose with long brown teeth. Terminal cone 4-5.6 cm.
Locally common in northern parts of Madeira up to c. 750 m, also in Porto Santo, on wet fields, roadsides, springs and stream beds. **MP**

Aberrant specimens have been found with a terminal cone on an otherwise normal vegetative shoot. Records of *E. ramosissimum* Desf., *Fl. atlant.* 2: 398 (1799) are based merely on depauperate plants of **1**; their branches show the characteristic bi-angled ridges with saw-like teeth.

IV. OPHIOGLOSSACEAE

1. Ophioglossum L. [Lingua de cobra]

Rhizome fleshy. Fronds not spirally coiled when young, consisting of sterile lamina and fertile spike; lamina simple, veins reticulate; fertile spike simple, sporangia in 2 rows, sunken, opening by a transverse slit. Spores tetrahedral.

1. Lamina 9-13 mm wide, broadly lanceolate to ovate, veins forming broad areoles with free veinlets . . . . **1. azoricum**
- Lamina less than 4 mm wide, narrowly lanceolate, veins forming narrow areoles parallel with lamina axis and without free veinlets . . . . **2. lusitanicum**

1. O. azoricum C. Presl, *Tent. pterid.* Suppl.: 49 (1845).
O. polyphyllum sensu auct. mad., non A. Braun (1844)
Rhizome subglobose. Fronds 1-3 per plant, appearing in summer (April - July), without old leaf-sheaths at base; lamina 20-48 mm long, 9-13 mm wide, often reflexed, broadly lanceolate to ovate, usually with a cuneate base and acute to obtuse apex, veins forming broad areoles, sometimes with free secondary veinlets, cells on underside of lamina short and broad, not parallel with lamina axis; fertile spike with 4-31 pairs of sporangia.
Rare, from Monte and Trapiche in the south of Madeira, and near Ponta Delgada in the north. **M**

True *O. polyphyllum* A. Braun is not recorded from Madeira but is found in the Canaries and Cape Verdes; it has persistent leaf-sheaths at base of plant, a longer rhizome and thick veins.

O. reticulatum L., *Sp. pl.* 2: 1063 (1753) has been reported from Madeira but no specimens have been seen. Plants are large, up to 25 cm, lamina broadly ovate with cordate to truncate base.

2. O. lusitanicum L., *Sp. pl.* 2: 1063 (1753).
Rhizome ± cylindrical. Fronds 1-3 per plant, appearing in winter (Nov. - Feb.); lamina up to 40 × 4 mm, reflexed or held upright, narrowly lanceolate, narrowing gradually to the base, apex slightly rounded, veins forming narrow areoles parallel with mid-vein, cells on underside of lamina narrow and parallel with long axis of lamina; fertile spike with 3-11 pairs of sporangia.
Rare, from Porto do Moniz and Ponta do Pargo in western Madeira; also recorded from Porto Santo but these records have not been confirmed. **M ?P**

2. Botrychium Sw. in Schrad.

Rhizome fleshy. Fronds not spirally coiled when young, consisting of sterile lamina and fertile spike; sterile lamina pinnate, veins dichotomous, free; fertile spike compound, sporangia large, subsessile, in 2 rows. Spores tetrahedral.

1. B. lunaria (L.) Sw. in Schrad. in *J. Bot. Göttingen* **1800**: 110 (1801).
Osmunda lunaria L. [basion.]; *Ophioglossum pennatum* Lam.
Fronds appearing in summer; lamina up to 15 cm, sessile, oblong; pinnae fan-shaped, without mid-vein, apex semicircular, irregularly dissected.
Very rare, known in Madeira from a single locality at Encumeada at c. 1000 m, under deep litter; possibly overlooked elsewhere. M

V. OSMUNDACEAE

1. Osmunda L.

Rhizome stout. Fronds large; veins free, forked. Terminal pinnae of fertile fronds bearing dense clusters of large, globular sporangia replacing the lamina. Spores green, trilete, globose.

1. O. regalis L., *Sp. pl.* **2**: 1062 (1753).
Fronds to 150 cm, 2-pinnate, dimorphic; pinnae and pinnules linear-oblong, entire; fertile fronds borne centrally. *Known for Madeira only from two old specimens (1850, 1914), for which no localities in Madeira are given. Possibly introduced.* M

VI. ADIANTACEAE

1. Cheilanthes Sw.

Rhizome short-creeping to erect, scaly. Fronds tufted; stipe rigid, dark and scaly, at least at base; lamina finely divided, 2-pinnate, basiscopic pinnules larger and more developed than acroscopic ones; ultimate segments with no obvious midrib; veins free; lamina glabrous or with sparse indumentum beneath. Margin of fertile segments deflexed and modified to form a continuous or interrupted membranous indusium; sori small, round and marginal on ends of veins; spores trilete, globose.

1. Lamina bearing scattered, red, 2- to 5-celled glandular hairs beneath, scales absent on axes .... **4. tinaei**
- Lamina lacking such glandular hairs, often with sparse narrow scales on axes beneath. 2

2. Sori well-spaced on rounded segment lobes, indusium not membranous, narrow with rounded, entire lobes .... **1. maderensis**
- Sori discontinuous to ± continuous along segment margin, indusium toothed or fimbriate 3

3. Scales mostly at base of stipe only; sori ± continuous along segment margin, indusium broad, subrectangular, somewhat toothed .... **2. guanchica**
- Scales throughout length of stipe and on rachis and costae beneath; sori discontinuous, indusium irregularly lobed with fimbriate margin .... **3. acrostica**

1. C. maderensis Lowe in *Trans. Camb. phil. Soc.* **6**: 528 (1838).
C. fragrans (L.) Sw., subsp. *maderensis* (Lowe) Benl; *C. pteridioides* (Reichard) C. Chr.
Fronds spreading, up to 15 cm, occasionally more; stipe usually shorter than lamina; rhizome

and stipe base clothed with long, narrow mid-brown scales. Lamina up to 3.5 cm wide, subtriangular to linear-lanceolate; ultimate segments of pinnules oblong to suborbicular, lobed. Lower surface of lamina bearing scattered inconspicuous, tiny, colourless to red or brown glands; stipe, rachis and costae bearing scales beneath. Sori usually well-spaced, indusium narrow, rounded, entire, with rounded lobes. $2n=60$.
Rare, on dry basalt walls above Funchal. M

2. **C. guanchica** Bolle in *Bonplandia, Hannover* **7**: 107 (1859).
C. sventenii Benl
Fronds ± erect, up to 12 cm; stipe usually shorter than lamina; rhizome and stipe base clothed with long narrow red-brown scales. Lamina up to 4.5 cm wide, triangular-ovate to ovate-lanceolate; ultimate segments of pinnules linear-oblong, lobed below, crenate above. Plant eglandular, fronds ± glabrous, sometimes with occasional narrow brown scales on stipe, rachis and costae beneath. Sori ± continuous around segments, indusium very broad, membranous, subrectangular, somewhat toothed.
Rare, on walls in the Funchal area. No recent records are known. M

Plants from Madeira are often only half the size of those known from the Canary Islands.

3. **C. acrostica** (Balb.) Tod. in *G. sci. nat. econ. Palermo* **1**: 215 (1866).
Fronds spreading, up to 9 cm; stipe slightly shorter or equal in length to lamina; rhizome and stipe base clothed with long, narrow, red-brown scales. Lamina up to 2.5 cm wide, ovate to lanceolate; pinnules oblong with suborbicular lobes. Plant eglandular, stipe, rachis and costae bearing scales beneath. Sori discontinuous, indusium broad, irregularly lobed with fimbriate margin.
Known from Madeira only from a single, old specimen (c. 1848) for which no precise locality is given. M

4. **C. tinaei** Tod. in *G. sci. nat. econ. Palermo* **1**: 217 (1866).
C. corsica Reichst. & Vida; *C. duriensis* Mendonça & Vasc.
Fronds spreading, up to 14 cm; stipe usually at least as long as lamina; rhizome and stipe base clothed with long, narrow, castaneous scales. Lamina up to 3 cm wide, ovate-lanceolate to triangular-ovate; pinnules ovate-lanceolate, sessile, decurrent, often pinnatisect with rounded, sometimes crenate lobes. Stipe and lower surface of lamina with scattered, red, 2- to 5-celled glandular hairs, scales lacking. Indusium discontinuous, narrow, with rounded lobes and a narrow, entire, membranous margin. $2n=120$.
Rare, on dry basalt walls above Funchal. M

C. pulchella Bory ex Willd., *Sp. pl.* **5**(1): 456 (1810) has been reported in the past as occurring in Madeira, possibly in error for **2**, but no definite records or specimens have been traced.

2. **Notholaena** R. Br.

Rhizome short-creeping and densely scaly. Fronds crowded; stipe rigid, dark and scaly; lamina pinnate-pinnatisect to 2-pinnate, ultimate segments with well-defined midribs; veins free; lamina densely covered beneath with scales. Margin of fertile segments slightly deflexed with a narrow, membranous border, forming a rudimentary indusium; sori not discrete, the sporangia submarginal, spreading inwards along veins; spores trilete, globose.

1. **N. marantae** (L.) Desv. in *J. Bot. Agric. Pharm. Med. Arts Paris* **1**: 92 (1813).
[*Maranta*]
Acrostichum marantae L. [basion.]; *A. subcordatum* Cav.; *Cheilanthes marantae* (L.) Domin, subsp. *subcordata* (Cav.) Benl & Poelt var. *cupripaleacea* Benl; *Paraceterach marantae* (L.) R.M. Tryon

Fronds erect, up to 45(-59) cm; stipe half to equal length of lamina; rhizome, stipe and rachis clothed with long, narrow, beige to pale brown to copper-coloured scales. Lamina linear-lanceolate to ovate-lanceolate, up to 6 cm wide, occasionally to 11.5 cm, ± 2-pinnate; pinnules mostly adnate to costae, ± rounded oblong, to 16 × 5 mm but normally half this size, often auricled at base, occasionally deeply lobed, upper surface glabrous, lower surface including costae densely covered with pale brown to rust brown or greyish scales of varying shapes, upper surface of costae with fewer scales. 2n=58. **Plate 1**
Locally common in crevices of rock faces and dry rocky slopes in open situations, mainly in coastal districts and particularly in the south-west of Madeira, but also in a few localities in central Madeira and on the north coast. **M**

All Madeiran plants are subsp. **subcordata** (Cav.) G. Kunkel in *Cuad. Bot. Canaria* **5**: 46 (1969), to which the above description and synonymy apply. Subspecies *marantae* from central Europe, the Mediterranean area, Ethiopia and across central Asia differs in being smaller, with the stipe longer than the lamina and the lamina narrower and having fewer pinnae and pinnules.

3. Cosentinia Tod.

Rhizome short-creeping to erect, densely scaly. Fronds tufted; stipe rigid, mid-brown, scaly at base, lanuginose above; lamina 2-pinnate, ultimate segments with no obvious midrib; veins free; lamina densely covered with a woolly indumentum of white to brown multicellular hairs beneath, more sparsely above. Margin of fertile segments usually deflexed, but not membranous or modified; sori not discrete, the sporangia mostly submarginal, spreading inwards along veins; spores trilete, tetrahedral.

1. C. vellea (Aiton) Tod. in *G. sci. nat. econ. Palermo* **1**: 220 (1866). [*Feto lanegero*]
Acrostichum velleum Aiton [basion.]; *Cheilanthes catanensis* (Cosent.) H.P. Fuchs; *C. vellea* (Aiton) F. Mueller; *Notholaena lanuginosa* (Desf.) Desv. ex Poir. in Lam.; *N. vellea* (Aiton) Desv.
Rhizome short-creeping. Fronds erect to spreading, up to 30 cm; stipe usually very short; rhizome and stipe base clothed with long, narrow, pale brown scales. Lamina linear-lanceolate, up to 3 cm wide; pinnae towards base of frond widely spaced and markedly reduced; pinnules up to 4 × 3 mm, rounded, often crenately lobed, lower short stalked, ± cordate, upper adnate to costa. Woolly indumentum may be lost from upper surface as frond ages. 2n=116.
Rare, now restricted to crevices in old, stone buildings in south-east Madeira. **M**

Madeiran plants are all subsp. **vellea**.

Pellaea viridis (Forssk.) Prantl in *Bot. Jb.* **3**: 420 (1882), a native of the Old World tropics, has been recorded for Madeira, but no further information is available. It has 2- or 3-pinnate, glabrous, coriaceous fronds, with sessile or shortly stalked, lanceolate to ovate pinnules, and ± continuous, marginal sori protected by reflexed, modified pinnule margins. **M**

4. Anogramma Link

Sporophyte annual, arising from tuberous gametophyte. Rhizome rudimentary. Fronds tufted, erect, thinly herbaceous, 2-pinnate-pinnatisect with decurrent pinnules, juvenile fronds less divided; veins free, forked, not reaching margin. Lamina margin flat, sori spread along veins, indusium absent. Spores tetrahedral.

1. A. leptophylla (L.) Link, *Fil. spec.*: 137 (1841).
Polypodium leptophyllum L. [basion.]; *Grammitis leptophylla* (L.) Sw.; *Gymnogramma leptophylla* (L.) Desv.
Fronds up to 22 cm. Stipe glossy red-brown; rhizome and stipe bases bearing whitish to red-brown, multicellular hairs, some arising from tiny brown scales. Lamina linear-lanceolate to ovate, up to 7 cm wide, usually much narrower; juvenile fronds (which may be fertile) ovate, much shorter with few pairs of flabellate pinnae, incised between veins, segments emarginate; successive fronds progressively larger and more dissected, with cuneate ultimate segments, emarginate, lobes rounded to acute. Lamina glabrous. Sporangia borne on veins but appearing to cover whole of ultimate segments when mature. 2n=52.
Locally common in shady, often damp places such as cracks in old walls or damp rock faces, among rocks above levadas or between the stones of terraces, mainly within 5 km of the coast in the eastern half of Madeira; also in Porto Santo and on the Desertas. **MDP**

Pityrogramma chrysophylla (Sw.) Link, *Handbuch* 3: 19 (1833), has been recorded from Câmara de Lobos in Madeira (as *P. calomelanos* (L.) Link) and may or may not be naturalized there. It is a native of tropical America; the lamina is 1- to 3-pinnatisect and yellow or white farinose beneath, with sori lacking indusia. **M**

5. Adiantum L.

Rhizome short-creeping, bearing narrow brown scales that may be of lattice appearance (*clathrate*). Stipe usually scaly only at base. Lamina simple or, usually, variously pinnately divided, ultimate segments with no obvious midrib and often unequal and undeveloped basiscopically (*dimidiate*); veins free. Margin of fertile segments deflexed and modified to form reniform, lunulate or oblong indusia, veins extending into indusia; sporangia borne on underside of indusium along and sometimes between veins; spores tetrahedral.

1. Fronds simple, reniform to orbicular . . . . **1. reniforme**
- Fronds variously branched 2

2. Fronds subpedately branched, subcoriaceous, hairy . . . . **4. hispidulum**
- Fronds flabellately branched, thin-textured, glabrous 3

3. Veins of sterile segments ending in teeth; indusium ± oblong, not around a sinus . . . .
. . . . **2. capillus-veneris**
- Veins of sterile segments ending in sinuses; indusium reniform around a sinus . . . . . .
. . . . **3. raddianum**

1. A. reniforme L., *Sp. pl.* **2**: 1094 (1753). [*Feto redondo*]
Rhizome very short-creeping; fronds numerous, tufted, up to 38 cm; rhizome and stipe base clothed with long, narrow, toothed, ± concolorous chestnut scales; stipe very much longer than lamina, shining red-black; scales becoming fewer and replaced by many long, multicellular brown hairs further up stipe. Lamina simple, subcoriaceous, up to 9.8 cm wide, reniform with cordate base to orbicular with rounded to truncate base, margin minutely scalloped, both surfaces with sparse long, pale-brown multicellular hairs, especially near junction with stipe. Indusia semi-lunulate to oblong, numerous, subcontinuous around margin. 2n=300.
Common, particularly in the north of Madeira up to c. 1000 m in rock crevices, on walls and very stony cliffs especially in the shade of overhangs above levadas, beside roads and in ravines; also on Deserta Grande. **MD**

Very small specimens from Macaronesia with stipes under 6.5 cm and laminas less than 1.9 × 2.5 cm have been called *A. reniforme* var. *pusillum* Bolle in *Z. allg. Erdk. Berl.* **14**: 300 (1863). Canary Island plants have a different chromosome number and further investigation into the *A. reniforme* complex in Africa and China is required.

2. A. capillus-veneris L., *Sp. pl.* **2**: 1096 (1753). [*Avenca, Avenca das fontes*]
Rhizome short-creeping; fronds erect to pendulous, up to 56 cm, glabrous. Rhizome and stipe base clothed with narrow, brown clathrate scales; stipe often shorter than lamina, shining red-black. Lamina triangular-ovate to ovate-lanceolate, up to 17 cm wide, small fronds pinnate, larger fronds 2- to 3-pinnate; rachis and costae red-black, shining; ultimate segments delicate, herbaceous, up to 3 cm wide on slender red-black stalks, wedge-shaped to flabellate but very variable, often dimidiate, often deeply incised, base cuneate, veins of sterile segments ending in teeth. Indusia lunulate to oblong, up to 10 per segment. $2n=60$.
Common mainly in coastal regions of Madeira in damp, shady places especially where water trickles, such as wet cliffs and cave mouths, rocks just above levadas and wet walls below terraces; also recorded from Porto Santo and Deserta Grande. **MDP**

3. A. raddianum C. Presl, *Tent. pterid.*: 158 (1836).
A. cuneatum Langsd. & Fisch.; *A. cuneipinnulum* N.C. Nair & S.R. Ghosh
Rhizome very short-creeping; fronds erect to pendulous, up to 53 cm, glabrous. Rhizome and stipe base clothed with brown semi-clathrate scales; stipe usually longer than lamina, shining red-black. Lamina triangular to ovate, up to 22 cm wide, commonly 3- to 4-(5-) pinnate; rachis and costae red-black, shining; ultimate segments delicate, herbaceous, up to *c*. 1 cm wide, on slender red-black stalks, wedge-shaped, scarcely dimidiate, often incised, base cuneate, veins of sterile segments ending in sinuses; sterile fronds, if present, with larger ultimate segments than fertile fronds. Indusia distinctly reniform around a deep sinus, up to 5 per segment.
Native to tropical America. Naturalized since 1911 and now common throughout Madeira, in damp, shady places such as wet rocks or walls beside levadas, damp roadside banks and terrace walls. **M**

4. A. hispidulum Sw. in Schrad. in *J. Bot. Göttingen* **1800**(2): 82 (1801).
Rhizome short-creeping, fronds tufted, erect, up to 53 cm. Rhizome and stipe base clothed with narrow, brown, semi-clathrate scales; stipe longer than lamina, shining red-brown; stipe and rachis (main axes) sparsely scaly with short, pale multicellular hairs and scattered, long, brown multicellular hairs. Lamina up to 18 cm wide, broadly ovate to deltate, subpedately branched with 5-9 linear pinnae; pinnae pinnate, pinnules firmly herbaceous to subcoriaceous, shortly stalked, 15 × 7 mm, ± trapeziform to lunulate, dimidiate, toothed, veins of sterile segments ending in teeth, with yellow or white hairs abundant on abaxial surface, less frequent above. Indusia reniform around sinus, pilose, up to 17 along acroscopic and outer margin of each pinnule.
Recent escape from cultivation, naturalized in a few damp, shady places in eastern Madeira. Native to Old World tropics and subtropics. **M**

VII. PTERIDACEAE

1. Pteris L.

Rhizome erect to short-creeping. Fronds tufted; lamina 1- to 2-pinnate, veins free. Sori marginal, oblong to linear; indusia comprising modified, deflexed frond margins. Spores tetrahedral.

1. Fronds pinnate, pinnae simple . . . . **1. vittata**
- Fronds more divided 2

2. Pinnae irregularly divided into linear segments . . . . **2. multifida**
-   Pinnae pinnatisect 3

3. Pinna segments contiguous, narrowly triangular; sori on pinna segment margins longer on basiscopic side than on acroscopic side . . . . **3. incompleta**
-   Pinna segments decurrent, lanceolate; sori on pinna segment margins equal in length . . . . **4. tremula**

1. P. vittata L., *Sp. pl.* **2**: 1074 (1753).
P. longifolia sensu auct. mad., non L. (1753)
Rhizome short, creeping. Fronds up to 40 cm; stipe short, less than ⅙ length of lamina, covered with hair-like, pale to mid-brown scales 3-4 mm long; lamina pinnate, markedly oblanceolate with a long terminal pinna, and lateral pinnae decreasing in size towards the frond base; pinnae linear, simple, cordate at the base, tapering towards the apex, serrate where sorus is absent. Sori continuous along all margins of upper pinnae, but not reaching to apex on lower pinnae.
Introduced from the Mediterranean region; on rocks and walls in the Funchal region. **M**

2. P. multifida Poir. in Lam., *Encycl.* **5**: 714 (1804).
P. serrulata L.f., non Forssk. (1775)
Rhizome short-creeping. Fronds up to 15 cm; stipe glabrous, equal in length to lamina; lamina pinnate, pinnae irregularly subdivided into linear segments with decurrent bases, segments of fertile fronds narrower than those of sterile fronds. Sori continuous along margins of segments excepting the apices.
Introduced; recorded only from Praia Formosa, Madeira. Native to China and Japan. **M**

3. P. incompleta Cav. in *An. Cien. nat. Madrid* **2**: 107 (1801). [*Feto de palma*]
P. arguta Aiton; *P. palustris* Poir. in Lam.; *P. serrulata* Forssk., non L.f. (1781)
Rhizome short-creeping. Fronds up to 150 cm; stipe glabrous, shiny mid-brown, equal in length to lamina; lamina mid-green, ovate, pinnate-pinnatisect, but lowest pinnae additionally pinnately divided for nearly half their length on the basiscopic side, and sometimes to a much lesser degree on the acroscopic side, and these pinnules pinnatisect; ultimate segments contiguous, narrow, triangular, toothed towards the apices. Sori along segment margins, occupying up to ¾ of length of the margin on the basiscopic side, shorter or even absent on acroscopic side.
Frequent in damp, shady localities in laurisilva in Madeira, along streams and levadas. **M**

4. P. tremula R. Br., *Prodr.* : 154 (1810).
Rhizome erect. Fronds up to 100 cm; stipe glabrous, shiny brown, equal in length to lamina; lamina light green, triangular-ovate, pinnate-pinnatisect but lowest pinnae additionally pinnately divided for nearly half their length on basiscopic side, and sometimes to a lesser degree on acroscopic side; in large specimens innermost pinnules on other lower pinnae sometimes pinnatisect; pinna segments decurrent at base, lanceolate, in fertile specimens curving towards apex of frond. Sori continuous along margins of ultimate segments except for toothed apex and decurrent base.
Introduced; recorded from Monte and Trapiche near Funchal, and occasionally beside levadas in laurisilva elsewhere in Madeira. Native to Australia, New Zealand and Fiji. **M**

VIII. HYMENOPHYLLACEAE

1. Hymenophyllum Sm.

Mat-forming perennial. Rhizomes wiry, branched. Fronds thin and translucent, pinnate, pinnae divided into segments with spinulose, serrate margins; lamina 1 cell thick except at the veins, veins free, 1 per segment. One sorus per pinna, borne on the innermost apiscopic segment, with 2 flattened indusial valves enclosing the sporangia. Spores tetrahedral, green.

A number of crucial diagnostic characters are clearly visible only with the aid of a low-power microscope.

1. Pinnae divided into 5-11 oblong segments; indusial valves toothed . . **1. tunbrigense**
- Pinnae divided into 3-6 oblanceolate segments; indusial valves with margins entire   2

2. Veins reaching the tips of ultimate segments; cells of the lamina rectangular, more than twice as long as broad and arranged in rows running obliquely to the vein; indusial valves tapering markedly at apex . . . . **2. wilsonii**
- Veins ending *c.* 4 cells short of ultimate segments; cells of lamina up to twice as long as broad and arranged haphazardly; indusial valves not tapering markedly at apex . . . . .
. . . . **3. maderense**

1. H. tunbrigense (L.) Sm. in Sowerby, *Engl. bot.* **3**: t. 162 (1794).
Trichomanes tunbrigense L. [basion.]
Fronds with a flattened appearance, lamina bluish green, up to 6 cm, broadly ovate to oblanceolate, pinnately divided; each pinna divided irregularly into 5-11 parallel-sided, oblong segments lying predominantly on the apiscopic side; segmental vein not reaching the apex; lamina cells not much longer than broad, haphazardly arranged, containing 30-40 chloroplasts per cell; chloroplasts *c.* 6-7 µm in diameter. Sori lying in the plane of the lamina; indusial valves ± round, only slightly swollen, apical margin irregularly toothed. Spores 30-35 µm in diameter. 2n=26.
Common in laurisilva and damp Erica *woodland throughout Madeira, beside levadas, on rocks or tree boles and even covering open ground.* **M**

2. H. wilsonii Hook., *Brit. fl*: 450 (1830).
H. peltatum sensu auct., non Desv. (1827); *H. unilaterale* sensu auct., non Bory (1810)
Fronds more erect and appearing less flattened than **1**, lamina dark olive green, up to 10 cm, oblong to lanceolate, pinnately divided; pinnae divided into 3-5 segments which are ± parallel-sided but tapering at each end, the vein usually reaching the apex of the segment; lamina cells more than twice as long as broad, arranged in rows lying at an angle to the vein, containing 60-80 chloroplasts per cell; chloroplasts *c.* 4 µm in diameter. Sori inserted perpendicularly to the plane of the lamina; indusial valves convex, pear-shaped, tapering towards the apex, margins entire; valves held open in ripe material. Spores 54-68 µm in diameter. 2n=36.
Less common than **1**, *recorded near Encumeada and from wooded valleys in the north-west of Madeira, on steep slopes, rocks and tree trunks, often growing above* **1** *on the tree boles.* **M**

3. H. maderense Gibby & Lovis in *Fern Gaz.* **13**: 285 (1989).
In gross morphology similar to **2** but with somewhat erect fronds, lamina up to 7 cm, dark olive-green, oblong to oblanceolate, pinnately divided, but sometimes with 1 or more lower pinna much longer and more divided than the rest giving a branched appearance to the frond; pinnae divided into 2-5 segments ± parallel-sided but tapering at each end, the vein ending *c.* 4 cells short of the segment apex; lamina cells up to twice as long as broad, arranged haphazardly, containing *c.* 80 chloroplasts per cell; chloroplasts *c.* 5 µm in diameter. Sori inserted perpendicularly to the plane of the lamina; indusial valves convex, ovoid, not tapering markedly at the apex, margins entire; valves held open in ripe material. Spores 53-57 µm in diameter. 2n=62.
Very rare endemic, known only from rocks and the side of a levada in northern Madeira. ● **M**

The hybrid **H. maderense** × **H. wilsonii** has been recorded from Madeira. It is very similar to **2** in morphology and can be distinguished only by its abortive spores, irregular meiosis, and chromosome number 2n=49. ● **M**

2. Trichomanes L.

Rhizome creeping, wiry, branched. Fronds thin, translucent; lamina 1 cell thick except at veins, veins free. Sori cylindrical, containing sporangia clustered on a bristle-like receptacle that becomes exerted from the flask-shaped indusium as the sporangia mature. Spores tetrahedral, green.

1. T. speciosum Willd., *Sp. pl.* ed. 4, **5**(1): 514 (1810). [*Feto frisado*]
T. brevisetum R. Br.; *T. radicans* sensu auct., non Sw. (1801)
Rhizomes creeping, slender, up to 4 mm in diameter, covered in fine, hair-like scales. Fronds up to 40 cm, arising singly; stipe about equal in length to lamina; lamina 2- to 3-pinnate, triangular to ovate. Sori borne on upper edges of pinnae, appearing green when young, brown after dehiscence; indurated cells of sporangia large, golden brown. $2n=144$.
Frequent in lush laurisilva in Madeira, in gullies by streams, flourishing on the forest floor or as an epiphyte. **M**

IX. POLYPODIACEAE

1. Polypodium L.

Epiphytes. Rhizome creeping, densely scaly; stipes articulated to the rhizome. Lamina pinnatisect, glabrous; veins free. Sori borne on the end of veins, round or oval, without indusia, forming a row on either side of the midrib; sterile hairs (*paraphyses*) may be present among sporangia. Spores ellipsoidal.

1. Lamina broadly triangular-ovate to oblanceolate, pinnae obviously serrate; sori with paraphyses
   .... **1. macaronesicum**
- Lamina oblanceolate to linear lanceolate, pinnae finely serrate; sori without paraphyses
   2

2. Lamina oblanceolate to lanceolate; sori (especially when immature) oval, annulus with *c*. 7-10 indurated cells, sporangium with 2-3 basal cells .... **2. interjectum**
- Lamina linear-lanceolate to lanceolate, sori round, annulus with *c*. 11-14 indurated cells, sporangium with (0-)1 basal cell .... **3. vulgare**

1. P. macaronesicum A.E. Bobrov in *Zh. russk. bot. Obschch. Akad. Nauk.* **49**: 540 (1964).
P. australe sensu auct. mad., non Fée (1852); *P. cambricum* L. subsp. *macaronesicum* (A.E. Bobrov) Fraser-Jenk. in Greuter, Burdet & G. Long; *P. serratum* sensu auct. mad., non (Willd.) Saut. (1882); *P. vulgare* sensu auct. mad., non L. (1753), subsp. *serratum* sensu auct. mad., non (Willd.) Christ (1900), var. *serratum* sensu auct. mad., non Willd. (1810)
Rhizome scales 5-14 × up to 4 mm, usually with a dark central stripe. Fronds up to 65 × 25 cm; lamina broadly triangular-ovate to oblanceolate; pinnae simply serrate. Sori, especially young ones, oval; paraphyses short with few or no branches, often sparse; indurated cells light brown in colour, *c*. 10-14 in each annulus; sporangium with 0-1(-2) basal cells. $2n=74$.
Very common throughout Madeira on rocks, cliffs, walls and trees; also in Porto Santo. **MDP**

2. P. vulgare L., *Sp. pl.* **2**: 1085 (1753).
Rhizome scales 3-6 mm long. Fronds up to 30 × 6 cm. Lamina linear-lanceolate to lanceolate; pinna margins with very small serrations. Sori round, lacking paraphyses; indurated cells reddish-brown in colour, *c*. 11-14 in each annulus; sporangium with (0-)1 basal cell. $2n=148$.
Rare in Madeira, on rocks above 850 m. **M**

3. P. interjectum Shivas in *Jour. Linn. Soc.* (Bot.) **58**: 28 (1961).
Rhizome scales up to 10 mm long. Fronds up to 40 × 10 cm; lamina oblanceolate to lanceolate;

pinnae margins with very small serrations. Sori, especially young ones, oval in shape, lacking paraphyses; indurated cells colourless or pale golden yellow, *c.* 7-10 in each annulus; sporangium with 2-3 basal cells. 2n=222.
Very rare in Madeira, known only from Pico Ruivo. M

Phlebodium aureum (L.) Sm. in *J. Bot.* **4**: 59 (1841) (*Polypodium aureum* L.) is similar to *Polypodium* but fronds up to 100 cm, glaucous, with veins anastomosing. Native of C. America, it sometimes escapes from gardens onto walls in Funchal. M

X. DICKSONIACEAE

1. Culcita C. Presl

Rhizome prostrate, densely covered with hairs. Fronds 3- to 4-pinnate, veins free, frond axes grooved adaxially. Sori marginal, 2-4 mm, ± spherical with 2-valved, cup-shaped indusium. Spores tetrahedral.

1. C. macrocarpa C. Presl, *Tent. pterid.*: 135 (1836). [*Feto abrum*]
Balantium culcita (L'Hér.) Kaulf.; *Dicksonia culcita* L'Hér.
Large plants with fronds up to 2 m. Rhizome thickly clad in long, silky, rich brown hairs. Stipe *c.* ⅓ length of lamina, stipe and rachis glabrous; lamina shiny mid-green above, coriaceous, triangular. Sori protected by recurved lobe of lamina and membranous inner indusium. 2n=136.
Rare in damp places in steep wooded valleys of north-west Madeira, except on Montado dos Pecegueiros where there is a population of about 200 plants. M

2. Dicksonia L'Hér.

Arborescent, rhizome clad in coarse bristles. Fronds large, 2- to 3-pinnate, veins free, frond axes raised adaxially. Sori marginal, 1-2 mm, globular with 2-valved, cup-shaped indusium. Spores tetrahedral.

1. D. antarctica Labill., *Nov. Holl. pl.* **2**: 100 (1806).
Tree-fern with erect stem up to 1 m in height and 25 cm diameter. Stipe short, base densely covered with shiny red-brown hairs. Fronds up to 130 cm, widest at the centre, narrowing towards the base. Sori protected by recurved lobe of lamina margin and membranous inner indusium.
Introduced; appearing naturalized in laurisilva in the mountains east of Encumeada at 1300m. Native to SE Australia. M

X1. DENNSTAEDTIACEAE

1. Pteridium Gled. ex Scop.

Rhizomes extensive, creeping underground. Fronds borne distantly; lamina 2- to 3-pinnate-pinnatifid; veins free. Sori linear, marginal, continuous along sides of ultimate segments, sporangia covered by revolute lamina margin and membranous indusium, a membranous inner indusium may also be present. Spores tetrahedral.

1. P. aquilinum (L.) Kuhn in Kerst., *Reis. Ost-Afr.* **3**(3): 11 (1879). [*Feiteiro*]
Pteris aquilinum L. [basion.]; *Asplenium aquilinum* (L.) Bernh. in Schrad.; *Polypodium austriacum* Jacq.
Fronds to 150 cm in length, 100 cm broad; stipe dark brown at base, glabrous, ¼-½ length

of lamina; lamina broadly triangular to ovate with many colourless and scattered red-brown multicellular hairs, especially below; lower pinnae broadly lanceolate, upper pinnae narrowly lanceolate, ultimate segments triangular to narrowly triangular with contiguous bases and rounded apices. Infrequently fertile; margins of fertile and sterile segments modified to form a ciliate, membranous indusium. $2n=104$.
Common in Madeira, usually in open habitats; also recorded also from Porto Santo, Deserta Grande and Selvegem Grande. **MDPS**

Microlepia platyphylla (D. Don) Sm. in *Lond. J. Bot.* **1**: 427 (1842) is a native of Asia has been recorded for Madeira, but no further information is available. Stipe glabrous; fronds 2-pinnate; pinna segments triangular, margins crenulate; sori circular, borne marginally; indusium attached basally. **M**

XII. THELYPTERIDACEAE

1. Oreopteris Holub

Rhizome erect; scales of rhizome and stipe thin, lacking hairs. Fronds 1-pinnate-pinnatisect, lanceolate, gradually narrowed to the base; pinnae deeply lobed, ± symmetrical; veins free. Unicellular, acicular hairs present on most axes; spherical glands on all parts of lower surface. Sori near margin, round; indusium ± reniform, often bearing glands. Spores ellipsoidal.

1. O. limbosperma (Bellardi ex All.) Holub in *Folia geobot. phytotax.* **4**: 46 (1969).
Polypodium limbospermum Bellardi ex All. [basion.]; *Lastrea oreopteris* (Ehrh.) Bory; *Nephrodium montanum* (J.A. Vogler) Baker; *N. oreopteris* (Ehrh.) Desv.; *Thelypteris limbosperma* (Bellardi ex All.) H.P. Fuchs; *T. oreopteris* (Ehrh.) Sloss. in Rydb.
Fronds arranged in a shuttlecock, up to 106 cm, lemon-scented when crushed. Stipe usually less than ¼ length of lamina, pale yellow-green; rhizome, stipe and rachis sparsely clothed with concolorous, glabrous, papery scales, whitish on croziers, becoming pale brown. Lamina bright yellow-green, up to 27 cm wide, pinnae lobed to 1-2 mm from costa, basal pinnae widely spaced and greatly reduced; pinna segments crowded, sometimes overlapping. Hairs sparsely covering rachis, both surfaces of costae, and scattered on lower surfaces of costules and veins; sessile, yellow glands abundant, mainly on lower surface of frond. Sori often restricted to sides of lobes; indusium inconspicuous, caducous, irregular in shape with lacerate margin, bearing spherical sessile yellow glands; sporangia glabrous. $2n=68$.
Locally common above c. 650 m in laurisilva and beside levadas and roads in wooded areas. **M**

2. Stegnogramma Blume

Rhizome short-creeping. Whole plant, including scales and sporangia hairy. Fronds 1-pinnate, ovate-lanceolate; pinnae ± entire to deeply lobed, ± symmetrical, upper pinnae adnate to rachis; veins free. Glands absent. Sori elongate along veins; indusium absent. Spores ellipsoidal.

1. S. pozoi (Lag.) K. Iwats. in *Acta phytotax. geobot. Kyoto* **19**: 124 (1963).
Hemionitis pozoi Lag. [basion.]; *Dryopteris africana* (Desv.) C. Chr.; *Gymnogramma lowei* Hook. & Grev.; *Lastrea africana* (Desv.) Copel.; *Leptogramma pilosiusculum* (Wikstr.) Alston; *L. totta* (Willd.) J. Sm.; *Thelypteris pozoi* (Lag.) C.V. Morton
Fronds ± solitary, spreading, up to 77 cm. Rhizome clothed with mid-brown scales, similar scales sparse on stipe. Stipe half to equal length of lamina, brownish green, dark brown or black towards base. Lamina up to 14 cm wide, pinnate-pinnatifid at base, pinnatisect towards apex; pinnae lobed *c.* mid-way to costa, lowest pinnae not or only slightly shorter than the rest and

not widely spaced. Whole plant covered with abundant brownish unicellular acicular hairs, giving a matt appearance. Sporangia setose. 2n=144.
Widespread and abundant in Madeira in damp, shady places, especially above 250 m on banks of streams and levadas, wooded slopes and ravines. **M**

3. Christella H. Lév.

Rhizome short-creeping. Whole plant, including scales on stipe and rhizome, hairy. Fronds 1-pinnate-pinnatifid, lanceolate, narrowed to the base; pinnae lobed, ± symmetrical; veins free except the lowermost pair of each segment. Glands absent on lamina and axes. Sori round, ± midway between costules and margins of lobes on veins; indusium reniform. Spores ellipsoidal.

1. C. dentata (Forssk.) Brownsey & Jermy in *Brit. Fern Gaz.* **10**: 338 (1973).
Polypodium dentatum Forssk. [basion.]; *Aspidium adultum* Wikstr.; *Cyclosorus dentatus* (Forssk.) Ching; *Lastrea dentata* (Forssk.) Romariz; *Nephrodium molle* (Sw.) R. Br.; *Thelypteris dentata* (Forssk.) E.P. St. John
Fronds tufted, up to 126 cm. Rhizome scales, mid-brown, long, narrow, stiff, shining, bearing short hairs. Stipe ⅐-½ length of lamina, brownish green, sparsely scaly at base. Lamina up to 29 cm wide; pinnae lobed *c.* mid-way to costa, lowest few pairs of pinnae usually very reduced and widely spaced. Whole plant covered with abundant, unicellular, colourless, acicular hairs, giving a matt appearance. Indusium bearing many unicellular, acicular hairs; sporangia glabrous except for unicellular yellow glands on sporangial stalks. 2n=144.
Widespread but not abundant throughout Madeira below c. 500 m in moderately damp, shady or exposed situations such as the foot of terrace-walls and rocky stream banks. **M**

XIII. ASPLENIACEAE

1. Asplenium L.
M. Gibby, J. Ormonde & A.M. Paul

Rhizome erect or creeping with dark, oblong-triangular to linear-lanceolate, sometimes filiform scales of lattice appearance (*clathrate*). Fronds in persistent tufts; stipe dark at least at the base, with 2 lateral, adaxial wings; lamina simple, entire to variously pinnate, often glabrous; veins free. Sori elliptical to linear, borne along one side of a vein; indusium attached to vein, usually opening towards the costal vein. Spores ± ellipsoidal.

1.	Lamina simple or forked	2
-	Lamina 1- to 4-pinnate	3
2.	Lamina palmate, subentire to lobate, cordate at the base	**10. hemionitis**
-	Lamina narrow, simple or irregularly forked	**9. septentrionale**
3.	Lamina 1-pinnate	4
-	Lamina 2- to 4-pinnate	7
4.	Rachis with membranous green wings; lamina coriaceous; scales concolorous	**8. marinum**
-	Rachis with scarious pale brown wings; lamina herbaceous; scales bicoloured	5
5.	Fertile pinnae strongly asymmetrical; usually a single sorus along the costa	**7. monanthes**
-	Fertile pinnae approximately symmetrical; usually several sori along the costa	6

6. Stipe and rachis with 3 wings, two adaxial, the third abaxial; median pinnae more than twice as long as wide .... **5. anceps**
- Stipe and rachis with only 2 lateral, adaxial wings; median pinnae usually less than twice as long as wide .... **6. trichomanes**

7. Lamina pinnate-pinnatisect to 2-pinnate with numerous small, reddish brown filiform scales .... **4. aethiopicum**
- Lamina 2- to 3-pinnate, ± glabrous or with a few hairs 8

8. Lamina ovate to lanceolate, lowest pair of pinnae shorter than adjacent pinnae; fully mature sori appearing ± round; stipe less than ⅔ length of lamina .... **3. obovatum**
- Lamina usually triangular (to triangular-ovate), the lowest pinnae usually the longest; fully mature sori appearing linear to oblong; stipe equal to or longer than lamina 9

9. Lamina not finely divided, ultimate segments broadly lanceolate with obtuse apices; pinnae not overlapping; exospore length greater than 34 µm .... **1. adiantum-nigrum**
- Lamina finely divided, ultimate segments narrowly lanceolate with acute apices; pinnae overlapping; exospore length less than 32 µm .... **2. onopteris**

1. A. adiantum-nigrum L., *Sp. pl.* **2**: 1081 (1753).
Rhizome short, creeping or decumbent, with dark brown, subulate scales. Fronds up to 40 cm; stipe equal in length to lamina, swollen at base, blackish; lamina (2-)3-pinnate, triangular-ovate to triangular-lanceolate, glossy green above, rachis blackish at base becoming green; pinnae ovate to triangular-lanceolate; pinnules ovate or elliptical, serrate. Sori linear oblong; indusium entire or sinuate. Mean exospore length 36-40 µm. 2n=144.
Rare, recently discovered in Madeiraabove Porto da Cruz on the north-east coast, in open situations in laurisilva. **M**

2. A. onopteris L., *Sp. pl.* **2**: 1081 (1753).
A. acutum Willd.; *A. adiantum-nigrum* subsp. *onopteris* (L.) Heufl.; *A. productum* Lowe
Rhizome short, creeping, sometimes branched, with brown, narrowly lanceolate scales. Fronds up to 60 × 17 cm; stipe often longer than lamina, swollen at base, shiny blackish-brown to reddish-black; lamina usually 3-pinnate, triangular-ovate to triangular-lanceolate, long attenuate-caudate at the apex, dark green; rachis blackish at base becoming green; pinnae triangular-ovate to triangular-lanceolate and tapering at apex; pinnules ovate-lanceolate to narrowly lanceolate; ultimate segments linear to ovate-lanceolate, with long acuminate teeth. Sori oblong or linear; indusium entire, lightly sinuate. Mean exospore length 26-32 µm. 2n=72.
Most frequent in northern Madeira from 300-1750 m, especially in cloud regions at 600-1000 m. Always in shady and damp places, on cliffs, stream banks and in roadside ditches, rarely on old stone walls. **M**

A. × ticinense D.E. Mey. in *Ber. dt. bot. Ges.* **73**: 391 (1961), the hybrid between **1** and **2**, was recently discovered near the Levada da Serra, inland from Porto da Cruz, growing together with both parents. The morphology is intermediate between the parents and it can be confirmed by the presence of abortive spores. 2n=108. **M**

3. A. obovatum Viv., *Fl. Lybyc. spec.*: 68 (1824).
A. billotii F.W. Schultz; *A. lanceolatum* Huds.; *A. obovatum* Viv. var. *billotii* (F.W. Schultz) Bech.; *A. rotundatum* Kaulf., nom. nud.
Rhizome short, erect or creeping, not branched, with dark brown, linear-lanceolate scales. Fronds up to 35 × 14 cm; stipe usually shorter than lamina, not swollen at base, usually shiny reddish-brown, with dark brown filiform scales at the extreme base; lamina 2(3)-pinnate,

ovate-lanceolate or oblong-lanceolate, bright green; rachis green; pinnae ovate-oblong to ovate-lanceolate, shortly acuminate; pinnules ovate to ovate-lanceolate, with acute mucronate teeth. Sori oblong-ovate; indusium entire, sinuate. Mean exospore length 35-43 μm. 2n=144.
In cloudy and moist regions in Madeira and frequent up to 1100 m, also in dry and rather sunny places on stone walls, sometimes in rocky roadside ditches, rarely on rocky cliffs; rare in Porto Santo and on Deserta Grande. **MDP**

All Madeiran plants are subsp. **lanceolatum** P. Silva in *Agronomia lusit.* **20**: 217 (1959), to which the above description and synonymy apply.

A. × joncheerei D.E. Mey. in *Willdenowia* **2**: 332 (1960), a presumed hybrid between **2** and **3**, is based on specimens from Pico da Lagoa, Madeira; it is intermediate in morphology and has abortive spores. **M**

4. A. aethiopicum (Burm.f.) Bech., *Candollea* **6**: 23 (1935).
Trichomanes aethiopicum Burm.f. [basion.]; *Asplenium canariense* sensu auct. mad., non Willd. (1810); *A. furcatum* sensu auct. mad., non Thunb. (1800); *A. maderense* Penny nom. nud.; *A. praemorsum* sensu auct. mad., non Sw. (1788)
Rhizome long, creeping, much branched, with blackish-brown, narrowly triangular scales. Fronds up to 50 × 15 cm; stipe at least half length of lamina, somewhat swollen at base, bright blackish-brown with scales similar to those on the rhizome and numerous smaller, reddish-brown scales; lamina pinnate-pinnatisect to 2-pinnate, oblong-lanceolate to triangular-lanceolate, usually acuminate at the apex, dark green above, pale green below, thinly coriaceous, with numerous small, reddish-brown filiform scales; rachis bright green, with scales similar to those on lamina, pinnae usually lanceolate-trapeziate, acute to caudate at apex; lower pinnules oblong, with 2-3 irregular toothed lobes; median and upper pinnules linear, adnate and decurrent, irregularly toothed at apex. Sori linear; indusium entire. Spores ellipsoidal with few ridges or wings anastomosing, mean exospore length 40-52 μm. 2n=432. **Plate 2**
Most frequent in eastern and northern Madeira, rare in the south, from 90-800 m, but usually between 300-500 m. Always in damp and shady places, on stone walls of cultivated terraces and roadsides, also on rocky fissures and on basaltic lava rocks near the coast. ■ **M** [*Canaries, Cape Verdes*]

All Madeiran plants are subsp. **braithwaitii** Ormonde in *Bolm Mus. munic. Funchal* **43**: 177-179 (1991), to which the above description and synonymy apply. Subsp. *aethiopicum*, from tropical and southern Africa, differs in being a sexual octoploid with more divided fronds and smaller spores. The *A. aethiopicum* complex in Macaronesia comprises two forms, the sexual dodecaploid, *A. aethiopicum* subsp. *braithwaitii*, and an apogamous hexaploid, *A. filare* subsp. *canariense* (Willd.) Ormonde from the Canary Islands that has a linear-lanceolate lamina with caudate apex, pinnae long acuminate at apices and spheroidal spores with dense ridges or wings anastomosing.

5. A. anceps Lowe ex Hook. & Grev., *Icon. filic.* **2**: t. 195 (1831).
Rhizome short, creeping to erect, bearing triangular-lanceolate, pale brown scales with blackish-brown central stripe. Fronds up to 40 × 2.5 cm; stipe short, bright blackish-brown, with few scales at the extreme base; lamina 1-pinnate, oblong-lanceolate to linear-lanceolate with acute apex, shiny dark green above, paler green below, herbaceous, glabrous; rachis shiny blackish-brown, glabrous, with 3 scarious, clathrate, blackish-brown wings, 2 lateral and adaxial, the other abaxial; most pinnae subsessile or shortly stalked, more than twice as long as broad, oblong-linear, crenulate, cuneate and usually acroscopically auriculate at the base, squarely inserted; veins mostly simple, clearly visible on abaxial surface. Sori short, oblong, nearer the margin than the costa and oblique to it; indusium subentire. Mean exospore length 27-32 μm. 2n=72. **Plate 1**

Frequent and abundant from 400-1400 m in shady and damp places on rocks, walls of levadas, streams and roadside ditches, and amongst bryophytes in laurisilva. Mostly in northen Madeira but also in some localities in the south in the cloud regions. ■ M [*Azores, Canaries*]

6. A. trichomanes L., *Sp. pl.*2: 1080 (1753).
Rhizome short, creeping to erect, sometimes branched, scaly. Fronds up to 22 × 2.5 cm; stipe shiny blackish-brown, scaly at extreme base; lamina 1-pinnate, linear-lanceolate, herbaceous to subcoriaceous, sometimes with short, glandular hairs on lower surface; rachis shiny blackish-brown; most pinnae subsessile or shortly stalked, less than twice as long as broad, suborbiculate, ovate or oblong, subentire to crenate-dentate round the apex and acroscopical margins, usually tapering towards the apex, cuneate and often slightly auriculate acroscopically at the base, squarely inserted; veins mostly simple. Sori oblong or linear, nearer the costa than the margin, oblique to both; indusium entire to irregularly crenulate. M

Two subspecies occur in Madeira.

a) subsp. **quadrivalens** D.E. Mey. emend. Lovis in *Brit. Fern Gaz.* **9**: 152 (1964).
A. trichomanes sensu auct. mad., non L. (1753)
Rhizome scales up to 5 mm, brown, ovate-lanceolate to linear-lanceolate with a narrow, blackish-brown central stripe. Lamina shiny dark green above, paler green below. Mean exospore length 32-38 μm. 2n=144.
Frequent but not abundant from 180-1750 m on rocks, walls, along levadas, streams and roadside ditches and on stones covered by bryophytes in laursilva. M

b) subsp. **maderense** Gibby & Lovis in *Fern Gaz.* **13**: 285 (1989).
Between **5** and **6a** in morphology, but much closer to the latter taxon, from which it can usually be distinguished as follows: rhizome scales up to 4.5 mm, narrowly lanceolate to subulate with broad, occluded central stripe; lamina in fresh specimens darker green, sub-shiny; occasionally a third abaxial wing can be detected on the rachis. Mean exospore length 35-40 μm. 2n=216.
From 1400-1800 m, in rock crevices. Rare endemic, recorded only from Pico de Arierio and Pico do Ferreiro, Madeira. ● M

A. trichomanes subsp. *maderense* is believed to be an allopolyploid, derived from a hybrid of **5** and **6a** by chromosome duplication.

7. A. monanthes L., *Mant. pl.*: 130 (1767). [*Feto de escoumas*]
Rhizome short, erect to suberect, with linear-lanceolate, brown scales with blackish central stripe and brown borders. Fronds up to 44 × 2.7 cm; stipe short, dark reddish brown; lamina 1-pinnate, linear-lanceolate, acute at the apex, darkish green above, paler green below, glabrous, herbaceous to subcoriaceous; rachis shiny dark reddish brown with proliferating buds at base; most pinnae subsessile or shortly stalked, oblong to rhombic, entire to crenate-dentate, with a very narrow basiscopic side, cuneate and acroscopically truncate-auriculate at base; veins mostly simple. Sori usually 1 per pinna along the costa near the basiscopic margin and parallel with it, oblong to linear; indusium entire. Mean exospore length 38-43 μm. 2n=108, apogamous.
Frequent throughout Madeira, but less abundant in the south, from 400-1400 m, usually from 700-1000 m. Always in damp places, usually in shade on cliffs, rocks and in lava fissures, in rocky roadside ditches and on stream banks, walls of levadas in the cloud regions, and amongst mosses in laurisilva. M

8. A. marinum L., *Sp pl* **2**: 1081 (1753). [*Feto maritimo*]
Rhizome short, thick, branched, with dense blackish-brown, linear-lanceolate scales. Fronds up to 50 × 10 cm; stipe from ¼ of length to as long as lamina, shiny blackish-brown to reddish-brown; lamina 1-pinnate, linear-lanceolate to oblong-lanceolate, usually acute at apex,

shiny dark green above, paler green below, usually coriaceous; rachis reddish-brown at base, becoming green towards the apex; most pinnae shortly stalked to sessile, ± oblong, crenate to lobate, usually rounded at apex, truncate to cuneate at base; often acroscopically truncate-auriculate at base and squarely inserted; veins clearly visible on both surfaces, 2- to 4-forked. Sori 2 to 8 on each side of costa, linear-oblong, along the terminal forked veins, oblique to margin and costa; indusium entire. Mean exospore length 26-30 μm. $2n=72$.

Up to 500 m, more frequent below 300 m, occurring in rock and lava fissures near the sea, usually in shade. Abundant on the north coast of Madeira, also on the south-eastern coast but less abundant there; rare in Porto Santo; very rare on Deserta Grande; recorded only once from the Salvages (c. 120 years ago), not found since and perhaps no longer present. **MDP ?S**

9. A. septentrionale (L.) Hoffm., *Deutschl. fl. Crypt.*: 12 (1795).
Acrostichum septentrionale L. [basion.]; *Asplenium bifurcatum* Opiz; *Scolopendrium septentrionale* (L.) Roth
Rhizome short, erect to creeping, branched into several crowns, with short, narrow linear-lanceolate, blackish-brown scales. Fronds up to 9 × 0.25 cm; stipe 2-3 × as long as lamina, erect, flexuose, green but blackish-brown at base; lamina simple or irregularly forked, linear-lanceolate to oblong-lanceolate with 2-3 long narrow teeth at apex, very decurrent at base; costa indistinct, veins dichotomous. Sori narrowly linear, along whole length of pinnae; indusium subentire. Mean exospore length 38-45 μm. $2n=144$.
Very rare in Madeira, recorded only at c. 1600 m in sunny rocky fissures on a mountainous cliff between Pico de Arranbementão and Pico Jorge. **M**

Madeiran plants are all subsp. **septentrionale**, to which the above description applies.

10. A. hemionitis L., *Sp. pl.* **2**: 1078 (1753). [*Feto de tres bicos*]
A. palmatum Lam.; *Phyllitis palmata* (Lam.) Samp.; *Scolopendrium palmatum* Samp.
Rhizome short, ascending to somewhat creeping, with dense, shiny black, ovate-lanceolate scales. Fronds up to 35 × 18 cm, persistent, in tufts; stipe usually twice as long as lamina, bright reddish-brown, somewhat swollen at base; lamina simple, deeply cordate at base, triangular to ovate, entire or 3- to 5(-7)-lobate, bright green, usually coriaceous, glabrous; lobes obtuse to acute, sometimes subcaudate; middle lobe usually longer than lateral ones, intermediate lobes patent, lower ones descendant; basal sinus rotund, usually very deep, mid-veins 3-5 palmate, 1 costa on each lobe, veins radiating dichotomously. Sori linear, on almost all veins, often extending from costa to margin; indusium entire, membranous, opening towards the apex of lamina or of lobes. Mean exospore length 23-28 μm. $2n=72$.
Shady, or sunny but damp places in humus-filled rocky fissures in laurisilva, on rocky cliffs and stream banks and in roadside ditches, sometimes on old, moss-covered levadas and stone walls; mostly from 300-900 m, rarely from 120-1300 m. More frequent in northern Madeira than in the south; rare in Porto Santo and on Ilhéu dos Garajaus off Deserta Grande. **MDP**

2. Ceterach DC.

Rhizome short. Fronds tufted, deeply pinnatisect, thick, glabrous above, densely covered with scales beneath, veins anastomosing at margins. Sori oblong to linear; indusium reduced or absent. Spores ellipsoidal.

1. C. lolegnamense Gibby & Lovis in *Fern Gaz.* **13**: 287 (1989). [*Deiradinha*]
C. aureum sensu auct. mad., non (Cav.) Buch (1819); *C. aureum* var. *madeirense* Ormonde nom. nud.; *C. officinarum* sensu auct. mad., non Willd. (1804); *Grammitis aurea* (Cav.) Sw.; *Gymnogramma ceterach* sensu auct. mad., non (L.) Spreng. (1827)
Rhizome ascending, covered densely with blackish-brown, narrow, lanceolate scales. Fronds

in tufts, up to 24 × 4 cm; stipe about half length of lamina, densely covered with ovate-lanceolate, brown scales with a broad black central stripe; lamina coriaceous, mid-green, linear-lanceolate, densely covered beneath with pale brown triangular scales; frond segments alternate, triangular or ovate, tapering at the apex, adnate. Sori lying in parallel rows diagonally on either side of midrib of frond segment, and midway between midrib and margin, up to *c.* 14 per frond segment. Mean exospore length 35-42 μm. 2n=216. **Plate 1**
Uncommon endemic on roadside walls in the hills around Funchal, and on some of the stream banks west of the town. ● **M**

Two varieties of *C. aureum* (Cav.) Buch are recorded in the Canary Islands, var. *aureum*, a tetraploid (2n=144), and var. *parvifolium* Benl & G. Kunkel, an octoploid (2n=288); *C. lolegnamense* differs from these varieties in being intermediate in size, spore size and chromosome number, and it may have evolved from them by hybridization.

3. Phyllitis Hill

Fronds in tufts. Lamina usually simple; veins free. Sori linear, borne in pairs along closely adjacent veins and appearing as 1; indusium linear, attached along 1 side and with pairs opening towards each other. Spores ellipsoidal.

1. P. scolopendrium (L.) Newman, *Hist. brit. ferns* ed. 2: 10 (1844). [*Lingua cervina*]
Asplenium scolopendrium L. [basion.]; *Scolopendrium officinarum* Sw. in Schrad.; *S. vulgare* Sm.
Rhizome ascending, densely covered with shiny black-brown, narrow, tapering scales with broad bases. Fronds up to 46 × 7 cm; stipe *c.* ⅓ length of lamina, purplish-brown with light brown, narrow, tapering scales; lamina coriaceous, linear, tapering at the apex, auriculate at the base, glossy mid-green on upper surface, lighter green and dull beneath; rachis and, to a lesser extent, underside of lamina with light brown, narrow, tapering scales. Sori linear, lying slightly diagonally to rachis and reaching almost from rachis to edge of lamina. 2n=72.
A rare fern in Madeira, recorded only from wet rocky places in the mountains in the north-west. **M**

XIV. WOODSIACEAE

1. Athyrium Roth

Rhizome upright or ascending, covered with broadly lanceolate, acuminate scales. Fronds forming a crown at apex of rhizome; veins free. Sori round to linear, indusium attached laterally. Spores ellipsoidal.

1. A. filix-femina (L.) Roth, *Tent. Fl. Germ.* 3(1): 65 (1799). [*Feto manso*]
Polypodium filix-femina L. [basion.]; *Athyrium filix-femina* var. *subincisum* Menezes
Fronds herbaceous, up to 90 cm. Stipe *c.* ¼ to equal length of lamina, pale brown, green or pale pink, scales dark brown, mostly confined to stipe base; lamina lanceolate, 2-pinnate-pinnatisect, gradually narrowed to the base; lowest pinnae more widely spaced than the rest, often bent downwards; pinnules oblong or oblanceolate, curving towards the pinna apex, sessile; pinnule segments toothed. Sori in a single row on either side of pinnule midrib; indusium J-shaped, attached along 1 side. 2n=80.
Frequent throughout Madeira in damp and shady conditions. **M**

2. Diplazium Sw. in Schrad.

Rhizome creeping, sparsely covered with ovate scales. Fronds arise singly or in groups of 2 or 3, lamina 2-pinnate-pinnatisect; groove of frond axis open at junction of costa; veins free. Sori oblong with flap-like indusium attached laterally along a vein. Spores ellipsoidal.

1. D. caudatum (Cav.) Jermy in *Brit. Fern Gaz.* **9**: 161 (1964). [*Feto de calvalto*]
Tectaria caudata Cav. [basion.]; *Allantodia umbrosa* R. Br.; *Athyrium umbrosum* sensu C. Presl, non Aiton (1789)
Fronds up to 150 × 50 cm. Stipe brown, *c.* half length of lamina, scales at base narrow-triangular to filiform, brown, less than 10 mm long; lamina triangular-ovate; pinnae lanceolate, widest near the middle, tapering at the apex; pinnules narrowly triangular, asymmetrical, being wider on acroscopic side, tapering at the apex, ultimate segments bluntly toothed. Sori alternately placed on either side of midrib of ultimate segments, on lateral veins. Indusium fimbriate.
Frequent but confined to dark, damp habitats in Madeira, in ravines and beside streams, usually in laurisilva. **M**

Deparia petersenii (Kunze) M. Kato in *Bot. Mag., Tokyo* **90**: 37 (1977) (*Asplenium petersenii* Kunze [basion.]; *Diplazium allorgei* Tardieu; *Lunathyrium petersenii* (Kunze) H. Ohba) is like *Diplazium* but has fronds up to 100 cm; lamina pinnate-pinnatisect, equal in length to stipe; groove of frond axis not open at junction of costa or costule, both surfaces of lamina bearing short, thick, brownish hairs. Native to tropical and subtropical Asia, it is known in Madeira only from one specimen found growing in *Calluna* near Passo, São Vicente. **M**

4. Cystopteris Bernh. in Schrad.

Rhizome shortly creeping. Fronds thinly herbaceous, pinnate-pinnatifid to 3-pinnate; veins free. Sori orbicular, indusium hood-like, attached at the base, sitting across a vein, toothed, becoming deflexed and shrivelled as sporangia ripen. Spores monolete, ellipsoidal or often orbicular.

1. C. diaphana (Bory) Blasdell in *Mem. Torrey bot. Club.* **21**: 47 (1963).
Polypodium diaphanum Bory [basion.]; *Cystopteris fragilis* sensu auct. mad., non (L.) Bernh. in Schrad. (1805); *C. fragilis* (L.) Bernh. in Schrad. subsp. *diaphana* (Bory) Litard.; *C. viridula* (Desv.) Desv.
Fronds up to 50 cm. Stipe brittle, *c.* half as long as lamina, black towards base, glabrous; lamina light green, linear-lanceolate, 2-pinnate-pinnatifid, glabrous; veins ending in a sinus. Spores with short dense spines. $2n=252$.
Frequent in damp shady habitats throughout Madeira. **M**

C. fragilis (L.) Bernh. in Schrad. and *C. diaphana* × *C. fragilis* are reported for Madeira, but no specimens of *C. fragilis* have been seen, nor any specimens with bad spores that could represent *C. diaphana* × *C. fragilis*.

XV. DRYOPTERIDACEAE

1. Polystichum Roth

Rhizome erect, densely covered with scales. Fronds pinnate to 2-pinnate; teeth of lamina segments with apical spine; veins free. Sori round; indusium peltate or lacking. Spores ellipsoidal.

1. Lamina triangular-ovate; indusium lacking . . . . **1. drepanum**
- Lamina oblanceolate to elliptic; indusium peltate 2

2. Lamina thin, soft textured, 2-pinnate; indusium pale brown . . . . **2. setiferum**
- Lamina thick, coriaceous, pinnate to pinnate-pinnatisect; indusium pale brown with central dark spot 3

3. Fronds pinnate; spores good (ellipsoidal, of uniform size) . . . . **3. falcinellum**
- Fronds pinnate-pinnatisect; spores abortive (irregular in size and shape) . . . .
. . . . **4. × maderense**

1. P. drepanum (Sw.) C. Presl, *Tent. pterid.*: 84 (1836).
Aspidium drepanum Sw. [basion.]; *Polypodium drepanum* (Sw.) Lowe
Rhizome scales dark brown, triangular with long tapering points. Fronds somewhat rigid, up to 140 × 95 cm; stipe *c.* ⅔ as long as lamina, green, scales at base similar to those of rhizome, narrower and paler towards the rachis; lamina triangular to triangular-ovate, usually 2-pinnate, large fronds 2-pinnate-pinnatifid, small fronds pinnate-pinnatifid; pinnules ovate to lanceolate, tapering, wider and sometimes with an auricle on the acroscopic side of pinna, coarsely serrate; scales on costae pale brown, triangular. Sori exindusiate. 2n=164. **Plate 3**
Very rare endemic; possibly now confined to a few steep forest areas in the north-west of Madeira.
● M

2. P. setiferum (Forssk.) Woyn. in *Mitt. naturw. Ver. Steierm.* **49**: 181 (1913).
Polypodium setiferum Forssk. [basion.]; *Polystichum aculeatum* sensu auct. mad., non (L.) Roth (1799); *P. angulare* (Kit. ex Willd.) C. Presl
Rhizome densely covered with triangular, gingery brown scales. Fronds herbaceous, up to 95 × 19 cm; stipe *c.* ⅓ length of lamina, green, with scales at base similar to those of rhizome, narrower with very long tapering points towards the rachis; lamina oblanceolate to elliptic, 2-pinnate; pinnules up to 11 mm, of uniform size except for innermost acroscopic pinnule on each pinna which is always larger than adjacent ones, pinnules with an auricle at the base on the apiscopic side, toothed, with a long apical spine; costae and costules bearing lanceolate scales with very long tapering points. Indusium 1 mm, peltate, concolorous, margin entire. 2n=82.
Frequent throughout Madeira in forest areas, along levadas and by shaded rocks and cliffs. M

A record of *P. aculeatum* (L.) Roth for Madeira has not been confirmed.

3. P. falcinellum (Sw.) C. Presl, *Tent. pterid.*: 83 (1836).
Aspidium falcinellum Sw. [basion.]
Rhizome densely covered with broadly triangular, tapering, dark brown scales. Fronds rigid, coriaceous, up to 70 × 18 cm; stipe *c.* ⅓ length of lamina, scales on base of stipe similar to those on rhizome, more sparse and ginger-brown with a darker central stripe towards rachis and those on upper rachis light brown. Lamina lanceolate, pinnate; pinnae falcate, narrow, tapering towards the apex, serrate, bearing blunt auricles at the base on the acroscopic side that often overlap rachis. Sori borne in 2 rows; indusium more than 1 mm, peltate, with central dark spot, margin bearing multicellular hairs. 2n=328. **Plate 3**
Locally common endemic in the north-western and central mountains of Madeira, in drier and more open habitats than other Polystichum *species.* ● M

4. P. × maderense J.Y. Johnson in *Ann. Mag. nat. Hist.*, III, **17**: 287 (1866).
(= *P. falcinellum* (Sw.) C. Presl × *P. setiferum* (Forssk.) Woyn.)
Rhizome covered with brown triangular scales with a dark central stripe. Fronds coriaceous, up to 80 × 13 cm; stipe *c.* half length of lamina, scales at base similar to those of rhizome, more sparse, concolorous, lanceolate with tapering points towards rachis; lamina oblanceolate-lanceolate, pinnate-pinnatifid, but 2-pinnate at base of lower pinnae; pinnae lanceolate

but with innermost acroscopic segment enlarged, pinna segments cut ¼-½ width of the pinnae, not overlapping, slightly toothed, tapering to a distinct apical spine. Indusium more than 1 mm, peltate with central dark spot, margin irregularly toothed with a few multicellular hairs. Spores abortive. 2n=205. **Plate 3**
Very rare endemic, recorded only from a few sites in central Madeira, from Pico deArieiro west to Rabaçal. ◓ **M**

A further *Polystichum* hybrid that probably has *P. falcinellum* as one parent was recorded last century as a single plant growing in a population of *P. falcinellum* at Camacha. It is similar to **4**, but can be distinguished by the more rounded and crowded pinnules, with innermost pinnules overlapping.

2. Cyrtomium C. Presl

Rhizome erect; rhizome and stipe bases with ovate to lanceolate scales. Fronds coriaceous, pinnate, veins anastomosing. Sori round with peltate indusium. Spores ellipsoidal.

1. C. falcatum (L.f.) C. Presl, *Tent. pterid.*: 86 (1836).
Polypodium falcatum L.f. [basion.]; *Polystichum falcatum* (L.f.) Diels in Engl. & Prantl
Fronds coriaceous, up to 75 × 18 cm; stipe up to *c.* half length of lamina, base densely covered with broadly lanceolate scales up to 25 × 7 mm, dark brown with paler margins; lamina oblanceolate, shiny dark green above, paler green below; pinnae ovate, acuminate with tapering apices that point towards frond apex; rachis with narrow, brown scales with very long tapering points at base of pinnae. Sori scattered; indusium peltate, round with undulate margins.
Cultivated in gardens in Madeira, sometimes escaping, possibly becoming naturalized; recently recorded as a new introduction on Selvagem Grande. **?M S**

3. Arachniodes Blume

Rhizome usually creeping. Fronds 2-pinnate-pinnatifid to 4-pinnate, never simply pinnate, broadly deltoid or pentagonal in outline, subcoriaceous to coriaceous in texture, the lowermost pinnae often basiscopically enlarged and the pinnules or their segments commonly spinose-serrulate; veins free. Indusium orbicular with a deep sinus and overlapping lobes. Spores ellipsoidal.

1. A. webbianum (A. Braun) Schelpe in *Bolm Soc. broteriana*, II, **41**: 203 (1967).
Aspidium webbianum A. Braun [basion.]; *A. frondosum* Lowe; *Polystichum frondosum* (Lowe) J. Sm.; *P. webbianum* (A. Braun) C. Chr.
Fronds up to 100 × 38 cm; stipe equal in length to lamina, dark brown at base and densely covered in narrow, lanceolate, dark brown scales up to 10 × 1 mm long, scales less dense and shorter further up the stipe and on rachis, upper stipe light brown; rachis green. Lamina bright green, 4-pinnate at base, elsewhere 3-pinnate or 2-pinnate-pinnatisect; lowest pinnae markedly asymmetrical with innermost basiscopic pinnule twice as long as innermost acroscopic pinnule; pinnule segments ovate to oblanceolate, tapering at apex; costae and costules bearing small brown scales. Sori more than 1 mm, crowded; indusium round. 2n=82. **Plate 2**
Endemic locally abundant in forest in the steep valleys of north-west Madeira from 10-1400 m, uncommon elsewhere. ● **M**

All Madeiran specimens are subsp. **webbianum** to which the above description applies; subsp. *foliosum* (C. Chr.) Gibby *et al.* is a tetraploid from Kenya.

4. Dryopteris Adans.

Rhizome usually erect and densely covered with scales. Fronds pinnate to 3-pinnate; veins free. Sori round with reniform indusium; spores ellipsoidal.

1. Fronds pinnate-pinnatisect to 2-pinnate 2
- Fronds 2-pinnate-pinnatisect to 3-pinnate 3

2. Fronds pinnate-pinnatisect; lamina oblanceolate to elliptic, 3-4 × length of stipe, lamina and indusia eglandular .... **1. affinis**
- Fronds 2-pinnate; lamina triangular-ovate to ovate, 1-2 × length of stipe **2. aitoniana**

3. Lamina and indusia glandular; lowest basal basiscopic pinnule longer than adjacent pinnule; lamina segments usually curling upwards .... **3. aemula**
- Lamina and indusia eglandular; lowest basal basiscopic pinnule shorter than adjacent pinnule; lamina segments usually deflexed .... **4. maderensis**

1. D. affinis (Lowe) Fraser-Jenk. in *Fern Gaz.* **12**: 56 (1979).
Nephrodium affine Lowe [basion.]; *Dryopteris borreri* sensu auct. mad., non (Newman) Newman ex Oberholzer & Tavel (1937); *D. pseudomas* sensu auct. mad., non Holub & Pouzar (1967)
Rhizome densely covered with narrow ginger-brown scales often with dark bases. Fronds arranged in a shuttlecock, up to 105 × 32 cm; stipe usually less than ⅓ length of lamina, green, stipe and rachis densely covered with scales similar to those on rhizome; lamina oblanceolate to elliptic, pinnate-pinnatisect, eglandular; pinnae ± symmetrical about axes, pinnules 5-15 mm long, adnate, crowded, rounded-truncate, ± parallel sided, ± entire with a few small apical teeth, lowest basal basiscopic pinnule usually longer than the adjacent one; costae bearing small narrow dark-based scales, denser towards rachis. Indusium rarely more than 1 mm. 2n=82, apomict.
Frequent in damp shade in woods and beside levadas; amongst rocks in the open in the central mountains of Madeira. **M**

All Madeiran plants are subsp. *affinis*, to which the above description and synonymy apply.

2. D. aitoniana Pic. Serm. in *Webbia* **8**: 154, f.3 (1951).
Dryopteris elongata (Aiton) Sim, non (Wall. ex Hook.) Kuntze (1891); *D. macaronesica* Romariz; *Nephrodium elongatum* (Aiton) Hook. & Grev.; *Polypodium elongatum* Aiton
Rhizome densely covered with shiny, dark brown, ovate-lanceolate scales, usually with a darker central stripe. Fronds spreading, up to 120 × 36 cm, thicker and more rigid than other *Dryopteris* species; stipe half to equal length of lamina, green, scales at base similar to those of rhizome; lamina triangular-ovate to ovate, 2-pinnate, with abundant stalked glands; pinnae asymmetrical, basiscopic pinnules markedly longer than acroscopic ones; pinnules 15-40 mm long, adnate to costa, often crenate, with rounded to pointed apices with well-developed teeth, lower ones deeply lobed, lowest basal basiscopic pinnule usually shorter than the adjacent one; costae and costules bearing scattered small, usually dark brown, ovate scales. Indusium more than 1 mm. 2n=82.
Plate 4
Endemic. Frequent in woods and along shady levadas throughout Madeira but in drier niches than other Dryopteris *species.* ● **M**

3. D. aemula (Aiton) Kuntze, *Revis. gen. pl.* **2**: 812 (1891).
Polypodium aemulum Aiton [basion.]; *Nephrodium aemulum* (Aiton) Baker; *N. foenisecii* Lowe
Rhizome sparsely scaly, scales dark brown, concolorous. Fronds spreading, up to 71 × 30 cm; stipe ± equal in length to lamina, lower half of stipe dark purplish brown, scales at base very sparse, narrow, dark brown, concolorous; lamina narrowly triangular, 2-pinnate-pinnatisect to 3-pinnate with abundant sessile globose glands; lowest pinnae asymmetrical, with longer and more developed basiscopic pinnules; lamina segments usually curled upwards. 2n=82.
Frequent amongst rocks, in woods and along levadas throughout Madeira. **M**

Madeiran specimens have more finely dissected fronds than most plants from the rest of Macaronesia, Europe and W. Asia.

4. D. maderensis Alston in *Bolm Soc. broteriana* **30**: 14 (1956).
D. austriaca sensu auct. mad., non (Jacq.) Woyn. (1915); *D. dilatata* sensu auct. mad., non (Hoffm.) A. Gray (1848); *D. intermedia* (Muhl. ex Willd.) Gray subsp. *maderensis* (Alston) Fraser-Jenk.; *D. spinulosa* sensu auct. mad., non (O.F. Müll.) Kuntze (1891)
Rhizome moderately scaly, scales ovate-lanceolate, brown with a dark central stripe. Fronds spreading, up to 90 × 40 cm; stipe ⅔ to equal length of lamina, green, dark brown towards the base, lower half of stipe with scales similar to those on rhizome; lamina triangular-ovate, 2-pinnate-pinnatisect to 3-pinnate, eglandular; lowest, basal, basiscopic pinnule distinctly shorter than its neighbour; lamina segments usually deflexed, especially when growing in exposed situations; scattered, pale brown, concolorous scales on rachis, costa and costules. 2n=82. **Plate 4**
Endemic. In the northern half of Madeira, at altitudes above 550 m, locally abundant in damp forest, infrequent beside levadas and streams. ● **M**

D. × furadensis Bennert *et al.* in *Fern Gaz.* **14**: 10 (1991), believed to be the endemic (●) hybrid **2 × 4**, was recently discovered above the Levada do Furado, west of Ribeiro Frio. The morphology is intermediate between the presumed parents, and the spores are abortive. 2n=82. **M**

XVI. LOMARIOPSIDACEAE

1. Elaphoglossum Schott ex Sm.

Epiphyte. Rhizome short, creeping with very dark, toothed scales. Fronds simple and entire, fertile fronds smaller than sterile ones; veins free. Sporangia occupy entire lamina surface of fertile frond. Spores ellipsoidal.

1. E. semicylindricum (Bowdich) Benl in *Botanica Macaronésia* **6**: 59 (1980).
[*Lingua de vaca*]
Lomaria semicylindrica Bowdich [basion.]; *Acrostichum paleaceum* Hook. & Grev.; *A. squamosum* Sw., nom. illegit. in Schrad.; *Elaphoglossum hirtum* sensu auct., non (Sw.) C. Chr. (1905); *E. paleaceum* (Hook. & Grev.) Sledge; *E. squamosum* (Sw.) J. Sm., comb. illegit.
Fronds to 45 cm in length; stipe short, less than ¼ length of lamina; lamina simple, lanceolate; stipe and both surfaces of lamina densely covered with lanceolate scales that bear many, very long-pointed teeth along the margins; fertile fronds narrower than sterile ones, and often with a relatively longer stipe. **Plate 2**
Epiphyte occasionally found on Laurus azorica *and* Erica arborea *in northern region of Madeira.* ■ **M** [*Azores*]

XVII. DAVALLIACEAE

1. Davallia Sm.

Epiphyte. Rhizome creeping. Fronds distant, articulated to rhizome; lamina deltate, coriaceous, glabrous, veins free. Sori terminal on veinlets, indusium pouch-shaped, fixed at base and sides, elongate, apex nearly reaching margin of segments. Spores ellipsoidal.

1. D. canariensis (L.) Sm. in *Mem. Acad. Sci. Turin* **5**: 414 (1793). [*Cabrinhas*]
Trichomanes canariensis L. [basion.]
Rhizome stout, covered with shiny brown, triangular, acuminate scales. Fronds up to 60 × 30 cm, stipe glabrous except for extreme base, equal in length to lamina; lamina ovate or deltate, 3- to 4-pinnate, ultimate segments ovate or lanceolate.
Common on trees, rocks and walls throughout Madeira, on rocks in Porto Santo and on the Desertas. **MDP**

The pantropical **Nephrolepis cordifolia** (L.) C. Presl, *Tent. pterid.*: 79 (1836), occurs occasionally in Madeira as a garden escape around Funchal. Fronds are tufted, herbaceous, 1-pinnate, with simple, sessile pinnae auricled at the base and articulated to the rachis, and reniform indusia. M

XVIII. BLECHNACEAE

1. Blechnum L.

Rhizome short ascending. Fronds dimorphic, ± pinnate; veins of sterile fronds free. Sori long, linear, continuous along both sides of the pinna midrib, indusium attached at pinna margin and opening inwards. Spores ellipsoidal.

1. B. spicant (L.) Roth in *Annln Bot.* **10**: 56 (1794). [*Feto pente*]
Osmunda spicant L. [basion.]; *Blechnum boreale* Sw.; *Lomaria spicant* (L.) Desv.
Rhizome covered in short (2-3 mm), lanceolate, acuminate, brown scales. Fronds coriaceous, glabrous. Sterile fronds up to 40 cm, held in a somewhat flattened rosette; stipe short, less than ¼ length of lamina; lamina oblong to lanceolate; pinnae entire or crenate, mostly contiguous, adnate at base, oblong with rounded tips. Fertile fronds erect, up to 60 cm, oblong to lanceolate, lamina twice as long as the stipe, pinnae linear, very narrow, widely spaced, adnate at base, curving towards apex. In ripe specimens almost the whole of the underside of the fertile pinnae appears to be covered in a mass of brown sporangia.
Very common in moist, shady habitats throughout Madeira. M

2. Doodia R. Br.

Rhizome short, ascending. Fronds dimorphic, pinnate; pinnae finely serrate. veins anastomosing except towards margin. Sori oblong, not contiguous, forming a row on each side of the mid-vein of the pinna, and parallel to the rachis at the base of the pinnae; indusium oblong, opening towards the mid-vein. Spores ellipsoidal.

1. D. caudata (Cav.) R. Br., *Prodr.*: 151 (1810).
Woodwardia caudata Cav. [basion.]
Rhizome and stipe bases bearing small, black or brown scales. Fronds lanceolate; pinnae stalked and auricled towards base of lamina, adnate towards the apex. Sterile fronds up to 16 cm, stipe less than ⅙ length of lamina; pinnae oblong. Fertile fronds longer, up to 30 cm, stipe *c.* ¼ length of lamina; lamina with long-tapering apex; pinnae linear-lanceolate, upper pinnae curving towards the frond apex. Sori 1.5-2.0 mm, may appear contiguous in mature specimens.
Native to Australia and New Zealand, naturalized in Madeira in several localities including Monte, Cruzinhas, Machico and Faial. M

3. Woodwardia Sm.

Large ferns; rhizome short, ascending; rhizome and stipe bases densely scaly; fronds pinnate-pinnatisect, veins anastomosing except towards margin. Sori oblong, borne in two rows parallel to and adjacent to the mid-vein of the pinna segments; indusium oblong, opening towards the mid-vein. Spores ellipsoidal.

1. W. radicans (L.) Sm. in *Mem. Acad. Sci. Turin* **5**: 412 (1793). [*Feto do botão*]
Blechnum radicans L. [basion.]
Fronds to over 300 cm, with a bulbil on the midrib towards the tip of the frond from which a young plant is produced; stipe *c.* ⅓ length of lamina, scales on rhizome and stipe bases 40 mm, lanceolate, acuminate, brown; lamina ovate-triangular; pinnae oblong; pinna segments lanceolate. Sori 2-3 mm.
Frequent, usually in shade or forest in Madeira, growing in wet gullies and where water-courses run down rocky slopes. M

SPERMATOPHYTA
GYMNOSPERMAE
XIX. PINACEAE
J.R. Press

Monoecious, resiniferous, trees, rarely shrubs. Branches in regular whorls. Leaves spirally arranged, needle-like, entire or minutely toothed, mostly evergreen. Flowers in cones with spiralled scales. Scales of male cones with 2 pollen sacs on the underside. Scales of female cones consisting of a lower scale (*bract*) subtending an ovuliferous scale (cone-scale) bearing 2 ovules on the upper surface. Fruiting cones woody. Seeds winged.

1. Pinus L.

Evergreen trees or shrubs with rough, scaling or furrowed bark. Branches regularly whorled. Shoots of 2 kinds, long shoots with brown scale-like leaves and short shoots (*bundles*) consisting of 2, 3 or 5 needle-like leaves surrounded by a common basal sheath, the whole eventually falling as a single unit. Male cones clustered at base of new growth. Female cones ovoid to cylindrical, ripening in second or third year. Exposed part of cone-scales (*apophysis*) thickened and with a prominent protuberance (*umbo*), often tipped with a spine.

1. Pinus pinaster Aiton, *Hort. kew.* 3: 367 (1789). [*Pinheiro*]
Tree up to 30 m or more with an open crown of wide-spreading branches. Bark cracking, reddish brown. Needles paired, up to 20 cm × 2 mm, coarse, stiff, pungent, the margins inrolled. Male cones purplish when young. Female cones in clusters of 3-5, conic-ovoid, light shiny brown, ripening in the second year but persisting on the tree; apophysis rhomboidal, keeled; umbo prominent, prickly.
Fl. III (male cones). *Extensively grown in Madeira as a plantation and timber tree, occasionally self-seeding.* **M**

Other species are planted on a very small or experimental scale, especially **P. radiata** D. Don which withstands sea winds and is planted around the summits of the higher peaks of Porto Santo. It has slender, grass-green needles in threes and oblique cones. **P**

XX. CUPRESSACEAE
J.R. Press

Monoecious or dioecious, resiniferous and evergreen trees or shrubs. Leaves opposite or whorled, usually scale-like, closely appressed and concealing the stem, sometimes needle-like. Flowers in cones with opposite or whorled scales. Scales of male cones with 3-5 pollen sacs on the underside. Scales of female cones with 2-many ovules on the upper surface, completely united with the subtending bract. Fruit woody or sometimes fleshy when ripe, indehiscent. Seeds winged or not.

1. Cone-scales woody . . . . **1. Cupressus**
- Cone-scales fleshy and coalescing, the fruit berry-like . . . . **2. Juniperus**

1. Cupressus L.

Monoecious trees. Leaves opposite and decussate, all scale-like, rarely needle-like on young plants. Cones woody, ± globose with 3-7 pairs of peltate scales, ripening in the second year. Seeds 6-20 on each scale, flattened, winged.

1. C. macrocarpa Hartw. in *J. hort. Soc. Lond.* **2**: 187 (1847).
Tree to 25 m, pyramidal when young, becoming broader with age, the branches erecto-patent. Leaves 1-2 mm. Female cones up to 25 mm diameter, reddish brown when ripe; scales 8-14, shortly mucronate.
Fl. ? *Planted on a small scale for shelter in Madeira.* **M**

C. lusitanica Mill. is also occasionally planted in Madeira. It has patent branches somewhat pendulous at the tips and female cones 10-15 mm diameter, glaucous when young, with 6-8 prominently mucronate scales. **M**

2. Juniperus L.

Monoecious or dioecious trees or shrubs with thin bark shedding in strips. Leaves of two kinds, needle-like and opposite or whorled, or scale-like and opposite. Cones berry-like, the 3-8 scales coalescent and fleshy in fruit, ripening in the first, second or third year. Seeds up to 12 in each cone.

1. Leaves all needle-like, with 2 white bands above . . . . **1. cedrus**
- Leaves of 2 kinds, in juvenile plants needle-like with 2 white bands above and below, in adult plants scale-like . . . . **2. phoenicea**

1. J. cedrus Webb & Berthel., *Hist. nat Isles Canaries.* **3** (2, sect. 3): 277 (1847).
[*Cedro, Cedro da Madeira*]
J. oxycedrus subsp. *maderensis* Menezes
Dioecious; shrub or small tree to *c.* 20 m, with pendulous twigs. Leaves all needle-like, erecto-patent, in whorls of three, 8-12 × 1-1.5 mm, acute to obtuse, sometimes mucronulate, with 2 white stomatiferous bands above, midrib prominent beneath. Female ones 9-10 mm, ± globose, bluish and pruinose when young, becoming yellow and later reddish when ripe. Seeds 3, coalesced to form a single large stone. **Plate 5**
Fl. I - III (male cones). *Once common in Madeira, now rare, occurring in laurisilva and other areas at high altitudes.* ■ **M** [*Canaries*]

The normal form of this species has a rounded habit but a few trees with a narrow, columnar form are known: all are female. This form has apparently arisen spontaneously.

2. J. phoenicea L., *Sp. pl.* **2**: 1040 (1753). [*Zimbreiro, Zimbro*]
Monoecious; shrub or small tree to 5 m. Twigs 1 mm diameter, terete. Juvenile leaves, if present, patent in whorls of 3, needle-like with 2 white stomatiferous bands above and below; adult leaves scale-like, 0.8-1.5 mm, ovate-rhombic with a gland in the centre of an oblong dorsal depression and very narrow scarious margins. Female cones *c.* 8 mm, globose, blackish and pruinose when young, ripening through green and yellow to dark red. Seeds 4-5, free.
Fl. II (male cones). *Very rare at a few sites in northeastern Madeira; rare in Porto Santo, on the seaward slopes of Pico Branco.* **MP**

XXI. TAXACEAE
J.R. Press

Usually dioecious evergreen trees or shrubs. Leaves needle-like, flattened, spirally arranged. Male flowers in catkin-like clusters. Female flowers axillary, solitary or paired, with 1 or more sterile scales and a single ovuliferous scale. Ripe seed surrounded by a fleshy aril.

1. Taxus L.

Trees or shrubs with irregularly whorled branches. Leaves spirally arranged but twisted and spreading to form 2 lateral ranks. Male flowers in axillary globose heads. Female flowers solitary, with several imbricate sterile scales and a single fertile scale. Seed erect, ovoid, partly surrounded by a scarlet, fleshy, cup-like aril.

1. T. baccata L., *Sp. pl.* **2**: 1040 (1753). [*Teixo*]
Shrub or small tree. Leaves 12-22 mm, shortly petiolate, mucronate, dark glossy green above, paler with 2 pale green stomatiferous bands beneath.
Fl. III - IV (male cones). *Very rare. Known from a few sites in eastern central Madeira and the north coast at Boa Ventura.* **M**

XXII. EPHEDRACEAE
J.R. Press

Dioecious shrubs with opposite or verticillate, scale-like leaves. Inflorescence axillary on short shoots, the flowers enclosed in 2 or more membranous bracts. Male flowers with pollen-sacs fused into groups borne on a peduncle and superficially resembling a stamen. Fruit a globose, fleshy syncarp with 1 or 2 seeds, surrounded by fleshy bracts.

1. Ephedra L.

Shrubs with green, opposite or verticillate, sometimes articulated twigs. Leaves green or brownish, fused at the base to form a sheath. Male inflorescences subsessile, with 4-8 flowers, the bracts in 1 or 2 pairs, free. Female flowers solitary, pedunculate, with 2-4 pairs of connate bracts. Fruit globose or ellipsoid, the bracts becoming fleshy, yellow or red. Seeds 1-2.

1. E. fragilis Desf., *Fl. atlant.* **2**: 327 (1799).
E. fragilis var. *dissoluta* Stapf
Dense, scrambling, ± erect shrub usually up to 1 m, sometimes more. Twigs 1-2 mm diameter, readily disarticulating. Leaves 1-2 mm, green on the back, caducous. Bracts of male flowers 1-2 mm, orbicular, those of female flowers ovate. Syncarp *c.* 9 mm, oblong-ellipsoid.
Fl. V - VI. *Very rare plant of the south coast of Madeira, from Praia Formosa and Ponta da Cruz east of Funchal, Cabo Girão and at Fajã de Ovelha; equally rare in Porto Santo; also recorded for Deserta Grande.* **MDP**

ANGIOSPERMAE

DICOTYLEDONES

XXIII. SALICACEAE
J.R. Press

Dioecious, deciduous trees or shrubs. Leaves alternate, rarely opposite, stipulate. Catkins appearing before or after the leaves. Flowers each subtended by a small membranous bract. Perianth reduced to a cup-like disc or 1 or more nectary-scales, rarely absent. Stamens 2-many, the filaments long and slender. Ovary unilocular. Fruit a 2-valved capsule. Seeds with a tuft of silky white hairs at the base.

1. Salix L.

Shrubs or trees. Buds with a single outer scale, terminal bud absent. Leaves generally much longer than wide, shortly petiolate. Catkins erect or spreading, rarely pendulous. Bracts subtending flowers entire. Perianth usually of 1-2 nectary-scales. Stamens usually 2, 3 or 5.

Several species of **Populus** L. are planted in Madeira, mainly as shade trees.

Literature: J.C. Rodriguez Pinero, W. Wildpret de la Torre & M. del Arco-Aguilar, Contribucion del estudio de *Salix canariensis* (Salicaceae), in *Vieraea* **17**: 121-142 (1987).

1. S. canariensis C. Sm. ex Link in Buch, *Phys. Beschr. Canar. Ins.*: 159 (1825).
[*Seixeiro, Seixo*]
S. pedicellata subsp. *canariensis* (Buch) A.K. Skvortsov
Small, slender tree up to 7 m but often less. Twigs weakly striate, densely hairy when young and only becoming glabrous by about the third or fourth year. Leaves 3-12 × 1-2.5 cm, lanceolate to narrowly elliptic, rarely obovate, margins entire or minutely undulate, yellowish green, sometimes glaucous beneath, densely hairy when young, becoming glabrous except for the midrib beneath; stipules very small, ovate, toothed, caducous. Male catkins 2-5 cm, cylindrical, dense; bracts dark chestnut-brown at the tip; stamens 2. Female catkins 3-4 cm, extending up to 7 cm in fruit; bracts with long white hairs; fruiting pedicels 1.5-2 (-3) mm, equalling or up to 1½ × as long as the bracts. Capsule glabrous.
Fl. I - II. *Common along streams and damp ravines in most parts of Madeira.* ■ M [*Canaries*]

Hybrid plants, given the local name *Vimeiro*, are widely cultivated for withies for the extensive basketry industry, the long, flexible twigs being cropped annually. There is some confusion as to the precise names of the parents involved but the plants are here referred to **S. fragilis** L. × **alba** L. Schrank (S. × **rubens** Schrank.). They have yellowish orange twigs, leaves up to 18 cm, lanceolate, glandular-toothed, dull green above, sometimes glaucous beneath, glabrous when mature, and glabrous, shortly pedicellate capsules. Only female plants are known on the island and although they readily reproduce vegetatively (from cuttings) they do not appear to have become naturalized.

XXIV. MYRICACEAE
M.J. Short

Shrubs or trees, often aromatic. Leaves alternate, simple, usually exstipulate. Flowers mostly unisexual, borne in axillary, bracteate, catkin-like spikes. Petals and sepals absent. Stamens 2-8(-16); filaments free or united at the base. Ovary superior, 1-locular; ovule 1; style short, 2-branched. Fruit usually a drupe.

1. Myrica L.

Usually dioecious. Leaves entire to irregularly toothed, glandular-punctate, exstipulate. Male flowers subtended by a single bract. Female flowers subtended by a bract and 2 or more small bracteoles. Fruit a small rough drupe with waxy papillae.

1. M. faya Aiton, *Hort. kew.* **3**: 397 (1789). [*Faia, Samouco*]
Evergreen shrub or small tree up to 8 m. Young twigs with small, brownish peltate glands. Leaves shortly petiolate, 4-10 × 1.3-3 cm, oblanceolate, acute to obtuse, attenuate, coriaceous, glabrous, inconspicuously dotted with glands, margins entire or shallowly and irregularly toothed towards the apex, often revolute. Male catkins branched, borne among the leaves on the current year's growth, the flowers yellow-green, usually with 4 stamens. Fruit 4-7.5 mm across, globose, papillose, slightly fleshy, red turning to black, glistening.
Fl. III - IV. *A locally abundant component of the laurisilva in Madeira, also frequent at lower altitudes in the northern half of the island, only very occasional in the south; 0-1000 m.* ?■ M [*Azores, Canaries*]

M. faya also occurs in the coastal regions of C. and SW Portugal, but it is uncertain whether these populations are native or naturalized.

ULMACEAE
J.R. Press

Celtis australis L., [*Lodão bastardo, Sementeira*] is a deciduous tree to 25 m with alternate leaves 7-10 × 3-5 cm, ovate to ovate-lanceolate, long-acuminate, rounded and often asymmetric at the base, sharply serrate, scabrous above, pilose to velutinous and brownish to greyish green beneath. Drupe 0.8 cm diameter, dark purple or brownish black when ripe. Introduced from S. Europe, it is planted as a shade tree in the lower regions of Madeira. **M**

XXV. MORACEAE
J.R. Press

Monoecious or dioecious trees or shrubs with milky latex. Leaves usually alternate, stipules 2 but often falling early. Flowers reduced, in spikes, heads or contained within a fleshy, urn-shaped receptacle. Male and female flowers 4-merous, the females with a 1- to 2-locular ovary and 1 or 2 styles. Fruit a collection of druplets forming a single syncarp.

1. Ficus L.

Monoecious trees or shrubs. Leaves usually lobed; buds covered by the connate stipules which fall early, leaving a distinctive circular scar. Flowers enclosed in the fleshy, urn-shaped receptacle.

Besides *Ficus carica* the edible fig, several species of *Ficus* are cultivated on the islands, notably **F. elastica** Roxb., the India-rubber tree, a tall evergreen tree with entire, glossy leaves, dark green above, frequently planted for shade and ornament.

1. F. carica L., *Sp. pl.* **2**:1059 (1753). [*Bebereira, Figueira*]
Small, deciduous tree up to 10m, sometimes a shrub. Bark pale or dark grey. Leaves palmately 3- to 5-lobed, the lamina up to 17 cm long, sometimes broader than long, base cordate, margins sinuate-toothed, bristly above, softer and more densely hairy beneath. Syncarps (*figs*) 5-8 cm when mature, green, sometimes ripening brownish or violet.
Fl. VI - VII. *Widely cultivated in gardens and on terraces in Madeira and Porto Santo and occasionally naturalized; also found on Deserta Grande.* **MDP**

Two species of **Morus** are cultivated on the islands.

M. nigra L., [*Amoreira*], is a deciduous, rough-barked tree with somewhat pendulous branches, heart-shaped, toothed leaves which are pubescent beneath and dark-red to purple fruits. Cultivated in the lower regions of Madeira and Porto Santo for its succulent fruit. **MP**

M. alba L., is similar to *M. nigra* but with bark smoother, leaves glabrous or pubescent on the veins only beneath and fruit white or pinkish. Originally introduced for the silkworm industry in Madeira. **M**

XXVI. URTICACEAE
J.R. Press

Herbs or shrubby perennials. Leaves opposite or alternate; stipules usually present. Flowers small, usually unisexual, in dense inflorescences. Perianth 4-merous, free or united and often enlarged in fruit. Fruit an achene.

Pilea microphylla (L.) Liebm. in *Danske Vid. Selsk.*, V, **2**: 296 (1851), from tropical America, is a procumbent, fleshy annual with opposite, entire leaves and globose, axillary inflorescences. It is cultivated in Maderia and may occur as an escape in a few places. **M**

1. Leaves opposite, toothed; plant with usually numerous stinging hairs . . . . **1. Urtica**
- Leaves alternate, entire; plant without stinging hairs 2

2. Flowers solitary; stems slender, creeping and rooting at nodes, forming a ± dense mat .... **3. Soleirolia**
-   Flowers in clusters; stems slender to robust, decumbent to erect, not creeping or rooting at nodes .... **2. Parietaria**

1. Urtica L. [*Urtiga*]

Annual or perennial herbs with stinging hairs. Leaves opposite, toothed, petiolate; stipules free. Inflorescence of slender, axillary, spike-like racemes. Flowers unisexual, 4-merous, the perianth-segments of the female flowers unequal, the inner two larger and enclosing the achene.

All the species present in Madeira are monoecious, but in species **3** and **4** occasional plants occur which are wholly male or female.

Literature: J.R. Press, The identity of *Urtica subincisa* Benth. var. *floribunda* Wedd. and the presence of *U. urens* L. in Madeira, in *Bocagiana* **123**: 1-6 (1988).

1. Stipules 4 at each node; male and female flowers borne in the same raceme 2
-   Stipules 2 at each node; male and female flowers borne in separate racemes 3

2. Ripe achenes more than 1.5 mm long; leaves usually coarsely dentate, the toothing extending to, but not around, the truncate leaf base .... **2. urens**
-   Ripe achenes less than 1.5 mm long; leaves dentate but never coarsely so, the toothing extending right around the subcordate base to the petiole ...... **1. portosanctana**

3. Axis of male racemes inflated, the flowers inserted unilaterally; inner perianth-segments of female flowers ± hispid but lacking stinging hairs .... **3. dubia**
-   Axis of male racemes not inflated, the flowers not inserted unilaterally; inner perianth-segments of female flowers ± glabrous but with 1 (rarely more) stinging hairs . . **4. morifolia**

1. U. portosanctana Press in *Bocagiana* **123**: 4 (1988).
U. subincisa Benth. var. *floribunda* Wedd.
Erect annual 20-40 cm high, ± hispid, sometimes densely so and usually with numerous stinging hairs. Leaves 2-5.5(-8) × 1-4(-6) cm, ovate, rounded to subcordate, acute, coarsely dentate to dentate-serrate, the straight or equally curved teeth extending around the base to the petiole; stipules 4 at each node. Racemes with few male and numerous female flowers, generally longer than the petioles in fruit. Inner perianth-segments of female flowers somewhat hispid on the back and bearing a single stinging hair. Achenes 1-1.2 mm. **Plate 5**
Fl. IV - VI. *Endemic confined to dry, rocky places, on cliffs and slopes near the sea, mainly in Porto Santo and the Desertas but also on the Ilhèu de Fora off the Ponta de São Lourenço in Madeira; probably introduced on Selvagem Grande.* ● **MDPS**

2. U. urens L., *Sp. pl.* **2**: 984 (1753).
Like **1** but leaves cuneate to rounded, often more coarsely dentate, the teeth more falcate or curved on the outer edge but not extending around the base to the petiole; achenes 1.6-1.7(-2.5) mm.
Fl. III - IV. *A weed of waste ground, apparently uncommon but easily confused with the preceeding species and probably overlooked.* **MP**

3. U. membranacea Poir. in Lam., *Encycl.* **4**: 638 (1798).
U. azorica Seub.; *U. dubia* Forssk.
Erect, sparsely hairy annual to 30 cm with few to numerous stinging hairs. Leaves (2-)3-4.5(-7) × (5-)5.5(-10) cm, ovate, rounded to sub-cordate, acute, coarsely dentate-serrate, the teeth

very prominent; stipules 2 at each node. Male and female flowers borne on separate racemes, the lower racemes usually female and shorter than the leaves, the upper male, the male flowers inserted laterally on an inflated axis as long as or longer than the leaves. Inner perianth-segments of the female flowers hispid, lacking stinging hairs. Achenes minutely verrucose.
Fl. all year. *Common in grassy and waste places, on banks and walls throughout the islands but not recorded for Bugío.* **MDPS**

4. U. morifolia Poir., *Encyc. Suppl.* **4**: 223 (1816).
U. elevata Banks ex Lowe; *U. morifolia* var. *elevata* (Banks ex Lowe) Menezes, var. *genuina* Menezes
Suffrutescent, sparsely hispid perennial with a few scattered, stinging hairs. Leaves (3.5-)4.5 (-7.5) × (4.5-)5(-10) cm, ovate to ovate-lanceolate, cordate to subcordate, acute, crenate to dentate-serrate, the teeth sometimes irregular; stipules 2 at each node. Male and female flowers borne in separate racemes, the females generally below and shorter than the leaves, the males above with the flowers inserted on an inflated axis as long as or longer than the leaves. Inner perianth-segments of the female flowers more or less glabrous on the back except for one (rarely more) stinging hair. Achenes verrucose. **Plate 5**
Fl. IV - IX. *Rather rare plant of ravines, cliffs and rocky places, mainly in the eastern mountains of Madeira but also along the north coast.* ■ **M** [*Canaries*]

2. Parietaria L.

Annual or perennial herbs, lacking stinging hairs. Leaves alternate, entire; stipules absent. Inflorescence of axillary, bracteate, few- to many-flowered cymes. Flowers unisexual or hermaphrodite; perianth 4-merous, in female flowers tubular, segments equal and enclosing the achene.

1. Perennial; bracts shorter than fruiting perianth; ripe achenes black . . . . . **1. judaica**
- Annual; bracts at least as long as fruiting perianth; ripe achenes olive-green to brown, asymmetrically apiculate . . . . **2. debilis**

1. P. judaica L., *Fl. palaest.*: 32 (1756). [*Alfavaca de cobra*]
P. diffusa Mert. & W.D.J. Koch in Röhling; *P. maderensis* Reich.; *P. officinalis* auct., non L., subsp. *judaica* (L.) Beg., var. *diffusa* (Mert. & W.D.J. Koch) Wedd.; *P. ramiflora* Moench
Pubescent perennial. Stem usually ascending, sometimes ± erect, usually branched. Leaves very variable, 1.4-4(-6.5) × 0.8-1.5(-2.2) cm, ovate-acuminate to lanceolate; petiole of the lower leaves less than half as long as the lamina. Bracts shorter than the fruiting perianth, connate towards the base. Fruiting perianth of hermaphrodite flowers 3-3.5 mm, tubular, clearly accrescent, that of the female flowers less obvious, c. 2 mm, subglobose, the segments long-acuminate and tightly closed. Achenes c. 1 mm or less, ovoid, shiny black.
Fl. III - XII. *Among rocks, on old walls and sea cliffs. Scattered in Madeira, locally common but absent from the higher regions; also in Porto Santo and on Deserta Grande.* **MDP**

2. P. debilis G. Forst., *Fl. ins. austr.*: 73 (1786).
P. gracilis Lowe; *P. lusitanica* auct. azor., non L. (1753); *P. micrantha* Ledeb.
Pubescent annual up to 50 cm but generally smaller. Stems much-branched, decumbent and spreading to ± erect, with long, soft hairs and short, stiffer hairs. Leaves (7-)10-16(-23) × (6-)8-12(-20) mm, usually broadly ovate, acute, subcordate; petioles of the lower leaves more than half as long as the lamina. Bracts about as long as the fruiting perianth. Female and hermaphrodite flowers similar, the fruiting perianth 1.2-1.3 mm, not accrescent. Achenes c. 0.8 mm, olive-green to brown, asymmetrically apiculate.

Fl. II - VI. *Infrequent on old walls, sea cliffs and rocky places in the south-east of Madeira around Garajau and Camacha and points all along the north coast. Very rare in Porto Santo; recorded from Deserta Grande and Ilhéu Chão.* **MDP**

Two varieties occur in the islands:

i) var. **micrantha** (Ledeb.) Wedd., in DC., *Prodr.* **16**: 235[45] (1869): Stems erect or ascending, sparsely branched. Leaves 2-6 × 1.5-5 cm, broadly ovate, cordate, bright green. *Sporadic in the north of Madeira.*

ii) var. **gracilis** (Lowe) Wedd., in DC., *Prodr.* **16**: 235[46] (1869): Stems and primary branches usually numerous, procumbent. Leaves *c.* 1 cm, sometimes larger, ovate to sub-rounded, pale green. *Throughout the range of the species.*

3. Soleirolia Gaudich.

Monoecious perennials lacking stinging hairs. Stems slender, creeping. Leaves alternate, entire; stipules absent. Flowers solitary, enclosed by an involucre of 1 bract and 2 bracteoles.

1. S. soleirolii (Req.) Dandy in *Feddes Reprium* **70**: 4 (1964).
Plant sparsely hispid. Stems reddish, prostrate and rooting at the nodes to form mats. Leaves 1.5-3 mm, suborbicular. Flowers sessile in the axils of the leaves, male flowers above, female flowers below.
Fl. III. *On damp, shady rocks, by levadas and in roadside ditches; sometimes cultivated in gardens. It spreads easily by vegetative means, rapidly invading new habitats. Native to the western Mediterranean islands; first recorded in 1968 from Ribeiro Frio, now naturalized in scattered sites throughout Madeira and spreading.* **M**

PROTEACEAE
J.R. Press

Hakea sericea Schrad., *Sert. hannov.*: 27 (1795), an erect shrub with narrowly cylindrical, pungent leaves up to 7 cm × *c.* 1 mm, axillary clusters of white flowers and woody capsules is sometimes grown in Madeira as a hedging plant and may perhaps become naturalized in a few places. **M**

Various species of this S. Hemisphere family are also cultivated as garden ornamentals.

XXVII. ARISTOLOCHIACEAE
J.R. Press

Woody climbers or occasionally herbs. Leaves alternate, simple, exstipulate. Flowers solitary or in racemes or clusters, hermaphrodite, actinomorphic or zygomorphic, 3-merous. Perianth petaloid, often tubular with a 1- to 3-lobed apex, luridly coloured and foetid. Stamens 6-many, free or united to the stylar column. Ovary inferior, 6-locular. Styles free or connate into a column with 3-many lobes. Fruit a capsule, rarely indehiscent.

1. Aristolochia L.

Perennial herbs or woody climbers. Flowers axillary. Perianth zygomorphic, deciduous, the tube with a swollen utricle at the base and a unilateral limb. Stamens usually 6, in a single whorl, connate with the stylar column. Fruit a septicidal capsule.

1. A. paucinervis Pomel in *Bull. Soc. Sci. Alge* **11**: 136 (1874).
A. longa auct. non L. (1753)
Glabrescent. Stems erect, usually branched, up to 75 cm long; tuber solitary, cylindrical. Leaves 2.5-7.5 × 2-6 cm, ovate-cordate with two rounded, backward projecting lobes and a broad sinus at the base, obtuse or emarginate, slightly sinuate, the lower surface puberulent and occasionally

papillose; petiole up to 1.5 cm. Flowers solitary, 3-5 cm, greenish brown with darker veins, utricle ± globular, tube straight with a conspicuous unilateral limb. Capsule up to 3 cm.
Fl. IV - X. *Infrequent on roadside walls and banks, mainly in the south-east of the island in the Funchal region.* M

XXVIII. POLYGONACEAE
J.R. Press

Herbs, shrubs or climbers. Leaves alternate, rarely opposite, stipules united to form a membranous sheath (*ochrea*). Flowers unisexual or hermaphrodite, perianth-segments 3-6, often accrescent and becoming membranous in fruit. Stamens 6-9. Styles 2-4. Fruit a trigonous or lenticular nut.

1.	Herbs or shrubs	2
-	Climbers or twining shrubs	4
2.	Perianth-segments 5	**1. Polygonum**
	Perianth-segments 6, in 2 whorls of 3	3
3.	Outer perianth-segments as large as inner, spiny in fruit	**3. Emex**
-	Outer perianth-segments smaller than inner, not spiny in fruit	**4. Rumex**
4.	Perianth fleshy and berry-like in fruit	**5. Muehlenbeckia**
-	Perianth keeled in fruit, not fleshy	**2. Fallopia**

1. Polygonum L.

Annual or perennial herbs, sometimes shrubby. Perianth-segments usually 5, free or united towards the base, ± petaloid. Styles 2-3. Nut trigonous or lenticular, black or dark brown, enclosed by the persistent perianth.

1.	Ochreae silvery-hyaline, sometimes brownish towards the base, eventually lacerate; flowers in small axillary clusters	2
-	Ochreae brown, entire or fimbriate; flowers in spikes or capitula	4
2.	Perennial, stems woody at least at the base; ochreae strongly veined, longer than the internodes	**1. maritimum**
-	Annual; ochreae obscurely veined, much shorter than the internodes	3
3.	At least the upper bracts short, scarious; flowers usually greenish	**4. patulum**
-	All bracts leafy; flowers often pink	4
4.	Leaves heterophyllous; perianth-segments united only at base	**2. aviculare**
-	Leaves all similar; perianth-segments united up to half-way	**3. arenastrum**
5.	Plant with glandular hairs; flowers in ± globular capitula	**5. capitatum**
-	Plant glabrous to sparsely hairy with eglandular hairs; flowers in spikes	6
6.	Spikes dense, stout, ± cylindrical, the flowers crowded	7
-	Spikes very lax, slender, the flowers separate, not crowded	8
7.	Leaves with sessile yellow glands beneath	**7. lapathifolium**
-	Leaves lacking glands beneath	**6. persicaria**

8. Perianth dotted with brown glands .... **8. hydropiper**
- Perianth lacking glands .... **9. salicifolium**

1. P. maritimum L., *Sp. pl.* **1**: 361 (1753).
Glabrous perennial or dwarf shrub. Ochreae silvery-hyaline, brownish towards the base with conspicuous veins, mostly longer than the internodes. Leaves elliptical, rather thick and glaucous with inrolled margins, subsessile. Flowers white or pink, solitary or in small axillary clusters. Nut *c.* 3 mm, trigonous, shiny, exceeding the perianth.
Fl. III - VI. *Confined to littoral regions, especially along the sandy shorelines of Ponta de São Lourenço and the southern coast of Porto Santo; also on Deserta Grande.* **MDP**

2. P. aviculare L., *Sp. pl.* **1**: 362 (1753). [*Sempre noiva*]
P. aviculare var. *commune* Menezes
Glabrous, much-branched annual, erect or spreading. Ochreae silvery-hyaline, sometimes brownish towards the base, deeply lacerate, shorter than the internodes. Leaves lanceolate, heterophyllous at least when young, those on the branches much smaller than those on the main stem; petioles very short and included in the ochreae. Bracts all leafy. Flowers bright or pale pink to whitish, in clusters of 3-6 in the axils. Perianth-segments united only at the base. Nut 2.5-3 mm, trigonous, dull, shorter than the perianth.
Fl. III - X. *Ruderal, common in the lower regions of Madeira, rare in Porto Santo; recorded only once from Selvagem Grande, possibly in error as to provenance.* **MP ?S**

3. P. arenastrum Boreau, *Fl. centre France* ed. 3, **2**: 559 (1857).
P. aviculare var. *confertum* Menezes
Like **2** but stems always procumbent, much-branched, often with very short internodes. Leaves all of similar size, subsessile. Flowers greenish white to pink, in clusters of 2-3. Perianth-segments united up to half-way. Nut 2 mm.
Fl. VII - IX. *Roadsides and waste places in the lower regions of Madeira.* **M**

P. arenastrum has only recently been recorded from Madeira. Sometimes difficult to distinguish from **2** and confused with that species, it has probably been previously overlooked and may prove to be quite common on the islands. Both species share similar habitats and often grow together but *P. arenastrum* prefers generally drier habitats such as paths and tracks where it is better able to withstand trampling.

4. P. patulum M.Bieb., *Fl. taur.-caucas.* **1**: 304 (1808).
Like **2** but leaves all the same size and often falling early; flowers greenish, axillary, forming a very loose, open spike; lower bracts leafy, upper ones scarious, shorter than the flowers.
Fl. ? *Apparently a very rare introduction from S. Europe.* **MP**

5. P. capitatum Buch.-Ham. ex D. Don, *Prodr. fl. nepal.*: 73 (1825).
Prostrate perennial with rooting stems, often trailing or forming mats. Stems and peduncles pubescent with glandular hairs, often strongly tinged dark- or reddish pink. Ochreae brown with a few short cilia. Leaves ovate, petiolate, with glandular and sometimes eglandular hairs. Flowers pink, capitate, in dense, ± globular heads on long peduncles.
Fl. III. *Introduced. Cultivated, frequently escaping and becoming naturalized. Common on waste ground and especially walls in and around towns and villages in Madeira. Native to the Himalayas, but naturalized in many areas.* **M**

6. P. persicaria L., *Sp. pl.* **1**: 361 (1753). [*Herva pecegueira*]
Glabrous to sparsely hairy annual. Stems ascending, swollen at the nodes. Ochreae brown, shortly

ciliate. Leaves lanceolate, sometimes with a black patch in the centre, shortly petiolate. Flowers pale or bright pink, in short, dense, cylindrical spikes. Nut 2 mm, glossy, lenticular.
Fl. III - X. *Ruderal. Common in most parts of Madeira except the mountains; also found in Porto Santo.* **MP**

The above description refers to var. **persicaria** (var. *genuinum* Gren. & Godr.); var. **elatum** Gren. & Godr. (var. *biforme* Wahlenb.) with erect stems, paler, narrower, more pointed leaves and elongated peduncles is said to be a rare plant in Madeira but no further information is available.

7. P. lapathifolium L., *Sp. pl.* **1**: 360 (1753).
Like **5** but stems procumbent to erect, sparsely hairy and sometimes reddish or spotted. Ochreae entire except for the uppermost which are shortly ciliate. Leaves broader, narrowly ovate-lanceolate, grey-tomentose or ± glabrous beneath, with numerous yellow, rarely colourless, glands beneath. Flowers greenish or dull pink, peduncle and perianth also usually with some yellow glands.
Fl. ? *Elsewhere this is a weed of cultivated and waste ground and marshy areas, and is presumably so in Madeira. It is recorded as being common in the lower regions, but this has not been confirmed and in fact the plant seems to be infrequent.* **M**

8. P. hydropiper L., *Sp. pl.* **1**: 361 (1753).
Ascending to erect annual, glabrous except for hairs on leaf veins and margins. Ochreae brown, truncate with a few short cilia. Leaves lanceolate to narrowly ovate, shortly petiolate or subsessile. Flowers in long, lax, leafless spikes. Perianth covered with brown glands. Nut 2.5-3.5 mm, dull, slightly trigonous.
Fl. VI - X. *Infrequent in damp places along levadas, streams and river-beds in Madeira.* **M**

9. P. salicifolium Brouss. ex Willd., *Enum. pl.* **1**: 428 (1809). [*Pulgueira*]
P. serrulatum Lag.
Perennial with spreading, rooting branches. Glabrous or sparsely hairy with hairs confined to the leaf margins and veins. Ochreae brown, strongly ciliate with long cilia. Leaves linear-lanceolate, shortly petiolate. Flowers pink, in long, lax, slender spikes. Nut *c.* 2 mm, glossy, trigonous. 2n=42.
Fl. IV - VIII. *A plant of wet places, stream sides and river-beds, formerly common in a few places in Madeira, such as around Funchal and Seixal, but now possibly rarer.* **M**

2. Fallopia Adans.

Annual or perennial climbers. Flowers in loose paniculate or spike-like inflorescences. Perianth-segments 5, rarely 6, the outer 3 becoming keeled or winged in fruit. Stamens 8. Styles 3. Nut triquetrous.

1. F. convolvulus (L.) Á. Löve in *Taxon* **19**: 300 (1970).
Polygonum convolvulus L.
Annual with climbing or spreading stems, often mealy. Leaves cordate to sagittate, acuminate, petiolate. Inflorescences narrow, unbranched. Pedicels up to 3 mm in fruit, articulated above the middle. Perianth-segments accrescent, the inner whitish, the outer greenish and keeled in fruit. Nut 4-5 mm, dull, finely granular.
Fl. VI - VIII. *An uncommon and sporadic field weed occurring on disturbed soil throughout cultivated regions in Madeira.* **M**.

F. baldshuanica (Regel) Holub is a vigorous woody climber with drooping panicles of pink or white flowers. Native to C. Asia, it is sometimes grown in gardens.

3. Emex Campd.

Monoecious annuals. Female flowers borne at the base of the inflorescence. Perianth-segments 6, free in male flowers, in female flowers connate with the outer 3 becoming spiny and hard in fruit. Stamens 4 or 6. Styles 3. Nut trigonous, concealed within the perianth.

1. E. spinosa (L.) Campd., *Monogr. Rumex*: 58 (1819). [*Labaça*]
Glabrous annual up to 50 cm, stems weakly erect to ascending, often reddish-tinged. Basal leaves up to 12 × 7 cm, bluntly ovate to oblong-cordate, petioles as long as or longer than the lamina. Male flowers in globular, pedunculate clusters. Female flowers sessile in axillary clusters. Tips of the outer 3 perianth-segments of female flowers wide-spreading and spinescent in fruit, tips of the inner 3 erect. Nuts 4-5(-6.5) mm, those produced at the root-collar often much larger. Fl. ± all year. *Common along shorelines, in waste places and on disturbed ground near the sea throughout the islands but not recorded from the Desertas.* **MPS**

4. Rumex L.

Perennials, sometimes woody, usually with long, stout roots, rarely annuals. Ochreae tubular. Flowers hermaphrodite or unisexual, in whorls, anemophilous. Outer 3 perianth-segments always small and thin, the inner 3 (*valves*) enlarging and often hardening in fruit. Valves with or without marginal teeth and dorsal tubercles which develop as they mature. Fruit a trigonous nut.

Positive identification is difficult without ripe fruits; size and shape of the valves, number and length of the teeth and size and shape of the tubercles are all important characters. In the following account measurements and descriptions of valve refer to those of ripe fruits

Literature: J.R. Press, Intraspecific variation in *Rumex bucephalophorus* L. In A.O. Chater (ed.), Flora Europaea: Notulae systematicae ad Floram Europaeam spectantes. Series 2. No. 2., in *Bot. J. Linn. Soc.* **97**: 335-355 (1988).

1. Tubercles absent, or, if present, hidden by the outwardly folded valves; at least the basal leaves usually hastate 2
- At least one valve with a clearly visible tubercle; leaves hastate 4

2. Plant dioecious; valves less than 2 mm long, ovate, not membranous, barely exceeding the nut . . . . **1. acetosella**
- Plant with hermaphrodite flowers; valves more than 2 mm long, ± orbicular, membranous, greatly exceeding the nut 3

3. Annual; valves folded longitudinally, completely hiding tubercle sand nut . . . .
 . . . . **2. simpliciflorus**
- Perennial; valves not folded to hide nut . . . . **3. maderensis**

4. Outer perianth-segments reflexed in fruit; basal leaves less than twice as long as broad; annual or slender perennial . . . . **8. bucephalophorus**
- Outer perianth-segments not reflexed in fruit; basal leaves at least twice as long as broad; robust perennials 5

5. Valves with distinct teeth more than 0.5 mm long 6
- Valves entire or with indistinct teeth or crenulations less than 0.5 mm long 7

6. Pedicels about as long as the valves, thick, persistent; valves with (5-)6-8 teeth on each side, the tubercles becoming verrucose when dry . . . . **4. pulcher**
- Pedicels twice as long as valves, slender, readily deciduous; valves with up to 5 teeth on each side, the tubercles not becoming verrucose . . . . **5. obtusifolius**

7. Valves more than 4 mm long, cordate; whorls of infructescence mostly crowded .... **6. crispus**
- Valves less than 4 mm long, oblong to oblong-lanceolate; all whorls of infructescence remote .... **7. conglomeratus**

1. R. acetosella L., *Sp. pl. 1: 338 (1753)* [*Azedinha*]
Slender, glabrous, dioecious perennial producing adventitious buds on long horizontal roots. Stems 10-40 cm, erect or occasionally ascending. Basal leaves (1.5-)2-3(-4) × (0.3-)0.5-1(-1.4) cm, hastate, acute to obtuse, entire, the mid-lobe lanceolate, the basal lobes usually forward-pointing, sometimes divided, sometimes very short; cauline leaves progressively smaller and narrower, hastate, rarely lanceolate; petioles about as long as lamina, rarely longer in the basal leaves. Inflorescence branched, the branches ascending, aphyllous; whorls with 3-6 flowers, all remote. Pedicels *c*. 1 mm, slender, articulated at the tip. Valves 1.5 × 1 mm, ovate, thin, entire, lacking tubercles, not easily separable from the nut. Nut *c*. 1 mm long and almost as broad.
Fl. III - VI. *Very common in pastures, cultivated ground, wayside and waste places in Madeira.*
M

Madeiran plants belong to subsp. **pyrenaicus** (Pourret ex Lapeyr.) Akeroyd in *Bot. J. Linn. Soc.* **106**: 99 (1991) (*R. angiocarpus* auct.). Elsewhere it occurs in SW Europe and N. Africa and is distinguished from other subspecies by the hastate leaves having a single (or rarely several) pairs of basal lobes and by the fruit with nut not readily separable from valves.

2. R. simpliciflorus Murb. in *Acta Univ. lund.* **35** Afd. II, no. 3: 11 (1899).
R. simpliciflorus var. *maderensis* Murb., var. *rubellianus* Menezes
Annual, 12-25 cm. Stems ascending or ± erect, branched. Basal leaves (2.5-)3-4 × (0.8-)1.5-2.5 cm, ovate-triangular to subhastate, acute, cuneate, the upper narrower, entire of minutely crenulate; petiole as long as or slightly longer than the lamina. Flowers in clusters of 3-4(-5); pedicels *c*. 5 mm, slender, articulated at about halfway or below. Outer perianth-segments reflexed in fruit. Valves 7.5-9 × 6.5-9 mm, ± orbicular, cordate, deeply emarginate, entire, thin and very membranous, reticulately veined, sometimes reddish when mature, folded longitudinally to completely hide the small, reflexed tubercles. Nut 3.5 mm, slightly winged on the angles.
Plate 6
Fl. III - V. *A rare plant recorded only from seashores and cliffs in the Funchal - Garajau region.*
M

In Madeira this species is represented by the endemic (●) subsp. **maderensis** (Murb.) Samuelson in *Bot. Notiser* **1939**: 522 (1939), to which the above description and synonymy apply. Subsp. *simpliciflorus* occurs in N. Africa and Arabia, from Morocco to the Red Sea, but is absent from Madeira. The closely related *R. vesicarius* L., with which Madeiran plants have been confused, occurs in the Canary Islands (as var. *rhodophysa* Ball), N. Africa and Asia.

3. R. maderensis Lowe, *Trans Camb. phil. Soc.* **4**: 534 (1834). [*Azeda*]
R. maderensis var. *glaucus* Lowe, var. *virescens* Lowe
Glabrous perennial 20-100 cm or more, woody towards the base, sometimes trailing, Stems flexuous, reddish. Leaves pale green or glaucous, 5-8 × 3-6.5(-8) cm, hastate to ± triangular, obtuse to acute, entire; petioles 4-7 cm, shorter than to a little longer than the lamina, Panicle many-branched, open and spreading; whorls all remote, each with 2-5(-7) flowers; pedicels 3-6 mm, filiform, articulated below the middle. Outer perianth-segments patent or reflexed in fruit. Valves 3-5.5 × 3-6 mm, orbicular, cordate, entire, thin and membranous, lacking tubercles, flushed pink or purplish. Nut *c*. 2 mm. 2n=20. **Plate 6**
Fl. VI - IX. *Common on banks, cliffs, old walls and rock faces throughout the island at heights of about 500-1000 m in Madeira.* ■ **M** [*Canaries*]

4. R. pulcher L., *Sp. pl.* **1**: 336 (1753). [*Coenha* (P), *Labaca* (M)]
Perennial to 40 cm high. Basal leaves 8-9.5 × 1.7-2.6 cm, oblong, occasionally panduriform, obtuse to acute, ± cordate, crenulate, somewhat fleshy; petiole 3-4.5 cm, shorter than the lamina, the petioles, ochreae and sometimes leaves white papillose, sometimes densely so. Inflorescence open, the branches at 90° or ascending; whorls many-flowered, remote, each subtended by a leafy bract; pedicels 2.5-4(-7) mm, somewhat thickened, articulated at or below the middle, not deciduous. Valves 4-5.5 × 2-3.5 mm, orbicular-ovate to triangular-ovate, ± obtuse, strongly reticulate, each valve with an oblong tubercle becoming verrucose to rugose when dry, 1 larger than the other 2, teeth 0.5-1.5 mm, less than half the width of the valve, (5-)6-8 on each side.
Fl. IV - VIII. *A weed of paths, roadsides and waste ground. Present on most of the islands, common in Madeira, rare elsewhere but not recorded for Bugío and Ilhéu Chão.* **MDPS**

Madeiran plants have been placed in subsp. **divaricata** (L.) Murb. in *Acta Univ. lund.* **27**(5): 46 (1891) but do not well fit the description, having smaller, narrower valves with longer teeth and occasionally panduriform leaves. They are better regarded as representing subsp. **pulcher**.

5. R. obtusifolius L., *Sp. pl.* **1**: 335 (1753).
Perennial up to 90 cm. Basal leaves up to 17 × 9.5 cm, oblong-ovate, obtuse to subacute, entire, sometimes slightly undulate, rather thin in texture, papillose beneath; petiole as long as or longer than the lamina; cauline leaves similar but smaller, generally narrower, acute and ± cuneate. Whorls all remote, each with numerous flowers; pedicels up to 10 mm, slender, articulated at or below the middle, easily deciduous. Outer perianth-segments somewhat patent in fruit. Valves 3-4.5 × 2-4 mm, ± triangular, acute or obtuse, with several (-5) well-developed teeth on each side, membranous, reticulate, at least one valve with a well-developed, oblong tubercle, the other two usually lacking tubercles or with very small ones. Nut 2.2-2.5 mm.
Fl. IV - X. *A weed of cultivated land and waste places, especially where damp; it requires freshly turned soil to establish successfully. Native to Europe and probably introduced.* **M**

Plants from Madeira are subsp. **obtusifolius**.

6. R. crispus L., *Sp. pl.* **1**: 335 (1753). [*Labaça*]
Glabrous perennial to 80 cm or more. Stems erect, reddish tinged. Lower cauline leaves up to 15 × 4 cm, narrowly lanceolate, acute, cuneate, crenulate, slightly undulate or crisped; petiole to 3 cm. Panicle sparingly branched, the branches erect and close to the main axis; upper whorls ± crowded, the lower remote, all but the uppermost subtended by a leafy bract; pedicels 5.5-12 mm, slender, articulated in the lower half. Outer perianth-segments patent in fruit. Valves 4 × 3.5-4 mm, ± cordate, membranous, slightly reticulate, entire or sometimes minutely toothed; all valves with a tubercle, at least one of them well-developed. Nut 2.3 mm.
Fl. III - VIII. *Waste places and cultivated land in Madeira and Porto Santo. Native to Europe and probably introduced.* **MP**

7. R. conglomeratus Murray, *Prodr. Stirp. Gotting.*: 52 (1770). [*Labaça*]
Glabrous perennial to 45 cm with erect, much-branched stems. Basal leaves up to 19 × 5 cm, oblong-lanceolate, obtuse to acute, rounded, entire, the cauline leaves progressively smaller and becoming lanceolate; petiole up to 4 cm. Whorls all remote, clearly separate on the stems, all except the uppermost subtended by a leafy bract, each with numerous flowers; pedicels 1.5-3(-4.5) mm, slender, articulated in the lower half. Outer perianth-segments pressed against the valves in fruit. Valves 2-2.5 × 1-1.5 mm, oblong to oblong-lanceolate, entire or rarely with minute teeth, each usually with a prominent oblong tubercle half the length of the valve or more. Nut 1.6 - 2 mm.
Fl. III - VII. *Damp and marshy places by watercourses, along sea-cliffs and shores in the lower regions of Madeira, mainly in the eastern parts. Native to Europe and probably introduced.* **M**

8. R. bucephalophorus L., *Sp. pl.* **1**: 336 (1753).
Annual or perennial. Stems few to many, 2-30 cm, arcuate-ascending to somewhat sprawling, slender or filiform to robust, usually much branched. Leaves with lamina 0.5-45 × 0.5-40 cm, broadly ovate-spathulate to ovate or rhombic, sometimes suborbicular, obtuse to acute, base rounded to cuneate and narrowing into the long petiole. Whorls remote, few-flowered; pedicels articulated in the lower half, the distal half becoming thickened and nodose. Outer perianth-segments reflexed in fruit. Fruits either homomorphic, all with valves 2-2.5 × 0.8-1 mm, narrowly triangular and with 4-5 slender, straight teeth on each margin, or heteromorphic with some fruits with valves 2.5-3.5 × 0.7-1.5, lingulate, entire. Heteromorphic plants have entire fruits at the lower nodes, entire and toothed fruits at the middle nodes and toothed fruits at the upper nodes.
Plate 6
Fl. III - VI. *Infrequent to rare. Disturbed soils, among rocks and by tracks and levadas, up to c. 1600 m in Madeira; in Porto Santo mostly on the high peaks; also from the Desertas.*
MDP

Madeiran plants all belong to subsp. **canariensis** (Steinh.) Rech. f. in *Bot. Notiser* **1939**: 502 (1939), to which the above description applies. Two varieties are recognized within this subspecies.

i) var. **canariensis**
Annual. Stems filiform to slender or robust.
Throughout the range of the subspecies but preferring dry, open situations. ■ [*Azores, Canaries*]

ii) var. **fruticescens** (Borm.) Press in *Bot. J. Linn. Soc.* **97**: 354 (1988).
R. bucephalophorus subsp. *fruticescens* Bornm.
Perennial forming dense tufts. Stems numerous, usually robust, rooting from the lower nodes. 2n=16.
Endemic to the high, central areas of Madeira, preferring damp situations. ●

5. Muehlenbeckia Meisn.

Woody, scandent perennials, dioecious or polygamous. Ochreae brown, opaque, entire, eventually becoming lacerate and disappearing. Perianth-segments 5, connate towards the base, accrescent. Stamens 8, represented by staminodes in female flowers. Styles 3, subsessile. Nut triquetrous, partly fused with the perianth.

1. M. sagittifolia (Ortega) Meisn., *Pl. vasc. gen.* **2**: 227 (1841).
Glabrous climber with twining branches. Leaves lanceolate, acuminate, hastate or truncate at the base. Flowers greenish, in lax, axillary racemes. Fruiting perianth fleshy, red at first, becoming white.
Fl. III. *Introduced, occasionally occurring as an escape from gardens and apparently naturalized on maritime rocks in a few places in Madeira. Native to temperate S. America, naturalized in the Azores.* **M**

M. complexa (A. Cunn.) Meisn., *Pl. vasc. gen.* **1**: 227 (1841) and **M. platyclados** Meisn. in *Bot. Ztg.* **22**: 313 (1865) (*Homalocladium platycladum* (Meisn.) Bailey), both from the southern hemisphere, are grown in gardens in Madeira and may occasionally occur as escapes. **M**

XXIX. CHENOPODIACEAE
J.R. Press

Annual or perennial herbs or shrubs, often succulent or farinose. Leaves alternate, rarely opposite; stipules absent. Flowers hermaphrodite or unisexual. Perianth-segments 1-5, often accrescent in fruit, or absent and replaced by 2 sepaloid bracteoles. Stigmas usually 2-3(-5). Fruit an achene. Seeds horizontal or vertical.

1. Leaves narrow, less than 5 mm wide, filiform or semi-cylindrical 2
 - Leaves broad, more than 5 mm wide, flat 3

2. Leaves spiny at the tips . . . . **6. Salsola**
 - Leaves not spiny . . . . **5. Suaeda**

3. Flowers all unisexual; perianth absent, the fruit enclosed by 2 sepaloid bracteoles . . . .
    . . . . **3. Atriplex**
 - Flowers mostly hermaphrodite; perianth-segments 2-5, even in unisexual flowers . 4

4. Perianth-segments and receptacle swollen in fruit; flowers often adhering in clusters . .
    . . . . **1. Beta**
 - Perianth-segments and receptacle not swollen in fruit; fruits single, not adhering . . 5

5. Plant densely white-tomentose; perianth-segments not keeled on the back in fruit . . . .
    . . . . **4. Bassia**
 - Plant glabrous to pubescent or farinose but never tomentose; perianth-segments often keeled on the back in fruit . . . . **2. Chenopodium**

1. Beta L.

Glabrous herbs. Leaves flat, ± entire. Flowers hermaphrodite, in few-flowered cymes, rarely solitary in the axils, in a branched, spicate inflorescence. Perianth-segments 5, keeled on the back, swelling along with the receptacle in fruit and causing the fruits to adhere. Stigmas 2-3. Seeds horizontal.

1. Inflorescence ebracteate, at least above, bracts if present, very small; fruits in clusters of 2-4(-5); receptacle basin-shaped 2
 - Inflorescence with large, leafy bracts ± to the apex; clusters with usually only one fruit developing; receptacle hemispherical 3

2. Leaves narrow, sessile . . . . **2. patula**
 - Leaves broad, clearly petiolate . . . . **1. vulgaris**

3. Procumbent annual; leaves ovate-cordate . . . . **4. patellaris**
 - Ascending perennial; leaves lanceolate, hastate to rhombic . . . . **3. procumbens**

1. B. vulgaris L., *Sp. pl.* **1**: 222 (1753).
Annual, biennial or perennial, sometimes with a conspicuously swollen root. Stems procumbent to erect, leafy. Leaves variable but usually ovate-cordate to rhomboid or lanceolate, often tinged dark red and the lower forming a basal rosette. Inflorescence dense, ebracteate or with narrow leafy bracts below. Perianth-segments incurved in fruit. Fruits adhering in clusters.
Fl. III - IX. **MDP**

a) subsp. **maritima** (L.) Arcang., *Comp. fl. ital.*: 593 (1882).
Stems usually procumbent or spreading. Leaves up to *c.* 10 cm, rhomboid. Cymes with (1-)2-3(-4) flowers.
Infrequent along shores and in fields near the sea along the southern coast of Madeira, in Porto Santo and Bugío.

b) subsp. **cicla** (L.) W.D.J. Koch, *Syn. deut. schweiz. Fl.* **2**: 606 (1837).   [*Acelga, Celga*]
B. vulgaris var. *cicla* L., var. *portosanctana* Gand. ex Menezes
Stems usually erect, robust. Leaves up to 20 cm, often ovate-cordate, the midrib sometimes fleshy. Cymes with 3-4 or more flowers.
Cultivated on Madeira and Porto Santo as a vegetable and sometimes occurring as an escape.

Subsp. **vulgaris** (*B. vulgaris* var. *crassa* Alef., var. *cruenta* Alef., var. *esculenta* Salisb.) [*Beterraba*], is also widely cultivated in Madeira and Porto Santo, both as a vegetable and as an ornamental foliage plant. It is similar to **b)** but with a usually conspicuously swollen root and undulate leaves lacking a fleshy midrib.

2. **B. patula** Aiton, *Hort. kew.* **1**: 352 (1789).
Biennial. Stems to 30 cm, branching freely from the base, the branches spreading or ascending. Leaves up to 8 × 0.5-1 cm, linear to lanceolate or oblanceolate, gradually narrowed at the base, sessile. Cymes 2- to 6(-9)-flowered, the lower subtended by leafy bracts, the upper usually ebracteate. Perianth-segments incurved in fruit. Fruits adhering in clusters. **Plate 7**
Fl. III - VII. *A rare endemic confined to the littoral of the Ponta de São Lourenço and islets, and Ilhéu Chão.* ● **MD**

Very similar to **1** except for the narrow, sessile leaves and 2- to 6-flowered cymes and perhaps only a subspecies of it.

3. **B. procumbens** C. Sm. ex Hornem., *Suppl. Hort. bot. hafn.*: 31 (1819).
Patellifolia procumbens (C. Sm. ex Hornem.) Scott, Ford-Lloyd & Williams
Ascending or procumbent perennial, often much-branched from the base. Leaves and bracts broadly lanceolate-hastate to narrowly rhombic, mostly acuminate. Inflorescence a lax spike, bracteate ± to the apex. Flowers in clusters of 1-3, but rarely more than 1 fruit developing. Perianth-segments erect or incurved in fruit. Receptacle hemisphaerical. $2n=18$. **Plate 7**
Fl. III - VII. *Rare littoral plant most common on the Salvages; recorded in Madeira from Praia Formosa, and in Porto Santo from Campo de Baixo and Serra de Fóra.* ■ **MPS** [*Canaries, Cape Verdes*]

4. **B. patellaris** Moq. in DC., *Prodr.* **13**(2): 57 (1849).
Patellifolia patellaris (Moq.) Scott, Ford-Lloyd & Williams
Like **1** but a procumbent annual; leaves and bracts broadly ovate-cordate, acute to obtuse. $2n=36$.
Fl. II - IV. *Littoral plant of stony and sandy shores, scattered and uncommon around the coasts of Madeira and Porto Santo but more common on Salvage Islands.* **MPS**

2. Chenopodium L.

Annual or perennial herbs or small shrubs, often farinose, sometimes glandular. Leaves flat, not fleshy. Flowers hermaphrodite or female, in small, ± globular cymes forming a branched inflorescence. Bracteoles absent. Perianth-segments 2-5, variously connate from the base nearly to the apex. Stigmas 2(-5). Seeds usually horizontal, sometimes vertical, especially in terminal flowers.

The surface markings of the testa are regarded as important specific characters but can only be clearly seen with a magnification of × 40, after removal of the membranous pericarp.

Literature: A.J. Scott, A review of the classification of *Chenopodium* and related genera (Chenopodiaceae) in *Bot. Jb. Syst.* **100**(2): 205-220 (1978).

1. Plant pubescent, not farinose; leaves aromatic, with numerous yellow glands . . . . . .
   . . . . **1. ambrosioides**
- Plant glabrous or farinose; leaves eglandular, not aromatic 2

2. Seeds with an acute margin; testa minutely pitted 3
- Seeds with an obtuse margin; testa smooth or with shallow, radial grooves 4

3. Leaves 1- to 2-pinnatifid, lobes cut halfway to the mid-rib . . . . **5. coronopus**
- Leaves sometimes coarsely toothed but not pinnatifid . . . . **4. murale**

4. Leaves generally 1½ × as long as wide, often toothed; plant white-farinose **2. album**
- Leaves less than 1½ × as long as wide, sometimes 3-lobed and the lobes toothed; plant grey-farinose . . . . **3. opulifolium**

1. C. ambrosioides L., *Sp. pl.* **1**: 219 (1753).   [*Formigueira, Lombrigueira*]
Very aromatic annual, rarely perennial. Stems up to 75 cm or more, sparsely hairy. Leaves up to 10 cm, lanceolate, entire or sinuate to irregularly toothed, the lower surface with numerous sessile, yellowish glands. Inflorescence a freely branching panicle composed of numerous sessile, bracteate cymes. Perianth-segments not keeled on the back. Seeds 0.6-0.8 mm.
Fl. I - IX. *A common weed on waste ground, along roadsides and tracks in Madeira and Porto Santo.* **MP**

A polymorphic species which has been divided into numerous subspecific taxa. Madeiran plants all appear to belong to the subsp. **ambrosioides.**

2. C. album L., *Sp. pl.* **1**: 219 (1753).   [*Fedegoso* (P)]
Annual, 30-60 cm, usually white-farinose. Stem erect, striped green and white, often tinged with red or purple. Leaves 1.5-6 × 1-1.5 cm, variable but usually at least 1½ × as long as wide, the middle and lower rhombic to ovate or ovate-lanceolate, bluntly dentate or entire, the upper leaves lanceolate, entire, often acuminate. Inflorescence a branched, spicate or cymose panicle. Perianth-segments 5, sometimes keeled on the back. Seeds 1.2-1.4 mm diameter, margin slightly obtuse, testa ± smooth or with shallow, widely spaced radial furrows.
Fl. III - X. *A common weed of roadsides, waste places and cultivated land, especially field margins and neglected terraces, throughout Madeira and Porto Santo.* **MP**

A polymorphic and taxonomically difficult species, sometimes regarded as forming an aggregate with several other, very similar, species. It has been divided into numerous subspecific taxa, many based on characters of the inflorescence, which is very variable. In Madeira it has possibly been confused with **3**.

3. C. opulifolium Schrad. ex W.D.J. Koch & Ziz, *Fl. Palat.*: 6 (1814).
Like **2** but usually grey-farinose and lacking reddish tinge on the stem; middle and lower leaves about as wide as long, triangular to 3-lobed, the lobes entire or with several teeth, upper leaves rhombic-ovate to 3-lobed, lobes acuminate.
Fl. VI - X. *A weed in similar habitats to* **2**. **MP**

Little is known of the distribution or frequency of this species in Madeira as it is difficult to distinguish from **2** with which it has been confused. It is apparently quite common around Vila Baleira in Porto Santo and has been reliably recorded from several sites in Madeira, mainly around Funchal and Machico; it is probably more widespread than has been supposed.

C. giganteum D. Don, *Prodr. fl. nepal*: 75 (1825), from northern India is like **2** but much larger, with leaves up to *c*. 14 cm long and wide, young leaves and inflorescence tinged bright reddish purple. It has been recorded for Madeira but no other data are available. **M**

4. C. murale L., *Sp. pl.* **1**: 219 (1753). [*Fedegoso* (P)]
Erect or spreading annual to 80 cm, often robust and much branched, glabrous, usually farinose, sometimes tinged deep red. Leaves 2.5-7 × 2-5 cm, triangular to rhombic-ovate, coarsely and irregularly dentate. Inflorescence axillary and terminal, divaricately branched, often dense. Perianth-segments bluntly keeled in the upper part of the back. Seed 1.2-1.5 mm diameter, margin acute, testa minutely pitted.
Fl. III - IV. *A widespread weed of disturbed soils on cultivated and waste land, also along roadsides and occasionally in rocky pastures. Common in Madeira and Porto Santo; also found on Ilhéu Chão and Bugío.* **MDP**

C. vulvaria L., *Sp. pl.* **1**: 220 (1753), a grey-farinose annual with a characteristic smell of bad fish has been recorded as a weed from the Serviços Florestais nursery in Porto Santo. It may occur only as a casual but has become naturalized in many areas outside its native Mediterranean region. **?M**

5. C. coronopus Moq. in DC., *Prodr.* **13**(2): 76 (1849).
Like **4** but stems more slender, up to 35 cm; leaves up to 5 × 2(-2.5) cm, 1- to 2-pinnatifid; lobes extending half-way to the mid-rib, linear to linear-ovate or narrowly triangular, slightly incurved; perianth-segments not keeled on the back. **Plate 7**
Fl. II. *On sandy shores and open ground on Selvagem Grande and Selvagem Pequena; possibly also present on Deserta Grande and Ilhéu Chão.* ■ **?D S** [*Canaries*]

This species appears to closely resemble plants from N. Africa and elsewhere and known as **C. murale** var. **spissidentatum** Murray in *Magy. bot. Lap.* **2**: 11 (1903). The exact relationship of these taxa is unknown. Records of **5** from the Desertas are based on poor specimens and require confirmation.

3. Atriplex L.

Annual herbs or small shrubs, often farinose. Leaves sometimes fleshy but always flat. Flowers unisexual, the males with 5 perianth-segments, the females lacking a perianth but enclosed by 2 persistent bracteoles, free or connate to the middle. Stigmas 2. Seeds usually vertical, often dimorphic.

1. Annual . . . . **3. prostrata**
- Shrubby perennial 2

2. Bracteoles orbicular to reniform, not lobed; leaves *c.* 4 cm, rhombic-ovate, shortly petiolate . . . . **2. halimus**
- Bracteoles triangular-ovate to deltate, often 3-lobed; leaves usually less than 3 cm, oblong-lanceolate, subsessile . . . . **1. glauca**

1. A glauca L., *Sp. pl.* ed. 2, **2**: 1493 (1763). [*Maçacota*]
Much-branched, shrubby perennial. Leaves greyish green to silvery, up to 3 cm long but often much shorter, oblong-lanceolate to obovate, slightly attenuate, entire or rarely dentate, slightly fleshy, 1-nerved, subsessile. Inflorescence long, spicate. Bracteoles 4-5 × 4-7 mm, connate to the middle, broadly triangular-ovate to deltate, ± entire or often 3-lobed, with numerous appendages on the back. Seeds brownish-black.
Fl. VI - VIII. *On roadsides, banks and dry gullies over much of Porto Santo; also present in Madeira on the Ponto de São Lourenço and Cabo Girão.* **MP**

Madeiran plants represent var. **ifniensis** (Caball.) Maire in *Bull. Soc. Hist. nat. Afr. N.* **28**: 377 (1937) (*A. parvifolia* Lowe), to which the above description applies. This variety is found elsewhere only in the Canaries and the Ifni region of Morocco and is distinguished from other subspecies by the concolorous leaves oblong, entire, attenuate at base and the sessile involucre with valves ovate-rhombic, entire and strongly tuberculate.

2. A. halimus L., *Sp. pl.* **2**: 1052 (1753). [*Salgueiro* (P)]
Shrubby perennial to 2 m or more. Leaves silvery, *c.* 4 cm, rhombic-ovate, the upper sometimes lanceolate, entire, obscurely 3-nerved at the base, shortly petiolate. Inflorescence paniculate. Bracteoles connate at the base, ± orbicular to reniform, entire or dentate but not lobed and usually lacking appendages on the back. Seeds brownish-black.
Fl. VIII. *Dry roadsides and waste ground around Vila Baleira, in Porto Santo.* **P**

Very variable; in the absence of fruiting material it is easily confused with **1**, which is usually smaller and more prostrate, with shorter and narrower leaves. Madeiran plants are var. **halimus**.

3. A. prostrata Boucher ex DC. in Lam. & DC., *Fl. franç.* ed. 3: 387 (1805).
A. hastata auct. mad. [as *hastatum*], & sensu Aellen in Flora Europaea 1: 97 (1964), non L. (1753); *A. triangularis* Willd.
Erect to ascending or prostrate annual, sometimes farinose. Lower leaves ± opposite, hastate or triangular with a truncate base, entire or slightly dentate, clearly petiolate. Upper leaves alternate, sometimes ovate. Inflorescence spicate. Bracteoles 3-4 mm broad and long, connate at the base, triangular to deltate, entire to dentate, with appendages on the back. Seeds of 2 kinds, few, large and dull brown, and numerous small, shiny black.
Fl. VI - IX. *Rare, confined to sea-cliffs along the north coast of Madeira from São Jorge westwards, and on the gravel shore at Madalena do Mar on the south coast.* **M**

4. Bassia All.

Annual or perennial herbs or small shrubs, usually villous. Leaves cylindrical (outside Madeira) or flat, entire, sessile. Flowers hermaphrodite or female, solitary or in few-flowered cymes forming a panicle. Bracteoles absent. Perianth-segments 5, sometimes accrescent and developing spines or tubercles on the back (outside Madeira). Stigmas 2-3. Seed horizontal.

1. B. tomentosa (Lowe) Maire & Weiller, *Fl. Af. N.* **8**: 53 (1962). [*Maçacota*]
Chenolea lanata (Masson) Moq. in DC.; *Chenoleoides tomentosa* (Lowe) Botsch.; *Salsola lanata* Masson; *Suaeda tomentosa* Lowe
Woody perennial, much-branched, densely lanate. Leaves up to 1 cm, ± semi-cylindrical, linear or lanceolate, the upper broader and overlapping. Flowers mostly solitary, sometimes 2-3 in the axils of the upper leaves. Perianth 2.5 mm, the segments fused for ⅔ of their length.
Fl. XII - I. *Littoral plant, in Madeira confined to the Ponta de São Lourenço and adjacent islets; on the south side of Porto Santo; Ilhéu de Cima; Ilhéu Chão and Deserta Grande; Selvagem Pequena and Ilhéu de Fora.* **MDPS**

5. Suaeda Forsk. ex Scop.

Annuals or shrubby perennials. Leaves usually semi-cylindrical. Flowers hermaphrodite or female, solitary or in few-flowered cymes. Bracteoles 2-3, minute. Perianth-segments 5. Stigmas 2-5. Seeds horizontal or vertical.

1. S. vera J.F. Gmel., *Syst. nat.* ed. 13, **2**: 503 (1792). [*Barrilha*]
S. fruticosa auct.; *S. laxifolia* Lowe, var. *crassifolia* Lowe, var. *tenuifolia* Lowe
Much-branched, shrubby perennial to about 1 m. Leaves 4-10 × *c.* 1 mm, fleshy. Flowers solitary or in 2- to 3-flowered cymes which are shorter than the bracts. Perianth-segments somewhat membranous, connate for about ⅓ of their length. Seeds usually vertical.

Fl. III - VI. *Maritime plant of rocky and sandy shores. Rare along the south coast of Madeira and the Ponta de São Lourenço, in Porto Santo and on Ilhéu Chão and Deserta Grande; common on the Salvages.* **MDPS**

6. Salsola L.

Annuals or small shrubby perennials. Leaves very narrow or semi-cylindrical, succulent. Flowers hermaphrodite, solitary or in clusters in the axils. Bracteoles 2, large. Perianth-segments 5, membranous, often winged on the back in fruit. Stigmas 2, rarely 3. Seeds horizontal.

1. S. kali L., *Sp. pl.* **1**: 222 (1753). [*Carqueja brava*]
Erect or spreading, hispid annual. Leaves, bracts and bracteoles spine-tipped. Leaves 1.5-3.5 cm, linear. Bracts and bracteoles triangular-ovate, patent. Perianth-segments ovate, acuminate. Fl. IV - V. *Common on Porto Santo along sandy shorelines, on dunes and banks near the sea.* **P**

Nucularia perrini Batt., *Bull. soc. bot. Fr.* **50**: 469 (1903). A single, sterile specimen of this small Saharan shrub has been recorded from Ilhéu de Fora in the Salvage Islands. It has opposite, fleshy, narow and spine-tipped leaves with tufts of woolly hairs in the axils. It is probably not established on the island. **S**

XXX. AMARANTHACEAE
N.J. Turland

Annual or perennial herbs, sometimes woody-based or sub-shrubby. Leaves alternate or opposite, simple, without stipules. Inflorescence spicate, capitate, or of axillary cymose clusters which are often condensed into spikes or panicles, at least above. Flowers hermaphrodite or unisexual, inconspicuous, subtended by small, often membranous bracteoles; perianth dry and membranous. Fruits dry, membranous.

Gomphrena celosioides Mart. in *Nova Acta physico-med.* **13**: 301 (1826) is a much-branched annual herb to 50 cm, with leaves opposite, oblanceolate to narrowly oblong, glabrous to sparsely hairy above and densely white-hairy at the margin and beneath; inflorescences are dense sessile spikes to 7 cm, coloured white or purple. Native to Central and South America, it has been recorded from Funchal (2n = 26), presumably as an escape from cultivation. **M**

Iresine herbstii Hook. in *Gdnrs Chron.* **1864**: 654, 1206 (1864) is a reddish-tinted perennial herb to 50 cm, with leaves opposite, broadly ovate to suborbicular, concave, deeply emarginate and flowers inconspicuous, whitish, borne in panicles 5-10 cm long. Native to Brazil, it is often cultivated for ornament in Madeira, in various colour variants, and has been recorded as an escape in a waste place in Funchal. **M**

1. Leaves alternate; inflorescence of axillary cymose clusters often condensed into spikes or panicles . . . . **1. Amaranthus**
- Leaves opposite; inflorescence spicate (but not as above) or capitate 2

2. Inflorescence spicate, elongate; fruits deflexed . . . . **2. Achyranthes**
- Inflorescence capitate, ovoid, dense; fruits not deflexed . . . . **3. Alternanthera**

1. Amaranthus L. [*Bredos*]

Annual or perennial herbs, monoecious or dioecious. Leaves usually alternate. Inflorescence of axillary cymose clusters, often condensed into spikes or panicles (at least above), green or sometimes purplish or red. Bracteoles 3-5, 0.5-6 mm, very small and herbaceous or membranous and spinescent. Flowers small and inconspicuous; perianth-segments (2-)3-5, 1.2-3 mm, linear or lanceolate to spathulate. Fruit 1.5-3 mm, dehiscing transversely (circumscissile), irregularly, or indehiscent.

Features of the perianth refer to the female flowers only. All of the species in the Madeiran islands are ruderals or weeds of cultivation.

Literature: A. Cavaco, Les *Amaranthus* de Madère et des Açores in *Bolm Soc. port. Ciênc. nat.* **16**: 79-89 (1976).

1. Leaf-axils with stout straight spines 8-15 mm . . . . **2. spinosus**
- Leaf-axils without spines 2

2. Perianth-segments (3-)5 3
- Perianth-segments (2-)3(-5) 5

3. Fruit indehiscent . . . . **4. muricatus**
- Fruit dehiscing transversely 4

4. Perianth-segments enlarged in upper part, ± spathulate, obtuse or truncate . . . .
. . . . **3. retroflexus**
- Perianth-segments lanceolate or narrowly ovate or elliptical, acute . . . . **1. hybridus**

5. Fruit dehiscing transversely . . . . **5. graecizans**
- Fruit indehiscent or dehiscing irregularly 6

6. Leaves obtuse; stems puberulent above; fruit inflated . . . . **6. deflexus**
- Leaves often conspicuously emarginate; stems usually glabrous; fruit not inflated . 7

7. Fruit feebly rugose, broadly ellipsoid; stems glabrous . . . . **7. blitum**
- Fruit strongly muricate, subglobose; stems sometimes puberulent above . . **8. viridis**

1. A. hybridus L., *Sp. pl.* **2**: 990 (1753) sensu lato. [*Crista de galo*]
A. hybridus var. *acicularis* Thell.; *A. paniculatus* L., var. *ambiguus* Mnzs, var. *purpurascens* Moq.; *A. patulus* Bertol.
Annual; stems to 100 cm, usually erect, sparsely to densely pubescent above. Leaves 1.5-9 × 1-5 cm, rhombic-ovate. Inflorescence of axillary cymose clusters condensed (at least above) into a narrowly cylindric spike or panicle, with branches ascending to erect, usually green or sometimes purplish. Bracteoles ovate, often long-aristate. Perianth-segments (3-)5, narrowly ovate, usually acute. Fruit rounded, wrinkled, dehiscing transversely.
Fl. V - X. *Introduced; naturalized on streamsides, roadsides, paths and pavements, cultivated ground and waste areas in south-eastern Madeira, from Funchal to Machico, and at the western end of the island. Probably widespread in the western and southern lowlands; also in Porto Santo.. Native to tropical & subtropical America.* **MP**

In view of the considerable taxonomic and nomenclatural confusion which surrounds *A. hybridus* and its relatives, the species is treated here in the broad sense, i.e. including *A. cruentus* L. nomen ambiguum, *A. hybridus* L. nomen ambiguum, *A. hypochondriacus* L., *A. paniculatus* L., *A. patulus* Bertol. and *A. powellii* S. Watson.

A. caudatus L., *Sp. pl.* **2**: 990 (1753) [*Crista de galo*] is like **1** but often larger, with inflorescence often very large and dense, paniculate, red (rarely green), with branches numerous, narrowly conical and often patent to pendent. Native to South America, it is commonly cultivated for ornament in gardens in Madeira and occasionally becomes established away from cultivation. **M**

A. palmeri S. Watson in *Proc. Am. Acad. Arts Sci.* **12**: 274 (1877) is native to the south-western U.S.A. A single female plant has been recorded from Madeira, growing in a waste area in Funchal harbour. It is probably only a casual. **M**

2. A. spinosus L., *Sp. pl.* **2**: 991 (1753).
Like **1** but each leaf-axil armed with 2 stout, straight spreading spines 8-15 mm long.
Fl. ?. *Introduced to Madeira; recorded from a roadside at Ponta da Cruz, west of Funchal. Probably native in tropical America.* **M**

3. A. retroflexus L., *Sp. pl.* **2**: 991 (1753).
Like **1** but stems lanate above; inflorescence usually short and dense; bracteoles stout and spinescent; perianth-segments linear-cuneate, wider apically, truncate or obtuse at apex, with short mucro; fruit feebly muricate.
Fl. VI - IX. *Introduced; naturalized in rocky places, as a weed of cultivation and on roadsides; scattered along both the northern and southern coasts of Madeira. Native to North America.* **M**

4. A. muricatus Gillies ex Moq. in A. DC., *Prodr.* **13**(2): 276 (1849).
Perennial; stems to 60 cm, decumbent, usually glabrous. Leaves 2-5 cm, linear to lanceolate. Inflorescence a long panicle, branched at base. Bracteoles ovate, acute. Perianth-segments 5, spathulate. Fruit strongly muricate, indehiscent.
Fl. ?. *Introduced; recorded from a nitrophilous community at the base of an old wall at Câmara de Lobos. Native to temperate S. America.* **M**

5. A. graecizans L., *Sp. pl.* **2**: 990 (1753).
A. graecizans subsp. *silvestris* (Vill.) Brenan
Annual; stems to 40 cm, usually erect and glabrous. Leaves 1.5-2.5 × 1-1.5 cm, ovate or elliptic-rhombic. Inflorescence of axillary cymose clusters. Bracteoles ovate, mucronulate. Perianth-segments 3, ovate-lanceolate, acute. Fruit somewhat muricate, with green longitudinal veins when young, dehiscing transversely.
Fl. VI - IX. *Roadsides in south-eastern coastal Madeira, from Machico to the base of the Ponta de São Lourenço; also at Arco da Calheta in south-western Madeira and in Porto Santo.* **MP**

6. A. deflexus L., *Mant. pl.*: 295 (1771).
Perennial; stems to 45 cm, procumbent, densely puberulent above. Leaves 0.5-3 × 0.3-1.5 cm, rhombic-ovate, obtuse, with finely undulate margin. Inflorescence interrupted and leafy towards base, spicate and usually dense towards apex. Bracteoles ovate, broad-based, mucronate. Perianth-segments 2 or 3, linear to oblong-spathulate. Fruit oblong-ovoid, inflated, smooth, with 3 dull green longitudinal veins, indehiscent; seeds much smaller than fruit.
Fl. VI - X. *Introduced; naturalized on cultivated ground and pavements; scattered along both the northern and southern coasts of Madeira and in Porto Santo. Native to S. America.* **MP**

7. A. blitum L., *Sp. pl.* **2**: 990 (1753).
A. lividus L., subsp. *polygonoides* (Moq.) Probst.
Annual; stems to 50 cm, procumbent to erect, glabrous. Leaves 1-8 × 0.8-6 cm, rhombic- to orbicular-ovate, light- or dark-spotted on upper surface, often conspicuously emarginate, often undulate at margin. Inflorescence of axillary cymose clusters, forming a dense ± leafless spike towards apex. Bracteoles ovate, broad-based, acute. Perianth-segments 3(-5), oblong-linear to spathulate. Fruit broadly ellipsoid, not inflated, feebly rugose, without green veins, indehiscent or dehiscing irregularly; seeds almost as large as fruit. 2n = 34.

Fl. IV - IX. *Cultivated areas, waste ground, roadsides and bases of walls; south-eastern Madeira, in the Funchal region and at Machico, and from Vila Baleira to Camacha in Porto Santo.* **MP**

8. A. viridis L., *Sp. pl.* ed.2, **2**: 1405 (1763).
Like **7** but stems sometimes puberulent above; fruit subglobose, strongly muricate.
Fl. ? *Introduced; recorded from Madeira without further data. Native to S. America.* **M**

A. viridis is superficially similar to **7** and has often been confused with that species in Europe.

2. Achyranthes L.

Perennial sub-shrubs. Leaves opposite. Inflorescence terminal, spicate. Bracteoles 3. Flowers hermaphrodite, erect, becoming deflexed; perianth-segments 4 or 5, narrowly lanceolate, becoming hardened. Fruits deflexed.

1. A. sicula (L.) All., *Auct. Syn. Stirp. Taurin.*: 41 (1773).
A. aspera sensu auct. mad., non L. sensu stricto (1753); *A. aspera* var. *sicula* L.
Stems to 1 m or more, sprawling or scrambling, 4-angled, often tinged reddish, sparsely pubescent to nearly glabrous. Leaves 2-11 × 1-7 cm, ovate to lanceolate (rarely rhombic), acuminate, sparsely silvery-sericeous above, usually densely so beneath. Inflorescence ascending, elongate, slender. Bracteoles 3-4.5 mm, narrowly linear, awned, broadly membranous-winged at base, pale yellowish, pink-flushed towards apex, glabrous. Perianth-segments 3.5-4.5 mm, narrowly lanceolate, pungent, pale green in lower part, dark green in upper part, pink-flushed at apex, pale yellowish at margin, glabrous. 2n = 84.
Fl. II - X. *Scrub, hillsides, roadsides, grassy, cultivated and waste areas up to 650 m; scattered around the coasts of Madeira, especially in the south-east, and at Serra de Agua; also on Selvagem Grande.* **MS**

Some authors do not separate *A. sicula* from *A. aspera* L., although the latter taxon, in the strict sense, differs in its densely pubescent stem and its broader leaves, which are green and sparsely hairy on both surfaces; it is native in the tropics and is locally naturalized elsewhere, but does not appear to occur in the Madeiran archipelago.

3. Alternanthera Forssk.

Somewhat woody-based perennial. Leaves opposite. Inflorescence axillary, sessile, 1 to 3 together, densely capitate. Bracteoles 2, membranous. Flowers hermaphrodite; perianth-segments 5, outer 3 longer than inner 2. Fruits not deflexed.

1. A. caracasana Kunth in Humb., Bonpl. & Kunth, *Nov. gen. sp.* **2**, ed. folio: 165; ed. 4°: 206 (1818).
A. achyrantha (L.) Sw., pro parte.
Stems to 30 cm, usually creeping, pubescent. Leaves 0.8-3 × 0.6-1.5 cm, elliptical to suborbicular, shortly petiolate, dark green above, paler beneath, sparsely hairy to almost glabrous. Inflorescence 5-10 mm, ovoid to narrowly so, creamy white. Bracteoles *c.* 3 mm, ovate, shortly aristate. Perianth-segments similar to but smaller than bracteoles, with tufts of minutely retrorsely barbed villous hairs.
Fl. III - XI. *Introduced; naturalized among cobblestones and in cracks in pavements in south-eastern coastal Madeira. Native to the West Indies, tropical America.* **MP**

XXXI. NYCTAGINACEAE
J.R. Press

Herbs, shrubs or (in Madeira) woody climbers. Leaves simple, entire, stipules absent. Flowers surrounded by brightly coloured bracts. Perianth petaloid, tubular, the lower part persistent, enclosing the indehiscent fruit.

Bougainvillea spectabilis Willd. [*Buganvilha*], a woody and prickly climber with brightly coloured involucral bracts, is cultivated in the towns, especially in Funchal where the courses of the main rivers are completely concealed by its prolific growth.

1. Mirabilis L.

Perennial herbs. Leaves opposite, the lower petiolate. Cymes branched, often dense. Bracts 5, joined at the base to form a calyx-like involucre surrounding 1-many flowers. Perianth-tube long, contracted over the ovary, the persistent portion coriaceous in fruit; limb spreading, 5-lobed.

1. M. jalapa L., *Sp. pl.* **1**: 177 (1753). [*Bonina*]
M. divaricata Lowe
Plant pubescent. Stems erect, 60-100 cm, with prominent nodes. Leaves ovate to acuminate, subcordate at the base. Flowers solitary in each involucre, often pinkish purple but very variable, opening in the late afternoon. Perianth-tube 2.5-3.5 mm, glabrous or thinly pubescent with long, multicellular hairs, limb *c*. 3.5 cm in diameter.
Fl. V - VII. *Common weed of waste ground, roadsides, walls and sea-cliffs in the lower regions of Madeira; also grown in gardens for ornament.* **M**

XXXII. PHYTOLACCACEAE
N.J. Turland

Trees, shrubs or perennial herbs. Leaves alternate, petiolate, simple, entire. Inflorescence an axillary, cylindrical raceme. Flowers regular, hermaphrodite or unisexual; perianth-segments 4-5, small, free, dull-coloured and appearing intermediate between petaloid and sepaloid, persistent in fruit; stamens 4, 10 or 20-30; ovary superior; carpels 7-10 and at least partly united, or 1, each containing a single ovule. Fruit a juicy berry.

1. Herb; perianth-segments 4; carpel 1 . . . . **2. Rivina**
- Herb, shrub or tree; perianth-segments 5; carpels 10 or 20-30 . . . . . . **1. Phytolacca**

1. Phytolacca L.

Herbs, shrubs or trees. Flowers hermaphrodite or unisexual; perianth-segments 5; stamens 10 or 20-30; carpels 7-10.

1. P. americana L., *Sp. pl.* **1**: 441 (1753).
Herbaceous perennial. Stems to 3 m, stout, fleshy, becoming woody at base, glaucous, becoming purple-tinged, glabrous. Leaves 8-20 × 3-6 cm (including petiole to 2.5 cm), narrowly elliptic to narrowly lanceolate. Racemes ascending, 6-20 cm, elongating in fruit, with axis and pedicels purple. Flowers hermaphrodite; perianth-segments white, green or tinged purplish; stamens 10; ovary prominent, green, with 10 carpels almost completely united. Fruit ripening glossy purple-black.
Fl. VI - IX. *Introduced; naturalized on terraces and in waste areas on the northern coast of Madeira, from Porto do Moniz to Seixal, and at Porto da Cruz. Native to N. and C. America.* **M**

P. dioica L., *Sp. pl.* ed. 2, **1**: 632 (1762) (*Pircunia dioica* (L.) Moq.) is a dioecious evergreen shrub or tree to 15 m, with trunk greatly thickened and buttressed at base, racemes pendent, perianth-segments greenish spotted with white, stamens 20-30 and carpels 7-10. Native to temperate and subtropical S. America, it is sometimes cultivated in parks and gardens in Madeira and may occasionally escape. It is not known if it successfully establishes. **M**

2. Rivina L.

Herb, becoming woody at base. Flowers hermaphrodite; perianth-segments 4; stamens 4; ovary with 1 carpel.

1. R. humilis L., *Sp. pl.* **1**: 121 (1753).
R. brasiliensis Nocca
Stems erect, to 60 cm, divaricately branched above, ± glabrous. Leaves 2-7 × 1-3.5 cm (excluding petiole to 3 cm), lanceolate to ovate-lanceolate. Racemes erect below, pendent above, 3-10 cm, elongating in fruit, slender; pedicels becoming recurved in fruit. Perianth-segments concave, whitish or pale pink, becoming greenish in fruit. Fruit 2.5-5 mm, globose, ripening brownish to red.
Fl. V - X. *Introduced; cultivated in Madeira and sometimes becoming naturalized on roadsides and in gardens and waste areas in the Funchal region. Native to southern U.S.A., West Indies and S. America.* **M**

XXXIII. AIZOACEAE
[Mesembryanthemaceae; Tetragoniaceae]
M.C. Tebbs

Herbs or small shrubs, usually succulent, often papillose. Leaves alternate or opposite, simple, entire, usually exstipulate. Flowers regular, hermaphrodite, terminal or axillary, solitary or in small cymes. Perianth segments 4-5, united at the base. Stamens 1-many; filaments sometimes basally connate in bundles. Staminodes 0-many, often petal-like and brightly coloured. Ovary inferior to superior, 1- to 5-locular. Fruit a capsule or fruit indehiscent, fleshy or dry.

1. Petal-like staminodes present 3
- Petal-like staminodes absent 2

2. Fruit a capsule dehiscing by 5 apical valves; ovary superior .... **1. Aizoon**
- Fruit indehiscent, with large tubercles near the top; ovary inferior ... **5. Tetragonia**

3. Flowers 5-10 cm across; leaves triquetrous .... **4. Carpobrotus**
- Flowers 0.5-3 cm across; leaves cylindrical or flat, never triquetrous 4

4. Stems red; leaves sessile; petal-like staminodes white or cream ....
   .... **2. Mesembryanthemum**
- Stems green; at least the lower leaves petiolate; petal-like staminodes deep pink to purple
   .... **3. Aptenia**

1. Aizoon L.

Annual or perennial herbs. Flowers solitary, in leaf axils or in forks of branches, sessile or subsessile. Perianth-segments (4-)5, triangular, united at the base into a short tube. Ovary superior. Fruit a 5-valved, loculicidal capsule, containing many kidney-shaped, tuberculate seeds.

1. Stems pilose, finely papillose; leaves suborbicular to spathulate; perianth segments 1-3 mm .... **1. canariense**
2. Stems densely papillose; leaves lanceolate; perianth segments 7-10 mm   **2. hispanicum**

1. A. canariense L., *Sp. pl.* **1**: 488 (1753).
Prostrate annual or perennial, stems often numerous, pilose and finely papillose. Leaves 8-70 × 4-45 mm, alternate, suborbicular to spathulate, base decurrent to petiole, both surfaces sparsely to densely pilose. Flowers sessile. Perianth segments 1-3 mm long, acute, pale yellow inside, green or red and pilose outside. Stamens 12-15, united in bundles. Fruit 5-9 mm across, red or pink, spongy, with a distinct central depression. Seeds dark chestnut-brown, shiny.
Fl. all year. *Frequent in dry sandy places, usually near the sea; present on all the islands, but in Madeira known only from Ponta de São Lourenço and its islets.* **MDPS**

2. A. hispanicum L., *Sp. pl.* **1**: 488 (1753).
Decumbent or semi-decumbent annual, stems branching, sometimes from the base, densely papillose. Leaves 20-50 × 4-8 mm, the upper opposite, the lower alternate, lanceolate, obtuse, sessile and partially clasping the stem. Flowers subsessile. Perianth segments 7-10 mm, lanceolate-linear, yellowish or white inside, green outside. Stamens 5-15, free or in bundles. Fruit 10-12 mm across, rigid, with a shallow central depression. Seeds black, dull.
Fl. II - V. *Introduced. Rare in Porto Santo on walls, along roadsides and field borders.* **P**

2. Mesembryanthemum L.

Succulent annuals with crystalline hairs. Flowers solitary, terminal, axillary or opposite the leaves, pedunculate. Perianth (4-)5-merous. Staminodes petaloid, united at the base into a tube. Ovary inferior. Fruit a 4- to 5-valved capsule.

1. Leaves linear; staminodes shorter than perianth-segments .... **1. nodiflorum**
- Leaves ovate-spathulate; staminodes longer than perianth-segments .. **2. crystallinum**

1. M. nodiflorum L., *Sp. pl.* **1**: 480 (1753). [*Barrilha*]
Stems 15-30 cm, prostrate, slender, red, spreading, moderately covered with crystalline hairs. Leaves 1-2(-3) cm × 1-3 mm, the lower opposite, the upper alternate, linear, semi-cylindrical, reddish, sessile. Flowers 1-1.5 cm across, solitary, axillary or leaf-opposed; peduncle 0.5 cm. Outer perianth-segments linear, inner segments triangular. Staminodes white or cream, much shorter than the perianth. Capsules 8-10 mm across. Seeds 1 mm, tuberculate, golden brown. $2n=36$.
Fl. IV - X. *Frequent along roadsides, on waste ground and cliffs on all the islands, always near the sea. Cultivated in the last century for burning to produce soda, especially on Ilhéu Chão.* **MDPS**

2. M. crystallinum L., *Sp. pl.* **1**: 480 (1753). [*Barrilha*]
Stems 30-40 cm, prostrate, thick, reddish, densely covered with crystalline hairs. Leaves (2-)5-8 × 2-4 cm, opposite, ovate-spathulate, reddish, glistening, the lower narrowing into a short petiole, the upper sessile. Flowers 1-1.5 cm across, axillary or in terminal groups of 3-5; peduncle 2-3 mm. Perianth-segments with broad red margins, outer segments large, leaf-like, inner segments smaller, ovate. Staminodes white or cream, much longer than the perianth. Capsules 10-14 mm across, dark purple-red or crimson, juicy. Seeds 0.6-0.8 mm, tuberculate, dark brown.
Fl. IV - VIII. *On tracks, dry waste ground, sandy places, cliffs and scree slopes near the sea.*

Common in Porto Santo, Ilhéu de Cima, the Salvages, Ilhéu Chão and Bugío, less common on Deserta Grande; in Madeira confined to Ponto de São Lourenço. **MDPS**

Dorotheanthus gramineus (Haw.) Schwantes in *Möllers Deutsche Gärtn.-Zeit.* **42**: 283 (1927) is a S. African species recorded from Madeira, but only as an occasional escape from cultivation. It differs from *Mesembryanthemum* by the compact habit, linear leaves often in a loose rosette, pink to white flowers up to 5 cm across, and concave capsules. **M**

Drosanthemum floribundum (Haw.) Schwantes in *Z. SukkulKde* **3**: 29 (1927) [*Arrozinho, Chorão baguinho de arroz*], also from S. Africa but widely cultivated elsewhere, has been recorded in areas close to hotel gardens. It is a low shrubby perennial about 12 cm high, with short, linear, fleshy leaves, and rose-pink flowers on a long, slender peduncle. It is undoubtably a garden escape, neither widespread nor truly naturalized in Madeira. **M**

3. Aptenia N.E. Br.

Prostrate or trailing, succulent, short-lived papillose perennials. Leaves opposite. Flowers solitary, pedunculate. Perianth-segments 4, free. Staminodes united at the base into a short tube. Ovary inferior. Fruit a 4-valved capsule.

1. A. cordifolia (L.f.) Schwantes in *Gartenflora* **77**: 69 (1928). [*Apténia*]
Mesembryanthemum cordifolium L.f.
Prostrate or trailing herb. Stems succulent, up to 60 cm long, papillose. Leaves petiolate, 10-40 × 5-20 mm, ovate-cordate, fleshy, bright green and glossy, minutely papillose, midrib prominent on the underside. Flowers 1-3 cm across, terminal or lateral; peduncle 8-20 mm. Perianth green, of 2 leaf-like, oblong-ovate segments 10-15 (-20) mm, alternating with 2 linear or awl-shaped segments 5-8 mm. Staminodes deep pink to purple; stamens many, erect, filaments white. Capsules 10 mm; seeds tuberculate, dark brown. $2n=18$.
Fl. ± all year. *Introduced. Locally common in dry places, on rocky cliffs and drystone walls in the lower regions of Madeira, particularly frequent along the north coast near Porto Moniz, 0-600 m; less common in Porto Santo.* **MP**

4. Carpobrotus N.E. Br.

Prostrate or decumbent perennial herbs or subshrubs with glabrous stems. Leaves opposite, sessile, partly clasping the stem, attached at the base, triquetrous. Flowers large, bracteate, terminal, with compressed peduncles. Perianth-segments (4-)5. Staminodes in 3-4 whorls, free. Ovary inferior. Fruit (6-)10- to 16-locular, indehiscent.

1. C. edulis (L.) N.E. Br. in E. Phillips, *Gen. S. Afr. fl. pl.*: 249 (1926).
[*Bálsamo, Chorão*]
Mesembryanthemum edule L.
Sturdy, succulent perennial herb, sometimes suffrutescent, stems branching, creeping. Leaves 4-10(-13) × 1-1.5 cm, oblong-lanceolate, fleshy, sometimes curving, with reddish margins. Flowers 5-10 cm across; peduncle 1-3 cm; perianth 2-3.5 cm, the 3 outer lobes oblong and larger than the 2 inner lobes. Staminodes cream, yellow or pink. Fruit 2-3 cm, fleshy. Seeds embedded in mucilage.
Fl. III - VIII. *Introduced. Locally common on sand dunes, banks, and on cliffs near the sea in Porto Santo; less common in Madeira, although occasionally locally abundant and forming extensive mats; sometimes planted.* **MP**

5. Tetragonia L.
M.J. Short

Sprawling or trailing, papillose herbs. Leaves alternate, flat. Flowers small, axillary, solitary, greenish or yellowish. Perianth segments 4-5, sepaloid externally but coloured within. Stamens 4-numerous, sometimes fasciculate. Staminodes absent. Ovary inferior; stigmas 3-8. Fruit indehiscent, hard, dry, horned.

1. T. tetragonoides (Pall.) Kuntze, *Revis. gen. pl.* 1: 264 (1891).   [*Espinafres*]
T. expansa Murray
Somewhat succulent annual or perennial herb; stems up to 75(-100) cm, much-branched from the base, procumbent or ascending. Leaves 2.5-8.5 × 2-6 cm, triangular-ovate to rhombic, dark green above, paler below, decurrent at the base into a short petiole. Flowers sessile or subsessile. Perianth segments 1.5-2 mm, unequal, triangular-lanceolate, green without, pale yellow within. Fruit 0.8-1.2 × 0.7-1.2 cm, subglobose to top-shaped, with several large, horn-like tubercles near the top, brown when ripe.
Fl. I - X. *Reported to have been introduced to Madeira c. 1825, cultivated as the vegetable known as 'New Zealand spinach'. It is now naturalized in several places in the coastal region, especially in and around Funchal and at Porto Moniz where it is found on waste ground, amongst rocks, and trailing down walls; 0-150 m; also known from Porto Santo. Native to New Zealand and Australia.* **MP**

XXXIV. PORTULACACEAE
M.J. Short

Annual or perennial, often fleshy, herbs or subshrubs. Leaves alternate or opposite, simple, entire, usually stipulate. Flowers regular, hermaphrodite. Sepals 2, imbricate, free or united below. Petals 4-6, free or connate at the base. Stamens 3 to many. Ovary superior or semi-inferior, 1-locular, with 1 to many ovules. Style usually divided above. Fruit a capsule, dehiscing loculicidally or transversely by a lid.

1. Portulaca L.

Glabrous, ± succulent, annual herbs. Leaves sessile, often forming involucre-like clusters beneath the flowers; stipules scarious or forming small bristles. Sepals shortly united at the base. Petals 4-6, falling early. Stamens 7-15. Ovary semi-inferior; style with 2-8 branches. Capsules thin-walled, dehiscing by a transverse lid. Seeds numerous, ovoid-reniform.

1. P. oleracea L., *Sp. pl.* 1: 445 (1753).
Stems up to 50 cm, branched, often reddish. Leaves alternate below but crowded and subopposite beneath the flowers, 1-2 cm, oblong-obovate to oblong-spathulate, obtuse, cuneate at the base, dark green and shiny. Stipules often reduced to a small tuft of bristles. Flowers sessile, 8-12 mm across, solitary or in small groups of 2-3, terminal or in the forks of the branches. Sepals 4-5 mm, keeled, unequal. Petals (4-)5(-6), ± free, slightly exceeding the sepals, obovate, deeply emarginate, delicate, yellow. Capsules 3-8 mm, ovoid-conical. Seeds 0.7-1.2 mm, bluntly tuberculate, black, shining. **MP**

Two subspecies have been recorded:

a) subsp. **oleracea**.
P. oleracea var. *sylvestris* DC.
Stems procumbent or decumbent.

Fl. all year. *A weed of cultivated land and open, waste places, particularly frequent along roads and paths in Madeira and Porto Santo; 0-900 m.*

b) subsp. **sativa** (Haw.) Čelak., *Prodr. F. Böhmen*: 484 (1875).
P. sativa Haw.
Stems ascending to erect; generally larger and more robust than subsp. *oleracea*.
Found in similar habitats to subsp. oleracea, but only recorded from Madeira.

Talinum paniculatum (Jacq.) Gaertn., *Fruct. sem. pl.* 2: 219 (1791), an erect perennial herb up to 1 m high from southern N. America to C. America, with succulent, elliptical to obovate leaves and a loose, terminal panicle of red, pink or yellow flowers, followed by thin-walled, 3-valved capsules, has been recorded from Madeira but without furthur data, presumably as a garden escape. **M**

XXXV. BASELLACEAE
J.R. Press

Perennial, herbaceous, often succulent vines, twining to the right. Leaves alternate, entire, sometimes fleshy; stipules absent. Flowers hermaphrodite, in spikes, racemes or clusters. Bracteoles 2 or 4. Perianth-segments 5. Ovary unilocular with 1 ovule; styles 3. Fruit fleshy, indehiscent, enclosed by the persistent perianth.

1. **Anredera** Juss.

Rhizomes with subterranean tubers. Bracteoles in 2 decussate pairs. Perianth-segments membranous, connate at the base and concealing the ovary. Fruit subglobose.

1. **A. cordifolia** (Ten.) Steenis, *Fl. Males.* I, **5**: 303 (1957).
Boussingaultia cordifolia Ten.; *B. baselloides* auct. mad., non Kunth (1825)
Slender plant with numerous tubers. Leaves ovate-cordate to lanceolate. Racemes up to 30 cm with numerous pedicellate flowers. Lower bracteoles 0.5 mm, broadly triangular, connate, remaining attached to the pedicel after the fruit has fallen; upper bracteoles 2 mm, petalloid, broadly ovate-elliptic. Perianth-segments spreading, 2.5-3 mm, ovate-oblong to elliptic, white. Fl. VII - X. *Originally introduced to gardens for its fragrant flowers but escaping and becoming naturalized in places in the lower regions of Madeira and Porto Santo. Native to tropical S. America.* **MP**

XXXVI. CARYOPHYLLACEAE
M.J. Short

Annual to perennial herbs. Leaves opposite and often decussate, rarely alternate or verticillate, simple, entire, with or without stipules. Flowers regular, generally hermaphrodite, solitary or more often in dichotomously branching, bracteate cymes. Sepals 4-5, free or fused into a tube. Petals (0)4-5, free, sessile or clawed, entire or lobed. Stamens usually 1-10, sometimes less. Ovary superior, 1-locular; styles (1-)2-5. Fruit a capsule, dehiscing by valves or teeth equalling in number or twice as many as the styles, or less commonly an indehiscent, 1-seeded achene.

1. Sepals joined to form a distinct calyx-tube 2
- Sepals free, or joined only at the extreme base 3

2. Styles 3; capsules dehiscing by 6 teeth . . . . **13. Silene**
- Styles 2; capsules dehiscing by 4 teeth . . . . **14. Petrorhagia**

3. Stipules absent 4
- Stipules present (mostly scarious, scale-like) 8

4. Fruit an indehiscent, 1-seeded achene, subsessile; styles 2 .... **5. Scleranthus**
- Fruit a capsule dehiscing by 4, 6 or 10 valves or teeth, distinctly pedicellate; styles 3-5 5

5. Leaves narrowly linear, less than 1 mm wide; sepals 4; seeds 0.2-0.3 mm   **4. Sagina**
- Leaves elliptic, lanceolate or ovate, more than 1 mm wide; sepals 5 (rarely 4); seeds 0.4-1.4 mm 6

6. Petals deeply bifid, divided almost to the base .... **2. Stellaria**
- Petals entire or emarginate and then divided less than ½-way to the base 7

7. Petals entire; styles 3; capsules 2-2.7 mm, dehiscing by 6 teeth ...... **1. Arenaria**
- Petals emarginate; styles 4-5; capsules 4-14 mm, dehiscing by 8 or 10 teeth .... **3. Cerastium**

8. Fruit 1-seeded, indehiscent or if opening, the valves remaining joined above 9
- Fruit a several-seeded capsule dehiscing by teeth or valves 12

9. Leaves all or mostly alternate 10
- Leaves opposite 11

10. Plant glabrous, mainly branched from the base .... **6. Corrigiola**
- Plant pubescent; stems with regularly alternating branches .... **8. Herniaria**

11. Calyx-lobes white and spongy, very conspicuous; stems glabrous .... **9. Illecebrum**
- Calyx-lobes not as above; stems pubescent .... **7. Paronychia**

12. Leaves obovate or elliptic; capsules c. 1.5 mm .... **10. Polycarpon**
- Leaves linear; capsules 3-6.5 mm 13

13. Leaves in whorls; stipules free; petals white .... **11. Spergula**
- Leaves opposite; stipules united at the base, surrounding the node; petals pink (white near base), rarely entirely white .... **12. Spergularia**

1. Arenaria L.

Low, caespitose annual herbs, occasionally overwintering. Leaves opposite, small, exstipulate. Flowers generally 5-merous, in terminal or axillary dichasial cymes. Bracts herbaceous. Sepals free. Petals white, entire. Stamens 10. Styles 3. Fruit a conical capsule, dehiscing by 6 narrow, acute teeth. Seeds numerous, orbicular-reniform, tuberculate.

1. A. serpyllifolia L., *Sp. pl.* **1**: 423 (1753).
Scabrid-puberulent, glandular-viscid above. Stems 2-15 cm, ascending or erect, usually branched, slender but rigid. Leaves 2.5-6 mm, ovate to ovate-lanceolate, acute to acuminate, the lower shortly petiolate, the upper sessile or nearly so. Bracts leaf-like, but smaller. Pedicels 1-8 mm, slender. Sepals 2.5-3 mm, lanceolate, acuminate, scarious-margined. Petals oblong, shorter than the sepals. Capsules 2-2.7 mm, thin-walled, shorter than or equalling the sepals. Seeds c. 0.4 mm, blackish brown, shallowly tuberculate.

Fl. IV - VIII. *Extremely rare in mountain turf in Porto Santo and Madeira; reported to be very common on Bugío in the last century, but its current status unknown; also known from Selvagem Grande.* **MDPS**

Represented in the Madeiran archipelago by subsp. **leptoclados** (Rchb.) Nyman, *Consp. fl. eur.*: 115 (1878) (*A. leptoclados* (Rchb.) Guss.), to which the above description applies. As in some other areas, plants are glandular-viscid, but varieties based on this character are not thought to warrant taxonomic recognition.

2. Stellaria L.

Annual or perennial, slender, weak-stemmed herbs. Leaves opposite, sessile or petiolate, exstipulate. Flowers generally 5-merous, in dichasial cymes, sometimes the flowers very few or solitary. Bracts scarious or herbaceous. Sepals free. Petals white, deeply bifid. Stamens 3-10. Styles 3. Fruit an oblong-ovoid, thin-walled capsule, dehiscing by 6 valves. Seeds numerous, orbicular-reniform, somewhat compressed, tuberculate.

1. Leaves ovate; stems terete; bracts herbaceous; pedicels with villous hairs . . **1. media**
- Leaves elliptic; stems 4-angled; bracts scarious; pedicels glabrous . . . . **2. alsine**

1. S. media (L.) Vill., *Hist. pl. Dauphiné* 3: 615 (1789). [*Marugem, Morugem*]
Variable sprawling annual; stems 5-40 cm, numerous, decumbent to ascending, loosely branched, leafy, terete, with a single line of villous hairs on alternating sides between the nodes. Leaves 4-18 × 3-11 mm, ovate, acute or shortly acuminate, glabrous, the lower petiolate, the upper mostly sessile; petioles ciliate. Bracts herbaceous. Pedicels up to 15 mm, with a longitudinal line of hairs, sometimes glandular. Sepals 3.5-5 mm, ovate-lanceolate, subobtuse, 1-veined, narrowly scarious, glabrous or long-pubescent, sometimes glandular. Petals divided almost to the base, ½-¾ as long as sepals. Stamens 3-8; anthers red-violet. Capsules 4.5-6 mm. Seeds 1-1.3 mm, reddish brown.
Fl. I - XII. *A very common weed of roadsides, grassy banks, cultivated ground, and among damp rocks in Madeira, 0-1000 m; fairly frequent in Porto Santo.* **MP**

2. S. uliginosa Murray, *Prodr. stirp. gott.*: 55 (1770).
S. alsine Grimm
Perennial, ± glabrous; stems 10-45 cm, numerous, decumbent to ascending, branched, slender, 4-angled, often rooting at the lower nodes. Leaves sessile, 7-22(-30) × 1.5-5(-6) mm, elliptic, acute, attenuate and slightly ciliate at the base. Bracts scarious except for a narrow green midrib. Pedicels up to 25 mm, glabrous. Sepals 2.2-3.5 mm, triangular-lanceolate, acute, 3-veined, scarious-margined. Petals divided almost to the base, less than ½ as long as sepals. Stamens 10; anthers yellow. Capsules *c.* 3 mm. Seeds 0.6-0.7 mm, reddish brown.
Fl. III - X. *Common in Madeira, often forming thick clumps in marshy ground by streams and pool margins, in wet flushes, and on rocks covered with running water; (30-)450-1000 m.* **M**

3. Cerastium L.

Pubescent annual or perennial herbs, the hairs often glandular. Leaves opposite, sessile or the lower shortly petiolate, exstipulate. Flowers 5- or 4-merous, mostly in terminal, dichasial cymes. Sepals free. Petals white, emarginate to bifid. Stamens 4, 5 or 10. Styles 4-5. Fruit a cylindrical or oblong capsule, sometimes curved, usually exceeding the calyx, dehiscing by twice as many short teeth as styles. Seeds numerous, subglobose, tuberculate.

Literature: J. Jalas, *Cerastium sventenii* Jalas, sp. nova, and the related Macaronesian taxa, in *Annls bot. fenn.* **3**: 129-139 (1966).

1. Perennial, often rather shrubby below or with short, sterile basal shoots 2
- Annual, not shrubby below and without sterile shoots 3

2. Petals *c.* 2 × sepals; capsules 6-8 × 5 mm, oblong-ovoid; leaves all sessile **1. vagans**
- Petals about equalling sepals; capsules 9-14 × 2.5-3.5 mm, cylindrical; lower leaves shortly petiolate . . . . **2. fontanum**

3. Flowers all 5-merous; sepals with eglandular hairs at and exceeding the apex; fruiting pedicels all shorter than or equalling the calyx . . . . **3. glomeratum**
- Flowers mostly 4-merous; sepals glabrous at the tips; at least some of the fruiting pedicels longer than the calyx . . . . **4. diffusum**

1. C. vagans Lowe in *Trans. Camb. phil. Soc.* **6**: 548 (1838).
C. vagans [var.] β *calva* Lowe, [var.] α *fulva* Lowe, [var.] β *subnuda* Lowe
Perennial; stems (6-)15-45(-60) cm, ± ascending, rather shrubby below, straggling and loosely branched. Leaves sessile, (7.5-)19-35 × (2.2-)4-7.5 mm, linear- to oblong-lanceolate, acute. Flowers 5-merous, in lax, densely glandular-pubescent cymes. Bracts herbaceous or the uppermost slightly scarious at the apex. Sepals 4.5-6.5 mm, lanceolate, obtuse, broadly scarious above, glabrous at the tip. Petals 9-10 mm; sinus 2-3 mm. Stamens 10; styles 5. Fruiting pedicels 9-12 mm, longer than the sepals. Capsules 6-8 × 5 mm, oblong-ovoid, straight; teeth erect, with revolute margins. Seeds 0.9-1.4 mm, unevenly tuberculate with both conical, acute and low, ± rounded tubercles. **Plate 8**
Fl. VI - VIII. *A very rare Madeiran plant with all known records confined to rocks of the central mountains east of Encumeada; 950-1800 m.* ■ M [*Azores*]

The above description refers to (●) var. **vagans** which is endemic to Madeira. Var. *ciliatum* Tutin & E.F. Warb., with ciliate margins to the leaves, occurs in the Azores.

2. C. fontanum Baumg., *Enum. stirp. Transsilv.* **1**: 425 (1816).
C. fontanum subsp. *triviale* (Spenn.) Jalas; *C. triviale* Link
Short-lived, straggling perennial, some hairs glandular above; stems (5-)10-45 cm, prostrate or trailing, with short, leafy sterile shoots at the base. Leaves 8-28 × 3.5-7(-11) mm, elliptic to oblong-lanceolate, mostly acute or subacute, the upper sessile, the lower attenuate into a short petiole. Flowers 5-merous, in rather lax, few-flowered cymes. At least the upper bracts scarious-margined. Sepals 6-7 mm, lanceolate, acute, broadly scarious, glabrous at the tip. Petals about equalling sepals; sinus 1.3-1.5 mm. Stamens 10; styles 5. Fruiting pedicels 7-14 mm, equalling or longer than the sepals. Capsules 9-12(-14) × 2.5-3.5 mm, cylindrical, straight or slightly curved above; teeth erect, with revolute margins. Seeds 0.7-0.8 mm, with broad, rounded tubercles.
Fl. III - X. *Common in Madeira along tracks, on banks and walls, and in mountain pastures; 300-1750 m. Also recorded from Porto Santo and Deserta Grande.* **MDP**

Plants from the Madeiran archipelago all belong to subsp. **vulgare** (Hartm.) Greuter & Burdet, *Willdenowia* **12**: 37 (1982), to which the above description and synonymy applies.

3. C. glomeratum Thuill., *Fl. env. Paris* ed. 2: 226 (1799).
Softly viscid-pubescent annual; stems 3-30(-45) cm, decumbent to erect, simple or branched from the base, with spreading glandular and eglandular hairs. Leaves 5-30(-37) × 3-15 mm,

broadly obovate-spathulate to elliptic, often obtuse, the lower attenuate into a short petiole. Flowers 5-merous, in dense cymose clusters. Bracts herbaceous. Sepals 3.5-6 mm, lanceolate, acute, the inner with a scarious margin, the outer not or only very narrowly scarious, glandular-pubescent and with long eglandular hairs at and exceeding the apex. Petals about equalling sepals; sinus 0.5-1.5 mm. Stamens 10; styles 5. Fruiting pedicels 1.5-4 mm, shorter than the sepals. Capsules 7-10 × 1.3-1.9 mm, cylindrical, curved; teeth erect, with revolute margins. Seeds $c.$ 0.5 mm, with low, blunt tubercles. $2n=72$.
Fl. II - X. *Common in Madeira on roadside banks and other grassy places, up to 850 m. Rare in lowland Porto Santo and also on the rocky summits of Pico do Facho and Pico do Castelo. Also known from Deserta Grande and Bugío.* **MDP**

4. **C. diffusum** Pers., *Syn. pl.* **1**: 520 (1805).
C. tetrandrum Curtis
Annual, the hairs glandular above; stems 2-15 cm, erect or ascending, usually much-branched from the base. Leaves 5-15 × 3-5 mm, acute, the lower oblanceolate, attenuate into the petiole, the upper elliptic to ovate, sessile. Flowers mostly 4-merous, a few 5-merous, in few-flowered cymes. Bracts herbaceous. Sepals 4-6 mm, lanceolate, acute, scarious, glabrous at the tip. Petals inconspicuous, $c.$ ⅔ the length of the sepals; sinus $c.$ 0.5 mm. Stamens and styles 4(-5). Fruiting pedicels up to 14 mm, shorter than to exceeding the sepals. Capsules 4-5.5 × 1.5-2 mm, cylindrical, nearly straight; teeth erect, with revolute margins. Seeds 0.7-0.8 mm, with low, blunt tubercles.
Fl. VII - VIII. *Very rare plant in Madeira recorded from Paúl da Serra and the summit of Pico Grande, its current status unknown.* **M**

4. Sagina L.

Small, often caespitose annual or perennial herbs with slender, decumbent to erect stems. Leaves opposite, linear-subulate, connate at the base, exstipulate. Flowers usually 4-merous, solitary or in few-flowered cymes, usually long-pedicellate. Sepals free. Petals white, entire, minute or sometimes absent. Stamens 4. Styles 4. Fruit an ovoid capsule dehiscing to the base by 4 valves. Seeds numerous, minute, orbicular-reniform, obscurely angled.

1. Mat-forming perennial, with short, non-flowering shoots, rooting towards the base; stems decumbent . . . . **1. procumbens**
- Tufted annual; stems all producing flowers, not rooting towards the base, decumbent to erect . . . . **2. apetala**

1. **S. procumbens** L., *Sp. pl.* **1**: 128 (1753).
S. procumbens var. *spinosa* (S. Gibson) Bab.
Mat-forming, moss-like perennial with dense central rosettes of leaves and numerous, decumbent lateral stems up to 20 cm, rooting at the nodes near the base. Leaves numerous and crowded, 5-16 mm, linear-subulate, mucronate, glabrous or minutely ciliate. Pedicels 3-15 mm, filiform, glabrous. Sepals 1.6-2 mm, ovate, obtuse, narrowly white-margined, glabrous, appressed or patent in fruit. Petals absent or minute and much shorter than the sepals. Capsules 2-3 mm. Seeds $c.$ 0.3 mm. brown, indistinctly papillose.
Fl. III - X. *Very common in Madeira among rocks, in mountain turf, on banks and along levadas in damp places, 150-1750 m; less frequent in Porto Santo; also reported from Deserta Grande.* **MDP**

2. **S. apetala** Ard., *Animadv. bot. spec. alt.*: 22 (1764).
S. apetala var. *glabra* Bab., [var.] α *glandulosa* Lowe
Tufted annual; stems 1.5-10 cm, diffusely branched from the base, decumbent to erect, not

rooting near the base; sterile stems absent. Leaves (2.5-)5-20 mm, narrowly linear-subulate, aristate, frequently sparsely ciliate near the base. Pedicels 3-17 mm, filiform, glabrous or shortly glandular-pubescent. Sepals 1.2-2.5 mm, ovate, obtuse to subacute, white-margined, glabrous or sparsely glandular, ± appressed or spreading in fruit. Petals minute, often falling early, sometimes absent.   Capsules 2-2.5 mm.   Seeds 0.2-0.3 mm, brown, minutely tuberculate.
Fl. III - VII. *Rare in Madeira, scattered but mainly found in and around Funchal; also rare in Porto Santo, recorded from the summits of Pico do Castelo and Pico Juliana and in the vicinity of Vila Baleira.* **MP**

5. Scleranthus L.

Low, annual to perennial herbs with procumbent to erect, usually diffusely branched stems. Leaves opposite, connate at the base, subulate, exstipulate. Flowers subsessile, small, greenish, in lax or dense terminal and axillary cymes. Bracts leaf-like. Sepals usually 5, inserted on the rim of the urceolate, perigynous zone, often scarious-margined. Petals absent. Fertile stamens 2-10. Styles 2, long, filiform. Fruit an indehiscent, 1-seeded achene, surrounded by the hardened wall of the perigynous zone and the persistent sepals which fall with it.

1. S. annuus L., *Sp. pl.* **1**: 406 (1753).
Annual or biennial herb; stems procumbent to erect, much-branched at the base, glabrous or minutely puberulent on one side. Leaves acute, ciliate at the base. Inflorescence usually lax, with axillary and terminal clusters of inconspicuous flowers. Sepals lanceolate, ± acute, with narrowly scarious margins scarcely extending around the apex. Stamens 2-5. Perigynous tube 10-furrowed. **MP**

Two subspecies have been recorded:

a) subsp. **annuus**.
Stems up to 15(-25) cm; leaves 3-8 mm; stamens almost as long as the sepals; achenes (including sepals) 3.2-4.5 mm, the sepals divergent.
Fl. V - VI.   *Collected only once, from an unknown locality in Porto Santo, in 1832.*

b) subsp. **polycarpos** (L.) Thell. in Schinz & R. Keller, *Fl. Schweiz* ed. 3, **2**: 109 (1914).
S. polycarpos L.
Like **1**, but smaller; stems 1-5 cm; leaves 3-8 mm; stamens almost as long as the sepals; achenes (including sepals) 1.8-3.2 mm, the sepals erect or slightly incurved.
Fl. VI - VII. *Reported to be frequent on the high peaks of Madeira; 1400-1800 m.*

The perennial species, *S. perennis* L., *Sp. pl.* **1**: 406 (1753), has also been recorded from Madeira. It may be distinguished from **1** by the oblong, obtuse sepals with broad, white, scarious margins extending around the apex. Available Madeiran specimens are scanty and poor, but all appear to be *S. annuus* subsp. *polycarpos*. No recent records are known for either species.

6. Corrigiola L.

Glabrous annual to biennial herbs with decumbent stems. Leaves alternate, sessile or subsessile; stipules small, scarious. Flowers small, 5-merous, in dense cymes. Bracts minute. Sepals free, white-margined. Petals membranous, whitish. Stamens 5. Stigmas 3, subsessile. Fruit an indehiscent, ± trigonous, 1-seeded achene, enclosed in the persistent calyx.

1. C. litoralis L., *Sp. pl.* **1**: 271 (1753).
Low, spreading, glaucous annual or biennial herb; stems up to 35 cm, much-branched from the base, slender, often reddish. Leaves sessile or nearly so, linear to oblanceolate, acute to subobtuse, slightly succulent; stipules acuminate, denticulate, semi-auriculate. Flowers

numerous, shortly pedicellate, crowded in small, axillary and terminal cymes. Sepals *c*. 1 mm, ovate, obtuse, slightly longer than the petals. Achenes 1-1.3 mm.
Fl. ? *An introduced weed of cultivated ground at Santa da Serra, Madeira, first recorded in 1957 and reported to be still present in the same area in 1973.* **M**

7. Paronychia Mill.

Annual herbs with procumbent to erect, often branched stems. Leaves opposite and somewhat unequal, sessile or subsessile, elliptical to linear; stipules conspicuous, scarious. Flowers 5-merous, small, in axillary and terminal, compact, rounded clusters. Bracts silvery-scarious, often very conspicuous. Calyx very deeply lobed, the lobes hooded, awned, with membranous margins. Petals filiform, minute, or absent. Stamens 5. Style 1; stigma bilobed. Fruit a 1-seeded, membranous-walled achene enclosed by the persistent calyx.

1. P. echinulata Chater in *Feddes Reprium Spec. nov. veg.* **69**: 52 (1964).
P. echinata sensu auct. mad., non Lam. (1779)
Stems 5-25 cm, ascending to erect, simple or branched, rigid, reddish, puberulous along one side. Leaves 4-10 × 1.3-3 mm, oblanceolate to linear-elliptic, glabrous, pale green (turning blackish on drying), margins pale, scabrid; stipules lanceolate, long-acuminate. Flower sessile or subsessile, in axillary clusters, 3-7 mm across. Bracts ovate, acuminate, ciliate. Calyx-lobes 2-2.5 mm, oblong-spathulate, prominently awned, thick and coriaceous, pubescent below with long, hooked hairs, glabrous above, margin broadly membranous. Achene *c*. 0.8 mm, globose, densely papillose above. Seed *c*. 0.7 mm, subglobose, brown, shiny.
Fl. IV - VII. *Dry, sunny, grassy places. Reported to be common in southern Madeira up to c. 900 m during the last century, but now believed to be very rare.* **M**

P. cymosa (L.) DC. in Lam, *Encycl.* **5**: 26 (1804) (*Chaetonychia cymosa* (L.) Sweet), a slender plant with narrowly linear, cuspidate leaves and small, dense, globose clusters of flowers with unequal, aristate calyx-lobes, has been recorded from Madeira, based on a single collection made in the late eighteenth century. No further records are known. **M**

8. Herniaria L.

Annual herbs; stems branched, prostrate. Leaves mostly alternate, sessile or subsessile, oblanceolate; stipules small, scarious. Flowers usually 5-merous, subsessile, small, greenish, in dense axillary clusters. Bracts similar to stipules but smaller. Sepals free. Petals filiform, inconspicuous. Stamens 2. Stigmas 2, subsessile. Fruit a 1-seeded, membranous-walled achene, ± enclosed by the persistent calyx.

1. H. hirsuta L., *Sp. pl.* **1**: 218 (1753).
Pale or yellowish green, low-growing plant covered with stiffly spreading hairs; stems 2-7(-10) cm, mat-forming, usually with distinctive regularly alternating branches. Leaves mostly alternate, up to 8 × 2.3 mm, narrowly elliptic to oblanceolate, attenuate to base, margins ciliate; stipules triangular-ovate, ciliate. Flowers in dense clusters opposite the leaves, usually contiguous into dense, leafy spikes on the short side branches. Sepals 1-1.7 mm, distinctly unequal at maturity, densely hispid, the basal hairs with hooked tips. Stamens 2. Achenes ovoid. Seed *c*. 1.6 mm, shiny brown.
Fl. IV - V. *On bare ground, in sparse pasture, and among rocks. Very rare in Madeira where it is only recorded from Ponta de São Lourenço; also rare in Porto Santo but known from several localities including Pico Branco, Vila Baleira, and Campo de Baixo; also found on Ilhéu de Cima, where it is quite common, and Ilhéu de Baixo. Reported from Bugío of the Desertas and also Selvagem Grande.* **MDPS**

Plants from the Madeiran archipelago all belong to subsp. **cinerea** (DC.) Cout., *Fl. Portugal*: 202 (1913) (*H. cinerea* DC.; *H. flavescens* Lowe), to which the above description applies.

9. Illecebrum L.

Annual herbs with prostrate to ascending, usually much-branched stems. Leaves opposite, subsessile; stipules scarious. Flowers 5-merous, sessile, small, in axillary clusters with small, scarious, silvery bracts. Calyx very deeply lobed, the lobes white, spongy, hooded, awned. Petals white, filiform, minute, or absent. Stamens 5. Stigmas 2, ± sessile. Fruit a 1-seeded, membranous-walled capsule, enclosed by the persistent calyx, dehiscing by 5 valves which remain joined above.

1. I. verticillatum L., *Sp. pl.* 1: 206 (1753).
Glabrous, often mat-forming; stems (2-)5-25 cm, prostrate or ascending, 4-angled, slender and often reddish, rooting at the nodes near the base. Leaves 1-5 × 0.7-3.5 mm, ovate, obovate or oblong-spathulate, obtuse, attenuate at the base. Stipules *c*. 1 mm, ovate. Flowers in small, crowded, axillary clusters of (2-)4-6, with 2 clusters at each node, forming conspicuous white whorls. Calyx-lobes 2-2.5 mm, thick and spongy, hooded, finely awned, shining white, occasionally tinged with purple. Petals much shorter than the sepals. Seed 0.7-0.9 mm, ellipsoid, shiny brown. $2n=10$.
Fl. IV - IX. *Reported to be extremely common in Madeira during the last century, growing in damp mountain pastures, cultivated fields, rock cracks, and among road cobblestones, 400-1400 m; now rather rare and seldom seen.* **M**

10. Polycarpon Loefl. ex L.

Small annual to perennial herbs; stems usually branched and rough at angles. Leaves opposite or in whorls of 4, ± petiolate, obovate to orbicular; stipules scarious, conspicuous. Flowers small, 5-merous, in terminal, dichasial, condensed cymes, with scarious bracts. Sepals free, keeled, hooded, scarious-margined. Petals white, narrow, shorter than the sepals. Stamens (1-)3-5. Style short, 3-lobed. Fruit a capsule dehiscing by 3 valves almost to the base. Seeds reniform, several.

1. P. tetraphyllum (L.) L., *Syst. nat.* ed. 10, **2**: 881 (1759). [*Saboneteira*]
Glabrous annual or rarely perennial; stems (2-)5-25 cm, prostrate to ascending, much-branched, sometimes mat-forming. Stipules ovate, long-acuminate, sometimes bifid. Leaves opposite or in whorls of 4, 4-15 × 2-8 mm, obovate or elliptic, obtuse, attenuate at base into a short petiole, rather bright green. Inflorescence many-flowered, often much-branched, and spreading. Pedicels 2-3 mm. Sepals 2-2.7 mm, oblong-lanceolate, broadly white-margined. Petals inconspicuous, linear-oblong, sometimes emarginate, *c*. ½ as long as sepals. Stamens 3-5. Capsules *c*. 1.5 mm, ovoid, shiny. Seeds 0.4-0.5 mm, pale brown, papillose.
Fl. I - XII. *Very common in both Madeira and Porto Santo, growing on cultivated and waste ground, drystone walls, rough pasture, and between the cobblestones of roads, up to 900 m in Madeira and 350 m in Porto Santo; also found on the Desertas (all islands), Selvagem Grande, and Ilhéu de Fora.* **MDPS**

The above description applies to subsp. **tetraphyllum**, to which most specimens from the Madeiran archipelago belong. However subsp. **diphyllum** (Cav.) O. Bolòs & Font Quer in *Collnea bot. Barcinone* **6**: 356 (1962) (*P. diphyllum* Cav.) has been recorded recently (1983) from Boca da Encumeada. It may be distinguished by the leaves which are mostly opposite and purplish tinged, at least near the base, and by the more condensed, fewer-flowered inflorescences, the flowers with 1-3 stamens. $2n=16$.

11. Spergula L.

Slender annual herbs; stems often much-branched at the base. Leaves appearing whorled with short, leafy, lateral branches borne on both sides at each node, linear; stipules small, scarious, free. Inflorescence a loose, terminal, cymose cluster with small scarious bracts. Sepals 5, free, scarious-margined. Petals 5, white, entire. Stamens 5-10. Styles 3-5. Capsules ovoid to subglobose, dehiscing by 3-5 valves. Seeds winged.

1. Leaves grooved beneath; pedicels and sepals glandular-pubescent; capsules 5-valved . . . . **1. arvensis**
- Leaves not grooved; pedicels and sepals glabrous; capsules 3-valved . . . . . **2. fallax**

1. S. arvensis L., *Sp. pl.* **1**: 440 (1753). [*Gorda, Orga*]
S. vulgaris Boenn.
Plant shortly glandular-pubescent above; stems 10-40 cm, ascending or sprawling, usually branched from the base. Leaves 1-3(-4.5) cm, narrowly linear, ± obtuse, fleshy, channelled beneath; stipules small, broadly triangular-ovate. Pedicels up to 2.7 cm, filiform, becoming deflexed after flowering. Sepals 2.5-4 mm, ovate-lanceolate, scarious-margined. Petals obovate, obtuse, slightly exceeding the sepals. Stamens 5-10. Styles 5. Capsules 3.5-4.5 mm, ovoid, 5-valved. Seeds 1-1.2 mm, subglobose, dull black, often with whitish or light brown papillae, with a pale keel or very narrow wing.
Fl. XII - VI. *An occasional weed of cultivated fields, roadsides, and gardens in Madeira, sometimes locally abundant; up to 450 m.* **M**

2. S. fallax (Lowe) E.H.L. Krause in Sturm, *Deutschl. Fl.* ed. 2, **5**: 19 (1901).
Spergularia fallax Lowe
Glabrous; stems 8-40 cm, ascending, usually much-branched from the base. Leaves 1-4 cm, narrowly linear, obtuse, fleshy, not channelled beneath; stipules small, triangular-ovate. Pedicels up to 1 cm, filiform. Sepals 3-5 mm, oblong-lanceolate, scarious-margined. Petals oblong-lanceolate, subacute to obtuse, ± equalling the sepals. Stamens 6-10. Styles 3. Capsules 3.5-4.5 mm, ovoid to subglobose, 3-valved. Seeds (0.8-)1-1.4 mm, bright black, minutely tuberculate, usually with a broad hyaline wing.
Fl. III - VIII. *Rare in south-east Madeira, growing on cliffs, in cultivated fields and pastures near the coast, from Funchal to Ponta de São Lourenço and Ilhéu de Agostinho. More widely distributed than* S. arvensis, *with records also from Porto Santo, Ilhéu Chão and Bugío of the Desertas, and Selvagem Grande.* **MDPS**

Some specimens from Selvagem Grande have seeds with atypically narrow, indistinct wings.

12. Spergularia (Pers.) J. & C. Presl

Annual to perennial herbs; stems often decumbent or ascending, dilated at the nodes, somewhat flattened. Leaves opposite, linear, often fleshy; stipules scarious, united to surround the node; short, leafy lateral branches (when present) borne on one side only at each node. Inflorescences cymose. Flowers 5-merous. Sepals free, scarious-margined. Petals white or pink, entire. Stamens 1-10. Styles 3. Fruit a capsule dehiscing by 3 valves. Seeds compressed, sometimes winged.

1. Capsules 4-6.5 mm, equalling or slightly exceeding the sepals; seeds 0.6-0.8 mm (excluding wing, if present) . . . . **1. marina**
- Capsules 3-4 mm, usually slightly shorter than or ± equalling the sepals; seeds 0.35-5 mm, unwinged . . . . **2. bocconii**

1. S. marina (L.) Griseb., *Spicil. fl. rumel.* **1**: 213 (1843).
S. salina J. & C. Presl
Annual to short-lived perennial herb; stems up to 25 cm, usually numerous and branched from the base, prostrate or ascending, sometimes densely matted. Leaves 6-30(-50) mm, linear, acute, not or very shortly mucronate; stipules broadly triangular-ovate, acuminate. Inflorescence glandular-pubescent. Pedicels 1-12 mm. Sepals 4-6 mm, lanceolate to narrowly ovate, obtuse. Petals 2-4 mm, ovate-oblong, obtuse, pink or purplish with a white base, rarely entirely white. Stamens 2-6(-8). Capsules 4-6.5 mm, ovoid, usually slightly exceeding the sepals. Seeds 0.6-0.8 mm (excluding wing), unwinged, winged or both, ovoid, usually finely papillose, light brown; wing (if present) with a fimbriate margin.
Fl. IV - VI. *Locally abundant in Porto Santo, for example in and around Vila Baleira and on roadside banks at Pedregal de Dentro. Recorded in Madeira from Porto do Moniz and the seashore near Caniçal.* **MP**

2. S. bocconii (Scheele) Graebn. in Asch. & Graebn., *Syn. mitteleur. Fl.* **5**(1): 849 (1919).
S. rubra sensu auct. mad., non (L.) J. & C. Presl (1819)
Annual or biennial herb, glandular-pubescent above and sometimes below; stems up to 25 cm, branched from the base, prostrate to ascending, sometimes forming large mats. Leaves (10-)15-40 mm, linear, distinctly mucronate to aristate; stipules triangular, acuminate. Pedicels 1-7 mm. Sepals 3-4 mm, oblong-lanceolate, obtuse. Petals 2-3 mm, oblong-lanceolate, obtuse, pink with a white base. Stamens 2-5(-8). Capsules 3-4 mm, ovoid, usually slightly shorter than or ± equalling the sepals. Seeds 0.35-0.5 mm, ovoid, brown, finely tuberculate, unwinged. 2n=36.
Fl. II - VIII. *Fairly frequent near the coast in Madeira, particularly in the south-east, growing along roads, tracks, in pastures, and on waste ground. Also recorded from Porto Santo and Deserta Grande.* **MDP**

This species has been reported as being common in Porto Santo, but a number of the records are based on misidentifications and are referable to *S. marina*.

13. Silene L.

Annual to perennial herbs. Leaves opposite, entire, exstipulate. Flowers 5-merous, in cymose inflorescences, sometimes raceme-like, or solitary. Calyx usually tubular, 10- or 20-veined, 5-toothed. Petals long-clawed, white, pink, yellow or greenish; limb often bifid or emarginate; coronal scales sometimes present at the base of the limb. Stamens 10. Styles 3. Fruit a capsule dehiscing by 6 teeth, borne on a carpophore. Seeds numerous, ± reniform.

1.	Flowers in raceme-like, usually simple, monochasial cymes	2
-	Flowers in branched dichasial cymes	3
2.	Calyx contracted at the mouth, pubescent with long, spreading, multicellular hairs and shorter glandular hairs; inflorescence ± 1-sided; capsules ovoid	**6. gallica**
-	Calyx not contracted at the mouth, appressed-pubescent with very short, ascending hairs; inflorescence not as above; capsules oblong	**5. nocturna**
3.	Perennial; calyx very inflated, 1 cm or more broad in fruit	4
-	Annual; calyx not or slightly inflated and then less than 1 cm broad in fruit	5
4.	Bracteoles scarious; capsule teeth erect or erecto-patent; plant ± erect	**1. vulgaris**
-	Bracteoles herbaceous; capsule teeth recurved; plant ± procumbent	**2. uniflora**
5.	Plant puberulent; stems very slender; leaves narrowly linear	**3. inaperta**
-	Plant glabrous; stems rather stout; leaves obovate or lanceolate	**4. behen**

1. S. vulgaris (Moench) Garcke, *Fl. N. Mitt.-Deutschland* ed. 9: 64 (1869). [*Orelha de boi*]
S. inflata Sm.; *S. venosa* Asch.
Perennial, ± glabrous, green or slightly glaucous; stems up to 60 cm, sometimes decumbent at the base, soon erect, simple or branched, all shoots flowering. Leaves sessile, up to 80 × 22 mm, oblong, lanceolate, oblanceolate or elliptic, acute, margins sometimes scabrid. Inflorescences usually many-flowered, the flowers slightly drooping; bracteoles scarious. Calyx 11-18 mm, oblong, 20-veined, very inflated and bladder-like, loosely enclosing the capsule, occasionally tinged purple; teeth 2-5 mm, margins minutely tomentulose. Petals usually white, rarely yellowish or pink, large, deeply bifid. Capsules 9-10 mm; teeth erect or erecto-patent; carpophore 2.5-3 mm. Seeds 1.4-1.5 mm, with low, blunt tubercles, brown.
Fl. II - VI. *Fairly common; grassy slopes and fields in Madeira, Porto Santo and Deserta Grande.* **MDP**

2. S. uniflora Roth in *Annln Bot. (Usteri)* **10**: 46 (1794). [*Chicharinha*]
S. maritima With.; *S. vulgaris* subsp. *maritima* (With.) Á. & D. Löve
Perennial, ± glabrous, often glaucous; stems up to 50 cm, prostrate, much-branched, trailing or forming mounds. Leaves sessile, 12-35 × 3-9(-12) mm, linear to oblanceolate, generally acute, slightly succulent, margins glabrous or scabrid. Inflorescences 1-3(-7)-flowered, the flowers slightly drooping; bracteoles herbaceous. Calyx 12-18 mm, oblong, 20-veined, very inflated and bladder-like, loosely enclosing the capsule, often flushed with purple; teeth 2-5 mm, the margins minutely tomentulose. Petals white, deeply bifid. Capsules 7-11 mm; teeth recurved; carpophore 2.5-3 mm. Seeds 1.4-1.5 mm, with low, blunt tubercles, blackish. $2n=24$.
Fl. III - XI. *Sea cliffs, rock crevices, roadsides, waste ground, rocky pastures. Locally common in Madeira and Porto Santo; also present on Ilhéu de Cima; recorded from all of the Desertas. Known from Selvagem Grande in the last century, but not recorded since.* **MDPS**

S. armeria L., *Sp. pl.* **1**: 420 (1753), a glabrous annual or biennial with simple, erect stems, glaucous, ovate to lanceolate, amplexicaul leaves, and bright pink flowers in a rather crowded, flat-topped cluster has been recorded growing on a rock-wall at Porto Moniz. Native to Europe and occasionally escaping from gardens where it is grown for ornament. **M**

S. noctiflora L., *Sp. pl.* **1**: 419 (1753) (*Melandrium noctiflorum* (L.) Fr.), an annual up to 60 cm, viscid-pubescent above, with large, sweetly-scented flowers (at night), *c*. 2 cm across, in a few-flowered dichasial cyme, the calyx 20-25 mm long, the petals deeply bifid, pink above, yellowish beneath, and inrolled during the day, has been recorded as a weed in Blandy's Garden at Palheiro Ferreiro in Madeira. **M**

3. S. inaperta L., *Sp. pl.* **1**: 419 (1753).
Puberulent annual; stems 15-50 cm, erect, divaricately branched and viscid above, rigid, very slender. Leaves narrowly linear, acute, the cauline mostly in fascicles. Flowers inconspicuous, in very loose, divaricately-branched dichasial cymes. Calyx 7-10 mm, ± glabrous, oblong, not contracted at the mouth; teeth *c*. 1 mm, triangular, acute, sparsely puberulent. Petals deeply emarginate, greenish brown, included within the calyx. Fruiting pedicels up to 2.5 cm, filiform. Capsules 5-8 mm, oblong; carpophore 2-3 mm, stout. Seeds 0.7-0.8 mm, dark brown, striate, lateral faces ± flat, dorsal face narrow, ± acutely grooved.
Fl. IV - VIII. *Extremely rare. Fields, waste ground and dry, rocky places in Madeira. Also recorded from Porto Santo but without further data.* **MP**

4. S. behen L., *Sp. pl.* **1**: 418 (1753).
S. ignobilis Lowe
Glabrous, glaucous annual; stems 10-30(-40) cm, erect, simple or branched from near the base,

rather stout. Lower cauline leaves obovate, acute, long-attenuate at the base into a petiole; upper cauline leaves lanceolate, sessile. Flowers erect or slightly nodding, in a lax, spreading, dichasial cyme. Calyx 9.5-16 mm, oblong-ovoid, somewhat contracted at the apex in fruit, membranous, whitish, with 10 reddish veins anastomosing above; teeth *c*. 1 mm, triangular, obtuse, margins minutely pubescent. Petals pale pink; limb deeply bifid. Fruiting pedicels up to 12(-15) mm, stout. Capsules 9-10 mm, ovoid. Seeds 1.2-1.4 mm, greyish brown; faces flattish with radiating ridges, topped by short, dark, conical papillae.
Fl. III - V. *Extremely rare in cornfields in Porto Santo; also known from Ponta de São Lourenço in Madeira, but not recorded recently.* **MP**

5. S. nocturna L., *Sp. pl.* **1**: 416 (1753).
Slender, pubescent annual, viscid above; stem (8-)15-60 cm, erect, simple or branched. Lower leaves oblong-spathulate to obovate, obtuse to subacute, sometimes mucronate, attenuate at base to a short petiole; upper leaves sessile or nearly so, narrower. Flowers erect or erecto-patent in raceme-like, monochasial cymes, occasionally forked below, crowded and sessile above, usually distant and pedicellate below. Calyx 8-11 mm, oblong, not contracted at the mouth, with 10 green veins anastomosing above, shortly appressed-pubescent with ascending hairs; teeth *c*. 1.5 mm, ciliate. Petals whitish above, blue-grey or greenish below, inrolled during the day; limb 2-6 mm, bifid. Fruiting pedicels up to 7(-20) mm. Capsules 7-11 mm, oblong; carpophore up to 1.5 mm. Seeds 0.6-0.9 mm, greyish brown; faces deeply concave; back very broad, with shallow groove.
Fl. III - V. *An inconspicuous and easily overlooked plant, common in Porto Santo in pastures and other grassy places; extremely rare in Madeira, recorded from Ponta de São Lourenço and near Funchal; also known from Deserta Grande.* **MDP ?S**

Two nineteenth century collections from Selvagem Grande have been seen, but the presence of this species in the Salvages has not been confirmed by any other records.

6. S. gallica L., *Sp. pl.* **1**: 417 (1753). [*Erva mel* (P)]
S. gallica var. *anglica* (L.) Mert. & Koch, var. *quinquevulnera* (L.) Mert. & Koch
Pubescent annual, glandular above; stems (2-)10-40 cm, usually erect, simple or branched from the base with often wide-spreading branches. Lower leaves oblong-spathulate to obovate, obtuse or mucronate, attenuate at base into an indistinct petiole; upper leaves sessile, narrower. Flowers shortly pedicellate, erect, in ± 1-sided, raceme-like monochasial cymes, sometimes forked below. Calyx 7-10 mm, cylindrical-ovoid, contracted at the mouth, 10-veined, the veins green or red, with mixed long, multicellular hairs and shorter glandular hairs; teeth 1.5-2.5 mm. Petals white to pale or occasionally deep pink, rarely with a crimson spot; limb 3-6 mm, entire or ocasionally emarginate. Fruiting pedicels up to 8(-20) mm. Capsules 7-9 mm, ovoid; carpophore up to 1 mm. Seeds 0.7-1 mm, blackish; faces deeply concave, ridged, back broad, flattish, papillate. $2n=24$.
Fl. III - VI. *Extremely common weed of cultivated and waste ground in Madeira and Porto Santo, also found in grassy places, on walls, banks, and cliffs, mainly below 600 m; also known from Deserta Grande, Bugío, and Ilhéu Chão; recorded from Selvagem Grande in the last century, but no recent records known.* **MDPS**

Gypsophila elegans M. Bieb., *Fl. taur.-caucas.* **1**: 319 (1808), a glabrous annual from the Caucasus with lanceolate, acute leaves and a lax inflorescence of white flowers, the petals purple-veined, has been recorded as a casual in Funchal. **M**

Saponaria officinalis L., *Sp. pl.* **1**: 408 (1753), a glabrous perennial from Europe with ascending to erect flowering shoots, ovate to elliptic, 3-veined leaves, and pink flowers in dense, branched clusters, the petals entire, has been reported as an occasional garden escape. **M**

Vaccaria hispanica (Mill.) Rauschert in *Wiss. Z.Martin-Luther-Univ. Halle-Wittenb.* **14**: 496 (1965) (*V. pyramidata* Medik.), a glabrous annual from Europe with an erect stem, glaucous, ovate to lanceolate leaves, and pink flowers with the calyx becoming inflated and winged on the angles in fruit, has been reported as a casual on waste ground near Funchal harbour. **M**

14. Petrorhagia (Ser. ex DC.) Link

Annual, wiry-stemmed herbs. Leaves opposite, narrow, united at the base, exstipulate. Flowers in capitate heads, opening 1 at a time. Bracts present, enclosing the calyx. Calyx distinctly 5-15-nerved; tube cylindrical or obconical with scarious intervals between the nerves, 5-toothed. Petals 5, pinkish, exceeding the calyx, clawed. Stamens 10. Styles 2. Fruit an ellipsoid capsule, dehiscing by 4 teeth. Seeds shield-shaped with an elevated rim.

1. P. nanteuilii (Burnat) P.W. Ball & Heywood in *Bull. Br. Mus. nat. Hist.* (Bot.) **3**: 164 (1964). [*Craveiro*]
Dianthus prolifer, Tunica prolifera, sensu auct. mad.
Stems 10-55 cm, solitary or much-branched from base, erect or sometimes procumbent at the base, simple or branched above, glabrous to somewhat scabrid-pubescent. Leaves 20-50(-70) mm, linear, acute, margins scabrid; leaf-sheaths 1.5-3.5 mm long. Inflorescence compact, often reduced to 1 flower. Bracts shiny brown, parchment-like, enclosing the calyces, the outermost mucronate, the inner obtuse. Calyx-tube 11-13 mm, cylindrical, puberulent and reddish above; teeth obtuse. Petals pink or lilac; limb 2-5 mm, obovate, deeply emarginate. Capsules *c*. 7 mm. Seeds 1.4-1.5 mm, finely tuberculate, black.
Fl. I - XII. *Fairly frequent in south-east Madeira along roadsides, levadas, in pastures and other grassy places, 50-250 m; extremely rare in Porto Santo. Recently recorded for the first time from Deserta Grande.* **MDP ?S**

One nineteenth century collection from Selvagem Grande has been seen, but the presence of this species on the island has not been confirmed by any other records.

XXXVII. RANUNCULACEAE
J.R. Press

Annual or perennial herbs. Leaves alternate, exstipulate, often lobed or dissected. Flowers hermaphrodite, actinomorphic or zygomorphic. Perianth all petaloid, or petaloid and sepaloid. Stamens numerous, spirally arranged, extrorse. Carpels 1-many, usually free and spirally arranged. Fruit consisting of a head of achenes or 1 or more follicles.

1.	Flowers spurred	2
-	Flowers lacking spurs	4
2.	Flowers actinomorphic	**4. Aquilegia**
-	Flowers zygomorphic	3
3.	Flowers with an upper and a lateral pair of petals; follicles (3-)5	**5. Delphinium**
-	Flowers lacking a pair of lateral petals; follicle 1	**6. Consolida**
4.	Flowers blue, the perianth segments in one whorl; fruit an inflated capsule	**1. Nigella**
-	Flowers yellow or red; perianth of 2 whorls; fruit of numerous achene	5
5.	Petals with nectaries at the base, concolorous	**3. Ranunculus**
-	Petals lacking nectaries, often with a dark patch at the base	**2. Adonis**

1. Nigella L.

Annuals. Leaves usually 1- to 3-pinnatisect. Flowers solitary, terminal. Perianth consisting of 5 outer segments (*sepals*) and 5 inner ones (*petals*), the inner nectariferous, bilabiate, the lower lip bilobed. Fruit a group of follicles forming a capsule.

Literature: M. Zohary, A revision of *Nigella*, in *Pl. Syst. Evol.* **142**: 71-107 (1983).

1. N. damascena L., *Sp. pl.* **1**: 534 (1753).
Stems erect, slender, usually little-branched. Leaves 2- to 3-pinnatisect with linear to filiform lobes, the uppermost leaves forming an involucre. Sepals petaloid, much longer than the petals, blue. Petals with the lobes of the lower lip rounded, 1-2 × as long as broad. Fruit a glabrous, somewhat inflated, membranous capsule, as long as or slightly shorter than the persistent styles. Seeds triquetrous, dull black.
Fl. mainly III - VI but irregularly throughout the year. *Probably introduced. Formerly a common weed of cornfields, cultivated and waste ground and roadsides on dry soils throughout the lower regions but now declining. Recorded only once from Selvagem Grande and probably no longer present there. Native from the Mediterranean eastwards to Iran but cultivated for ornament in many areas and often naturalized.* **M ?S**

2. Adonis L.

Annuals or perrenials. Leaves 3-pinnatisect, lobes linear. Flowers solitary, terminal. Petals lacking nectaries. Achenes reticulate, shortly beaked, in an elongated, ± cylindrical or rounded head.

Literature; H. Riedl, Revision der einjahrigen Arten von *Adonis* L., in *Annln. naturh. Mus. Wien* **66**: 51-90 (1963).

1. A. microcarpa DC., *Syst. nat.* **1**: 223 (1817).
Annual. Stems ± erect but often widely branched. Sepals 5, spreading, purple-tinged. Petals 5-8, bright red or sometimes yellow, black towards the base. Anthers blackish or dark purple. Achenes somewhat reticulate or ribbed, sometimes with a transverse, toothed ridge and a short dorsal projection near the upturned, bluish beak.
Fl. I - II. *Probably introduced. A sporadic and rare ephemeral weed of dry soils, especially in cornfields and waste places in Porto Santo. Native from the Mediterranean eastwards to Iran.* **P**

3. Ranunculus L.

Annual or perennial herbs (in Madeira all terrestrial). Leaves often palmately lobed or divided. Flowers solitary and terminal or in corymbose panicles. Sepals (3-)5. Petals (0-)5, yellow, sometimes fading to white, with a nectary at the base. Stamens and achenes numerous. Fruit a head of achenes.

1. All leaves entire, or the upper sometimes toothed . . . . **9. flammula**
- At least some leaves toothed or divided 2

2. Annual; achenes spiny, muricate or tuberculate 3
- Perennial; achenes smooth or pitted 6

3. Upper leaves 3-lobed, the middle lobe distinctly stalked; receptacle pubescent . . . 4
- Upper leaves simple or with short, sessile lobes; receptacle glabrous 5

4.	Sepals reflexed; achenes tuberculate .... **5. trilobus**
-	Sepals patent; achenes spiny .... **6. arvensis**

5.	Flowers 12-16 mm across; achenes *c.* 7 mm .... **4. muricatus**
-	Flowers 3-6 mm across; achenes *c.* 3 mm .... **7. parviflorus**

6.	Sepals reflexed .... **1. bulbosus**
-	Sepals not reflexed 7

7.	Stems creeping, rooting at nodes; pedicels sulcate .... **3. repens**
-	Stems not creeping and rooting at nodes; pedicels smooth 8

8.	Large, fusiform, tubers present; flowers more than 25 mm across . **8. cortusifolius**
-	Tubers absent, the roots all fibrous; flowers less than 25 mm across ..... **2. acris**

1. R. bulbosus L., *Sp. pl.* **1**: 554 (1753).
More or less hairy perennial. Basal and lower leaves 3-lobed, the middle lobe cuneate at base and sessile or narrowed into a stalk, all the lobes toothed or divided; upper leaves sessile, with narrow lobes. Flowers 20-30 mm diameter. Sepals about as long as petals, reflexed. Pedicels sulcate. Receptacle pubescent. Achenes *c.* 2 mm, the margin keeled and grooved, beak short, curved.
Fl. IV. *Probably introduced. Very localized, in a small area around Pico do Infante, between Monte and Quinta Palheiro Ferreiro. Native to Europe.* **M**

Madeiran plants are most similar to subsp. **aleae** (Willk.) Rouy & Foucaud, *Fl. France* **1**: 106 (1893), in which the basal stock is barely swollen but the roots are fusiform and tuberous. However, there is some resemblance to subsp. **adscendens** (Brot.) Neves, *Contrib. Portug. Ranunc.*: 84 (1944) which has similar roots but inner basal leaves narrower and more acutely lobed than the outer ones. *R. bulbosus* subsp. *aleae* closely resembles the upland form of *R. cortusifolius*, especially in the presence of tuberous roots. *R. bulbosus* is easily distinguished by the reflexed sepals and stipitately-lobed leaves.

2. R. acris L., *Sp. pl.* **1**: 554 (1753).
Hairy perennial with creeping rhizome. Lower leaves divided into 3-7, toothed lobes; cauline leaves similar but smaller. Flowers 15-25 mm diameter. Sepals ± equaling petals, not reflexed. Receptacle glabrous. Fruiting head globose. Achenes 2-3.5 mm, elliptic or sub-orbicular, glabrous, bordered, beak short, hooked.
Fl. IV - VI. *Very rare, only occurring in woods in the south-east of Madeira around Funchal and Camacha.* **M**

3. R. repens L., *Sp. pl.* **1**: 554 (1753).
Pubescent to strigose perennial with long stolons rooting at the nodes. Basal and lower leaves triangular-ovate, 3-lobed, the middle lobe long-stalked, all lobes divided into 3-toothed segments; upper cauline leaves similar but smaller. Flowers 18-25 mm diameter. Sepals patent, shorter than petals. Pedicels sulcate. Receptacle pubescent. Fruiting head globose. Achenes 2-3 mm, suborbicular, smooth, compressed, with a narrow, thickened border, beak 1 mm, hooked.
Fl. ± throughout the year. *Very common in damp, grassy places and woods almost everywhere in Madeira.* **M**

4. R. muricatus L., *Sp. pl.* **1**: 555 (1753).
Sparsely hairy to ciliate annual with fibrous roots. Leaves all similar, suborbicular, coarsely toothed or at most shallowly lobed. Flowers axillary, solitary, 12-16 mm diameter. Sepals

equalling or shorter than petals, reflexed. Receptacle pubescent. Achenes 4-6 mm, ovate, strongly compressed, spiny, with a thickened border, beak 2-3 mm, hooked.
Fl. II - V. *In levadas, gutters and ditches along roadsides and in damp patches of waste ground. Common in the lower regions of Madeira; rare in Porto Santo.* **MP**

5. R. trilobus Desf., *Fl. atlant.* **1**: 437 (1798).
R. sardous subsp. *trilobus* (Desf.) Rouy & Fouc.
Glabrescent ot densely pubescent annual with fibrous roots. Lowest leaves simple; upper leaves 3-lobed, the middle lobe stalked, all lobes shallowly to deeply toothed, those of uppermost leaves deeply divided. Flowers 10-12 mm diameter. Sepals ± equalling or slightly shorter than petals, reflexed. Receptacle pubescent. Achenes *c.* 2 mm, ovate, strongly compressed, tuberculate, with a thickened border, beak short, triangular.
Fl. IV - VI. *Uncommon and perhaps rare in Madeira. Generally occurring in the lower regions, often near the sea but occasionally inland and on higher ground.* **M**

6. R. arvensis L., *Sp. pl.* **1**: 555 (1753).
More or less pubescent annual. Lower leaves simple; upper leaves divided into 2-3 narrow, toothed lobes. Flowers 4-12 mm diameter. Sepals shorter than petals, patent. Pedicels terete. Receptacle pubescent. Achenes 4-8 in a single whorl, 5-8 mm, ovate, spiny, with a broad sulcate border, beak 3-4 mm, straight or slightly curved.
Fl. ? *Apparently recently introduced but very rare and perhaps only a casual. Occurring in the Funchal region only. Native to Europe.* **M**

7. R. parviflorus L., *Syst. nat.* ed. 10 **2**: 1087 (1759).
Softly hairy annual. Lower leaves 3(-5)-lobed, the lobes toothed; upper leaves simple or with entire lobes. Flowers 3-6 mm diameter. Sepals equalling petals, reflexed. Receptacle glabrous. Achenes 2-3 mm, tuberculate, the tubercles each with a short, hooked spine, border thickened, beak short, hooked.
Fl. V - VII. *In damp ravines and wet places. Scattered and rare in Madeira, from the northeast of the island, Paúl da Serra to Ribeiro Frio and Machico.* **M**

8. R. cortusifolius Willd., *Enum. pl.*: 588 (1809). [*Douradhina, Rainúnculo*]
Erect, sparsely to densely villous perennial up to 1 m, with fleshy, fusiform tubers. Basal leaves coriaceous, up to 21 × 30 cm, orbicular-cordate, lobed, the usually 3-5 main segments themselves lobed and toothed at the apex, long petiolate; cauline leaves similar but progressively smaller, narrower and bract-like, shortly petiolate to subsessile. Inflorescence corymbosely branched. Flowers shining, golden yellow. Sepals horizontally spreading in flower, ovate, glabrous, margins scarious. Petals 12-27 mm. Receptacle hairy, at least at base. Achenes 3-3.5 mm, glabrescent, smooth, beak 0.5-1 mm, curved. **Plate 8**
Fl. III - VI. *Found mainly in central and northern Madeira.* ■ **M** [*Azores, Canaries*]

Two forms of this plant have been recognized (as varieties of *R. grandifolius*) in Madeira. Although the extremes of variation are well marked, a number of plants are intermediate. The exact status and relationship of these and plants from the Azores and Canaries is uncertain. The two forms are:

[*R. grandifolius* var. *major* Lowe]. Robust, usually somewhat sparsely villous plants more than 50 cm high. Basal leaves 10-21 × 14-30 cm, lobed to half-way or less. Inflorescence much-branched and spreading, many-flowered. Flowers (24-)30-60 mm diameter. Fruiting heads ovoid, about as long as broad. *Damp, shady places among rocks and in ravines up to c. 1000 m, typical of laurisilva. Mainly in central Madeira, around Fajã da Nogueira, Ribeiro Frio, Ribeira da Metade and Pico das Pedras.*

[*R. grandifolius* var. *minor* Lowe]. Smaller, more slender plants up to 55 cm but usually much less, rather densely villous. Basal leaves 2-10(-13) × 2-11(-14) cm, lobed to half-way or more, the segments often narrow. Inflorescence little-branched, few-flowered (flowers usually 1-15). Flowers 25-30(-60) mm diameter. Fruiting heads cylindrical, about 2 × as long as broad. *In dry, more exposed situations on rock ledges, grassy banks and in* Vaccinium *scrub. An apparently endemic upland form found above c. 1300 m around Pico de Arieiro, Pico Ruivo and Paúl da Serra and occasionally elsewhere such as at Cabo Girão.* ?●

9. R. flammula L., *Sp. pl.* **1**: 548 (1753).
Perennial with fibrous roots. Stems erect or creeping and rooting at the nodes. Leaves variable in shape but all simple, orbicular to subulate, entire or serrate, the lower stalked, the upper sessile. Flowers 1-many, 7-20 mm diameter. Pedicels sulcate. Receptacle glabrous. Achenes glabrous, weakly bordered, beak short.
Fl. ? *Extremely rare, known only from a single locality in Madeira.* **M**

4. Aquilegia L.

Perennial herbs with erect, woody stock. Leaves ternate. Flowers hermaphrodite, regular. Outer perianth-segments (*sepals*) 5, petaloid; inner perianth segments (*petals*) 5, each tubular with a broad limb and backward projecting, nectariferous spur. Stamens numerous, the innermost reduced to scarious staminodes. Follicles several, free.

1. A. vulgaris L., *Sp. pl.* **1**: 533 (1753).   [*Luvas de Nossa Senhora, Viuvas*]
Stem branched above, puberulent to pubescent. Basal leaves 2-ternate, the leaflets with 3 bluntly toothed lobes, glabrous above, pubescent and somewhat glaucous beneath, long-stalked; cauline leaves smaller, short-stalked, the uppermost sessile. Flowers nodding, bluish-violet. Sepals 16-20 × 10-12 mm, ovate. Petal-limb 8-10 mm, oblong-truncate; spur 7-10 mm, hooked. Stamens barely exceeding the petals. Follicles up to 20 mm, hairy.
Fl. V - VI. *A rare plant from ravines above Funchal, Ribeiro Frio and Porto do Moniz. Various cultivars are also grown for ornament in Madeira.* **M**

5. Delphinium L.

Annuals. Leaves palmately divided. Flowers zygomorphic with both outer perianth-segments (*sepals*) and inner ones (*petals*) petaloid. Sepals 5, the upper one spurred. Petals 4, in 2 pairs, the upper pair nectariferous and spurred, the spurs inserted within the sepal spur, those of the lateral pair with a distinct limb narrowing to a claw. Stamens in 8 spiralled series. Follicles several, free.

Literature: C. Blanché, A new species of *Delphinium* L. (Ranunculaceae) from Madeira, in *Bocagiana* **154**: 1-5 (1992).

1. D. maderense C. Blanché in *Bocagiana* **154**: 1 (1992).
D. peregrinum auct. mad., non L.(1753)
Pubescent annual 20-35 cm. Basal leaves palmately 3-5 × divided into narrow lobes, often withered at anthesis; cauline leaves 2.5-5.0 cm × 1-3 mm, progressively less divided, the upper linear, entire. Racemes (3-)5- to 8(-15)- flowered. Bracteoles short, linear, in the upper ¼ of the pedicel. Flowers pale blue, (22-)24-26 mm including the horizontal or ascending spur of 16-18 mm. Follicles 3-5, 10-11 mm, linear-oblong, pubescent, divergent. **Plate 8**
Fl. VI - VIII. *Very rare endemic confined to cornfields and roadside banks in the valley at Alegría, north of São Roque in Madeira.* ● **M**

6. Consolida (DC.) Gray

Similar to *Delphinium* but perianth with the upper pair of petals forming a single structure with a nectariferous spur inserted within the sepal spur; lateral petals absent. Stamens in 5 spiralled series. Follicles solitary.

1. C. ajacis (L.) Schur in *Ver. Mitt. siebenb. Ver. Naturw.* **4**: 47 (1853). [*Ciumes*]
C. ambigua sensu P.W. Ball & Heywood; *Delpinium ajacis* L.; *D. ambiguum* auct., non L. (1753)
Stiffly erect annual with few, ascending branches. Basal leaves long-stalked, with numerous oblong segments, cauline leaves sessile, lobes linear. Racemes 3- to 5-flowered; at least the lower bracts dissected; bracteoles shorter than pedicels, not reaching the base of the flower. Flowers 21-27 mm long, including the straight or slightly curved spur of 12-15 mm. Follicle pubescent, narrowing gradually to the apex.
Fl. IV - IX. *Cornfield weed, common in the lower regions of Madeira; rare in Porto Santo.* **MP**

XXXVIII. BERBERIDACEAE

Perennial herbs or shrubs. Leaves alternate, simple or compound. Flowers hermaphrodite, (2-)3(-4)-merous. Perianth-segments in several whorls, at least the inner ones petaloid, the innermost also nectariferous. Stamens opposite the nectariferous segments; anthers opening by valves or slits. Carpel solitary. Fruit a berry or capsule.

1. Berberis L.
J.R. Press

Evergreen shrubs (in Madeira) with spiny long shoots and leafy short shoots in the axils of the spines. Flowers in umbel-like clusters. Perianth-segments in whorls of 3, the outermost small and usually sepaloid, the 2 middle whorls petaloid and equalling or longer than the 2 inner, petaloid and nectariferous whorls. Fruit a berry.

1. B. maderensis Lowe in *Hookers J. Bot.* **8**: 289 (1856). [*Ameixieira de espinho, Fustete*]
Evergreen shrub 1-2 m, stems grooved, glabrous to sparsely hairy, dark purplish red when young. Spines tripartite, sometimes absent. Leaves in fasicles, 2.5-5 cm, obovate, obtuse with a mucronate tip, attenuate at the base, entire, denticulate or with occasional slender spines, rather coriaceous, glabrous but with scattered, sessile glands beneath. Inflorescence 6- to 10(-14)-flowered. Flowers *c.* 5 mm diameter, globose, bright yellow. Berries *c.* 12 mm, lacking a style, yellowish red, ripening black with a blue bloom. **Plate 8**
Fl. V - VI. *A very rare endemic, occurring in only a few places on less accessible rock faces and in ravines in the eastern mountains of Madeira.* ● **M**

Epimedium pinnatum Fisch. in DC., *Veg. syst.* **2**: 29 (1821) has been recorded from Madeira, presumably as an escape from cultivation; it does not appear to be fully naturalized. It is a rhizomatous herb with basal, ternately divided leaves, leafless flowering stems and yellow inner perianth-segments. **M**

XXXIX. LAURACEAE
M.J. Short

Trees or shrubs, mostly evergreen. Leaves usually alternate, simple, entire, coriaceous, exstipulate. Inflorescences mostly axillary, paniculate or pseudo-umbellate. Flowers small, usually greenish, yellowish or whitish, regular, hermaphrodite or unisexual. Perianth-segments 4-6, in 2 whorls, sometimes persistent in fruit. Stamens usually in 4 whorls, the innermost sterile or sometimes lacking; anthers 2- or 4-locular, dehiscing by valves. Ovary superior, 1-locular; style simple. Fruit a 1-seeded berry.

1. Twigs pubescent when young 2
- Twigs glabrous when young 3

2. Leaves with tiny, dark gland-like depressions in the axils of the midrib and main veins on the lower surface; petioles green; flowers in pseudo-umbels, surrounded by an involucre of bracts; perianth-segments 4 .... **4. Laurus**
- Leaves without gland-like depressions; petioles reddish; flowers in panicles, without involucral bracts; perianth-segments 6 .... **1. Persea**

3. Lower leaf surface with 2 large gland-like depressions at the lower vein axils covered by hairs; berry surrounded at the base by the hardened, cupule-like perianth-tube, the segments deciduous .... **3. Ocotea**
- Lower leaf surface without gland-like depressions; berry clasped at the base by the persistent perianth-segments .... **2. Apollonias**

1. Persea Mill.

Leaves alternate. Inflorescences paniculate, long-pedunculate. Flowers hermaphrodite. Perianth-segments 6, equal, persistent in fruit. Fertile stamens 9; anthers 4-locular; staminodes 3, large, conspicuous, sagittate, stipitate.

P. americana Mill. (*P. gratissima* Gaertn.) [*Abacate, Abacateira, Pereira abacate*] is commonly cultivated in the lower regions of Madeira for its edible fruit, the avocado, produced from October to December.

1. P. indica (L.) Spreng., *Syst. veg.* **2**: 268 (1825). [*Vinhático*]
Phoebe indica (L.) Pax
Tree up to 15(-25) m with a broad, rounded crown. Twigs finely pubescent when young. Leaves 10-20 × 3-8 cm, elliptic, acute, sparsely pubescent below with very fine, appressed hairs not easily visible to the naked eye, ± glabrous above except for a few hairs on the midrib, pale green, becoming reddish with age; petioles 1.5-3 cm, reddish, pubescent or glabrous. Peduncles 4-12 cm, shorter than the leaves, pubescent. Perianth densely covered with whitish, appressed, silky hairs; segments 4-6 mm, ovate-lanceolate, acute. Berry 1.5-2 cm, ovoid-ellipsoid, green, turning to black when ripe. **Plate 9**
Fl. VIII - XI. *Rare in Madeira. A constituent of the laurisilva found in shady, humid valleys and ravines; 400-1300 m.* ■ M [*Azores, Canaries*]

2. Apollonias Nees

Leaves alternate or sometimes subopposite. Inflorescences paniculate. Flowers hermaphrodite. perianth-segments 6, subequal. Fertile stamens 9; anthers 2-locular; staminodes 3, conspicuous, sagittate, stipitate. Berry clasped at the base by the persistent perianth-segments.

1. A. barbujana (Cav.) Bornm. in Engl. in *Bot. Jb.* **33**: 420 (1903).
[*Barbusano, Barbuzano*]
A. canariensis (Willd.) Nees
Tree up to 25 m, with a dense, rounded crown. Twigs glabrous. Leaves 5-15 × 2-5 cm, elliptic, acute to obtuse, margins sometimes revolute, dark glossy green, glabrous; petioles 0.5-1.3 cm, glabrous. Peduncles up to 3 cm, glabrous. Perianth-segments 3-4.5 mm, oblong-ovate, acute to obtuse, greenish white, glabrous without, densely pubescent within. Berry 1.5-2 cm, ellipsoid, dull green, turning to black when ripe. **Plate 9**
Fl. II - V. *Rare in Madeira, occurring in laurisilva and on rocky hillsides near the coast; up to 1000 m. Recently found on Deserta Grande. Extinct in Porto Santo.* ■ MDP [*Canaries*]

Plants from the Madeiran archipelago belong to subsp. **barbujana** which otherwise occurs in the Canaries and to which the above description and synonymy apply. Subsp. *ceballosi* (Svent.) G. Kunkel is endemic to Gomera in the Canaries and is distinguished by its broadly ovate, cuspidate leaves.

The leaves are frequently marked with wart-like protuberances caused by a mite (*Eriophyes barbujanae* Carmona) which is specific to this tree.

3. Ocotea Aubl.

Leaves alternate. Inflorescences paniculate. Flowers polygamous or unisexual. Perianth-segments 6, equal, not persistent in fruit. Fertile stamens 9; anthers 4-locular; staminodes 3. Berry surrounded at the base by the hardened, cupule-like perianth-tube.

1. O. foetens (Aiton) Baill., *Hist. pl.* **2**: 438 (1870). [*Til*]
Tree up to 30(-40) m, with a dense, pyramidal to rounded crown. Twigs glabrous. Leaves 6-13(-18) × 2-7 cm, elliptic to elliptic-ovate, mostly acute, glossy and glabrous above, glabrous below except for two gland-like depressions covered by long, tangled hairs at the lower and sometimes subsequent vein axils; petioles 1-1.5 cm, glabrous. Peduncles 1-7 cm, glabrous, much shorter than the leaves. Perianth-segments 2-3 mm, oblong-ovate, obtuse, cream-coloured, glabrous without, densely pubescent within. Berry 1.8-2(-2.5) cm, ellipsoid, black, the lower third surrounded by the cupule-like perianth-tube. **Plate 9**
Fl. VII - XII. *An important component of the Madeiran laurisilva, preferring moist, slightly exposed sites, formerly widespread, but now a rather rare tree. Its black wood is highly valued for use in furniture and building; (0-)600-1500 m.* ■ M [*?Azores, Canaries*]

4. Laurus L.

Leaves alternate. Inflorescences of shortly pedunculate, pseudo-umbels borne on a short shoot, each usually comprising five flowers, surrounded by a globose involucre of large, persistent bracts. Flowers unisexual. Perianth-segments 4, equal, not persisting in fruit. Male flowers with 8-30 fertile stamens; anthers 2-locular; staminodes absent. Female flowers with 2-4 staminodes. Berry seated on a small, knob-like cupule.

1. L. azorica (Seub.) Franco in *Anais. Inst. Sup. Agron. (Lisboa)* **23**: 96 (1960).
[*Loireiro, Loureiro, Louro*]
L. azorica var. *longifolia* (Kuntze) G. Kunkel, var. *lutea* (Menezes) A. Hansen; *L. canariensis* Webb & Berthel., non Willd.
Tree up to (10-)20 m with a dense crown. Twigs densely brown-tomentose when young. Leaves variable, 5-17 × 3-6 cm, elliptic, oblong, obovate, ovate or suborbicular, acute to obtuse, dark glossy green, strongly aromatic when crushed, glabrous above, usually tomentose below when young, at least on the midrib, with tiny, dark gland-like projections in the axils of the midrib and main veins; petioles 1.5-2.5 cm, pubescent or glabrous. Inflorescence densely tomentulose. Peduncles 0.5-1 cm. Flowers creamy yellow. Perianth-segments 3.5-5 mm, oblong-obovate, obtuse. Berry 1-2 cm, broadly ellipsoid, green turning to black when ripe, very rarely yellow. 2n=48. **Plate 9**
Fl. XI - IV. *Abundant in the laurisilva of Madeira, where it is usually the dominant tree, (200-)600-1200(-1500) m. It has become scarce above 1200 m due to the effects of overgrazing, burning and cutting.* M

The parasitic fungus *Laurobasidium laurii* (Geyl.) Jülich (*Exobasidium lauri* Geyl.) is locally frequent on the trunks and branches of this species.

XL. PAPAVERACEAE
J.R. Press

Herbs, mostly exuding latex when cut. Leaves spirally arranged, lobed or dissected; exstipulate. Flowers hermaphrodite, hypogynous, actinomorphic or zygomorphic. Sepals 2(-3). Petals 4-6. Stamens 2 or numerous. Ovary superior, often unilocular, of 2 or more united carpels. Fruit a capsule opening by valves or pores, or a 1-seeded, indehiscent nut.

1. Flowers strongly zygomorphic; fruit a small, 1-seeded nut . . . . **5. Fumaria**
- Flowers actinomorphic; fruit a many-seeded capsule 2

2. Capsule ellipsoid to globular, less than 2 × as long as wide; plant often setose or prickly 3
- Capsule linear, many × longer than wide; plant sometimes pubescent but not setose 5

3. Plant prickly . . . . **2. Argemone**
- Plant not prickly 4

4. Flowers yellow; plant with orange latex . . . . **4. Chelidonium**
- Flowers red, pink, white or mauve; plant with white latex . . . . **1. Papaver**

5. Leaves pinnatifid with broad unequal segments; capsule opening from apex to base. . . . . **3. Glaucium**
- Leaves 1- to many-ternate with ± linear segments; capsule opening from base to apex . . . . **4. Eschscholtzia**

Subfam. PAPAVEROIDEAE

Latex present. Flowers usually solitary, large, actinomorphic. Sepals falling before the flower opens fully. Stamens numerous. Fruit a dehiscent capsule.

1. Papaver L.

Annuals with white latex. Leaves usually toothed or pinnately divided. Flowers nodding in bud. Sepals 2, free, falling as flower opens. Petals 2+2, entire, crumpled in bud. Stamens many. Stigmas 4-15, sessile and connate, forming a broad persistent disc. Capsule opening by pores just beneath the stigmatic disc. Seeds small, numerous.

Literature: J. W. Kadereit, *Papaver somniferum* L. (Papaveraceae): a triploid hybrid?, in *Bot. Jb.* **106**: 221-244 (1986); A revision of *Papaver* L. sect. *Papaver* (Papaveraceae), in *op. cit.* **108**: 1-16 (1986); A revision of *Papaver* L. section *Rhoeadium* Spach, in *Notes R. bot. Gdn Edinb.* **45**: 225-286 (1989).

1. Flowers white or mauve; upper leaves amplexicaule . . . . **4. somniferum**
- Flowers red, rarely pink or white; upper leaves not amplexicaule 2

2. Capsule more than 2 × as long as wide, broadly obovoid to ± globose; stigmatic rays 10-13 . . . . **1. rhoeas**
- Capsule less than 2 × as long as wide, narrowly obovoid or clavate; stigmatic rays 4-11 3

3. Anthers yellow; leaves lanceolate to elliptic in outline . . . . **2. pinnatifidum**
- Anthers brown to greenish; leaves obovate to ovate in outline . . . . **3. dubium**

1. P. rhoeas L., *Sp. pl.* **1**: 507 (1753). [*Papoila vermelha*]
Setose annual. Stems 20-80 cm, erect or ascending to almost decumbent, mostly branched from base. Leaves 2-18 × 0.5-6 cm, obovate to ovate in outline, pinnatifid with toothed segments to 2-pinnatisect, terminal segment more than 2 × as long as broad; lower leaves petiolate, petiole up to 8 cm, upper leaves mostly sessile. Pedicels and calyces with patent or loosely appressed setae. Petals 2.3-3.3 × 2-6 cm, outer pair larger than the inner, broadly obovate, often much wider than long, dull red to crimson, rarely pink or white, with or without a black basal spot. Stamens numerous, filaments slender, red, black or rarely white to yellow, anthers brown. Capsule 10-14 × 7-10 cm, broadly obovoid to almost globose, sessile or short-stalked; stigmatic disc with 10-13 rays, mostly flat and broader than capsule.
Fl. III - VI. *A common weed of cultivated and waste ground, grassy areas and roadsides throughout the lower regions of Madeira and Porto Santo.* **MP**

Two varieties occur in the islands:

i) var. **rhoeas**
Stems and pedicels with spreading setae.
Throughout the range of the species.

ii) var. **strigosum** Boenn., *Prodr. Fl. Monast. Westphal.*: 157 (1824).
P. rhoeas subsp. *strigosum* (Boenn.) Menezes
Stems and pedicels with appressed setae.
Also throughout the range of the species, but rare.

2. P. pinnatifidum Moris, *Fl. sardoa* **1**: 74 (1837).
Erect, setose annual, 45 cm, branching from base. Leaves held very erect, up to 7 × 1.5 cm, lanceolate to elliptical in outline, simple and toothed to pinnatifid with appressed, ± dense setae above and below; lobes mostly triangular, toothed or entire; lower leaves petiolate, upper sessile with a pair of large, triangular basal lobes. Pedicels appressed setose above. Petals obovate, orange-red, with or without dark basal spot. Filaments slender, black; anthers yellow. Capsule 12-16 × 4-5 mm, more than 2 × as long as wide, clavate, ribbed, contracted below pores; stigmatic disc narrower than capsule, the lobes often with dark margins, rays 4-9.
Fl. V. *A weed of cultivated fields and roadsides around Funchal, apparently rare.* **M**

3. P. dubium L., *Sp. pl.* **2**: 1196 (1753).
Erect, setose annual 10-90 cm. Leaves 3-11 × 0.5-5 cm, obovate to ovate in outline, pinnatifid to mostly pinnatisect, with ± dense, patent setae, or glabrous above, sparsely setose on veins beneath; lobes erect to spreading, ovate to oblong, pinnatisect, toothed or entire; lower leaves petiolate, upper shortly petiole to sessile. Pedicels with closely appressed setae. Petals 1.8 × 1.4 cm, obovate, mostly orange-red, with or without a black spot. Filaments filiform, violet to black; anthers brown to green. Capsule 10-20 × 5-8 mm, usually more than 2 × as long as wide, narrowly obovoid to clavate; stigmatic disc narrower or equalling capsule width, the lobes usually broadening towards the tips; rays 4-11. Seeds brown, glaucous.
Fl. III - VI. *Occasional weed of cultivated and waste ground on disturbed soils.* **M**

Madeiran plants all appear to belong to subsp. **dubium**, distinguished from other subspecies principally by the plant being green with a dense indumentum and by the colour of the latex (white or cream).

4. P. somniferum L., *Sp. pl.* **1**: 508 (1753). [*Papoila*]
Glaucous or green annual, 30-60 cm, stem erect, simple or branched above. Leaves simple or shallowly pinnate-lobed, 6-13(-16) × 2.5-7(-9) cm, oblong to ovate, acute to obtuse, irregularly

dentate; basal and lower cauline leaves narrowing into a petiole, upper cauline amplexicaule. Petals 3-5 cm, suborbicular to obovate-oblong, white or mauve with a purple spot at base. Stamens with clavate filaments, anthers yellow. Capsule 1-1.5 cm, subglobose, shortly stalked, dehiscent or not; stigmatic disc ± flat, deeply lobed, lobes shallowly toothed, rays 8-12.
Fl. IV - V. *Common weed of crops, waste ground and roadsides throughout the lower regions of all the islands except the Salvages.* **MDP**

Two subspecies occur in the islands:

a) subsp. **somniferum** [*Papoila branca*]
P. somniferum subsp. *nigrum* (Garsault) Thell.
Plant usually glaucous, ± glabrous. Petals white with a purple basal spot. Capsule dehiscent.
Throughout the range of the species.

b) subsp. **setigerum** (DC.) Arcang., *Comp. fl. ital.*: 25 (1882).
Plant usually green, setose on stems, veins of leaves beneath and on sepals. Leaf lobes and teeth tipped with a single seta. Petals mauve with a dark purple basal spot. Capsule usually indehiscent.
Porto Santo; formerly present on Madeira, perhaps no longer so.

2. Argemone L.

Glaucous annuals with yellow latex. Leaves dentate to deeply lobed. Sepals usually 3, free, ending in a hollow, spine-tipped horn. Petals usually 3+3. Stigma 3- to 7-lobed. Ovary unilocular, carpels 3-7. Capsule narrow, glabrous or spiny, opening from above by means of valves. Seeds numerous, blackish brown.

Literature: G.B. Ownby, Monograph of the genus *Argemone* for North America and the West Indies, in *Mem. Torrey bot. Club* **21**: 1-159 (1958).

1. A. mexicana L., *Sp. pl.* **1**: 508 (1753).
Erect, branched annual with prickly stem and bright yellow latex. Leaves oblanceolate to elliptic in outine, pinnatifid, segments unequal and with spine-tipped teeth, main veins spiny and conspicuously glaucous above; lower leaves petiolate, upper sessile and sometimes amplexicaule. Flowers shortly pedicellate, subtended by a leafy bract. Petals 2-2.5 cm, yellow. Capsule 2.5-3.5 cm, ellipsoid, prickly. Seeds 1.5 mm, globular, black, reticulate.
Fl. ± throughout the year. *A weed of waste ground around Funchal.* **M**

3. Glaucium Mill.

Glaucous annuals or biennials, with yellow latex. Leaves pinnately lobed or divided. Sepals 2, free. Petals 4. Stigma 2-lobed, sessile. Capsule linear, 2-celled, opening from apex by 2 valves reaching almost to the base. Seeds embedded in septum, pitted.

1. G. corniculatum (L.) Rudolph, *Fl. jen.*: 13 (1781).
Plant pubescent. Stems up to 18 cm, erect. Leaves obovate-oblong in outline, pinnatifid, the segments unequal, irregularly toothed; basal and lower cauline petiolate, upper cauline sessile. Petals obovate, orange-red, often with a dark spot at the base. Capsule up to 18 cm, pubescent.
Fl. IV. *A rare plant of very dry, light soils in Porto Santo.* **P**

4. Chelidonium L.

Glaucous perennials, with orange latex. Leaves pinnate. Inflorescence umbellate. Sepals 2, free. Petals 4. Stigma 2-lobed, style very short. Capsule linear, opening from the base upwards by 2 valves, septum absent. Seeds with a crested, white aril.

1. C. majus L., *Sp. pl.* **1**: 505 (1753). [*Celidonia, Cedronha, Ceredonha*]
Plant sparsely pubescent. Stems up to 50 cm. Basal leaves up to 40 cm including the long petiole, cauline 13-19 cm, all with 5-7 ovate to oblong, bluntly crenate leaflets, the terminal usually 3-lobed, the lateral usually with a stipule-like basal lobe on the lower side. Umbels 5- to 11-flowered. Sepals pubescent. Petals *c.* 1 cm, obovate, yellow. Capsule 4-5 cm × *c.* 3 mm, glabrous; style 2-4 mm.
Fl. III - VIII. *Common weed of waste ground, grassy banks and verges, rocks and walls up to 800 m in Madeira, mainly near habitation.* M

4. Eschscholzia Cham.

Glaucous annuals or perennials with watery latex. Leaves finely divided. Sepals 2, connate, forming a cap and falling as flower opens. Petals 4. Stigma 4- to 6-lobed, style very short. Capsule linear, unilocular, opening from base by 2 valves which separate from the placentae. Seeds numerous.

1. E. californica Cham. in Nees, *Horae phys. berol.*: 74 (1820).
Erect annual. Leaves 1- to many-ternate, segments linear, the ultimate sometimes spathulate. Petals 2-3.2 cm, yellow to orange, obovate, broadly cuneate, entire, inrolled longitudinally in dull weather. Cap formed by connate sepals with a long, slender tip, the persistent base forming a broad collar below petals. Capsule up to 8 cm, strongly ribbed. Seeds black, reticulate.
Fl. VI. *Cultivated as an ornamental; naturalized in considerable numbers on dry slopes by the road near Santa Cruz and occasionally elsewhere.* M

Subfam. FUMARIOIDEAE

Latex absent. Flowers in bracteate racemes, zygomorphic. Stamens 2. Fruit a small, indehiscent nut.

5. Fumaria L.

Erect, diffuse or scandent annuals. Leaves 2- to 4-pinnate. Flowers strongly zygomorphic, in ± sessile or pedunculate, bracteate racemes, with pedicels often becoming thickened in fruit. Sepals 2, lateral. Corolla white or pink, the wings of the spurred upper petal and the tips of the 2 inner petals often dark purple. Fruit a 1-seeded nut, smooth to rugose, with 2 apical pits.

Fumaria shows considerable environmentally-induced variation in growth. Floral characters are difficult to observe except in fresh material and are diagnostically not as reliable as is often claimed; the rugosity and apical pits of the fruits are only apparent when the mature fruit is dry. Flower colour is a useful diagnostic character but in a number of species flowers which are initially white become suffused with pink after fertilization. In low light intensities or late in the season, many species produce 'shade forms' with cleistogamous flowers which are smaller, paler and simpler than normal. Plants may also be found in which the fruits greatly exceed normal size. These monstrous forms are the result of infection by the larvae of the small dipter *Ailax*.

Literature: M. Lidén, Synopsis of Fumarioideae (Papaveraceae) with a monograph of the tribe Fumarieae, in *Op. bot. Soc. bot. Lund* **88**: 5-133 (1986). H.W. Pugsley, The genus *Fumaria* in Britain, in *J. Bot., Lond.* **50**, Suppl. 1: 1-76 (1912); A revision of the genera *Fumaria* and *Rupicapnos*, in *Bot. J. Linn. Soc.* **44**: 233-354 (1919).

1. Flowers less than 6 mm; sepals 1 mm or less; leaf-segments linear and channeled .... **6. parviflora**
- Flowers more than 7 mm long, sepals more than 2 mm long; leaf-segments flat, broad 2

2. Fruiting pedicels recurved .... **1. capreolata**
- Fruiting pedicels erect to patent 3

3. Pedicels slender; fruit almost or completely smooth, with small, rounded apical pits 4
- Pedicels stout; fruit rugulose to rugose, with squarish apical pits 5

4. Corolla 12-14 mm, initially white, becoming purplish red; sepals dentate at base only .... **2. sepium**
- Corolla 9-12 mm, pink; sepals usually dentate throughout or entire but not dentate at base only .... **3. muralis**

5. Bracts ⅔ or more as long as pedicels; corolla initially white .... **5. montana**
- Bracts ⅓-½ as long as pedicels; corolla pale pink .... **4. bastardii**

1. F. capreolata L., *Sp. pl.* **2**: 701 (1753).
Diffuse, often scandent by means of twining petioles. Racemes 14- to 18-flowered, shorter than the peduncles. Bracts ½-1 × as long as the rigidly arcuate-recurved fruiting pedicels. Sepals broader than and more than half as long as the corolla, ± entire or toothed towards the base. Corolla 10-11 mm, creamy white, the wings of the upper petal and the tips of the inner dark purple. Fruit 2-2.5 mm, subglobose, obscurely keeled, slightly rugulose towards the keel, otherwise smooth, with small apical pits. 2n=64.
Fl. III. *Introduced and first recorded in 1928 at Santo Antonio, near Funchal. Infrequent on disturbed soils on waste and cultivated ground. Native to the Mediterranean region and C. and W. Europe.* **M**

2. F. sepium Boiss. & Reut. in Boiss., *Diagn. pl. Orient.* **3**(1): 16 (1854).
Diffuse, often scandent, with long branches. Racemes 8- to 14-flowered, about as long as the peduncles. Bracts up to ⅔ as long as the slender, erecto-patent fruiting pedicels. Sepals 4-5 mm, less than half as long as the corolla, toothed towards the base. Corolla 12-14 mm, white but tinged dark purplish red after fertilization, wings of the upper petals and tip of the inner dark purple. Fruit *c.* 2 mm, subglobose, smooth or slightly rugulose, obscurely keeled, with small apical pits.
Fl. ? *Recently introduced as a field weed at Santo de Serra. Native to southern Iberia and Morocco.* **M**

Madeiran plants are subsp. **sepium**.

3. F. muralis Sonder ex W.D.J. Koch, *Syn. fl. germ.* ed. 2,: 1017 (1845).
[*Herva pombinha, Molarhina*]
Slender, much-branched, often diffuse and glaucous. Racemes 10- to 17-flowered, usually a little longer than the peduncles. Bracts ½-1 × as long as the slender, erect to patent or rarely slightly recurved fruiting pedicels. Sepals 2-3.5 × 1.5-2.5 mm, orbicular-ovate, dentate, rarely entire. Corolla 8-11 mm, pink, the wings of the upper petal and tips of the inner blackish red. Fruit 2 mm, rarely longer, ± globular, obtuse or subacute, slightly keeled, ± smooth, with small apical pits.
Fl. XI - VII but sporadically ± throughout the year. *Very common everywhere except the highest*

regions, among rocks, on walls, waste and cultivated ground, fields and roadsides. Not recorded from the Desertas. MPS

Plants from the islands all appear to belong to subsp. **muralis**, to which the above description applies. Within this subspecies two endemic varieties can be distinguished; both differ from var. *muralis* in having the upper petal blunt and fruits which are ± globular and therefore obtuse or, at most, subacute rather than subrotund-ovate and apiculate.

i) var. **lowei** Pugsley in *J. Bot., Lond.* **50**, Suppl 1: 23 (1912).
F. muralis L. var. *vulgaris* Lowe
Sepals dentate. **Plate 10**
Throughout the range of the subspecies and the common variety in Madeira. ■ [*Canaries*]

ii) var. **laeta** Lowe, *Man. fl. Mad.* **1**: 15 (1868).
F. laeta Lowe
Sepals entire, corolla bright pink. **Plate 10**
Confined to Pico do Facho in Porto Santo. ●

The subsp. **boraei** (Jord.) Pugsley, *J. Bot., Lond.* **40**: 15 (1902) is similar but larger in all its parts than subsp. *muralis*. It has been recorded for both Madeira and Porto Santo but no material definately assignable to this taxon has been seen and the records require confirmation.

4. F. bastardii Boreau in Duch., *Rev. bot.* **2**: 359 (1847).
F. muralis L. var. *gussonei* (Boiss.) Menezes, nom. illegit., ?var. *pallida* Menezes; *F. vulgaris* Sond. var. *pustulosa* Lowe
Diffuse, glaucous. Racemes 11- to 15-flowered, longer than the peduncles. Bracts ⅓-½ × as long as the erecto-patent fruiting pedicels. Sepals 2 mm, toothed, rather persistent. Corolla 8.5-10 mm, pale pink, tips of the upper and inner petals, or of the inner only, dark purple. Fruit 2-2.2 mm, ovoid or somewhat square in outline, obscurely keeled, rugulose, with shallow apical pits.
Fl. XII - V. *Common in Porto Santo on dry soils as a weed of fields and waste ground; also on Ilhéu Chão; possibly present in Madeira.* ?M DP

5. F. montana J.A. Schmidt, *Beitr. Fl. Cap Verd. Ins.*: 263 (1852).
F. praetermissa Pugsley
Diffuse, branched, rarely erect. Racemes 8- to 19-flowered, equalling or longer than the peduncles. Bracts ⅔-1⅓ × as long as the thickened, erecto-patent pedicels. Sepals dentate, 2.5-3.5 × 1.5-2 mm. Corolla 7-9 mm, initially white, soon flushed red, the tips of the inner, and sometimes of the upper, petals dark purple. Fruit 2.25-2.5 × 2.25-2.5 mm, rugose and conspicuously keeled. **Plate 10**
Fl. ? *A plant of dry slopes, among rocks or scrub on the north coast of Madeira, Ilhéu Chão and Selvagem Pequena.* ■ MDS [*Canaries, Cape Verdes*]

6. F. parviflora Lam., *Encycl.* **2**: 567 (1788).
Diffuse, much-branched, glaucous. Leaflets less than 1.5 mm wide, slightly channeled. Racemes 8- to 11(-20)-flowered, subsessile. Bracts equalling or slightly longer than the thickened, ± erect pedicels. Sepals inconspicuous, deeply incised. Corolla 4-4.5 mm, white, the wings of the upper petal and tips of the inner with a minute dark purple blotch. Fruit *c.* 2 mm, ± globose to subacute, obscurely keeled, rugulose with broad apical pits.
Fl. IV - V. *Common in vineyards and on cultivated and waste ground in the lower parts of Porto Santo.* P

XLI. CRUCIFERAE
[BRASSICACEAE]
M.J. Short

Annual to perennial herbs, rarely small shrubs. Leaves mostly alternate, exstipulate. Inflorescences racemose, usually ebracteate, often elongating greatly in fruit. Flowers usually hermaphrodite, regular. Sepals 4, free, the inner pair often saccate at the base. Petals 4, free, clawed, limb spreading. Stamens usually 6, in two whorls, rarely 4 or 2; filaments often winged or appendaged. Nectaries present at base of stamens. Ovary with 2 united carpels, usually 2-locular. Style simple or 0; stigma entire to 2-lobed. Fruit a capsule, dehiscing by 2 valves from below, rarely indehiscent or lomentaceous, separating transversely into 1-seeded sections.

A fruit is called a *siliqua* when it is clearly more than 3 times longer than broad, and a *silicula* when clearly less than 3 times longer than broad. Measurements are based on mature fruits and, unless stated otherwise, do not include the style or beak.

1. Petals white, cream, lilac, pink or purple, or petals absent 2
 - Petals yellow 18

2. Petals 0.5-5 mm (including claw) 3
 - Petals 7-22 mm (including claw) 13

3. Fruit with two segments, the upper globose, indehiscent, the lower stalk-like . . . . . .
   . . . . **26. Crambe**
 - Fruit dehiscent, not as above 4

4. Fruit more than 3 × as long as wide 5
 - Fruit up to 3 × as long as wide 7

5. Leaves simple, entire to shallowly dentate . . . . **2. Arabidopsis**
 - Leaves pinnate 6

6. Stems erect; lower leaves forming a rosette; seeds in 1 row in each locule . . . .
   . . . . **8. Cardamine**
 - Stems ± procumbent; leaves not forming rosettes; seeds in 2 rows in each locule . . .
   . . . . **7. Nasturtium**

7. Fruit triangular-obcordate . . . . **13. Capsella**
 - Fruit not as above 8

8. Plant glabrous or with simple hairs 9
 - Plant pubescent, with at least some hairs bifid or stellate 12

9. Stems ± procumbent; at least some inflorescences leaf-opposed; fruit reticulate-pitted or ridged . . . . **17. Coronopus**
 - Stems erect or ascending; inflorescences terminal; fruit smooth 10

10. Cauline leaves sagittate-amplexicaul; fruit 11-16 mm wide; seeds 3-8 per locule . . . . **15. Thlaspi**
 - Cauline leaves not sagittate-amplexicaul or stems leafless; fruit 1.6-5 mm wide; seeds 1-2 per locule 11

11. Flowering stems leafless or nearly so; seeds 2 per locule . . . . **14. Teesdalia**
- Flowering stems leafy; seeds 1 per locule . . . . **16. Lepidium**

12. Cauline leaves ovate, the margins serrate-dentate . . . . **11. Draba**
- Cauline leaves linear to linear-oblanceolate, entire . . . . **12. Lobularia**

13. Leaves succulent, glabrous; fruit with an ovoid, 4-angled, 1-seeded segment separating from the stout lower joint . . . . **24. Cakile**
- Leaves not succulent, pubescent; fruit not as above 14

14. Petals with conspicuous dark veins; fruit strongly constricted between the seeds . . . . **27. Raphanus**
- Petals without conspicuous dark veins; fruit not strongly constricted between the seeds 15

15. Leaves triangular-ovate, with simple hairs; fruit an oblong-elliptic to suborbicular silicula . . . . **10. Lunaria**
- Leaves not as above, with branched hairs; fruit a linear siliqua 16

16. Seeds in 2 rows in each locule . . . . **4. Erysimum**
- Seeds in 1 row in each locule 17

17. Petals white; fruit glabrous . . . . **9. Arabis**
- Petals purple (rarely white); fruit pubescent . . . . **5. Matthiola**

18. Fruit pendulous, flattened and winged . . . . **3. Isatis**
    Fruit not as above 19

19. Fruit strongly constricted between the seeds, forming bead-like segments . . . . **27. Raphanus**
- Fruit not or only very weakly constricted between the seeds 20

20. Fruit without valves, comprising two indehiscent segments, the upper globose, ribbed, the lower stalk-like . . . . **25. Rapistrum**
- Fruit with valves, not as above 21

21. Seeds in 2 rows in each locule 22
- Seeds in 1 row in each locule 24

22. Dwarf shrub; leaves all linear, entire . . . . **4. Erysimum**
- Annual herb; basal leaves pinnately lobed 23

23. Petals 6-8 mm, plain yellow; fruiting pedicels 7-10(-16) mm, slender; fruit 1.3-1.9 mm wide . . . . **18. Diplotaxis**
- Petals 15-20 mm, whitish or yellowish with dark violet veins; fruiting pedicels 3-4 mm, thick; fruit 3-4 mm wide . . . . **22. Eruca**

24. Fruit stiffly erect and closely appressed to the stem 25
- Fruit spreading, not closely appressed to the stem 27

25. Petals 2-4 mm . . . . **1. Sisymbrium**
- Petals 6-9 mm 26

26. Stem glabrous or sparsely hispid below; fruit 2.5-3 mm wide; beak slender, seedless . . . . **19. Brassica**
- Stem densely hispid below; fruit 1-1.5 mm wide; beak swollen at the base, 1- to 2-seeded . . . . **23. Hirschfeldia**

27. Cauline leaves auriculate, clasping the stem at their base . . . . **6. Barbarea**
- Cauline leaves not clasping, petiolate or narrowed at the base 28

28. Lower leaves ± simple, entire or toothed, sometimes with 1-2 small lobes at the base . . . . **20. Sinapidendron**
- Lower leaves pinnately divided 29

29. Siliquae 18-80 × 0.6-1 mm, terminating in a ± sessile stigma or a short style up to 3 mm long . . . . **1. Sisymbrium**
- Siliquae 12-22 × 2-3 mm, terminating in a narrowly conical beak, 8-16 mm long . . . . **21. Sinapis**

1. Sisymbrium L.

Annual to biennial herbs, glabrous or with simple hairs. Lower leaves pinnately lobed. Sepals erecto-patent. Petals indistinctly to shortly clawed, yellow (often drying white). Fruit a dehiscent, linear siliqua; valves convex, with a prominent midvein and usually 2 weaker lateral veins; style very short; stigma capitate, emarginate or 2-lobed. Seeds numerous, in 1 row in each locule.

1. Siliquae broadest at the base, stiffly erect and appressed to stem, 8-15 mm long . . . . **4. officinale**
- Siliquae linear, patent to ascending, not appressed to stem, 18-80 mm long 2

2. Petals 7-9 mm; leaves pubescent; siliquae pubescent when young . . . . . **2. orientale**
- Petals 2-4 mm; leaves glabrous or nearly so; siliquae glabrous 3

3. Pedicels 1-2.5 mm in flower; in fruit thick, about as wide as the siliqua . . . . **3. erysimoides**
- Pedicels 3-6 mm in flower; in fruit slender, distinctly narrower than the siliqua . . . . **1. irio**

1. S. irio L., *Sp. pl.* 2: 659 (1753).
More or less glabrous annual; stem up to 60 cm, erect, simple or sparingly branched above. Lower leaves not rosette-forming, petiolate, pinnatisect, lateral lobes distant, terminal lobe hastate, all somewhat toothed; upper leaves with fewer, narrower lobes. Sepals 2-3 mm. Petals 2.5-3.5 mm, pale yellow. Fruiting pedicels 4-10 mm, erecto-patent, slender. Siliquae 25-45 × 0.6-0.8 mm, usually curved, ascending; valves thin-walled, slightly torulose; style very short. Seeds 0.8-1 mm, light brown.
Fl. IV. *A rare and local casual from S. Europe found in Porto Santo on waste ground around Vila Baleira.* **P**

2. S. orientale L., *Cent. pl. II*: 24 (1756).
S. columnae Jacq.
Pubescent annual with long, spreading hairs; stem up to 70 cm, erect, simple or branched above. Lower leaves often rosette-forming but soon withering, petiolate, pinnatisect, lateral lobes distant, terminal lobe hastate, all subentire to shallowly toothed; upper leaves with fewer,

narrower lobes, the uppermost often entire. Sepals 3.5-4.5 mm. Petals 7-9 mm, pale yellow. Fruiting pedicels 4-7 mm, erecto-patent, thick. Siliquae 60-80 × c. 1 mm, straight, patent to erecto-patent; valves pubescent when young; style 1.5-3 mm. Seeds 0.8-1.2 mm, brown.
Fl. I (-VI?). *A rare and local casual of waste places in Madeira, recorded from São Gonçalo and Gorgulho near Funchal, and more recently (1973) from Funchal harbour. Native to S. and SE Europe, N. Africa.* **M**

3. S. erysimoides Desf., *Fl. atlant.* **2**: 84 (1798).
Annual, glabrous or shortly and sparsely pubescent; stem 10-60 cm, erect, simple or sparingly branched. Leaves mostly glabrous, the lower petiolate, pinnatifid, the terminal lobe ± ovate, all lobes subentire to irregularly dentate; upper leaves smaller, with fewer lobes. Pedicels 1-2.5 mm in flower, pubescent. Sepals 2-3 mm. Petals 2-3 mm, pale yellow. Fruiting pedicels 2-6 mm, patent, thick. Siliquae 18-35 × c. 1 mm, linear, straight, patent to erecto-patent; style 0.7-1 mm. Seeds 0.8-1 mm, yellow-brown.
Fl. XII - VI. *Common along walls and in waste places in the lower coastal region of southern Madeira; rare in Porto Santo, recorded from Fontinha, Pedras Pretas, and Fajã Pequena.* **MP**

4. S. officinale (L.) Scop., *Fl. carniol.* ed. 2, **2**: 26 (1772). [*Fedorento manso* (P)]
Annual or biennial, glabrous or with short, bristly hairs; stems 10-80 cm, stiffly erect, branched above. Leaves variable, the lower rosette-forming, petiolate, pinnatisect, terminal lobe large, ± ovate, all lobes irregularly toothed or lobed; lower stem leaves with a long hastate terminal lobe; uppermost leaves simple, linear. Sepals 1.8-2.5 mm. Petals 2-4 mm, pale yellow. Fruiting pedicels 1.5-3 mm, erect, thick. Siliquae 8-15 × 1-1.5 mm, conical-cylindrical, straight, closely appressed to stem; valves pubescent or glabrous; style c. 0.5 mm. Seeds 1-1.5 mm, reddish brown.
Fl. III - VI. *Occasional by roadsides, in cultivated and waste ground in the lower regions of Madeira; also fairly uncommon in Porto Santo; collected from Selvagem Grande in 1868 but not recorded since.* **MPS**

Specimens of *S. officinale* with glabrous fruits are common and have been recognized as var. **leiocarpum** DC., *Syst. nat.* **2**: 460 (1821).

2. Arabidopsis (DC.) Heynh.

Annual herbs with simple and branched hairs. Leaves entire to dentate. Sepals erect. Petals white, sometimes absent. Fruit a slender, linear, dehiscent siliqua; valves convex, 1-veined; style short; stigma subcapitate. Seeds in 1 row in each locule, numerous, small.

1. A. thaliana (L.) Heynh. in Holl & Heynh., *Fl. Sachsen* **1**: 538 (1842).
Sisymbrium thalianum (L.) J. Gay; *Stenophragma thalianum* (L.) Čelak.
Stems (1-)6-45 cm, solitary or several in a small tuft, erect, simple or sparingly branched, slender, pubescent below, glabrous above. Leaves pubescent with simple, bifid and stellate hairs; basal leaves forming a rosette, elliptic or spathulate, attenuate into a short petiole, sparingly and shallowly dentate; cauline leaves few, sessile, linear-lanceolate, ± entire. Sepals 1.2-2 mm. Petals 2-3.5 mm. Fruiting pedicels 4-12 mm, erecto-patent. Siliquae 7-22 × 0.5-1 mm, glabrous; style 0.2-0.4 mm. Seeds c. 0.5 mm, pale brown.
Fl. III - VIII. *Apparently rare, but a short-lived and inconspicuous plant which is possibly often overlooked. Rocks, slopes, and paths on high mountain peaks and valleys in Madeira, for example at Fajã Nogueira and between Pico Ruivo and Pico do Arieiro; also known from Deserta Grande.* **MD**

Represented in the Madeiran archipelago by var. **thaliana**, to which the above description applies. Var. *apetala* O.E.Schulz is found in the Canaries.

3. Isatis L.

Biennial or short-lived perennial herbs, glabrous or with simple hairs. Basal leaves simple, ± entire; upper leaves sessile, amplexicaul. Inflorescence corymbose-paniculate. Sepals erecto-patent. Petals bright yellow. Pedicels deflexed in fruit, often thickened above. Fruit 1-seeded, indehiscent, flattened, pendulous, with a thick, broad wing; valves with a slender midrib. Style 0; stigma weakly bilobed. Seeds large.

1. I. tinctoria L., *Sp. pl.* **2**: 670 (1753). [*Pastel*]
I. praecox sensu auct. mad., non Kit. ex Tratt. (1812).
Stem 40-100 cm, erect, branched, leafy, glabrous or with scattered long hairs. Leaves glabrous or pubescent, usually very glaucous, thin; the basal oblong-obovate, entire or weakly sinuate-dentate, attenuate into the petiole; upper leaves sessile, with acute, triangular auricles. Inflorescence branched, spreading, many-flowered. Petals *c.* 3.5 mm, bright yellow, *c.* 2 × as long as the sepals. Fruiting pedicels 6-9 mm, very slender. Fruit 9-20 × 3-6 mm, pendulous, truncate to rounded at apex, cuneate or tapering at the base, dark brown to purplish black when ripe, glabrous. Seeds 3-4 mm, oblong.
Fl. III - IX. *Naturalized on roadsides, fields, slopes and waste ground in Madeira, especially near the sea. Reported to be very common in the last century, but now only occasional. Native to ?C. and S. Europe, W. Asia, and widespread as a relic of cultivation.* **M**

4. Erysimum L.
N.J. Turland

Small or dwarf shrubs, strigose with branched appressed hairs. Leaves borne in rosettes at the stem-tips, linear to linear-lanceolate, patent to erecto-patent. Inflorescence terminal, racemose, simple or branched from below. Sepals erect, the inner saccate at the base. Petals yellow, white or lilac. Fruit a stiff, straight, dorsally compressed siliqua, erecto-patent to erect; valves 1-veined; style distinct, not more than one third as long as rest of siliqua; stigma weakly 2-lobed. Seeds in 2 rows in each locule.

Hairs are either 2-fid, when medifixed, or 3- or 4-fid, when stellate. Pedicels and styles should be measured only if attached to ripe siliquae.

Literature: A. Polatschek, Die Gattung *Erysimum* auf den Kapverden, Kanaren und Madeira, in *Annln naturh. Mus. Wien* **80**: 93-103 (1976).

1. Leaves linear-lanceolate, always serrate (sometimes minutely and remotely so), with mostly 3- or 4-fid hairs; flowers changing colour with age . . . . **3. bicolor**
- Leaves linear, always entire, with mostly 2-fid hairs; flowers usually not changing colour
. . . . 2

2. Leaves 1-4 mm wide, persistent when withered; flowers lilac, rarely white; siliquae 2.5-3 mm wide . . . . **2. arbuscula**
- Leaves 0.5-2 mm wide, not persistent; flowers yellow; siliquae 1-2 mm wide . . . . .
. . . . **1. maderense**

1. E. maderense Polatschek in *Annln naturh. Mus. Wien* **80**: 100 (1976).
Cheiranthus tenuifolius L'Hér.
Much-branched dwarf shrub to 60 cm, with very slender ascending stems. Leaves 10-50 × 0.5-2 mm, linear, entire, not persistent when withered, densely covered with 2-fid hairs. Inflorescence usually simple but sometimes branched from below, erect. Pedicels 3-8 mm. Sepals 4.5-7 mm. Petals yellow, not changing colour; limb 4-6 mm. Siliquae 20-65 × 1-2 mm, densely covered with 2-fid hairs; style 2-3 mm. Seeds *c.* 2.5 mm. 2n= *c.* 28. **Plate 10**

Fl. XI - VI. *A rare Madeiran endemic of rock ledges on cliffs west of Funchal, between Cabo Girão and Câmara de Lobos, 350-500 m.* ● **M**

Records from Porto Santo are referable to **2**.

2. E. arbuscula (Lowe) Snogerup in *Op. bot. Soc. bot. Lund* **13**: 9 (1967).
[*Goivos, Queiranto*]
Cheiranthus arbuscula Lowe; *E. bicolor* sensu auct. mad., non (Hornem.) DC. (1821); *E. maderense* sensu A. Hansen & Sunding (1985), non Polatschek (1976); *E. scoparium* sensu auct. mad., non (Brouss. ex Willd.) Wettst. (1889).
Dwarf shrub to 30 cm, with ascending but often crowded and twisted stems. Leaves 13-75 × 1-4 mm, linear, entire, persistent when withered, densely covered with mostly 2-fid hairs, rarely also with 3-fid hairs. Inflorescence unbranched, erect. Pedicels 6-12 mm. Sepals 5.5-7 mm. Petals lilac, rarely white initially; limb 4.5-7 mm. Siliquae (13-)20-65 × 2.5-3 mm, densely covered with 2-fid and sometimes 3-fid hairs; style 2-4 mm. Seeds 2-3 mm. 2n=28. **Plate 10**
Fl. IX - V. *An endemic of rock-crevices on the peaks and higher slopes of Porto Santo from 300-450 m. Formerly locally abundant but now being reduced by overgrazing in places accessible to sheep and goats. A single collection from a Funchal garden originated from seed gathered on the plateau of Selvagem Grande.* ● **PS**

3. E. bicolor (Hornem.) DC., *Syst. nat.* **2**: 509 (1821). [*Goivo da serra, Goivos, Queiranto*]
Cheiranthus mutabilis L'Hér., non Brouss. ex Spreng. (1825); *C. scoparius* sensu Menezes (1927), non Brouss. ex Willd. (1809); *E. scoparium* sensu A. Hansen (1969), non (Brouss. ex Willd.) Wettst. (1889).
Very variable branched shrub to 1.8 m, with ascending stems. Leaves 20-150 × (1-)1.5-15 mm, linear-lanceolate, serrate but sometimes minutely or remotely so, not persistent when withered, sparsely covered with mostly 3- and 4-fid hairs; midrib with a high density of 2-fid hairs. Inflorescence simple or branched from below, ascending. Pedicels (3-)4-14 mm. Sepals (5-)6-11 mm. Petals white in bud and on opening, usually becoming lilac; limb (5-)6-12 mm. Siliquae 20-100 × 1.5-3 mm, densely covered with mostly 3- and 4-fid hairs; style (2-)3.5-7 mm. Seeds 2-3 mm. 2n=42. **Plate 10**
Fl. XI - VIII. *Crevices of rock faces in ravines, laurisilva and other shady places; widely distributed in Madeira but not common, from near sea-level to 1750 m.* ■ **M** [*Canaries, Cape Verdes*]

Madeiran plants are hexaploid, whereas plants in the Canaries are tetraploid (2n=28). Dwarf plants with small narrow leaves, often found in exposed situations or at high altitudes in Madeira, tend to have a more dense indumentum with a higher percentage of 2-fid hairs. Plants from north-western coastal areas of Madeira have been determined as *E. scoparium* (Brouss. ex Willd.) Wettst. in *Öst. bot. Z.* **39**: 283 (1889) (*Cheiranthus scoparius* Brouss. ex Willd.) but in fact belong to **3**; they all have narrow, minutely and remotely serrate leaves covered with mostly 2- and some 3-fid hairs, inflorescences much-branched from below and flowers apparently opening yellow. Genuine *E. scoparium* is endemic to the Canaries. Records of *E. bicolor* from the Salvage Islands are erroneous; they are based on an incorrect determination of the collection of **2**, mentioned above, which originated on Selvagem Grande. A record of *E. scoparium* from the same island is almost certainly also referable to **2**.

5. Matthiola R. Br.

Densely pubescent annual or biennial herbs with branched hairs. Leaves entire to sinuate-dentate or pinnatisect. Sepals erect. Petals long-clawed, purple or rarely white. Fruit a dehiscent, pubescent siliqua, compressed or ± cylindrical; valves 1-veined; style absent; stigma deeply 2-lobed, each lobe with a dorsal swelling or horn. Seeds in 1 row in each locule.

1. Robust biennial or perennial; leaves usually entire; sepals 10-14 mm; siliquae compressed, not torulose and without conspicuous horns . . . . **1. maderensis**
- Annual; leaves sinuate-dentate, sometimes pinnatisect; sepals 5-6.5 mm; siliquae ± cylindrical, torulose, with conspicuous horns 2-3.5 mm long . . . . **2. parviflora**

1. M. maderensis Lowe in *Trans. Camb. phil. Soc.* **6**: 551 (1838).
[*Bofe de Burro, Cravo de burro, Goivo, Goivo da rocha*]
M. maderensis [var.] γ *albiflora* Lowe, [var.] β *mitis* Lowe, [var.] α *muricata* Lowe
Robust, densely greyish white tomentose biennial, (15-)30-50(-90) cm high, sparsely to densely covered with yellow, stipitate glands; stem stout, woody at the base, simple or usually with spreading branches. Leaves elliptic to oblong-oblanceolate, acute, attenuate at the base into a short petiole, entire or very rarely sparingly sinuate-dentate, the lower leaves 5-25 × 0.7-5 cm, the upper becoming smaller. Flowers fragrant. Sepals 10-14 mm, linear, with scarious margins. Petals 18-28 × (6-)8-12 mm, obtuse or shallowly emarginate, lilac to purple (rarely white), paler towards the base, the claw greenish. Fruiting pedicels 0.5-1.5 cm. Siliquae (50-)80-150 × 3-4.5 mm, erecto-patent, compressed, usually with yellow or blackish glands; stigma lacking conspicuous horns. Seeds 3-3.5 mm, compressed, suborbicular, brown with a paler wing. 2n=14. **Plate 12**
Fl. II - X. *Common in Madeira, mainly on coastal rocks and cliffs, 0-100 m. Fairly frequent in Porto Santo and sometimes found inland, recorded growing at altitudes of up to 200 m on Pico do Castelo; also occurring on Ilhéu de Cima. Known from all of the Desertas, but reported to be rare on Bugío.* ● **MDP**

2. M. parviflora (Schousb.) R. Br. in W.T. Aiton, *Hortus kew.* ed. 2, **4**: 121 (1812).
[*Goivo da rocha*]
Greyish, ± densely pubescent annual, 5-25 cm, sometimes woody at the base, stems spreading and often much-branched. Leaves sinuate-dentate, the lower in a basal rosette, oblanceolate, obtuse, sometimes pinnatisect, the upper elliptic, smaller. Sepals 5-6.5 mm. Petals 7-10 mm, purple. Fruiting pedicels *c.* 1 mm. Siliquae 40-60 × 1-1.5 mm, patent to somewhat reflexed, ± cylindrical, torulose; horns 2-3.5 mm, spreading to erect, straight. Seeds 0.9-1.2 mm, oblong, light brown, indistinctly winged.
Fl. I - IV. *Tracks and roadsides, in dry, rocky places. Rare in Porto Santo; recorded from the south-west of the island at Ponta, near the peak of Ana Ferreira and at Espigão. An early record was from Ponta da Malhada to the east of Vila Baleira. 0-150 m.* **P**

Malcolmia maritima (L.) R. Br. in W.T. Aiton, *Hortus kew.* ed. 2, **4**: 121 (1812), an appressed-pubescent annual with grey-green, ovate to broadly elliptic leaves and a lax inflorescence of pink or purple-petalled flowers, followed by slender, linear siliquae, 20-70 mm long, has been recorded as a casual from cultivated ground in Madeira. **M**

6. Barbarea R. Br.

Annual or biennial herbs, glabrous or with simple hairs. Basal leaves pinnately lobed; cauline leaves sessile, clasping the stem at the base. Sepals erect. Petals yellow (often drying white). Fruit a linear, terete to 4-angled, dehiscent siliqua; valves with a strong median vein and a network of weak, indistinct lateral veins. Style short; stigma capitate. Seeds in 1 row in each locule.

1. B. verna (Mill.) Asch., *Fl. Brandenburg* **1**: 36 (1860).
B. praecox (Sm.) R. Br.
Glabrous or with a few scattered long, simple hairs; stem 6-70 cm, erect, usually branched above. Basal leaves in a rosette, petiolate, up to 27 cm long, pinnatisect, with 4-10 pairs of oblong-suborbicular lateral lobes and a larger, elliptic-ovate terminal lobe, the lobes usually irregularly and shallowly sinuate-toothed; upper leaves sessile, auriculate, deeply pinnatifid, the lobes linear-oblong, entire. Sepals *c.* 3.5 mm. Petals 4.5-7 mm, bright yellow. Fruiting pedicels 3-7 mm, stout. Siliquae (26-)35-55 × 1.5-2 mm, erect or ascending; style 1-2 mm. Seeds 1.7-2.2 mm, oblong, blackish.
Fl. IV - VII. *Rare casual from SW Europe found on waste ground and roadsides in the lower regions of southern Madeira.* **M**

7. Nasturtium R. Br.

Glabrous perennial herbs. Leaves pinnate. Sepals erect. Petals with a short claw, white. Fruit a narrowly oblong, ± terete siliqua; valves with a weak median vein; style short; stigma entire or indistinctly 2-lobed. Seeds numerous, in 2 rows in each locule, their surfaces marked with polygonal depressions.

1. N. officinale R. Br. in W.T. Aiton, *Hortus kew.* ed. 2, **4**: 110 (1812). [*Agrião*]
N. officinale var. *genuinum* Gren. & Godr., var. *siifolium* (Rchb.) Steud.
Stems 8-70 cm, weak, simple or branched, leafy, often purplish, often rooting at the nodes, procumbent below, then ascending or floating, or occasionally ± erect. Leaves petiolate, with oblong-elliptic lateral leaflets and a larger, suborbicular terminal leaflet, all weakly sinuate, dark green. Sepals 2-2.5 mm. Petals 4-5 mm. Fruiting pedicels 4-10(-15)mm, patent, ascending or recurved. Siliquae (6-)10-16 × 1.8-3 mm, straight or incurved. Seeds *c.* 1 mm, ovoid, brown.
Fl. II - X. *A semi-aquatic, very common in ditches, streams, wet flushes, waste places with running water up to 1000 m in Madeira; infrequent in Porto Santo.* **MP**

8. Cardamine L.

Annual herbs, glabrous or with simple hairs. Leaves pinnate. Sepals erect. Petals white. Stamens 4(-6). Fruit a strongly compressed, linear siliqua, dehiscing explosively, veinless, often weakly torulose. Style usually present; stigma ± entire. Seeds numerous, in 1 row in each locule.

1. C. hirsuta L., *Sp. pl.* **2**: 655 (1753).
Plant 7-20(-30) cm; stems erect, usually branched from the base and sparingly above, glabrous or with scattered, long, spreading hairs. Lower leaves petiolate, forming a compact rosette, with 1-4(-5) pairs of ovate to orbicular leaflets and a larger, orbicular-reniform terminal leaflet; leaflets shortly petiolulate, margins entire to weakly sinuate, sparsely pubescent above and around the margins. Upper leaves few, with oblanceolate or linear leaflets. Sepals 1.5-2 mm. Petals 2-4 mm. Fruiting pedicels 2-5 mm, erect. Siliquae 10-25 × 1-1.2 mm, slender, the young fruits overtopping the flowers. Seeds 1-1.2 mm, oblong, light brown.
Fl. XI - VIII. *Fairly common in Madeira along tracks, roads, levada paths, wall and cliffs, often in moist shady places, up to 1720 m; less frequent in Porto Santo. Recently found on Deserta Grande.* **MDP**

9. Arabis L.

Perennial herbs with simple and stellate hairs. Leaves simple, the lower forming rosettes, the upper amplexicaul. Sepals erect. Petals white. Fruit a linear siliqua; valves strongly compressed, sometimes with a weak midrib. Style short; stigma capitate or emarginate. Seeds in one row in each locule.

1. A. alpina L., *Sp. pl.* **2**: 664 (1753). [*Agrião da rocha*]
A. albida Steven ex M. Bieb.; *A. caucasica* Willd. ex Schltdl.
Softly, and sometimes densely, pubescent perennial herb with stellate and simple hairs. Flowering stems 12-35 cm, ascending or procumbent, arising from a slender horizontal stock with many vegetative leaf-rosettes. Leaves rather hoary, those at the base oblong-obovate to oblong-oblanceolate, sinuate-dentate, long attenuate into the petiole; cauline leaves sessile, oblong, amplexicaul-sagittate. Sepals 3-6 mm. Petals 9-16 mm. Fruiting pedicels 10-20 mm, patent to erect. Siliquae 40-65 mm. Seeds 1.2-1.7 mm, dark brown, narrowly winged. $2n=16$.
Fl. III-VII. *Very common in Madeira, trailing down rock walls of levadas, ravines, and high peaks, (200-)600-1650 m.* **M**

Plants from the Madeiran archipelago all belong to subsp. **caucasica** (Willd.) Briq., *Prodr. fl. Corse* **2**(1): 48 (1913), to which the above description and synonymy applies.

10. Lunaria L.

Biennial herbs with simple hairs. Leaves simple, toothed. Sepals erect. Petals long-clawed, purple, rarely white. Stamens 6. Style long; stigma weakly 2-lobed. Fruit a large, strongly compressed, dehiscent, oblong-elliptical to orbicular silicula; valves thin-walled, translucent, veinless; septum silvery-white, shiny, persistent. Seeds large, strongly compressed, in 2 rows in each locule.

1. L. annua L., *Sp. pl.* **2**: 653 (1753).
Plant up to 1 m, pubescent; stem erect, simple or sparingly branched above. Leaves up to 11 × 8 cm, triangular-ovate, acuminate, shallowly cordate, coarsely and irregularly dentate, the lower long-petiolate, becoming progressively shorter-stalked above, the uppermost ± sessile. Sepals 6-8 mm, pubescent, often purplish tinged. Petals 12-20 mm, reddish purple. Fruiting pedicels 9-25 mm, slender. Silicula 25-45 × 15-25 mm; style 5-11 mm. Seeds 5-9 mm, orbicular, dark brown, winged.
Fl. III - VI. *Grown in gardens as an ornamental and sometimes escaping. Naturalized in Madeira along roadsides at Monte and between Terreiro da Luta and Monte. Native to SE Europe.* **M**

11. Draba L.

Annual herb with simple, forked and stellate hairs. Leaves simple, toothed, the basal in a rosette. Sepals erecto-patent. Petals white. Stamens 4(-6). Fruit a dehiscent silicula; valves ± flat, without conspicuous veins; style very short; stigma entire. Seeds in 2 rows in each locule.

1. D. muralis L., *Sp. pl.* **2**: 642 (1753).
Plant ± densely pubescent. Stems 8-30(-45) cm, erect, simple or sparingly branched. Basal leaves forming a rosette, shortly petiolate, obovate or oblanceolate, margins serrate-dentate, ciliate. Cauline leaves sessile, broadly ovate, semi-amplexicaul. Sepals 0.7-1 mm, usually pubescent, often purplish. Petals 1.2-2.5 mm. Fruiting pedicels 2-7 mm, patent, slender. Silicula 3-6 × 1.5-2 mm, oblong-elliptic, glabrous; style *c.* 0.1 mm. Seeds 4-6(-8) per locule, 0.8-0.9 mm, ellipsoid, brown.
Fl. III - VI. *Rare mountain plant in Madeira, found on rocky walls, banks of ravines and high peaks; 600-1800 m.* **M**

12. Lobularia Desv.

Low, annual herbs or suffruticose perennials, with appressed, medifixed hairs; stems usually much-branched from the base. Leaves sessile, simple, narrow, entire. Flowers in short, crowded, terminal racemes, sometimes elongating in fruit. Sepals spreading. Petals with a short, distinct claw, white, sometimes tinged with purple. Stigma capitate. Fruit a dehiscent, orbicular or elliptic silicula; valves convex or flat, with or without a distinct midvein. Seeds with or without a wing.

Literature: L. Borgen, *Lobularia* (Cruciferae). A biosystematic study with special reference to the Macaronesian region, in *Op. bot. Soc. bot. Lund* **91**: 1-96 (1987).

1. Petals less than 1.2 mm wide; pedicels shorter than to ± equalling the silicula; seeds 4-5 per locule . . . . **2. libyca**
- Petals more than 1.5 mm wide; pedicels longer than the silicula; seeds 1-3 per locule 2

2. Seeds 1 per locule . . . . **3. maritima**
- Seeds 2-3 per locule . . . . **1. canariensis**

1. L. canariensis (DC.) L. Borgen in *Op. bot. Soc. bot. Lund* **91**: 66 (1987).
L. maritima var. *canariensis* (DC.) Cout.
Annual herb or suffruticose perennial; stems branched from the base. Leaves alternate or in rosettes, linear to obtrullate, acute to obtuse. Racemes sometimes elongating in fruit. Sepals 1.6-2.3 mm. Petals 2.7-4 mm, white, sometimes tinged with purple. Fruiting pedicels 4-8 mm, patent or ascending. Siliculas 2.5-4 × 2.2-2.8 mm, elliptic or suborbicular; valves pubescent. Seeds(1-)2-3 per locule, dark brown, sometimes narrowly winged. **Plate 12** ● S

Two subspecies have been recorded from the Salvage Islands:

a) subsp. **rosula-venti** (Svent.) L. Borgen in *Op. bot. Soc. bot. Lund* **91**: 80 (1987).
L. maritima var. *rosula-venti* Svent.
Suffruticose perennial. Leaves alternate in rosette-like clusters at the tips of the branches, 4-15 × 0.9-1.6 mm, linear, acute, silvery-pubescent, recurved. Sepals 1.6-1.8 mm. Petals 2.7-3.3 mm, white. Fruiting pedicels 4-8 mm, patent to erecto-patent. Siliculas 2.5-4.5 × 2-2.5 mm, suborbicular to elliptic-obovate. Seeds (1)2(3) per locule, 1-1.5 mm, ovate, very narrowly winged.
Fl. ± all year. *Common on the Salvage Islands among rocks and in sandy soil; 0-150 m.*

b) subsp. **succulenta** Borgen in *Op. bot. Soc. bot. Lund* **91**: 82 (1987).
Annual with procumbent to ascending branches. Leaves alternate, 23-42 × 3.5-5 mm, linear-oblanceolate to obtrullate, obtuse, green to greyish pubescent, succulent. Sepals 1.6-2.3 mm, purplish. Petals 3.2-4 mm, white, sometimes tinged with purple. Fruiting pedicels 6-8 mm, ascending. Siliculas 3.3-4 × 2.2-2.8 mm, elliptic. Seeds (1)2-3 per locule, 1.1-1.3 mm, ovate, ± wingless.
Fl. II - VI. *Very common on Selvagem Pequena amongst stones and in sandy soil; 0-50 m.*

This account follows Borgen (1987), who described only two taxa from the Salvage Islands, *L. canariensis* subsp. *rosula-venti* and *L. canariensis* subsp. *succulenta*. However, there are quite a few specimens, mostly from Selvagem Pequena and Ilhéu de Fora, which are apparently intermediate and cannot readily be placed in either one of these subspecies.

2. L. libyca (Viv.) Meisn., *Pl. vasc. gen.* **2**: 11 (1837).
Appressed, greyish pubescent annual; stems up to 20 cm, much-branched from the base,

prostrate or ascending. Leaves alternate, 10-20 × 1-2.5 mm, linear, obtuse to ± acute. Racemes ± elongating in fruit. Sepals 1.2-1.5 mm. Petals 1.6-2 mm, white. Fruiting pedicels 2.5-5.5 mm, ascending. Siliculas 3-7 × 2.7-3.3 mm, ovate to oblong-elliptic; valves flat, pubescent. Seeds 4-5 per locule, 0.9-1.1 mm, compressed, suborbicular, narrowly winged.
Fl. X - IV. *Infrequent; pastures and waste ground in dry, sunny places in Porto Santo.* **P**

3. L. maritima (L.) Desv. in *J. Bot. Agric. Pharm. Med. Arts Paris* **3**: 162 (1814).
Koniga maritima (L.) R. Br.
Greyish green to silvery-pubescent suffruticose perennial; stems up to 25 cm, branched from the base, decumbent to ascending. Leaves alternate, 12-30 × 1.3-3.5 mm, linear to linear-oblanceolate, usually acute. Racemes greatly elongating in fruit. Sepals 2-3 mm, often tinged with purple. Petals 2.5-4 mm, white, often tinged with purple. Fruiting pedicels 4.5-8 mm, ± patent to ascending. Siliculas 2-3 × 1.6-2.5 mm, elliptic to orbicular; valves convex, pubescent when young, becoming glabrous. Seeds 1 in each locule, 1.2-1.4 mm, ovate, wingless or with an extremely narrow wing.
L. VIII - XII. *Garden escape naturalized in Madeira on cultivated and waste ground in and around Funchal; infrequent. Native to SW Europe and N. Africa.* **M**

13. Capsella Medik.

Annual or biennial herbs, glabrous or with simple and branched hairs. Basal leaves pinnatifid or entire; cauline leaves sagittate-amplexicaul. Sepals erect. Petals white, pink or reddish. Fruit a strongly compressed, dehiscent, triangular-obcordate silicula; valves keeled, weakly reticulately veined, several seeded; style short; stigma capitate.

1. C. bursa-pastoris (L.) Medik., *Pfl.-Gatt.*: 85 (1792).
Plant up to 40 cm, usually pubescent; stems erect, simple or branched. Basal leaves forming a rosette, oblanceolate, pinnatifid to entire, attenuate into a broad petiole; upper leaves narrow, entire to remotely and shallowly denticulate, clasping the stem with acute basal auricles. Sepals sometimes reddish. Petals 2-3 mm, white, up to *c.* 2 × as long as the sepals. Fruiting pedicels 4-10 mm, erecto-patent. Silicula 5.5-7.5 × 5-6.5 mm, slightly emarginate. Seeds 0.8-1 mm, ellipsoid, brown.
Fl. XII - V. *Ruderal. Fairly common in Madeira along paths, roadsides, and in cultivated and waste ground, mainly in the lower regions, but also occurring up to 1800 m; very rare in Porto Santo, recorded from Camacha and Portela.* **MP**

A variable species with many named varieties that scarcely warrant taxonomic recognition.

C. rubella Reut., *Compt. Rend. Soc. Hallér.* **2**: 18 (1854), distinguished from **1** by its smaller, pink or red-tinged petals, not or scarcely exceeding the sepals, has been recorded from Bico da Cana in Madeira. However, the taxonomic status of this species is uncertain and due to frequent confusion with **1**, its presence in Madeira is questionable.

14. Teesdalia R. Br.

Annuals, glabrous or with simple hairs. Leaves mostly in a basal rosette, usually pinnatifid; flowering stems almost leafless. Flowers small. Sepals erecto-patent. Petals white, subequal or unequal. Stamens 4 or 6; filaments with a broad, white basal scale-like appendage. Fruit a dehiscent silicula, strongly compressed, emarginate above; valves thin-walled, narrowly winged in the upper part; stigma subsessile, entire. Seeds usually 2 in each locule.

1. Petals unequal, the two outer 1.5-2 × as long as the sepals; stamens 6 . **1. nudicaulis**
- Petals subequal, all ± equalling the sepals; stamens 4 . . . . **2. coronopifolia**

1. T. nudicaulis (L.) R. Br. in W.T. Aiton, *Hortus kew.* ed. 2, **4**: 83 (1812).
Glabrous or occasionally with simple hairs on the stems and pedicels. Stems (1.5-)3.5-30 cm, branched at the base, ascending to erect. Basal leaves long-petiolate, obovate to spathulate, lyrate-pinnatifid or sometimes entire, the lobes usually rounded; petioles winged to the base. Flowers minute. Petals unequal, the outer two larger, 1.5-2 × as long as the sepals. Stamens 6. Fruiting pedicels 2.5-4(-6) mm, patent. Silicula 2.5-4 mm, broadly elliptic-obovate. Seeds 0.9-1.3 mm, ovoid-orbicular, light brown.
Fl. III - VIII. *Very common in mountain pastures in Madeira, especially on Paúl da Serra and Pico Ruivo; mainly occurring between 1000-1860 m.* **M**

2. T. coronopifolia (J.P. Bergeret) Thell. in *Reprium Spec. nov. Regni veg.* **10**: 289 (1912).
T. lepidium DC.
Glabrous. Stems 3-10 cm, usually branched from the base, slender, ascending or erect. Basal leaves oblanceolate to spathulate, attenuate into the petiole, entire to deeply pinnatifid, the lobes usually ± acute. Flowers minute. Petals 0.7-1 mm, subequal, ± equalling the sepals. Stamens 4. Fruiting pedicels 2.5-5 mm, patent. Silicula 2-3 mm, broadly elliptic to suborbicular. Seeds 0.5-0.8 mm, ovoid, pale brown.
Fl. III - IV. *Rare. Only known from south-east Porto Santo and Ilhéu de Cima.* **P**

15. Thlaspi L.

Glabrous annual herbs. Leaves simple, entire or dentate, petiolate or sessile, the cauline amplexicaul. Sepals erect. Petals shortly clawed, white. Fruit a strongly compressed, dehiscent silicula; valves keeled and winged; stigma subsessile, entire. Seeds in 2 rows in each locule.

1. T. arvense L., *Sp. pl.* **2**: 646 (1753).
Plant somewhat foetid when crushed; stem 10-60 cm, erect, leafy, simple or sparingly branched above. Leaves entire to sinuate-dentate; the basal, attenuate into a short petiole, oblanceolate to obovate; upper leaves sessile, oblong to lanceolate, sagittate-amplexicaul. Sepals 1.5-2 mm. Petals 3-4 mm. Fruiting pedicels 7-15 mm, erecto-patent. Silicula 12-16 × 11-15 mm, almost orbicular; valves broadly winged; apical sinus 2.5-4.5 mm. Seeds 1.3-2 mm, 3-8 per locule, brownish black.
Fl. V - VIII. *Rare casual of roadsides, waste and cultivated ground, recorded from São Vicente and Serra d'Água in the last century, but no recent records known from Madeira; also reported from Vila Baleira in Porto Santo.* **MP**

16. Lepidium L.

Annual or biennial herbs, glabrous or with simple hairs. Leaves entire to pinnately divided. Flowers small. Sepals erect. Petals white, pink or purplish, sometimes absent. Stamens 2, 4 or 6. Fruit a dehiscent silicula, strongly compressed; valves winged; style short or 0; stigma entire. Seeds 1 per locule.

1. Silicula 5-6.5 mm; pedicels erect, distinctly shorter than the fruit, glabrous; sepals pubescent .... **1. sativum**
- Silicula 1.8-3.5 mm; pedicels patent or erecto-patent, ± equalling or longer than the fruit, puberulent; sepals glabrous 2

2. Upper stem leaves deeply pinnately lobed .... **3. bonariense**
- Upper stem leaves entire to serrate 3

3. Silicula 2.7-4 mm wide; seeds 1.6-1.8 mm; stems usually minutely puberulous . . . . .
    . . . . **2. virginicum**
-   Silicula 1.6-2 mm wide; seeds 1.1-1.3 mm; stems glabrous or with a few short spreading hairs . . . . **4. ruderale**

1. L. sativum L., *Sp. pl.* **2**: 644 (1753). [*Mastruço*]
Annual, glabrous or with scattered simple hairs below; stem 20-50 cm, erect, branched above. Lower leaves long-petiolate, pinnate to bipinnate, the lobes obovate, irregularly toothed; upper leaves ± sessile, linear, simple or with a few lobes, the margins entire. Sepals 1-1.5 mm, pubescent. Petals 2-3 mm, white or sometimes pink or purplish. Stamens 6. Fruiting pedicels 2-3.5 mm, erect, glabrous. Silicula 5-6.5 × 4-5 mm, broadly elliptic, winged above; style not projecting beyond the apical notch. Seeds 3-3.5 mm, oblong-ovoid, dark brown.
Fl. III - IV. *A rare casual on waste ground in Madeira. Native to Egypt and W. Asia, but widely cultivated elsewhere.* **M**

2. L. virginicum L., *Sp. pl.* **2**: 645 (1753). [*Mastruço*]
Annual or biennial; stem (3.5-)12-50 cm, erect, branched above and occasionally below, usually minutely puberulous. Leaves glabrous or with scattered simple hairs; the basal lyrate or pinnate, rough with short bristles; middle and upper leaves simple, incised-serrate, the lower petiolate, broadly elliptic, the uppermost sessile, linear-lanceolate. Sepals 0.7-1 mm, glabrous. Petals white, shortly exceeding the sepals Stamens 2(-4). Fruiting pedicels 3-5 mm, patent or erecto-patent, puberulous. Silicula 2.7-3.5 × 2.7-4 mm, orbicular, narrowly winged above, style not exceeding apical notch. Seeds 1.6-1.8 mm, oblong-ovoid, narrowly winged on one side, brown.
Fl. III - VII. *Roadsides and waste ground; a very common weed everywhere in the lower regions of Madeira. Native to N. America, but widely introduced elsewhere.* **M**

3. L. bonariense L., *Sp. pl.* **2**: 645 (1753).
Annual or biennial; stem 10-45 cm, ± erect, branched above and sometimes below, with simple hairs. Leaves all pinnatisect, the lobes irregularly incised or entire, with scattered unbranched hairs. Sepals 0.5-1 mm, glabrous. Petals white, shorter than sepals. Stamens 2(-4). Fruiting pedicels (1.5-)2-4 mm, patent or erecto-patent, puberulous. Silicula 2.3-3.5 × 2.1-2.5 mm, elliptic-ovate to suborbicular, narrowly winged above, style not exceeding apical notch. Seeds 1.1-1.3 mm, ovoid, unwinged, brown.
Fl. I - VII. *A S. American species occasionally found on roadsides and waste places in the lower regions of Madeira in both the north and south of the island.* **M**

4. L. ruderale L., *Sp. pl.* **2**: 645 (1753).
Foetid annual or biennial; stem 10-35 cm, erect or ascending, branched above and sometimes below, glabrous or occasionally with a few short spreading hairs. Basal leaves petiolate, pinnatisect, the lobes entire or irregularly incised, glabrous or minutely pubescent along veins below. Cauline leaves becoming less divided, the upper sessile, linear, entire or with very small lobes. Sepals 0.6-0.7 mm, glabrous. Petals usually absent. Stamens usually 2. Fruiting pedicels 2-3.5 mm, erecto-patent or patent, puberulous. Silicula 1.8-2.4 × 1.6-2 mm, ovate or broadly elliptical, narrowly winged above, style not exceeding apical notch. Seeds 1.1-1.3 mm, oblong-ovoid, unwinged or narrowly winged, brown.
Fl. V - XII. *Roadsides and waste places. Occasional along the south coast of Madeira, from Funchal to Calheta. Native to Europe and SE Asia, but widely introduced elsewhere.* **M**

17. Coronopus Zinn
[*Senebiera* DC.]

Annual or biennial herbs, glabrous or with simple hairs. Leaves deeply pinnatisect. Racemes short, terminal, leaf-opposed or axillary, the flowers small. Sepals patent. Petals white, sometimes absent. Fertile stamens 2-6. Fruit a silicula, indehiscent or splitting into 2 indehiscent, nutlet-like halves; valves 1-seeded, reticulate-pitted or ridged; style very short or 0; stigma entire.

1. Petals equalling to exceeding the sepals; pedicels stout, glabrous, shorter than the fruit . . . . **1. squamatus**
- Petals shorter than the sepals or absent; pedicels slender, pubescent, longer than the fruit . . . . **2. didymus**

1. C. squamatus (Forssk.) Asch., *Fl. Brandenburg* 1: 62 (1860).
C. procumbens Gilib.; *Senebiera coronopus* (L.) Poir.
Glabrous annual or biennial, much-branched from the base; stems up to 30 cm, procumbent to ascending, leafy. Leaves petiolate, somewhat fleshy, deeply pinnatisect, the segments oblanceolate to oblong, entire or infrequently incised. Inflorescence dense, not or scarcely elongating in fruit. Sepals 1-1.5 mm, oblong, persistent. Petals 1-2 mm, equalling or exceeding the sepals. Fertile stamens 6. Fruiting pedicels up to 2 mm, stout. Silicula 2.5-3 × 3-4 mm, subreniform, apiculate above with a short style; valves reticulate-pitted, strongly and irregularly ridged and verrucose, indehiscent.
Fl. I - IV. *Very seldom in streets and waste ground; occurring in the Funchal and Machico regions of Madeira and also reported from Tanque in Porto Santo.* **MP**

2. C. didymus (L.) Sm., *Fl. brit.* 2: 691 (1800).
Senebiera didyma (L.) Pers.; *S. pinnatifida* DC.
Foetid annual or biennial, much-branched from the base; stems 2-40 cm, procumbent and often mat-forming, sparsely pilose, leafy. Lower leaves petiolate, deeply pinnatisect, the lobes oblanceolate, entire to pinnatifid; upper leaves sessile. Inflorescence dense in flower, elongating in fruit. Sepals 0.5-1 mm, triangular-ovate. Petals *c.* 0.5 mm, shorter than the sepals, or often absent. Fertile stamens 2(-4). Fruiting pedicels up to 3 mm, patent, slender, pubescent. Silicula *c.* 1.5 × 2.5 mm, deeply 2-lobed; valves subglobose, indehiscent, reticulate-pitted, readily separating at maturity; style 0. $2n = 32$.
Fl. I - VI. *Roadsides, walls, banks, pasture, and waste ground. Extremely common in the lower regions of Madeira and also frequent in Porto Santo. Recorded from Deserta Grande and Bugío, also Selvagem Grande. Native to S. America, but widely introduced elsewhere.* **MDPS**

18. Diplotaxis DC.

Annual herbs, glabrous or with simple hairs. Leaves entire to pinnately divided. Sepals spreading. Petals yellow. Style short; stigma 2-lobed. Fruit a long, slender, dehiscent, linear siliqua, shortly beaked; valves compressed, prominently 1-veined. Seeds numerous, in 2 rows in each locule.

1. D. catholica (L.) DC., *Syst. nat.* 2: 632 (1821).
Stems up to 90 cm, erect, usually branched from the base and above, with ascending branches, glabrous or sparsely pubescent below. Leaves variable, ± glabrous to very sparsely and shortly hispid; the basal usually pinnatisect, lobes narrow, toothed to pinnatifid; the upper smaller with narrower lobes or entire. Sepals 3-4 mm. Petals 6-8 mm, pale yellow. Fruiting pedicels 7-10(-16) mm, erecto-patent to patent, slender. Siliqua 15-25 × 1.3-1.9 mm, spreading; valves

sometimes weakly torulose; beak 2-5 mm, 0-2-seeded. Seeds 0.6-0.8 mm, subglobose, brown. Fl. III - V. *An introduced species from SW Europe, recorded in 1973 from waste ground near the stadium in Funchal, Madeira.* **M**

Records of *D. siifolia* Kunze from the same locality are referable to this species.

19. Brassica L.

Annual to perennial herbs with simple, often bristly hairs, rarely glabrous. Leaves simple or lyrate-pinnatifid. Sepals erect or patent. Petals yellow or rarely white. Fruit a linear or oblong dehiscent siliqua with a conspicuous long or short, conical or filiform beak, 0-3-seeded; valves convex, with a prominent midvein; style long; stigma capitate or weakly 2-lobed. Seeds in 1(-2) rows in each locule.

1. B. nigra (L.) W.D.J. Koch in Röhl., *Deutschl. Fl.* ed. 3, **4**: 713 (1833).

[*Mostarda, Rinchões* (PS)]

Annual; stem up to 1 m, erect, sometimes purplish tinged, branched above, glabrous or sparsely hispid below. Leaves petiolate, the lower lyrate-pinnatisect, with 1-3 pairs of oblong-ovate lateral lobes and a large, ovate terminal lobe, all irregularly serrate-dentate, hispid; upper leaves linear to oblong, entire or toothed, glabrous. Sepals 3-5 mm, spreading. Petals 7-9 mm, bright lemon yellow. Fruiting pedicels 3-5 mm, erect, appressed to stem. Siliqua 10-17 × 2.5-3 mm, erect, ± 4-angled, attenuate into a slender, seedless beak, 2-4 mm long. Seeds 1.1-1.3 mm, globose, dark brown, in 1 row in each locule.
Fl. I - VI. *Very common in cultivated and waste ground in the lower regions of Madeira; infrequent in Porto Santo.* **MP**

B. oleracea L. is extensively cultivated as a vegetable in numerous forms which are variously treated as varieties or subspecies, including cabbage [*Couve*] kale, broccoli [*Broculos*] and cauliflower [*Couve flôr*], and though these may occasionally escape and appear as casuals, apparently not persisting or becoming naturalized. Swede, **B. napus** subsp. **rapifera** Metzger (*Nabo*) and turnip, **B. rapa** L. (*B. campestris* L.) are also grown.

20. Sinapidendron Lowe

Perennial herbs or small shrubs, glabrous or with simple hairs. Leaves simple, entire or toothed, sometimes rather fleshy. Sepals erecto-patent, usually purple-spotted. Petals yellow, the claw often purplish. Fruit a straight or curved, dehiscent siliqua; valves thick, convex with a prominent midvein; beak cylindrical; stigma capitate or weakly 2-lobed. Seeds in 1 row in each locule.

1. All leaves linear; siliquae 16-30 mm . . . . **1. angustifolium**
- Lower leaves elliptic, oblong, ovate or suborbicular; siliquae 35-80 mm 2

2. Lower leaves elliptic to elliptic-oblanceolate, entire to regularly crenate-serrate; fruit 1-1.4 mm wide . . . . **2. frutescens**
- Lower leaves oblong to ovate or suborbicular, irregularly dentate; fruit 1.4-2 mm wide 3

3. Stems hispid; leaves oblong to ovate; pedicels and sepals pubescent . . . . **3. rupestre**
- Stems ± glabrous; leaves broadly ovate to suborbicular; pedicels and sepals glabrous . . . . **4. gymnocalyx**

1. S. angustifolium (DC.) Lowe, *Man. fl. Madeira* **1**: 30 (1857).
[*Sinapidendro de folha estreita*]
S. salicifolium Lowe
Glabrous shrub; stems 30-110 cm, slender, branched. Leaves (25-)50-110 × 3-11 mm, linear, acute, attenuate into petiole, entire or rarely sparsely and shallowly toothed towards the apex, rather fleshy. Sepals 4.5-5.5 mm, purple-spotted, with a few scattered hairs. Petals 8-9.5 mm, the claw often purplish. Filaments often purple. Fruiting pedicels 4-10 mm, erecto-patent. Siliquae 16-30 × 1.6-2.8 mm, usually somewhat curved, spreading; beak 3.5-7 mm. Seeds 1-1.3 mm, dark brown. $2n=20$. **Plate 11**
Fl. IV - VI (X). *Locally common on sea cliffs along the south coast of Madeira between Madalena do Mar and Praia Formosa to the west of Funchal, occasionally also occurring a little inland.* ● M

A hybrid between species **1** and **4** has been recorded from Madeira, but this seems unlikely in view of their distributions on the island and needs confirmation.

2. S. frutescens (Sol. in Aiton) Lowe in *Trans. Camb. phil. Soc.* **4**: 37 (1831).
Shrub; stems up to 50(-120) cm, slender, erect or hanging, usually appressed-pubescent below. Leaves variable, the lower 11-75 × 5-25(-30) mm, elliptic to elliptic-oblanceolate, acute to obtuse; petiole 5-20 mm. Upper leaves linear, entire, attenuate to base. Sepals 3.5-6 mm, blotched purple, with occasional appressed hairs. Petals 7-10 mm, claw often purplish. Fruiting pedicels 7-15 mm, erecto-patent. Siliquae 35-70 × 1-1.4 mm, ± straight and erect; beak 3-4.5 mm. Seeds 1-1.2 mm, reddish brown. $2n=20$. **Plate 11** ● M

Two varieties have been recorded in Madeira:

i) var. frutescens
S. frutescens [var.] α *diffusa* Lowe
Leaves usually distinctly crenate-serrate, with scattered, short, appressed hairs, at least when young, somewhat fleshy.
Fl. III - VII. *Locally frequent in crevices of often vertical rock-faces, usually in shade. Ravines and peaks of central Madeira, 750-1800 m.*

ii) var. succulentum Lowe, *Man. fl. Madeira* **1**: 29 (1857).
S. frutescens subsp. *succulentum* (Lowe) Rustan
Leaves entire or very shallowly crenate, glabrous, very fleshy.
Rare; sea cliffs along the north coast of Madeira.

S. sempervivifolium Menezes in *Broteria* (Bot.) **20**: 113 (1922), is a glabrous plant, becoming shrubby, with erect, branched stems; lower leaves 5-7 × 2-3 cm, obovate to obovate-oblong, crenulate at the apex, crowded; upper leaves linear; sepals 5-6 mm, glabrous to subglabrous, purplish tinged; petals 9-10 mm; siliquae torulose; beak short, sometimes 1-seeded. $n=10$. Very rare and no specimens have been seen. Only known from Deserta Grande. ● D

3. S. rupestre Lowe in *Trans. Camb. phil. Soc.* **4**: 37 (1831). [*Couve da rocha*]
S. rupestre [var.] α *chaetocalyx* Lowe
Perennial herb, becoming woody below, hispid, especially below; stems erect, sparingly branched. Leaves coarse and scabrid, the lower 7-22 × 3.5-13 cm, oblong to ovate, acute to obtuse, often unequal at the base, occasionally with a pair of small lobes, irregularly dentate; petioles 15-35 mm; uppermost leaves linear, entire. Sepals 4-5.5 mm, purple-spotted. Petals 7.5-10 mm, claw purple. Fruiting pedicels 9-14 mm, erecto-patent. Siliquae (35-)50-70 × 1.4-2 mm, ± straight, spreading; beak 4-6 mm. Seeds 1.4-1.6 mm, dark brown. **Plate 11**
Fl. VI - VIII. *Generally rare in ravines and mountains of northern Madeira, 850-1500 m.* ● M

4. S. gymnocalyx (Lowe) Rustan in *Sommerfeltia* **17**: 8 (1993). [*Couve da rocha*]
S. rupestre [var.] β *gymnocalyx* Lowe
Shrubby perennial up to 1.5 m; stems ± glabrous. Leaves ± glabrous to sparsely scabrid, rather succulent, the lower 4-21 × 3.5-15 cm, broadly ovate to suborbicular, obtuse, sometimes unequal at the base, occasionally with 2 or more small lobes, irregularly dentate; petioles 2.5-9 cm; upper leaves linear-oblanceolate, toothed towards the apex to ± entire. Pedicels and sepals glabrous. Sepals 3-6 mm, horned dorsally. Petals 6.5-9 mm, the claw purple. Fruiting pedicels 8-15 mm, erecto-patent. Siliquae 35-50 × 1.8-2 mm; beak 3-4 mm. Seeds 1-1.2 mm, dark brown. **Plate 11**
Fl. (III) VI - IX. *Locally common in Madeira, mainly on cliffs along the north coast between Ribeira do Tristão and Ponta de São Jorge, 0-30 m, but also inland in Ribeiro do Tristão, 100 m, and Montado dos Peseguiros, 1000 m.* ● **M**

21. Sinapis L.

Annual herbs with simple hairs. Lower leaves pinnately lobed. Sepals patent. Petals clawed, yellow. Fruit a dehiscent siliqua with a long indehiscent beak; valves convex, distinctly 3(-7)-veined. Stigma sessile, entire or emarginate. Seeds in 1 row in each locule.

1. S. arvensis L., *Sp. pl.* **2**: 668 (1753). [*Mostarda, Saramago* (P)]
Brassica sinapistrum subsp. *vulgaris* Cout.
Plant ± glabrous to hispid with deflexed hairs; stems up to 75 cm, erect, simple or with spreading branches. Lower leaves petiolate, simple or pinnatifid, with a large terminal lobe and a few small lateral lobes, coarsely toothed; upper leaves sessile, usually simple, lanceolate, toothed. Sepals 4-5.5 mm. Petals 7.5-10 mm, yellow with a paler claw. Fruiting pedicels 3-6 mm, erect, stout. Siliquae 12-22 × 2-3 mm, erect; valves 3(-5)-veined, glabrous or rarely hispid (var. *orientalis* (L.) W.D.J. Koch & Ziz), sometimes torulose; beak 8-16 mm, conical, straight, 0-1-seeded. Seeds 1.2-1.6 mm, subglobose, brownish black.
Fl. XI - VI. *Very common weed of cultivated and waste ground in Madeira and Porto Santo; also known from Deserta Grande and Ilhéu Chão.* **MDP**

S. alba L., *Sp. pl.* **2**: 668 (1753) [*Mostarda*], with the uppermost leaves pinnately lobed and densely hispid fruits with a strongly compressed, sword-shaped beak, has occasionally been recorded as an escape from cultivation in and around Funchal. **M**

22. Eruca Mill.

Annual herbs, usually with stiff, white, simple hairs. Leaves pinnatifid. Sepals erect. Petals long-clawed, whitish or yellowish with dark violet veins. Stigma 2-lobed. Fruit a short, dehiscent siliqua with a triangular, flat, seedless beak; valves convex, with a prominent, slightly keeled midvein. Seeds in 2 rows in each locule.

1. E. vesicaria (L.) Cav., *Descr. pl.*: 426 (1802). [*Fedorênte, Fedorento*]
E. sativa Mill.
Stems up to 50 cm, erect or ascending, stiff, sparsely hispid below, usually branched near the base. Basal leaves sometimes rosette-forming, petiolate, lyrate-pinatifid, with ovate-oblong, ± acute lateral lobes and a larger, ovate to suborbicular terminal lobe, lobes entire to dentate, glabrous or with a few scattered hairs; upper leaves smaller, becoming sessile. Sepals 6.5-10 mm, sometimes purplish. Petals 15-20 mm, *c.* 2 × as long as the sepals. Fruiting pedicels 3-4 mm, erect, thick. Siliquae 8-14 × 3-4 mm, erect, glabrous or with scattered hairs; beak 4-9 mm. Seeds 1.3-1.7 mm, ovoid, light brown.
Fl. XII - VI. *Common in Porto Santo on slopes, roadsides, cultivated and waste ground; also*

found on Ilhéu de Baixo. In Madeira only known from Ponta de São Lourenço and the adjacent islet of Ilhéu de Fora. **MP**

Plants from the Madeiran archipelago all belong to subsp. **sativa** (Mill.) Thell., *Ill. fl. Mitt. Eur.* **4**(1): 201 (1918), to which the above description refers.

23. Hirschfeldia Moench

Annual or biennial herbs, densely hispid with simple hairs. Lower leaves pinnatifid; uppermost leaves linear. Sepals ± erect. Petals yellow. Stigma capitate. Fruit a short, linear, dehiscent siliqua with a 0-2-seeded conical, indehiscent beak, swollen at the base; valves 1-3-veined. Seeds in 1 row in each locule.

1. H. incana (L.) Lagr.-Foss., *Fl. Tarn Garonne*: 19 (1847).
Erucastrum incanum (L.) W.D.J. Koch
Plant densely white hispid below, less pubescent or glabrous above. Stem up to 60(-90) cm, erect, usually with spreading branches. Basal leaves rosette-forming, petiolate, lyrate, with ± oblong lateral lobes and a broadly ovate, obtuse, terminal lobe, all lobes shallowly toothed; uppermost leaves nearly sessile, linear-lanceolate, toothed. Petals 6-8 mm, pale lemon yellow, *c.* 2 × as long as the sepals. Fruiting pedicels 3-4 mm, club-shaped; pedicels and fruit erect and closely appressed to stem. Siliquae 9-11 × 1-1.5 mm (including beak), glabrous or pubescent; valves with a weak central vein when mature (3-veined when young); beak *c.* ½ as long as valves, 1- to 2-seeded. Seeds *c.* 1 mm, oblong-ovoid, brown, 3-6 per locule.
Fl. V - VI. *Extremely rare in Madeira; recorded along the road between Madalena do Mar and Arco da Calheta; also found on Deserta Grande.* **MD**

Varieties based on whether fruits are hairy or glabrous do not merit taxonomic recognition.

24. Cakile Mill.

Annual, succulent, ± glabrous, and often glaucous herbs. Leaves simple to pinnately divided. Sepals erect. Petals pale lilac or white. Stigma capitate. Fruit with 2 indehiscent segments, the upper ovoid, ± 4-angled, 1-seeded, readily separating from the lower obconical segment, also usually 1-seeded, remaining attached to the plant.

1. C. maritima Scop., *Fl. carniol.* ed. 2, **2**: 35 (1772). [*Carqueja mansa*]
C. aegyptiaca Willd.
Stems up to 50 cm, much-branched from the base, the branches prostrate to ascending. Leaves succulent, variable, entire, linear to spathulate, entire to pinnatifid, the lobes obtuse. Sepals 4-5 mm, oblong, narrowly scarious towards the apex and with scattered, long, simple hairs. Petals 8-12 mm. Fruiting pedicels 4-7 mm, stout, patent to ascending. Mature fruit 16-24 × 4-7 mm; lower segment ± obconical; upper segment conical, 4-angled.
Fl. IV - VI. *Common on dunes and along the sandy beach in Porto Santo. Also recorded from Selvagem pequena.* **PS**

Plants from Porto Santo have been referred to subsp. **maritima**. Specimens examined differ from typical material in having fruits which lack the two projecting lateral horns on the lower segment.

25. Rapistrum Crantz

Annual herbs with simple hairs. Leaves pinnatifid. Sepals erecto-patent. Petals very shortly clawed, yellow. Style long; stigma capitate. Fruit with 2 indehiscent segments, the lower segment narrower than the upper, 0-3-seeded, the upper segment ovoid to globose, 1-seeded, readily separating from the lower segment when mature.

1. R. rugosum (L.) All., *Fl. pedem.* 1: 257 (1785). [*Rinchão*]
Variable annual, hispid below; stem 18-60 cm, erect, simple or more usually with spreading branches, sometimes purplish tinged. Lower leaves in a rosette, petiolate, deeply lyrate-pinnatifid, the lobes toothed or rarely subentire; upper leaves smaller, subsessile, linear-oblong or oblanceolate, toothed. Sepals 2.5-3.5 mm. Petals 5-8 mm, pale yellow. Fruiting pedicels 2-4 mm, erect, appressed to stem. Silicula glabrous to densely hispid; upper segment 2.5-3.5 × 1.7-4 mm, longitudinally rugose-sulcate, abruptly contracted into the persistent style *c.* 2 mm long; lower segment 1.5-3 × 1-2 mm. $2n=16$.
Fl. III - VIII. *Very common in pastures, grassland, waste places and on roadside banks and cliff slopes in Madeira and Porto Santo; also known from Deserta Grande.* **MDP**

Varieties based on whether the fruit is glabrous or pubescent do not merit taxonomic recognition. Plants with an ellipsoidal or cylindrical lower fruit segment have been recognized as subsp. **orientale** (L.) Arcang., *Comp. fl. ital.*: 49 (1882), and this is the form most frequently encountered in the Madeiran archipelago. Plants belonging to subsp. **rugosum** have an obconical lower fruit segment.

26. Crambe L.

Perennial herbs, glabrous or with unbranched hairs. Lower leaves pinnatifid or ± entire with a large terminal lobe and 2 small lateral lobes. Inflorescence paniculate. Sepals erecto-patent. Petals white. Stigma capitate, mostly sessile. Fruit a 2-membered, indehiscent silicula; lower segment short, sterile, forming a stalk; upper segment 1-seeded, globose, readily separating from the lower.

1. C. fruticosa L.f., *Suppl. pl.*: 299 (1781). [*Couve da rocha* (P)]
Small, shrubby perennial up to 90 cm high; stems glabrous or hispid below, usually branched. Leaves sometimes glaucous beneath, glabrous to sparsely or densely strigose, occasionally densely white pubescent; the lower petiolate, with 0-2 pairs of very small lateral lobes and a large, oblong-ovate to suborbicular, obtuse, dentate terminal lobe 1.8-6 × 1.5-5 cm, to shallowly to deeply pinnatifid, lobes acute to obtuse, dentate. Inflorescence often large and much-branched, the flowers in dense racemes at the ends of the branches. Sepals 2-2.5 mm. Petals 4-4.5 mm. Fruiting pedicels 4-11 mm, erect, slender. Silicula dark brown when mature; lower segment 0.5-1 mm, sharply 4-angled; upper segment *c.* 2.5 mm, globose, apiculate, reticulate-rugose, 4-angled. **Plate 12**
Fl. III - VII. *Rare in Madeira; ravines, rocks and cliffs, mainly in the west of the island, usually near the coast, up to 100 m. Locally common in Porto Santo on the peaks of Ana Ferreira, Branco and Castelo and on Ilhéu de Cima; also found on Deserta Grande, Bugío and Ilhéu Chão.* ● **MDP**

A very variable species with regard to leaf shape and indumentum. Plants having thick, fleshy leaves with a large, terminal lobe and a pair of very small lateral lobes, mainly occurring on the Desertas and Porto Santo (var. **brevifolia** Lowe) have been distinguished from pinnatifid-leaved plants (var. **pinnatifida** Lowe), but a wide range of intermediates may be found.

27. Raphanus L.

Annual herbs with stiff, white, simple hairs. Lower leaves pinnate. Sepals erect. Petals long clawed, white, yellow or violet, usually clearly veined. Stigma capitate. Fruit indehiscent, with

2 segments, the lower very small, narrow and stalk-like, seedless; upper segment cylindrical, constricted between the seeds sometimes breaking into 1-seeded segments; beak narrow, seedless.

R. sativus L., *Sp. pl.* **2**: 669 (1753) [*Rabanete, Rabão*], the cultivated radish, is occasionally found as an escape from cultivation in Madeira and Porto Santo, but apparently not persisting. It is a herb with a thick, swollen tap-root, white to pink or purple flowers, and a ± oblong siliqua, not or scarcely constricted between the seeds. **MP**

1. R. raphanistrum L., *Sp. pl.* **2**: 669 (1753). [*Saramago*]
Plant sparsely hispid below, glabrescent above; stems up to 1 m, erect with ascending branches. Basal leaves long-petiolate, with oblong, obtuse, sinuate to coarsely dentate lateral lobes and a large, obovate terminal lobe; lower cauline leaves fewer lobed, with a lanceolate, acute terminal lobe, uppermost leaves subsessile, lanceolate, acute. Sepals 8-11 mm, often purplish. Petals 15-22 mm, white, cream or pale yellow with conspicuous dark veins. Fruiting pedicels 10-16 mm, erect to patent. Siliqua usually glabrous; lower segment 1-2 mm; upper segment 15-35 × 2-3.5 mm, with 2-8 segments, prominently longitudinally veined; beak (6-)10-14 mm, narrowly conical.
Fl. I - VI. *Very common in the lower regions of Madeira by roadsides, banks and walls, on waste and cultivated ground; rare in Porto Santo.* **MP**

Represented in the Madeiran archipelago by subsp. **raphanistrum**, to which the above description applies. Varieties based on petal colour do not merit taxonomic recognition, white and yellow colour forms often growing intermixed.

XLII. RESEDACEAE
M.J. Short

Annual to perennial herbs. Leaves alternate, entire to pinnately-lobed; stipules minute, glandular. Flowers usually hermaphrodite, irregular, solitary in the axils of the bracts, forming terminal racemes or spikes. Sepals 4-7. Petals 4-7, free, frequently lobed or laciniate. Stamens 7-numerous, inserted on a nectar-secreting disc. Ovary superior, of 3-7 carpels connate into a unilocular ovary or 4-7, ± free, carpels. Stigmas borne on apical lobe of each carpel. Fruit a many-seeded capsule open at the top or of 4-7, spreading, 1-seeded carpels.

Literature: M.S. Abdallah & H.C.D. De Wit, The Resedaceae. A taxonomical revision of the family, in *Belmontia* II, **8**: 1-416 (1978).

1. Fruit a single, bottle-like capsule containing several seeds . . . . **1. Reseda**
- Fruit of 4-7 spreading, 1-seeded carpels . . . . **2. Sesamoides**

1. Reseda L.

Annual to perennial herbs. Leaves sessile, entire, toothed or pinnatifid. Petals white, yellow, or greenish, unequal, the upper and lateral variously dissected. Stamens connate at the base and merged with the fleshy disc which is usually extended dorsally as a crescent-shaped rim. Carpels 3-4, fused marginally; ovary unilocular, open above. Capsules thin-walled, opening more widely but not splitting at maturity. Seeds reniform.

1. All leaves entire; sepals and petals usually 4 . . . . **1. luteola**
- Some leaves ternate or pinnatifid; sepals and petals usually 6 2

2. Petals white; capsules nodding . . . . **2. media**
- Petals yellow; capsules erect . . . . **3. lutea**

1. R. luteola L., *Sp. pl.* **1**: 448 (1753). [*Lirio*]
R. luteola var. *australis* (Webb) Walp., var. *crispata* (Link) Müll. Arg., var. *gussonii* (Boiss. & Reut.) Müll. Arg.
Robust, glabrous, usually biennial, tap-rooted herbs; stems up to 100(-150) cm, erect, simple or sparingly branched, leafy. Leaves 5-10(-16) × 0.5-1.5(-2) cm, linear-oblong, entire, margins usually strongly undulate, the lower forming a rosette. Pedicels 1-2 mm. Sepals 4, c. 2 mm long, linear-lanceolate. Petals 4, creamy yellow, the upper one clawed, 3-8-lobed, the lateral and lower ones smaller, clawed or not, fewer-lobed, the lower often entire. Capsules 4-5 mm, erect, subglobose, rugose, 3-lobed. Seeds 0.9-1 mm, dark brown, shiny.
Fl. II - VII. *Common on grassy slopes and pastures in Madeira (up to c. 1000 m) and Porto Santo; also occurring on Deserta Grande and Ilhéu Chão.* **MDP**

2. R. media Lag., *Gen. sp. pl.*: 17 (1816).
Sparsely pubescent, sprawling annual or biennial herb; stems up to 50(-75) cm, decumbent to ascending, wiry. Leaves variable; the lower 2-6(-8) × 0.3-1(-2) cm, linear-elliptic, entire; upper leaves entire, deeply 3-lobed or pinnatisect with usually 1-4 pairs of lobes and an elliptic terminal lobe. Pedicels 1-5 mm in flower. Sepals 6, 2-5 mm long in flower, linear-spathulate, accrescent and reflexed in fruit. Petals white, the upper 3-lobed, with the lateral lobes laciniate, lateral and lower petals smaller, fewer lobed, the lower usually entire. Capsules 9-12 mm, nodding, obovoid-cylindrical. Seeds c. 2.5 mm, rugose, cream.
Fl. III - X. *On grassy banks, roadside verges, and along levadas. First recorded from Madeira in 1925 and now well-established in several places, mainly in the south-east of the island and particularly around Santo da Serra and Terreiro da Luta. Also recorded occasionally from the north. Native to SE Europe and N. Africa.* **M**

3. R. lutea L., *Sp. pl.* **1**: 449 (1753).
Bushy annual to perennial, tap-rooted herbs; stems numerous, up to 70(-100) cm high, erect or ascending, branched, scabrid, leafy. Leaves 2.5-8(-15) × 1.5-2 cm, usually pinnatifid, with 1-4 pairs of narrow, distant, entire or pinnatifid lobes, glabrous, margins somewhat undulate, the lower forming a rosette, withering early. Bracts usually caducous. Pedicels 4-8 mm in flower. Sepals 6, c. 3 mm, linear. Petals 6, yellow, clawed, the upper 3-lobed, the lateral smaller, 2-3-lobed, the lower usually entire. Capsules 8-16 mm, usually erect, oblong to obovoid, 3-lobed. Seeds 1.6-2 mm, smooth, black.
Fl. ? *Doubtfully still present in Madeira. Collected in 1839, but no recent records known.* **M**

2. Sesamoides Ortega

Usually perennial herbs. Leaves simple. Flowers solitary, forming dense, terminal, bracteate, spike-like racemes. Calyx 5- or 6-lobed, the lobes unequal. Petals 5 or 6, clawed, unequal, the upper two multilaciniate, the lower ones less divided. Stamens 7-15, inserted on an urn-shaped, hypogynous disc. Carpels 4-7, ± free, stellate-patent in fruit, dehiscing irregularly, 1-seeded. Seeds ovoid.

1. S. clusii (Spreng.) Greuter & Burdet in *Willdenowia* **19**: 47 (1989).
Glabrous, perennial herbs; stems 5-15(-35) cm high, spreading, numerous, slender, woody below. Leaves 10-40 × 0.5-3 mm, linear to linear-oblanceolate, entire, the lower forming dense rosettes, caducous. Pedicels c. 1 mm. Calyx-lobes usually 5, oblong, obtuse, pale-margined, with narrow sinuses, equalling or exceeding the calyx-tube. Petals usually 5, white, the upper cut into 5-7 narrow lobes. Carpels 4-7, often papillose basally, obovoid; style subterminal, overtopping the dorsal swelling of the carpel. Seeds c. 1 mm, papillose, brown.
Fl. IV. *Occasionally recorded as naturalized in grassy areas in Madeira, for example in lawns at Santo da Serra. Native to SW Europe and NW Africa.* **M**

Specimens from Madeira have been recorded under two names: *S. canescens* var. *suffruticosa* (Lange) Abdallah & de Wit and *S. pygmaea* (Scheele) Kuntze. This account follows the treatment in *Flora europaea* 1, 2nd ed., where both taxa are included in the synonymy of *S. clusii*.

XLIII. CRASSULACEAE
M.J.Short

Herbs or small shrubs, usually succulent. Leaves alternate, opposite or whorled, generally simple and entire, rarely pinnate, exstipulate. Flowers regular, hermaphrodite, 3- to *c.* 20-merous, in bracteate cymes or racemes, or solitary in the leaf-axils. Sepals free or fused towards the base. Petals free or united into a tube. Stamens equalling petals in number or twice as many. Carpels as many as petals, free or nearly so, subtended by a nectariferous scale at the base. Fruit a group of follicles, usually many-seeded.

1. Petals fused into a tube for more than half their length 2
- Petals free or fused only at the very base 3

2. Basal and lower leaves orbicular, peltate; calyx and corolla 5-lobed . . **3. Umbilicus**
- Leaves neither orbicular nor peltate; calyx and corolla 4-lobed . . . . . . **2. Kalanchoe**

3. Leaves opposite; stamens equal in number to petals . . . . **1. Crassula**
- Leaves alternate, sometimes crowded towards the ends of the branches, or in crowded, terminal rosettes; stamens twice as many as petals 4

4. Petals 5; leaves not rosette-forming . . . . **6. Sedum**
- Petals 6-18; leaves rosette-forming 5

5. Inflorescence generally 3- to 4-flowered; petals minutely denticulate at the apex (Salvage Is. only) . . . . **7. Monanthes**
- Inflorescence usually many-flowered; petals entire 6

6. Annual herbs; leaves not ciliate, forming a lax rosette, but soon falling **5. Aichryson**
- Perennial herbs or shrubs, woody at least at the base; leaves ciliate, forming dense, persistent terminal rosettes . . . . **4. Aeonium**

1. Crassula L.

Delicate annuals to perennial subshrubs. Leaves opposite, sessile or petiolate, ± connate at the base. Flowers 3- to 5-merous, small, axillary or in a compact, panicle-like cluster. Sepals free, or nearly so. Petals exceedingly shortly fused at the base, pink or white. Stamens equal in number to the petals, alternating with them.

1. Annual; stems up to 5 cm long, slender . . . . **1. tillaea**
- Perennial; stems more than 5 cm long, thick 2

2. Leaves broadly oblong to suborbicular; petals white tinged with pink . . **2. multicava**
- Leaves lanceolate; petals cream . . . . **3. tetragona**

1. C. tillaea Lest.-Garl. *Fl. Jersey*: 87 (1903).
Tillaea muscosa L.
Minute, glabrous, moss-like annual, 1-5 cm high, often yellow or red-tinged; stems prostrate or ascending, branched. Leaves 1-2.5 mm, oblong-lanceolate, mostly acute, aristate, concave,

crowded. Flowers sessile, axillary, 3(-4)-merous. Sepals ovate, aristate-acuminate. Petals *c*. 1 mm, shorter than sepals, lanceolate, acute, white or pale pink. Follicles usually 2-seeded.
Fl. II - VII. *Walls, bare patches on the ground; up to 1000 m. Generally rare in Madeira, Porto Santo and Selvagem Grande, although occasionally locally abundant. A little-recorded, inconspicuous plant which is probably often overlooked.* **MPS**

2. C. multicava Lem. in *Rev. Hort. Paris* **11**: 97 (1862). [*Crássula*]
Succulent, glabrous perennial herb; stems up to 40 cm, decumbent, thick, sparingly branched, often rooting at the nodes. Leaves 15-40(-65) × 15-25(-40) mm, broadly oblong to suborbicular, obtuse, sometimes shallowly emarginate; petioles 3-20 mm. Inflorescence a rounded or elongate panicle. Flowers 4-merous, stellate. Sepals 1-1.5 mm, triangular, acute. Petals 5-6 mm, lanceolate, acute, white, tinged pink. Anthers purple.
Fl. I - VI. *A species native to Natal and Cape Province which is commonly cultivated in parks and gardens, often escaping and becoming naturalized, usually in shady areas; up to 700 m.* **M**

3. C. tetragona L., *Sp. pl.* **1**: 283 (1753).
Succulent perennial up to 1 m; stems branched. Leaves lanceolate, acute. Flowers 5-merous, shortly pedicellate. Calyx-lobes 1-2 mm, triangular. Corolla tubular, cream; lobes 1-3 mm, oblanceolate to elliptic, ridged dorsally, ± recurved. Anthers brown.
Fl. ? *A S. African species naturalized on rocks by tunnels on the Machico-Caniçal road, where it was reported to be growing abundantly.* **M**

Records of *Sedum sediforme* (Jacq.) Pau from Madeira are referable to this species.

2. Kalanchoe Adans.

Succulent, glabrous, perennial herbs. Leaves opposite, in whorls of 3 or alternate, sessile or petiolate, simple or pinnate, crenate or dentate, sometimes producing plantlets on the margins. Flowers 4-merous, often showy, pendent, in lax cymes. Sepals ± free or fused into a tube, fleshy. Corolla tubular, reddish, purplish or orange; lobes short, spreading. Stamens 8, inserted near base of the corolla, exceeding the tube.

1.	Leaves narrowly linear, subterete	**2. delagonensis**
-	Leaves obovate or divided with elliptic to obovate leaflets	2
2.	Upper leaves ternate or pinnate; corolla-tube 3.5-4.5 cm	**1. pinnata**
-	Leaves all simple; corolla-tube up to 1.8 cm	**3. fedtschenkoi**

1. K. pinnata (Lam.) Pers., *Syn. pl.* **1**: 446 (1805).
Bryophyllum pinnatum (Lam.) Oken
Stems up to 1 m or more, erect, stout. Leaves opposite, petiolate, the lower simple, the upper ternate or pinnate; leaflets broadly elliptic to obovate, obtuse, deeply crenate, often red-tinged. Inflorescence lax. Calyx 3-3.5 cm, tubular, inflated, green with reddish markings; lobes short, triangular. Corolla green, tinged with red; tube 3.4-4.5 cm, constricted near the base; lobes oblong-ovate, acuminate.
Fl. I - IV. *Naturalized and common in Madeira along Levada Machico between the valleys of Ribeiras Funda and Seca. Native to Madagascar.* **M**

2. K. delagonensis Eckl. & Zeyh., *Enum. pl. afric. austral.*: 305 (1837).
Bryophyllum delagonensis (Eckl. & Zeyh.) Schinz; *B. tubiflorum* Harv.; *Kalanchoe tubiflora* Raym.-Hamet
Stems up to 1 m or more high, erect, solitary or branched from the base. Leaves sessile, in

whorls of 3 or sometimes alternate, up to 8 × 0.5 cm, narrowly linear, subterete, with several sharp teeth at the apex, marked with dark brown or dull purplish splashes. Inflorescence often large, dense. Calyx 5-10 mm; lobes lanceolate, acute. Corolla red to purplish brown or orange; tube 2.5-3.5 cm; lobes oblong-obovate, ± obtuse.
Fl. ? *Established as a garden escape in southern Madeira, recorded from waste ground near Funchal and on a road slope near Calheta. Native to Madagascar.* **M**

3. K. fedtschenkoi Raym.-Hamet & Perrier in *Annls Mus. colon. Marseille* III, **3**: 75 (1915).
Bryophyllum fedtschenkoi (Raym.-Hamet & Perrier) Lauz.-March.
Stems up to 50 cm, decumbent at the base. Leaves opposite, up to 6 × 4 cm, obovate, dentate towards the apex, attenuate into a very short petiole at the base, glaucous, sometimes reddish tinged, crowded near the base of the stems. Inflorescence small, lax. Calyx up to 1.8 cm; lobes triangular, acute. Corolla purplish red; tube up to 1.8 cm, campanulate; lobes obovate, obtuse.
Fl. ? *Recorded in Madeira from a dry river-bed near Faial. Native to Madagascar.* **M**

3. Umbilicus DC.

Perennial herbs with tuberous or rhizomatous rootstock. Leaves alternate, the basal suborbicular, peltate, petiolate. Flowers 5-merous, in a long, terminal, bracteate racemes. Calyx divided almost to the base. Corolla tubular, greenish; lobes erect. Stamens 10.

1. U. rupestris (Salisb.) Dandy in Ridd., Hedley & W.R. Price, *Fl. Gloucestershire*: 611 (1948).   [*Inhame de galatixa, Inhame de lagartixa, Lagartixa*]
Cotyledon umbilicus sensu Menezes, non L.; *Umbilicus pendulinus* DC.
Up to 30(-50) cm, erect, glabrous. Leaves mostly basal, up to 4(-7) cm across, orbicular, peltate, concave above, sinuate-crenate; cauline leaves progressively smaller and shorter stalked. Bracts linear, equalling or exceeding pedicels, the lower often leaf-like. Pedicels up to *c.* 4 mm. Raceme dense, occupying more than half the stem, sometimes branched from the base. Flowers pendulous. Sepals 1.5-5 mm, ovate, acute. Corolla 6.5-8(-10) mm, pale yellowish green; lobes ovate, acute, mucronate, *c.* ¼ the length of tube.
Fl. IV - VI. *Extremely common in all regions of Madeira on walls and rocks; very rare in Porto Santo on mountain peaks; also known from Deserta Grande.* **MDP**

2. Umbilicus horizontalis (Guss.) DC., *Prodr.* **3**: 400 (1828).   [*Inhame de lagartixa*]
Umbilicus pendulinus [var.] β Lowe
Like **1**, but with the raceme usually occupying half or less of the stem; flowers subsessile, usually horizontal.
Fl. ? *Recorded infrequently from Madeira, Porto Santo, and Selvagem Grande.* **MPS**

4. Aeonium Webb & Berthel.

Evergreen, biennial or perennial herbs or subshrubs. Stems woody at the base. Leaves ± sessile, in crowded terminal rosettes, succulent, glabrous to puberulent, margins ciliate. Flowers 6-16-merous, in a terminal, bracteate panicle. Petals free, yellow, equalling sepals in number. Stamens twice as many as petals.

Literature: Ho-Yih Liu, Systematics of *Aeonium* (Crassulaceae), *Nat. Mus. nat. Sci., Taiwan*, special publ. no. 3: 1-102 (1989).

1. Leaves pubescent; plant subsessile, the stem very short and usually concealed by the leaves . . . . **1. glandulosum**
-   Leaves ciliate but otherwise glabrous; plant with a distinct stem 2

2. Leaf rosette flattened in the centre, the young leaves tightly compressed; inflorescence dense .... **2. arboreum**
- Leaf rosette lax, not flattened in the centre, the young leaves suberect; inflorescence lax .... **3. glutinosum**

1. A. glandulosum (Aiton) Webb & Berthel., *Hist. nat. Iles Canaries* 3(2, sect. 1): 185 (1840). [*Ensaião, Ensaião de pasta, Pastinha, Saião*]
Sempervivum glandulosum Aiton
Glandular-pubescent biennial to perennial subshrub, strongly smelling of balsam. Stem very short, concealed by the leaves, occasionally stoloniferous. Leaf rosette up to 30(-45) cm across, closely imbricate, ± flat and plate-like, but becoming dome-shaped centrally approaching flowering time. Leaves up to 10(-20) × 5(-8) cm, obovate- to rhombic-spathulate, obtuse, mucronulate, long attenuate, bright green, puberulent, margins ciliate with thick, bead-like unicellular and multicellular hairs. Inflorescence lax, up to 30 × 45 cm, branches spreading. Pedicels 3-19 mm, becoming recurved distally. Sepals 4-10 mm, triangular-lanceolate, acuminate. Petals 7-10 mm, pale yellow, occasionally tinged with red without. Carpels 3.5-5 mm; styles 3-3.5 mm. **Plate 13**
Fl. VI - VIII. *Cliffs, ravines and rocks, 0-300(-700) m. Very common on sea cliffs in Madeira, particularly along the north coast; frequent on the north coast and rocky peaks of Porto Santo; also found on Deserta Grande and Bugío.* ● **MDP**

Smaller, branched, caespitose populations from mountainous areas of Porto Santo have been regarded as distinct by some botanists but Liu (1989) does not consider they merit taxonomic recognition.

A. glandulosum × glutinosum
Like **1**, but with slightly sticky, more rigid leaves marked with brownish stripes.
● **M**

2. A. arboreum (L.) Webb & Berthel., *Hist. nat. Iles Canaries* 3(2, sect. 1): 185 (1840).
[*Ensaião, Ensaião da festa* (P), *Saião*]
Sempervivum arboreum L.
Perennial subshrub up to 2 m; stem erect, stout, branches suberect with conspicuous leaf-scars. Leaf rosettes up to *c.* 25 cm across, flattened, the young leaves tightly compressed. Leaves up to 6(-15) × 2(-4.5) cm, obovate, obtuse, mucronate, attenuate, margins ciliate with slender, unicellular hairs, otherwise glabrous, bright green. Inflorescence dense, puberulent. Pedicels 2-12 mm. Sepals 2-3.5 mm, lanceolate, acuminate. Petals 5-7 mm, bright yellow. Carpels 2-3.5 mm; styles 1.5-2.5 mm.
Fl. XII - VI. *Gardens in Madeira and Porto Santo, especially in and around Funchal, sometimes becoming naturalized on walls. Native to Morocco.* **MP**

3. A. glutinosum (Aiton) Webb & Berthel., *Hist. nat. Iles Canaries* 3(2, sect. 1): 185 (1840).
[*Ensaião, Ensayão, Farrobo, Saião*]
Sempervivum glutinosum Aiton
Very viscid, perennial subshrub; stem ± erect, stout, branches up to 60 cm, decumbent to ascending, occasionally plants subsessile. Leaf rosettes up to 22 cm across, lax, young leaves suberect. Leaves up to 12 × 5 cm, obovate-spathulate, obtuse, attenuate, yellowish green, often brown-striped near the apex, ± glabrous, margins ciliate with sparse to numerous, slender, unicellular hairs. Inflorescence very lax, remotely branched, up to 40 × 45 cm, branches spreading. Pedicels 2-6(-10) mm. Sepals 1.1-3.4 mm, triangular-ovate. Petals 5-7 mm, golden yellow, usually red striped abaxially. Carpels 2.5-3 mm; styles *c.* 1.5 mm. $2n=36$. **Plate 13**

Fl. VI - X. *Very common on cliffs, ravines and rocks in Madeira; 0-300(-1700) m. Also found on Deserta Grande.* ● **MD**

5. Aichryson Webb & Berthel.

Annuals. Leaves alternate, entire, petiolate, caducous, somewhat fleshy, glabrous to densely villous. Flowers 6- to 10-merous, in a dichotomous cyme. Petals free, yellow, equalling sepals in number. Stamens twice as many as petals.

1. Leaves densely villous; petals 5-8.5 mm . . . . **1. villosum**
- Leaves glabrous or sparsely puberulent; petals 3-5 mm 2

2. Leaves rhomboid-orbicular, abruptly cuneate at the base; petal midrib greenish dorsally
 . . . . **2. divaricatum**
- Leaves spathulate to narrowly obovate, attenuate at the base; petal midrib reddish dorsally
 . . . . **3. dumosum**

1. A. villosum (Aiton) Webb & Berthel., *Hist. nat. Iles Canaries* 3(2, sect. 1): 181 (1840).
[*Ensaião*]
Sempervivum villosum Aiton
Densely and softly glandular-villous, 3.5-15(-20) cm high, bushy, often tinged reddish purple; stem ascending then erect, branched above, soon naked. Leaves initially forming a lax rosette but soon falling, (7-)12-30 × (4-)8-26 mm, rhomboid-spathulate, obtuse, abruptly cuneate at the base; petioles (3-)10-25 mm. Pedicels 3-16 mm. Flowers 10-15 mm across, mostly 8-merous, in dense cymes; pedicels 3-16 mm. Sepals 2.5-5 mm, linear-lanceolate, acute. Petals 5-8.5 mm, oblanceolate, aristate, golden-yellow, spreading. Carpels 2-2.5 mm; style 1.3-2 mm.
Plate 14
Fl. III - VIII. *Walls, rocks above levadas, ravines; common in Madeira especially in the north, up to 1300 m; on Pico do Facho and the peak of Ana Ferreira in Porto Santo, up to 500 m; Deserta Grande; Bugío, up to 350 m.* ■ **MDP** [Azores]

2. A. divaricatum (Aiton) Praeger, *Acc. Sempervivum*: 125 (1932). [*Ensaião*]
Sempervivum divaricatum Aiton
Usually smooth and glabrous, 4-30(-60) cm high, bushy; stem ascending then erect, dichotomously branched, dark green, later becoming reddish, soon naked. Leaves initially forming a lax rosette but soon falling, 10-30 × 9-23 mm, rhomboid-orbicular, obtuse, minutely retuse, abruptly cuneate; petioles 6-27 mm. Inflorescence glabrous or rarely minutely glandular-puberulous. Flowers 6-10 mm across, mostly 7-merous, in dense cymes; pedicels 2-10 mm, very slender. Sepals 1.5-3 mm, linear to linear-lanceolate, acute. Petals 3-5 mm, ovate, shortly aristate, pale golden-yellow with a greenish midrib dorsally, spreading. Carpels 2-3 mm; style 1-1.5 mm. 2n=30. **Plate 14**
Fl. III - XI. *Very common in Madeira on rocks above levadas, cliffs, walls, ravines, sometimes epiphytic on tree trunks; (100-)500-1000 m. Also known from near Ponta do Pedregal on Deserta Grande.* ● **MD**

Plants with a glandular-puberulent inflorescence have been distinguished as a distinct variety, var. *pubescens* Lowe, but are not thought to merit taxonomic recognition.

3. A. dumosum (Lowe) Praeger, *Acc. Sempervivum*: 127 (1932).
Sempervivum dumosum Lowe
Glandular-puberulous at least above, up to 38 cm high; stem erect, branched, naked nelow, reddish-purple. Leaves not forming a rosette, 12-20 × 3-5 mm, spathulate to narrowly obovate, obtuse, attenuate into the petiole, becoming red-tinged, caducous. Flowers 8-10 mm

across, mostly 7-merous, in lax cymes; pedicels up to 9 mm. Sepals 2.5-3 mm, linear-lanceolate, acute. Petals 4-5 mm, lanceolate, bright golden-yellow with a reddish midrib dorsally, spreading. Carpels *c.* 3 mm; style *c.* 1 mm. **Plate 14**
Fl. III - V. *Rocks and walls; extremely rare and known only from south-west Madeira near Madalena do Mar, 300-400 m.* ● M

6. Sedum L.

Succulent, evergreen, perennial herbs. Leaves alternate, often crowded toward the stem tips, sessile. Inflorescences terminal, cymose, sometimes paniculate; flowers generally 5-merous, stellate. Sepals somewhat fleshy, unequal. Petals free, patent, yellow or white. Stamens usually twice as many as the petals. Carpels equal in number to the petals, free or shortly connate at the base.

1. Leaves 35-65 mm, flat; inflorescence large, many-flowered, up to 18 × 14 cm . . . . **1. praealtum**
- Leaves 3-17 mm, subterete; inflorescence few-flowered, small 2

2. Flowers white; plant farinose . . . . **5. farinosum**
- Flowers yellow; plant not farinose 3

3. Stems smooth; leaves fusiform, acute; sepals acute . . . . **4. fusiforme**
- Stems papillate; leaves linear to oblong or clavate, obtuse; sepals obtuse 4

4. Sepals 3.5-5 mm, oblong; petals 5-7 mm; follicles 4.5-7 mm . . . . **3. brissemoretii**
- Sepals 1.5-2 mm, ovate; petals 3.5-4 mm; follicles 3.5-4 mm . . . . **2. nudum**

1. S. praealtum A.DC. in DC. & A.DC. in *Not. pl. rar. bot. jard. Genève* **10**: 21 (1847).
[*Ensaião, Saião*]
Glabrous, evergreen, shrubby perennial, often forming large clumps; stems terete, stout, woody below, green above, branched; sterile branches up to 1.5 m, trailing to ascending; flowering branches erect. Leaves 3.5-6.5 cm, oblong-spathulate, obtuse to subacute, thick but flattened, crowded and ascending above. Flowers subsessile, in large, lax, much-branched paniculate cymes. Sepals 1-2.5 mm, ovate, obtuse. Petals 5-8 mm, acute, bright yellow. Follicles 5-6 mm, suberect.
Fl. XII - V. *A Mexican species with showy flowers which is commonly grown in Madeiran gardens for ornament, often escaping and becoming naturalized, hanging over walls and banks or covering the ground; up to 700 m.* M

S. forsterianum Sm. in Sowerby, *Engl. bot.* **26**: t. 1802 (1808), a species from W. Europe with bright yellow flowers has been recorded from a roadside in Funchal, presumably as a garden escape. M

2. S. nudum Aiton, *Hort. kew.* **2**: 112 (1789).
[*Arroz da rocha, Erva arroz, Lagartixa, Uva de galatixa, Uva de rato*]
Compact, bushy, glabrous perennial up to 15(-25) cm; stems ± erect, numerous, tortuous, much-branched, papillate, brittle and somewhat woody below, the older shoots reddish brown. Leaves crowded towards the ends of the branches, spreading, 3-6(-9) mm, linear to oblong or clavate, obtuse, subterete, pale green. Flowers subsessile to shortly pedicellate, in few-flowered cymes. Sepals 1.5-2 mm, ovate, obtuse. Petals 3.5-4 mm, greenish yellow. Scales linear-spathulate, *c.* 2 × longer than wide. Follicles 3.5-4 mm, connate *c.* 1 mm at the base, substellate. 2n=26. **Plate 15**

Fl. VII - X. *Frequent on exposed rocks and sea cliffs in Madeira, occurring along the south coast and at the eastern and western ends of the north coast, up to 300 m; sporadic on mountain peaks of Porto Santo, up to 330 m; very rare on Deserta Grande.* ● **MDP**

Plants of *S. lancerottense* R.P. Murray from Lanzarote (Canaries), similar to **2** but with spurred sepals, have sometimes been treated as a variety or subspecies of *S. nudum*.

3. S. brissemoretii Raym.-Hamet in *Bull. soc. bot. France* **72**: (1925).

[*Arroz da rocha, Erva arroz*]

Compact, bushy, glabrous perennial up to 20 cm; stems ± erect, much-branched, tortuous, naked and somewhat woody below, papillate, the young shoots smooth. Leaves crowded towards the ends of the branches, (3.5-)5-11 mm, oblong to subobovate, obtuse, terete, fresh green. Flowers subsessile, in few-flowered inflorescences. Sepals 3-5 mm, oblong to subobovate, obtuse. Petals 4.5-7 mm, pale yellow with a greenish keel. Scales broadly spathulate, wider than long. Follicles 4.5-7 mm, stellate. 2n=22. **Plate 15**

Fl. VI - VII. *Very local, on cliffs, steep rocks and walls along the north coast of Madeira between Ribeira da Janela and São Vicente, sometimes occurring up to 500 m inland; 0-100 m.* ● **M**

4. S. fusiforme Lowe in *Trans. Camb. phil. Soc.* **4**: 31 (1831).

Bushy perennial up to 30 cm; stems ± erect, often numerous, tortuous, branched, brittle and slightly woody below, the young shoots smooth. Leaves spreading, crowded towards the ends of the branches, (6-)8-17 mm, fusiform, acute, subterete, glaucous with a brownish red central stripe. Flowers subsessile, in few-flowered cymes. Sepals 2-3.5 mm, triangular-ovate, acute. Petals 7-8 mm, acuminate, yellow, mottled with red within. Follicles 5-6 mm, stellate. 2n=24. **Plate 15**

Fl. VII - X. *Sea cliffs, rocks and walls along the south coast of Madeira between Cabo Girão and Garajau; 100-400 m. Very rare.* ● **M**

5. S. farinosum Lowe in *Trans. Camb. phil. Soc.* **4**: 31 (1831). [*Erva arroz*]

Farinose, mat-forming, glabrous perennial; stems creeping, woody and naked at the base, sparingly branched, with short, loosely-tufted sterile shoots and erect flowering stems 4-14 cm high. Leaves spreading to suberect, crowded towards the ends of the branches, alternate-imbricate, 3.5-7 mm, oblong to obovate, obtuse, subterete, glaucous, often becoming tinged with red. Flowers *c.* 12 mm across, subsessile, in few-flowered cymes. Sepals 2-4 mm, oblong-ovate, obtuse to subacute. Petals 5-8 mm, acute, mucronate, white, often red-nerved without. Anthers dark purple. Follicles *c.* 5 mm, suberect. 2n=*c.* 382. **Plate 15**

Fl. VI - X. *Rather rare; rocks of mountains and the highest peaks of Madeira; 900-1800 m.* ● **M**

7. Monanthes Haw.

Small, papillose, herbaceous, short-stemmed perennials. Leaves in rosettes, petiolate, succulent. Inflorescences racemose, few-flowered, axillary. Flowers 6-8-merous. Calyx divided *c.* ½-way to the base. Petals free. Stamens twice as many as the petals, unequal, those opposite the petals longer than those opposite the sepals. Nectariferous scales large, conspicuous. Carpels 6-8.

1. M. lowei (A. Paiva) P. Pérez & Acebes in *Vieraea* **14**(1-2): 153 (1985).

M. brachycaulon sensu auct. mad., non (Webb in Webb & Berthel.) Lowe

Stem usually less than 1 cm, thick. Leaves up to 12 × 5 mm, obovate-spathulate to ± rhomboid, obtuse, attenuate at the base into the petiole, green, often with purple streaks or blotches. Peduncles *c.* 2 cm. Inflorescences generally 3- to 4-flowered, glandular-pubescent. Pedicels up to 1 cm, filiform. Flowers 0.8-1 cm across. Calyx 2-3.5 mm; lobes lanceolate,

acute to obtuse, sometimes marked with purple lines. Petals lanceolate, lightly denticulate at the apex, margins subrevolute, greenish or yellowish, sometimes streaked with red. Scales pedicellate, weakly bilobed, yellowish green. Seeds longitudinally ribbed. **Plate 13**
Fl. II - X. *An inconspicuous little plant of damp rock crevices which is fairly frequent and in places abundant on Selvagem Pequena and Selvagem Grande, preferring north facing sites.*
● S

XLIV. SAXIFRAGACEAE
J.R. Press

Herbs, usually perennial, sometimes slightly succulent. Leaves alternate, exstipulate. Inflorescence usually a cyme or panicle. Flowers 5-merous. Ovary usually with 2 carpels united below, diverging above. Styles free. Fruit a capsule with numerous small seeds.

1. Saxifraga L.

Perennials, sometimes woody at the base. Leaves alternate, basal, simple, often deeply lobed; sessile or stalked glands often present. Flowers in cymes or panicles. Sepals 5. Petals 5. Ovary superior or semi-inferior.

Literature: D.A. Webb & J.R. Press, The genus *Saxifraga* L. in the Madeiran archipelago, in *Bocagiana* **105**: 1-4 (1987).

1.	Plant with slender stolons; flowers zygomorphic	**3. stolonifera**
-	Plant without stolons; flowers actinomorphic	2
2.	Plant completely glabrous though sometimes subsessile glands on petioles and pedicels	**2. portosanctana**
-	Plant with at least a few hairs on petioles, pedicels or hypanthium ..	**1. maderensis**

1. S. maderensis D. Don in *Trans. Linn. Soc. Lond.* **13**: 414 (1821).
Perennial with woody, much-branched stems forming large cushions. Leaf-lamina up to 3 × 4 cm, lobes usually 5 or more, each with several shallow, apical segments, coriaceous, viscid on the upper surface with numerous sessile glands; petiole up to 8 cm, usually much longer than the lamina, glandular-hairy. Cymes axillary, borne on previous year's growth, lax, leafy. Pedicels usually with at least some hairs. Petals (4-)6-12 mm, up to 2 × as long as broad, white.
Fl. IV - VI. *Locally frequent endemic growing on damp and shady vertical rock faces in Madeira above c. 600 m.* ● M

Two varieties occur in Madeira:

i) var. **maderensis**
Basal leaves with a reniform to semi-circular lamina, base cordate-truncate to very shortly cuneate, segments obtuse to acute; petioles 1 mm wide in the upper part, sometimes dilated at base, sparsely to moderately hairy. Cymes with 6-13 flowers. Petals (4-)6-10 mm. **Plate 16**
Throughout the range of the species.

ii) var. **pickeringii** (C. Simon) D.A. Webb & Press in *Bocagiana* **105**: 3 (1987).
S. pickeringii C. Simon
Basal leaves with a rhombic to fan-shaped lamina, distinctly cuneate at base, lobes acute to apiculate; petioles 2 mm wide, base dilated and densely hairy. Cymes with up to 6 flowers. Petals 8-12 mm. **Plate 16**
Confined to the high peaks of central Madeira, around Pico de Arieiro, Pico Ruivo and Pico do Cidrão.

2. S. portosanctana Boiss., *Diagn. pl. orient.* **3**: 68 (1856).
Like **1** but completely glabrous or with a few subsessile glands on petioles and pedicels. Leaves fleshy, lamina up to 2 × 2 cm, rhombic to fan-shaped, base long cuneate, ± ternately lobed, the apical segments 3-9, all obtuse. Petals 10-11 mm, 3 × as long as broad. **Plate 16**
Fl. IV - V. *Endemic, growing in clefts and crevices among rocks in Porto Santo. Declining and now rare, confined to rocky outcrops on the summits of the high peaks of Porto Santo.* ● **P**

3. S. stolonifera Meerb., *Afb.zeldz. gew.*: t. 23 (1775).
Perennial with long, thread-like, red stolons rooting at intervals to produce new plants. Leaves up to 5 cm, orbicular with sinuate-dentate margins, dark green with greyish veins and hairy above, reddish and glabrous beneath. Flowers in panicles, strongly zygomorphic, the upper 3 petals *c*. 3 mm, ovate, spotted with yellow and red or tinged pink, the lower two petals 10-20 mm, elliptical to lanceolate, pure white.
Fl. VII. *Frequently cultivated in gardens and as a pot plant; naturalized among shady rocks in at least one site at Ribeiro Frio and possibly elsewhere.* **M**

XLV. HYDRANGEACEAE
J.R. Press

Mostly shrubs. Leaves simple; stipules absent. Calyx tube ± adnate to the ovary. Stamens numerous, in several series. Fruit a loculicidal capsule.

1. Hydrangea L.

Shrubs. Leaves opposite, entire or lobed. Flowers in terminal corymbs, at least the outer sterile and lacking petals but with showy, petaloid sepals.

1. H. macrophylla (Thunb.) Ser. in DC., *Prodr.* **4**: 15 (1830). [*Hortensias, Novelos*]
Plant with numerous erect or ascending stems forming a more or less domed bush. Leaves thick, somewhat coriaceous, glabrous, bright green. Flowers numerous, all sterile, in large heads. Petaloid sepals 4, variable in size and shape, predominantly blue (in Madeira), sometimes pink or white.
Fl. VI - IX. *Very common in parks, gardens and especially in dense borders along roadsides from 300-1100 m, apparently naturalized in both Madeira and Porto Santo.* **MP**

Although widespread and growing freely on the islands this species is planted in most areas where it occurs and, in the Madeiran islands at least, it is probably not truly naturalized on more than a very local scale.

XLVI. PITTOSPORACEAE
J.R. Press

Trees or shrubs. Leaves alternate, simple, stipules absent. Flowers actinomorphic and usually hermaphrodite, the parts in fives. Petals clawed and united below. Fruit a capsule or berry.

1. Pittosporum Banks ex Gaertn.

Small, evergreen trees or shrubs. Flowers in few- to many-flowered clusters. Petals with the limb spreading or reflexed. Capsule woody, dehiscing by means of valves. Seeds immersed in a sticky, resinous mucilage.

Literature: R.C. Cooper. The Australian and New Zealand species of *Pittosporum*. In *Ann. Mo. bot. Gdn* **43**: 87-188 (1956).

1. Leaves thin-textured, lanceolate to elliptic, narrowly pointed with undulate margins . . .
  . . . . **1. undulatum**
- Leaves thick, coriaceous, oblong to oblong-ovate, blunt, the margins not undulate . . . .
  . . . . **2. coriaceum**

1. P. undulatum Vent., *Descr. pl. nouv.*: t.76 (1802).   [*Arvore do incenso*]
Evergreen tree with pale, grey-brown bark. Leaves 12-19 × 3.5-4 cm, lanceolate to oblanceolate or elliptic, acute, somewhat coriaceous but thin, the margins entire, undulate; petioles 1.5-3 cm glabrescent, slightly flattened. Flowers white, fragrant, in loose umbellate cymes. Peduncles and calyces thinly tomentose. Calyx 8-10 mm, 5-lobed but splitting to form 2 lips, one with 2 lobes, the other with 3, falling before the petals. Corolla campanulate, petals 10-15 mm, the limb a little broader than the claw, reflexed and spreading towards the tip. Ovary villous. Capsule orange when ripe, 2-valved, tipped with the persistent base of the style. Fl. III - IV. *Naturalized in lowland woods and along levadas; also planted in many parts of Madeira. Native to SE Australia and originally introduced as an ornamental and shade tree, it is now widespread .* **M**

2. P. coriaceum Dryander ex Aiton, *Hort. kew.*: 3: 488 (1789).   [*Mocano, Pitosporo*]
Evergreen tree 5-8 m high. Bark pale grey, smooth. Smallest branches in umbellate clusters. Leaves generally crowded at the tips of the branches, 6.5-8.5 × 3-5 cm, oblong to obovate-oblong, obtuse, glabrous, entire, thick and coriaceous, dark green above, paler below; petioles glabrous. Flowers creamy, fragrant, in short umbellate cymes. Bracts, peduncles and calyces densely rusty-villous. Calyx *c.* 4 mm, 5-lobed, regular. Petals *c.* 10 mm, thinly villous on the outside, the limb little broader than the claw, reflexed. Ovary densely villous. Capsule up to 2 cm long, broadly ovoid, slightly compressed, rather finely rugose, 2-valved, the valves *c.* 1.5 mm thick, tipped with the persistent base of the style. **Plate 16**
Fl. V - VI. *Endemic tree, formerly scattered in various parts of Madeira, including the cliffs of the north coast, now rare and confined to rocky areas and ravines in the laurisilva in the central mountains. It was formerly cultivated in private gardens for its attractive foliage and flowers and it may still be so.* ● **M**

P. tobira (Thunb.) W.T. Aiton, a tall shrub from China and Japan and similar to **2**, is grown in gardens.

PLATANACEAE
J.R. Press

Platanus × hispanica Mill. ex Münch (*P. acerifolia* (Aiton) Willd.; *P. hybrida* Brot.; *P. occidentalis* sensu auct. mad., non L (1753)) [*Platano*], is widely planted along roads and in gardens in coastal regions. It is a spreading, deciduous tree up to 20 m with bark flaking to reveal large buff and yellow patches. Leaves pinnately 5-lobed to half-way, lobes coarsely dentate. Flowers in globose heads, male heads 2-6, female heads usually 2.

XLVII. ROSACEAE
N.J. Turland

Trees, shrubs or herbs. Leaves alternate, petiolate, stipulate. Flowers regular, usually hermaphrodite. Sepals 4-5. Petals usually 5, sometimes absent. Fruit a pome, drupe, or cluster of drupelets or of 1 or more achenes.

1. Trees, shrubs or dwarf shrubs 2
- Herbs 8

2. Plant with at least a few prickles 3
- Plant unarmed 5

3. Plant with spines; leaves simple, pinnatifid . . . . **13. Crataegus**
- Plant with prickles; leaves pedate or pinnate 4

4. Leaves usually ternate or pedate with 3-5 leaflets; stipules basally attached to petiole; inflorescence paniculate; fruit a head of drupelets, usually ripening black . **2. Rubus**
- Leaves usually pinnate with 5-7 leaflets; stipules attached to petiole for much of their length; inflorescence a few-flowered terminal corymb; fruit not a head of drupelets, berry-like, ripening red, containing numerous achenes . . . . **10. Rosa**

5. Leaves simple 6
- Leaves pinnate 7

6. Evergreen shrub with leathery subentire leaves 1-4.5 cm, gradually tapered into petiole; fruit a small pome . . . . **12. Chamaemeles**
- Evergreen or deciduous trees or shrubs; leaves larger, with characters not as above; fruit a drupe . . . . **1. Prunus**

7. Inflorescence a flat-topped terminal compound corymb; petals 5, white; fruits fleshy, ripening red . . . . **11. Sorbus**
- Inflorescence a slender axillary raceme; sepals 4, green; petals absent; fruit dry, ripening rusty brown . . . . **9. Marcetella**

8. Leaves pinnate 9
- Leaves trifoliolate, digitate or digitately lobed 10

9. Leaves with both large and small leaflets; inflorescence spike-like; petals 5, yellow; fruit with hooked bristles . . . . **7. Agrimonia**
- Leaves with uniformly sized leaflets; inflorescence capitulate; sepals 4, green; petals absent; fruit without bristles . . . . **8. Sanguisorba**

10. Leaves trifoliolate; receptacle swollen in fruit, berry-like, ripening red 11
- Leaves trifoliolate, digitate or digitately lobed; receptacle not swollen in fruit . . . 12

11. Epicalyx-segments 3-toothed at apex; petals yellow; mature fruiting receptacle spongy, tasteless . . . . **4. Duchesnea**
- Epicalyx-segments entire at apex; petals white; mature fruiting receptacle fleshy, sweet . . . . **3. Fragaria**

12. Perennials with procumbent, wiry, rooting flowering stems; leaves trifoliolate or digitate; petals yellow . . . . **5. Potentilla**
- Small annual plant not as above; leaves digitately lobed; petals absent . . **6. Aphanes**

1. Prunus L.

Evergreen or deciduous trees or shrubs. Leaves simple, subentire to serrate. Flowers solitary or in clusters, umbels or racemes. Flowers 5-merous; epicalyx absent; petals deep pink to white. Fruit a fleshy drupe, often succulent and sweet-tasting when ripe.

The genus is represented in the Madeiran archipelago mainly by cultivated species, most of which are fruit trees. The single native species is extremely rare as a wild plant.

1. P. lusitanica L., *Sp. pl.* **1**: 473 (1753). [*Ginjeira brava*]
Cerasus lusitanica (L.) Loisel.
Evergreen tree to 18 m, glabrous throughout; trunk to 50 cm in diameter. Young twigs slender, dark red. Leaves 6-15 × 2.5-5 cm, narrowly elliptic-oblong, broadly aristate, serrate, leathery, glossy dark green above, paler beneath; petioles dark red; stipules very small, membranous. Inflorescence an elongate axillary raceme 4-18 cm. Petals *c.* 4 mm, ± orbicular, white. Fruit ovoid to subglobose, ripening through red to purplish black.
Fl. VI - VIII. *A laurisilva tree formerly scattered in central Madeira and the Ribeira da Janela region, but now known only from the Ribeira Seca valley north of Ribeiro Frio; recorded also as cultivated at Camacha.* **M**

The above description refers to the Macaronesian endemic (■) subsp. **hixa** (Brouss. ex Willd.) Franco in *Bolm Soc. port. Ciênc. nat.* II, **10**: 81 (1964) **Plate 17**, which otherwise occurs in the Canaries, and to which Madeiran plants belong.

P. laurocerasus L. [*Louro inglês, Louro cerejo*] is an evergreen shrub or small tree similar to **1** but with young twigs and petioles pale green, and leaves sometimes subentire. It is native to the E. Balkan peninsula and SW Asia, and is occasionally cultivated for ornament in Madeira.

The following additional species are all deciduous trees or shrubs cultivated for their edible fruits or seeds. Some may be locally established away from cultivation.

P. persica (L.) Batsch (*Amygdalus persica* L.) [*Pecegueiro*] is a tree with leaves oblong-lanceolate, serrulate and glabrescent. Mostly solitary, subsessile, usually deep pink flowers appear as the leaf-buds open, and are followed by large, globose, velvety fruits, which ripen yellow or pale green, tinged reddish, with succulent flesh. It is native to China and is cultivated in Madeira and Porto Santo. Plants with glabrous fruits [*Pecegueiro calvo*] are also cultivated in Madeira.

P. dulcis (Mill.) D.A. Webb (*Amygdalus communis* L.) [*Amendoeira*] is a tree or shrub similar to *P. persica*, but with leaves glabrous, and flowers mainly in pairs, appearing before the leaves, fading with age from bright pink to pale pink or almost white. Fruits have leathery flesh which eventually splits and separates away from the hard seed. It is native to SW and C. Asia and is cultivated in Madeira and Porto Santo for its seeds.

P. armeniaca L. [*Damasqueiro*] is a small tree or shrub with leaves broadly ovate to suborbicular, serrate and glabrous. Solitary or paired, subsessile, white or very pale pink flowers appear before the leaves, and are followed by fruits similar to those of *P. persica*. It is native to C. Asia and N. China and is cultivated in Madeira.

P. domestica L. [*Ameixieira*] is a tree or shrub with leaves obovate to elliptic, crenate-serrate, glabrous above and densely pubescent to subglabrous beneath. White flowers, usually in clusters of 2-3, appear with the leaves, and are followed by globose to oblong, glabrous fruits, which are often pruinose, ripening purple, red, yellow or green, with succulent flesh. It is of hybrid origin and is cultivated in Madeira.

P. avium (L.) L. (*Cerasus avium* (L.) Moench) [*Cerejeira*] is a tree with leaves obovate-oblong, deeply crenate-serrate, dull and glabrous above and usually with some persistent hairiness beneath. Sessile umbels of 2-6 white flowers appear just before the leaves, and are followed by small globose glabrous fruits, which ripen creamy yellow, red, dark red or black, with succulent flesh. It is native to Europe and SW Asia and is cultivated in Madeira.

P. cerasus L. (*Cerasus vulgaris* Mill.) [*Ginjeira*] is a tree or shrub similar to *P. avium*, but with suckering habit, indistinct trunk, leaves glossy above, glabrescent beneath, and fruits ripening bright red, with acidic flesh. It is native to SW Asia and is cultivated in Madeira.

2. Rubus L. [*Silvado*]

Evergreen, prickly shrubs usually with long, arching or procumbent stems. Leaves usually ternate or pedate, with 3-5 leaflets; stipules basally attached to petiole. Inflorescence paniculate. Flowers 5-merous; epicalyx absent; petals pink to white. Fruit a head of fleshy 1-seeded drupelets, usually ripening black.

Literature: C. A. Menezes, 'Rubus' madeirenses, in *Jorn. Sci. math. phys. nat.* II, **7**(28): 309-314 (1910). W.O. Focke, Species ruborum, pars III. In *Biblthca bot.* **83**: 113-120, 206 - 207 (1914).

R. pinnatus Willd. has very prickly stems and pinnate leaves, usually with 5-7 ovate to lanceolate leaflets. It is native to the mountains of tropical and southern Africa, Ascension, St Helena and São Tomé, and was apparently subspontaneous near some of the gardens of Funchal, but by 1914 had already disappeared from Madeira.

R. idaeus L. has a distinctive suckering habit, erect stems, pinnate or ternate leaves, and drupelets ripening red, juicy and sweet. It is a circumboreal species which has been recorded as occasionally cultivated in Madeira, but it apparently does not grow well on the island.

1. Leaflets 3; stipules ovate-lanceolate; drupelets ripening black with a bluish, pruinose surface . . . . **6. R. caesius**
- Leaflets 3-5; stipules narrowly lanceolate to filiform; drupelets ripening glossy black, not pruinose 2

2. Plant with long-stipitate glands often densely covering axes of inflorescences (occasionally only a few on pedicels and sepals) . . . . **5. grandifolius**
- Plant with subsessile glands but no long-stipitate glands 3

3. Terminal leaflets on flowering shoots 2-6 × 1-4 cm, obovate to oblanceolate, rhombic, obtuse to cuspidate, densely and minutely white-tomentose beneath; inflorescence usually narrow; petals 5-9 × 4-6 mm . . . . **1. ulmifolius**
- Terminal leaflets on flowering shoots generally larger, (4-)5-14 × (2.5-)3-8 cm, usually broadly obovate or broadly elliptic to orbicular, cuspidate to aristate, variably hairy beneath (very sparsely hirsute only on veins to densely white-tomentose throughout); inflorescence usually broader, pyramidal; petals often larger, 7-23 × 3-16 mm . . **2.-4. bollei group**

1. R. ulmifolius Schott in *Isis, Jena* **1818**: 821 (1818).
R. discolor Weihe & Nees; *R. inermis* sensu Hansen & Sunding (1985), non Pourr. (1788); *R. ulmifolius* subsp. *rusticanus* var. *communis* Menezes, var. *dalmatinus* (Tratt. ex Focke) Menezes, var. *neglectus* Menezes, var. *nutritus* Menezes
Stems arching or procumbent, rooting at tips, angled, often furrowed, minutely white-tomentose, prickly; prickles to 8 mm, recurved. Leaves pedate with 3-5 leaflets; terminal leaflets on flowering shoots 2-6 × 1-4 cm, obovate to oblanceolate, rhombic, unevenly serrate, obtuse to cuspidate, dark green and glabrous above, densely and minutely white-tomentose beneath; stipules 3-14 mm, narrowly lanceolate to filiform. Inflorescence usually a long narrow panicle, leafy at base, with erecto-patent, shortly cymose primary branches; all axes minutely white-tomentose, prickly, with minute reddish, subsessile glands. Sepals deflexed during and after flowering, 4-9 × 3-3.5 mm, ovate, cuspidate to long-acuminate, densely and minutely white-tomentose, with minute reddish, subsessile glands; petals 5-9 × 4-6 mm, obovate, pink to white. Fruiting head 8-10 mm, subglobose, with *c.* 10-20 drupelets ripening glossy black and juicy, not pruinose.
Fl. III - X. *Woodland, scrub, rocky and grassy places and terrace walls in southern Madeira, westward to Encumeada, from sea level to 1300 m; also at the western end of Vila Baleira in Porto Santo and on Deserta Grande.* **MDP**

2-4. R. bollei group
Like **1** but stems ± glabrous to sparsely hirsute; leaves generally larger; terminal leaflets on flowering shoots (4-)5-14 × (2.5-)3-8 cm, usually broadly obovate or broadly elliptic to orbicular, cuspidate to aristate, variably hairy beneath (very sparsely hirsute only on veins to densely white-tomentose throughout); stipules generally larger, 7-20 mm; inflorescence usually a broader pyramidal panicle with primary branches patent to erecto-patent, and shortly cymose, dichasial, racemose or paniculate; axes sparsely to densely tomentose, or villous; sepals 4-15 × 2.5-5 mm, densely white-tomentose to villous, sometimes with a few minute prickles on outer face; petals generally larger, 7-23 × 3-16 mm, obovate to oblanceolate; fruiting heads 5.5-13 mm, hemispherical to globose or ovoid.
Fl. VI - XI. *Woodland, scrub, ravines, humid gullies, rock-faces, steep banks and levadas mainly in northern Madeira, from 50-950 m.* **M**

The species within this group are imperfectly known; the features said to distinguish them show a wide range of overlap among Madeiran specimens and cannot satisfactorily be used for identification purposes. Further detailed study of these and related species is required throughout Macaronesia.

2. R. bollei Focke in *Abh. naturw. Ver. Bremen* **9**: 405 (1889), **12**: 338, t.3 (1892) has been recorded from Madeira, and is said to be frequent in many places along the northern coast. ■ **M** [*Canaries*]

3. R. canariensis Focke in *Abh. naturw. Ver. Bremen* **9**: 405 (1889), **12**: 338, t.4 (1892) has been recorded from Madeira. ■ **M** [*Canaries*]

4. R. vahlii Frid. in *Bot. Tidsskr.* **27**: 108 (1906) is endemic to Madeira. It was described from pinewoods north of Funchal, at 700 m, and has also been recorded from Santana. ● **M**

R. concolor Lowe, *Man. fl. Madeira* **1**: 249 (1862), non Ley in *Bot. Cent. DtLd* **21**: 434 (1846) (*R. ulmifolius* var. *concolor* (Lowe) Frid.) was described from the area north of the Paúl da Serra, but is possibly not distinct from **4**. If it were to be maintained as a distinct taxon, a new name would be required, since it is a later homonym of *R. concolor* Ley. (● **M**)

The hybrid **R. ulmifolius** × **R. vahlii** has been recorded from Madeira, at Caminho do Arrebentão north of Monte. It was described as **R.** × **suspiciosus** Menezes in *Jorn. Sci. math. phys. nat.* II, **7**: 313 (1910), where the parentage was given as *R. ulmifolius* × *R. bollei*; this was later amended to *R. ulmifolius* × *R. vahlii* by Menezes (*Fl. Madeira*: 243 (1914)). ● **M**

5. R. grandifolius Lowe in *Trans. Camb. phil. Soc.* **4**: 32 (1831). [*Silvado da serra*]
R. grandifolius var. *dissimulatus* Menezes
Stems robust, arching, angled, furrowed, prickly, glabrous but often covered with long-stipitate glands; prickles small, to 3 mm, recurved. Leaves pedate with 5 leaflets; terminal leaflets on flowering shoots 4-11 × 2-6 cm, obovate to oblanceolate or suborbicular, unevenly serrate, obtuse to cuspidate, green and glabrous above, slightly paler and occasionally very sparsely hairy on veins beneath; stipules 7-25 mm, narrowly lanceolate to linear. Inflorescence a pyramidal panicle; axes with small prickles, usually densely covered with long-stipitate glands (at least a few on pedicels). Sepals deflexed during and after flowering, 4-14 × 1-7 mm, ovate to lanceolate, cuspidate to long-acuminate, sparsely white-hairy to densely tomentose, with long-stipitate glands on outer face. Petals 10-18 × 5-12.5 mm, obovate, white. Fruiting head 5-11 × 5-9 mm, subglobose to subcylindrical, with *c.* 50 drupelets ripening glossy black, firm-textured, not very juicy, not pruinose. **Plate 17**
Fl. VI - IX. *A widespread but uncommon endemic of laurisilva, scrub, rocky and moist shady places in Madeira, mainly restricted to ravines and river valleys from 50-1500 m.* ● **M**

6. R. caesius L., *Sp. pl.* **1**: 493 (1753).
Stems slender, branched, cylindrical, prickly, glabrous, pruinose, rarely with a few glands; prickles small. Leaves ternate; terminal leaflet sometimes 3-lobed; lateral leaflets often 2-lobed; stipules ovate-lanceolate. Inflorescence short, with a few 2- to 5-flowered corymbs. Sepals appressed to developing fruit. Petals large, white. Fruiting heads with 2-20 large and loosely coherent drupelets ripening black with a bluish pruinose surface.
Fl. ? *Recorded as occurring in Madeira but without further data.* M

3. Fragaria L.

Hairy perennial herbs with basal leaf-rosette and long, wiry, rooting epigaeal stolons. Leaves trifoliolate. Flowers in scapose, laxly cymose inflorescences, 5-merous; epicalyx present, with segments entire at apex; petals white. Fruiting receptacle swollen, fleshy, berry-like, sweet-tasting, with brown achenes scattered over the surface.

1. F. vesca L., *Sp. pl.* **1**: 494 (1753). [*Morangueiro*]
Stolons to 65 cm. Petioles 2-20 cm; leaflets subsessile, and ovate, broadly elliptic or obovate, coarsely serrate with obtuse teeth; central leaflets 2-8.5 × 1.5-6 cm. Scape 3-25 cm. Epicalyx-segments lanceolate to narrowly lanceolate, equalling sepals but slightly shorter than petals; sepals ovate to lanceolate; petals 4-9 × 3-6 mm, obovate. Fruit 9-15 mm, glabrous, ripening red.
Fl. I - XII. *Woodland,* Vaccinium *scrub, wet shady rocky places, mountain slopes, banks, levadas, tracks and paths in eastern Madeira, westward to Encumeada, from 450-1200 m.* M

4. Duchesnea Sm.

Like *Fragaria* but flowers solitary, axillary on the stolons; epicalyx-segments 3-toothed at apex; petals yellow. Mature fruiting receptacle spongy, tasteless; achenes red.

1. D. indica (Andrews) Focke in *Natürl. PflFam.* **3**(3): 33 (1888).
Fragaria indica Andrews
Leaflets usually shortly petiolulate. Pedicels 3.5-14 cm. Epicalyx-segments obovate, equalling sepals and equalling to exceeding petals; petals 6-9 × c. 4 mm, oblong-elliptic to oblanceolate.
Fl. ± all year. *Introduced and naturalized on banks, tracksides and along levadas, in eastern Madeira and at Chão da Ribeira in the north, from 400-1100 m. Probably native in S. and E. Asia.* M

5. Potentilla L.

Sparsely hairy perennial herbs with basal leaf-rosettes and procumbent, wiry, rooting flowering stems. Leaves trifoliolate or digitate. Flowers solitary in leaf-axils, or the upper sometimes in a few-flowered cyme; 4- to 5-merous; epicalyx present; petals yellow. Fruit a dry head of achenes.

1. Leaves trifoliolate, or digitate with 4-5 leaflets; leaflets with acute to obtuse teeth; flowers 4- to 5-merous, the upper sometimes in a few-flowered cyme . . . . **1. anglica**
- All leaves digitate with 5(-7) leaflets; leaflets with obtuse to rounded teeth; all flowers 5-merous, solitary, axillary . . . . **2. reptans**

1. P. anglica Laichard., *Veg. europ.* **1**: 475 (1790). [*Solda*]
P. procumbens Sibth.
Flowering stems to 90 cm. Leaves trifoliolate, or digitate with 4-5 leaflets; leaflets broadly obovate to oblanceolate, cuneate towards base, coarsely and often deeply serrate with acute to obtuse

teeth; central leaflets 7-30 × 5-22 mm. Flowers 4- to 5-merous, solitary, axillary, or the upper sometimes in a few-flowered cyme; epicalyx-segments and sepals similar, shorter than petals, lanceolate; petals 6-10 × 4-10 mm, obcordate, yellow with an orange base.
Fl. I - XII. *Woodland, scrub, grassland, rocks, banks, levadas and pathsides in eastern Madeira, westward to São Vicente, from 450-1500 m.* M

2. P. reptans L., *Sp. pl.* **1**: 499 (1753).
Like **1** but all leaves digitate with 5(-7) leaflets; leaflets crenate-dentate with obtuse to rounded teeth; all flowers 5-merous, solitary, axillary.
Fl. IV - VIII. *Rare on roads and roadsides in western Madeira, at Fajã da Ovelha, Lugar de Baixo, São Vicente, and the area between Canhas and Ribeira Brava.* M

6. Aphanes L.

Small, sparsely hairy annual herb. Leaves digitately lobed; stipules connate. Inforescence a contracted, leaf-opposed cyme, ± sessile in the stipular cup. Sepals 4(-5); epicalyx present; petals absent. Fruit dry, with a single achene.

1. A. microcarpa (Boiss. & Reut.) Rothm. in *Reprium Spec. nov. Regni veg.* **42**: 172 (1937).
?*A. arvensis* sensu auct. mad., non L. (1753).
Plant usually yellowish green. Stems to 10 cm, decumbent to erect, much branched. Leaves 3-8 × 3-10 mm, fan-shaped, deeply divided into 3 segments, each of which is further divided into up to 5 lobes. Stipules forming an involucre and enclosing the inflorescence; stipule-lobes oblong-triangular. Calyx with an urn-shaped tube and spreading triangular lobes. Fruit 1.2-1.8 mm including the persistent sepals.
Fl. III - VI, XI - XII. *Short turf, mossy areas, rocky banks, tracks, paths and among cobblestones, sporadically distributed in Madeira, mainly in the eastern half.* M

A. arvensis L. (*Alchemilla arvensis* (L.) Scop.) has been recorded from Madeira, but most probably in error for **1**. It differs in usually being more greyish-green, with stipule-lobes triangular, sepals more oblong-triangular, and fruit 2.2-2.8 mm including the sepals.

7. Agrimonia L.

Pubescent or villous, rhizomatous perennial herb with erect stems. Leaves pinnate with both large and small leaflets, the basal often in a rosette. Inflorescence a spike-like terminal raceme. Flowers 5-merous; epicalyx absent; petals yellow. Fruit dry, with hooked bristles at upper end, containing 1-2 achenes.

1. A. eupatoria L., *Sp. pl.* **1**: 448 (1753). [*Amoricos*]
Stems to 80 cm. Leaves 6-25 cm, with 3-6 pairs of main leaflets with smaller, unequally sized leaflets in between; main leaflets 8-75 × 8-35 mm, elliptic to broadly obovate, coarsely crenate to serrate, dark green above, whitish- or greyish-tomentose beneath. Petals 4-6 × 1.5-3 mm, obovate to oblanceolate. Mature fruits 6-7.5 × 5-6 mm including bristles, obconical to turbinate, deeply concave, grooved, with long subappressed hairs; inner bristles erect, the outer patent to ascending; pedicel deflexed. $2n = 28$.
Fl. IV - XI. *Woodland, grassland, ravines, levadas, roadsides and field margins, mainly in eastern Madeira, sporadic elsewhere, from 150 m upwards.* M

Most or all Madeiran material appears to belong to subsp. **eupatoria**, which is widespread in Europe. Large plants with villous hairs may be referable to subsp. **grandis** (Andrz. ex Asch. & Graebn.) Bornm. in *Beih. Repert. nov. Spec. Regni veg.* **89**: 244 (1940), which occurs in E., EC and S. Europe.

8. Sanguisorba L.

More or less glabrous to sparsely hairy perennial herb with erect leafy stems. Leaves pinnate with leaflets uniformly sized, the basal in a rosette. Inflorescence terminal, capitulate; upper flowers of capitula female. Sepals 4, green; epicalyx and petals absent. Fruit dry, without bristles, containing 1(-2) achenes.

1. S. minor Scop., *Fl. carniol.* ed. 2, **1**: 110 (1772).
Stems to 70 cm. Leaves 3-35 cm, with 7-17 leaflets; leaflets 3-30 × 3-30 mm, elliptic to orbicular, coarsely serrate. Sepals 3-4 × 2-3 mm, broadly elliptic, with pale scarious margins. Mature fruiting capitula 7-23 × 7-14 mm, globose to cylindric; fruits 3-5 mm, subglobose, slightly compressed laterally or angled, strongly verrucose, hard, ripening pale brown. $2n = 28$.
Fl. IV - VI. *Pine woodland, grassy and rocky places, roadsides and as a weed of cultivation in the area from Funchal and Caniço northwards to Queimadas and Porto da Cruz.* **M**

The above description refers to subsp. **verrucosa** (Ehrenb. ex Decne.) Holmboe in *Bergens Mus. Skr.* II, **1**(2): 100 (1914) (*Poterium verrucosum* Ehrenb. ex Decne., *S. minor* subsp. *magnolii* (Spach) Briq.), which otherwise occurs in S. Europe, N. Africa and SW Asia, and to which Madeiran plants belong.

9. Marcetella Svent.

Evergreen, unarmed dioecious shrubs. Leaves pinnate. Inflorescence a slender axillary raceme. Sepals 4, green; epicalyx and petals absent. Fruit dry, ripening rusty brown, containing 2-3 achenes.

Literature: E. Sventenius, Estudio taxonómico en el género *Bencomia*, in *Boln Inst. nac. Invest. agron., Madr.* **18**: 253-272 (1948). D. Bramwell, The endemic genera of Rosaceae (*Poterieae*) in Macaronesia, in *Botanica Macaronésica* **6**: 67-73 (1980).

1. M. maderensis (Bornm.) Svent. in *Boln Inst. nac. Invest. agron., Madr.* no. **18**: 267 (1948).
Bencomia caudata sensu auct. mad., non Webb & Berthel. (1846); *B. maderensis* Bornm.; *Sanguisorba maderensis* (Bornm.) Nordborg
Plant to *c*. 1 m. Twigs slender, leafy and very sparsely hairy when young, later becoming bare, glabrous and covered with flaky greyish-brown bark. Leaves 4-15 cm, with 7-13 leaflets; leaflets 1-3.5 × 0.5-2 cm, oblong-lanceolate, serrate, green and glabrous above, slightly paler and sparsely pubescent beneath; petiolules up to ⅓ as long as leaflets; stipules to 14 mm, lower part fused to base of petiole, upper part free, pinnatifid. Male and female inflorescences similar, 3.5-17 cm, spicate, flexuous, often branching and leafy near base; flowers sessile, each subtended by 3 ovate to lanceolate scarious bracts to 3 mm. Sepals of male flowers *c*. 3 mm, obovate, strongly revolute from apex, light green with pale scarious margin; sepals of female flowers similar but smaller, *c*. 2 mm, not or scarcely revolute, darker green. Mature fruit 3-6 mm, compressed-obovoid with a marginal rim and a central ridge on both faces (thus appearing weakly 4-angled), spongy, smooth. $2n = 28$. **Plate 18**
Fl. IV - V. *A rare Madeiran endemic, growing on rocks in the area from Pico do Cedro southward to the coast at Câmara de Lobos; also above Porto do Moniz.* ● **M**

Bencomia caudata (Aiton) Webb & Berthel., *Hist. nat. Iles Canaries* **3**(2, sect. 2): 10 (1842) is endemic to the Canaries, although a single male plant occurred in Madeira during the nineteenth century, at Monte north of Funchal, where it was apparently planted. It is like *Marcetella maderensis* but generally larger, with young twigs, petioles, midribs and undersides of leaflets all densely white-hairy; leaflets larger, 2.5-5.5 × 1-2.5 cm, lanceolate; petiolules generally shorter, to one thirtieth as long as leaflets, and inflorescences denser, with white-hairy bracts and larger flowers.

10. Rosa L. [*Roseira*]

More or less deciduous prickly shrubs with ascending stems. Leaves usually pinnate, with 5-7 leaflets; stipules attached to petiole for much of their length. Inflorescence a few-flowered terminal corymb. Flowers usually 5-merous; epicalyx absent; petals usually white. Fruit fleshy, berry-like, ripening red, containing numerous achenes.

1. Leaflets eglandular beneath; pedicels eglandular, unarmed; sepals eglandular on outer face, rarely with stipitate glands at margins . . . . **1. canina**
- Leaflets glandular beneath; pedicels with long-stipitate glands and fine straight prickles to 3 mm; sepals with stipitate glands on outer face and at margins . . . **2. rubiginosa**

1. R. canina L., *Sp. pl.* **1**: 491 (1753). [*Rosa brava*]
Stems short and much-branched, or longer and arching to 4 m or more, sparsely prickly to ± unarmed; prickles to 7 mm, recurved. Leaves pinnate with 5-7 leaflets; leaflets 6.5-37 × 5-23 mm, elliptic to orbicular, simply serrate or rarely slightly biserrate, glabrous and eglandular on both surfaces; teeth eglandular (rarely a few with ± sessile glands); petiole and midrib glabrous, with a few small prickles and stipitate glands; stipules sometimes with stipitate glands at margins. Inflorescence terminal with 1-9 flowers; pedicels glabrous, eglandular, unarmed; sepals not persistent on ripe fruit, 15-30 mm, lanceolate, caudate, often pinnatifid, pale-pubescent towards margins, eglandular (rarely with stipitate glands at margins); petals 13-33 × 12-25 mm, cuneate-obcordate, white, sometimes very slightly pink-tinged; ovary glabrous, eglandular, unarmed. Fruit 13-25 × 8-18 mm, narrowly ovoid to subglobose, ripening orange-red.
Fl. IV - VIII. *Widespread but sporadic in laurisilva, scrub, ravines, and on cliffs and roadsides, from 600-1600 m.* **M**

The above description applies to almost all Madeiran plants, which have been treated as an endemic (●) taxon, **R. mandonii** Déségl. in *Mém. Soc. Acad. Angers* **28**: 111 (1873) (*R. canina* var. *mandonii* (Déségl.) Menezes, var. *glabra* sensu Lowe (1868)) **Plate 17**. Plants with more numerous glands, pubescent petioles and midribs, and leaflets sometimes pubescent on the mid-vein beneath, while appearing slightly biserrate because of glands on the marginal teeth, have been called **R. canina** var. **pubescens** Menezes in *Bull. Soc. port. Sci. nat.* **4**(1): 4 (1910). Such plants are very rare in Madeira, having been recorded only from the levada below Bica da Cana on the eastern edge of Paúl da Serra, at 1370 m, and from Santo da Serra in the east of the island.

2. R. rubiginosa L., *Mant. pl. alt.*: 564 (1771).
R. [cf.] *wilsoni* sensu Lowe (1868).
Like **1** but usually a low shrub with stems more prickly; prickles stouter and longer, to 10 mm, sometimes straight; leaflets with ± sessile glands on teeth (therefore appearing ± biserrate) and on underside; petioles and midribs with many more stipitate glands; stipules with many more glands at margins; pedicels with long-stipitate glands and fine straight prickles to 3 mm, the latter sometimes extending onto ovary; sepals persistent on ripe fruit, with stipitate glands on outer face and at margins.
Fl. VI. *Known with certainty only from the area east of Ribeiro Frio, above Porto da Cruz in eastern Madeira, growing along levadas.* **M**

Recent, unconfirmed, records of *R. stylosa* Desv. from Madeira may be referable to **2**, in which case the distribution of the species would extend from Serra de Água in the west to Santo da Serra in the east.

Several additional species and numerous hybrids are cultivated for ornament in Madeira; among these, the following may be locally established away from cultivation.

R. bracteata J.C. Wendl., *Bot. Beob.*: 150 (1798) has pinnate leaves with 5-7 obovate-oblong leaflets and few-flowered terminal inflorescences. It is native to China. **M**

R. laevigata Michx., *Fl. bor.-amer.* 1: 295 (1803) [*Rosa mosquêta*] is a robust, shrubby climber with large stout prickles and ternate leaves with leaflets lanceolate, leathery, mid-green and glabrous. Large white flowers, each with a single whorl of petals, have pedicels and ovaries densely covered with fine straight prickles to 5 mm. It is native to N. America. **M**

R. multiflora Thunb., *Fl. jap.*: 214 (1784) [*Rosa de toucar*] is a shrub with small, slender prickles and pinnate leaves with leaflets elliptic and pubescent. Numerous small pink flowers, each with several whorls of petals, have pedicels and sepals densely pubescent. Fruits are never formed. The species is native to China and Japan, and is commonly planted in Madeira for hedges. **M**

Cydonia oblonga Mill., *Gard. dict.* ed.8, no.1 (1768) (*C. oblonga* subsp. *maliformis* (Mill.) Thell.; *C. vulgaris* Pers., var. *oblonga* (Mill.) DC.) [*Marmeleiro*] is a deciduous tree or shrub with leaves simple, ovate and entire. Large pink or white flowers are followed by large globose or pyriform tomentose pomes, which ripen yellow and fragrant, with firm flesh. Native to SW and C. Asia, it is cultivated in Madeira for its edible fruits and may also be locally established away from cultivation. **M**

Pyrus communis L. [*Pereira*] is a deciduous tree with leaves simple, ovate or elliptic and crenulate-serrulate. White flowers are followed by oblong pyriform turbinate or subglobose pomes, which are variously coloured when ripe, with firm flesh. It is of hybrid origin and is cultivated in Madeira and Porto Santo for its edible fruits.

P. pyraster Burgsd., *Vers. Gesch. Holzarten* 2: 193 (1787) has been recorded from Pináculo east of Funchal as a likely escape from cultivation and, therefore, as a likely error for *P. communis*. It differs mainly in having smaller fruits, only 1.3-3.5 cm.

Malus domestica Borkh., *Theor. prakt. Handb. Forstbot.* 2: 1272 (1803) (*Pyrus malus* L.) [*Macieira, Pereiro*] is a deciduous tree or shrub with leaves simple, ovate-elliptic and serrate. White or pink flowers are followed by more or less globose pomes, which are variously coloured when ripe, with firm flesh. Of hybrid origin, it is cultivated in Madeira and Porto Santo for its edible fruits and may also be locally established away from cultivation. **M**

11. Sorbus L.

Deciduous, unarmed shrubs or small trees. Leaves pinnate; stipules soon falling. Inflorescence a flat-topped terminal compound corymb. Flowers 5-merous; epicalyx absent; petals creamy white. Fruit a small fleshy pome, ripening red.

1. S. maderensis Dode in *Bull. Soc. dendr. France*: 206 (1907).
Pyrus aucuparia var. *maderensis* Lowe; *P. maderensis* (Dode) Menezes
Twigs stout, smooth, dark reddish-brown, with scattered pale dots. Leaves 5-14 cm, with (11-)13-17 leaflets; leaflets 1-4.5 × 0.5-2 cm, elliptic, oblong or lanceolate, crenate, green and glabrous above, pale green and sparsely hairy mainly along mid-vein beneath. Inflorescence 2-7 × 3-9.5 cm. Petals 3-4 × 3-4 mm, obovate. Fruiting heads erect, dense, showy; each fruit 8-13 mm, globose. **Plate 18**
Fl. VI - VII. *A very rare Madeiran endemic of scrub, rock-ledges and screes, known only from the Pico de Arieiro and Pico Ruivo area, from 1500-1750 m, and Paúl da Serra, at 1525 m.*
● **M**

Eriobotrya japonica (Thunb.) Lindl. [*Nespera de Japão*] is a small, evergreen, brownish hairy tree with leaves simple, obovate to elliptic-oblong, wrinkled, leathery and dark green. Compact terminal panicles of white flowers are followed by pyriform or ellipsoid pomes, which ripen yellow, with succulent flesh surrounding one or more large seeds. It is native to China and Japan and is cultivated in Madeira for its edible fruits.

Mespilus germanica L. [*Nespereira*] is a deciduous shrub or small tree with leaves simple and lanceolate or oblanceolate to obovate. Solitary white flowers are followed by pyriform to depressed-globose pomes, which ripen brown with persistent narrow leafy sepals. It is native to SE Europe and SW and C. Asia and is cultivated locally in Madeira for its fruits, which become soft and edible as they begin to decay.

12. Chamaemeles Lindl.

Evergreen, unarmed shrub. Leaves simple, gradually tapered into petiole, subentire, leathery; stipules soon falling. Inflorescence a small axillary raceme or panicle; flowers 5-merous; epicalyx absent; petals white, tinged with pink or with red markings. Fruit a small fleshy 1-seeded pome.

An endemic monospecific genus not obviously closely related to any other extant genus.

1. C. coriacea Lindl. in *Trans. Linn. Soc. Lond.* **13**: 104 (1821). [*Buxo da rocha*]
Plant to 4 m, much-branched, completely glabrous except for inflorescences and dormant growth buds. Twigs slender, dark reddish-brown with scattered pale dots. Young growth tinted bright reddish. Leaves 1-4.5 × 0.5-2 cm, obovate to oblanceolate, glossy dark green above, paler beneath. Inflorescence 1-4.5 cm, elongate to pyramidal; all axes and pedicels sparsely pubescent. Sepals persistent and ± fleshy in fruit. Petals 2-2.5 mm. Fruit 7-12 × 5-9 mm, ± ovoid, with grit cells, ripening creamy yellow to white approximately 1 year after flowering. **Plate 18**
Fl. X - VI, also VIII. *An endemic of coastal cliffs and rocks, as well as cliffs in ravines to c. 3 km inland, reaching 400 m; occurring along the southern coast of Madeira, from Fajã da Ovelha to Santa Cruz, as well as at Santana on the northern coast; also on Pico Juliana in Porto Santo and on Deserta Grande and the adjacent islet of Doca.* ● **MDP**

13. Crataegus L.

More or less deciduous, spiny shrubs or small trees with slender twigs. Leaves simple, ovate-deltate, pinnatifid; stipules persistent, leaf-like. Inflorescence corymbose; flowers 5-merous; epicalyx absent; petals white. Fruit a small globose fleshy pome.

1. C. monogyna Jacq., *Fl. austriac.* **3**: 50 (1775).
Spines 7-14 mm, ± patent. Leaves 1-6 × 1-5 cm; lobes oblong, dentate at apex; stipules 5-14 mm, compoundly dentate. Fruit ripening red.
Fl. ?. *Probably introduced to Madeira from Europe; in scrub and laurisilva in the upper Ribeira da Janela valley by the road to Rabaçal and at Vinte-cinco Fontes; also recorded as naturalized at Monte north of Funchal.* **M**

Madeiran individuals have been compared with subsp. **brevispina** (Kunze) Franco in *Collnea bot. Barcinone* **7**: 463 (1968), which occurs in the Iberian peninsula and the Balearics. However, fertile parts are not yet known from Madeira, and subspecific identity cannot properly be ascertained in their absence.

XLVIII. LEGUMINOSAE
[FABACEAE]
N.J. Turland

Trees, shrubs or herbs. Leaves usually alternate, simple to bipinnate, stipulate. Flowers usually hermaphrodite and 5-merous, very often zygomorphic; sepals usually united into a tube; petals free or somewhat united; stamens usually 10. Fruit a dehiscent 2-valved or indehiscent 1-valved, occasionally segmented pod (*legume*); seeds often with a small, hard appendage (*strophiole*).

1. Trees or shrubs 2
- Annual or perennial herbs, sometimes woody-based 12

2. Plant spiny 3
- Plant unarmed 5

3. Leaves of adult plants reduced to persistent phyllodes terminating in spines .. **7. Ulex**
- Leaves not reduced to spinose phyllodes 4

4. Leaves simple .... **16. Ononis**
- Leaves bipinnate .... **2. Acacia**

5. Leaves bipinnate or reduced to simple phyllodes; flowers regular, small, crowded into cylindrical spikes or globose capitula 6
- Leaves simple, trifoliolate or pinnate; corolla weakly to strongly zygomorphic, not as above 7

6. Trees or shrubs to 30 m; flowers yellow; stamens numerous (more than 10); legume slightly to strongly compressed .... **2. Acacia**
- Shrub to *c.* 2 m; flowers white or off-white; stamens 10; legume flat . **3. Leucaena**

7. Leaves pinnate 8
- Leaves simple or trifoliolate 9

8. Leaves imparipinnate; flowers strongly zygomorphic, with petals dissimilar; legume disarticulating into segments at maturity .... **23. Coronilla**
- Leaves paripinnate; flowers weakly zygomorphic, with petals similar; legume not disarticulating at maturity .... **1. Senna**

9. All leaves simple .... **6. Genista**
- At least some leaves trifoliolate 10

10. Twigs often rush-like, with leaves falling early; usually at least some leaves simple and sessile .... **4. Cytisus**
- Twigs never rush-like, persistently leafy; all leaves trifoliolate and petiolate 11

11. Calyx and legume prominently covered with dark glandular papillae **8. Adenocarpus**
- Calyx and legume without dark glandular papillae .... **5. Teline**

12. Leaves consisting of a tendril only (but with large leaf-like stipules) .. **15. Lathyrus**
- Leaves not consisting of a tendril only 13

13. Leaves simple 14
- Leaves compound 15

14. Spiny perennial; corolla pink and white .... **16. Ononis**
- Unarmed annual; corolla yellow or orange .... **25. Scorpiurus**

15. Leaves digitate or appearing so, with 5 or more leaflets	16
- Leaves trifoliolate or pinnate	17
16. Leaves long-petiolate	**9. Lupinus**
- Leaves sessile	**20. Lotus**
17. All leaves trifoliolate or appearing so, sometimes with leaflet-like stipules or with genuine leaflets at base of petiole	18
- Leaves obviously pinnate	26
18. Legume spirally coiled	**18. Medicago**
- Legume not spirally coiled	19
19. Plant glandular-hairy, at least above	**16. Ononis**
- Plant not glandular-hairy	20
20. Plant dark-glandular-punctate, at least on basal part of legume	21
- Plant not dark-glandular-punctate	22
21. Leaflets entire; inflorescence a subglobose head; legume beaked, with beak exserted from calyx	**11. Bituminaria**
- Leaflets sinuate-crenate; inflorescence a cylindrical raceme; legume not beaked, completely enclosed within calyx	**12. Cullen**
22. Corolla sometimes persistent in fruit; filaments of all or 5 of the stamens dilated at apex; legume usually enclosed within calyx	**19. Trifolium**
- Corolla not persistent in fruit; filaments not dilated at apex; legume nearly always longer than calyx	23
23. Inflorescence a 5- to 90-flowered raceme, elongating after flowering	**17. Melilotus**
- Inflorescence a compact 1- to 20-flowered head, not elongating after flowering	24
24. Legume not or only slightly longer than calyx	**20. Lotus**
- Legume obviously longer than calyx	25
25. Corolla 2-3 mm, yellow; legume reniform, biconvex, less than twice as long as wide	**18. Medicago**
- Corolla larger, variously coloured; legume cylindrical or linear, sometimes curved, many times longer than wide	**20. Lotus**
26. Leaves paripinnate, often terminating in a tendril	27
- Leaves imparipinnate, without a tendril	31
27. Stems winged	**15. Lathyrus**
- Stems angled but not winged	28
28. Leaflets parallel-veined	**15. Lathyrus**
- Leaflets pinnately veined	29
29. Calyx-teeth all equal and at least twice as long as tube	**14. Lens**
- At least 2 calyx-teeth less than twice as long as tube	30

30.	Style pubescent all round or on lower side only, or glabrous	**13. Vicia**
-	Style pubescent on upper side only	**15. Lathyrus**

31.	Legume segmented, the segments disarticulating at maturity	32
-	Legume not segmented	33

32. Legume flat, the segments with a central ± circular sinus opening on convex edge . . . . **24. Hippocrepis**
- Legume compressed or cylindrical, the segments entire . . . . **22. Ornithopus**

33. Perennial; inflorescence terminal; legume enclosed within persistent the membranous and somewhat inflated calyx . . . . **21. Anthyllis**
- Annuals; inflorescence axillary; legume much longer than calyx . . . . **10. Astragalus**

Subfam. CAESALPINIOIDEAE

Flowers somewhat zygomorphic; sepals almost free; petals similar, free, the adaxial petal innermost, overlapped by lateral petals; stamens usually not more than 10, free.

1. Senna Mill.

Evergreen shrubs or small trees. Leaves paripinnate; stipules often falling early. Inflorescence of axillary racemes together forming a terminal, leafy panicle. Sepals equal or the inner larger. Petals usually yellow, becoming darker-veined on drying. Stamens usually 9 or 10, the 3 adaxial usually sterile, the others fertile (usually 4 median and 2 or 3 abaxial). Legume variable in shape, usually indehiscent.

Literature: H.S. Irwin & R.C. Barneby, The American *Cassiinae*. A synoptical revision of Leguminosae tribe *Cassieae* subtribe *Cassiinae* in the New World, in *Mem. N. Y. bot. Gdn* **35** (1982).

1. Leaflets 30-90 × 10-30 mm, ovate to lanceolate, acute to acuminate at apex . . . . **3. septemtrionalis**
- Leaflets 8-43 × 5-20 mm, elliptic to broadly so or obovate to oblanceolate, obtuse to rounded at apex 2

2. Leaflets in 2-3 pairs; inflorescence to 7 cm (including peduncle to 3.5 cm); pedicels 1-5 mm . . . . **1. bicapsularis**
- Leaflets in 4-5 pairs; inflorescence to 12 cm (including peduncle to 6 cm); pedicels 13-30 mm . . . . **2. pendula**

1. S. bicapsularis (L.) Roxb., *Fl. ind.* ed. 1832, **2**: 342 (1832).
Cassia bicapsularis L.
Glabrous shrub to 9 m, often climbing, scrambling or diffuse. Leaflets in 2-3 pairs, 8-25 × 5-13 mm, elliptic to broadly so or obovate, obtuse to rounded at apex, finely mucronate. Inflorescence with racemes to 7 cm (including peduncle to 3.5 cm), with 1-several flowers; pedicels 1-5 mm. Sepals unequal, 6-10 mm, narrowly to broadly elliptic or obovate, membranous at margin. Petals ± equal, 10-12 mm, obovate. Legume 9-11 × 1.2-1.4 cm, narrowly oblong, compressed, straight or curved, ripening dark brown.
Fl. all year. *Introduced; naturalized among* Opuntia *on the sea cliffs and on roadsides along the southern coast of eastern Madeira, from Câmara de Lobos to São Gonçalo; also cultivated.*
M

The above description refers to var. **bicapsularis**, which is native to the Caribbean region and to which Madeiran plants belong.

2. S. pendula (Humb. & Bonpl. ex Willd.) Irwin & Barneby in *Mem. N. Y. bot. Gdn* **35**: 378 (1982).
Cassia bicapsularis sensu auct. mad. pro parte, non L. (1753); *C. pendula* Humb. & Bonpl. ex Willd.
Like **1** but leaflets in 4-5 pairs, larger, 10-43 × 7-20 mm, obovate to oblanceolate, rounded at apex, not mucronate; racemes longer, to 12 cm (including peduncle to 6 cm); pedicels much longer, 13-30 mm; flowers larger, the sepals 7-14 mm, the petals 13-20 mm.
Fl. probably all year. *Introduced; cultivated for ornament in Madeira in Funchal and naturalized at Campanàrio to the west.* **M**

The above description refers to var. **glabrata** (Vogel) Irwin & Barneby in *Mem. N. Y. bot. Gdn* **35**: 382 (1982), which is native to Paraguay and Brazil and to which Madeiran plants belong. *Senna pendula* is similar and related to **1** and has been misidentified as that species in Madeira.

3. S. septemtrionalis (Viv.) Irwin & Barneby in *Mem. N. Y. bot. Gdn* **35**: 365 (1982).
Cassia laevigata Willd.
Like **1** but shrub or small tree to 6 m; leaflets in 3-5 pairs, much larger, 30-90 × 10-30 mm, ovate to lanceolate, acute to acuminate at apex; racemes longer, to 12 cm (including peduncle to 6 cm); pedicels much longer, 13-25 mm; legume 4-8 × 1-1.5 cm.
Fl. all year. *Introduced; cultivated for ornament and naturalized on roadsides, in vineyards and waste areas in the Funchal region, westwards to Campanário, and at São Vicente on the northern coast. Native from Mexico to Costa Rica.* **M**

Several additional species are cultivated for ornament in Madeira, including the following:

S. × floribunda (Cav.) Irwin & Barneby (*Cassia floribunda* Cav.) is the hybrid *S. multiglandulosa* × *S. septemtrionalis*. Native to Mexico, it is like **3** but with leaflets oblong-elliptic and very sparsely hairy.

S. multiglandulosa (Jacq.) Irwin & Barneby (*Cassia tomentosa* L. f.), native from Mexico to Bolivia, is densely tomentose, with narrowly oblong leaflets smaller than those of **3**, in 5-7 pairs.

S. multijuga (Rich.) Irwin & Barneby (*Cassia multijuga* Rich.), native from Mexico to Bolivia and Brazil, is a large shrub or small tree with leaves long and narrow, the leaflets similar to those of *S. multiglandulosa* but in 9-14 pairs and only sparsely hairy; the inflorescence is a large and showy, pyramidal panicle.

S. didymobotrya (Fresen.) Irwin & Barneby (*Cassia didymobotrya* Fres.), native to tropical Africa, is like *S. multijuga* but has inflorescences consisting of erect, elongate racemes.

Ceratonia siliqua L., *Sp. pl.* **2**: 1026 (1753) [*Alfarrôba, Alfarrobeira*] is an evergreen tree with leaves paripinnate, leaflets in 2-5 pairs, broadly elliptic to suborbicular, coriaceous, dark green, glabrous, flowers without petals, borne in short catkin-like racemes from late summer to winter, and legumes to 20 cm, linear-oblong, thick, compressed, pendent, initially green, ripening brownish-violet. Native to the Mediterranean region, it is planted in Madeira and Porto Santo and may also be locally established away from cultivation. **MP**

Various other species within this subfamily are cultivated for ornament in Madeira, including: **Caesalpinia decapetala** (Roth) Alston (*C. sepiaria* Roxb.), **Cercis siliquastrum** L. [*Arvore de Judas, Olaia*] and **Gleditsia triacanthos** L. (*G. triacanthos* var. *armata* Lowe; *G. triacanthos* var. *inermis* DC.) [*Alfarrôba, Alfarrobeira*].

Subfam. MIMOSOIDEAE

Flowers regular; sepals and petals united into tubes; stamens 10-numerous, usually free.

2. Acacia Mill.

Evergreen or deciduous trees or shrubs. Leaves bipinnate in juvenile state, remaining so in adult state or reduced to simple, entire phyllodes; stipules spinescent or rudimentary. Flowers small, orange-yellow to pale yellow, crowded into cylindrical spikes or globose capitula themselves arranged in racemes, panicles or axillary clusters; stamens numerous (more than 10), usually free, conspicuous. Legume variable in shape, usually dehiscent.

1. Stipules on older branches spinescent; inflorescence of 1-3 capitula in axils of the older leaves .... **1. farnesiana**
- Stipules inconspicuous, not spinescent; inflorescence a profuse panicle of capitula .... **2. mearnsii**

1. A. farnesiana (L.) Willd., *Sp. pl.* **4**: 1083 (1806).
Vachellia farnesiana (L.) Wight & Arn. [*Arôma amarelo*]
Deciduous shrub or small tree to 7 m. Leaves bipinnate, glabrous; pinnae in 2-8 pairs; pinnules in 10-25 pairs, linear-oblong; stipules on older branches forming straight spines to 3 cm. Inflorescence of 1-3 capitula in axils of older leaves; capitula 10-20 mm in diameter; peduncles 1-3.5 cm, pubescent. Flowers bright orange-yellow, fragrant. Legume 50-90 × 9-16 mm, linear, slightly compressed, straight or curved, not constricted between seeds, ripening black or dark brown, glabrous.
Fl. all year. *Introduced from the Dominican Republic; cultivated for ornament in gardens in Madeira and naturalized at Gorgulho west of Funchal; recorded also as occurring in Porto Santo but without further data.* **MP**

2. A. mearnsii De Wild., *Pl. Bequaert.* **3**: 61 (1925).
Unarmed, evergreen tree to 15 m. Leaves bipinnate, villous when young; pinnae in 8-20 pairs; pinnules in 20-60 pairs, 1-3 mm, linear; stipules inconspicuous, not spinescent. Inflorescence a profuse panicle, consisting of capitula each 5-7 mm in diameter. Flowers pale yellow, fragrant. Legume 30-90 × 5-8 mm, linear, strongly compressed, straight or curved, constricted between seeds and appearing segmented, ripening blackish-brown, minutely white-pubescent.
Fl. IV - X. *Introduced and extensively planted for forestry and now frequently naturalized in eastern Madeira, south and east of the high peaks, between c. 500 and 900 m. Native to SE Australia and Tasmania.* **M**

A. dealbata Link, *Enum. hort. berol. alt.* **2**: 445 (1822) [*Acacia branca*] is like **2** but taller, to 30 m, with pinnules longer, 3-4 mm, and legumes 10-12 mm wide, not or scarcely constricted between seeds. Native to SE Australia and Tasmania, it is planted for ornament in south-eastern Madeira, from 300-1100 m, and is possibly locally naturalized. The foliage is used as cattle fodder. **M**

Several additional species are planted in Madeiran parks, gardens and as street trees or for forestry, including **A. longifolia** (Andrews) Willd., **A. melanoxylon** R. Br. [*Acacia*], **A. retinodes** Schldl. [*Acacia*] and **A. sophorae** (Labill.) R. Br., all of which have adult leaves reduced to simple, entire, flattened phyllodes. A further species, **A. verticillata** (L'Hér.) Willd., has phyllodes narrow and needle-like, sharply pointed at tips, resembling the leaves of a conifer.

3. Leucaena Benth.

Like *Acacia* but evergreen shrubs to *c.* 2 m; leaves always bipinnate; stipules inconspicuous, not spinescent, deciduous; inflorescence a terminal, often leafy raceme with 1-4 globose capitula at each node; flowers white or off-white; stamens 10, free; legume 8-15 × 1-1.5 cm, linear, flat, straight, not constricted between seeds, dehiscent.

1. L. leucocephala (Lam.) de Wit in *Taxon* **10**: 54 (1961). [*Arôma branco*]
Acacia leucocephala (Lam.) Link
Plant suckering, sprawling. Pinnae in 4-8 pairs; pinnules in 10-20 pairs, narrowly oblong, acute. Capitula 10-18 mm in diameter; flowers white or off-white. Legume ripening dark brown. Fl. all year. *Introduced and cultivated for ornament in gardens in Madeira; naturalized and sometimes forming thickets along levadas, in rocky places and on waste ground on the western side of Funchal, up to* c. *150 m. Native to southern U.S.A.* **M**

Albizia lophantha (Willd.) Benth. in *Lond. J. Bot.* **3**: 86 (1844) (*Acacia lophantha* Willd.; *Albizia distachya* (Vent.) Macbr.) [*Acacia*] is an evergreen, suckering shrub or small tree to 5 m with leaves bipinnate, inflorescences axillary, racemose, shortly cylindrical, stamens very long, forming a dense mass, yellowish or yellowish white and legumes oblong, straight, coriaceous with ribbed margins, ripening brown. Native to Australia, it is planted in Madeira and Porto Santo and may also be locally established away from cultivation. **MP**

The somewhat similar **A. julibrissin** Durazz. (*Acacia julibrissin* (Durazz.) Willd.), with pink stamens, is cultivated for ornament in Madeira.

Subfam. LOTOIDEAE

Flowers strongly zygomorphic; sepals united into a tube; petals dissimilar, the adaxial petal (*standard*) outermost, the 2 lateral petals (*wings*) free, the 2 lower petals innermost and usually partly adhering to each other with interlocking marginal hairs to form the *keel*; stamens 10, rarely 5, with all of the filaments united (*monadelphous*), 9 united and 1 free (*diadelphous*), or all free.

Various members of this subfamily are cultivated in Madeira, either for ornament or for their edible legumes and seeds, including: **Dipogon lignosus** (L.) Verdc. (*Dolichos lignosus* L.), **Erythrina crista-galli** L., **Glycine max** (L.) Merr., **Phaseolus coccineus** L. (*P. multiflorus* Lam.) [*Feijoa*], **P. vulgaris** L. (including numerous infraspecific taxa) [*Feijoeiro*], **Pisum sativum** L. (*P. arvense* L.; *P. sativum* var. *saccharatum* Ser.) [*Ervilha*], **Robinia hispida** L., **R. pseudoacacia** L. [*Acacia*] and **Wisteria sinensis** (Sims) Sweet [*Cacho roxo, Lilaz*].

4. Cytisus L.
M.J. Cannon & N.J. Turland

Unarmed shrubs. Twigs often rush-like with leaves falling early. Leaves sessile or petiolate, simple or trifoliolate; stipules very small or absent. Flowers axillary, forming leafy racemes; calyx bilabiate; teeth short; corolla usually yellow. Legume oblong-ovate to oblong or narrowly so, or elliptic, dehiscent; seeds strophiolate.

1. Young twigs 8- to 10-angled; calyx minutely appressed-sericeous; legume compressed but somewhat inflated, with valves densely villous . . . . **1. striatus**
- Young twigs 5-angled; calyx glabrous; legume strongly compressed, with valves glabrous except on margins . . . . **2. scoparius**

1. C. striatus (Hill) Rothm. in *Feddes Reprium Spec. nov. veg..* **53**: 149 (1944).
C. pendulinus L. f.
Shrub or small tree to 3 m. Young twigs cylindric, striate, 8- to 10-angled, pubescent, later glabrescent. Lowest leaves trifoliolate, petiolate; upper leaves trifoliolate or simple, sessile; leaflets 2.5-10 × 1-2 mm, narrowly oblanceolate to narrowly lanceolate, glabrous above, sericeous or

villous beneath. Flowers solitary or paired, rarely in threes; calyx minutely appressed-sericeous; corolla 15-20 mm, yellow. Legume 15-35 × 8-12 mm, oblong-ovate to oblong or elliptic, cuspidate at apex, compressed but somewhat inflated, with valves densely villous.

Fl. IV - VII. *Introduced; locally established in gullies and on the margins of woodland in south-eastern Madeira, from São Martinho to Poiso and Camacha, and at Porto do Moniz, up to 1400 m; also cultivated. Native to Portugal, Spain and Morocco.* **M**

2. C. scoparius (L.) Link, *Enum. hort. berol. alt.* **2**: 241 (1822). [*Giesta*]
Sarothamnus scoparius (L.) Wimmer ex W.D.J. Koch
Shrub to 2.5 m. Young twigs 5-angled, pubescent, later glabrous. Lowest leaves trifoliolate, petiolate or subsessile, uppermost leaves sometimes simple, sessile; leaflets 2.5-10 × 1-5 mm, narrowly elliptic to obovate, glabrous or slightly pubescent beneath. Flowers solitary or paired; calyx glabrous; corolla *c.* 20 mm, yellow, occasionally white or with wings crimson. Legume 30-65 × 8-12 mm, narrowly oblong, cuspidate at apex, strongly compressed, with valves glabrous except on margins, ripening black.

Fl. I - VIII. *Probably an early introduction (before 1800); now very common in Madeira, especially in the central and higher regions, up to 1850 m; rare in Porto Santo and Deserta Grande. Used for fuel, stakes, fertilizing the soil and as a source of fine withies.* **MDP**

The above description refers to subsp. **scoparius**, which is widely distributed in Europe and to which Madeiran plants belong.

The hybrid *C. balansae* subsp. *europaeus* (G. López & Jarvis) Muñoz Garmendia (*C. purgans* auct.) × *C. multiflorus* (L'Hér.) Sweet has been recorded from Madeira as an introduction, from the Paúl da Serra and Camacha, but its presence on the island should be regarded as doubtful. The record from the Paúl da Serra is possibly referable to *C. multiflorus*, native to Portugal and Spain, which was once recorded (in 1926) as evidently introduced at Porto do Moniz in north-western Madeira (as *C. lusitanicus* Quer. ex Willk.). However, further collections are required before the exact identity of these plants can be ascertained. The record from Camacha most probably belongs to **1**, which is known with certainty from that area.

Chamaecytisus proliferus (L.f.) Link (*Cytisus proliferus* L.f.) is a shrub with trifoliolate leaves and white flowers, borne in rounded heads of up to 12. Endemic to the Canaries, it is occasionally cultivated for ornament in Madeira.

5. Teline Medik.
M.J. Cannon & N.J. Turland

Unarmed, evergreen shrubs or small tree. Leaves petiolate, trifoliolate, persistent; stipules conspicuous and persistent or inconspicuous and deciduous. Inflorescence racemose, usually terminal. Calyx tubular-campanulate, bilabiate; upper lip bifid; lower lip with 3 distinct or indistinct teeth. Corolla yellow. Stamens monadelphous, 5 long and 5 short. Legume narrowly oblong, strongly compressed, dehiscent; seeds 2-8, strophiolate.

Literature: P.E. Gibbs, & I. Dingwall, A revision of the genus *Teline*, in *Bolm Soc. broteriana* **45**: 269 - 316 (1971).

1. T. maderensis Webb & Berthel., *Hist. nat. Iles Canaries* **3**(2,2): 37 (1842). [*Piorno*]
Genista maderensis (Webb & Berthel.) Lowe; *G. paivae* Lowe; *T. maderensis* var. *paivae* (Lowe) Arco; *T. paivae* (Lowe) Gibbs & Dingwall
Plant to 6 m. Indumentum of whitish- to brownish-villous, ± patent hairs, or of silvery-sericeous, ± appressed hairs. Leaflets 5-25 × 2.5-11 mm, obovate to oblanceolate or elliptic, obtuse to rounded or retuse at apex, often mucronate; mucro to 1.5 mm; petiole usually at least half as long as leaflets; stipules 1.5-6 mm, persisting after leaves have fallen (making twigs appear scaly), or deciduous. Inflorescence terminal, 1.5-7 cm, 3- to 20-flowered. Calyx 4-8 mm. Corolla 9-16

mm. Legume 20-40 × 5-8 mm, narrowed at base, cuspidate at apex, often widely and irregularly sinuate at margins. 2n = 48. **Plate 19**
Fl. IV - X. *An endemic frequent in laurisilva, rocky wooded ravines and maritime cliffs in the northern half of Madeira, up to 1500 m, from Santo da Serra westwards to Seixal, extending around the coast to Fajã da Ovelha; also at Curral das Freiras and on Deserta Grande; recorded in error from the Salvage Islands.* ● **MD**

The plants of the Madeiran interior have an indumentum of whitish- to brownish-villous, ± patent hairs, leaflets obovate to oblanceolate or elliptic, the apex obtuse, with a mucro 0.3-1.5 mm, stipules 2.5-6 mm, persistent, and the lower calyx-lip with 3 usually distinct teeth. The plants on maritime cliffs along the coast from Fajã da Ovelha to São Vicente, as well as on Deserta Grande, have an indumentum of silvery-sericeous, ± appressed hairs, leaflets mostly obovate, the apex rounded to retuse, not or only very shortly mucronate, stipules only 1.5-3 mm, deciduous, and the lower calyx-lip often only minutely and indistinctly toothed. These coastal plants have been called *T. paivae* (Lowe) Gibbs & Dingwall in *Bolm Soc. broteriana* **45**: 288 (1971), but do not seem worthy of separation, since the plants occupying the northern coastal strip from Seixal eastwards to São Jorge constitute a range of intermediates.

T. monspessulana (L.) K. Koch, *Dendrologie* **1**: 30 (1869) has been collected at Monte, north of Funchal. It closely resembles **1**, with whitish, ± patent indumentum, but differs in its short petioles, much less than half as long as leaflets, its inflorescences, which are borne on short lateral shoots, and its smaller legumes, 20-30 × 4-5 mm, which are closely and regularly sinuate at margins. One collection is from a park, and all were made during the last 25 years, which suggests the species is probably a recent introduction to Madeira, planted as an ornamental. It is native in the Azores and the Mediterranean region. **M**

6. Genista L.
M.J. Cannon & N.J. Turland

Unarmed shrubs. Twigs sometimes rush-like with leaves falling early. Leaves subsessile, simple; stipules minute or absent. Inflorescence terminal, shortly racemose. Calyx tubular-campanulate, bilabiate; upper lip bifid; lower lip 3-toothed. Corolla yellow. Stamens monadelphous, 5 long and 5 short. Legume narrowly oblong, strongly compressed, dehiscent; seeds 3-5, with a minute but distinct strophiole.

Literature: P.E. Gibbs, Taxonomic notes on some Canary Island and North African species of *Cytisus* and *Genista*, in *Lagascalia* **4**: 33 - 41 (1974).

3. G. tenera (Jacq. ex Murray) Kuntze, *Revis. gen. pl.* **1**: 190 (1891).
[*Giesta de piorno. Piorno*]
G. virgata (Aiton) DC., non Lam. (1788).
Plant to 2.5 m. Twigs erect or spreading. Indumentum silvery-sericeous, with hairs short, appressed to ± patent. Leaves 2.5-14 × 0.5-4.5 mm, linear-lanceolate to obovate, acute to rounded at apex, mucronulate; stipules minute or absent. Inflorescence 1-3(-5) cm, 1- to 9-flowered. Calyx 4-7 mm. Corolla 10-15 mm. Legume 15-35 × 4-6.5 mm, narrowed at base, acuminate at apex, often sinuate at margins. 2n = 48. **Plate 19**
Fl. all year but mainly III - VII. *A common endemic of dry sunny cliffs and ravines, especially in southern Madeira, from sea-level to 1700 m.* ● **M**

Spartium junceum L., *Sp. pl.* **2**: 708 (1753) is an unarmed shrub to 3 m, with branches cylindrical, glabrous, leaves simple, subsessile, glabrous above, sericeous beneath, falling early, flowers yellow, fragrant, with membranous calyces, borne in lax terminal racemes, legumes linear or lanceolate, flattened, glabrous at least at tip, ripening blackish. Native in SW Europe and the Mediterranean region, it is frequently planted in Madeira and is perhaps becoming naturalized. **M**

7. Ulex L.

M.J. Cannon & N.J. Turland

Spiny shrubs. Leaves of young plants trifoliolate, without stipules, those of mature branches and twigs reduced to persistent phyllodes terminating in spines. Flowers solitary or clustered in axils of phyllodes, with 2 small bracteoles at base of calyx; corolla yellow; calyx divided to base into 2 lips, the upper lip 2-toothed, the lower 3-toothed; stamens monadelphous. Legume scarcely exserted from calyx, dehiscent; seeds 1-several, strophiolate.

1. Spines stout, to 25 mm; bracteoles 1.5-3.5 mm wide; calyx 10-17 mm, with long, $\pm$ patent hairs .... **1. europaeus**
- Spines slender, to 15 mm; bracteoles c. 0.5 mm wide; calyx 6-10 mm, with short appressed hairs .... **2. minor**

1. U. europaeus L., *Sp. pl.* **2**: 741 (1753). [*Carqueja*]
Plant 60-150 cm, with main stems ascending or erect, becoming bare at base. Young twigs and spines $\pm$ glaucous; spines stout, to 25 mm. Bracteoles 1.5-3.5 mm wide. Calyx 10-17 mm, yellowish, with long, $\pm$ patent hairs. Corolla 12-18 mm, with wings longer than keel. Legume 14-19 mm, villous; seeds 4-6.
Fl. all year but mainly I - VI. *Introduced in the early 19th century; now common in Madeira along roadsides, in forestry plantations, steep gullies and mountain pastures, up to c. 1300 m. Native mainly to W. Europe.* **M**

The above description refers to subsp. **europaeus**, which occurs mainly in western Europe and to which Madeiran plants belong.

2. U. minor Roth, *Catal. bot.* **1**: 83 (1797).
Plant 10-100 cm, with main stems often procumbent. Young twigs and spines not glaucous; spines slender, to 15 mm. Bracteoles c. 0.5 mm wide. Calyx 6-10 mm, yellowish, with short, appressed hairs. Corolla 7-10 mm, with wings and keel equal. Legume c. 10 mm, sparsely villous; seeds 2-6.
Fl. IX - X. *Recently introduced; cultivated and locally established; occurring in southeast Madeira along the levada between Monte and Camacha and at Poiso to the north; recorded also as occurring in Porto Santo but without further data. Native from Britain to Portugal, introduced to the Azores.*
MP

8. Adenocarpus DC.

M.J. Cannon & N.J. Turland

Unarmed shrub. Leaves petiolate, trifoliolate, persistent; axils with dense clusters of small leaves (undeveloped lateral shoots). Inflorescence a lax panicle of many-flowered, lax, terminal racemes. Calyx tubular, bilabiate, with dark glandular papillae; upper lip with 2 prominent teeth; lower lip longer that upper, with 3 smaller teeth. Corolla yellow. Stamens monadelphous. Legume narrowly oblong, strongly compressed, prominently covered with dark glandular papillae but otherwise glabrous; seeds usually 2-8.

1. A. complicatus (L.) Gay in Durieu, *Pl. hispano-lusit.* Sect. 1, *Astur.*: no. 350 (1836) [in sched.]. [*Codeço, Codeso, Tudesco*]
A. divaricatus Sweet
Plant to 4 m. Twigs interlacing, stiff, slender, pale. Indumentum of short, $\pm$ patent hairs. Main leaflets 6-18 $\times$ 2-7 mm, often folded inwards along midrib. Inflorescence with racemes to 20 cm. Calyx 5-9 mm, often pubescent on tube. Corolla 10-15 mm, pubescent. Legume (10-)20-40 $\times$ 3.5-6 mm.

Fl. III - XII. *Ravines, roadsides, sometimes forming thickets; restricted to the area above Funchal, ascending to Curral das Freiras and Poiso; recorded also from Santo da Serra, as possibly planted.* **M**

The above description refers to subsp. **complicatus**, which occurs in SW Europe and the Mediterranean region and to which Madeiran plants belong.

9. Lupinus L.
M.J. Cannon & N.J. Turland

Hairy annual herbs. Leaves long-petiolate, digitate, with several oblanceolate to linear leaflets; stipules ± linear. Inflorescence a terminal raceme, with flowers arranged in whorls. Calyx bilabiate, deeply divided; upper lip 2-partite to shallowly 2-dentate; lower lip 3-dentate to subentire. Corolla yellow, blue or ± white. Stamens monadelphous. Legume ± compressed, somewhat contracted between seeds, villous, dehiscent.

1. Leaflets 6-11 × as long as broad, linear; corolla blue . . . . **2. angustifolius**
- Leaflets 2-7 × as long as broad, oblanceolate to narrowly so; corolla white tinged with blue, or yellow 2

2. Leaflets hairy on both surfaces; corolla yellow . . . . **1. luteus**
- Leaflets glabrous above, hairy beneath; corolla white tinged with blue . . . . **3. albus**

1. L. luteus L., *Sp. pl.* **2**: 722 (1753). [*Tremoço amarello*]
Plant to 60 cm. Leaflets of upper leaves 20-40 × 4-8 mm, 5-7 times as long as broad, narrowly oblanceolate, obtuse to acute at apex, mucronate, hairy on both surfaces; stipules of upper leaves linear. Calyx 7-9 mm. Corolla 12-14 mm, yellow. Legume 40-45 × 10-12 mm, oblong, aristate.
Fl. III - VI. *Introduced; cultivated for fodder and its seeds and naturalized in a few places to the east of Funchal. Native to the western Mediterranean region; cultivated and naturalized elsewhere.* **M**

2. L. angustifolius L., *Sp. pl.* **2**: 721 (1753). [*Tremoço*]
Plant to 60 cm. Leaflets 10-40 × 2-3.5 mm, 6-11 times as long as broad, linear, rounded to truncate or retuse at apex, minutely mucronate, sparsely hairy to subglabrous; stipules linear. Calyx *c*. 6 mm. Corolla 11-13 mm, blue. Legume 25-40 × 10-14 mm, narrowly elliptic to oblong, cuspidate.
Fl. III - VI. *Introduced; rarely naturalized on field-margins and in grassy places between Câmara de Lobos and Funchal. Native to the Mediterranean region.* **M**

3. L. albus L., *Sp. pl.* **2**: 721 (1753). [*Tremoço*]
Plant to 60 cm. Leaflets 15-50 × 5-20 mm, 2-3.5 times as long as broad, oblanceolate to narrowly so, obtuse to rounded at apex, mucronate, glabrous above, hairy beneath; stipules linear-lanceolate, long-acuminate. Calyx 8-10 mm. Corolla 14-16 mm, white tinged with blue. Legume at least 50 × 10 mm, oblong.
Fl. III - VII. *Introduced; widely grown in the lowlands of Madeira as a crop or for green manure; sometimes escaping or persisting as a relic of cultivation on field-margins. The seeds are used as an ingredient of soups.* **M**

The above description refers to subsp. **albus** (*L. albus* subsp. *termis* (Forssk.) Cout.; *L. termis* Forssk.), which occurs in central Europe and the Mediterranean region and to which Madeiran plants belong.

10. Astragalus L.
M.J. Cannon & N.J. Turland

Annual herbs. Leaves imparipinnate, stipulate; stipules entire, usually membranous. Inflorescence axillary, pedunculate, shortly racemose. Calyx tubular or campanulate, with 5 teeth. Stamens 10, diadelphous, sometimes only 5 fertile. Legume much longer than calyx, variously shaped, glabrous or hairy, dehiscent or not.

1. Calyx 2.5-4 mm; corolla 4-6 mm, bluish white; legume flat, with margins prominently serrate . . . . **3. pelecinus**
- Calyx 4-9 mm; corolla 8-11 mm, cream to yellow or greenish yellow; legume trigonous in cross-section, the edges entire 2

2. Plant appearing green, with leaflets glabrous above, sparsely hairy beneath; peduncles much shorter than to almost equalling leaves; legume oblong, straight . . **1. boeticus**
- Plant appearing silvery, with leaflets almost glabrous above, densely appressed-white-hairy beneath; peduncles usually equalling or exceeding leaves; legume sickle-shaped to semicircular . . . . **2. solandri**

1. A. boeticus L., *Sp. pl.* **2**: 758 (1753).
Plant 10-70 cm, procumbent or diffusely straggling, appearing green. Leaves 3.5-16 cm; leaflets in 8-16 pairs, 4-23 × 1.5-9 mm, oblong, truncate to emarginate at apex, glabrous above, sparsely hairy beneath; stipules 4-13 mm. Peduncles 0.5-10 cm, much shorter than to almost equalling leaves; racemes dense, with up to *c*. 10 flowers. Calyx 6.5-9 mm, black-hairy. Corolla *c*. 10 mm, cream to yellow. Legume (12-)20-40 × 6-10 mm, oblong, straight, trigonous in cross-section, with a broad channel with rounded and thickened edges on dorsal side, sparsely hairy; beak stoutly hooked.
Fl. IV - V. *Probably introduced, at least in Madeira, where it is found only on dry sunny slopes at Gorgulho and Garajau either side of Funchal; locally common in Porto Santo, growing in pastures, along roadsides and on waste ground. Native from the Mediterranean region to Iran.*
MP

2. A. solandri Lowe in *Hooker's J. Bot.* **8**: 294 (1856).
?*A. hamosus* sensu auct. mad., non L. (1753)
Plant 4-30(-45) cm, procumbent to diffuse, appearing silvery. Leaves 3-10 cm; leaflets in 8-15 pairs, 3-14 × 1.5-7 mm, elliptic-oblong, slightly retuse to distinctly emarginate at apex, almost glabrous above, densely appressed-white-hairy beneath; stipules 2-9 mm. Peduncles 2-8.5 cm, usually equalling or exceeding leaves; racemes lax, at least below, with up to 13 flowers. Calyx 4-6 mm, densely black-hairy. Corolla 8-11 mm, yellow or greenish yellow. Legume 15-30 × 2-3 mm, sickle-shaped to semicircular, trigonous in cross-section, with a broad channel with raised but not thickened edges on dorsal side, shortly and ± densely appressed-white-hairy; beak finely hooked. **Plate 19**
Fl. III - VI. *In Madeira a very rare plant confined to the Ilhéus de Ponta de São Lourenço; much more common and widespread in Porto Santo, growing in stony pastures, on sunny rocky slopes and sandy coastal ground.* **MP**

A. solandri is otherwise known only from the Canaries and the Atlantic coast of Morocco.

A. hamosus L. has been recorded from Porto Santo, but almost certainly in error for **2**. It differs in having peduncles much shorter than the subtending leaves.

3. A. pelecinus (L.) Barneby in *Mem. N. Y. bot. Gdn* **13**: 26 (1964).
Biserrula pelecinus L.; *B. pelecinus* var. *glabra* Lowe, var. *pubescens* Lowe
Plant 4-30(-45) cm, prostrate or procumbent, appearing green to greyish green. Leaves 1.5-8

cm; leaflets in 7-13 pairs, 2-9 × 1-4 mm, elliptic-oblong, emarginate at apex, glabrous to white-hairy on both surfaces; stipules 1-4 mm. Peduncles 1-4.5 cm, shorter than to equalling leaves; racemes moderately dense, with up to 6 flowers. Calyx 2.5-4 mm, black-hairy. Corolla 4-6 mm, bluish white. Legume 10-25 × 4-8 mm, oblong, straight, flat, shortly appressed-white-hairy to glabrous, the margins prominently serrate.
Fl. III - VI. *Dry pastures, cliffs and open rocky areas; in Madeira only around Garajau east of Funchal and on the Ponta de São Lourenço and its associated islets; much more common and widespread in Porto Santo; also on Ilhéu Chão and Bugío in the Desertas and on Selvagem Grande.* **MDPS**

11. Bituminaria Fabr.
M.J. Cannon & N.J. Turland

Pubescent perennial herbs, often becoming woody at base. Leaves long-petiolate, trifoliolate; leaflets entire, pinnately veined; stipules free, ± linear. Inflorescence an axillary pedunculate subglobose head. Calyx campanulate, with 5 unequal teeth longer than tube. Stamens monadelphous. Legume indehiscent, consisting of 2 portions; basal portion black glandular-punctate, containing the single seed and enclosed within calyx; apical portion a flattened and slightly curved beak exserted from calyx.

1. B. bituminosa (L.) C.H. Stirt. in *Bothalia* **13**: 318 (1981).   [*Fedegoso*]
Aspalthium bituminosum (L.) Fourr.; *Psoralea bituminosa* L.
Plant smelling or bitumen or naptha when rubbed, especially in hot weather. Stems to 1(-1.5) m, sprawling to erect. Leaflets 1.5-5.5 × 0.8-3 cm, becoming narrower towards upper part of stem, those of lowermost leaves ovate-orbicular, those of uppermost leaves linear-lanceolate. Peduncles 3.5-17 cm. Heads 2-3 cm wide, 4- to 17-flowered. Calyx 10-15 mm, covered with long white hairs and shorter black hairs; teeth acuminate, awned. Corolla 13-18 mm, bluish violet and white. Legume 13-17 mm including beak, the basal portion, with long white and black hairs and a few yellowish prickles, the beak shortly white- and black-hairy.
Fl. all year. *Maritime cliffs, grassy, rocky and dry open places, roadsides and waste areas; common throughout the lower regions of Madeira, up to c. 650 m; in Porto Santo only in the south-west around Ponta.* **MP**

12. Cullen Medik.
M.J. Cannon & N.J. Turland

Like *Bituminaria* but whole plant prominently darkly glandular-punctate; leaflets sinuate-crenate; inflorescence a cylindrical raceme; legume compressed-ovoid, not beaked, completely enclosed within calyx.

1. C. americanum (L.) Rydb. in Britton, *N. Amer. fl.* **24**: 3 (1919).
Psoralea americana L., var. *polystachya* (Poir.) Cout.
Stems to 1 m, procumbent or diffuse. Leaflets 1.5-4.5 × 1-3.5 cm, ovate- or obovate-rhombic or suborbicular. Peduncles 3-10 cm. Racemes 4-10 cm, 15- to 40-flowered. Calyx 5-7 mm, covered with long white hairs; teeth acute, not awned. Corolla 6-8 mm, white tinged violet, the keel violet at tip. Legume 3.5-5 mm, blackish brown, muricate, glabrous.
Fl. all year but mainly III - VI. *Coastal slopes and cliffs, rocky places, fields and field-margins, roadsides and waste areas; a rare plant restricted to south-eastern Madeira, known only from the western side of Funchal and from Machico.* **M**

Lablab purpureus (L.) Sweet, *Hort. brit.* ed. 1: 481 (1826) (*Dolichos lablab* L.; *L. vulgaris* Savi), native to tropical Africa, is a slender twining perennial similar to the commonly cultivated *Phaseolus vulgaris* L. It has leaves trifoliolate, leaflets ovate-rhombic or ovate-deltate, the terminal long-petiolulate, flowers purple or white, borne in pedunculate racemes, and legumes oblong, flat, glabrous. It is cultivated in Madeira and has also been recorded, possibly as an escape, from waste ground in the Funchal region. **M**

Cicer arietinum L., *Sp. pl.* **2**: 738 (1753) [*Grão de bico*] is an erect annual with leaves imparipinnate, with 3-8 pairs of dentate leaflets, inflorescence axillary, pedunculate, 1-flowered, corolla purple or white and legumes containing 2 large seeds. Possibly native in SW Asia, it is commonly cultivated for its edible seeds in both Madeira and Porto Santo and occurs also as an escape on field-margins and waste ground. **MP**

13. Vicia L.
D.A. Goyder

Annual or perennial herbs. Stems climbing or scrambling, occasionally erect, angled but not winged. Leaves paripinnate, with a tendril or rarely mucronate; leaflets in 1-16 pairs, pinnately veined; stipules small, herbaceous. Flowers axillary, solitary or in racemes; calyx regular or the lower teeth longer than the upper, with at least 2 teeth less than twice as long as tube; stamens diadelphous; style dorsally or laterally compressed, pubescent all round or on lower side only, or glabrous. Legume oblong, usually compressed, dehiscent with 2 or more seeds.

1. Stipules a with nectariferous spot on abaxial surface; inflorescence sessile or with peduncle shorter than flowers, rarely longer [subgen. *Vicia*] 2
- Stipules without a nectariferous spot; inflorescence pedunculate, the peduncle much longer than flowers [subgen. *Vicilla*] 5

2. Mouth of calyx-tube oblique, or calyx-teeth unequal 3
- Mouth of calyx-tube not oblique, the teeth equal or subequal .... **1. sativa**

3. All leaves without tendrils, the rachis terminated by a short mucro .... **4. faba**
- At least upper leaves with tendrils 4

4. Leaflets in 1-3 pairs, at least 10 mm wide, ovate, elliptic or oblong . **3. narbonensis**
- Leaflets in 3-10 pairs, 1-6 mm wide, linear to narrowly elliptic or oblong . . **2. lutea**

5. Stipules of each leaf markedly dimorphic, one entire, the other palmatifid ....
 .... **14. articulata**
- Stipules of each leaf ± identical 6

6. Calyx-teeth equal, longer than tube 7
- Calyx-teeth unequal, at least upper teeth shorter than tube 8

7. Leaves without tendrils; corolla 6-9 mm .... **15. ervilia**
- At least the upper leaves with tendrils; corolla 2-5 mm .... **9. hirsuta**

8. Corolla 9-18 mm, rarely shorter; legume 20-45 mm 9
- Corolla 4-9 mm; legume less than 20 mm 12

9. Calyx strongly gibbous at base; legume pubescent or glabrous 10
- Calyx not or only weakly gibbous; legume glabrous 11

10. Racemes shorter than or equalling leaves; corolla reddish, usually dark purple at the tip; legume pubescent .... **5. benghalensis**
- Racemes usually longer than leaves; corolla violet or purple, sometimes with blue or white wings; legume glabrous .... **6. villosa**

11. Racemes 7- to 15-flowered, about equalling leaves . . . . **7. capreolata**
- Racemes 2- to 4-flowered, much shorter than leaves . . . . **8. ferreirensis**

12. Racemes shorter than leaves; seeds usually 2 . . . . **10. disperma**
- Racemes equalling or longer than leaves; seeds 3-6 13

13. Leaflets 2-5 mm wide; lower calyx-teeth equalling or longer than tube   **13. pubescens**
- Leaflets 1-3 mm wide; lower calyx-teeth shorter than tube 14

14. Racemes longer than leaves, (1-)2- to 5- flowered . . . . **11. parviflora**
- Racemes about equalling leaves, 1- or 2-flowered . . . . **12. tetrasperma**

Subgen. VICIA. Stipules with a glandular nectary on abaxial surface. Inflorescence 1- to several-flowered, shorter than subtending leaf, the flowers sometimes sessile in leaf-axils. Legume not stipitate, linear or rhomboid with woolly tissue between seeds.

Sect. VICIA. Leaves with tendrils. Flowers 1-2(-4), usually sessile in leaf-axils; calyx-teeth equal; claw of standard narrower than limb. Legume with parallel sutures.

1. V. sativa L., *Sp. pl.* **2**: 736 (1753). [*Ervilhaca*]
Pubescent annual to 80 cm. Leaflets in 3-8 pairs, 5-20(-25) × 1-8(-10) mm, linear to obcordate, the apex acute to emarginate, mucronate; stipules dentate, usually with a dark spot. Flowers 1-2(-4), subsessile or very rarely borne on a peduncle of 15-35 mm; mouth of calyx-tube not oblique; calyx-teeth equal, longer or shorter than tube; corolla 10-30 mm, reddish purple, rarely white. Legume 25-55 × (3-)4-6 mm, not contracted between seeds (except in subsp. *sativa*), yellowish brown to black, glabrous or pubescent; seeds 6-12, with hilum c. $^1/_5$ of circumference. Fl. all year. *Coastal cliffs, scrub, grassy places, cornfields and waste areas; common up to c. 1000 m in Madeira and Porto Santo; also on Deserta Grande.* **MDP**

V. sativa is extremely variable and intermediates exist between the two subspecies given below. Subsp. *devia* J.G. Costa in *Bolm Mus. munic. Funchal* **3**: 62 (1948), described from Porto Santo, merely represents a form in which the flowers are borne on a distinct peduncle and is not recognized here. Such plants occur occasionally in Madeira as well as in Porto Santo.

a) subsp. **nigra** (L.) Ehrh. in *Hannover. Mag.* **18**: 229 (1780).
V. conspicua Lowe, var. *dumetorum* Lowe, var. *lactea* Lowe, var. *laeta* Lowe; *V. pectinata* Lowe; *V. sativa* var. *bobartii* (E. Forst.) Koch, var. *segetalis* (Thuill.) Burnat
Leaflets linear to oblong-cuneate, the apex acute, obtuse or truncate. Calyx-teeth shorter than tube. Corolla 10-18(-29) mm; standard light reddish purple; wings similar or somewhat darker. Legume 3-6 mm wide, usually glabrous; seeds 2-4 mm.
Common in both Madeira and Porto Santo.

b) subsp. **cordata** (Wulfen ex Hoppe) Batt. in Batt. & Trab., *Fl. Algérie* **1**: 267 (1889).
V. cordata Wulfen ex Hoppe; *V. sativa* subsp. *sativa* sensu A. Hansen & Sunding (1985), var. *maculata* sensu Menezes (1914), non (C. Presl) Burnat
Leaflets oblong- to obovate-cuneate, the apex truncate to emarginate. Calyx-teeth longer than tube. Corolla 18-30 mm; standard reddish-purple; wings dark red. Legume 4.5-6 mm wide, usually glabrous; seeds 3-4.5 mm.
Occasional, often mixed with subsp. a. In Madeira at Serra de Água, around Funchal and eastwards towards Machico; also on the slopes of Pico de Facho in Porto Santo and on Deserta Grande.

Sect. HYPECHUSA (Alef.) Asch. & Graebn. Leaves with tendrils. Flowers 1-many, pedunculate or sessile in leaf-axils; calyx irregular; standard oblong, or with claw narrower than limb. Legume rhomboid.

2. V. lutea L., *Sp. pl.* **2**: 736 (1753).
V. lutea var. *pallidiflora* DC., var. *purpurascens* Lowe; *V. peregrina* sensu auct. mad., non L. (1753); *V. portosanctana* Gand.
Subglabrous or pubescent annual to 60 cm. Leaflets in 3-10 pairs, 5-25 × 1-6 mm, linear to narrowly elliptic or oblong, those of lower leaves obovate to suborbicular; stipules entire or dentate. Flowers subsessile, 1-3 together; mouth of calyx-tube oblique; calyx-teeth unequal, the lower equalling or longer than tube; corolla (15-)20-25 mm, pale yellow, often purple-tinged. Legume 20-40 × 8-12 mm, yellowish brown to black, pubescent, with hairs tuberculate at base, or occasionally glabrous; seeds 3-9, with hilum ⅓-½ of circumference.
Fl. III - VII. *Cliffs, waste ground, roadsides and dry river-beds up to c. 800 m. Uncommon in Madeira, occurring on the northern coast between Porto do Moniz and Faial, at Serra de Agua, Curral das Freiras and around Funchal; also on Pico do Castelo in Porto Santo.* **MP**

Plants from the Madeiran archipelago belong to subsp. **lutea** (subsp. *genuina* Cout.), which has 4- to 10-seeded, white-hairy legumes, each hair with a small tubercle at its base. Subsp. **vestita** (Boiss.) Rouy, *Fl. France* **5**: 219 (1899) (subsp. *muricata* (Ser.) Guinea) has been doubtfully recorded from Madeira and its presence requires confirmation. It has 3- to 4-seeded, reddish or brownish hairy legumes, each hair with a large tubercle at its base. *V. peregrina* L. has been erroneously recorded from Porto Santo, based on a misidentified specimen of *V. lutea*, which also forms the basis of *V. portosantana* Gand.

Sect. FABA (Mill.) Ledeb. Leaves with tendrils or mucronate. Flowers 1-6, pedunculate or sessile in leaf-axils; claw of standard narrower than limb. Legume with parallel sutures.

3. V. narbonensis L., *Sp. pl.* **2**: 737 (1753).
Pubescent annual; stems 20-60 cm, erect, much-branched. Upper leaves with branched tendrils; leaflets in 1-3 pairs, 10-50 × 8-30 mm, ovate or obovate to elliptic, the apex obtuse or emarginate, the middle and upper leaflets with serrate margins; stipules *c*. 10 mm, entire or dentate. Flowers 4-6, subsessile; mouth of calyx-tube oblique; calyx-teeth unequal, the lower longer than tube; corolla 20-25 mm, purple, the wings with a darker tip. Legume 40-70 × 10-15 mm, brown, glabrous with tuberculate-dentate pubescent margin; seeds 4-8, 4-6 mm, with hilum ⅛ of circumference.
Fl. VI - VIII. *Introduced; cultivated and occasionally naturalized in vineyards around Funchal. Native to S. Europe and the Mediterranean region.* **M**

Only var. **serratifolia** (Jacq.) Ser. in DC., *Prodr.* **2**: 365 (1825) has been recorded from Madeira.

4. V. faba L., *Sp. pl.* **2**: 737 (1753). [*Fava, Faveira*]
Faba vulgaris Moench; *V. faba* var. *major* Cout.
Like **3** but more robust; leaves without tendrils, the rachis terminated by a short mucro; leaflets 40-80(-100) × 10-20(-40) mm; flowers 1-6; corolla usually white with black tips to the wings; legume 80-200 × 10-20 mm, densely pubescent but becoming sparsely pubescent when mature; seeds 20-30 mm, ovoid-oblong, compressed.
Fl. II - III. *Introduced; widely cultivated for its edible seeds in both Madeira and Porto Santo and occasionally naturalized in vineyards, fields and gardens around Funchal. Native origin unknown but cultivated particularly in the Mediterranean region and S. Asia.* **MP**

Subgen. VICILLA (Schur) Rouy. Stipules without a nectariferous spot. Inflorescence 1- to many-flowered, usually equalling or longer than subtending leaf. Legume often stipitate; sutures rarely parallel; woolly tissue not present between seeds.

Sect. CRACCA Dumort. Leaves with tendrils. Racemes few- to many-flowered. Calyx irregular, sometimes gibbous at base. Standard of corolla waisted. Style laterally compressed.

5. V. benghalensis L., *Sp. pl.* **2**: 736 (1753).
V. albicans Lowe; *V. atlantica* J.G. Costa, non Pomel (1874); *V. atropurpurea* Desf.; *V. costae* A. Hansen
Villous annual or short-lived perennial, 20-80 cm. Leaves with tendrils; leaflets in 5-9 pairs, (6-)10-20(-30) × 2-5(-10) mm, linear, oblong or elliptic; stipules entire or dentate. Racemes 2- to 13-flowered, usually shorter than leaves. Calyx strongly gibbous at base; teeth unequal, the lower longer or shorter than tube. Corolla 10-18 mm, reddish, usually dark purple at tip; limb of standard about as long as claw. Legume 25-40 × 7-11 mm, shortly stipitate, brown, pubescent; seeds 3-5, with hilum ⅕ of the circumference. $2n = 14$.
Fl. I - VI. *Along levadas, roadsides and on rocky ground around Funchal, rare; also in the eastern half of Porto Santo.* **MP**

Plants from Porto Santo (*V. atlantica*; *V. costae*) have few-flowered peduncles and unusually short calyx-teeth.

6. V. villosa Roth, *Tent. fl. Germ.* **2**(2): 182 (1793).
Like **5** but glabrous or appressed-pubescent annual; stipules entire; racemes 10- to 30-flowered, usually longer than leaves; calyx-teeth all shorter than tube; corolla violet or purple, sometimes with blue or white wings; limb of standard about half as long as claw; legume glabrous.
Fl. ? *Recorded as occurring in Madeira but without further data.* **M**

The above description refers to subsp. **varia** (Host) Corb., *Nouv. Fl. Normandie*: 181 (1894), to which Madeiran plants have been referred.

7. V. capreolata Lowe in *Trans. Camb. phil. Soc.* **4**(3): 545 (1838).
Ervum capreolatum (Lowe) Lowe
Pubescent annual, scrambling to 300(-400) cm. Leaves with tendrils; leaflets in 2-6 pairs, often scattered along the rachis, 5-15(-20) × 1.5-3(-6) mm, linear-oblong to elliptic or suborbicular; stipules entire, semi-sagittate. Racemes 7- to 15- flowered, about equalling leaves at anthesis, longer in fruit. Calyx weakly gibbous at base; teeth unequal, shorter than tube. Corolla 9-13 mm, cream or lilac, often with blue veins. Legume (25-)30-45 × (4-)6-8 mm, brown, glabrous; seeds 3-5, sub-cuboid, with hilum *c*. ⅙ of circumference. **Plate 20**
Fl. III - VIII. *A rare endemic of cliffs and wooded banks in Madeira up to c. 1600 m, especially in ravines; found above Seixal and in the mountains between Serra de Agua and Ribeiro Frio; also in Porto Santo, growing in pastures at lower altitudes, and on Ilhéu Chão and Deserta Grande.*
● **MDP**

8. V. ferreirensis Goyder in *Bocagiana* **113**: 1 (1987).
V. portosanctana Menezes, non *V. portosantana* Gand. (1912)
Glabrous or subglabrous annual, to *c*. 100 cm. Leaves with tendrils; leaflets in 4-6 pairs, 5-20 × 1-3 mm, linear or oblong, the apex truncate; stipules semi-sagittate. Racemes 2- to 4-flowered, shorter than leaves, slender, often ending in a short arista. Calyx weakly gibbous at base; teeth unequal, shorter than tube. Corolla 8-11 mm, pale blue. Legume 25-40 × 4-6 mm, linear, glabrous; seeds 4-6. $2n = 14$.
Fl. ?. *An endemic of shady places on Ana Ferreira and Pico de Castelo in Porto Santo.* ● **P**

9. V. hirsuta (L.) Gray, *Nat. arr. Brit. pl.* **2**: 614 (1821). [*Cigerão*]
Ervum hirsutum L.
Pubescent annual, 20-70 cm. Leaves with tendrils; leaflets in 4-10 pairs, 5-20 × 1-3 mm, linear or oblong; stipules entire, the lower linear-lanceolate, often with 2-4 setaceous teeth. Racemes 1- to 8- flowered, almost equalling leaves. Calyx-teeth subequal, longer than tube. Corolla 2-4(-5) mm, dirty white with a bluish tinge. Legume 6-11 × 3-5 mm, black, pubescent; seeds usually 2, with hilum ⅓ of circumference.

Fl. all year. *A common weed of waste ground and grassy banks up to c. 800 m in Madeira; also in Porto Santo and on Ilhéu Chão and Deserta Grande.* **MDP**

10. V. disperma DC., *Cat. pl. horti monsp.*: 154 (1813).
Ervum parviflorum (Loisel.) Bertol.
Sparsely pubescent annual, 10-50 cm. Leaves with tendrils; leaflets in 5-10 pairs, 8-16 × 1.5-4 mm, linear to elliptic; stipules semi-hastate, entire. Racemes (1-)2- to 6-flowered, shorter than leaves. Calyx-teeth unequal, the lower slightly longer than tube. Corolla 4-5 mm, pale blue. Legume 12-20 × 5-8 mm, brown, glabrous; seeds usually 2, with hilum $1/7$- $1/6$ of circumference.
Fl. III - VI. *Rocky ground, roadsides and waste areas; rare in Madeira, occurring at Serra de Agua, Curral das Freiras, along the levadas around Funchal and at Caniçal; recorded once from Selvagem Grande, over a century ago.* **MS**

Sect. ERVUM (L.) Taub. Leaves with tendrils. Flowers 1-6; calyx ± regular; standard of corolla oblong to ovate; style circular in cross-section to dorsally compressed. Legume less than 20 mm, linear.

11. V. parviflora Cav. in *An. Cienc. nat. Madrid* **4**: 73 (1801).   [*Cigerão*]
Ervum gracile (Loisel.) DC.; *V. gracilis* Loisel., non Banks & Sol. in Russell (1794); *V. laxiflora* Brot.; *V. tenuissima* Schinz & Thell., non *Ervum tenuissimum* M. Bieb. (1798).
Subglabrous annual, 15-60 cm. Leaves with a short simple tendril; leaflets in 2-5 pairs, 6-25 × 1-3 mm, linear-lanceolate; stipules semi-sagittate or semi-hastate, entire. Racemes (1-)2- to 5-flowered, longer than leaves, usually with a short awn. Calyx-teeth unequal, shorter than tube. Corolla 5-9 mm, pale bluish pink. Legume (10-)12-17 × 3-4 mm, brown, glabrous or pubescent; seeds 4-6, with hilum $1/8$-$1/2$ of circumference.
Fl. III - VII. *A common weed of grassy and dry places, roadsides and waste areas up to c. 900 m; in Madeira occurring in the south and around Porto do Moniz and Faial on the northern coast; also in Porto Santo and on Deserta Grande and Bugío.* **MDP**

12. V. tetrasperma (L.) Schreb., *Spic. fl. lips.*: 26 (1771).
Ervum tetraspermum L.
Like **11** but leaflets in 3-6 pairs; racemes 1- or 2-flowered, about equalling leaves, not terminating in an awn; corolla 4-8 mm; legume 9-16 × 3-5 mm, usually glabrous; seeds 3-5, with hilum $1/5$ of circumference.
Fl. ? *Known only from a single collection made in 1777 and possibly no longer present in Madeira.* **?M**

13. V. pubescens (DC.) Link, *Handbuch* **2**: 190 (1831).
Ervum pubescens DC., var. *glabrescens* Lowe, var. *subpilosa* Lowe
Like **11** but sparsely pubescent; leaves with a well developed and sometimes branched tendril; leaflets in 3-5 pairs, 10-20 × 2-5 mm, elliptic to ovate-oblong; racemes 1- to 6-flowered, about equalling or longer than leaves; lower calyx-teeth equalling or longer than tube; legume usually pubescent; hilum of seeds $1/16$- $1/12$ of circumference.
Fl. III - VII. *Cornfields and waste ground up to c. 800 m; frequent in the southern half of Madeira from Calheta to Machico.* **M**

Sect. ERVOIDES (Godr.) Kupicha. Leaves with tendrils; stipules dimorphic. Flowers 1-2; calyx slightly irregular; standard of corolla oblong; style dorsally compressed. Legume with slight bulges and constrictions.

14. V. articulata Hornem., *Enum. pl. hort. hafn.* ed.2: 41 (1807). [*Lentilha*]
Ervum monanthos L.; *Vicia monanthos* (L.) Desf., non Retz. (1783)
Glabrous annual, 20-70 cm. Leaf-rachis terminating in a branched tendril; leaflets in 5-9 pairs, 6-18 × 1-4 mm, linear, oblong, the apex usually emarginate with a mucro and 2 acute lobes; stipules dimorphic, 1 of each pair simple, linear, the other laciniate-palmatifid, stalked. Racemes 1- to 2-flowered. Calyx-teeth slightly unequal, longer than tube. Corolla 8-14 mm, white or pale blue. Legume 15-35 × 6-10 mm, glabrous, yellow; seeds 2-4, with hilum $1/10$ of circumference.
Fl. IV - V. *Introduced; cultivated during the 19th century in south-eastern Madeira; rarely naturalized along the Ribeira de Santa Luzia and on the hills around Caniço. Native to the Mediterranean region.* **M**

Sect. ERVILIA (Link) W.D.J. Koch. Leaves mucronate. Flowers 1-4; calyx ± regular; standard of corolla ovate; style dorsally compressed. Legume with slight bulges and constrictions.

15. V. ervilia (L.) Willd., *Sp. pl.* 3: 1103 (1802). [*Marroios, Marroiso, Marruiço*]
Ervum ervilia L.
Glabrous or pubescent annual, 15-30(-50) cm. Leaves without tendrils, the rachis terminating in a short mucro; leaflets in 8-16 pairs, 5-15 × 1-4 mm, oblong or linear, the apex emarginate-mucronate; stipules entire or palmatifid. Racemes 1- to 4-flowered. Calyx-teeth equal, longer than tube. Corolla 6-9 mm, white with violet or purple veins. Legume 10-30 × 4-6 mm, glabrous, yellow; seeds 2-4, subglobose, with hilum $1/12$ of circumference.
Fl. IV - VI. *Introduced; cultivated in cornfields in southern Madeira, at Cabo Girão and Caniço; rarely also naturalized. Native to the Mediterranean region.* **M**

14. Lens Mill.
D.A. Goyder

Annual herbs. Stems angled but not winged. Leaves paripinnate, ending in a mucro or tendril; leaflets in 4-6 pairs, pinnately veined; stipules small, herbaceous. Flowers axillary, solitary or in racemes; calyx-teeth all equal and at least twice as long as tube; stamens diadelphous; style filiform, dorsally compressed and pubescent on upper side. Legume broadly rhomboid, strongly compressed, dehiscent; seeds 1-2(-3), flattened, orbicular.

1. L. culinaris Medik. in *Vorles. Churpfälz. phys.-ökon. Ges.* 2: 361 (1787).
[*Ervilha* (P), *Lentilha* (M)]
Ervum lens L.
Erect pilose annual, 10-30 cm. Lower leaves mucronate, upper leaves ending in a simple tendril; leaflets 4-14 × 1-3.5 mm, oblong to obovate or narrowly elliptic; stipules 1.5-5 mm, lanceolate, without a basal appendage. Racemes 1- to 2(-3)-flowered, approximately equal to the leaves. Calyx 4-7 mm. Corolla pale blue or lilac, included within or slightly exceeding the calyx. Legume 9-13 × 5-6 mm, yellowish, glabrous; seeds *c*. 4 mm.
Fl. III - VI. *A field weed, probably a remnant of earlier cultivation; in Madeira restricted to the Ponta de São Lourenço but widespread in Porto Santo. Widely cultivated in the Mediterranean region and the Middle East.* **MP**

15. Lathyrus L.
D.A. Goyder

Annual or perennial herbs. Stems often climbing by means of tendrils, sometimes winged. Leaves paripinnate, ending in a tendril, or the leaf occasionally reduced to a tendril; leaflets in 1-4(-5) pairs, often only 1 pair, mostly parallel-veined; stipules herbaceous. Flowers axillary, solitary or in racemes; calyx regular or the lower teeth longer than the upper; stamens diadelphous; style dorsally compressed, pubescent on the upper side only. Legume usually oblong, compressed, dehiscent, containing 2-many seeds.

1. All leaves with 1 pair of leaflets, the rachis never laminate 2
- Leaflets absent (although stipules sometimes leaf-like), or leaf-rachis laminate . . . 9

2. Stem winged, at least in upper part 3
- Stem angled but not winged 8

3. Flowers always solitary; calyx-teeth 2-3 × as long as tube 4
- Flowers in 1- to 12-flowered racemes; calyx-teeth shorter or only slightly longer than tube 5

4. Corolla brick-red; legume oblong, with 2 keels on dorsal suture . . . . **2. cicera**
- Corolla blue, violet or white; legume broadly oblong, with 2 broad wings on dorsal suture . . . . **3. sativus**

5. Corolla yellow . . . . **1. annuus**
- Corolla variously coloured, but never yellow 6

6. Racemes 3- to 12-flowered; calyx-teeth distinctly unequal . . . . **6. sylvestris**
- Racemes 1- to 3-flowered; calyx-teeth subequal 7

7. Plant glabrous . . . . **5. tingitanus**
- Plant pubescent, at least on legume . . . . **4. odoratus**

8. Peduncles 5-20 mm; legume 3-7 mm wide, with prominent longitudinal veins . . . . **10. sphaericus**
- Peduncles 20-70 mm; legume 2-4 mm wide, with indistinct reticulate veins . . . . **11. angulatus**

9. All leaves without leaflets; rachis forming tendril . . . . **7. aphaca**
- Upper leaves with 1-4 pairs of leaflets 10

10. Upper leaves with 1-2 pairs of leaflets; corolla yellow; dorsal suture of legume broadly 2-winged . . . . **9. ochrus**
- Upper leaves with 2-4 pairs of leaflets; corolla purple, with wings violet, lilac, pink or white; dorsal suture of legume narrowly keeled or channelled . . . . . . **8. clymenum**

Sect. LATHYRUS. Stems winged. Leaves with tendrils and 1 pair of pinnate- or parallel-veined leaflets: stipules semi-sagittate. Flowers 1-12; calyx-teeth equal; standard of corolla with a very wide limb and a short claw; style contorted.

1. L. annuus L., *Demonstr. pl.*: 20 (1753).
Glabrous annual; stems 40-150 cm, winged. Leaflets in 1 pair, 50-130 × 3-10 mm, linear or linear-lanceolate; stipules 10-25 × 0.3-1.5 mm, linear, semi-sagittate. Racemes 1- to 3-flowered; peduncles (10-)50-90 mm. Calyx-teeth equal, shorter or slightly longer than tube. Corolla 12-18 mm, yellow or orange-yellow. Legume 40-60 × 7-12 mm, pale brown, glandular when young, glabrescent; seeds 7-8, tuberculate or papillose.
Fl. III - VII. *Occasional in fields and hedges around Funchal, up to 500 m.* **M**

2. L. cicera L., *Sp. pl.* **2**: 730 (1753). [*Chícharo branco* (P), *Chícaros*]
L. cicera subvar. *caerulea* Lowe, subvar. *purpurea* Lowe
Glabrous annual; stems 20-100 cm, winged. Leaflets in 1 pair, 10-70 × 1-9 mm, linear to lanceolate; stipules 10-30 × 1-7 mm lanceolate, semi-sagittate. Flowers solitary; peduncles 10-30 mm, articulated near middle or apex; calyx-teeth equal, 2-3 × as long as tube; corolla 8-12 mm, brick-red. Legume 20-40 × 6-10 mm, brown, glabrous, narrowly keeled on dorsal suture; seeds 2-6, smooth.
Fl. IV - VII. *Introduced; cultivated as a vegetable and for fodder during the 19th century at São Vicente and in the valley of the Ribeira de Santa Luzia above Funchal in Madeira, as well as on the hills of Porto Santo. Native to the Mediterranean region and Middle East.* **MP**

3. L. sativus L., *Sp. pl.* **2**: 730 (1753). [*Chícharos*]
Like **2** but leaflets 25-100 mm; peduncles 20-60 mm; corolla 12-16 mm, white, violet or blue; legume 10-15 mm wide, with 2 wings on dorsal suture.
Fl. II - V. *Introduced; a field weed occurring around Campanário and Funchal up to c. 700 m; cultivated, at least formerly, as a vegetable and for fodder. Native origin unknown; widely cultivated in the Mediterranean region and S. Asia.* **M**

4. L. odoratus L., *Sp. pl.* **2**: 732 (1753).
Tuberculate-hairy annual; stems 50-200 cm, winged. Leaflets in 1 pair, 20-60 × 7-30 mm, ovate-oblong or elliptic; stipules 15-25 × 2-4 mm, lanceolate, semi-sagittate. Racemes 1- to 3-flowered; peduncles 120-200 mm. Calyx-teeth subequal, longer than tube. Corolla 20-35, purple, pink or white. Legume 50-70 × 10-12 mm, brown, densely covered with shortly tuberculate-based hairs; seeds *c*. 8, smooth.
Fl. VII. *Introduced; cultivated in Madeira as an ornamental and occasionally naturalized. Native to S. Italy and Sicily.* **M**

5. L. tingitanus L., *Sp. pl.* **2**: 732 (1753).
Glabrous annual; stems 60-120 cm, winged. Leaflets in 1 pair, 20-80 × 4-30 mm, linear-lanceolate to broadly elliptic or ovate; stipules 12-20 × 3-10 mm, lanceolate to ovate, semi-sagittate or semi-hastate. Racemes 1- to 3-flowered; peduncles 20-80 mm. Calyx-teeth subequal, shorter than tube. Corolla 20-30 mm, bright purple. Legume 60-100 × 8-10 mm, brown, glabrous; seeds 6-8, smooth.
Fl. IV - VI. *Probably introduced and almost certainly now extinct in Madeira, having been recorded only from rocky ground along the Levada de Santa Luzia above Funchal. Native to the Azores, Canaries and W. Mediterranean region.* **?M**

6. L. sylvestris L., *Sp. pl.* **2**: 733 (1753).
Glabrous or pubescent perennial; stems 60-200 cm, winged. Leaflets in 1 pair, 40-110 × 5-20 mm, linear to lanceolate; stipules 10-30 × 2-5 mm, linear to lanceolate, semi-sagittate. Racemes 3- to 12-flowered; peduncles 60-200 mm. Calyx-teeth unequal, the lower about equalling tube. Corolla 13-30 mm, purplish pink. Legume 40-70 × 5-13 mm, brown, glabrous; seeds 10-15, reticulate-rugose.
Fl. VIII. *Known in Madeira only from a single collection from near the sea at Santana.* **M**

Sect. APHACA (Mill.) Dumort. Stems not winged. Leaves of mature plants lacking leaflets, but with a tendril and large, leaf-like hastate stipules. Calyx-teeth equal. Corolla with standard bossed and limb of wings waisted. Style linear, not contorted.

7. L. aphaca L., *Sp. pl.* **2**: 729 (1753).
Glabrous annual; stems trailing or scrambling to 100 cm, angled. Seedling leaves with 1 pair of small leaflets; mature leaves reduced to a tendril, with leaflets absent; stipules 10-35 × 5-25

mm, broadly ovate, hastate. Flowers 1(-2); peduncles 20-60 mm; calyx-teeth equal, 2-3 × as long as tube; corolla 6-11 mm, yellow. Seeds 6-8, 2-3.5 mm, smooth.
Fl. III - VII. *Occasional in lowland pastures, on borders of cornfields and waste ground in both Madeira and Porto Santo.* **MP**

Sect. CLYMENUM (Mill.) DC. ex Ser. Stems winged. Juvenile leaves reduced to phyllodes, later leaves with tendrils and several pairs of pinnate-veined leaflets; stipules semi-sagittate. Flowers 1-3; standard of corolla with 2 prominent pouches; style spathulate with 2 stigmas separated by a sterile mucro.

8. L. clymenum L., *Sp. pl.* **2**: 732 (1753).
L. articulatus L., var. *latifolius* Rouy, subvar. *atropurpurea* Lowe, subvar. *rosea* Lowe; *L. clymenum* subvar. *albiflora* Lowe, subvar. *atropurpurea* Lowe, subvar. *roseopurpurea* Lowe
Glabrous annual; stems 30-100 cm, winged. Leaves with broad, leaf-like petiole and rachis, the lower linear-lanceolate, without leaflets, the upper with 2-4(-5) pairs of leaflets; leaflets 10-60 × 1-10 mm, linear to ovate, semi-hastate. Racemes 1- to 3-flowered. Calyx-teeth equal, shorter than tube. Corolla 15-20 mm, crimson, with wings violet, lilac, blue, pink or white. Style aristate or obtuse. Legume 30-70 × 5-12 mm, with or without bulges and constrictions, narrowly keeled or channelled on the dorsal suture, brown, glabrous; seeds 4-7, smooth.
Fl. XII - VII. *Common in hedges, on field-margins, roadsides and waste ground in lowland areas of Madeira and Porto Santo, from 100-500 m; also on Ilhéu Chão and in the central valley of Deserta Grande; cultivated in Porto Santo during the 19th century.* **MDP**

L. clymenum and *L. articulatus* are not clearly distinguishable from each other and, although plants from Porto Santo and Deserta Grande are closer to 'typical' *L. articulatus*, the two taxa are not separated here.

9. L. ochrus (L.) DC. in Lam. & DC., *Fl. franç.* ed. 3, **4**: 578 (1805). [*Chícharo*]
Glabrous annual; stems 30-60 cm, winged. Leaves with broad, leaf-like petiole and rachis, the lower ovate-oblong, the upper with 1-2 pairs of lealets; leaflets 20-30 × 10-15 mm, ovate; stipules 6-12 mm. Racemes 1- to 2-flowered. Calyx-teeth slightly unequal, as long as tube. Corolla pale yellow. Legume 40-60 × 10-12 mm with 2 wings on the dorsal suture; seeds 5-7, 4-5 mm.
Fl. IV - V. *Introduced; very rare in Madeira and known only from two records in the Funchal region; known in Porto Santo from a single collection from the summit of Pico Branco. Native to Europe and W. Asia.* **MP**

Sect. LINEARICARPUS Kupicha. Stems angled or narrowly 2-winged. Leaves mucronate or with a simple tendril, with 1 pair of parallel-veined leaflets; stipules semi-sagittate. Flowers solitary; calyx-teeth equal; standard of corolla bossed; style not contorted.

10. L. sphaericus Retz., *Observ. bot.* **3**: 39 (1783).
Glabrous annual; stems 10-50 cm, angled or weakly 2-winged. Leaflets in 1 pair, 15-60(-80) × 1-4(-10) mm, linear to linear-lanceolate; stipules 6-20 × 0.5-1 mm, linear, semi-sagittate. Flowers solitary; peduncles 5-20 mm, aristate, articulated near middle or apex; calyx-teeth equal; corolla 6-13 mm, brick-red. Legume 30-70 × 3-7 mm, brown, glabrous, with prominent longitudinal veins; seeds 8-15, smooth or slightly rugose.
Fl. I - VII. *Occasional in deep ravines and grassy and stony places in the lowlands of south-eastern Madeira, up to Curral das Freiras, Monte, Fajã da Nogueira and Ribeira Seca above Machico.* **M**

11. L. angulatus L., *Sp. pl.* **2**: 731 (1753).
Like **10** but stipules 1-15 mm long and up to 2 mm wide; peduncles 20-70 mm, articulated near

apex; corolla purple or pale blue; legume 25-50 × 2-4 mm with indistinct reticulate venation; seeds rugose to finely tuberculate.
Fl. VI. *Known from a single collection from Fajã da Nogueira in eastern Madeira.* **M**

16. Ononis L.
M.J. Cannon & N.J. Turland

Annual, rarely perennial herbs, glandular-hairy, at least above. Stems prostrate to erect, branching, rarely spiny and becoming woody. Leaves usually trifoliolate, rarely simple; leaflets serrate; stipules joined to petiole, leaf-like or membranous. Inflorescence racemose or paniculate. Calyx campanulate, with 5 equal teeth. Corolla pink or white, or both. Stamens monadelphous. Legume ovoid or broadly ellipsoid to narrowly oblong-ellipsoid, compressed, dehiscent; seeds 1-several.

1. Perennial; stems spiny, becoming woody; leaves mostly simple . . . . **2. spinosa**
- Annual; stems unarmed, not becoming woody; leaves mostly trifoliolate 2

2. Stems whitish or pale yellow; inflorescence densely ovoid to narrowly ellipsoid, stipules conspicuous, pale, membranous . . . . **5. mitissima**
- Stems green or reddish; inflorescence laxer, stipules not conspicuous as above . . . 3

3. Pedicels deflexed after flowering; calyx-teeth entire or 3-toothed at apex; legume 8-13 mm, narrowly oblong-ellipsoid . . . . **1. dentata**
- Pedicels erecto-patent to erect after flowering; calyx-teeth always entire; legume 5-6.5 mm, ovoid to ellipsoid 4

4. Leaflets obovate to oblanceolate, finely serrate, 7-14 teeth on each side; inflorescence leafy; calyx-teeth lanceolate to narrowly so . . . . **3. diffusa**
- Leaflets narrowly oblong-lanceolate, coarsely serrate, 4-7 teeth on each side; inflorescence ± leafless, at least above; calyx-teeth subulate to narrowly linear . . . . . **4. serrata**

1. O. dentata Sol. ex Lowe in *Trans. Camb. phil. Soc.* **4**: 34 (1831).
O. reclinata sensu auct. mad., non L. (1763), var. *simplex* Lowe, var. *tridentata* Lowe
Annual; stems 6-20 cm, diffuse to erect, green. Leaves trifoliolate; leaflets 5-17 × 3-10 mm, obovate to oblanceolate; stipules leaf-like. Pedicels deflexed after flowering. Corolla 7-14 mm, pink or white. Calyx 6-12 mm; teeth much longer than tube, narrowly oblong, entire or 3-toothed at apex. Legume 8-13 × 2.5-4 mm, scarcely longer than calyx, narrowly oblong-ellipsoid; seeds 5-12.
Fl. III - VII. *A rare plant of rough pastures, dry ground and rocky places; in Madeira restricted to the Ponta de São Lourenço and its associated islets; also in the higher parts of Porto Santo and on all three Desertas.* **MDP**

2. O. spinosa L., *Sp. pl.* **2**: 716 (1753).
Perennial; stems to 60 cm, ascending to erect, spiny, becoming woody. Leaves mostly simple; leaflets 6-20 × 3.5-8 mm, oblong-obovate to oblong-lanceolate; stipules leaf-like. Pedicels erecto-patent after flowering. Calyx 7-12 mm; teeth much longer than tube, narrowly lanceolate, entire. Corolla 8-15 mm; standard pink on outer face, white-striate on inner face; wings white. Legume *c.* 5 × 3.5 mm, shorter than calyx, broadly ellipsoid; seeds 1-2.
Fl. VI. *Grassy banks at c. 450 m; known only from the Achadas da Cruz area in north-western Madeira.* **M**

Madeiran plants were originally described as the endemic *O. costae* Menezes in *Broteria* (Bot.) **20**: 114 (1922), which is here included within the synonymy of **O. spinosa** subsp. **maritima**

(Dumort.) P. Fourn., *Quatre Fl. france*: 540 (1936) (*O. repens* L.), to which the above description refers.

3. O. diffusa Ten., *Fl. napol.* **1**: xli (1811-1815).
O. micrantha Lowe
Annual; stems 5-30 cm, prostrate to ascending, green or reddish. Leaves trifoliolate, occasionally simple directly below inflorescence; leaflets 5-16 × 3-9 mm, obovate to oblanceolate, finely serrate, with 7-14 teeth on each side; stipules leaf-like or membranous. Inflorescence leafy; pedicels erecto-patent to erect after flowering. Calyx 6-10 mm; teeth longer than tube, lanceolate to narrowly so, entire. Corolla 8-13 mm; standard pinkish-purple; wings paler; keel white, tipped with purple. Legume 5-6.5 × 3-3.5 mm, shorter than to equalling calyx, ovoid to ellipsoid; seeds usually 2.
Fl. III - V. *Dry stony and grassy places; in Madeira restricted to the south-east, occurring at Garajau and on the Ponta de São Lourenço and its associated islets; more common in Porto Santo; also on Ilhéu Chão in the Desertas.* **MDP**

4. O. serrata Forssk., *Fl. aegypt.-arab.*: 130 (1775).
Like **3** but leaflets narrower, 2.5-5 mm wide, narrowly oblong-lanceolate, coarsely serrate, with only 4-7 teeth on each side; stipules membranous; inflorescence ± leafless, at least above; calyx-teeth much longer than tube, subulate to narrowly linear; legume containing up to 5 seeds.
Fl. XII - IV. *A rare plant of dry stony pastures in Porto Santo.* **P**

5. O. mitissima L., *Sp. pl.* **2**: 717 (1753).   [Trevo branco (P)]
Annual; stems 15-90 cm, usually erect, whitish or pale yellow. Leaves trifoliolate, sometimes simple on upper part of stem; leaflets 5-17 × 3-12 mm, broadly obovate to oblanceolate; stipules membranous. Inflorescence densely ovoid to narrowly ellipsoid, with conspicuous pale membranous stipules; pedicels erecto-patent to erect after flowering. Calyx 5-9 mm; teeth about equalling tube, triangular, entire. Corolla 8-10 mm; standard pink; keel white. Legume 5.5-7.5 × 3.5-4 mm, shorter than calyx, ellipsoid to oblong; seeds 2-3.
Fl. IV - VI. *Rough pastures, rocky places, dry ground and field-margins; frequent along the southern coast of Madeira from Funchal eastwards; common in Porto Santo; common on Deserta Grande, rare on Ilhéu Chão, also recorded from Bugío.* **MDP**

17. Melilotus Mill. [Trevo]
M.J. Cannon & N.J. Turland

Annual, sometimes biennial herbs. Stems diffuse to erect, branching, glabrous, sometimes sparsely pubescent at tips. Leaves trifoliolate; leaflets serrate, usually becoming narrower towards stem-tips; stipules joined to petiole, linear-setaceous to subulate, entire or dentate. Inflorescence an axillary 5- to 90-flowered slender raceme, elongating after flowering. Calyx with 5 subequal teeth. Corolla yellow or white, not persistent in fruit. Stamens diadelphous; filaments not dilated at apex. Legume longer than calyx, globose to obovoid-oblong, apiculate, veined, glabrous, usually indehiscent; seeds usually 1-2.

1. Corolla white . . . . **1. albus**
-   Corolla yellow 2

2. Stipules of middle and upper cauline leaves entire or ± so; veins of legume reticulate or transverse and parallel, not concentric 3
-   Stipules of middle and upper cauline leaves dentate; veins of legume concentric . . 4

3. Corolla 2-3 mm; legume 2-3 mm, reticulate-veined . . . . **2. indicus**
-   Corolla 3.5-5 mm; legume 3-4 mm, with transverse parallel veins . . . . . **3. elegans**

4. Racemes ± equalling subtending leaves at flowering, 5- to 15-flowered, the axis extended into a distinct point; corolla 3-4 mm . . . . **4. sulcatus**
- Racemes 2-3½ × as long as subtending leaves at flowering, 25- to 80-flowered, the axis scarcely extended and forming only a minute point; corolla 5.5-6.5 mm . **5. segetalis**

1. M. albus Medik. in *Vorles. Churpfälz. Phys.-ökon. Ges.* **2**: 382 (1787).
Annual or biennial. Stems to 60 cm or more, erect. Leaflets 20-35 × 4-13 mm, narrowly elliptic or oblanceolate; stipules entire. Corolla 4-6 mm, white; wings and keel subequal, shorter than standard. Legume 3-6 mm, obovoid to obovoid-oblong, reticulate-veined.
Fl. I - VIII. *Recently introduced; occurring as a weed in south-eastern Madeira, in the Funchal region and near Água da Pena to the east. Native to Eurasia.* **M**

2. M. indicus (L.) All., *Fl. pedem.* **1**: 308 (1785). [*Trevo de namorado* (P)]
M. parviflorus Desf.
Annual. Stems to 75 cm, ascending or erect. Leaflets 4-30 × 2-14 mm, obovate to narrowly cuneate-oblanceolate; stipules ± entire. Corolla 2-3 mm, yellow; wings and keel equal, shorter than standard. Legume 2-3 mm, subglobose, reticulate-veined.
Fl. I - VII. *Cultivated and open sunny ground, roadsides and dry waste areas; common in lowland Madeira, usually near the sea; common in Porto Santo; rarer in the Desertas, occurring on all three islands.* **MDP**

3. M. elegans Ser. in DC., *Prodr.* **2**: 198 (1825).
M. elegans subsp. *lippoldianus* (Lowe) Menezes; *M. lippoldianus* Lowe
Annual. Stems to 60 cm, erect. Leaflets 8-30 × 3-20 mm, broadly obovate to narrowly cuneate-oblanceolate; stipules entire. Corolla 3.5-5 mm, yellow; standard and wings subequal, shorter than keel. Legume 3-4 mm, subglobose to obovoid, with transverse parallel veins.
Fl. III - VIII. *Infrequent weed of coastal cliffs, fields and waste areas around Funchal.* **M**

4. M. sulcatus Desf., *Fl. atlant.* **2**: 193 (1799). [*Trevo de seara* (P)]
Annual. Stems to 30 cm, diffuse to erect. Leaflets 9-25 × 2-8 mm, cuneate-obovate to narrowly cuneate-oblanceolate or narrowly elliptic to ± linear; stipules dentate. Racemes ± equalling subtending leaves at flowering, 5- to 15-flowered, the axis extended into a distinct point. Corolla 3-4 mm, yellow; standard longer than wings but shorter than keel. Legume 3-4 mm, globose, with raised concentric veins.
Fl. III - V. *Cornfield weed; in Madeira only in the south-east, in the Funchal region and on the Ponta de São Lourenço; common in Porto Santo.* **MP**

5. M. segetalis (Brot.) Ser. in DC., *Prodr.* **2**: 187 (1825).
Annual. Stems to 60 cm, erect. Leaflets 10-27 × 5-14 mm, oblanceolate; stipules of lower cauline leaves entire, those of middle and upper cauline leaves dentate. Racemes 2-3.5 × as long as subtending leaves at flowering, 25- to 80-flowered, the axis scarcely extended and forming only a minute point. Corolla 5.5-6.5 mm, yellow; standard longer than wings but shorter than keel. Legume 2.5-5 mm, subglobose, with raised concentric veins.
Fl. III - IV. *Recently introduced; occurring on waste ground and as a garden weed in Funchal. Native to the Mediterranean region.* **M**

18. Medicago L. [*Trevo*]
M.J. Cannon & N.J. Turland

Annual, rarely perennial herbs. Leaves trifoliolate; leaflets usually denticulate or serrulate, at least at apex, mucronate; stipules entire, dentate or laciniate. Inflorescence an axillary, pedunculate and usually few-flowered raceme. Calyx campanulate, with 5 ± equal teeth. Corolla yellow, rarely blue to violet, not persistent in fruit. Stamens diadelphous; filaments not dilated at apex. Legume longer than calyx, usually spirally coiled, spiny or unarmed, indehiscent; seeds 1-several.

Features of legumes refer only to mature examples; diameters include any spines which may be present.

Literature: K.A. Lesins & I. Lesins, *Genus* Medicago *(Leguminosae): a taxogenetic study.* (1979) The Hague.

M. sativa L., *Sp. pl.* **2**: 778 (1753) is a perennial with blue to violet flowers. Of uncertain origin, it is occasionally cultivated as a fodder plant in the Funchal region where it may also occur as an escape. M

1. Legume unarmed 2
- Legume spiny, at least with a few very short spines 3

2. Racemes 1- to 4-flowered; legume 10-17 mm in diameter, spirally coiled ....
.... **2. orbicularis**
- Racemes (3-)10- to 15-flowered; legume 1.5-3 mm, not coiled .... **1. lupulina**

3. Legume glabrous 4
- Legume with at least a few hairs 6

4. Legume 6.5-14 mm in diameter, densely spiny; spines 1-4.5 mm, spreading, hooked at tips .... **8. polymorpha**
- Legume 3.5-8 mm in diameter, almost unarmed to sparsely spiny; spines to 1.5 mm, spreading or appressed, not hooked at tips 5

5. Legume 5.5-8 mm in diameter, with a distinct submarginal vein running parallel with and near to marginal vein; spines spreading .... **3. italica**
- Legume 3.5-5.5 mm in diameter, the submarginal vein often obscure and confluent with marginal vein; spines appressed .... **4. littoralis**

6. Legume 12-18 mm in diameter, the spiral with 6-10 coils .... **9. ciliaris**
- Legume 2.5-13 mm in diameter, the spiral with (2-)3-6 coils 7

7. Leaflets 4.5-16 × 4-14 mm, obovate to obdeltate; legume 6-13 mm in diameter, shortly cylindric .... **5. truncatula**
- Leaflets generally smaller, 2.5-11 × 1.5-9 mm, often similar in shape but narrower (to narrowly oblanceolate); legume generally smaller, 2.5-10 mm in diameter, ± globose .
.... 8

8. Leaflets sometimes irregularly incise-dentate to laciniate, subglabrous to pubescent; stipules dentate to deeply laciniate; peduncle of raceme equalling or longer than subtending leaf .... **7. laciniata**
- to Leaflets never as deeply cut as above, though often sharply 3-toothed at apex, pubescent densely villous; stipules entire or dentate; peduncle of raceme much shorter than subtending leaf .... **6. minima**

1. M. lupulina L., *Sp. pl.* **2**: 779 (1753).
Annual or short-lived perennial, subglabrous to pubescent. Stems to 45 cm, usually procumbent. Leaflets 5-13 × 4.5-12 mm, orbicular to obovate; stipules lanceolate, acuminate, entire or slightly denticulate. Racemes (3-)10- to 15-flowered, very dense; peduncles usually longer than subtending

leaves. Corolla 2-3 mm, yellow. Legume 1.5-3 mm, not coiled, reniform, biconvex, glabrous or pubescent, unarmed; seed 1.
Fl. ± all year. *Frequent in grassy places and waste areas along the southern coast of Madeira, from Paúl do Mar to the Funchal region; also at São Vicente.* **M**

2. M. orbicularis (L.) Bartal., *Cat. piante Siena*: 60 (1776).
Annual, glabrous or sparsely pubescent. Stems to 45 cm, prostrate. Leaflets 4.5-17 × 4-14 mm, obovate to obdeltate; stipules laciniate. Racemes 1- to 4-flowered, lax; peduncle usually shorter than subtending leaves. Corolla 4-6 mm, yellow. Legume 10-17 mm in diameter, spirally coiled with 4-5 turns, discoid to lenticular, glabrous, unarmed; seeds several.
Fl. III - V. *Frequent in grassy places, waste areas and on roadsides along the southern coast of Madeira, from Ribeira Brava to Caniço.* **M**

3. M. italica (Mill.) Fiori in Fiori & Paol., *Iconogr. fl. ital.* ed. 2: 237 (1921).
M. helix Willd., var. *calcarata* Lowe, var. *inermis* sensu Lowe; *M. obscura* subsp. *helix* (Willd.) Batt., subsp. *helix* var. *aculeata* sensu Menezes, subsp. *helix* var. *inermis* sensu Menezes
Annual, sparsely pubescent to ± villous. Stems to 40 cm, prostrate. Leaflets 2.5-11 × 2-8 mm, obovate to rhombic; stipules laciniate. Racemes 1- to 6-flowered, dense; peduncles slightly shorter to longer than subtending leaves. Corolla 4-6 mm, yellow. Legume 5.5-8 mm in diameter, spirally coiled with 2-4.5 turns, discoid to very shortly cylindric, glabrous, almost unarmed to sparsely spiny, with a distinct submarginal vein running parallel with and near to marginal vein; spines to 1.5 mm, spreading, conical, not hooked at tips; seeds several.
Fl. III - V. *A rare plant of sunny slopes, cliffs and walls in Porto Santo.* **P**

The above description refers to subsp. **tornata** (L.) Emberger & Maire, *Cat. pl. Maroc.*: 1038 (1941) (*M. tornata* (L.) Mill.), which occurs in the western half of the Mediterranean region and to which plants from Porto Santo belong.

4. M. littoralis Rhode ex Loisel., *Not. fl. France*: 118 (1810).
M. littoralis var. *breviseta* DC., var. *inermis* Moris; *M. tribuloides* var. γ. Lowe
Like **3** but legume smaller, 3.5-5.5 mm in diameter; submarginal vein often obscure and confluent with marginal vein; spines appressed.
Fl. XII - VI. *A rare plant of rocky places and rough pastures in Porto Santo; recorded also as occurring in Madeira but without further data.* **MP**

5. M. truncatula Gaertn., *Fruct. sem. pl.* 2: 350 (1791).
M. tribuloides Desr. (excl. var. γ. Lowe), var. α. Lowe, var. β. Lowe, var. *muricata* Menezes
Annual, sparsely pubescent to ± villous. Stems to 45 cm, prostrate. Leaflets 4.5-16 × 4-14 mm, obovate to obdeltate; stipules laciniate. Racemes 1- to 3-flowered, dense; peduncles usually shorter than subtending leaves. Corolla 4.5-7 mm, yellow. Legume 6-13 mm in diameter, spirally coiled with (2-)4-6 turns, shortly cylindric, sparsely villous, with at least a few hairs, ± densely spiny; spines 1.5-4 mm, spreading to appressed, usually interlacing, slender to stout, usually curved, sometimes also hooked at tips; seeds several.
Fl. III - VII. *Cliffs, rocky places, sunny pastures, fields and waste areas; in Madeira frequent in the Funchal region, from Praia Formosa to Caniço, and on the Ponta de São Lourenço; very rare on Bugío in the Desertas; recorded also as occurring in Porto Santo but without further data.* **MDP**

Madeiran authors recognized two varieties distiguished by the armament of the legumes: *M. tribuloides* var. α. Lowe, with long spreading spines, and *M. tribuloides* var. β. Lowe (*M. tribuloides* var. *muricata* Menezes), with short appressed spines. However, intermediates occur and the species is not divided here. *M. tribuloides* var. γ. Lowe is synonymous with **4**.

6. M. minima (L.) L., *Fl. angl.*: 21 (1754).
Annual, pubescent to densely villous. Stems to 30 cm, prostrate, often densely branched and mat-forming. Leaflets 2.5-11 × 1.5-9 mm, obovate to narrowly oblanceolate, often sharply 3-toothed at apex; stipules lanceolate, acuminate, entire or dentate. Racemes 1- to 4-flowered, dense; peduncles much shorter than subtending leaves. Corolla 3-4.5 mm, yellow. Legume 2.5-10 mm in diameter, spirally coiled with 3-4.5 turns, ± globose, sparsely villous, with at least a few hairs, densely spiny; spines 0.2-4 mm, spreading to appressed, interlacing or scarcely so, slender to moderately stout, straight or curved, often but not always hooked at tips; seeds several.
Fl. I - VIII. *Cliffs, rocky and sandy places, pastures, roadsides and waste areas; in Madeira apparently restricted to the south-east, recorded from Ribeira Brava, Curral das Freiras, the Funchal region and the Ponta de São Lourenço, though probably elsewhere too; widespread in Porto Santo; also on Deserta Grande and Bugío.* **MDP**

Madeiran authors have recognized the following three varieties which appear to lack intermediates. Some authors include var. **iii** in the synonymy of var. **ii**, although these two taxa as they occur within the Madeiran archipelago seem sufficiently distinct to be kept separate.

i) var. **minima**
M. minima var. *mollissima* (Roth) Cout., var. *longispina* Benth.
Spines 1.5-4 mm, spreading, scarcely interlacing, slender, straight or slightly curved, mostly hooked at tips.
Throughout the range of the species; by far the commonest of the three varieties.

ii) var. **brevispina** Benth. in Sm. & Sowerby, *Engl. bot. Suppl.* **1**: t. 2635 (1831).
M. minima var. *vulgaris* Urban
Spines 0.5-1.5 mm, ± spreading, ± interlacing, straight or slightly curved, not hooked at tips.
Scattered in Porto Santo.

iii) var. **pulchella** (Lowe) Lowe, *Man. fl. Madeira* **1**: 166 (1862).
M. minima subsp. *pulchella* (Lowe) Menezes; *M. pulchella* Lowe
Spines 0.2-0.75 mm, appressed, interlacing (when sufficiently long), straight or hooked.
In Madeira known only from the eastern side of Funchal; rare in Porto Santo.

7. M. laciniata (L.) Mill., *Gard. dict.* ed. 8, no.5 (1768).
Like **6 i)** but less hairy, subglabrous to pubescent; leaflets sometimes irregularly incise-dentate to laciniate; stipules dentate to laciniate, often divided to their base into narrowly linear segments; racemes 1- to 2-flowered, the peduncle equalling or longer than subtending leaf.
Fl. ?. *Recorded as occurring in Madeira but without further data.* **M**

8. M. polymorpha L., *Sp. pl.* **2**: 779 (1753). [*Trevo preto* (P)]
M. hispida Gaertn., subsp. *lappacea* var. *longispina* sensu Menezes, subsp. *pentacycla* var. *nigra* (Willd.) Cout., subsp. *polymorpha* var. *apiculata* sensu Menezes; *M. lappacea* Desr., var. *brachycantha* Lowe, var. *macracantha* (Webb & Berthel.) Lowe
Annual, glabrous or subglabrous, rarely pubescent. Stems to 60 cm, prostrate or ascending through other plants. Leaflets 4-23 × 3-19 mm, obovate to obdeltate, sometimes narrowly so; stipules laciniate. Racemes 1- to 5-flowered, dense; peduncles shorter than or equalling subtending leaves. Corolla 3-5 mm, yellow. Legume 6.5-14 mm in diameter, spirally coiled with 1-5.5 turns, discoid to shortly subcylindric, glabrous, densely spiny; spines 1-4.5 mm, spreading, not interlacing, slender, straight or slightly curved, all hooked at tips; seeds several.
Fl. XI - VIII. *A weed of grassy places, cliff slopes, cultivated ground, roadsides and waste areas; in Madeira common along the southern coast, from Madalena do Mar to the Ponta de São Lourenço; common in Porto Santo; rarer on Deserta Grande and Bugío.* **MDP**

9. **M. ciliaris** (L.) All., *Fl. pedem.* **1**: 315 (1785).
Annual, glabrous to sparsely pubescent. Stems to 60 cm, prostrate. Leaflets 8-24 × 6-16 mm, obovate to obdeltate or rhombic; stipules laciniate. Racemes 1- to 5-flowered, moderately lax; peduncles shorter than subtending leaves. Corolla 5-7 mm, yellow. Legume 12-18 mm in diameter, spirally coiled with 6-10 turns, globose to shortly cylindric, villous, densely spiny; spines 2.5-4 mm, subappressed, interlacing, slender, straight, sometimes slightly curved at tips, rarely a few hooked; seeds several.
Fl. III - VII. *A rare weed of fields and waste areas; known only from Praia Formosa.* M

19. Trifolium L. [*Trevo*]
J.R. Press

Annual or perennial herbs. Leaves trifoliolate; leaflets toothed or entire; stipules partly joined to each other and to the petioles. Inflorescence terminal or axillary, capitate, usually many-flowered. Calyx 5- to 20(-30)-nerved; tube sometimes inflated in fruit; throat naked and open or closed by a callosity or ring of hairs; teeth 5, equal or unequal. Corolla persistent in fruit or not. Stamens diadelphous, all or 5 of the filaments dilated at apex. Legume usually membranous with 1-2(-9) seeds, enclosed within calyx, rarely exserted.

Literature: M. Zohary & D. Heller (1984), *The Genus* Trifolium. Jerusalem.

1. Terminal leaflets with a distinct stalk, longer than that of lateral leaflets; calyx 5-nerved 2
- Leaflets all sessile or at most very shortly stalked; calyx usually 10- to 20-nerved ... 3

2. Standard of corolla with a smooth ovate limb .... **11. dubium**
- Standard of corolla with a suborbicular sulcate limb .... **10. campestre**

3. Calyx bilabiate, the upper lip becoming conspicuously inflated and bladder-like in fruit and concealing lower lip 4
- Calyx not bilabiate, not inflated in fruit or if so only the tube expanding 6

4. Perennial; corolla not turned upside-down .... **7. fragiferum**
- Annual; corolla turned upside-down 5

5. Fruiting heads globose-stellate, the glabrescent calyces projecting ... **8. resupinatum**
- Fruiting heads smoothly globose, the pilose calyces not projecting .... **9. tomentosum**

6. Flowers usually pedicellate, subtended by small bracts; throat of fruiting calyx naked and open 7
- Flowers sessile, ebracteate; throat of fruiting calyx usually with a callosity or ring of hairs 12

7. Heads with 1-5 flowers; legume curved .... **1. ornithopodioides**
- Heads with 6-numerous flowers; legume ± straight 8

8. Perennial; stems creeping and rooting .... **3. repens**
- Annual; stems neither creeping nor rooting 9

9. Heads pedunculate; flowers shortly pedicellate 10
- Heads sessile (rarely with a very short peduncle); flowers sessile or subsessile 11

10. Heads lax; peduncles shorter than petioles, the upper very short . . . . . . **2. cernuum**
- Heads dense; peduncles longer than petioles . . . . **4. isthmocarpum**

11. Heads ± remote; corolla longer than calyx . . . . **5. glomeratum**
- Heads crowded, often ± contiguous; corolla much shorter than calyx   **6. suffocatum**

12. Heads with 2(-4) fertile outer flowers and numerous sterile inner calyces . . . .
. . . . **25. subterraneum**
- Heads with usually numerous fertile flowers only 13

13. Calyx 20-nerved 14
- Calyx 10-nerved 15

14. Heads sessile, backed by an involucre formed from expanded stipules of upper leaves .
. . . . **19. cherleri**
- Heads shortly pedunculate, involucre absent . . . . **20. lappaceum**

15. Uppermost leaves opposite 16
- All leaves alternate 17

16. Calyx-tube densely hairy; upper 2 calyx-teeth fused for *c.* ⅓ of their length . . . .
. . . . **24. squarrosum**
- Calyx-tube glabrescent; upper 2 calyx-teeth not fused . . . . **23. squamosum**

17. Heads sessile, involucrate 18
- Heads pedunculate, without involucres 21

18. Perennial . . . . **12. pratense**
- Annual 19

19. Leaflets with lateral veins curved towards margin . . . . **17. scabrum**
- Leaflets with lateral veins ± straight 20

20. Throat of calyx naked within; calyx-teeth ± equal . . . . **15. striatum**
- Throat of calyx with a ring of hairs within; lowest calyx-tooth longest . **16. bocconei**

21. Throat of calyx narrowed or closed in fruit by an annular or bilabiate callosity . . . 22
- Throat of calyx remaining open in fruit, without a callosity 24

22. Leaflets linear-lanceolate; throat of calyx completely closed in fruit by a bilabiate callosity
. . . . **22. angustifolium**
- Leaflets obovate to suborbicular; throat of calyx narrowed but not closed in fruit by an
annular callosity 23

23. Corolla longer than calyx . . . . **14. incarnatum**
- Corolla much shorter than calyx . . . . **18. ligusticum**

24. Fruiting heads globose; calyx-teeth stellately spreading in fruit . . . . . **13. stellatum**
- Fruiting heads cylindrical; calyx-teeth erect in fruit . . . . **21. arvense**

Sect. LOTOIDEA Crantz. Bracts entire, bifid or crenulate, rarely absent. Calyx 5- to 30- but usually 10-nerved; throat open, naked; teeth usually equal. Legume 2- to 9-seeded.

1. T. ornithopodioides L., *Sp. pl.* **2**: 766 (1753).
Trigonella ornithopodioides (L.) DC.
Glabrous annual. Stems procumbent. Leaflets 2-6 × 1-3 mm, obovate-cuneate, mucronate, serrate above. Stipules lanceolate, acuminate, ± membranous. Heads axillary, subsessile to shortly pedunculate, 1- to 3(-5)-flowered; flowers shortly pedicellate. Calyx 10-nerved; tube cylindric to campanulate; teeth equal, longer than tube, narrowly triangular at base, subulate. Corolla 6-8 mm, white, persistent. Legume clearly exserted from calyx, curved, 4- to 6(-9)-seeded.
Fl. VI. *Short turf in mountain pastures; very rare and known only from one locality in the high peaks of eastern Madeira, between Pico de Arieiro and Pico do Cidrão, but possibly overlooked elsewhere.* **M**

2. T. cernuum Brot., *Phytogr. Lusit. select.* **1**: 150 (1816).
Glabrous annual. Stems procumbent or ascending, much-branched, slender. Leaflets 3-10 × 2-7 mm, obovate to obovate-cuneate, ± rounded or emarginate, serrulate. Stipules triangular, acuminate, membranous. Heads axillary, pedunculate, *c.* 8 mm, hemispherical, lax; peduncles shorter than petioles, the upper very short; flowers with pedicels *c.* 2 mm, strongly deflexed in fruit. Calyx tubular, 10-nerved; 2 upper teeth very slightly longer than 3 lower, all recurved in fruit. Corolla 4 mm, slightly longer than calyx, whitish or pale pink, persistent.
Fl. V - VII. *A rare plant occurring in a few places in the high central mountains of Madeira and in scattered localities in the south-east, between Monte and Caniço; also on Pico do Castelo in Porto Santo.* **MP**

3. T. repens L., *Sp. pl.* **2**: 767 (1753).
More or less glabrous perennial. Stems far-creeping, rooting. Petioles long, to 27 cm, ribbed. Leaflets 15-25(-50) × 10-20(-45) mm, ovate-elliptic to obcordate, shallowly emarginate or entire at apex, serrulate at margins, with light or dark markings. Stipules membranous, sheathing, the free upper portion subulate. Heads axillary, pedunculate, (15-)20-30 mm, globose, lax; peduncles exceeding petioles; flowers fragrant, with pedicels 2-4 mm, strongly deflexed after flowering. Calyx campanulate, (6-)10-nerved; 2 upper teeth longer than 3 lower. Corolla 6-10 mm, longer than calyx, white to pale pink or deep pink, persistent.
Fl. II - VII, but often at other times. *Roadsides, grassy places and waste areas; common almost everywhere in Madeira, up to 1700 m.* **M**

The above description refers to var. **repens**, to which Madeiran plants belong.

4. T. isthmocarpum Brot., *Phytogr. Lusit. select.* **1**: 148 (1816).
Like **3** but annual; stems procumbent to ascending, not rooting at nodes; heads 9-20 mm; pedicels only weakly deflexed after flowering; calyx-teeth subequal.
Fl. IV - V. *Introduced; recorded from waste ground in Funchal harbour. Native to Portugal and the western Mediterranean region.* **M**

5. T. glomeratum L., *Sp. pl.* **2**: 770 (1753).
Glabrous annual. Stems procumbent or ascending, much branched, slender. Petioles mostly short. Leaflets 6-15 × 5-11 mm, obovate to obcordate, mucronate, serrulate. Stipules ovate, membranous, the free upper portion subulate. Heads axillary, sessile or rarely very shortly pedunculate, 6-14 mm, ± globose, dense; flowers subsessile. Calyx ± obconic, 10(-12)-nerved; teeth spreading or deflexed, subequal, triangular-ovate, auriculate at base. Corolla 4-6 mm, longer than calyx, pale pink to rose pink, persistent.
Fl. II - VI. *Rocky places, cornfields and waste areas; common in all but the highest regions of Madeira; less common in Porto Santo where it occurs on the higher slopes and summits; also on Ilhéu Chão and Deserta Grande.* **MDP**

6. T. suffocatum L., *Mant. pl.*: 276 (1771).
Glabrous, tufted annual. Stems procumbent, slender. Leaflets (4-)5-10 × (2-)3-7 mm, obovate-cuneate, truncate, serrulate above. Stipules ovate-acuminate, membranous. Heads axillary, sessile and often crowded on stems, 5-8 mm, ± globose; flowers sessile. Calyx ovoid, 10-nerved; teeth subequal, lanceolate, recurved in fruit. Corolla *c*. 3 mm, much shorter than calyx, pale rose-pink, persistent.
Fl. IV - VI. *Rare, in a few localities in the central mountains of Madeira and on or near the south-eastern coast; also recorded from the western half of Porto Santo and the highest parts of Deserta Grande.* **MDP**

Sect. VESICARIA Crantz. Bracts entire or toothed, sometimes united to form a small involucre. Calyx bilabiate; upper lip inflated and vesiculous in fruit; lower lip remaining unchanged; throat narrowed or closed. Legume 1- to 2-seeded.

7. T. fragiferum L., *Sp. pl.* **2**: 772 (1753).
Hairy perennial. Stems creeping, often rooting at nodes, woody. Leaflets 6-13 × 4-9 mm, ovate to suborbicular but mostly obovate-cuneate, obtuse or notched at apex, minutely toothed to spinulose at margins, glabrous or hairy on veins; lateral veins somewhat curved. Stipules lanceolate, membranous, sheathing stems in plants with short internodes. Heads axillary, pedunculate, 8-13 mm, hemispherical in flower, ± globose in fruit; peduncles usually longer than petioles, hairy; lowest bracts entire or toothed, united to form an irregular involucre; flowers ± sessile. Calyx tubular, bilabiate; 2 upper teeth longer and narrower than 3 lower, pilose on upper side; upper lip becoming deflexed, inflated, membranous, reticulately veined and often pink or reddish in fruit. Corolla 4-6 mm, longer than the calyx at anthesis, pink, persistent, included within or just projecting from fruiting calyx.
Fl. V - IX. *Common in grassy places and waste areas near the sea in Madeira.* **M**

Madeiran plants appear to belong to var. **pulchellum** Lange, *Meddel. Nat. Floren.* **2**, Aart. 7: 169 (1865), distinguished by its short stems, short internodes covered by long membranous stipules and relatively small fruiting calyces.

8. T. resupinatum L., *Sp. pl.* **2**: 771 (1753).
Glabrous annual. Stems erect or ascending and spreading, much-branched from base, grooved. Leaflets very shortly stalked, 6-16 × 4-8 mm, obovate-cuneate, minutely toothed to spinulose. Stipules membranous, fused in lower half, the free upper portion subulate. Heads axillary, pedunculate, 6-10 mm, hemispherical, expanding and becoming globose-stellate in fruit; peduncles shorter than to slightly exceeding petioles; lowest bracts reduced to an involucre of scales; flowers subsessile. Calyx bilabiate; 2 upper teeth much longer and narrower than 3 lower, pale green, pilose above but glabrescent later; upper lip becoming inflated, ovoid to flask-shaped, membranous, reticulately veined and projecting but not deflexed in fruit. Corolla turned upside-down, 3-4 mm, longer than calyx, pink to purplish, persistent, hidden by enlarged upper lip of fruiting calyx.
Fl. IV - VIII. *Rare in dry open places near the sea in Madeira; also in pastures and on rocky slopes in Porto Santo.* **MP**

9. T. tomentosum L., *Sp. pl.* **2**: 771 (1753).
Like **8** but smaller and procumbent; heads 5-7 mm, becoming smoothly globose in fruit; peduncles often very short and always shorter than petioles; upper lip of fruiting calyx ovoid, whitish or pinkish, pilose, sometimes densely so; corolla not turned upside-down, white or pale pink.
Fl. III - V. *Rough pastures, roads, paths and waste areas; in Madeira frequent in dry open sunny places near the sea; rarer in Porto Santo, occurring on the higher slopes and peaks.* **MP**

Sect. **CHRONOSEMIUM** Ser. Bracts minute or absent. Calyx 5-nerved, throat open, naked, teeth unequal. Legume 1- to 2-seeded.

10. T. campestre Schreb. in Sturm, *Deutschl. Fl.* Abt. 1, Band 4, Heft 16 (1804).
T. agrarium L., pro parte; *T. procumbens* L., pro parte
Hairy annual. Stems prostrate to ascending, sometimes erect, much-branched, the branches spreading. Leaflets (7-)10-16 × 4-9 mm, obovate, very shortly toothed, the terminal leaflet with a long stalk. Stipules ovate, green. Heads axillary, pedunculate, *c.* 10 mm, ovoid to subglobose, dense; peduncules as long as or exceeding petioles; flowers pedicellate, becoming deflexed and closely overlapping in fruit. Calyx 5-nerved, membranous, sparsely hairy or glabrous; teeth slender, the 2 upper short and not exceeding length of tube, the 3 lower 1-3 × as long as tube. Corolla 3-5 mm, much longer than calyx, yellow, becoming brown in fruit; limb of standard suborbicular, grooved.
Fl. III - VI. *Pastures, rocky ground and waste areas; very common throughout Madeira and Porto Santo, up to 1700 m; also in the higher parts of Deserta Grande.* **MDP**

11. T. dubium Sibth., *Fl. oxon.*: 231 (1794).
T. minus Sm.
Glabrous or sparsely hairy annual. Stems prostrate to erect, slender. Leaflets 5-10 × 3-5 mm, obovate-cuneate, emarginate, toothed above, the terminal leaflet stalked. Stipules ovate, green. Heads axillary, pedunculate, to 10 mm, hemispherical to ovoid; peduncles generally twice as long as the leaves, slender; flowers pedicellate, becoming brown and slightly deflexed in fruit. Calyx 5-nerved, membranous, glabrous; 2 upper teeth shorter than tube; 3 lower teeth 1-2 × as long as tube. Corolla 3 mm, yellow, persistent; limb of standard ovate, smooth.
Fl. IV - VIII. *Mountain pastures, open ground and levada banks in woodland; rare in the higher parts of Madeira and the peaks of Porto Santo.* **MP**

Sect. **TRIFOLIUM**. Bracts absent or few at the base of each head. Calyx 10- to 20-nerved; throat usually closed by a bilabiate or annular callosity or a ring of hairs; teeth equal or unequal. Legume 1(-2)-seeded.

12. T. pratense L., *Sp. pl.* **2**: 768 (1753).
Patent- or appressed-pubescent to glabrescent perennial. Stems ascending or erect. Leaflets shortly stalked, 15-40 mm, ovate to elliptic or obovate, usually obtuse at apex, entire or minutely toothed at margins, hairy on both surfaces or glabrescent above, often spotted; lateral veins slightly recurved towards margin. Stipules ovate, abruptly acuminate. Heads terminal, usually sessile, 20-25 mm, ± globose, backed by an involucre of stipules. Calyx 10-nerved, appressed-hairy; throat with a ring of hairs within; teeth erect, unequal, subulate, sparsely hairy, the 4 upper about equalling tube, the lowest twice as long as tube. Corolla up to twice as long as the calyx, reddish purple.
Fl. V - VII. *Introduced from Britain as a pasture plant; now established in Madeira but rare, not spreading and ± confined to cultivated land. Native to Europe & W. Asia but cultivated throughout the N. hemisphere.* **M**

13. T. stellatum L., *Sp. pl.* **2**: 769 (1753).
Densely patent-pubescent annual. Stems erect to ascending, branched only at base. Leaflets 8-12 mm, obcordate-cuneate to obovate-cuneate, minutely toothed above. Stipules ovate, toothed above, with prominent green veins and margins. Heads terminal, pedunculate, 15-20 mm, hemispherical to ovoid, expanding and becoming globose in fruit. Calyx 10-nerved, densely tuberculate-hairy; throat white-villous within and turning a contrasting reddish purple in fruit; teeth equal, 2-3 × as long as tube, narrowly triangular to subulate, fused at base, stellately spreading in fruit and remaining soft. Corolla shorter than or about equalling calyx, white or creamy to rose-pink.

Fl. III - VII. *Infrequent in dry sunny places in the hills near the south-eastern coast of Madeira, as well as on the Ponta de São Lourenço.* **M**

14. T. incarnatum L., *Sp. pl.* **2**: 769 (1753).
Patent- or appressed-pubescent annual. Stems erect. Leaflets obovate-cuneate to suborbicular, minutely toothed above. Stipules ovate, blunt, minutely toothed and often green or purplish towards apex. Heads terminal, pedunculate, oblong-cylindric, elongating in fruit. Calyx 10-nerved, hairy; throat villous within and narrowed in fruit; teeth ± equal, 1-2 × as long as the tube, setaceous, spreading in fruit. Corolla longer than calyx, deep red, sometimes pink.
Fl. IV - VI. *Introduced to Madeira as a forage crop and apparently naturalized in a few places. Native and widely cultivated in Europe and Turkey.* **M**

15. T. striatum L., *Sp. pl.* **2**: 770 (1753).
T. striatum subsp. *genuinum* (Lan.) Cout.
Sparsely to densely patent-villous annual. Stems numerous, ascending to erect. Upper leaflets 9-20 mm, obovate-cuneate, obtuse or acute, minutely toothed. Stipules connate in lower half, ovate with a long subulate tip. Heads sessile, 12-16 × 8-12 mm in fruit, ovoid, becoming cylindrical, backed by an involucre formed from expanded stipules of upper leaf. Calyx 10-nerved; tube ovoid, becoming inflated in fruit and readily falling; throat open, naked; teeth equal, about equalling or shorter than tube, subulate, erect or slightly spreading in fruit. Corolla about as long as calyx, pink, persistent.
Fl. III - V. *Very common in open rocky places up to c. 1000 m in Madeira; known only from Pico do Facho in Porto Santo.* **MP**

16. T. bocconei Savi in *Atti Accad. Firenze* **1**: 191 (1808).
Patent- or appressed-pubescent annual. Stems ascending, sparsely branched. Leaflets of upper leaves 8-18 mm, narrowly obovate-cuneate, toothed above; lateral veins ± straight. Stipules oblong-lanceolate with a subulate upper portion. Heads sessile, 10-14 mm, ovoid to cylindrical, subtended by 1-2 leaves. Calyx 10-nerved, pubescent; throat open, with a ring of hairs within; teeth erect, unequal, subulate, the lowermost longest and equalling tube. Corolla about as long as calyx, pink, persistent.
Fl. IV - VI. *Possibly recently introduced on waste ground in Funchal, but recorded once before, over a century ago, on Pico da Silva north-east of Funchal. Native to S. and W. Europe.* **M**

The above description applies to var. **bocconei**, to which Madeiran plants belong.

17. T. scabrum L., *Sp. pl.* **2**: 770 (1753).
Sparsely to densely pubescent annual. Stems prostrate, flexuous, rather thick. Leaflets 6-10 mm, obovate-cuneate, sometimes broadly so, rounded or emarginate, entire or minutely toothed, coriaceous; lateral veins thickened and recured at margins. Stipules oblong with a triangular-acuminate upper portion, slightly membranous, pale. Heads all axillary, sessile, 5-8 mm, ovoid, becoming more cylindric in fruit, backed by an involucre formed from stipules of subtending leaf, the whole head falling when ripe. Calyx-tube cylindrical, 10-nerved, hairy; throat narrowed by an annular callosity; teeth unequal, lanceolate to narrowly triangular, hairy, erect or slightly spreading and very stiff in fruit, the lowest tooth slightly longer than tube, sometimes recurved in fruit. Corolla shorter than or equalling the calyx, white, persistent.
Fl. III - VI. *Dry ground in open sunny places, usually near the sea; common in Madeira and throughout Porto Santo; rather rare on Ilhéu Chão and present only in the higher parts of Deserta Grande and Bugío.* **MDP**

18. T. ligusticum Balb. ex Loisel., *Fl. gall.*: 731 (1807).
Patent-pubescent annual. Stems decumbent to ascending or erect, sparsely branched from base.

Leaflets 8-14 mm, obovate, minutely toothed above. Stipules oblong, the free upper portion with a long setaceous point. Heads pedunculate, often paired, 10-20 mm, ovoid to oblong. Calyx 10-nerved; tube campanulate, hairy or somewhat glabrescent; throat narrowed in fruit by an annular calosity; teeth erect or slightly spreading, ± equal, 1-2 × as long as tube, setaceous with a triangular base, ciliate. Corolla much shorter than the calyx, pale pink.
Fl. VI - VII (-VIII). *Common in pastures and rocky places in the mountains of Madeira, sometimes also in the lower regions and along maritime cliffs; very common along the ridges of Deserta Grande.* **MD**

19. T. cherleri L., *Demonstr. pl.*: 21 (1753).
Patent-pubescent annual. Stems prostrate to ascending. Leaflets 6-12 mm, obovate-cuneate, minutely toothed above. Stipules united in lower half, membranous, pale with green veins, the free upper portion ovate-lanceolate. Heads sessile, 16-20 mm in fruit, many-flowered, ± globose, subtended by a membranous involucre formed from expanded stipules of upper leaves, the whole head readily falling when ripe. Calyx 20-nerved, densely hairy; teeth equalling or longer than tube, plumose; throat open, with a dense ring of hairs within. Corolla shorter than to slightly longer than calyx, white, persistent. Legume 1-seeded.
Fl. IV - V. *Rather rare on dry stony ground in the hills around Caniço and Garajau to the east of Funchal.* **M**

20. T. lappaceum L., *Sp. pl.* **2**: 768 (1753).
Sparsely pubescent annual. Stems prostrate to ascending. Leaflets 7-12 mm, obovate-cuneate to obcordate-cuneate, minutely toothed, dull blackish green. Stipules oblong, membranous with conspicuous green veins, the free upper portion subulate. Heads ± sessile at first, becoming pedunculate, 12-20 mm in fruit, globose. Calyx 20-nerved, glabrous; teeth equal, longer than tube, hairy, spinescent in fruit; throat open, with a ring of hairs within. Corolla equalling calyx at anthesis but shorter in fruit, white or pinkish, persistent. Legume 1-seeded.
Fl. III - IV. *Rare in fields and undisturbed waste ground in the south-eastern coastal lowlands of Madeira, from Funchal to the islets off Ponta de São Lourenço; also in Porto Santo and on Ilhéu Chão in the Desertas.* **MDP**

The above description applies to var. **lappaceum**, to which Madeiran plants belong.

21. T. arvense L., *Sp. pl.* **2**: 769 (1753).
Appressed- and greyish-pubescent annual. Stems erect, much branched above, the branches spreading. Leaves shortly petiolate, the upper subsessile. Leaflets shortly stalked, 9-20 mm, linear-oblong to oblanceolate or elliptic, mucronulate. Stipules ovate, prominently nerved; upper portion much longer than lower part, setaceous. Heads numerous, pedunculate, 10-20 × 8-10 mm, ovoid to cylindrical. Calyx 10-nerved; tube campanulate, becoming globose in fruit, densely hairy; throat open and slightly hairy within; teeth erect, ± equal, 2-3 × as long as tube, setaceous, often pinkish, plumose. Corolla much shorter than calyx, white or pink, persistent.
Fl. III - VI. *Rocky ground, roadsides and waste places in the lowlands; frequent but scattered in Madeira; very rare in Porto Santo; also on Ilhéu Chão and on the islet of Doca off the western coast of Deserta Grande.* **MDP**

22. T. angustifolium L., *Sp. pl.* **2**: 769 (1753). [*Trevo massaroco* (P)]
Appressed-pubescent annual. Stems erect, branching at base. Leaflets 2-5 cm, linear-lanceolate. Stipules prominently nerved; upper portion equalling or longer than lower part, subulate or lanceolate. Heads terminal, pedunculate, to 8 cm, cylindric or conical, elongating in fruit. Calyx 10-nerved, densely tuberculate-hairy; throat closed by a bilabiate callosity in fruit; teeth narrowly triangular to subulate, spinescent and stellately spreading in fruit, the 4 upper teeth about equalling tube, the lower tooth longer. Corolla equalling or slightly longer than calyx, pale to deep rose-pink.

Fl. III - V. *Common in grassy places, waste areas and along roadsides throughout the lower coastal parts of Madeira and Porto Santo; common in the higher parts of Deserta Grande; also present on Ilhéu Chão and on Selvagem Grande.* **MDPS**

23. T. squamosum L., *Amoen. acad.* **4**: 105 (1759).   [*Trevo de pé de passaro* (P)]
T. maritimum Hudson.
Sparsely hairy to pubescent annual. Stems erect or ascending, branched. Leaves alternate except the 2 uppermost which are opposite, dark green. Leaflets of upper leaves 10-17 mm, narrowly obovate-cuneate, minutely toothed above. Stiples oblong, membranous; upper free portion longer than lower part, linear, green. Heads shortly pedunculate, 10-12 mm, ovoid. Calyx 10-nerved; tube glabrescent, leathery; throat closed by a bilabiate callosity; teeth narrowly triangular, ciliate, stiff and spreading in fruit, all $\pm$ equal and shorter than tube or lowest tooth longest and about as long as tube. Corolla longer than calyx, pink.
Fl. IV - V. *Rocky slopes and pastures near the sea; rare on the south coast of Madeira; more frequent in Porto Santo, though still rare.* **MP**

24. T. squarrosum L., *Sp. pl.* **2**: 768 (1753).
Glabrous or hairy annual. Stems robust, procumbent to erect. Leaves alternate except the 2 uppermost which are opposite. Leaflets oblanceolate to narrowly elliptic, acute, rounded or notched, mucronate. Stipules with prominent often purplish veins; free upper portion much longer than lower part, linear, green. Heads pedunculate or shortly so, ovoid; flowers $\pm$ sessile. Calyx 10-nerved, densely hairy; tube ovoid, becoming constricted at mouth; throat closed in fruit by a hairy bilabiate callosity; teeth somewhat spreading, narrowly triangular, ciliate, the 4 upper shorter than or about equalling tube, the lower 1½-2 × as long, the 2 uppermost connate for *c.* ⅓ of their length. Corolla equalling or slightly longer than calyx, white or pink.
Fl. III - V. *Grassy slopes by streams; restricted to Porto Santo, where it occurs near the northern coast, north of Pico Juliana.* **P**

Sect. TRICOCEPHALUM W.D.J. Koch. Bracts 9. Inner flowers sterile, consisting of a solid calyx-tube and 5 long bristly teeth, developing during or after flowering of fertile outer flowers and eventually concealing them in fruit. Legume 1-seeded.

25. T. subterraneum L., *Sp. pl.* **2**: 767 (1753).
Procumbent patent-hairy annual. Leaves usually long-petiolate; leaflets obcordate, minutely toothed. Stipules ovate, acute. Outer fertile flowers (2-)4, becoming deflexed after fertilisation; corolla 10 mm, about twice as long as calyx, white. Sterile flowers numerous, eventually strongly deflexed over fertile fruiting calyces; calyx-teeth twice as long as tube; corolla absent. Fruiting heads globose, becoming buried in the ground by the growth and change in orientation of the peduncles and sterile flowers, ripening underground. Legume membranous, completely enclosed by or just protruding from the accrescent calyx.
Fl. III - VI. *Rare in woods, mountain pastures and other grassy places in Madeira, up to 800 m.* **M**

The above description applies to subsp. **subterraneum** var. **subterraneum**, although Madeiran plants appear to have fewer fertile flowers than is usual in this variety. Subsp. **brachycalycinum** var. **flagelliforme** Guss., *Enum. pl. Ins. Inarime*: 50 (1854) is recorded from Madeira but possibly in error. It differs in having long-acuminate stipules, sterile calyces with teeth twice as long as the tube, and only the lower ⅓ of the legume concealed by the calyx.

20. Lotus L.

Glabrous to densely hairy annual, biennial or perennial herbs. Stems prostrate to erect, usually slender, sometimes becoming woody below. Leaves imparipinnate, 5-foliolate; lower pair of leaflets borne at base of petiole and simulating stipules, often smaller than upper leaflets and of a different shape; rachis (if present) simulating a petiole; stipules minute. Inflorescence axillary, sessile or pedunculate, 1- to several-flowered. Bracts trifoliolate, like the upper leaflets. Calyx campanulate or tubular-campanulate, regular or bilabiate. Corolla pale yellow to orange, white to pink, or purple to blackish, sometimes streaked and tinged with red or purple, not persistent in fruit; keel beaked. Stamens diadelphous; filaments not dilated at apex. Style toothed or not. Legume usually longer than calyx, cylindric, rarely fusiform or linear and laterally compressed, straight or curved, often with bulges and constrictions, usually glabrous, dehiscent; valves usually twisting on dehiscing; seeds 2-numerous.

1. Style toothed [Sect. Pedrosia] 7
- Style not toothed 2

2. Calyx bilabiate, with lateral teeth very short; legume laterally compressed, distinctly constricted between seeds, appearing segmented . . . . **6. ornithopodioides**
- Calyx ± regular, with teeth all ± equal but sometimes curved; legume cylindric or fusiform, not or weakly constricted between seeds, not appearing segmented 3

3. Perennial, with persistent woody stock; peduncles 5- to 20-flowered   **1. pedunculatus**
- Annuals, with slender roots; peduncles 1- to 6-flowered 4

4. Peduncles shorter than leaves; corolla white or pale pink, with violet keel; legume normally exceeding 25 mm, strongly curved upwards . . . . **5. conimbricensis**
- Peduncles usually equalling or longer than leaves; corolla yellow or orange-yellow, sometimes streaked and tinged with red; legume up to 25 mm, straight or only slightly curved upward 5

5. Peduncles (2-)4- to 6-flowered; legume not or only slightly longer than calyx, the valves not twisting on dehiscing . . . . **2. parviflorus**
- Peduncles 1- to 3(-5)-flowered; legume at least twice as long as calyx, the valves twisting on dehiscing 6

6. Legume 8-16 × 1.2-2 mm, not more than 3 × as long as calyx . . . . . **3. suaveolens**
- Legume 15-25 × 1-1.7 mm, at least 3 × as long as calyx . . . . . . **4. angustissimus**

7. Rachis of leaves 0-5 mm long 8
- Rachis of leaves 2-15 mm long 9

8. Upper leaflets obovate to oblanceolate, rarely narrower; corolla yellow to orange, rarely paler, streaked with red or purple, often darkening on fading; legume glabrous . . . . **8. glaucus**
- Upper leaflets narrowly oblanceolate to linear; corolla deep mauve to blackish purple, or white with standard tinged pink; legume densely pubescent . . . . **11. loweanus**

9. Peduncles 20-90 mm, 1- to 5-flowered; corolla yellow . . . . **7. lancerottensis**
- Peduncles 2-14 mm, 1(-2)-flowered; corolla entirely pale yellowish or dark purplish, or a combination of both colours 10

10. Stems becoming stout and woody below; leaflets densely sericeous, appearing silvery or silvery green; legume 2.5-4 mm wide . . . . **10. argyrodes**
- Stems remaining slender below; leaflets pubescent, appearing greenish or slightly glaucous; legume 1.5-3 mm wide . . . . **9. macranthus**

Sect. LOTUS. Calyx ± regular; teeth all ± equal but sometimes curved; corolla yellow or orange-yellow, sometimes streaked or tinged with red; style not toothed.

1. L. pedunculatus Cav., *Icon.* **2**: 52 (1793).
L. uliginosus Schkuhr, subsp. *pisifolius* (Lowe) Menezes, var. *glabriusculus* Bab., var. *pisifolius* (Lowe) Lowe
Glabrous perennial with persistent woody stock. Stems to 90 cm, sometimes stout. Leaves with rachis; upper leaflets 7-40 × 3.5-20 mm, obovate- to oblanceolate-rhombic; lower leaflets shorter than but about as wide as the upper, 4-25 × 3-20 mm, ovate. Peduncles 45-180 mm, much longer than leaves, 5- to 20-flowered. Bracts usually longer than calyx. Calyx 5-7 mm. Corolla 10-12 mm, yellow; standard streaked reddish at base, tinged reddish on reverse. Legume 10-35 × 1-2.5 mm, cylindric, straight, weakly constricted between seeds, glabrous; valves twisting on dehiscing.
Fl. V - IX. *Moist banks and gullies, dripping rocks, stream-sides, damp maritime cliffs and in other wet places; widespread in Madeira, up to 1400 m.* **M**

2. L. parviflorus Desf., *Fl. atlant.* **2**: 206 (1799).
L. parviflorus var. *robustus* Lowe, var. *tenuis* Lowe
Villous annual, with hairs long, rather dense and ± patent. Stems to 30 cm. Leaves with rachis; upper leaflets 4-11 × 2-5 mm, obovate- to oblanceolate-rhombic; lower leaflets about as long as but wider than the upper, 3.5-10 × 2.5-6 mm, ovate. Peduncles 10-30 mm, usually equalling or longer than leaves, (2-)4- to 6-flowered, becoming recurved in fruit. Bracts about as long as calyx. Calyx 4-6 mm. Corolla 4.5-6 mm, yellow to orange-yellow. Legume not or only slightly longer than calyx, fusiform, glabrous; valves not twisting on dehiscing.
Fl. III - VII. *Mountain pastures and other grassy places, dry rocky ground, roadside banks, cultivated and waste areas; widespread in eastern Madeira, up to 950 m.* **M**

3. L. suaveolens Pers., *Syn. pl.* **2**: 354 (1807).
L. hispidus sensu auct. mad., non DC. (1805).
Villous annual, with hairs weakly appressed to patent. Stems to 45 cm. Leaves with rachis; upper leaflets 3-15 × 1.5-7 mm, obovate- to oblanceolate-rhombic; lower leaflets about as long and as broad as the upper, 3-10 × 1.5-6 mm, ovate. Peduncles 15-60 mm, longer than leaves, (1-)3(-5)-flowered, not recurved in fruit. Bracts equalling or shorter than than calyx. Calyx 3-6 mm. Corolla 5-8 mm, orange-yellow; standard streaked with red, tinged red on reverse. Legume 8-16 × 1.2-2 mm, up to 3 × as long as calyx, cylindric, straight or slightly curved upwards, weakly constricted between seeds, glabrous; valves twisting on dehiscing.
Fl. IV - X. *Cliffs, grassy, rocky and sandy places; widespread in Madeira up to 850 m, but more common in the east; also on Pico do Castelo in Porto Santo and on Deserta Grande.* **MDP**

4. L. angustissimus L., *Sp. pl.* **2**: 774 (1753).
Like **3** but peduncles 1- to 2(-3)-flowered; corolla usually paler (base colour yellow); legume 15-25 × 1-1.7 mm, at least 3 × as long as calyx.
Fl. VI - IX. *Cleared woodland, levadas, waysides and similar habitats to 3 (the two species often growing together); eastern Madeira, from Funchal up to Curral das Freiras, Fajã da Nogueira and Camacha.* **M**

Sect. ERYTHROLOTUS Brand. Calyx ± regular; teeth all ± equal but sometimes curved; corolla white or pale pink, with violet keel; style not toothed.

5. L. conimbricensis Brot., *Phytogr. Lusit. select.*: 59 (1800).
Glabrous or sparsely villous annual. Stems to 25 cm. Leaves with rachis; upper leaflets 5-10 × 3-5 mm, obovate- to oblanceolate-rhombic; lower leaflets longer and broader than the upper,

6-10 × 3.5-9 mm, ovate-deltoid. Peduncles 3-10 mm, shorter than leaves, 1- to 2-flowered. Bracts equalling or slightly exceeding calyx. Calyx 5-6 mm. Corolla 6-7 mm. Legume 40-65 × 1-1.5 mm, cylindrical, strongly curved upwards, weakly constricted between seeds, glabrous; valves twisting on dehiscing.
Fl. ?. *Introduced; recorded from waste ground in Funchal harbour. Native to the Mediterranean region.* **M**

Sect. LOTEA (Medik.) Willk. Calyx bilabiate; lateral teeth very short; corolla yellow; style not toothed.

6. L. ornithopodioides L., *Sp. pl.* **2**: 775 (1753).
Pubescent annual, with hairs weakly appressed to patent. Stems to 45 cm. Leaves with rachis; upper leaflets 10-20 × 5-12 mm, obovate-rhombic; lower leaflets shorter and narrower than the upper, 4-10 × 3-8 mm, ovate-rhombic. Peduncles 20-50 mm, longer than leaves, (2-)3- to 5-flowered. Bracts longer than calyx. Calyx *c*. 6 mm. Corolla 7-8 mm, yellow. Legume 30-55 × *c*. 2.5 mm, linear, laterally compressed, curved upwards, distinctly constricted between seeds, appearing segmented, glabrous; valves twisting on dehiscing.
Fl. III - V. *Introduced; a rare plant occurring in southern Madeira, from Ribeira Brava and Quinta Grande up to Serra de Água. Native to the Mediterranean region.* **M**

Sect. PEDROSIA (Lowe) Brand. Calyx ± bilabiate; lateral teeth usually shorter than others; upper teeth usually curved upwards; corolla variously coloured; style toothed.

7. L. lancerottensis Webb & Berthel., *Hist. nat. Iles Canaries* **3**(2,2): 84 (1842).
L. neglectus (Lowe) Masf.; *Pedrosia neglecta* Lowe, var. *cinerea* Lowe, var. *virescens* Lowe
Pubescent tap-rooted perennial with persistent woody stock. Stems and leaves appearing greenish, with hairs short and closely appressed. Stems to 60 cm, much-branched. Leaves with rachis 3-5 mm; upper leaflets 4-10 × 2.5-7 mm, obovate; lower leaflets shorter and narrower than the upper, 2.5-5.5 × 2-5.5 mm, ovate to broadly so. Peduncles 20-90 mm, longer than leaves, 1- to 5-flowered. Bracts shorter than calyx. Calyx 5.5-7 mm. Corolla 10-12 mm, yellow. Legume 15-25 × 1.8-2.3 mm, cylindric, straight, weakly constricted between seeds, glabrous; valves twisting on dehiscing. 2n=28. **Plate 21**
Fl. III - VIII. *Rare on maritime cliffs on the southern coast of Madeira, known from only a few localities between Ribeira Brava and Cabo Garajau and apparently not seen recently.* ■ **M** [*Canaries*]

Madeiran plants are tetraploid, whereas those from the Canaries are diploid (2n = 14).

8. L. glaucus Aiton, *Hort. kew.* **3**: 92 (1789). [*Bacaira* (P), *Trevina*]
L. glaucus var. *angustifolius* R.P. Murray; *L. salvagensis* R.P. Murray; *Pedrosia florida* Lowe, var. *aurantiaca* Lowe, var. *sulphurea* Lowe; *P. glauca* (Aiton) Lowe, var. *dubia* Lowe, var. *intricata* Lowe; *P. paivae* Lowe
Variable, minutely pubescent, tap-rooted perennial with persistent woody stock. Stems and leaves appearing glaucous to silvery, with hairs sparse to dense, appressed. Stems to 45 cm, much-branched, usually forming dense mats. Leaves sometimes fleshy, without rachis, or with short rachis to 2(-3) mm; upper leaflets 2-8(-13.5) × 0.5-3.5(-4) mm, obovate to oblanceolate, rarely narrower; lower leaflets shorter than but about as wide as the upper, 2-4.5(-12.5) × 0.5-4 mm, ovate or obovate to oblanceolate, rarely narrower. Peduncles 5-30(-40) mm, usually longer than leaves, 1- to 3(-6)-flowered. Bracts usually shorter than calyx. Calyx 5-8(-9) mm. Corolla 10-15 mm, yellow to orange, rarely paler, streaked with red or purple, often darkening to reddish brown on fading. Legume 15-35 × 1.5-2.5 mm, cylindric, straight, weakly constricted between

seeds, often also with irregular deep constrictions, glabrous; valves twisting on dehiscing. 2n=14.
Plate 21
Fl. all year. *Maritime cliffs, rocks, stony and sandy ground, coastal hills and dry roadside banks, up to 100 m; common around the coasts of Madeira; rarer but also occurring inland in Porto Santo; common on Ilhéu Chão and Deserta Grande, rarer on Bugío; common on Selvagem Grande, also on Selvagem Pequena and Ilhéu de Fora.* ■ **MDPS** [*Canaries*]

Some plants from the Salvage Islands have very large leaflets, the upper to 13.5 mm, the lower to 12.5 mm, often unusually narrow in relation to their length, with a rachis to 3 mm. They closely resemble plants from the Canarias described as *L. glaucus* var. *angustifolius* R.P. Murray in *J. Bot., Lond.* **35**: 382 (1897).

9. L. macranthus Lowe in *Trans. Camb. phil. Soc.* **6**: 546 (1838).   [*Cabelleira* (P)]
Pedrosia macrantha (Lowe) Lowe
Pubescent tap-rooted annual, biennial or short-lived perennial. Stems and leaves appearing greenish or slightly glaucous, with hairs appressed. Stems to 90 cm, much-branched, remaining slender below. Leaves sometimes slightly fleshy, with rachis 3-15 mm; upper leaflets 2.5-12 × 1.5-8 mm, obovate; lower leaflets shorter and usually narrower than the upper, 1-8 × 1-8 mm, broadly ovate to suborbicular. Peduncles 2.5-10 mm, shorter than leaves, 1-flowered. Bracts shorter than calyx. Calyx 7-14 mm. Corolla 14-25 mm, dull purple or pale yellow, streaked or tinged with brownish-purple, darkening with age; base of standard and tip of keel both dark purple. Legume 20-60 × 1.5-3 mm, cylindric, straight, not constricted between seeds, glabrous; valves twisting on dehiscing. 2n=14. **Plate 21**
Fl. XII - VI. *An endemic of stony pastures, hillsides, mountain summits, cornfields and other open sunny and rocky places; rare in Madeira, only on the coast at Cabo Garajau east of Funchal; more common and widespread in Porto Santo, up to 500 m; also on Deserta Grande and in the southern part of Bugío.* ● **MDP**

10. L. argyrodes R.P. Murray in *J. Bot., Lond.* **35**: 386 (1897).
L. argenteus (Lowe) Masferrer, non Salisb. (1796); *Pedrosia argentea* Lowe
Densely sericeous tap-rooted biennial or perennial. Stems and leaves appearing silvery or silvery green, with hairs usually appressed. Stems to 55 cm, much-branched, becoming stout and woody below. Leaves with rachis 2-10 mm; upper leaflets 4.5-18 × 3.5-11.5 mm, obovate; lower leaflets shorter and slightly narrower than the upper, 3-11 × 3-9 mm, broadly ovate. Peduncles 2-14 mm, shorter than leaves, 1(-2)-flowered. Bracts shorter than calyx. Calyx 6.5-15 mm. Corolla 13-25 mm, dark red or dark purple, rarely pale greenish yellow. Legume 20-55 × 2.5-4 mm, cylindric, straight, not constricted between seeds, glabrous; valves twisting on dehiscing. 2n=14.
Plate 21
Fl. XI - VII. *A rare endemic of open rocky places, cliffs, ledges and mountain summits; in Madeira known from the northern coast, between Porto do Moniz and Ribeira da Janella and at Boa Ventura, and in the south-east on the Ponta de São Lourenço and its associated islets; in Porto Santo known from Pico do Facho and Pico do Conçelho; also on all three Desertas.* ● **MDP**

Records of *L. argyrodes* and *L. macranthus* from the Azores are referable to *L. azoricus* P.W. Ball in *Feddes Reprium* **79**: 40 (1968).

11. L. loweanus Webb & Berthel., *Nat. Hist. Iles Canaries* **3**(2,2): 87 (1842).
[*Cabelleira de coquinho*]
Pedrosia loweana (Webb & Berthel.) Lowe
Densely sericeous tap-rooted perennial with a persistent woody stock. Stems and leaves appearing silvery green, hairs appressed. Stems to 45 cm, much-branched, often forming dense mats. Leaves with rachis ± absent or up to 5 mm; upper leaflets 4.5-13 × 1-3 mm, narrowly oblanceolate to linear; lower leaflets slightly shorter and narrower than upper, 4-9 × 1-2 mm, narrowly lanceolate

to linear. Peduncles 0-3 mm, shorter than leaves, 1-flowered. Bracts shorter than calyx. Calyx 8-12 mm, densely hirsute. Corolla 10-13 mm, deep mauve to blackish purple, or white with standard tinged pink. Legume 8-16 × 2.5-3 mm, cylindric, straight, distinctly constricted between seeds, appearing segmented, densely pubescent; valves not twisting on dehiscing. **Plate 21**
Fl. III - VI. *A widespread endemic of maritime sands and cliffs, rocky places, coastal hillsides and dry stony pastures in Porto Santo, up to 150 m.* ● **P**

21. Anthyllis L.
M.J. Cannon & N.J. Turland

Tufted, white-sericeous, woody-based perennial. Leaves imparipinnate; stipules small, falling early. Inflorescence terminal, of 1-4 subsessile, rounded heads each backed by an involucre of 2 palmatisect bracts. Calyx tubular, somewhat inflated, constricted at apex, with an oblique mouth and 5 unequal teeth, membranous. Stamens monadelphous. Legume enclosed within persistent calyx, indehiscent, 1-seeded.

1. A. lemanniana Lowe in *Hooker's J. Bot.* **8**: 291 (1856).
Stems 15-45 cm, decumbent to ascending, much-branched from base. Leaflets in 5-6 pairs, 5-20 × 1-6 mm, narrowly elliptic, glabrescent above. Heads 20-35 mm wide, 5- to 20-flowered, dense; bracts about as long as calyces, with several linear segments, glabrescent above. Calyx 8-11 mm, pale yellow. Corolla 12-15 mm, cream slightly tinged with pink, the keel dark reddish at tip, the colours darkening with age. Legume 4.5-5.5 mm, oblong, with a winged margin surrounding the seed on three sides, yellowish to light brown, glabrous. **Plate 20**
Fl. V - VII. *A rare endemic of cliffs, steep rocks and ledges in the high peaks of eastern Madeira, from Serra de Água eastwards to Pico Ruivo and Pico de Arieiro, from 1200-1800 m.* ● **M**

Possibly merely a subspecies of *A. vulneraria* L., *Sp. pl.* **2**: 719 (1753), which occurs almost throughout Europe and the Mediterranean region. *A. vulneraria* subsp. *iberica* (W. Becker) Jalas in *Bull. Jard. bot. natn. Belg.* **27**: 409 (1957) differs from *A. lemanniana* only in the colour of its calyx, which is red at the apex, and in its wholly red corolla. It occcurs along the W. coast of Europe from Belgium to Portugal.

22. Ornithopus L.
M.J. Cannon & N.J. Turland

Annual herbs. Stems prostrate to ascending, branched from base. Leaves imparipinnate; stipules free, small, linear. Inflorescence an axillary pedunculate head, with minute membranous bracts or a leaf-like imparipinnate bract at top of peduncle. Calyx tubular or campanulate, with 5 equal teeth. Stamens diadelphous. Legume compressed or cylindric, segmented; segments entire, 1-seeded, the terminal beaked, all disarticulating at maturity.

1. Plant glabrous or very sparsely pubescent; heads with minute membranous bracts; legume cylindrical .... **4. pinnatus**
- Plant pubescent; heads with a leaf-like imparipinnate bract; legume compressed .. 2

2. Corolla yellow; legume with segments only slightly narrowed at joints, the terminal segment with a large beak 5-15 mm .... **1. compressus**
- Corolla lilac and white, or white with red or yellow veins; legume with segments distinctly narrowed at joints, the terminal segment with a small beak 1-3 mm 3

3. Calyx-teeth not more than half as long as tube, triangular; corolla 4-7 mm, white with red or yellow veins .... **3. perpusillus**
- Calyx-teeth ± equalling tube, linear-subulate; corolla 7-9 mm, lilac and white .... **2. sativus**

1. O. compressus L., *Sp. pl.* **2**: 744 (1753).
Plant to 45 cm, pubescent. Lower leaves petiolate, the upper sessile; leaflets in 5-20 pairs, 2-8 × 1-4 mm, elliptic-oblong, mucronate. Heads (1-)3- to 4-flowered; bract leaf-like, imparipinnate, with 4-7 pairs of pinnae. Calyx 3.5-5.5 mm; teeth half to nearly as long as tube, linear-subulate. Corolla 6-8 mm, yellow. Legume 30-55 × 2-3 mm, compressed, curved, often strongly so; segments 6-12, 3-4.5 mm, oblong, only slightly narrowed at joints; beak of terminal segment 5-15 mm.
Fl. III - VIII. *Common in grassy, open and waste areas in all but the highest regions of Madeira; recorded also as occurring in Porto Santo but without further data.* **MP**

2. O. sativus Brot., *Fl. lusit.* **2**: 160 (1804).
Plant to 60 cm, pubescent. Leaves often all sessile; leaflets in 6-12 pairs, 6-14 × 2-4 mm, narrowly elliptic-oblong, mucronate. Heads 3- to 5-flowered; bract leaf-like, imparipinnate, with 2-3 pairs of pinnae. Calyx 4.5-6.5 mm; teeth ± equalling tube, linear-subulate. Corolla 7-9 mm; standard lilac with darker veins; wings and keel white or flushed lilac. Legume 10-20 × *c*. 2.5 mm, compressed, straight or very slightly curved; segments 3-7, *c*. 3 mm, broadly elliptic-oblong, distinctly narrowed at joints; beak of terminal segment 2.5-3 mm.
Fl. VI. *Introduced; frequently cultivated for fodder in Madeira and sometimes naturalized, as at Queimadas. Native in the Azores, Portugal, Spain, France, Morocco and Algeria; cultivated and naturalized elsewhere.* **M**

3. O. perpusillus L., *Sp. pl.* **2**: 743 (1753).
Plant to 30(-45) cm, pubescent. Lower leaves petiolate, the upper sessile; leaflets in 4-15 pairs, 2-10 × 0.75-4 mm, broadly to narrowly elliptic-oblong, mucronate. Heads (1-)2- to 4(-5)-flowered; bract leaf-like, imparipinnate, with 2-4 pairs of pinnae. Calyx 1.5-4 mm; teeth not more than half as long as tube, triangular. Corolla 4-7 mm, white with red or yellow veins. Legume 10-23 × 1.5-2 mm, compressed, curved, sometimes ± straight; segments 2-10, 2-2.5 mm, broadly elliptic-oblong, distinctly narrowed at joints; beak of terminal segment 1-3 mm.
Fl. V - VIII. *Common in grassy areas in the higher regions of Madeira, forming an important constituent of the mountain pastures; recorded also as occurring in Porto Santo but without further data.* **MP**

4. O. pinnatus (Mill.) Druce in *J. Bot., Lond.* **45**: 420 (1907).
Arthrolobium ebracteatum (Brot.) Lowe; *O. exstipulatus* Thore
Plant to 90 cm, glabrous or very sparsely pubescent. All leaves petiolate; leaflets in 3-8 pairs, 3-20 × 1-8 mm, narrowly elliptic-oblong to oblanceolate, mucronate. Heads 1- to 4(-5)-flowered; bracts minute, membranous. Calyx 3-5 mm; teeth *c*. ¼ as long as tube, triangular. Corolla 6-8 mm, yellow. Legume 15-30 × 1-1.5 mm, cylindric, curved, often strongly so; segments 5-17, 2-2.5 mm, cylindric, not narrowed at joints; beak of terminal segment 1-3.5 mm.
Fl. III - VII. *Common in grassy and rocky places, scrub and on roadsides throughout Madeira; rare in Porto Santo, occurring on Pico do Castelo.* **MP**

23. Coronilla L.
M.J. Cannon & N.J. Turland

Evergreen, unarmed, glabrous shrub. Leaves imparipinnate; stipules inconspicuous, falling early. Inflorescence an axillary pedunculate head. Calyx campanulate, with 5 subequal teeth shorter than tube. Corolla yellow. Stamens diadelphous. Legume straight, consisting of slightly compressed ellipsoid segments which disarticulate at maturity.

1. C. valentina L., *Sp. pl.* **2**: 742 (1753).
C. glauca L.; *C. valentina* subsp. *glauca* (L.) Batt.
Plant to 1.5 m, erect, much-branched; twigs slender. Leaflets in 2-4 pairs, 8-22 × 6-12 mm,

broadly to narrowly cuneate or obovate-oblong, truncate to emarginate at apex, finely mucronate, green above, glaucous beneath. Heads 5- to 11-flowered, flattened-globose, the flowers in a single whorl. Calyx 2-3 mm. Corolla 7-12 mm. Legume deflexed, 7-40 mm, with 1-7 segments, each segment 5-8 × 2-3 mm.
Fl. I - VIII. *Introduced; cultivated for ornament in gardens in Madeira; sometimes escaping and becoming naturalized in the Funchal region. Native to the Mediterranean region.* **M**

24. Hippocrepis L.
M.J. Cannon & N.J. Turland

Almost glabrous annual herb. Leaves imparipinnate; stipules inconspicuous. Inflorescence an axillary, pedunculate head. Calyx campanulate with 5 subequal teeth equalling or slightly shorter than tube. Stamens diadelphous. Legumes almost straight to curved into a full circle, flat, disarticulating at maturity into ± square segments, each with a central ± circular sinus opening on the convex edge.

1. H. multisiliquosa L., *Sp. pl.* **2**: 744 (1753).
Stems to 55 cm, prostrate to diffuse, branched from base, slender. Leaflets in 3-8 pairs, 3.5-15 × 0.6-6 mm, narrowly elliptic-oblong or narrowly cuneate to linear, truncate to emarginate at apex, mucronate. Heads 2- to 6-flowered. Calyx 3-4 mm. Corolla 6-8 mm, yellow, the standard sometimes streaked with reddish-purple on reverse. Legume (10-)20-50 × 3-5.5 mm; segments with margin of sinus thickened, glabrous or papillose-hairy, containing the single horseshoe-shaped seed.
Fl. III - V. *Dry slopes, stony pastures and sand dunes; rare in Madeira, known only from Garajau to the east of Funchal; scattered throughout Porto Santo.* **MP**

25. Scorpiurus L.

Annual herbs. Leaves simple, petiolate, elliptic to narrowly so; stipules free, linear. Inflorescence an axillary pedunculate few-flowered head. Calyx campanulate, with 5 equal teeth slightly shorter to slightly longer than tube. Corolla yellow or orange. Stamens diadelphous. Legume coiled or irregularly contorted, segmented; segments 1-seeded, longitudinally ridged, the ridges smooth, spiny or densely capitate-tuberculate.

1. Heads 2- to 4-flowered; segments of legume lunate, the ridges of convex side spiny, those of concave side smooth . . . . **1. muricatus**
- Heads 1(-2)-flowered; segments of legume cuneate-cuboid, with the broader end curved and all ridges densely covered with stout capitate tubercles . . . . . . **2. vermiculatus**

1. S. muricatus L., *Sp. pl.* **2**: 745 (1753). [*Azeda* (P), *Cabreira* (M)]
S. muricatus var. *sulcatus* (L.) Fiori; *S. sulcatus* L.
Plant very sparsely white-villous, with hairs ± appressed. Stems to 35 cm, procumbent, branched from base. Leaves 4-16 × 0.7-3 cm (including petiole 1.5-7 cm), obtuse to rounded at apex, mucronate, glossy. Heads 2- to 4-flowered. Calyx 3.5-5.5 mm. Corolla 6-10 mm, yellow or orange. Segments of legume 4-7 × 1.5-3.5 mm (excluding spines), lunate, narrowed towards ends, the ridges of convex side spiny, those of concave side smooth.
Fl. II - V. *Rocky places, stony pastures, maritime sands, cultivated and waste ground; coastal south-eastern Madeira, from the Funchal region to the Ponta de São Lourenço; scattered in Porto Santo.* **MP**

2. S. vermiculatus L., *Sp. pl.* **2**: 744 (1753).
Like **1** but more densely hairy, the hairs ± patent; stems to 60 cm; heads 1(-2)-flowered; corolla *c.* 10 mm, orange; segments of legume 6-7 × 5-6.5 mm (including tubercles), cuneate-cuboid,

with the broader end curved and all ridges densely covered with stout capitate tubercles.
Fl. III - V. *Coastal cliffs, rocks, cultivated and abandoned fields and roadsides; coastal south-eastern Madeira, from the Funchal region to the Ponta de São Lourenço; also at Santana on the northern coast.* **M**

XLIX. OXALIDACEAE
J.M. Mullin

Mostly perennial herbs. Leaves compound, alternate or mostly basal; stipules present or absent. Flowers solitary or in cymes, hermaphrodite and regular. Sepals 5, imbricate. Petals 5, hypogynous, connate at the base. Stamens 10 (5+5), connate at base; anthers ovoid, 2-locular, dehiscing by longitudinal slits. Ovary 5-locular; styles 5, free.

1. Oxalis L.

Perennial, rarely annual herbs often producing bulbs and bulbils. Stems aerial or subterranean, erect or with creeping branches. Leaves 3-foliolate, alternate or mostly basal; stipules adnate to the petiole. Flowers solitary, or in axillary inflorescences consisting of two umbellate cymes on a common peduncle with two small bracts. Fruit an oblong to cylindrical capsule.

At night and in bad weather the leaves and flowers close up. Most species show varying degrees of heterostyly. Species which produce bulbils often spread solely vegetatively, and can become troublesome weeds.

1. Flowers predominantly yellow 2
- Flowers predominantly pink to purple 3

2. Plant creeping .... **1. corniculata**
- Plant not creeping .... **4. pes-caprae**

3. Flowers in umbellate cymes; plant not creeping 4
- Flowers solitary; plant creeping .... **5. purpurea**

4. Leaflets sub-triangular, apical sinus wide .... **3. latifolia**
- Leaflets orbicular, apical sinus narrow .... **2. debilis**

1. O. corniculata L., *Sp. pl.* **1**: 435 (1753). [*Bolsa de pastor*]
Annual or perennial. Stems to 20 cm, branched, creeping and often rooting at nodes, ± densely pubescent. Leaves sometimes subopposite; petioles 1-4 cm; stipules 1-2 mm; leaflets obcordate, emarginate, entire, pilose on margins and sometimes beneath. Cymes with (1-)2-7 flowers, the peduncle up to 6 cm, normally exceeding the leaves, bracts linear to lanceolate, 1-2 mm. Pedicels *c.* 1 cm, with ± appressed hairs and usually deflexed in fruit. Sepals lanceolate, acute. Petals 7-10 mm, yellow, narrow, cuneate at base. Stamens all with fertile anthers. Capsule 10-15 mm, cylindrical, pubescent, acute at apex.
Fl. all year. *Introduced; a very common weed throughout the lower regions of Madeira and Porto Santo; also recorded from Selvagem Grande. Cosmopolitan weed of unknown origin.* **MPS**

O. exilis A. Cunn., in *Ann. nat. Hist.* **3**: 316 (1839) (*O. corniculata* var. *microphylla* Hook.f.), a native of New Zealand and Tasmania, is like **1** but generally finer and much smaller, with stems less than 1.5 mm, flowers solitary on peduncles shorter than the leaves and with the five longer stamens lacking anthers. It is recorded from Madeira but its status and distribution there are unknown. **M**

2. O. debilis Kunth in Humb., Bonpl. & Kunth, *Nov. gen. sp.* **5**: 236 (1821).
O. corymbosa DC.
Perennial with leaves and flowers arising from an underground bulb covered in ovate scales, later developing numerous bulbils between these scales. Petioles 5-15 cm; leaflets 2-5 cm, orbicular, deeply emarginate with a narrow apical sinus, ± hairy and with small reddish glands near the margins below. Flowers in short, dense cymes on a glabrous peduncle up to 30 cm long. Sepals acute with two glands at tip. Petals up to 20 mm, purplish pink. Fruit unknown.
Fl. X - III. *Introduced; the plant spreads vegetatively by means of bulbils and has become a common weed in Madeira, especially in plantations. Native to S. America.* **M**

3. O. latifolia Kunth in Humb., Bonpl., Kunth, *Nov. gen. sp.* **5**: 237 (1821).
O. intermedia A. Rich.
Like 3 but bulbils on short stolons up to 2 cm long. Petiole 15-30 cm; leaflets usually much broader than long, subtriangular, the broad, shallow apical sinus with straight sides forming an obtuse angle, glabrous except for the ciliate margins, glands absent. Peduncle 20 cm, ± pubescent. Petals, 10-15 mm, pink. Fruit not seen.
Fl. XII - V. *Introduced; naturalized at São Martinho near Funchal. Cosmopolitan weed.* **M**

4. O. pes-caprae L., *Sp. pl.* **1**: 434 (1753).
O. cernua Thunb.
Bulbous perennial with erect underground stem bearing bulbils. Leaves numerous; petiole up to 20 cm; leaflets up to 2 cm, obcordate, ± hairy below. Cyme umbellate, peduncle up to 30 cm, longer than leaves. Petals up to 20 mm, bright yellow, sometimes pale pink on the outside, often double. Fruit not seen.
Fl. XII - IV. *Introduced; commonly cultivated as an ornamental and widespread as an escape by waysides and on waste ground around houses in Madeira and Porto Santo. Cosmopolitan weed.* **MP**

The common, low-styled heteromorphous plants originate from Europe and the Mediterranean; occasional mid-styled plants are introductions from S. Africa.

5. O purpurea L., *Sp. pl.* **1**: 443 (1753).
O. variabilis Jacq.; *O. venusta* Lowe
Pubescent, bulbous perennial with underground, generally wide-creeping stems producing rosettes of leaves at soil level. Leaves often spreading or prostrate; petiole 2-8 cm, pilose; leaflets 1-2 cm, rhombic, upper surface usually glabrous or glaucescent, with orange or brown cilia, lower surface with many pellucid glands and streaks which turn black on drying. Flowers solitary, peduncles shorter to longer than the leaves, with 2 alternate, linear bracts below the middle. Sepals lanceolate. Corolla 1-3 cm long, campanulate, deep rose-purple with cream to yellow tube.
Fl. (X-) XI - IV. *Common in the lower regions of Madeira, in open woods, and especially as a weed of cultivated land. Native to S. Africa.* **M**

L. GERANIACEAE
M.J.Short

Herbs or small shrubs. Leaves alternate or opposite, lobed or compound, stipulate. Flowers hermaphrodite, regular or irregular, 5-merous, in terminal or axillary cymes, often umbellate. Sepals free, persistent, the dorsal one sometimes spuured. Petals free. Stamens mostly 10 (or some reduced to staminodes), in 2 alternating whorls. Ovary superior, usually 5-locular. Fruit splitting into five 1-seeded mericarps; styles united in flower, often splitting longitudinally in fruit, forming a slender awn.

1. Dorsal sepal with a spur fused to the pedicel; corolla irregular, the upper 2 petals wider .... **2. Pelargonium**
- Sepals not spurred; corolla not or only weakly irregular 2

2. Leaves longer than broad, pinnately divided or lobed; fertile stamens 5, alternating with 5 scale-like staminodes; awn of mericarp spirally twisted below at maturity .... **3. Erodium**
- Leaves not longer than broad, palmately divided or lobed; fertile stamens usually 10; awn of mericarp curving upwards, but never spirally twisted .... **1. Geranium**

1. Geranium L.

Annual to perennial herbs. Leaves alternate or the uppermost opposite, orbicular to reniform in outline, palmately lobed or divided, appressed-pubescent, the basal long-petiolate. Flowers regular. Inflorescence cymose, the ultimate peduncles usually 2-flowered. Stamens 10, usually all fertile. Stigmas 5, filiform. Mericarps 1-seeded, usually dehiscent, without apical pits, usually remaining attached to outer part of the style which forms a long awn, separating and curving from the base upwards but remaining attached for a while by the apex to the top of the central axis.

Literature: P.F. Yeo, The biology and systematics of *Geranium*, sections *Anemonifolia* Knuth and *Ruberta* Dum., in *Bot. J. Linn. Soc.* **67**: 285-346 (1973).

1. Leaves dull green; sepals ± spreading; petals very shortly clawed, entire or notched 2
- Leaves bright green; sepals erect, often connivent near apex; petals distinctly and often long-clawed, entire 4

2. Upper leaves alternate; mericarps wrinkled, glabrous; beak 5-8 mm ..... **3. molle**
- Upper leaves opposite; mericarps smooth, pubescent; beak 10-15 mm 3

3. Leaves deeply divided, almost to base, the lobes cut into linear segments; petals emarginate .... **1. dissectum**
- Leaves divided *c.* ½-way to base, the lobes crenate to lobulate at apex; petals entire . ... **2. rotundifolium**

4. Sepals keeled; leaves 5-lobed up to ½-way to base .... **4. lucidum**
- Sepals not keeled; leaves dissected to the base, appearing compound 5

5. Throat of flower whitish, paler than the rest of the flower; mericarps with a strand of fibres attached near the apex 6
- Throat of flower crimson to purplish, darker than the rest of the flower; mericarps without a strand of attached fibres 7

6. Petals 7-9.5 × 2-2.5 mm; anthers yellow .... **5. purpureum**
- Petals 11-14 × 4.5-7 mm; anthers red or orange-pink .... **6. robertianum**

7. Biennial; hairs of the inflorescence with ± colourless stalks .... **7. rubescens**
- Rather robust perennial; hairs of the inflorescence purple, or rarely inflorescence glabrous 8

9. Filaments 14-17 mm long; anthers orange or creamy yellow .... **8. palmatum**
- Filaments 8-10 mm long; anthers dark red .... **9. maderense**

1. G. dissectum L., *Cent. pl. I*: 21 (1755).
Appressed-pubescent annual; stems 8-50 cm, procumbent to erect, branched and often rather straggling. Leaves orbicular or reniform, palmatisect, dull or grey-green; the lower 2-6 cm wide, segments 5-7, rhombic-cuneate, deeply pinnatifid or incised, the ultimate lobes linear, mostly acute; petioles up to 13 cm; upper leaves opposite, more shortly petiolate. Peduncles 5-25(-33) mm. Pedicels 5-15 mm, usually glandular-pubescent, spreading in fruit. Sepals 5-7 mm; awn 1-1.5 mm. Petals 5-7 mm, emarginate, very shortly clawed, deep pink. Mericarps smooth, pubescent; beak 10-13 mm. Seeds pitted.
Fl. II - X. *Very common in Madeira in pastures, on roadside banks, along paths and levadas, and in other rocky or grassy places, up to 1000 m; less frequent in Porto Santo where it is found both on high peaks as well as in lowland areas.* **MP**

2. G. rotundifolium L., *Sp. pl.* **2**: 683 (1753).
Softly pubescent annual; stems 5-35 cm, usually branched from the base, erect or ascending, with spreading hairs. Leaves orbicular-reniform, palmatifid up to *c.* ½-way to base, dull or grey-green; the lower 2-6 cm wide; segments 5-7, cuneate, crenate-lobulate at apex; petioles up to 25 cm; upper leaves opposite, shortly petiolate, lobes more acute. Peduncles 7-35(-50) mm. Pedicels 8-18 mm, usually glandular-pubescent, spreading or deflexed in fruit. Sepals 5-6 mm; awn 0.4-0.7 mm. Petals 5-6.5 mm, entire, shortly clawed, rose pink or whitish. Mericarps smooth, pubescent; beak 10-15 mm. Seeds reticulate.
Fl. II-VII. *Common in Madeira in cultivated and waste ground, along paths and roadsides, and on walls, up to 650 m; less frequent in Porto Santo, and reported to be very rare on Bugío.* **MDP**

3. G. molle L., *Sp. pl.* **2**: 682 (1753).
Softly pubescent annual; stems 8-30(-40) cm, branched from the base, procumbent to ascending, with spreading hairs. Leaves orbicular-reniform, palmatifid to *c.* ⅔-way to base, dull or grey-green; the lower 1-4 cm wide; segments 5-9, obovate-cuneate, mostly 3-lobed at apex, the lobes obtuse to subacute; petioles up to 14 cm; uppermost leaves alternate, becoming subsessile. Peduncles 4-20(-30) mm. Pedicels 5-15 mm, with very long eglandular and very short glandular hairs intermixed, deflexed in fruit. Sepals 4-5 mm, mucronulate. Petals 3-6 mm, very shortly clawed, bifid, rose-purple. Mericarps wrinkled, glabrous; beak 5-8 mm. Seeds smooth.
Fl. II - VII. *Fairly frequent in Madeira along levadas, roadsides, and on waste ground, up to 1000(-1500) m; rare in Porto Santo.* **MP**

4. G. lucidum L., *Sp. pl.* **2**: 682 (1753).
Subglabrous annual, often tinged red; stems 10-46 cm, procumbent to erect, usually branched from base. Leaves orbicular-reniform, palmatifid ½-¾-way to base, shining green, sparsely pubescent; the lower 1.5-6 cm wide; segments 5-7, obovate-cuneate, 3-lobed at apex, the lobes obtuse, mucronulate; petioles up to 14 cm; upper leaves opposite, smaller, shortly petiolate. Peduncles 10-33 mm. Pedicels 5-17 mm, sparsely pubescent, spreading in fruit. Sepals 4.5-7 mm; awn 0.3-0.8 mm, keeled, transversely wrinkled, connivent towards apex. Petals 8.5-10 mm, long clawed, entire, bright pink. Mericarps ridged above, reticulate-rugose below, apex pubescent; beak 8-11 mm. Seeds smooth.
Fl. IV - VII. *Locally frequent in Madeira, mainly in the ribeiras of the central mountains, growing on shady banks, rocky and grassy places, along levadas, for example along Ribeiro Frio.* **M**

5. G. purpureum Vill. in L., *Syst. pl. eur. 1, Fl. Delph.*: 72 (1786).
G. robertianum var. *purpureum* (Vill.) DC.
Sparsely pubescent, red-tinged, foetid annual up to 35 cm; stems procumbent to erect,

branched. Leaves 2.5-7 cm wide, palmately 5-lobed, dark green, segments irregularly pinnately to bipinnately cut; petiole up to 10(-19) cm; lower leaves forming a basal rosette. Inflorescence glandular-pubescent, the hairs mostly with colourless stalks, red-tipped. Peduncles 6-40 mm. Pedicels 2-15 mm. Sepals 4-7 mm, connivent; awn 0.7-1 mm. Petals 7-9.5 × 2-2.5 mm, long-clawed, apex rounded, pink, claw whitish above. Anthers bright yellow. Mericarps 2.5-3 mm, with 3-5 collar-like ridges at apex, reticulate-ribbed below, glabrous, with a strand of white fibres attached near the apex; beak 14-20 mm. 2n=32.
Fl. III - VIII. *Widespread in Madeira on grassy banks and slopes, in woods, and along paths and levadas, usually in shade, 0-1000 m; less frequent in Porto Santo.* **MP**

6. G. robertianum L., *Sp. pl.* **2**: 681 (1753).
Sparsely pubescent, foetid biennial, often red-tinged, up to 60 cm; stems prostrate to erect, branched. Leaves up to 7 cm wide, palmately 5-lobed, dark green, segments irregularly pinnately to bipinnately cut; petiole up to 15 cm; lower leaves forming a dense basal rosette before flowering. Inflorescence glandular-pubescent, the hairs with colourless stalks, red-tipped. Peduncles 20-50 mm. Pedicels 3-20 mm. Sepals 5-9 mm, connivent; awn 1.2-2 mm. Petals 11-14 × 4.5-7 mm, long-clawed, apex rounded, pink, claw whitish above. Anthers red to orange-pink. Mericarps 2.5-3.5(-4.2) mm, with 1-2(-3) collar-like ridges at apex, reticulate below, glabrous, with a strand of white fibres attached near the apex; beak 17-20 mm.
Fl. III - VIII. *Fairly frequent in Madeira, but less widespread than G. purpureum; along roadsides, paths, and levadas, mainly 800-1000 m.* **M**

Records of this species from Porto Santo are probably all referable to *G. purpureum*.

7. G. rubescens Yeo in *Bolm Mus. munic. Funchal* **23**: 26 (1969).
Sparsely pubescent biennial, sometimes red-tinged, up to 75 cm; stems ± erect, branched. Leaves up to 9 cm wide, palmately 5-lobed, dark green, segments irregularly bipinnately cut; petiole up to 14 cm. Inflorescence glandular-pubescent, the hairs with ± colourless stalks, red-tipped. Peduncles 3-8 cm. Pedicels 4-15(-25) mm. Sepals 7.5-8.5 mm, connivent; awn 1.6-2.3 mm. Petals 18-21 × 8-11 mm, long-clawed, rounded to truncate at apex, mauve-pink, the claw dark red above. Anthers orange-pink. Mericarps *c.* 4 mm, with 1-2 collar-like ridges at apex, closely reticulate below, glabrous; beak 22-27 mm. **Plate 22**
Fl. III - X. *In woodland, on banks, and along levadas in central Madeira, where it is locally frequent.* ● **M**

8. G. palmatum Cav., *Diss.* **4**: 216 (1787). [*Gerânio folha-de-anémona*]
G. anemonifolium L'Hér.
Short-lived perennial up to 1 m with a short, ± woody caudex, subglabrous to sparsely pubescent below. Leaves mostly crowded at the top of the short stem, up to 30 cm wide, bright green, rather shiny, palmately 5-lobed, segments irregularly bipinnately cut; petioles up to 38 cm, reflexed in age, the bases dilated, contiguous, and densely crowded. Stipules up to 25 mm, not appressed, velutinous, margin sometimes villous. Inflorescence large, wide-spreading, densely glandular-pubescent, with purple hairs, or less commonly glabrous. Peduncles 30-50 mm. Pedicels 8-30 mm. Sepals 6-8 mm; awn 2-3(-4) mm. Petals 20-25 × (8-)11-15 mm, long-clawed, rounded to truncate at apex, lilac, deep crimson towards the base, claw crimson. Filaments 14-17 mm long; anthers orange or creamy yellow. Mericarps 3.2-4 mm, with 1-2 collar-like keels at apex and minutely glandular, reticulate below, glabrous; beak 20-25 mm. 2n=68. **Plate 22**
Fl. III - XII. *Fairly common in Madeira on rocky cliffs and along levadas, usually in moist, shady places, mainly 700-1000(-1500) m, but almost down to sea-level on the north coast.* ● **M**

9. G. maderense Yeo in *Bolm Mus. munic. Funchal* **23**: 26 (1969). [*Pássaras*]
Massive monocarpic or perennial herb up to 1 m or more high with a ± woody caudex, subglabrous to sparsely pubescent below. Leaves mostly crowded at the top of the stem, up to *c.* 60 cm wide, light green, palmately 5-lobed, segments irregularly bipinnately cut; petioles up to *c.* 60 cm, reflexed in age, the bases swollen amd contiguous. Stipules up to *c.* 12 mm, appressed, sparsely and minutely pubescent, margin sparsely ciliate when young. Inflorescence very large, wide-spreading and much-branched, densely glandular-pubescent with purple hairs. Peduncles 5-15 cm. Pedicels 1-14 cm. Sepals 8-10 mm; awn *c.* 1 mm. Petals (14-)19-21 × (10-)13-18 mm, shortly but distinctly clawed, rounded to subtruncate at apex, purplish pink, darker towards the base, claw blackish purple. Filaments 8-10 mm long; anthers dark red. Mericarps 4-4.5 mm, with 1-3 collar-like keels at apex, coarsely reticulate below, sparsely and minutely puberulent; beak 22-26(-30) mm. **Plate 22**
Fl. II - IX. *Restricted to a small number of sites in the mountains of central and northern Madeira, where it is locally abundant. Reported growing on moist, north-facing rocks. Also cultivated in gardens as an ornamental.* ● M

2. Pelargonium L'Hér.
M. Gibby

Flowers irregular, with a nectar spur; fertile stamens 7.

1. Petals bright red . . . . **1. inquinans**
- Petals white or pink 2

2. Decumbent or erect, branched shrub, becoming woody with age; petals pale pink to purple 3
- Short stemmed, often prostrate shrub; petals white with crimson markings . . . . **2. odoratissimum**

3. Viscid, balm-scented shrub; leaves 3-palmatilobate to pinnatisect . . . . **5. glutinosum**
- Aromatic shrub; leaves pilose to villous, cordiform, shallowly lobed 4

4. Erect shrub with unpleasant smell; leaves not crisped; pseudo-umbels with 3-12 flowers . . . . **3. vitifolium**
- Decumbent, rose-scented shrub; leaves crisped; pseudo-umbels with 8-20 flowers . . . . **4. capitatum**

1. P. inquinans (L.) L'Hér. in Aiton, *Hort. kew.* **2**: 424 (1789). [*Malva*]
Branched shrub with soft, somewhat succulent stems, becoming hardened with age. Leaves orbicular with cordate base, very shallowly lobed and with crenate margin. Pseudo-umbel with 5-20 flowers, not markedly zygomorphic, petals bright red.
Fl. all year. *A native of S. Africa, commonly naturalized on cliff-faces, ledges, and among rocks in Madeira; at Serra de Foro in Porto Santo.* **MP**

2. P. odoratissimum (L.) L'Hér. in Aiton, *Hort. kew.* **2**: 419 (1789).
Perennial, often prostrate shrub with a short, rough main stem. Leaves round to ovate-cordate, light green and covered with fine short hairs, emitting a strong, sweet minty smell when bruised. Flowers borne on 3-10-flowered umbels, relatively small, petals white with crimson markings.
Fl. IV - VIII. *A native of S. Africa, naturalized near Funchal.* **M**

3. P. vitifolium (L.) L'Hér. in Aiton, *Hort. kew.* **2**: 425 (1789).
Erect, branched, strongly aromatic shrub. Stems herbaceous, becoming woody with age, villous

with glandular hairs. Leaves pilose to villous with glandular hairs, lamina cordiform, shallowly lobed. Pseudo-umbels with 3-12 flowers, petals pale pink to purple-pink, posterior two petals with purple markings.
Fl. IV - VIII. *A native of S. Africa, naturalized in the south of Madeira near Funchal and Machico; grown in gardens as clipped hedges.* **M**

According to Lowe *P. graveolens* L'Hér. is widely grown in gardens and as clipped hedges, but a specimen of Lowe incorrectly labelled *P. graveolens* from Machico is *P. vitifolium*.

4. P. capitatum (L.) L'Hér. in Aiton, *Hort. kew.* **2**: 425 (1789).
Decumbent, branched rose-scented shrub. Stems herbaceous, becoming woody with age, pilose to densely villous with glandular hairs. Leaves villous to densely villous with glandular hairs interspersed, lamina 3-5 palmatilobate to 3-5 palmatipartite, crisped, base cordate. Pseudo-umbels with 8-20 flowers, petals pale to dark pink, posterior two petals with purple markings.
Fl. IV - VIII. *A native of S. Africa, possibly naturalized in the south of Madeira.* **M**

5. P. glutinosum (Jacq.) L'Hér. in Aiton, *Hort. kew.* **2**: 426 (1789).   [*Malva*]
Erect, branched, viscid, balm-scented shrub. Stems herbaceous when young but soon becoming woody, with glandular hairs. Leaves with glandular hairs, 3-palmatilobate to pinnatisect, base cordate, margins dentate with sharp pointed hairs. Pseudo-umbels with 1-8 flowers, petals pale to dark pink, upper two petals with darker markings.
Fl. IV - VIII. *A native of S. Africa, frequently escaped from gardens. Recorded forming dense thickets at Larano, to the east of Porto da Cruz, in Madeira.* **MP**

Other *Pelargonium* species are cultivated and occasionally found as garden escapes, including **P. alchemilloides** (L.) L'Hér., **P. zonale** (L.) L'Hér., **P. peltatum** (L.) L'Hér. and **P. cucullatum** (L.) L'Hér.

3. Erodium L'Hér.

Annual to perennial herbs. Leaves longer than broad, pinnate or pinnately lobed, usually appressed-pubescent. Flowers regular or the corolla slightly irregular, borne in umbels (rarely reduced to 1 flower), subtended by 2 or more usually scarious bracts. Stamens 5, antesepalous; staminodes 5, scale-like. Stigmas 5, filiform. Mericarps indehiscent, with 2 apical pits on either side of apex, usually remaining attached to outer part of style which forms a long awn, spirally twisted below at maturity, separating from the base upwards.

1. Leaves lobed, pinnatifid or pinnatisect (occasionally compound at the base, but then only with 1(-2) pairs of distinct leaflets) 2
- Leaves pinnate 4

2. Sepals 11-15 mm in fruit; mericarps 9-11 mm; awn 7-11 cm; petals 10-15 mm . . . . **3. botrys**
- Sepals up to 7 mm in fruit; mericarps up to 5 mm; awn up to 4 cm; petals up to 8 mm 3

3. Apical pit of mericarps with a furrow beneath; awn (18-)22-26 mm; sepals glandular-pubescent; basal leaves ovate to oblong . . . . **2. malacoides**
- Apical pit of mericarps without a furrow beneath; awn 30-40 mm; sepals eglandular; basal leaves suborbicular . . . . **1. chium**

4. Cauline leaves with most leaflets divided more than ½-way to midrib; apical pits of mericarps eglandular . . . . **4. cicutarium**
-   Cauline leaves with most leaflets divided less than ½-way to midrib; apical pits of mericarps glandular . . . . **5. moschatum**

1. E. chium (L.) Willd., *Phytographia*: 10 (1794). [*Agulha* (P), *Alfinête* (P)]
Softly pubescent and glandular annual or biennial. Stems 15-70 cm, procumbent or ascending. Basal and lower cauline leaves 4-7.5 cm, ± suborbicular, variably lobed, crenate-dentate; petioles up to 14 cm. Upper cauline leaves ovate, smaller. Umbels 4-8-flowered; bracts ovate, acute, membranous, whitish, ciliate. Pedicels eglandular-pubescent. Sepals 4-6 mm in flower, up to 7 mm in fruit, mucronate, membranous-margined, eglandular-pubescent, veins prominent. Petals 5-8 mm, pinkish purple. Mericarps 3.5-4.5 mm, appressed white-setose; apical pits small, deep, eglandular, without a furrow beneath; awn 30-40 mm.
Fl. II - VI. *Grassy banks, walls, roadsides, and waste ground. Reported to be rare in Madeira in the last century, but now a fairly common species at low altitudes in the south of the island. Frequent in Porto Santo and also occurring on Ilhéu de Cima. Also recorded from Bugío.* **MDP**

Represented in the Madeiran archipelago by subsp. **chium**, to which the above description applies.

2. E. malacoides (L.) L'Hér. in Aiton, *Hort. kew.* **2**: 415 (1789).
Softly pubescent, often glandular annual or biennial. Stems 6-40(-60) cm, procmbent or ascending. Basal and cauline leaves ovate to oblong, 2-4.5(-6) × 1.5-3(-4) cm, crenate-dentate, sometimes shallowly lobed; petioles up to 8 cm. Umbels (2-)3-8-flowered; bracts ovate, acute, membranous, whitish, ciliate. Pedicels glandular-pubescent. Sepals 3-4.5 mm in flower, up to 7 mm in fruit, mucronate, the mucro often purplish, membranous-margined, glandular-pubescent, veins prominent. Petals 4-7 mm, mauvish pink. Mericarps 4-5 mm, appressed white or brownish-setose; apical pits deep, sparsely glandular, with a single wide furrow beneath, also usually glandular; awn (18-)22-26 mm.
Fl. XI - V. *Track and roadsides, waste ground. Fairly frequent at low altitudes in the south of Madeira. Very rare in Porto Santo with no known recent records. Also recorded from the Desertas but without further data.* **MDP**

3. E. botrys (Cav.) Bertol., *Amoen. ital.*: 35 (1819). [*Agulheta*]
Hispid annual. Stems 5-30(-40) cm, erect or ascending, often reddish. Basal leaves 2-6(-9) × 0.8-4 cm, oblong in outline, pinnately lobed to pinnatifid, the lobes crenate-dentate to pinnatifid; petioles up to 6 cm. Cauline leaves pinnatisect, the lobes acutely incised or pinnatifid; petioles 0-1 cm. Umbels (1)2-3(-5)-flowered; bracts ovate, acuminate, membranous, brown, densely ciliate. Pedicels glandular-pubescent, sharply reflexed in fruit. Sepals 5-8 mm in flower, 11-15 mm in fruit, mucronate, membranous-margined, glandular-pubescent. Petals 10-15 mm, mauvish pink. Mericarps 9-11 mm, appressed white-setose; apical pits deep, eglandular, with 2(3) deep furrows beneath; awn 7-11 cm.
Fl. I - VI. *Grassy banks, rocky ground, pastures and rough grassland. Fairly frequent in Madeira; rather rare in Porto Santo, recorded recently (1986) from the peak of Ana Ferreira.* **MP**

4. E. cicutarium (L.) L'Hér. in Aiton, *Hort. kew.* **2**: 414 (1789).
Very variable, often reddish tinged annual, sparsely or densely pubescent with coarse white hairs. At first stemless, later usually branching from the base, stems up to 45 cm, prostrate. Leaves up to 8 cm long, pinnate; leaflets sessile, 3-17 mm, ovate-oblong, deeply pinnatifid, usually divided more than ½-way to midrib. Umbels with 2-6 flowers; bracts ovate, mucronate,

membranous, brownish, sparsely ciliate. Pedicels eglandular or glandular-pubescent. Sepals 3-5 mm in flower, 6-7 mm in fruit, mucronate, membranous-margined, eglandular or sometimes glandular-pubescent. Petals 4-8 mm, bright purplish pink, rarely white, the upper 2 often slightly larger than the others. Mericarps 4.5-6.5 mm, with brownish erecto-patent hairs; apical pits large, eglandular, with a shallow furrow beneath; awn 20-35 mm.
Fl. III - X. *Grassy mountain pastures, levada paths, roadsides. At altitudes up to 1300 m in Madeira on Paúl da Serra, where it is common in the heavily grazed turf. Locally frequent in Porto Santo, often growing with* E. moschatum; *white-flowered plants have been recorded from Pico Branco and Camacha. Also known from Bugío and Selvagem Grande.* **MDPS**

5. E. moschatum (L.) L'Hér. in Aiton, *Hort. kew.* **2**: 414 (1789). [*Malvas*]
Pubescent, annual or biennial, glandular, smelling of musk. Stems 6-40(-50) cm, usually procumbent, often forming dense clumps. Basal and cauline leaves up to 20 cm, pinnate; leaflets subsessile, somewhat remote, 8-35(-45) × 4-25(-30) mm, ovate, serrate or sometimes pinnatifid, most divided less than ½-way to midrib. Umbels (2-)4-10-flowered; bracts united at base, ovate-orbicular, subacute to obtuse, membranous, whitish, subglabrous. Pedicels usually with white eglandular and longer glandular hairs. Sepals 5-6 mm in flower, 7-9 mm in fruit, mucronate, membranous-margined, glandular-pubescent. Petals 7-8 mm, rose-pink. Mericarps 5-6 mm, pubescent, hairs patent, brown or whitish; apical pits broad, glandular, with a single broad, deep furrow beneath; awn 25-40(-45) cm.
Fl. XI - VI. *Track and roadsides, waste ground, walls. Very common in Madeira and also frequent in Porto Santo. Also recorded from the Desertas but without further data, and Selvagem Grande.* **MDPS**

LI. TROPAEOLACEAE
M.J. Short

Prostrate or climbing herbs with acrid sap. Leaves alternate, peltate, simple (in Madeira), stipulate. Flowers hermaphrodite, irregular. Sepals 5, ± free, the abaxial one produced into a long spur. Petals 5, free, often clawed, unequal, the upper two different from lower three. Stamens 8. Ovary superior, 3-locular. Style 1; stigma 3-lobed. Carpels separating from the central axis at maturity, forming three, 1-seeded, indehiscent segments.

1. Tropaeolum L.

Somewhat succulent, annual herbs, trailing or climbing by means of coiling petioles. Leaves long-petiolate; stipules small, caducous. Flowers solitary in the leaf-axils, long-pedicellate. Carpels slightly fleshy, unwinged.

1. T. majus L., *Sp. pl.* **1**: 345 (1753). [*Chagas*]
Glabrous; stems long, fleshy, prostrate or climbing. Leaves up to 8(-15) cm across, ± orbicular, subentire, glaucescent; petioles up to 20 cm. Sepals lanceolate, acute; spur up to 4 cm. Flowers showy, 4-6 cm across, mainly bright orange, but also yellow or red, and often multi-coloured. Petal limb suborbicular; upper two petals darker veined towards the base, lower three petals smaller, the claw long-ciliate. Carpels c. 10 mm, ribbed dorsally.
Fl. I - IX. *An introduced species only known in cultivation. Very common in the lower regions of Madeira, especially near towns and villages, along roadsides, on waste and cultivated land, often covering the ground in large masses or trailing over walls and banks; infrequent in Porto Santo.* **MP**

LII. ZYGOPHYLLACEAE
M.J. Short

Herbs or small shrubs (in ours). Leaves usually opposite, simple, 2-3-foliolate, or compound, often fleshy, stipulate. Flowers hermaphrodite, usually regular, solitary, paired or in cymes.

Disc usually present. Sepals and petals (4-)5, free. Stamens usually twice as many as petals. Ovary superior, 4-5-locular, usually angled or winged. Fruit a capsule, rarely a berry.

1. Leaves 3-foliolate, coriaceous; stipules spiny . . . . **1. Fagonia**
- Leaves 2-foliolate, very fleshy; stipules membranous . . . . **2. Zygophyllum**

1. Fagonia L.

Small, shrubby perennial herbs, with spinose stipules. Leaves opposite, mostly 3-foliolate. Flowers solitary in the axils of the leaves, 5-merous. Sepals deciduous. Petals clawed. Disc inconspicuous. Stamens 10. Ovary 5-locular. Fruit a deeply 5-lobed capsule dehiscing along the axis; carpels dehiscent, 1-seeded; style persistent. Seeds compressed.

1. F. cretica L., *Sp. pl.*: 386 (1753).
Low, sprawling plant with stems 10-35 cm, procumbent, much-branched, striate, ± glabrous. Leaves petiolate, 3-foliolate; leaflets 5-22 mm, linear-lanceolate to narrowly elliptic, with a short, spiny tip, slightly asymmetrical at the base, somewhat coriaceous. Pedicels 3-6 mm, glandular-pubescent, reflexed in fruit. Sepals 3-4 mm, ovate, acuminate, glandular-pubescent. Petals 8-10 mm, purple, caducous. Capsules 4-5 × 5-8 mm, pubescent. Seeds *c.* 4 mm, ovoid, dark brown.
Fl. II - IV. *Dry, stony and sandy places on the Salvage Islands.* S

2. Zygophyllum L.

Small fleshy shrubs with stems jointed at the nodes. Leaves opposite, 2-foliolate. Flowers axillary, solitary, 5-merous. Petals clawed. Disc fleshy, angled. Stamens 10. Ovary 5-locular. Fruit a deeply 5-lobed capsule.

1. Z. fontanesii Webb & Berthel., *Hist. nat. Iles Canaries* 3(2, sect. 1): 17 (1836).
Low, pubescent shrub; stems up to 30 cm or more, ascending, much-branched, rather brittle. Leaves very fleshy, glaucous, turning yellow before falling; leaflets 3-20 mm, cylindrical to clavate, obtuse, mucronate; petioles thick and fleshy, resembling the leaflets. Stipules small, membranous. Pedicels 3-5 mm, thickened above. Sepals *c.* 3 mm, margins scarious. Petals 3.5-4 mm, white or pale mauve, the apex denticulate. Capsules 4-5 × 4-6 mm, globose when young, becoming thick and corky. Seeds *c.* 2 mm, ellipsoid, greenish black.
Fl. III - VII. *Frequent in dry, sandy ground on Selvagem Pequena; also recorded from Ilhéu de Fora, but no recent records known. A single record of a collection made in the last century from Selvagem Grande has not been confirmed.* S

LIII. LINACEAE
M.J. Short

Herbs or small shrubs. Leaves usually alternate, simple, entire. Inflorescence cymose or rarely flowers solitary or in corymbs. Flowers 4- or 5-merous, hermaphrodite, regular. Sepals free. Petals free or rarely joined at the base, caducous. Fertile stamens 4-5, fused at their base, often alternating with staminodes. Ovary superior, 4- to 5-locular, each locule partially divided by a false septum; styles 3-5. Fruit a capsule.

Reinwardtia indica Dumort., *Comment. bot.*: 19 (1822) (*Linum trigynum* Roxb.), native to SE Asia, is a low, glabrous shrub up to *c.* 60 cm with subsessile, obovate-elliptic leaves and large, bright yellow, trumpet-shaped flowers, the 5 petals twisted and basally united into a tube. It is cultivated as an ornamental in Madeiran gardens and may occasionally become naturalized. **M**

1. Leaves alternate; sepals 5, entire .... **1. Linum**
- Leaves opposite; sepals 4, toothed .... **2. Radiola**

1. Linum L.

Annual or biennial herbs. Leaves sessile, alternate, exstipulate, narrow, 1- or parallel-veined. Flowers in cymes, 5-merous. Sepals imbricate, entire. Petals blue or yellow, clawed, longer than the sepals. Stamens 5, alternating with 5 filiform staminodes. Capsules shortly beaked, dehiscing loculicidally and septicidally into 10 valves. Seeds 2 per locule, elliptic, flat, shiny.

Capsule measurements in the following species descriptions exclude the beak.

1. Petals blue; capsules 4-9 mm 2
- Petals yellow; capsules 1.5-3 mm 3

2. Stems usually several, branched from the base, procumbent, ascending or erect; capsules 4-6 mm, fully dehiscent .... **1. bienne**
- Stem usually solitary, unbranched at base, erect; capsules 6-9 mm, partially dehiscent . ... **2. usitatissimum**

3. Leaf margins smooth; inflorescence lax; pedicels filiform, up to 10 mm .... **3. trigynum**
- Leaf margins scabrid; inflorescence dense; pedicels stout, 0.5-1(-2.5) mm .... **4. strictum**

1. L. bienne Mill., *Gard. dict.* ed. 8, no. 8 (1768). [*Linho bravo*]
L. angustifolium Huds.
Glabrous annual or biennial. Stems (7-)15-40 cm, several, usually branched from the base, procumbent, ascending or erect, slender. Leaves 5-20 × 0.8-2 mm, linear, acute, 1- to 3-veined, somewhat glaucous. Pedicels up to 2.5 cm, slender. Sepals 5-6 × 1.7-3.5 mm, elliptic-ovate, acuminate, with a conspicuous midvein, margin scarious. Petals 9-12 mm long, pale blue with darker veins. Capsules 4-6 mm, subglobose, fully dehiscent. Seeds 2-2.3 mm.
Fl. III - XI. *Common in Madeira and Porto Santo along roadsides, on cultivated and waste ground, and in grassy places; 0-300 m.* **MP**

2. L. usitatissimum L., *Sp. pl.* **1**: 277 (1753). [*Linho*]
Like **1**, but more robust with an erect, usually solitary stem, generally unbranched at the base; leaves 1.5-3(-5) mm wide, 3-veined; sepals 6-9 mm; capsules 7-9 mm, partially dehiscent.
Fl. XII - VI. *A species of cultivated origin grown for its fibre (flax) and its seeds (source of linseed oil) and recorded as becoming locally naturalized. Formerly much planted in the north of Madeira, now common around Ponta do Pargo in the south-west.* **M**

3. L. trigynum L., *Sp. pl.* **1**: 279 (1753).
L. gallicum L.
Slender, glabrous annual; stems 10-40(-50) cm, erect or ascending. Leaves 5-20 × 0.6-3(-5.5) mm, linear-elliptic, acute, margins smooth. Inflorescence lax, spreading, delicate. Pedicels up to 10 mm, filiform. Sepals 3-4 mm, ovate, long and narrowly acuminate, margins glandular-ciliate. Petals 4-6 mm, yellow. Capsules 1.5-2 mm, subglobose. Seeds 0.9-1 mm.
Fl. IV - IX. *Fairly frequent in Madeira along paths, roadsides, and in pastures; reported to be extremely rare in Porto Santo.* **MP**

4. L. strictum L., *Sp. pl.* **1**: 279 (1753).
L. strictum var. *cymosum* Gren. & Godr.
Rather rigid, scabrous annual; stems 6-25 cm, several, ascending to erect, leafy. Leaves crowded, 8-23 × 1-3.5 mm, linear-lanceolate, acute, margins scabrid. Inflorescence dense, leafy. Pedicels 0.5-1(-2.5) mm, stout. Sepals 4-6 mm, lanceolate, acuminate, scabrid, margins glandular-ciliate. Petals 5-8 mm, yellow. Capsules 2-3 mm, subglobose. Seeds 1.3-1.5 mm. Fl. III - VI. *Roadside banks, cliff slopes, pastures and cultivated ground. Very common in Porto Santo; less frequent in Madeira, where it is restricted to areas near the south coast between Ponta de São Lourenço and Calheta.* **MP**

2. Radiola Hill

Like *Linum* but leaves opposite; flowers 4-merous; sepals toothed; petals white, ± equalling sepals; capsules dehiscing by 8 valves; seeds ovoid, 2 per locule.

1. R. linoides Roth, *Tent. fl. Germ.* **1**: 71 (1788).
R. millegrana Sm.
Low, delicate, glabrous annual; stems 1-6(-10) cm, erect, filiform, dichotomously branched. Leaves 0.5-3 mm, ovate-elliptic, acute, 1-veined. Flowers minute, numerous, in dichasial cymes, shortly pedicellate. Sepals *c.* 0.7 mm, deeply (2-)3-toothed at apex. Petals *c.* 1 mm, white. Capsules *c.* 1 mm, globose. Seeds 0.3-0.4 mm.
Fl. IV - VI. *Mountain pastures and along paths and tracks on bare ground and in grassy places above 300 m. ?Abundant. An inconspicuous little plant and probably often overlooked.* **M**

LIV. EUPHORBIACEAE
M.J. Short

Herbs, shrubs or trees, often with milky sap. Leaves alternate or occasionally opposite, simple, entire to pinnately lobed, usually stipulate. Flowers small, unisexual, usually regular, often without petals and sometimes without sepals. Stamens one to numerous. Ovary superior, 2-3-locular; styles 2-3. Disc often present. Fruit a capsule. Seeds 1-2 per locule, often carunculate.

1. Plants with milky sap; perianth absent; several male flowers and a single female flower surrounded by an involucre . . . . **4. Euphorbia**
- Plants with watery sap; perianth present; flowers not arranged as above 2

2. Leaves entire, appearing like the leaflets of a compound leaf . . . . . . **1. Phyllanthus**
- Leaves not entire, not as above 3

3. Small, slender, annual herb up to 45 cm; leaves opposite, ovate, crenate-serrate . . . . **2. Mercurialis**
- Large, stout, shrub-like herb up to 2(-4) m; leaves alternate, palmately lobed . . . . **3. Ricinus**

1. Phyllanthus L.

Monoecious herbs with watery sap. Leaves alternate, stipulate, simple, entire, appearing like the leaflets of a compound leaf. Flowers small, axillary, solitary or in clusters. Perianth-segments 5. Stamens 5. Ovary 3-locular; styles 3. Capsules with 3, 2-seeded locules, explosively dehiscent. Seeds trigonous.

1. P. tenellus Roxb., *Fl. ind. ed. 1832* **3**: 668 (1832).
Glabrous annual or short-lived perennial herb, up to 60 cm high; stems erect, branched, often rather woody at the base. Leaves 5-20 × 3.5-9 mm, broadly elliptic to obovate, acute to obtuse, thin, paler beneath, very shortly petiolate. Perianth-segments 0.8-1 mm, lanceolate to ovate, broadly scarious. Styles bifid, persistent. Fruiting pedicels 3-7 mm, filiform. Capsules 1.7-2.2 mm wide, compressed-globose, smooth, green. Seeds 0.9-1 mm, pale brown, minutely papillose.
Fl. ? all year. *A weed of gardens and waste places in southern Madeira. Native to the Mascarene Is.* **M**

Records of *P. niruri* L. from Madeira are referable to this species.

2. Mercurialis L.

Small, dioecious or monoecious herbs with watery sap. Leaves opposite, stipulate, simple. Flowers with 3 perianth-segments; the male flowers borne in clusters on long-pedunculate, slender, axillary and terminal spikes; stamens 8-15; the female flowers axillary, solitary or in small clusters, shortly pedunculate or subsessile. Ovary 2-locular; styles 2. Capsules with 2, 1-seeded locules, septicidally dehiscent. Seeds carunculate.

1. M. annua L., *Sp. pl.* **2**: 1035 (1753).   [*Urtiga morta*]
Subglabrous annual. Stem 5-45 cm, erect, often much-branched. Leaves 1-5.5 × 0.5-3 cm, ovate to elliptic-lanceolate, crenate-serrate, sparsely ciliate; petioles 1-25 mm. Male inflorescences up to 7 cm; perianth-segments *c*. 1-1.5 mm, ovate, acute. Female flowers 1-4 per axil. Capsules 2.5-4 mm wide, tuberculate, the tubercles bristle-tipped. Seeds 1.5-2 mm, ovoid, rugulose.
Fl. all year. *Walls, grassy banks, paths, pastures, cultivated and waste ground. Very common in Madeira at altitudes up to 800 m; also frequent in Porto Santo, and found on both Ilhéu de Cima and Ilhéu de Baixo. Known on Deserta Grande and also recorded from Ilhéu Chão. Very rare on Selvagem Grande.* **MDPS**

Madeiran plants are frequently regularly monoecious (var. **ambigua** (L.f.) Duby), whilst typically dioecious plants (var. **annua**) may also exhibit a tendency to develop male flowers on female plants.

Acalypha virginica L., *Sp. pl.* **2**: 1003 (1753), a monoecious annual with erect, puberulent stems, 10-50 cm high, alternate, ovate, crenate leaves, and a pubescent capsule with 3, 1-seeded locules, was reported from Madeira in 1927 as a recently introduced weed of cultivated ground at Seixal and also between Seixal and São Vicente. No recent records are known and it is possibly no longer present. Native to N. America. **M**

3. Ricinus L.

Large monoecious herbs with watery sap. Leaves alternate, peltate, palmately-lobed, stipulate. Inflorescence paniculate, pistillate flowers above, staminate flowers below. Male flowers with 3-5-lobed perianth; filaments repeatedly branched. Female flowers with 5-lobed perianth, lobes caducous; ovary 3-locular; styles 3, bifid. Capsules with 3, 1-seeded locules, usually spinose. Seeds carunculate.

1. R. communis L., *Sp. pl.* **2**: 1007 (1753).   [*Carrapateira*]
Glabrous, often stout and shrub-like, annual herb up to 2(-4) m, tinged with reddish purple, especially when young; stem erect, branched above, stout. Leaves petiolate, up to 30(-60) cm broad, 5-9(-11)-lobed; lobes lanceolate to ovate-lanceolate, acuminate, glandular-serrate. Inflorescences axillary and terminal, up to 30 cm long, erect. Male flowers globose;

perianth-segments ovate-lanceolate, acute; stamens numerous; anthers globose. Female flowers *c.* 1 cm long; styles dark red. Capsules 1-2 cm, oblong-ellipsoid. Seeds 8-15 mm, compressed-ovoid, reddish brown, variously mottled.
Fl. III - X. *Commonly naturalized along roadsides and on waste ground in the lower regions of Madeira and in Porto Santo; also known from Selvagem Grande. Native to tropical Africa.*
MPS

4. Euphorbia L.

Monoecious herbs, shrubs or small trees with milky sap. Leaves alternate or opposite, simple, mostly entire, usually exstipulate. Flowers small, greenish, with several male flowers and one female flower arranged in a small group, surrounded by a 4-5-lobed, calyx-like involucre, the lobes alternating with 4-5 conspicuous glands, the whole forming a *cyathium*. Cyathia axillary, solitary or more usually forming umbel-like cymes, the primary branches (*rays*) subtended by *ray-leaves*, the ultimate branches subtended by *raylet-leaves*; axillary rays often present below the umbel. Male flowers comprising a single stamen on a jointed pedicel. Female flower pedicellate; ovary 3-locular; ovules 1 per locule; styles 3; stigmas often bifid. Capsules shallowly to deeply 3-sulcate. Seeds sometimes carunculate.

Several species are cultivated in Madeira, including the shrub **E. pulcherrima** Willd. ex Klotzsch [*Manhãs de páscoa, Poinsétia*], a native of Mexico, which is frequently grown as an ornamental for its brilliant red bracts.

1. Annual to perennial herb 2
- Shrub or small tree 12

2. Leaves opposite, stipulate 3
- Leaves alternate, exstipulate 5

3. Stems glabrous, fleshy; seeds smooth . . . . **2. peplis**
- Stems pubescent, at least when young, not fleshy; seeds wrinkled 4

4. Capsules pubescent along keels; leaves 2-7 mm . . . . **3. prostrata**
- Capsules glabrous; leaves 10-25(-35) mm . . . . **1. nutans**

5. Involucral glands with horns, or if horns absent, with the outer margin emarginate . 8
- Involucral glands without horns, the outer margin rounded 6

6. Raylet-leaves similar in shape to cauline leaves; capsules smooth, unwinged . . . . . .
. . . . **9. helioscopia**
- Raylet-leaves different in shape to cauline leaves; capsules tuberculate or winged . . 7

7. Leaves obtuse; capsules valves winged, otherwise smooth; seeds reticulate . . . .
. . . . **8. pteroccocca**
- Leaves acute; capsules valves unwinged, warty; seeds smooth . . . . . **7. platyphyllos**

8. Seeds tuberculate or pitted 10
- Seeds smooth 9

9. Cauline leaves oblong-obovate to ovate, the midrib obscure; glands emarginate or shortly horned . . . . **13. paralias**
- Cauline leaves linear, the midrib distinct; glands with very long, filiform horns . . . .
. . . . **14. terracina**

10. Cauline leaves broadly obovate to suborbicular, shortly petiolate . . . . **10. peplus**
- Cauline leaves linear, sessile 11

11. Raylet-leaves triangular-lanceolate; capsules c. 1.3 mm wide; seeds 1-1.3 mm, tuberculate
. . . . **10. exigua**
- Raylet-leaves deltate-rhombic; capsules 3-4 mm wide; seeds 1.5-2.2 mm, pitted . . . . .
. . . . **12. segetalis**

12. Leaves 60-175 × 10-28 mm, dark green with pale midrib; inflorescence paniculate; capsules 7-9 × 10-11 mm . . . . **4. mellifera**
- Leaves 25-70 × 3.5-12 mm, pale glaucous green; inflorescence umbellate; capsules 4-6 × 5-8 mm 13

13. Leaves 8-11 mm wide, elliptic to oblanceolate, obtuse or retuse; glands usually 4; seeds smooth (Salvage Is. only) . . . . **5. desfoliata**
- Leaves 3.5-8(-12) mm wide, linear, subacute to obtuse; glands usually 5; seeds rugulose
. . . . **6. piscatoria**

1. E. nutans Lag., *Gen. sp. pl.*: 17 (1816).
E. preslii Guss.
Annual; stems up to 50 cm, ascending to erect, branched, slender, pubescent when young, later becoming glabrous. Leaves opposite, 10-25(-35) × 3-13 mm, oblong, subacute to obtuse, somewhat asymmetrical at the base, serrate, sparsely pilose above, often with a reddish spot, glabrous or glabrescent below; petioles 1-2 mm. Stipules triangular, fringed. Cyathia axillary, solitary. Glands 4, transversely ovate, pale yellow, with small appendage. Capsules 1.8-2.3 × 2.2-2.5 mm, smooth, glabrous. Seeds 1.2-1.3 mm, ovoid-quadrangular, transversely wrinkled, blackish under a grey coat, ecarunculate. $2n = 12$.
Fl. VII - X. *Naturalized along roadsides and on cultivated and waste ground, often occurring as a garden weed, mainly in the south of Madeira and especially in and around Funchal. Native to N. America.* **M**

2. E. peplis L., *Sp. pl.* 1: 455 (1753). [*Trovisco*]
Glabrous annual, fleshy and often purplish tinged, usually 4-branched from base; branches up to 20 cm long, prostrate, branching dichotomously. Leaves opposite, 5-15 × 2.5-9 mm, oblong, obtuse or retuse, markedly asymmetrical at the base, entire or denticulate towards the base; petioles 1-2 mm. Stipules membranous, laciniate. Cyathia solitary in axils and in forks of the stems. Glands 4, transversely elliptic, reddish brown, with narrow appendages. Capsules 3-4 × 4.5-5 mm, smooth, glabrous. Seeds 2.5-2.8 mm, ovoid, smooth, pale grey, mottled brown, ecarunculate.
Fl. V - XII. *Rare, on the sandy beach of Porto Santo.* **P**

3. E. prostrata Aiton, *Hort. kew.* 2: 139 (1789).
Annual; stems up to 20 cm, prostrate, much-branched, forming small mats, purplish tinged, puberulous along upper side. Leaves opposite, 2-7 × 1.5-4 mm, oblong-elliptic to oblong-obovate, obtuse, slightly asymmetrical at the base, minutely serrulate towards the base, glabrous to sparsely pubescent; petioles up to 1 mm. Stipules minute, triangular, fringed. Cyathia in short, axillary clusters. Glands 4, transversely elliptic, dark purple, with minute appendages. Capsules 1-1.5 × 1.3-1.5 mm, smooth, pilose along keels only. Seeds 0.9-1 mm, ovoid-quadrangular, transversely wrinkled, reddish beneath a grey coat, ecarunculate.
Fl. I - IX. *Very common in the lower regions of Madeira and Porto Santo, growing on waste ground and between the pavements and cobblestones of streets and paths in towns and villages, especially Funchal and Vila Baleira; 0-300 m. Native to N. America.* **MP**

4. E. mellifera Aiton, *Hort. kew.* **3**: 493 (1789). [*Alindres, Figueira do inferno*]
Shrub or small tree up to 8 m with spreading branches; bark grey, smooth. Leaves subsessile, alternate, crowded towards the ends of the branches on the current year's growth, 60-175 × 10-28 mm, oblong-elliptic to oblong-lanceolate, acute to subacute, long mucronate, dark green with prominent pale midrib, often sparsely to densely pilose beneath along lower part of midrib, otherwise glabrous. Inflorescence terminal, paniculate, very dense in flower, lax in fruit. Raylet-leaves 4-6.5 mm, oblong-obovate, mucronate, sparsely to densely villous-ciliate. Glands usually 5, transversely ovate, reddish purple with paler margin. Capsules 7-9 × 10-11 mm, hard, tuberculate, glabrous. Seeds 3.5-5 mm, ovoid, smooth, dark brown, carunculate. $2n=40$. **Plate 23**
Fl. II - VII. *A rare but characteristic species of the Madeiran laurisilva, growing in moist, shady places in sheltered ravines, scattered throughout the island, mainly 400-1100 m.* ■ M [*Canaries*]

5. E. desfoliata (Menezes) Monod in *Bol. Mus. Mun. Funchal,* suppl. 1: 34 (1990).
[*Tabaiba*]
E. anachoreta Svent.; *E. obtusifolia* var. *desfoliata* Menezes
Low, glabrous shrub, 20-40(-60) cm high; stemsthick and succulent, much-branched, with numerous leaf-scars, lower branches whitish grey, the upper branches subverticillate, greenish. Leaves sessile, in crowded rosettes at the ends of the branches, *c.* 4-7 × 0.8-1.6 cm, linear-elliptic to oblanceolate, obtuse and often retuse, somewhat coriaceous, glaucous. Inflorescence umbellate, simple. Rays 2-7. Raylet-leaves *c.* 3 mm, ovate-oblong, acute or obtuse. Glands usually 4, transversly orbicular. Capsules *c.* 4-5 × 5-6.5 mm, hard, weakly muricate. Seeds *c.* 3 mm, ovoid, smooth, glabrous, grey-black, mottled white, carunculate. **Plate 23**
Fl. III - VI. *Very rare; only found among rocks and stones, and in rock fissures on Ilhéu de Fora.* ● S

6. E. piscatoria Aiton, *Hort. kew.* **2**: 137 (1789). [*Figueira do inferno*]
Variable, glabrous, rounded shrub up to 1.75 m; stems much-branched, leafless below. Leaves alternate, sessile, crowded towards the ends of the branches, 25-70 × 3.5-8(-12) mm, linear to linear-elliptic, subacute to obtuse, glaucous, the margin yellowish. Inflorescence umbellate, simple or compound. Rays 5-6(-8). Raylet-leaves 3-6.5 mm, oblong-obovate to suborbicular, obtuse, mucronate. Glands usually 5, transversely ovate, emarginate or with 2 short horns, yellowish green or deep purplish red. Capsules 6 × 7-8 mm, hard, smooth, glabrous. Seeds 3-3.5 mm, oblong-ovoid, rugulose, dark brown to black, carunculate. $2n=20$. **Plate 23**
Fl. I - VIII. *Fairly common and locally abundant in Madeira on cliffs and rocky slopes, mainly in the lower regions near the coast, 0-300(-550) m; often growing with the introduced* Opuntia *in the south. Frequent in Porto Santo, on both high peaks and in lower regions. Also found on Ilhéu de Cima, Deserta Grande and Ilhéu Chão.* ● MDP

7. E. platyphyllos L., *Sp. pl.* **1**: 460 (1753). [*Trovisco*]
Annual, 12-50(-80) cm high; stems erect, simple or occasionally branched from the base, glabrous or pilose. Leaves sessile, alternate, cauline and ray-leaves 15-45 × 5-13 mm, oblong-oblanceolate, acute, base auriculate, serrulate, sparsely pilose, especially beneath. Raylet-leaves 5-17 mm, deltate, mucronate, serrulate. Inflorescence umbellate. Rays 5, 3-4-chotomous, then 2-3-chotomous; axillary rays usually numerous. Glands 4, transversely ovate. Capsules 2.5-3 × 2.7-3.5 mm, hard, warty, glabrous. Seeds 1.9-2.3 mm, ovoid, smooth, dark grey-brown, carunculate.
Fl. III - VI. *Infrequent in the lower coastal region of Madeira, along roadsides and in waste places.* M

8. E. pterococca Brot., *Fl. lusit.* **2**: 312 (1804).
Annual, 15-45 cm high; stems ascending to erect, simple or branched from the base, sparsely pilose. Leaves alternate; cauline and ray-leaves 20-42 × 7-12 mm, obovate-spathulate, obtuse, base attenuate, serrulate, sparsely pilose, the lower shortly petiolate. Raylet-leaves 3-17 mm, rhombic, serrulate. Inflorescence umbellate. Rays 5, 3-5-chotomous, then 2-3-chotomous; axillary rays usually numerous. Glands 4, transversely ovate. Capsules *c.* 1.5 × 2-2.2 mm, hard, glabrous, with 2 undulate wings on each valve. Seeds 1.2-1.3 mm, ± spherical, reticulate, dark reddish brown; caruncle minute.
Fl. IV - V. *Very rare in the southern coastal region of Madeira, along roadsides and in waste places.* **M**

E. pterococca is typically glabrous, but the leaves and stems of Madeiran specimens are sparsely pubescent.

9. E. helioscopia L., *Sp. pl.* **1**: 459 (1753). [*Trovisco*]
Erect annual, 14-45 cm high. Stems simple or with a few branches below, glabrous or sparsely pilose. Leaves alternate, sessile; cauline and ray-leaves 18-40 × 10-25 mm, obovate to obovate-spathulate, obtuse, base attenuate, serrulate in the upper half, more or less glabrous, the cauline caducous. Raylet-leaves similar but smaller and less tapered at the base. Inflorescence umbellate. Rays 5, 3-chotomous, then usually 1-chotomous; axillary rays absent. Glands transversely ovate, entire. Capsules 2.5-3.3 × 3.5-4 mm, hard, smooth, glabrous. Seeds 2-2.3 mm, ovoid, reticulate, dark brown, carunculate.
Fl. I - VI. *Infrequent in Madeira and Porto Santo along roadsides and on cultivated and waste ground.* **MP**

E. lathyris L., *Sp. pl.* **1**: 457 (1753), a glabrous and glaucous biennial up to 1.5 m high with a solitary, erect stem, opposite and decussate, linear-lanceolate leaves in 4 rows, and smooth, spongy capsules, 8-20 mm long, has occasionally been reported from Madeira, but these records possibly refer only to garden plants. **M**

10. E. exigua L., *Sp. pl.* **1**: 456 (1753). [*Trovisco*]
E. exigua var. *retusa* L.
Glabrous, glaucous annual 10-23 cm. Stem solitary, erect, often much-branched from the base. Leaves alternate, sessile, 3-25 × 0.5-2 mm, linear, acute, obtuse, truncate or retuse, mucronate. Ray-leaves similar to upper cauline; raylet-leaves triangular-lanceolate, acute, obliquely subcordate, rarely 1-2-toothed on one side near the base. Inflorescence umbellate. Rays 3-5. Glands with 2 long slender horns, red or orange. Capsules *c.* 1.5 × 1.3 mm, hard, smooth or slightly granulate on keels, glabrous. Seeds 1-1.3 mm, ovoid-quadrangular, pale grey, white tuberculate, carunculate.
Fl. VI - VIII. *Extremely rare weed of cultivated fields and roadsides in Madeira.* **M**

11. E. peplus L., *Sp. pl.* **1**: 456 (1753). [*Sarmento, Trovisco*]
E. peplus var. *genuina* Cout., var. *peploides* (Gouan) Vis.
Glabrous annual, 5-32 cm. Stem solitary, erect, usually branched, often ± leafless below, sometimes reddish. Leaves alternate, 5-30 × 4-14 mm, obovate or suborbicular, obtuse; petiole 2-8 mm. Ray-leaves like the cauline, subsessile. Raylet-leaves smaller, somewhat obliquely ovate. Inflorescence umbellate. Rays 3. Glands with long slender horns, greenish. Capsules 2 × 2-2.2 mm, glabrous, each valve with 2 narrow dorsal wings. Seeds 1.4-1.6 mm, oblong-hexagonal, with a longitudinal groove on 2 ventral faces and (2)3-4 pits on the other 4, pale grey, darker in the depressions, carunculate.
Fl. II - X. *Very common in Madeira and Porto Santo on waste and cultivated ground, on walls and cliffs, along paths and roads, up to 550 m; also found on Ilhéu de Cima and Deserta Grande.* **MDP**

12. E. segetalis L., *Sp. pl.* **1**: 458 (1753). [*Trovisco*]
Glabrous annual to perennial; stems 27-41 cm, erect, simple or branched from the base. Leaves alternate, sessile, 10-45 × 1.5-5 mm, linear to linear-oblanceolate, acute to obtuse, mucronate. Ray-leaves elliptic-oblong. Raylet-leaves deltate-rhombic, obtuse, base cuneate to subcordate. Inflorescence umbellate. Rays 5-6. Glands with 2 horns. Capsules 2.5-3.5 × 3-4 mm, hard, muricate on the keels, otherwise smooth, glabrous. Seeds 1.5-2.2 mm, ovoid, distinctly and coarsely pitted, pale grey, carunculate.
Fl. I - VII. *Very rare in the coastal region of southern Madeira, growing in fields and cultivated land, with no recent records known.* **M**

Two varieties have been recorded from Madeira, var. **segetalis**, an annual, and var. **pinea** (L.) Lange in Willk. & Lange, *Prodr. Fl. Hisp.* **3**: 499 (1877) (*E. pinea* L.), a perennial with more crowded and generally shorter and broader leaves.

13. E. paralias L., *Sp. pl.* **1**: 458 (1753). [*Eufórbia marítima*]
Glabrous, glaucous, fleshy, caespitose perennial up to 50(-70) cm; stems numerous, erect, usually simple, often reddish at the base, densely leafy. Leaves alternate, sessile; cauline 5-20 × 2-10 mm, the lower oblong-obovate to elliptic-oblong, acute to obtuse, the uppermost ovate; ray-leaves similar to the upper cauline. Raylet-leaves 5-12 mm, suborbicular-rhombic to reniform, mucronulate. Inflorescence umbellate. Rays 3-6, up to 3 times dichotomous; axillary rays 0-10. Glands emarginate or shortly horned, yellow-orange. Capsules 3-5 × 4.5-6.5 mm, hard, granulate-rugulose. Seeds 2.5-3.5 mm, ovoid, smooth, pale grey; caruncle minute.
Fl. III - VII. *Rather common on the sandy beach and cliffs of Porto Santo, up to c. 100 m; also found on Ilhéu de Baixo.* **P**

14. E. terracina L., *Sp. pl.* ed. 2: 654 (1762). [*Figueirinha* (P), *Trovisco*]
E. heterophylla Desf.; *E. juncea* Aiton
Glabrous perennial 10-50 cm high; stems procumbent to erect, simple to much-branched. Leaves alternate, sessile, the cauline 10-50 × 1.5-6(-10) mm, linear, acute, obtuse or tricuspidate, entire to very minutely serrulate; ray-leaves lanceolate to oblong-lanceolate. Raylet-leaves 4-15 mm, ovate-rhombic, sometimes slightly asymmetrical, acute. Inflorescence umbellate. Rays 3-5, up to 5 times dichotomous; axillary rays 0-5. Glands with 2 very long, slender horns, often tinged red. Capsules 2.5-3 × 3.3-4.2 mm, hard, smooth, glabrous. Seeds 1.9-2 mm, ovoid, smooth, pale grey; caruncle large, boat-shaped. 2n=18.
Fl. I - VI. *Roadsides and on cultivated and waste ground. Common in the lower regions of southern Madeira, including Ponta de São Lourenço; also frequent in Porto Santo, up to c. 200 m.* **MP**

LV. RUTACEAE
M.J. Short

Trees, shrubs or rarely herbs. Leaves alternate or opposite, simple or pinnately compound, often gland-dotted and aromatic, sometimes evergreen, exstipulate. Flowers usually regular and hermaphrodite, solitary or in clusters, cymes, racemes or panicles. Sepals 4-5, free or connate below. Petals 4-5, usually free. Disc present. Stamens equalling or twice the number of petals, rarely more numerous. Ovary superior, usually 4- to 5-locular, carpels occasionally more or less free. Ovules mostly 1-2 per locule. Fruit a capsule, berry or drupe.

Several important commercial fruit trees belonging to the genus **Citrus** L. are cultivated in the lower regions of Madeira, principally citron, **C. medica** L. [*Cidreira*, the fruit *Cidra*]; lemon, **C. limon** (L.) Burm.f. [*Limoeiro*, the fruit *Limão*]; sweet lemon, **C. limetta** Risso [*Limão doce, Lima, Lima de chêiro*]; lime, **C. aurantifolia** (Christm.) Swingle [*Limão de gallinha*]; Seville orange, **C. aurantium** L. [*Laranjeira azeda*, the fruit *Laranja azeda*]; sweet orange,

C. sinensis (L.) Osbeck [*Laranjeira*, the fruit *Laranja*]; and tangerine, **C. reticulata** Blanco [*Tangerineira, Laranja tangerina*].

The heath-like, white-flowered shrub **Diosma ericoides** L. [*Urze de cheiro*] from S. Africa is reported to be commonly grown in Madeiran gardens. **Choisya ternata** Humb., Bonpl. & Kunth [*Laranjeira do México*], an evergreen shrub native to Mexico with glossy, ternate leaves and white, 5-petalled, sweetly-scented flowers, 2-3 cm across, is also frequently cultivated.

1. Ruta L.

Glandular-punctate, perennial herbs, usually woody below. Leaves alternate, pinnately divided. Inflorescence corymbose, bracteate. Sepals and petals 4 (except the terminal flower which usually has 5). Petals hooded, yellow, the margin fimbriate. Stamens twice as many as the petals. Ovary 4(-5)-locular; style 1. Capsules 4(-5)-lobed; lobes dehiscent inwardly at apex.

1. R. chalepensis L., *Mant. pl.*: 69 (1767). [*Arruda*]
R. bracteosa DC.
Glabrous herb with pungent odour, up to 50(-70) cm. Stems prostrate to ascending, branched, densely leafy. Leaves 1.5-10 cm, 2- to 3-pinnatisect, glaucous; ultimate segments 5-15(-20) × 1.5-4(-6) mm, oblong-elliptic to oblanceolate, obtuse. Bracts ovate-lanceolate to cordate-ovate. Flowers *c.* 1-1.5 cm across, pedicellate. Sepals deltate-ovate, acute. Petals greenish yellow, oblong, fringed with long cilia, contracted at the base into a short claw. Capsules 6-8 mm, globose; lobes acuminate, erect. Seeds *c.* 2 mm.
Fl. III-VII. *Rocky ground, pastures, and dry banks. Sporadic along the south coast of Madeira, mainly on Ponta de São Lourenço; locally common in Porto Santo and Ilhéu de Cima; also found on Deserta Grande.* **MDP**

LVI. POLYGALACEAE
M.J. Short

Polygala myrtifolia L., *Sp. pl.* **2**: 703 (1753), an erect, much-branched shrub up to 2.5 m, with elliptic to narrowly oblong-obovate leaves, irregular flowers in short, few-flowered, leafy racemes, the inner two sepals wing-like, petaloid, violet-purple, the corolla 13-18 mm long, lilac to deep violet, is grown for ornament in Madeira and may occasionally escape. Native to S. Africa. **M**

LVII. ANACARDIACEAE
M.J. Short

Trees or shrubs, often with resinous bark. Leaves alternate, pinnately compound (in Madeira), exstipulate. Flowers small, hermaphrodite or unisexual, regular, in large panicles. Sepals (4-)5, fused at the base. Petals (4-)5, free. Stamens 5-10, inserted on a disk. Ovary superior, 1-locular. Styles 3. Fruit a 1-seeded drupe.

Schinus molle L., *Sp. pl.* **1**: 388 (1753) [*Pimenteira bastarda*], a graceful, evergreen tree up to 10 m with slender, pendulous branches, leaves up to 30 cm long, divided into 20-40 linear-lanceolate leaflets, small greenish yellow flowers, and pendulous clusters of small, globose, rose-pink fruits, is a commonly planted ornamental tree in Madeira. Native to S. America. **M**

1. Rhus L.

Shrubs or small trees. Leaves imparipinnate with toothed leaflets, the rachis often slightly winged. Flowers numerous, in branched, axillary or terminal panicles. Calyx deeply 5-lobed.

Petals 5, imbricate, exceeding the calyx. Stamens 5. Fruit a globose drupe. Seed reniform, flattish.

1. R. coriaria L., *Sp. pl.* **1**: 265 (1753). [*Sumagre*]
Deciduous shrub or small tree up to 3 m. Young branches densely pubescent. Leaves 10-20 cm, divided into 4-8 pairs of sessile leaflets, each 1.5-5.5 cm long, broadly lanceolate to oblong-elliptic, crenate-serrate, dark green above, paler beneath, softly pubescent to subglabrous, often turning purplish red in the autumn; rachis narrowly winged in the uppermost part. Flowers greenish white, in very dense panicles up to 15 cm long. Pedicels up to 2.5 mm, stout. Petals 3-4.5 mm, oblong-ovate. Drupe 4-6 mm, globose, villous, brownish purple when ripe.
Fl. V - IX. *Naturalized along roadsides, on open sunny banks, rough grassland, and in dry, rocky waste places. Reported as common in the last century but apparently now only occasional in lowland regions of Madeira, principally in the south-east; 0-600 m. Native to S. Europe.* **M**

LVIII. ACERACEAE
J.R. Press

Trees or shrubs. Leaves opposite, exstipulate. Inflorescence a raceme, corymb or (in Madeira) a panicle. Flowers actinomorphic, dioecious or functionally monoecious. Sepals and petals 4-5. Stamens (4-)8(-10), inserted on the disc. Ovary 2-lobed. Styles 2, free or united below. Fruit of 2 single-seeded, winged mericarps (*samaras*).

1. Acer L.

Deciduous trees or occasionally shrubs. Leaves palmately lobed. Samaras winged on one side only.

1. A. pseudoplatanus L., *Sp. pl.* **2**: 1054 (1753).
Deciduous tree to *c*. 30 m. Leaves up to 15 x 17 cm, palmately 5-lobed to half-way with the 2 basal lobes often smaller than the others, coarsely toothed. Flowers monoecious, appearing with the leaves in narrow, drooping panicles. Petals greenish yellow. Fruits 4-5 cm, diverging at 90° or slightly more.
Fl. IV. *Introduced and widely planted in Madeira, readily self-seeding and sometimes becoming a pest in a few areas such as Ribeiro Frio. Native to Europe and W. Asia.* **M**

A. campestre L., *Sp. pl.* **2**: 1055 (1753), also from Eurasia, has small, coriaceous leaves and fruits with horizontally spreading wings. It is also occasionally planted in Madeira, at least in the Santana area, and may perhaps self-seed. **M**

LIX. SAPINDACEAE
M.J. Short

Climbers (in Madeira). Leaves alternate, compound, stipulate (in Madeira). Flowers often unisexual or functionally so, irregular and 4-merous (in Madeira), in racemose or paniculate cymes, often small and numerous. Sepals free, unequal. Petals free, imbricate, usually with a scale-like appendage. Disk usually present. Stamens often 8. Ovary superior, 3-locular. Style 1. Fruit a capsule (in Madeira).

1. Cardiospermum L.

Climbing herbs. Leaves biternate, the leaflets toothed. Inflorescences axillary, of panicle-like clusters with a pair of opposite, curled tendrils at the peduncle apex. Flowers bracteate, white. Sepals persistent, the outer two much shorter than the inner. Style 3-lobed. Fruit an inflated, bladder-like, membranous capsule. Seeds globose, black with a white hilum, 1 per locule.

1. C. grandiflorum Sw., *Prodr.*: 64 (1788). [*Corriola de balões*]
C. grandiflorum forma *hirsutum* (Willd.) Radlk.
Vigorous, slightly woody annual vine up to 8 m, pubescent with long, yellowish brown hairs; stems sulcate. Leaves petiolate; leaflets divided into ovate, acute, coarsely serrate segments; stipules minute, caducous. Peduncles 4.5-8(-15) cm. Flowers 6-9 mm long. Two outer sepals 1.5-2.5 mm, suborbicular, two inner sepals 5.5-7 mm, obovate, petaloid. Petals obovate, obtuse, creamy-white. Ovary pubescent. Capsules 4.5-6.5 × 2-3 cm, broadly ellipsoid, apiculate, green turning to pale brown. Seeds 4-5 mm across; hilum triangular in outline. 2n=22.
Fl. I - XII. *Commonly grown as an ornamental in the lower regions of Madeira and quite often becoming naturalized, sprawling down walls, cliff faces, and over waste ground; 0-400 m. Native to tropical Africa and America.* **M**

Records of *C. halicacabum* L. from Madeira are probably referable to this species.

LX. BALSAMINACEAE
M.J. Short

Herbs with translucent stems. Leaves simple, exstipulate. Flowers irregular, hermaphrodite, solitary or in racemes. Sepals 3, free. Petals 5, the lower 4 fused into 2 lateral pairs. Stamens 5; anthers united around ovary. Ovary superior, 5-locular. Stigma sessile, 5-toothed. Fruit a 5-valved capsule, dehiscing by elastically coiling valves which shoot out the seeds explosively.

1. Impatiens L.

Annual herbs. Sepals 3, very unequal, the lowest large and with a sac-like spur, the 2 lateral small, ovate, usually green. Corolla 2-lipped, with a large, hood-like upper petal and 2 pairs of lateral petals, united except for 2 apical lobes and together forming a divided lower lip.

1. I. balsamina L., *Sp. pl.* **2**: 938 (1753). [*Maravilhas*]
Stems up to 60 cm, ± erect, simple or sparingly branched, glabrous or pubescent when young. Leaves mostly alternate, shortly petiolate, 3-10 × 1-2.5 cm, narrowly lanceolate to elliptic or oblanceolate, acute, cuneate, margin serrate and with several dark, almost stalkless glands near the leaf-base, glabrous or finely pubescent. Flowers 1(-3) in axils of the leaves, up to 3.5 cm, pink or white; pedicels 1-1.5 cm, slender, pendent in fruit. Lower sepal abruptly contracted into a filiform, curved spur. Capsules 0.8-1.5 cm, broadly ellipsoid, densely pubescent. 2n=14.
Fl. IX. *Mainly in gardens, but naturalized at Faial and perhaps elsewhere. Native to India and SE Asia.* **M**

LXI. AQUIFOLIACEAE
S. Andrews

Trees or shrubs. Leaves simple; exstipulate. Flowers actinomorphic, dioecious, usually 4- to 8-merous. Petals imbricate. Stamens sometimes adnate to petals. Ovary superior. Fruit a berry.

1. Ilex L.

Evergreen trees or shrubs. Leaves alternate, rarely opposite. Flowers axillary, solitary, in fascicles or cymes. Petals fused at base. Stamens 4-5, adnate to petals. Fruit globose or obovoid, containing 3-5 pyrenes.

Literature: S. Andrews. A reappraisal of *Ilex aquifolium* and *I. perado* (Aquifoliaceae). In *Kew Bull.* **39**: 141-155 (1984); Report on the *Ilex perado* complex of the North Atlantic Islands. (Part 1). In *Internat. Dendrol. Soc. Yearb.* **1982**: 69-72 (1983); G. Kunkel. The *Ilex perado* complex in the Canary Islands and Madeira. In *Cuad. Bot. canaria* **28**: 17-29 (1977); K. Lems. Hollies of the Canary and Madeira Islands. In *Am. hort. Mag.* **47**: 290-295 (1968).

1. Leaves glossy green, apex acute or mucronate, petioles winged; flowers in fascicles . . . . **1. perado**
- Leaves dull green, apex obtuse, rarely acute, petioles winged; flowers in cymes . . . . **2. canariensis**

1. I. perado Aiton, *Hort. kew.* **1**: 169 (1789). [*Perado*]
I. maderensis Lam.; *I. perado* Aiton var. *maderensis* (Lam.) Loes. subvar. *spinulosa-serrata* Loes.
Tree up to 5 m, with smooth, pale grey bark. Leaves 6.2-8(-9.7) × 4-5.6(-6.4) cm, obovate, oblong-ovate, ovate or oblong, acute or mucronate, glabrous, entire, thick and coriaceous, dark glossy green above, paler below; leaves of seedlings and sucker growth elliptic with closely spaced, forward-pointing spines 1-1.5(-2) mm; petioles 7-17 mm, usually winged. Flowers white or tinged pink on outer lobes, fragrant, in axillary fascicles. Peduncles and calyces often pubescent. Calyx 1-1.5(-2.2) mm, 4-lobed. Corolla 3-4.5 mm, 4-lobed, the lobes all ciliate. Stamens 4. Ovary 2-3 mm, stigma not protruding. Fruiting pedicels 6.5-12(-16) mm, pubescent or glabrous. Fruit 6-10 × 5-10 mm, red, globose; pyrenes (3-)4(-5). **Plate 24**
Fl. IV - V. *Sporadic but locally abundant in laurisilva, sometimes on exposed crests but more often on the deeper soils of shaded groves mainly in the central and northern parts of Madeira from 700-1200m. Often burnt for firewood.* ■ M [*Azores, Canaries*]

Represented in Madeira by the endemic (●) subsp. **perado**, to which the above description and synonymy apply. Of the other Macaronesian taxa, subsp. *platyphylla* (Webb & Berthel) Tutin is endemic to the Canaries and distinguished by its larger leaves and longer fruiting pedicels; subsp. *azorica* (Loes.) Tutin is endemic to the Azores and has smaller leaves.

I. aquifolium L. has strongly spinose leaves with undulate margins and is frequently cultivated in gardens in Madeira.

2. I. canariensis Poir. in *Encycl. Suppl.* **3**: 67 (1813). [*Azevim, Azevinho*]
I. azevinho Sol. ex Lowe; *I. aestivalis* Buch.; *I. canariensis* subsp. *azevinho* (Sol. ex Lowe) Kunkel
Tree to 6.5 m with greyish white bark. Leaves 4-9.5(-11) × 2-4.5 cm, ovate or ovate-lanceolate, obtuse, rarely acute, glabrous, entire, those of seedlings and suckers elliptic or ovate with widely spaced, spreading spines 1-2.5 mm, semi-coriaceous, dull blackish green above, paler below; petiole 1-1.8 cm, pubescent of glabrous. Flowers white or tinged with purple on outer lobes, strongly scented, in terminal and axillary cymes. Peduncles and calyces often pubescent. Both calyx and corolla 4- to 5-lobed, ciliate, calyx 1-1.5 mm, corolla 4-4.5 mm. Stamens 4-5. Ovary 3-3.3 mm, 4- to 5-lobed, stigma protruding. Fruit 9.5-11 x 7-9mm, sphaerical or oblong, red but darkening to almost black; pedicels 1.2-2.4 mm, pubescent or glabrous. Pyrenes 4-5, 6.5-7.5 × 3.5 mm, deeply sulcate. **Plate 24**
Fl. IV - VI. *In laurisilva and heath forests on dry, exposed soil, mainly in central and northern Madeira from 300-880 m. Becoming rare.* ■ M [*Canaries*]

LXII. CELASTRACEAE
N.J. Turland

Evergreen, glabrous, unarmed shrubs. Leaves alternate, simple; stipules small, deciduous. Flowers clustered in leaf-axils, small, regular, hermaphrodite, 5-merous; ovary superior, 3-celled, with 1 ovule in each cell. Fruit a 3-celled capsule dehiscing loculicidally; seeds arillate.

1. Maytenus Molina

Plant densely branched and leafy. Older twigs with short, woody spurs bearing leaves and flowers. Peduncles very short, nearly always 1-flowered. Calyx 5-lobed. Petals 5, alternate with calyx-lobes. Stamens 5, alternate with petals and opposite calyx-lobes. Ovary ± included in and coherent with a fleshy, 5-crenate hypogynous disc; stigma 3-fid.

1. M. umbellata (R. Br.) Mabb. in *Taxon* **30**: 486 (1981). [*Buxo da rocha*]
Catha dryandri Lowe
Plant to 2 m. Leaves 2-7 × 1-3.5 cm, elliptic to narrowly so, obtuse to rounded at apex, weakly crenate at margins, gradually narrowed at base into a short petiole, coriaceous, glossy. Peduncles and pedicels together 3-20 mm, wiry, articulate half-way or more from base, with 1-3 minute bracteoles below joint of peduncle and pedicel. Flowers 5-7 mm across; sepals ovate, obtuse; petals oblong-ovate to oblong-lanceolate, somewhat fleshy, pale greenish yellow, tending to become reddish. Fruit *c.* 10 mm across, globose, ± angular, usually containing only 2 seeds (1 cell being abortive), splitting open before fully ripe to reveal the separate cells and seeds within; cells becoming reflexed and pale yellowish as fruit ripens; seeds glossy light brown, each with a conspicuous white aril at base. **Plate 24**
Fl. XI - I. *An endemic scattered around the coast of Madeira, growing on maritime cliffs, very steep slopes and rocks in ravines near the sea, from sea-level to 400 m; also at two localities on Deserta Grande; very rare in Porto Santo, known only from Pico de Ana Ferreira.* ● **MDP**

BUXACEAE
J.R. Press

Buxus sempervirens L. (including cv. '**Arborescens**') [*Buxo*] is frequently planted in gardens and near houses, even in remote areas. It is a slender evergreen shrub or small tree to 5 m, with leaves ovate to lanceolate or elliptic, dark glossy green above, and inflorescence axillary, consisting of a single female flower surrounded by several males, all lacking petals. The fruit is a woody capsule with three persistent, spreading styles.

LXIII. RHAMNACEAE
M.J. Short

Evergreen or deciduous trees or shrubs. Leaves alternate or subopposite, simple, usually stipulate. Flowers small and inconspicuous, perigynous, regular, hermaphrodite or unisexual, usually in axillary cymes. Sepals 4-5, attached to the hypanthium. Petals (0)4-5, free, often incurved. Stamens 4-5, opposite the petals. Ovary superior to semi-inferior, 2-4-locular. Style simple or divided above. Fruit usually a fleshy drupe.

1. Leaves serrate, evergreen; flower clusters stalked; fruit glabrous . . . . . **1. Rhamnus**
- Leaves entire, deciduous; flower clusters sessile; fruit pubescent . . . . . **2. Frangula**

1. Rhamnus L.

Evergreen trees. Winter buds with scales. Leaves alternate or opposite, petiolate, serrate; stipules caducous. Flowers greenish or yellowish, 5-merous. Ovary superior; styles 2-3-fid. Fruit a globose drupe containing 2-4, 1-seeded pyrenes.

1. R. glandulosa Aiton, *Hort. kew.* **1**: 265 (1789). [*Sanguinho*]
Small tree up to 8(-10) m with a dense crown; trunk cylindrical, slender, bark greyish; young twigs minutely pubescent. Leaves 3-10 × 2-4.5 cm, broadly ovate to oblong-ovate, acute to obtuse, margins bluntly serrate, often revolute, dark green, coriaceous, the lower axils with small, gland-like, pubescent depressions, otherwise glabrous; petioles 10-23 mm. Stipules

subulate. Flowers small, yellow-green, pedicellate, in short, erect, stalked clusters in the axils of the leaves. Pedicels densely pubescent. Calyx glabrous; lobes 2.5-3.5 mm, triangular-ovate, erect or spreading. Fruit 5-9 mm across, glabrous, shiny purplish black when mature. **Plate 25**
Fl. III - IV (-VII). *A very rare tree of the laurisilva in high mountain valleys of Madeira; 800-1200 m.* ■ M [*Canaries*]

2. Frangula Mill.

Like *Rhamnus* but always deciduous. Winter buds without scales. Leaves alternate, entire; stipules caducous or persistent. Style simple. Fruit containing 2-3 pyrenes.

1. F. azorica Grubov in *Trudy bot. Inst. Akad. Nauk SSSR* I, **8**: 259 (1949).
[*Gingeira brava, Tintureira*]
Frangula azorica Tutin, nom. superfl.; *Rhamnus latifolia* L'Hér.
Large shrub or small tree up to 10 m with a wide-spreading crown; branches little-divided, leafy only towards the end; bark reddish brown; young twigs brown-puberulent. Leaves 9-14 × 5-7 cm, broadly elliptic, abruptly acuminate, bright shining green, with parallel lateral veins, brownish pubescent on veins below; petioles 10-20(-35) mm. Stipules lanceolate, brown-pubescent, persistent. Flowers small, pale yellow, pedicellate, 3-5 in small, axillary, sessile clusters. Pedicels pubescent. Calyx pubescent; lobes 3.5-4.5 mm, triangular-lanceolate. Petals ± equalling sepals. Fruit 0.8-1.2 cm across, sparsely pilose, bright red turning to purplish black. **Plate 25**
Fl. V - VII. *Based on fossil evidence, this species is believed to be native to Madeira, but is extinct in the wild. May still occasionally be found in gardens.* ■ M [*Azores*]

VITACEAE
N.J. Turland

Shrubs with slender stems, to 35 m if left unpruned, becoming woody, climbing by means of axillary tendrils; bark usually peeling off in long shreds. Leaves alternate, petiolate, stipulate, rounded in outline, usually palmately 3- to 5-lobed, cordate at base, ± glabrous above, tomentose or lanate beneath, at least on veins; lobes coarsely dentate or compoundly dentate. Inflorescence an axillary panicle. Flowers small, 5-merous, regular, hermaphrodite, greenish, sometimes fragrant; sepals united; petals united at apex and falling without separating; ovary superior, 2-celled; style and stigma 1. Fruit a juicy berry, ripening to various shades of green, yellowish, red or dark purple, often with a pruinose surface, edible, sweet- or sour-tasting, containing 2-4 seeds.

Many cultivars of species and hybrids of **Vitis** L. are grown for wine production and dessert grapes in both Madeira and Porto Santo, on sunny slopes from sea-level to *c.* 800 m. They may occasionally become established as escapes. **V. vinifera** L., *Sp. pl.* **1**: 202 (1753) [*Vidêira, Vinha*] was introduced from the eastern Mediterranean region in the early years of settlement, but was devastated by the root aphis *Phylloxera* in Madeira in the 1870s (about 20 years later in Porto Santo). These vines have since been largely replaced by resistant American cultivars, some of which are apparently of hybrid origin. The exact taxonomic status of modern vines in the archipelago is, therefore, very difficult to determine. The species introduced from N. America include the following: **V. aestivalis** Michx., *Fl. bor.-amer.* **2**: 230 (1803) (*V. monticola* Buckl.), **V. berlanderi** Planch., in *C. r. hebd. Séanc. Acad. Sci., Paris* **91**: 425 (1880), **V. labrusca** L., *Sp. pl.* **1**: 203 (1753) [*Isabella, Vinha americana*], and **V. riparia** Michx., *Fl. bor.-amer.* **2**: 231 (1803) [*Riparia, Vinha silvado*]. **MP**

216 FLORA

LXIV. MALVACEAE
J.R. Press

Herbaceous to shrubby perennials, usually stellate-hairy. Leaves alternate, stipulate, simple or palmately divided. Flowers regular, hermaphrodite. Epicalyx usually present. Calyx-segments 5, united towards the base. Petals 5. Stamens numerous, the filaments united to form a tube surrounding the styles and ovary. Fruit a schizocarp.

A number of species, especially members of the genus **Hibiscus** L., including **H. rosa-sinensis** L. [*Cardeal vermelho*] and **H. syriacus** L. [*Cardeal violeta*], are widely grown as ornamentals in gardens, parks and roadside plantings.

1. Epicalyx absent 2
- Epicalyx present 3

2. Mericarps dehiscent, each containing several seeds; leaves cordate at the base ......
  .... **1. Abutilon**
- Mericarps indehiscent, 1-seeded; leaves not cordate at the base .... **2. Sida**

3. Epicalyx-segments 6 .... **6. Alcea**
- Epicalyx-segments 3 4

4. Epicalyx-segments ovate to oblong-lanceolate, united at least at the base . **5. Lavatera**
- Epicalyx-segments linear to narrowly ovate, free 5

5. Leaves palmately divided to more than half-way, the lobes pinnately toothed; mericarps 2-seeded .... **7. Modiola**
- Leaves simple or shallowly lobed to less than half-way; mericarps 1-seeded 6

6. Mericarps with 1 apical awn and 2 short laterally spreading dorsal spines; flowers yellow
  .... **3. Malvastrum**
- Mericarps lacking awns or spines; flowers lilac to purplish pink .... **4. Malva**

1. Abutilon Mill.

Herbaceous or shrubby perennials. Flowers solitary or few in clusters in the leaf axils; peduncles articulated in the upper half. Epicalyx absent. Mericarps compressed, ± reniform to oblong, with a ventral tooth, dehiscent, each containing several seeds.

1. A. sonneratianum (Cav.) Sweet, *Hort. brit.* ed.1,: 54 (1826).
A. indicum sensu auct. mad., non (L.) Sweet (1826); *A. populifolium* sensu auct. mad., non (Lam.) Sweet (1826).
Slender, shrubby perennial, ± shortly and densely stellate-tomentose with a few scattered, long, simple, patent hairs. Stems and branches purple-tinged. Leaves 3.5-4.5 × 2-3 cm, cordate to narrowly so, sometimes very shallowly 3-lobed, obtuse, irregularly dentate, the teeth rounded, dull green above, grey-green below with a velvety texture. Flowers solitary. Calyx-lobes triangular, acuminate; petals 14-16 mm, yellow. Mericarps 10-12, dorsally stellate-pubescent, the outer dorsal angle with a point up to 1.5 mm long.
Fl. X - XI. *An ornamental from S. Africa, apparently occurring as a weed of cultivated or waste ground in and around Funchal but not recorded in recent years.* **M**

2. A. grandifolium (Willd.) Sweet, *Hort. brit.* ed.1,: 53 (1826).
A. permolle (Willd.) Lowe
Shrubby, tomentose perennial with short stellate hairs and long, simple, patent hairs, the latter

especially on stems and petioles. Leaves 6-16 × 4-12 cm, cordate, acuminate, crenate-serrate; petioles 2-9 cm. Flowers 1 or 2 on a common peduncle shorter than the subtending leaf. Calyx-segments broadly ovate to triangular-ovate. Petals *c.* 16 mm, yellow. Mericarps 8-10, papery when ripe, dorsally pubescent with stellate hairs and long, simple hairs, the outer dorsal angle with a point up to 1 mm long.
Fl. X - VIII. *Introduced. On waste ground and in gardens in and around Funchal. Sporadic and rather rare. Native to tropical America and Africa, cultivated elsewhere in the tropics for fibre and ornament.* **M**

A. megapotamicum (K. Spreng.) A. St-Hil. & Naudin in *Ann. Sci. nat.*, II, **28**: 49 (1842), native to Brazil, is a slender shrub with pendulous flowers, red calyx and yellow petals. It is a popular and commonly cultivated garden plant and may perhaps occur as an escape. **M**

A. striatum J. Dicks. ex Lindl. in *Bot. Reg.* **25**: 39 (1839), a widely cultivated species from Central America with palmately divided leaves and large reddish flowers has been recorded twice from Madeira, most likely as a garden escape. **M**

2. Sida L.

Low shrubs. Flowers 1-several in axillary clusters. Pedicels articulated. Epicalyx absent. Mericarps awned or muticous at the tip, indehiscent, 1-seeded.

1. S. rhombifolia L., *Sp. pl.* **2**: 684 (1753). [*Cha bravo*]
Plant with short, often dense stellate hairs, especially on the underside of the leaves which have a mealy appearance. Stems 30-60 cm, slender, straight and rather stiff. Leaves 3-6 × 1-2 cm, rhombic-ovate to lanceolate, crenate-serrate, dull green above, greyish green beneath; petioles short. Flowers solitary, pedicels equalling or exceeding the subtending leaves. Sepals broadly rhombic, apiculate; petals *c.* 6 mm, dull yellow. Mericarps 9-11, with 1-2 apical awns up to 2 mm long.
Fl. all year. *In dry, sunny positions along roadsides and in waste places, mainly along the south coast from Funchal west to Tábua.* **M**

Plants with 1-awned mericarps, the common form in Madeira, have been called var. **maderensis** Lowe, *Man. fl. Mad.*: 68 (1857) and those with 2-awned mericarps var. **canariensis** Lowe, *Man. fl. Mad.*: 68 (1857).

3. Malvastrum A. Gray

Annuals or perennials. Flowers axillary, solitary or in few-flowered clusters, sessile or shortly pedunculate; peduncles not articulated. Epicalyx-segments 3, free. Petals entire or emarginate. Mericarps laterally compressed, awned or not, the lateral faces radially veined, indehiscent, 1-seeded.

1. M. coromandelianum (L.) Garcke in *Bonplandia* **5**: 295 (1857).
Sida carpinifolia auct. mad., non L.f.
Pubescent annual or perennial to about 1 m or more, with 4-armed hairs, the arms in 2 opposing pairs or somewhat forked. Stems erect, little branched, tough and rather woody. Leaves 3-6 × 3(-4) cm, rather variable but usually ovate to rhombic-ovate, simple or slightly 3-lobed, crenate-serrate; petiole 1-3 cm. Flowers solitary, peduncles very short, later accrescent. Epicalyx-segments linear-lanceolate. Calyx-segments triangular-ovate. Petals *c.* 6 mm, rounded or emarginate, yellow. Mericarps 12, sharply angled, dorsally setose, with 2

short but prominently laterally spreading spines and an apical awn.
Fl. all year. *Introduced; a weed of waste ground, the fringes of cultivation and, especially, of roadsides. Common in the lowlands of the south-east, especially around Funchal. Native to tropical America but now pantropical.* **M**

4. **Malva** L. [*Malva*]

Annual, biennial or perennial herbs. Flowers usually several in axillary clusters. Epicalyx-segments 3, free. Petals emarginate or bilobed. Mericarps indehiscent, 1-seeded.

1. Calyx accrescent in fruit; mericarps narrowly winged on the angles . . . **1. parviflora**
- Calyx not accrescent; mericarps sharply angled but not winged 2

2. Petals more than 12 mm, bright purple or pink with darker veins, the claw bearded . . . . **2. sylvestris**
- Petals less than 12 mm, pale pink, lacking dark veins, glabrous . . . . . **3. nicaeensis**

1. M. parviflora L., *Sp. pl.* **1**: 18 (1753).
Prostrate to ascending annual, indumentum variable but with both stellate and simple hairs. Leaves long-petiolate, up to 7 cm, suborbicular-cordate, usually with 5-7 shallow, crenate lobes. Flowers sessile or shortly pedicellate, in clusters of 2-4. Epicalyx-segments linear to linear-lanceolate. Calyx ± glabrous to stellate-hairy beneath, accrescent. Petals 5-7 mm, slightly longer than the calyx, pale pink, the claw glabrous. Mericarps 10, glabrous or hairy, reticulate, the angles sharply toothed to narrowly winged.
Fl. III - VII. *A common and widespread ruderal weed occurring in lower regions throughout the archipelagos.* **MDPS**

2. M. sylvestris L., *Sp. pl.* **1**: 689 (1753).
M. mauritiana L.
Very variable, especially in habit, leaf-size and indumentum. Erect to decumbent biennial or perennial with stellate and simple hairs. Leaves long-petiolate, reniform to suborbicular cordate, palmately 5- to 7-lobed, the lobes shallow, crenate. Epicalyx-segments narrowly ovate to oblong-lanceolate. Calyx densely stellate-hairy beneath. Petals 15-22 mm, 3-4 × as long as the calyx, pink to purplish with darker veins, the claw bearded at the base. Mericarps 10, pubescent, rugose, with sharp angles.
Fl. III - VIII. *Sporadic on roadsides, waste ground and other ruderal sites, mainly in the Funchal region.* **M**

3. M. nicaeensis All., *Fl. pedem.* **2**: 40 (1785).
Like **2** but annual or biennial, setose, lacking stellate hairs; calyx with a few setae beneath; petals 6-11 mm, 1-2 × as long as the calyx, pale lilac, lacking darker veins, not bearded; mericarps 8, glabrous.
Fl. III. *Probably introduced. Rare weed of fields near the sea, only recorded near Funchal and Santa Cruz but perhaps overlooked elsewhere. Native to S. Europe, N. Africa, W. Asia.* **M**

5. **Lavatera** L. [*Malva*]

Annual or biennial herbs. Flowers axillary, solitary or in few-flowered clusters. Epicalyx-segments 3, united at least at the base. Petals emarginate or shortly bifid. Mericarps indehiscent, 1-seeded.

1. Epicalyx-segments shorter than the calyx; mericarps with rounded angles; annual or herbaceous perennial . . . . **1. cretica**
- Epicalyx-segments longer than the calyx, at least in fruit; mericarps sharply angled; biennial, woody towards the base . . . . **2. arborea**

1. L. cretica L., *Sp. pl.* **2**: 691 (1753).
Annual or biennial up to 1.5 m, erect to procumbent, stellate-pubescent. Leaves long-petiolate, up to 10 cm, reniform to suborbicular-cordate, usually with 5-7 shallow, crenate lobes. Flowers in clusters of 1-5, pedicellate. Epicalyx-segments ovate to ovate-oblong. Calyx densely pubescent beneath, the lobes triangular-ovate. Petals 12-20 mm, 3-4 × as long as calyx, emarginate, bearded at the base of the claw, purplish lilac with darker veins. Mericarps (8-)9(-11), smooth or slightly ridged, with rounded angles.
Fl. III - VII. *On roadsides, waste ground, field margins and other ruderal sites. Very common throughout the lower regions of Madeira, Porto Santo and Ilhéu de Fora, Ilhéu Chão and Selvagem Grande.* **MDPS**

2. L. arborea L., *Sp. pl.* **2**: 690 (1753).
Shrubby biennial to 1 m or more. Stems erect, branched, woody towards the base. Flowers and leaves similar to **1** but larger. Epicalyx-segments longer than the calyx, at least in fruit; petals deep pinkish purple with darker veins and base. Mericarps 6, ridged, with sharp angles.
Fl. III - VI. *Cultivated in gardens near the sea, especially on Porto Santo, and sometimes surviving in abandoned plots or escaping; doubtfully completely naturalized. Native to the Mediterranean region and W. Europe.* **MP**

6. Alcea L.

Tall, erect perennial herbs. Flowers ± sessile in long, spike-like racemes. Epicalyx of 6 segments, united at the base. Mericarps indehiscent, each divided into a sterile upper cell and a fertile lower cell with a single seed.

1. A. rosea L., *Sp. pl.* **2**: 687 (1753).
Althea rosea (L.) Cav.
Plant stellate-tomentose, especially on the younger parts. Leaves cordate, weakly 5-lobed, finely crenate. Epicalyx shorter than the calyx. Corolla large, variously coloured.
Fl. V - IX. *Widely cultivated in gardens, escaping and sometimes becoming naturalized. Native to the E. Mediterranean region.* **M**

Numerous forms of this species are cultivated for ornament in the islands.

7. Modiola Moench

Annual herbs. Flowers solitary, axillary, the peduncles not articulated. Epicalyx-segments 3, free. Petals entire. Mericarps numerous, dehiscent, divided by an internal septum into an upper and lower chamber, each containing one seed.

1. M. caroliniana (L.) G.Don, *Gen. hist.* **1**: 465 (1831).
Setose annual. Stems prostrate, rooting, ascending at the tips. Leaves 1-4 cm, palmately divided into 3-7 pinnately lobed or toothed segments; petioles up to 8 cm. Peduncles about as long as, or shorter than, the leaves. Epicalyx-segments lanceolate. Calyx-segments triangular-ovate. Petals *c.* 4 mm, dull scarlet with a dark patch at the base. Mericarps 20-22, the lower half of the lateral faces ribbed, the dorsal face setose and with 2 short, pronounced spines.
Fl. III - V. *Rather rare introduction, confined to waste places in and around Funchal. Native to tropical and N. America; cultivated and naturalized elsewhere.* **M**

LXV. THYMELAEACEAE
R. Khan

Shrubs and (outside Madeira) toxic trees, lianes or herbs. Leaves alternate, exstipulate, simple, entire. Flowers hermaphrodite, 4-merous, usually in small heads or clusters. Petals inserted at the mouth of the tubular to urceolate hypanthium (*calyx-tube*). Stamens in 2 whorls; filaments short. Ovary superior, at the base of the hypanthium but free from it; ovule solitary. Fruit a nut or drupe, indehiscent.

1. Gnidia L.

Heath-like shrubs. Leaves small and narrow. Flowers hermaphrodite, in terminal bracteate heads. Bracts ovate-elliptic with silky hairs at the base. Calyx-tube cylindrical, the 4 lobes imbricate, opening into a funnel-shape, cut across above the ovary. Petals 8, scale-like, smaller than the calyx-lobes. Stamens 8, dorsifixed. Ovary sessile, 1-celled; style filiform.

1. G. polystachya Bergius *Descr. pl. Cap.*: 123 (1767).
G. carinata Thunb.
Small shrub 1-2(-3.5) m with small but prominent leaf-scars. Leaves spirally arranged, sessile, linear-lanceolate, subacute, 5-10(-15) × 0.5-1 mm, 1-nerved, often slightly keeled, glandular-pitted, glabrescent or glabrous. Bracts 4-8 mm ovate, obtuse and glaucous. Flowers in small clusters often terminating short branches, pale yellow, fragrant and sessile. Calyx hirsute, the tube 5-9 mm, narrowly ovoid, the lobes 2(-3) mm. Petals half as long as the calyx-lobes. Ovary 5 mm, oblong, with silky hairs arising from base; stigma capitate. Fruit *c.* 2.5 mm.
Fl. III - VII. *Often cultivated for ornament and naturalized in rocky ravines, on banks and along levadas, mainly in the region of Funchal, also at São Roque near Faial. Native to S. Africa.* **M**

LXVI. THEACEAE
C. Whitefoord

Trees and shrubs. Leaves spirally arranged, usually 2-ranked, without stipules. Flowers usually large, bisexual, regular and solitary. Bracteoles present. Sepals and petals 5, imbricate. Stamens numerous. Ovary superior. Fruit a loculicidal capsule or indehiscent, sometimes fleshy.

1. Visnea L. f.

Evergreen trees with glossy, leathery leaves. Flowers solitary or in clusters. Stamens *c.* 20. Ovary densely hairy, usually 3-locular. Styles usually 3, free. Seeds 2 per loculus.

1. V. mocanera L. f., *Suppl. pl.* **1**: 251 (1782). [*Mocano*]
Slender tree 6-8 m, with short branches. Bark grey. Young twigs reddish, angular, warty and sparsely hairy. Leaves 4-6 × 1.5-2.5 cm. elliptic, lanceolate or oblanceolate, blunt or emarginate, entire to broadly and shallowly toothed, sparsely long-hairy beneath at first, midrib reddish beneath; petiole short. Flowers 6-10 mm in diameter, nodding, fragrant; peduncles 4-15 mm, slender, with 2 minute, widely-spaced bracteoles. Sepals ovate, fleshy, sparsely hairy on the outer surface, erect in fruit. Petals creamy white, broadly obovate, united at the base. Fruit red, ripening purplish black, the receptacle and sepals becoming greatly enlarged, subglobose and obscurely angular. **Plate 25**
Fl. XII - III. *Very rare and probably decreasing; confined to banks and steep rock faces in the deep ravines of the north coast of Madeira, from São Vicente westwards. Few, if any, large trees are now known in the wild, but it is occasionally planted in gardens.* ■ **M** [*Canaries*]

Camellia japonica L. [*Camélia, Japonesa*] is the most common species of **Camellia** L. grown for ornament in Madeira. Native to China and Japan, it is much cultivated in slightly shaded sites up to *c.* 1000 m. It is evergreen, frequently forming a tree, with ovate, serrate, coriaceous leaves and abundant flowers ranging in colour from white to pink or red; single forms are most common but double forms also occur.

LXVII. GUTTIFERAE
[HYPERICACEAE]
B.R. Tebbs

Trees, shrubs or herbs with translucent glands containing essential oil, and often with black or red glands containing hypericin. Leaves simple, opposite, alternate or rarely whorled. Flowers regular. Sepals often overlapping in bud. Petals free. Stamens many, often in distinct bundles. Styles up to 5. Ovary superior, 1- to many-chambered. Fruit usually a dry capsule, sometimes fleshy and berry-like.

1. Hypericum L.

Perennial herbs, shrubs, or rarely small trees. Leaves opposite. Flowers usually 5-merous, in terminal, cymose inflorescences. Petals yellow, often tinged red. Stamens in 3-5 bundles, alternating with sterile bundles. Ovary 1- to 5-locular, ovules many Fruit a dry capsule or fleshy and berry-like.

1. Black glands absent; shrubs or small trees 2
- Black glands present; herbaceous or rarely shrubby 4

2. Petals persistent; stamens in 3 bundles; leaves narrowly elliptic to oblong-elliptic . . . .
. . . . **8. canariense**
- Petals deciduous; stamens in 5 bundles; leaves oblong-ovate, lanceolate or broadly ovate
. . . . 3

3. Petals 1.5-2 × as long as sepals; outer sepals broadly ovate; fruit red when ripening . .
. . . . **2. × inodorum**
- Petals 3-4 × as long as sepals; sepals lanceolate to narrowly oblong; fruit not red when ripening . . . . **1. grandifolium**

4. Shrub to 1 m; leaves crowded at ends of branches . . . . **9. glandulosum**
- Herb; leaves not crowded at ends of branches 5

5. Plant procumbent, often mat-forming; sepals unequal . . . . **7. humifusum**
- Plant erect, never mat-forming; sepals ± equal 6

6. Sepals denticulate; leaves usually amplexicaul . . . . **5. perfoliatum**
- Sepals entire, leaves not amplexicaul 7

7. Leaf-margins undulate; petals tinged red . . . . **4. undulatum**
- Leaf-margins not undulate; petals not tinged red 8

8. Leaves narrowly oblong to linear; sepals with numerous black streaks **6. linarifolium**
- Leaves oval to linear; sepals not streaked . . . . **3. perforatum**

1. H. grandifolium Choisy, *Prodr. monogr. Hypéric.*: 38 (1821). [*Malfurada*]
Evergreen shrub to 1.8 m, branches erect or ascending. Leaves 4-9 × 2.5-4.5 cm, opposite, decussate, oblong-ovate, blunt at apex, sessile and clasping the stem at base. Sepals lanceolate to narrowly obovate, imbricate in bud. Petals 3-4 × as long as sepals, spreading. Capsule ellipsoid, dehiscent, not turning red on ripening. **Plate 26**

Fl. all year. *Common on cliffs and dry stony hillsides, and the edges of laurisilva, up to 800 m in Madeira.* ■ M [*Canaries*]

2. H. × inodorum Mill., *Gard. Dict.* ed. 8, no. 6 (1768). [*Malfurada*]
H. hircinum L. × *H. androsaemum* L.
Evergreen shrub to 2 m. Leaves 3.5-9 cm, ovate to oblong-lanceolate, sessile. Sepals unequal, ovate. Petals 1.5-2 × as long as sepals. Capsule ellipsoid to sub-cylindrical, red at first, ripening to brown. **Plate 26**
Fl. VI - VIII. *A much cultivated hybrid, often naturalized in Madeira.* M

H. × *inodorum* is frequently confused with the **1** and probably over-recorded as a consequence. It is not, as is often stated, a Macaronesian endemic.

3. H. perforatum L., *Sp. pl.* **2**: 785 (1753). [*Herva de São João, Malfurada*]
Perennial herb 10-100 cm, stems erect from a decumbent, rooting base. Leaves 8-30 mm, linear to ovate, ± sessile, with numerous large, translucent dots. Sepals ± equal, lanceolate or oblong-linear, acute to shortly aristate, not streaked. Petals with a few black dots on the margins.
Fl. V - X. *Common up to 1000 m throughout Madeira.* **M**

4. H. undulatum Schousboe ex Willd., *Enum. pl.*: 810 (1809).
H. acutum Moench
Perennial herb 15-100 cm, with erect stems from a decumbent, rooting base. Leaves 7-40 mm, narrowly ovate or elliptic to oblong, sessile, with small or medium-sized, translucent dots, the margins markedly undulate. Sepals ± equal, lanceolate, acute to acuminate. Petals 7.5-10 mm, red-tinged. Sepals and petals persistent.
Fl. VI - VIII. *Frequent in damp places, mainly in eastern Madeira.* **M**

5. H. perfoliatum L., *Syst. nat.* ed. 12, **2**: 510 (1767).
Perennial herb 15-75 cm, erect or with a decumbent base. Leaves 13-60 mm, ovate to triangular-lanceolate, usually amplexicaule. Sepals ± equal ovate-lanceolate, obtuse or subacute, dotted with sessile black glands, denticulate-fringed with cilia or stalked black glands. Petals glandular with black streaks.
Fl. V - VII. *A rare plant of meadows and rocky, shady areas in south-eastern Madeira and at São Vicente on the northern coast.* **M**

6. H. linarifolium Vahl, *Symb. bot.* **1**: 65 (1790).
Perennial herb 5-70 cm, stem erect or decumbent, occasionally branched, rooting at the base. Leaves 5-35 mm, narrowly oblong or linear, patent, often without translucent glands. Sepals ± equal, ovate to lanceolate, with numerous black streaks, dots and black glands. Petals at least 2 × as long as sepals. Sepals and petals persistent.
Fl. VI - VIII. *A rare plant of dry rocky areas in the higher parts of central Madeira, including the Paúl da Serra.* **M**

7. H. humifusum L., *Sp. pl.* **2**: 785 (1753). [*Pelicão*]
Perennial herb 5-40 cm, often procumbent and forming a dense mat, with rooting stems. Leaves 3-20 mm, oblong to lanceolate, obovate at the base of stems, usually with translucent glands. Sepals markedly unequl, lanceolate to ovate, with black glands. Petals 1-2 × as long as sepals. Sepals and petals persistent.
Fl. V - X. *Common in open habitats up to 2000 m throughout Madeira.* **M**

8. H. canariense L., *Sp. pl.* **2**: 784 (1753). [*Hipericão*]
H. floribundum Aiton
Deciduous shrub or small tree 1-4 m, with many erect branches. Leaves 26-70 × 5-15mm, sessile, narrowly elliptic to oblong-elliptic, acute or acuminate, cuneate at base, with translucent glands. Sepals ± equal, to 4.5 mm, ovate to lanceolate, acute to obtuse. Petals usually 3 × as long as sepals, persistent. Stamens in 3 distinct bundles. **Plate 26**
Fl. V - IX. *Cliffs, ravines, open rocky slopes and in laurisilva, from 900-1200 m.* ■ M [*Canaries*]

H. canariense is a variable species which has been separated into a number of varieties, none of which is recognized here.

9. H. glandulosum Aiton, *Hort. kew.* **3**: 107 (1789). [*Malfurada*]
Small evergreen shrub to 1 m. Leaves 3-5 cm, opposite, decussate, crowded towards ends of branches, elliptic-lanceolate, sessile, with translucent dots and black glands around the margins. Sepals unequal, lanceolate, acute, with black glands. Petals *c.* 1 cm, pale yellow, persistent in fruit. **Plate 26**
Fl. IV - VII. *Rocky areas, especially in ravines in eastern Madeira; on Pico de Ana Ferreira in Porto Santo.* ■ MP [*Canaries*]

LXVIII. VIOLACEAE
M.J. Short

Herbs (in ours). Leaves alternate or sometimes opposite, simple, stipulate. Flowers regular or irregular, hermaphrodite, sometimes cleistogamous. Sepals 5, free. Petals 5, the lowermost spurred when corolla irregular. Stamens 5; anthers erect, connivent in a ring around the ovary. Ovary superior, 1-locular; style simple, often curved or thickened above. Fruit a capsule, dehiscing loculicidally, or rarely a berry.

1. Viola L.

Annual or perennial herbs. Leaves alternate, sometimes in basal tufts, petiolate; stipules persistent, often prominent. Flowers irregular, solitary, axillary or arising from the basal tuft. Sepals prolonged into short appendages below their insertion point. Petals unequal, the lower with a basal spur. Lower 2 stamens with basal appendages extending back into the petal spur. Style thickened above, sometimes hooked. Fruit a 3-valved capsule, dehiscing explosively. Seeds numerous, smooth, with a sometimes conspicuous elaiosome.

Species **1** and **2** may produce cleistogamous flowers. Descriptions refer only to open flowers and are either measured from tip of spur to apex of lower petal (length) or vertically between apices of upper and lower petals. Sepal measurements include appendages.

1. Leaves cuneate, truncate or tapering at the base; corolla flat-faced, the lateral petals directed upwards; style globose-capitate 3
- Leaves deeply cordate at the base; corolla not as above, the lateral petals directed downwards; style hooked 2

2. Sepals acuminate, glabrous; stolons absent; capsules glabrous, on erect peduncles; flowers scentless . . . . **2. riviniana**
- Sepals obtuse to acute, ciliate towards the base; stolons present; capsules pubescent, on drooping peduncles; flowers sweetly scented . . . . **1. odorata**

3. Stipules entire, linear to spathulate; sepals not exceeding the petals; corolla bright yellow .... **3. paradoxa**
- Stipules pinnatifid, the mid-lobe ovate-lanceolate, leaf-like; sepals equalling or exceeding the petals; corolla whitish yellow .... **4. arvensis**

1. V. odorata L., *Sp. pl.* **2**: 934 (1753).   [*Violeta*]
V. odorata subsp. *maderensis* (Lowe) G. Kunkel
Appressed-pubescent perennial up to 18 cm, often becoming shortly stemmed and slightly woody at the base; long trailing and rooting stolons present. Leaves up to 5(-8) × 5(-7.5) cm, ovate-orbicular, mostly obtuse, deeply cordate, crenate; petioles (3.5-)5-15(-25) cm. Stipules lanceolate, acuminate, sparsely and shortly glandular-fimbriate, pale green or straw-coloured when dry. Sepals 8-9 mm, oblong-ovate, obtuse to acute, shortly ciliate towards the base. Corolla 16-18 mm long, pale violet, very sweetly scented; lateral petals spreading ± horizontally; spur 5-6 mm, exceeding calycine appendages. Style hooked apically. Capsules 6-8 mm, globose, pubescent, on drooping peduncles.
Fl. X - VII. *Very common in Madeira on damp, grassy banks and in woodland, 300-850 m.*
M

2. V. riviniana Rchb., *Iconogr. bot. pl. crit.* **1**: 81 (1823).   [*Violeta*]
V. sylvatica Fr. ex Hartm.; *V. sylvestris* var. *riviniana* (Rchb.) W.D.J. Koch
Glabrous, shortly caulescent perennial up to 20 cm with central rosette and ascending or procumbent axillary flowering branches. Leaves 1-5(-6.5) × 1-5(-6) cm, ovate-orbicular, mostly subacute, deeply cordate, crenate; petioles 0.5-5(-8) cm. Stipules lanceolate, acuminate, long glandular-fimbriate, reddish brown when dry. Sepals 5-12 mm, lanceolate, acuminate, glabrous; appendages conspicuous. Corolla 17-23 mm long, blue to purple with paler spur, scentless; lateral petals spreading ± horizontally; spur 3-5 mm, exceeding calycine appendages, often upturned. Style hooked apically. Capsules 9-13 mm, ellipsoid, trigonous, glabrous, on erect peduncles.
Fl. I - XII. *Common along paths, on banks and rock walls in mountainous areas of Madeira, 600-1300 m.* **M**

3. V. paradoxa Lowe in *Trans. Camb. phil. Soc.* **6**: 550 (1838).
                              [*Violeta amarela da Madeira, Violeta da Madeira*]
Perennial, glabrous below, puberulent above; stems up to 30 cm, prostrate to suberect, branched, becoming woody at the base. Leaves ternate, up to 1.5 × 1.5 cm; the lower ovate to ovate-orbicular, acute to obtuse, base cuneate to truncate, margin crenate; the upper linear-elliptic to linear-spathulate. Stipules entire, narrowly linear to linear-lanceolate. Sepals 6-10 mm, linear-oblanceolate, shortly acuminate, thickly ciliate, much shorter than corolla. Corolla 2-2.5 cm vertically, bright yellow, scentless; lateral petals directed upwards; spur 2.5-3.5 mm, exceeding calycine appendages. Style capitate. Capsules 5-7 mm, ellipsoid, shallowly 6-angled, glabrous. 2n=34. **Plate 27**
Fl. V - VII. *Very rare Madeiran endemic, growing among rocks and on ledges of high peaks, 1600-1800 m.* ● **M**

4. V. arvensis Murray, *Prodr. stirp. gott.*: 73 (1770).   [*Amor perfeito*]
V. tricolor sensu auct. mad., non L. (1753).
Shortly pubescent annual; stems up to 40 cm, ± erect, branched. Leaves up to 5 × 1.5 cm, ovate-oblong to elliptic, becoming narrower above, obtuse to acute, base shallowly cordate, truncate or attenuate, margin crenate. Stipules pinnatifid, the terminal segment large, ovate-lanceolate, leaf-like. Sepals 8-10.5 mm, linear-lanceolate, acute, equalling or exceeding corolla, glabrous or very sparsely ciliate. Corolla 8-15(-20) mm vertically, whitish yellow, the lateral and upper petals occasionally tinged with blue, scentless; lateral petals directed upwards;

spur 2-3 mm, about equalling calycine appendages. Style capitate. Capsules 6-8 mm, ellipsoid, glabrous.
Fl. III - IX. *Occasional on roadsides, cultivated and waste ground in the lower regions of Madeira.* M

LXIX. PASSIFLORACEAE
N.J. Turland

Herbaceous or woody perennials climbing by means of axillary tendrils. Leaves alternate, petiolate, entire or digitately lobed, usually stipulate. Flowers axillary, solitary, pedunculate, regular, hermaphrodite.

1. Passiflora L.

Flowers subtended by 3 bracts at apex of peduncle, often concealing at least the base of cup-shaped to elongate-tubular hypanthial tube. Sepals usually 5, free or joined at base, often scarcely distinguishable from petals and together with them inserted at rim of tube. Petals 5, usually free, rarely absent. *Corona* often present on rim of tube, consisting of 1-5 series of filaments, though sometimes rudimentary. *Androgynophore* inserted at base of tube and passing through its length, consisting of filaments fused to each other and to a *gynophore* (stalk of ovary). Stamens free above apex of androgynophore, usually 5. Styles and stigmas 3. Fruit a fleshy, often large berry containing numerous seeds.

Several species are cultivated in Madeira for ornament or for their edible fruits, including **P. antioquiensis** H. Karst. (*P. vanvolxemii* (Hook.) Triana & Planch.; *Tacsonia vanvolxemii* Hook.), **P. edulis** Sims [*Maracujá roxo*], **P. ligularis** Juss. (*P. lowei* Heer) [*Maracujá amarelo, Maracujá inglez*], **P. manicata** (Juss.) Pers. (*Tacsonia manicata* Juss.), and **P. quadrangularis** L. [*Maracujá*].

1. Plant glabrous; stipules conspicuous, ovate-lanceolate to lanceolate; hypanthial tube short, cup-shaped; sepals and petals ± white, at least above; corona-filaments well-developed 2
- Plant shortly pubescent; stipules inconspicuous, linear; hypanthial tube long and slender; sepals and petals pink; corona-filaments reduced to a rim of tubercles 3

2. Leaves 3- to 7(-9)-lobed, cut almost to base, the lobes oblanceolate to linear; flowers 5-8 cm across; corona-filaments blue, white in middle part .... **1. caerulea**
- Leaves 3-lobed, cut ½-³/₅ of way to base, the lobes oblong-lanceolate; flowers smaller, 4-5 cm across; corona-filaments white .... **2. subpeltata**

3. Leaves cut ⅔-⅞ of way to base, the lobes lanceolate, shortly acuminate; petiole-glands usually several .... **3. mollissima**
- Leaves very deeply cut, ± to base, the lobes narrowly lanceolate, acuminate; petiole-glands usually 2 .... **4. × exoniensis**

1. P. caerulea L., *Sp. pl.* **2**: 959 (1753). [*Flôr da paixão, Martyrio*]
Plant glabrous. Stems to 10 m. Leaves to 6 cm, deeply 3- to 7(-9)-lobed, cut almost to base, ± glaucous; lobes oblanceolate to linear, entire at margins; petiole with 2-4(-6) glands at apex; stipules conspicuous, ovate-lanceolate. Bracts free, ovate, entire. Flowers 5-8 cm across; hypanthial tube very short, cup-shaped; sepals and petals 5 each, ± alike, *c.* 3 cm, oblong, ± white; corona-filaments in 4 series, well-developed, blue towards base and apex, white in middle part. Fruit pendent, *c.* 5 cm, ovoid, ripening bright orange, edible.
Fl. almost all year. *Introduced; cultivated for ornament in Madeira and naturalized in vineyards,*

waste areas, on roadsides and among buildings in the Funchal region. Native to Brazil and Argentina. **M**

2. P. subpeltata Ortega, *Nov. pl. descr. dec.* **6**: 78 (1798).
P. alba Link & Otto
Plant glabrous. Leaves 5-7.5 cm, 3-lobed, cut ½-⅗ of way to base; lobes oblong-lanceolate, obtuse to rounded at apex, mucronate, entire at margins, but often with a few gland-tipped teeth in the 2 sinuses; petiole with 2-4 glands in middle part; stipules conspicuous, 2.5-3 cm, lanceolate. Bracts 1-1.5 cm, free, ovate, entire. Flowers small, 4-5 cm across; hypanthial tube short and cup-shaped; petals and sepals 5 each, ± alike, *c.* 2 × 0.5 cm, oblong, white, the sepals greenish on reverse; corona-filaments in 5 series, well-developed, white. Fruit ovoid or subglobose, ripening yellowish with a powerful, unpleasant odour.
Fl. IX. *Introduced; naturalized along the Levada do Bom Successo near Funchal. Native to C. and S. America, from C. Mexico to Colombia and Venezuela.* **M**

3. P. mollissima (Kunth) L.H. Bailey in *Rhodora* **18**: 156 (1916).
Tacsonia mollissima Kunth
Plant shortly pubescent. Stems to 4 m or more. Leaves 5.5-11 cm, 3-lobed, cut ⅔-⅞ of way to base; lobes lanceolate, shortly acuminate at apex, shallowly dentate at margins, glabrescent above, pubescent beneath; petiole usually with several glands; stipules inconspicuous, linear. Flowers pendent. Bracts 3-4 cm, fused for ⅓-½ of way from base, the lobes lanceolate to triangular-ovate. Hypanthial tube 6.5-7 × *c.* 0.7 cm, bulbous at base, parallel-sided above, green, sometimes tinged pink. Sepals and petals 5 each, ± alike, patent, 4.5-5.5 × 1.5-2 cm, narrowly oblong-lanceolate, mucronate at apex, pink. Corona-filaments reduced to a single series of pinkish tubercles on a purple rim. Androgynophore exserted 2-2.5 cm from tube; free section of filaments *c.* 15 mm; anthers 9-10 mm, medifixed, versatile, narrowly oblong, bright yellow or orange; ovary 8-13 mm, narrowly ellipsoid, densely pubescent; styles 7-9 mm; stigmas almost half as long as anthers, capitate, green. Fruit pendent, *c.* 10 cm, elongate, ripening yellow, pubescent, resembling a straight banana.
Fl. all year. *Introduced; cultivated for ornament and its edible fruits in Madeira; naturalized in the valleys of the Ribeira da Janela valley and Fajã da Nogueira, also in the Funchal region and probably elsewhere, climbing through shrubs and trees. Native to Venezuela, Colombia, Peru and Bolivia.* **M**

4. P. × exoniensis hort. ex L.H. Bailey, *Stand. cycl. hort.* **5**: 2485 (1916).
P. antioquiensis × *P. mollissima*; *Tacsonia* × *exoniensis* hort.   [*Maracujá banana*]
Like **3** but with leaves very deeply divided, cut ± to base; lobes 5-10 × 0.8-2.5 cm, narrowly lanceolate, acuminate; petiole-glands usually 2.
Fl. ? *Recorded as naturalized in Ribeira da Janela.* **M**

There has been some confusion between this hybrid and *P. mollissima*, with several records of the former referable to the latter and, in one instance, vice versa. Therefore, only records which could be confirmed have been used for this account.

LXX. CISTACEAE
M.J. Short

Shrubs (in Madeira), often stellate-pubescent. Leaves usually opposite, simple, stipulate or not. Flowers hermaphrodite, regular, solitary or in cymes. Sepals 5 or 3, free or fused at the base, often unequal, persistent. Petals 5, free, often caducous. Stamens numerous. Ovary superior, 1-locular or partially septate at base. Style simple; stigma capitate. Fruit a capsule, dehiscing loculicidally. Seeds numerous.

1. Cistus L.

Leaves opposite, entire, exstipulate. Flowers large, showy, pedicellate, in cymes or solitary. Sepals 5 or 3. Petals 5, caducous, white. Ovary partially 5(10)-locular. Capsules woody, dehiscing to the middle into 5(10) valves. Seeds small, angular.

1. Leaves linear to linear-lanceolate, 3-8 mm wide, margins revolute; outer sepals 5-7 mm cuneate at the base .... **1. monspeliensis**
- Leaves oblong to oblong-ovate, 5-23 mm wide, margins flat; outer sepals 9-18 mm, cordate at the base .... **2. psilosepalus**

1. C. monspeliensis L., *Sp. pl.* **1**: 524 (1753). [*Alecrim de fora*]
Erect, much-branched dwarf shrubs up to 1 m, viscid, aromatic, the young shoots pubescent. Leaves sessile, 1.5-5.5 × 0.3-0.8 cm, linear to linear-lanceolate, margins strongly revolute, upper surface rugose, dark green and shining, subglabrous, lower surface with prominent reticulate venation, densely stellate-pubescent and with long simple hairs on the midrib. Flowers 2-3 cm across, white, 5-10 in a densely villous, 1-sided inflorescence. Sepals 5, the outer 5-7 mm long, ovate, shortly acuminate, cuneate at base. Capsules 4-5 mm across, subglobose, 5-valved, stellate-pubescent towards the apex.
Fl. III - V. *Recorded from waste areas near Santo Antonio in the last century but probably no longer present in this locality. Its present status in Madeira is unknown, but presumed to be very rare; it has also been recorded from Porto Santo.* **MP**

2. C. psilosepalus Sweet, *Cistineae*: t. 33 (1826).
Straggling dwarf shrubs up to 1 m, the stems pubescent. Leaves sessile, 2-6 × 0.5-2.3 cm, oblong to oblong-ovate, 3-veined at least towards the base, bright green, sparsely pubescent above and below with stellate and long simple hairs, margins flat. Flowers up to 4 cm across, white, 1-5 in somewhat villous, terminal, flat-topped clusters. Sepals 5, the outer 9-18 mm long, ovate, acuminate, cordate at base, much larger than the inner. Capsules 6-7 mm long, ovoid, 5-valved, stellate-pubescent. $2n=18$.
Fl. III - IX. *Track and roadsides. First collected in Madeira at Terreiro da Luta in 1953 and presumed to be an introduction. It has since been recorded from several scattered localities on the island and also north of Pico do Castelo in Porto Santo. Native to the Iberian Peninsula.* **MP**

C. ladanifer L., *Sp. pl.* **1**: 523 (1753) is a distinctive, very viscid species from SW Europe, with glabrous branches and large, usually solitary flowers, 6-10 cm across, with 3 sepals and white petals, often purple-blotched at the base. It was collected near Monte in the last century but has not been recorded since. **M**

C. salvifolius L., *Sp. pl.* **1**: 524 (1753) is a Mediterranean species recorded twice from Madeira at Fanal and Ribeira da Janela, but probably introduced and apparently not persisting. Also white-flowered, it may be distinguished from species **1** and **2** by its distinctly petiolate leaves. **M**

LXXI. TAMARICACEAE
M.J. Short

Shrubs or small trees. Leaves alternate, simple, often scale-like, exstipulate. Flowers hermaphrodite, regular, in spike-like racemes or rarely solitary. Sepals 4-5, free or slightly fused at the base. Petals 4-5, free. Stamens 4 to numerous. Disc present. Ovary superior, usually of 3-4 carpels, unilocular. Fruit a capsule. Seeds numerous, usually with long hairs.

1. Tamarix L.

Deciduous, the smaller twigs falling with the leaves. Leaves small, scale-like, sessile, with deep salt-secreting glands. Flowers small, white or pink, in bracteate, spike-like racemes. Sepals and petals 5, imbricate. Stamens 5, attached to a fleshy, lobed, nectariferous disc. Styles 3. Capsules pyramidal. Seeds with an apical tuft of long, unicellular hairs.

1. T. gallica L., *Sp. pl.* **1**: 270 (1753).   [*Cedro* (P), *Tamargueira*]
Glabrous, much-branched, feathery-looking shrub up to 4 m; bark dark brown or purplish; branches slender. Leaves densely imbricate on young branches, up to 2 mm long, triangular-lanceolate, acute, adpressed to and clasping the stem at the base, margins scarious. Racemes 10-35 × 3-6 mm in flower, catkin-like, forming large, terminal panicles. Bracts triangular-lanceolate, acuminate. Flowers subsessile. Sepals 0.7-1.3 mm, ovate, margins scarious. Petals 1.5-2(-2.5) mm, elliptic to elliptic-obovate, pink or less commonly white, caducous. Stamens 5. Styles 3. Capsules 3-4 mm, trigonous, acuminate.
Fl. III - VIII. *A species from SW Europe planted around the coast of Madeira, especially in the north, for example near Porto Moniz and Porto da Cruz, where it may occasionally become naturalized. Introduced to Porto Santo in 1834 and used for fencing and firewood, now extremely widespread.* **MP**

LXXII. FRANKENIACEAE
M.J. Short

Herbs or dwarf shrubs, often encrusted or dotted wtih salt on the stem and leaves. Leaves opposite, simple, entire, often heath-like, exstipulate. Flowers regular, usually hermaphrodite, solitary or in cymes. Calyx tubular, 4-5(-7)-toothed, persistent. Petals 4-5(-7), usually free, clawed, with a scale-like appendage on the claw. Stamens 4-6, in two whorls. Ovary superior, 1-locular. Style 1; stigmas 3-4. Fruit a many-seeded capsule dehiscing by 2-4 valves.

1. Frankenia L.

Annual or perennial herbs. Leaves small, often clustered on short, lateral branches. Flowers sessile, subtended by a pair of bracts and bracteoles. Calyx-tube prominently ribbed, (4-)5-toothed. Petals (4-)5, pink. Stamens usually 6, unequal. Capsules 3-angled, dehiscing by 3 valves, enclosed by the persistent calyx. Seeds numerous, minute.

1. Leaves obovate to oblong-spathulate, flat or margins very weakly revolute . . . .
   . . . . **1. pulverulenta**
- Leaves appearing narrowly linear, margins strongly revolute . . . . **2. laevis**

1. F. pulverulenta L., *Sp. pl.* **1**: 332 (1753).
Mat-forming annual; stems 3.5-30 cm, usually prostrate, numerous, much-branched, leafy, puberulent. Leaves 2-7 × 1-3 mm, obovate to oblong-spathulate, obtuse or retuse, tapering at the base into a short petiole with membranous, ciliate margin, flat or very weakly revolute, ± glabrous above, pubescent below, often becoming reddish. Flowers axillary, solitary or in short clusters. Bracts and bracteoles leaf-like. Calyx 3-4 mm, subglabrous. Petals 3.5-5 mm, obovate, minutely denticulate at the apex, very pale pink.
Fl. IV - VII. *In Madeira recorded only from Porta da Cruz and S. Jorge on the north-east coast and presumed to be extremely rare, although its current status unknown. Reported to be fairly common in Porto Santo.* **MP**

2. F. laevis L., *Sp. pl.* **1**: 331 (1753).   [*Rasteira, Rasteyro, Resteira*]
F. hirsuta var. *intermedia* auct. mad., non (DC.) Boiss. (1867); *F. laevis* [var.] α *hebecaulon* Lowe

Low-domed perennial, usually white-encrusted; stems procumbent, much-branched, rather tough and wiry, densely puberulent. Leaves mostly densely crowded on short, lateral shoots, 3-8(-12) mm, appearing linear, heather-like, margins strongly revolute, ciliate at the base, otherwise glabrous. Flowers solitary or in small clusters. Calyx 2.8-4.5 mm, sparsely puberulent, somewhat fleshy, twisted in fruit. Petals 4.5-6 mm, obovate, minutely denticulate at the apex, pale or deep pink.

Fl. IV - VII. *In dry pasture and grassland, slopes and rocks, and roadside banks. Very rare in Madeira where it is only recorded from Ponta de São Lourenço. Common in Porto Santo, mainly in lower areas near the sea, 0-35(-125) m; also found on Ilhéu de Baixo. Recorded from all of the Savage Islands.* **MPS**

Plants from the Salvage Islands and one of two collections seen from Ilhéu de Baixo, differ from other material in having densely puberulent leaves and calyces and need critical assessment.

CUCURBITACEAE
J.R. Press

Two members of this widely cultivated family have been recorded from the Madeiran archipelago, although neither species appears to be truly naturalized.

Citrullus lanatus (Thunb.) Matsum. & Nakai, *Cat. Sem. Spor. hort. Univ. Tokyo*: 30 (1916), from tropical S. Africa, is a monoecious, creeping and hairy vine with branched tendrils, leaves pinnately lobed, flowers yellow, fruits large, globose to cylindrical, dark or pale green, spotted or striped. It has been recorded from Porto Santo, probabbly as an escape. **P**

Sechium edule (Jacq.) Sw., *Fl. Ind. occid.* 2(2): 1150 (1800) [*Caiota, Pepinella*], from the W. Indies, is widely cultivated as a vegetable in Madeira and is occasionally found as a crop remnant growing on the fringes of cultivated terraces and along levadas. It is a tall annual vine, sprawling or climbing by means of branched tendrils. Leaves cordate, weakly 3- to 5-lobed, bristly. Flowers greenish yellow. Fruits pear-shaped, pale green, ribbed and prickly. **M**

LXXIII. CACTACEAE
N.J. Turland

Perennials, almost always with succulent stems. Nodes regularly scattered over surface of stems, appearing as small cushion-like structures (*areoles*), often bearing spines. Leaves absent, or very small and present only on young growth, subulate, soon falling, very rarely larger and persistent. Flowers hermaphrodite, borne singly at areoles, sessile; sepals and petals numerous, not differentiated from each other, overlapping in several rows; stamens numerous; style 1; stigmas 2-many; ovary inferior, 1-celled. Fruit fleshy, with seeds scattered through pulp.

Almost all members of the family are native to the New World. Numerous exotic species and hybrids are grown as ornamentals in parks and gardens in Madeira, including **Hylocereus triangularis** (L.) Britton & Rose (*Cereus triangularis* (L.) Mill.), native to the West Indies, with stems creeping or climbing by means of aerial roots, and flowers large, white, nocturnal, and **Pereskia aculeata** Mill., native to tropical America, with stems scrambling, non-succulent, leaves persistent, ovate to lanceolate or elliptic and flowers greenish white.

1. Opuntia Mill.

Shrubs with strongly succulent jointed stems. Leaves tiny, subulate, succulent, soon falling. Flowers borne towards apical margins of uppermost mature joints, cup-shaped to saucer-shaped; ovary obovoid-cylindric, succulent, scattered with areoles, with a depression (*umbilicus*) at apex.

Opuntia species are characterized by the presence of fine barbed bristles (*glochids*) at the areoles. These readily detach and penetrate the skin when touched. Larger spines may also be present.

1. O. tuna (L.) Mill., *Gard. Dict.* ed. 8, no. 3 (1768). [*Tabaibeira*]
Plant to 2 m. Joints to *c*. 45 × 15 × 2.5 cm, obovate to oblong or oblanceolate to narrowly elliptic, strongly compressed, dull green, readily detachable; oldest joints becoming greyish brown and woody. Spines 0-5(-6) at each areole, 3-20(-25) mm, ± spreading, slender, initially pale yellowish, soon becoming white; glochids numerous, crowded, yellowish. Flowers *c*. 5 cm across, cup-shaped, dull orange to dull red. Fruit 5-7.5 × 3.5-5 cm, obovoid-cylindric, broadly umbilicate at apex, ripening pale green or yellowish.
Fl. V - IX. *Introduced; commonly naturalized and widespread along the southern coast of Madeira but rare on the northern coast, growing on maritime cliffs and in rocky and waste areas from sea-level to 500 m, often dominating other vegetation; also on the islet of Doca, off the western coast of Deserta Grande, and scattered in Porto Santo. Native to Jamaica.* **MDP**

O. ficus-barbarica A. Berger in *Monatsschr. Kakteenk.* **22**: 181 (1912) (*O. ficus-indica* auct. eur.) differs from **1** in being taller, to 5 m, eventually tree-like, with joints not readily detachable and spines absent (rarely 1-2 at each areole and then less than 10 mm); flowers are bright yellow and fruit ripens yellow, red or parti-coloured. Native to tropical America, it has been recorded from Madeira, the Desertas and Porto Santo but without further data. **MDP**

LXXIV. LYTHRACEAE
N.J. Turland
Annual or perennial herbs. Leaves simple, entire; stipules minute or absent. Flowers axillary, regular, hermaphrodite, (4-)6-merous; petals free; ovary superior; style 1; stigma capitate. Fruit a dehiscent capsule containing numerous seeds.

1. Lythrum L.
Stems slender, 4-angled. Leaves alternate or mostly so, sessile. Flowers axillary, solitary, shortly pedicellate, with 2 bracteoles; hypanthial tube and epicalyx present as well as sepals and petals; petals pink to purple; stamens 4-6 in 1 series, or 12 in 2 series of different length.

1. Petals 4.5-7 mm; stamens 12, in 2 series of different length . . . . **1. junceum**
- Petals 2-3.5 mm; stamens usually 4-6, in 1 series . . . . **2. hyssopifolia**

1. L. junceum Banks & Sol. in Russell, *Nat. hist. Aleppo* ed. 2, **2**: 253 (1794).
L. flexuosum sensu auct. mad., non Lag. (1816); *L. graefferi* Ten.
Plant usually perennial, glabrous. Stems to 120 cm, usually creeping to decumbent, branched and rooting near base, sometimes reddish-tinged. Leaves 5-21 × 1.5-8 mm, oblong-elliptic to linear, rounded to acute at apex. Flowers borne usually on upper half of stem; hypanthial tube obconic in flower, cylindric in fruit; epicalyx-segments *c*. 0.75 mm, triangular-subulate, equalling or slightly longer than sepals; sepals broadly deltate, membranous; petals 4.5-7 mm, bright pinkish purple; stamens 12, in 2 series of different length; style trimorphic, with stigma held below, between or above the 2 series of anthers.
Fl. IV - IX. *Streams, springs, wet flushes and hollows, bases of cliffs and other wet places, from near sea-level to 900 m; widespread and common in Madeira; rare in Porto Santo, known from the Fonte das Pombas and beyond Camacha.* **MP**

2. L. hyssopifolia L., *Sp. pl.* **1**: 447 (1753). [*Peceguaia*]
L. hyssopifolia var. *acutifolium* DC.; *L. hyssopifolia* forma *typicum* Cout.
Like **1** but annual, ± glabrous; stems to 80 cm, sprawling or ascending to erect, branched

towards base; leaves 10-25(-35) × 1.5-5(-11) mm, lanceolate to linear-oblong, obtuse to acute at apex; flowers borne ± from base of stem upwards; epicalyx-segments c. 1 mm, subulate, c. twice as long as sepals; sepals deltate; petals 2-3.5 mm, pink to violet; stamens usually 4-6, in 1 series; style monomorphic.

Fl. V - VIII. *Streams, wet flushes, damp patches in cornfields and other wet places, up to c. 600 m; somewhat rare in Madeira, mainly in the north from Ponta do Pargo to Santo da Serra, also in the Funchal region; abundant on the margins of a pool in the middle of Ilhéu Chão in the Desertas; also in Porto Santo, near Serra de Dentro.* **MDP**

LXXV. MYRTACEAE

N.J. Turland

Evergreen trees or shrubs. Leaves alternate or opposite, simple, entire, with aromatic oil-glands; stipules absent. Flowers regular, hermaphrodite; calyx and corolla 4- or 5-merous; stamens numerous; ovary inferior. Fruit a fleshy berry or woody capsule.

1. Leaves alternate, at least those of adult growth; fruit a rigid woody capsule 2
- All leaves opposite; fruit a fleshy, large or small berry 3

2. Trees; leaves of juvenile growth opposite; leaves of adult growth alternate, large, strap-like; flowers large, seemingly apetalous .... **4. Eucalyptus**
- Much-branched shrub; all leaves alternate, small, ± needle-like; flowers small, with 5 obvious petals .... **5. Leptospermum**

3. Densely branched shrub, rarely a tree; leaves to 5 cm, glabrous; fruit c. 6 mm excluding persistent sepals .... **2. Myrtus**
- Trees or shrubs; leaves 8-20 cm, glabrous or pubescent beneath; fruit large, 3-6 cm . 4

4. Leaves with a prominent submarginal vein running parallel to and c. 2 mm from margin, glabrous; inflorescence terminal; calyx 4-lobulate or with 4 segments; petals 4 .... **3. Syzygium**
- Leaves without a submarginal vein, usually pubescent beneath; inflorescence axillary; calyx entire before flowering, later splitting irregularly; petals 5 .... **1. Psidium**

1. Psidium L.

Trees or shrubs. Leaves opposite, usually large, glabrous or pubescent beneath; lateral veins reaching margin, with no submarginal vein present. Inflorescence axillary, 1- to few-flowered. Calyx entire before flowering, later splitting irregularly. Petals 5, white. Fruit berry-like, usually large, containing numerous seeds.

1. P. guajava L., *Sp. pl.* **1**: 470 (1753). [*Goiaba, Goiabeira*]
P. pomiferum L.; *P. pyriferum* L.
Shrub or small tree to 10 m. Twigs 4-angled, oblong in cross-section, shortly pubescent. Leaves 8-12.5 × 3.5-7 cm, oblong-elliptic, rounded to obtuse at apex, sometimes mucronate, ± coriaceous, ± glabrous above with parallel prominently impressed lateral veins, pubescent beneath; petiole short, c. 5 mm, decurrent at base into 2 narrow ridges tapering down internode. Flowers usually large, shortly pedicellate; pedicels c. 10 mm; calyx thick-textured, pubescent, especially on outer surface; petals slightly longer than calyx, to c. 13 mm, ovate, sparsely pubescent, densely glandular-punctulate. Fruit variable in size, shape, colour and flavour, 3.5-6 cm, ovoid to pear-shaped or ± globose, ripening yellow or reddish, with a strong musky odour.

Fl. VI - VII. *Introduced; cultivated for its edible fruits in the lowlands of Madeira, though much less common nowadays than formerly; sometimes naturalized as an escape on roadsides, in gardens and rocky and waste areas, especially in the Funchal region. Native to tropical America.* **M**

The similar **P. littorale** Raddi (*P. cattleianum* Sabine; *P. littorale* var. *globosum* Heer.) [*Araça amarelo, Araça roxo*], native to Brazil, differs in having glabrous leaves, while **P. guineense** Sw., native to tropical America, has twigs rounded in cross-section. Both are cultivated in parks and gardens in Madeira.

2. Myrtus L.

Much-branched, densely leafy shrub, rarely a small tree. Leaves opposite, small, glabrous, with a sweet resinous fragrance when rubbed. Flowers axillary, borne singly on slender pedicels; calyx with 5 segments, prominently persistent in fruit; petals 5, white, tinged reddish on reverse. Fruit a small fleshy berry.

1. M. communis L., *Sp. pl.* **1**: 471 (1753). [*Murta*]
M. communis var. *latifolia* sensu Lowe, var. *lusitanica* L.
Plant usually to *c.* 2 m, sometimes taller, to 9 m. Twigs brown. Leaves 20-50 × 5-25 mm, lanceolate to narrowly so, acute at apex, coriaceous, glossy, punctulate. Pedicels 7-30 mm. Flowers 15-20 mm across; sepals ovate-deltate; petals *c.* 10 mm, much longer than sepals, suborbicular. Fruit *c.* 6 mm (excluding sepals), ± globose, usually ripening bluish black.
Fl. III - X. *Dry sunny cliffs, rocks and slopes, sometimes forming thickets with* Erica *and Lauraceae; mainly in eastern Madeira, west to Funchal and São Jorge, from 100-1000 m; also at Serra de Agua and Paúl do Mar. Much cultivated in parks and gardens in the Funchal region.* **M**

The above description refers to subsp. **communis**, which is widespread in the Mediterranean region and to which wild Madeiran plants belong. Among the cultivated plants in Madeira occurs subsp. **tarentina** (L.) Nyman (*M. communis* var. *parvifolia* sensu Lowe) [*Murta da India*], which differs in its smaller, much more crowded leaves, not more than 12 mm long. It is native mainly in the W. half of the Mediterranean region.

3. Syzygium Gaertn.

Trees or shrubs. Leaves opposite, usually large, glabrous; lateral veins ending at a prominent submarginal vein running parallel to and *c.* 2 mm from margin. Inflorescence terminal, few-flowered. Calyx 4-lobulate or with 4 segments. Petals 4, white. Fruit berry-like, usually large, with 1-4 segments.

1. S. jambos (L.) Alston in Trimen, *Handb. fl. Ceylon* **6**: 115 (1931).
Eugenia jambos L.; *Jambosa vulgaris* DC. [*Jamboeiro*]
Shrub or tree to 12(-15) m. Twigs angled or compressed. Leaves 10-20 cm, narrowly elliptic, acuminate at apex, with a short petiole *c.* 5 mm. Flowers shortly pedicellate, fragrant; pedicels 3-8 mm; petals reflexed at flowering, orbicular, concave; stamens creamy white, the filaments long. Fruit *c.* 3 cm, depressed-globose, green, yellowish or orange-yellow, tinged with red or purple, fragrant.
Fl. III - V. *Introduced; cultivated in parks and gardens in the lowlands of Madeira, mainly in the Funchal region; occasionally ± naturalized on roadsides, in hedges and waste areas. Native from S. China and SE Asia to Australia.* **M**

Other, somewhat similar species are cultivated in gardens in Funchal, including **S. malaccense** (L.) Merr. & Perry (*Eugenia malaccensis* L.; *Jambosa malaccensis* (L.) DC.), native to Australia and Peninsular Malaysia, and **Eugenia uniflora** L. (*E. brasiliana* (L.) Aubl.) [*Pitangueira*], native to tropical S. America. The latter has globose, cherry-like, 8-ribbed fruits, which ripen bright red with a spicy, acidic flavour.

4. Eucalyptus L'Hér.

Trees. Leaves glabrous, dimorphic; those on juvenile growth opposite, sessile or shortly petiolate; those on adult growth alternate, petiolate, pendent, large and strap-like, coriaceous. Inflorescence axillary, 1- to several-flowered. Flowers large, seemingly apetalous, closed in bud by a cone or dome of fused perianth-segments (*operculum*) which falls off as the flower opens. Fruit a rigid woody capsule, dehiscing by apical pores.

1. E. globulus Labill., *Voy. rech. Pérouse* 1: 153 (1800).   [*Eucalypto*]
Large tree to 45 m or more. Bark usually smooth, pale, peeling off in strips. Juvenile leaves 8-11 × 3.5-6 cm, ± oblong or ovate to broadly lanceolate, cuspidate at apex, perfoliate or clasping at base; mature leaves 10-20 × 1.5-3 cm (excluding recurved petiole 2-4 cm), lanceolate to very narrowly so, often curved (sickle-shaped), acuminate at apex, greyish green. Flowers solitary, ± sessile to shortly pedicellate, warty and silvery green in bud, with hemispherical operculum; pedicel to *c*. 7 mm, stout; stamens cream. Fruit *c*. 23 mm across, obconic-hemispherical to subglobose, with 4 prominent ribs in lower part below a prominent circular rim, usually glaucous; pores 4-5, set in a circular depression.
Fl. X - III. *Introduced; commonly planted for its timber in south-eastern Madeira, up to 1250 m, often together with* Acacia *species and* Pinus pinaster, *and apparently naturalized. Native to Tasmania.* **M**

Further species are cultivated for ornament in parks and gardens in Madeira, including **E. citriodora** Hook., with lemon-scented foliage, **E. ficifolia** F. Muell., with bright red stamens, and **E. robusta** Sm.

5. Leptospermum J.R. & G. Forst.

Much-branched heath-like shrub. All leaves alternate, small, crowded, ± needle-like. Flowers small, borne at the tips of short leafy lateral spurs; petals obvious, 5, white or pale pink, sometimes dark red, surrounding a red centre. Fruit a rigid woody capsule, dehiscing by valves in upper half.

1. L. scoparium J.R. & G. Forst., *Char. gen. pl.* ed. 2: 72 ["48"] (1776).
Plant to *c*. 2 m, occasionally to 5 m or more. Leaves 5-12 × 1.5-2.5 mm, narrowly elliptic-oblong to linear, cuspidate at apex. Flowers 10-13 mm across; sepals ovate, brownish, membranous, deciduous; petals spreading. Fruit 4-5 mm, ± globose, with a rim around equator; valves forming a 5-pointed star when viewed from above.
Fl. XI - VI. *Introduced; often planted for wind-breaks, hedges and ornament in Madeira, particularly in cooler wetter areas; also naturalized in the west, between Prazeres and Porto do Moniz, and in the east between Camacha and Santo da Serra, growing in* Erica *scrub and along roadsides and levadas from 350-900 m. Native to Australia (New South Wales, Victoria and Tasmania) and New Zealand.* **M**

LXXVI. ONAGRACEAE
N.J. Turland
Annual, biennial or perennial herbs or shrubs. Leaves alternate, opposite or whorled. Inflorescence a spike, raceme or panicle, or flowers axillary. Flowers hermaphrodite, regular or nearly so, with a hypanthial tube below the 4 sepals and 4 petals; stamens 8 in 2 whorls; ovary inferior, 4-locular; style 1; stigma entire or lobed. Fruit an elongate capsule or a fleshy berry, containing numerous small seeds.

1. Shrubs; fruit a fleshy berry . . . . **1. Fuchsia**
- Herbs; fruit an elongate capsule 2

2. Leaves alternate; sepals strongly reflexed in flower; seeds without a plume of hairs . . . . **2. Oenothera**
- Leaves opposite or whorled, at least below, sepals ± erect in flower; seeds with a plume of hairs . . . . **3. Epilobium**

1. Fuchsia L. [*Mimos*]

Shrubs. Leaves usually opposite or in whorls of 3, sometimes alternate near stem-tips. Flowers regular, axillary or borne in racemes or panicles; pedicels slender, wiry; ovary narrowly obovoid to fusiform; hypanthial tube often long and brightly coloured together with sepals; stamens and style exserted; stigma entire or 4-lobed. Fruit a fleshy berry; seeds without a plume of hairs.

1. Stems slender, often scrambling or trailing; flowers axillary, 1-6 per node, pendent; sepals 15-24 mm, 2-3 times as long as hypanthial tube, erecto-patent . . . . . **1. magellanica**
- Stems stout, erect, straight; flowers borne in a terminal panicle, erect; sepals 5-6 mm, only slightly longer than hypanthial tube, spreading or recurved . . . . **2. arborescens**

1. F. magellanica Lam., *Encycl.* **2**: 564 (1788).
F. coccinea sensu auct. mad., non Aiton (1789)
Plant 0.6-10 m, sparsely hairy. Stems slender, often scrambling or trailing, often tinged reddish. Leaves 2-9 × 1-3.5 cm (including petiole to 1 cm), lanceolate, remotely serrate, shortly acuminate, often with veins and petiole tinged reddish. Flowers axillary, 1-6 per node, pendent; pedicels 2-6 cm; ovary 3.5-9 mm; hypanthial tube 6-12 × 2-5 mm, widest below middle, red; sepals 15-24 mm, 2-3 × as long as hypanthial tube, erecto-patent, narrowly lanceolate, acuminate, red; petals 7-15 mm, shorter than sepals, erect, obovate, violet or purple, overlapping to form a campanulate tube; filaments red, glabrous; anthers 2-2.5 mm, oblong, glabrous; style red, sparsely patent-hairy except towards apex; stigma 1-2 mm, capitate to cylindric or clavate, glabrous. Fruit shortly cylindric, ripening black, juicy, edible.
Fl. all year. *Introduced; naturalized in woodland, scrub, on slopes, steep banks, walls, roadsides, waste ground and along levadas, often forming hedges or scrambling over other shrubs and trees. Mainly in southern Madeira, but recorded elsewhere and probably widespread in the lowlands below 600 m. Native to Chile and Argentina.* **M**

2. F. arborescens Sims in *Curtis's bot. Mag.* **53**: t. 2620 (1826).
Schufia arborescens (Sims) Spach
Plant to 3 m, glabrous. Stems stout, erect, straight. Leaves 6-23 × 1.5-6 cm (including petiole 1-4 cm), narrowly elliptic, ± entire, acuminate. Inflorescence a many-flowered terminal panicle. Flowers erect; pedicels 1-1.5 cm; ovary *c.* 3.5 mm; hypanthial tube 4-5 mm, obconic, bright pink or reddish purple; sepals 5-6 mm, spreading or recurved, oblong-lanceolate, acute, concolorous with hypanthial tube; petals *c.* 4 mm, erect to spreading, oblanceolate, concolorous with sepals, not overlapping to form a tube; filaments glabrous; anthers 1.5-2 mm, oblong, glabrous; style glabrous, at least above; stigma *c.* 1.5 mm, 4-lobed. Fruit *c.* 10 mm, shortly cylindric to subglobose.
Fl. V - X. *Introduced; escaping from cultivation and locally naturalized along levadas, walls and rivers in south-eastern Madeira, from Jardim da Serra to Santo da Serra. Native to Mexico.* **M**

Other species are cultivated for ornament in Madeira. One of these is perhaps locally established away from cultivation:

F. boliviana Carr. in *Revue hort.*: **1876**: 150 (1876) is a shortly tomentose shrub, with leaves elliptic, glabrescent above, densely pale-hairy beneath; flowers in terminal racemes, shortly pedicellate, pale pink to red, with hypanthial tube very long, gradually and uniformly tapered to base, sepals spreading, linear-lanceolate and petals erect, oblanceolate, not overlapping to form a tube. Native to S. America, from Peru to northern Argentina, it has been recorded from Madeira but its present status is unknown. **M**

2. Oenothera L.

M.J. Short

Annual to perennial herbs. Leaves alternate. Flowers regular, usually opening in the evening, borne in the axils of the upper leaves, forming elongated spike-like clusters; hypanthial tube often conspicuous. Sepals strongly reflexed in flower. Petals broad, yellow or white, sometimes aging reddish or pink. Stigma deeply 4-lobed. Fruit a capsule splitting into 4 valves; seeds smooth or sharply angled, not plumed.

Literature: K. Rostanski, The representatives of the genus *Oenothera* L. in Portugal, in *Bolm Soc. broteriana* II, **64**: 5-33 (1919).

1. Petals white, becoming pink; capsules obovoid, winged, narrowed into a ribbed pedicel .... **1. tetraptera**
- Petals yellow, sometimes becoming reddish; capsules ± cylindrical, unwinged, sessile 2

2. Hypanthial tube 6-8(-10) cm; leaves densely pubescent .... **2. longiflora**
- Hypanthial tube *c.* 1 cm; leaves with ciliate margins, otherwise subglabrous .... **3. stricta**

1. O. tetraptera Cav., *Icon.* **3**: 40 (1796).
Perennial herbs covered with short, ± appressed hairs and longer, spreading hairs; stem 15-50 cm, decumbent to ascending, simple or branched. Basal leaves petiolate, 3-10 × 1-3 cm, oblanceolate, irregularly sinuate-dentate to sinuate-pinnatifid, or occasionally subentire; cauline leaves smaller, lanceolate-ovate. Bracts lanceolate. Hypanthial tube *c.* 1 cm. Petals 2-3.5 cm, broadly obovate, white, becoming pink with age. Filaments 1-1.5 cm; anthers 5-6 mm. Fruiting pedicels 0.5-2.5 cm, ribbed, hollow. Capsules 1-1.5 cm, obovoid, densely pubescent, prominently 4-winged, the wings 2-3 mm wide. Seeds 1.1-1.3 mm, obovoid, smooth.
Fl. V - VII. *Occasionally naturalized in vineyards and on waste ground in the Funchal and Monte areas but rare, with no recent records known. Native from Mexico to Columbia and Venezuela.* **M**

2. O. longiflora L., *Mant. pl. alt.*: 227 (1771).
Annual or biennial herbs, densely hispid and also with sparse, shorter, glandular hairs; stem 40-100 cm, erect, simple or sparingly branched, stout. Leaves irregularly and shallowly toothed, the basal forming a rosette, 6-15 × 1.5-2.5 cm, oblanceolate; cauline leaves smaller, oblong to oblanceolate, denticulate. Bracts ovate, usually red-margined. Inflorescence branched. Hypanthial tube 6-8(-10) cm long, often red-streaked. Petals 2-3.5 cm, broadly obovate, golden yellow, withering to orange-brown. Filaments 15-20 mm; anthers 8-10 mm. Capsules 3-4.5 cm, ± cylindrical but slightly enlarged above, densely hispid. Seeds 1.8-2 mm, ellipsoid, smooth.
Fl. V - VI. *Extremely rare. Naturalized in Madeira on cultivated ground in the Monte and São Roque areas, but no recent records known. Native to Brazil, Uruguay, and Argentina.* **M**

Represented in Madeira by subsp. **longiflora**.

3. O. stricta Ledeb. ex Link, *Enum. hort. berol. alt.* **1**: 377 (1821).
O. odorata sensu auct. mad., non Jacq. (1789)
Annual or biennial herbs; stem up to 60(-150) cm, erect, simple or sparingly branched, slender,

subglabrous or with sparse, short, appressed hairs below, glandular-pubescent and also with long, spreading haris above. Leaves with a pale midrib, glabrescent, margins ciliate, remotely and shallowly toothed, sometimes undulate; lower leaves forming a rosette, 5-10 × 0.8-1.3 cm, linear to narrowly oblanceolate; cauline leaves smaller, lanceolate. Bracts lanceolate, usually red-margined. Inflorescence simple or branched. Hypanthial tube 1.5-4 cm. Petals 2-3.5 cm, broadly obovate, yellow, later becoming reddish. Filaments 10-20 mm; anthers 7-8 mm. Capsules 2.5-4 cm, ± cylindrical but enlarged in upper half, puberulent and with very sparse, spreading hairs. Seeds 1.5-1.8 mm, ellipsoid, smooth.

Fl. V - VI. *The most frequently encountered species of* Oenothera *in Madeira, which is locally naturalized in fields, on cultivated ground, and along roadsides at Monte, Terreiro da Luta, and São Roque. Native to Chile and Argentina.* **M**

A plant with subglabrous sepals and basally pubescent petals collected at São Vicente under the name of *O. suaveolens* Pers. (= *O. biennis* L.) has been tentatively identified by Rostanski as a variety of *O. grandiflora* L'Hér. ex Aiton.

O. biennis L. has also been reported from Madeira but records of this species lack confirmation.

3. Epilobium L.

Perennial herbs overwintering by epigaeal leafy stolons or rosettes. Leaves opposite, at least below. Flowers regular or nearly so, borne in a leafy raceme; ovary long and slender, similar to and sometimes exceeding pedicel; hypanthial tube very short; sepals erecto-patent to erect; petals purplish pink to white, never yellow; stamens and style shorter than petals; stigma entire or deeply 4-lobed. Fruit a long slender capsule, splitting into 4 backward-curving segments when ripe; seeds tiny, with a plume of hairs.

The very short hypanthial tube together with the sepals give the impression of calyx-lobes united at the base. For convenience, they are here referred to as the 'calyx', with the sepals 'calyx-lobes'.

1. Stems rounded in cross-section, sparsely white-villous; stigma distinctly 4-lobed . . . . **1. parviflorum**
- Stems appearing angular in cross-section because of raised lines, minutely white-hairy, at least above; stigma clavate, entire 2

2. Plant with leafy epigaeal stolons; calyx-tube sometimes with a few patent glandular hairs; capsules regularly dispersed along stems . . . . **3. obscurum**
- Plant without stolons, overwintering by rosettes; calyx-tube eglandular; capsules clustered towards stem-tips . . . . **2. tetragonum**

1. E. parviflorum Schreb., *Spic. fl. lips.*: 146 (1771).
E. parviflorum var. *menezesi* Leveil.; *E. parviflorum* var. *subglabrum* Koch
Plant sparsely white-villous, overwintering by leaf-rosettes or short leafy stolons. Stems to 70 cm, stout, erect, sometimes procumbent and rooting at base, rounded in cross-section, without ridges. Leaves 3-14 × 0.7-3 cm, lanceolate to linear-lanceolate, subsessile, serrulate, obtuse. Calyx 4.5-5.5 mm; lobes oblong-lanceolate. Petals 8-10 mm, obovate, deeply notched at apex. Stigma distinctly 4-lobed. Capsules 3-7.5 cm.
Fl. III - IX. *Damp places on and beneath maritime cliffs, damp banks and rocks, ravines and along levadas; widespread in Madeira, from near sea-level to 900 m.* **M**

2. E. tetragonum L., *Sp. pl.* **1**: 348 (1753).
Plant glabrous below, minutely white-hairy above, overwintering by subsessile lax leaf-rosettes.

Stems to 45 cm, slender, erect, appearing quadrangular in cross-section because of ridges. Leaves 2-10 × 0.5-1.4 cm, lanceolate to linear-lanceolate, shortly petiolate or with margins slightly extended down stem-ridges, serrulate, obtuse. Calyx 3-5.5 mm; tube eglandular; lobes lanceolate. Petals 4-6.5 mm. Stigma clavate, entire. Capsules 5-9 cm, crowded towards stem-tips. M

The following two subspecies occur in Madeira; they are very similar and sometimes grow together:

a) subsp. **tetragonum**.
Leaves linear-lanceolate; margins slightly extended down stem-ridges. 2n = 36.
Fl. IV - VII. *Cliffs, especially by the sea, wet rocks and river-beds; scattered along both the northern and southern coasts of Madeira, occasionally inland.*

b) subsp. **lamyi** (F.W. Schultz) Nyman, *Consp. fl. Eur.*: 247 (1879).
Leaves lanceolate, shortly petiolate; margins not extended down stem-ridges.
Fl. VI - IX. *Wet rocks and river-beds; less frequent than subsp.* tetragonum *and mainly found at inland localities; also on the southern coast from Ponta do Sol to Ribeira Brava.*

3. E. obscurum Schreb., *Spic. fl. lips.*: 147 (1771).
E. lanceolatum sensu auct. mad., non Sebast. & Mauri (1818); *E. lanceolatum* var. *maderense* (Hausskn.) Leveil.; *E. maderense* Hausskn.
Plant sparsely and minutely white-hairy below, more densely so above, overwintering by ± elongate leafy epigaeal stolons. Stems to 50 cm, moderately stout, erect, decumbent below, appearing angular in cross-section because of raised lines. Leaves 2-5 × 0.5-1 cm, lanceolate to linear-lanceolate, very shortly petiolate or with margins slightly extended down stem-ridges, serrulate, obtuse to acute. Calyx 3.5-4.5 mm; tube sometimes with a few patent glandular hairs; lobes lanceolate. Petals 4-5.5 mm. Stigma clavate, entire. Capsules 3-5 cm, not clustered towards stem-tips but regularly dispersed along them.
Fl. XII - IX. *Moist banks and rocks in ravines, up to 1300 m; widely scattered in eastern Madeira.* M

Species **2** and **3** are very similar and can be confused if their method of overwintering is not evident. All three species are known to form hybrids with each other, although no such plants have yet been recorded from Madeira, even though mixed populations occur.

LXXVII. ARALIACEAE
N.J. Turland

Evergreen trailing or climbing perennial. Old stems becoming woody. Leaves alternate, petiolate, palmately lobed or entire, without stipules. Inflorescence terminal, umbellate. Flowers small, 5-merous, regular, hermaphrodite; ovary inferior, 5-locular. Fruit a fleshy berry.

1. Hedera L.

Juvenile growth climbing or trailing, clinging to support by means of aerial roots; adult growth shrubby, not clinging, present only on climbing plants. Young shoots and inflorescence covered with minute stellate-peltate hairs (*trichomes*). Umbels solitary or several in a panicle, simple, globose. Flowers yellowish green; ovary tipped with a conspicuous domed disc terminating in a short style. Fruit globose, containing 2-3 seeds.

1. H. helix L., *Sp. pl.* **1**: 202 (1753). [*Hera*]
Trichomes 12- to 20-rayed, reddish; rays all in one horizontal plane, united at base. Leaves ± coriaceous, glossy dark green, of 2 sorts; those of juvenile growth 1.5-8.5 × 1.5-9 cm

(excluding petiole 1-14 cm), somewhat reniform, truncate or cordate at base, with 3-5 short, broadly triangular lobes; those of adult growth 2.5-9 × 1.5-8 cm, broadly ovate to lanceolate, distinctly rhombic, obtuse to rounded at base, entire. Umbels 1.5-4.5 cm across, many-flowered. Sepals 2.5-3.5 × 1.5-2.5 mm, deltate, free, greenish. Fruits 5-8 mm, ripening black.

Fl. IX - X. *Trailing over ground or climbing cliffs, walls, banks and trees; common and widespread in Madeira, up to 1100 m.* **M**

The above description refers to subsp. **canariensis** (Willd.) Cout., *Fl. Portugal*: 428 (1913) (*H. canariensis* Willd.; *H. maderensis* K. Koch ex A. Rutherf.), which otherwise occurs in the Azores, Canaries, Portugal and NW Africa and to which Madeiran plants belong. Plants from Madeira have been treated as *H. maderensis* K. Koch ex A. Rutherf. subsp. *maderensis* in *Plantsman* **15**: 120 (1993), while in the same journal plants from elsewhere in Macaronesia and from the Mediterranean region have been referred to several other taxa. However, this treatment is not accepted here, since it is incomplete and appears to be based on very poor and scanty evidence. A satisfactory taxonomic treatment of the genus would require much detailed study in all areas of its distribution.

LXXVIII. UMBELLIFERAE
[APIACEAE]
M.J. Cannon

Herbs or small shrubs. Leaves alternate, rarely opposite, blade usually divided, rarely entire; petiole usually sheathing. Flowers hermaphrodite or male, in simple or compound umbels; umbels sometimes subtended by bracts to form an involucel. Sepals 5 or absent; petals 5; stamens 5; ovary inferior, 2-locular, styles 2, usually swollen at the base (*stylopodium*). Fruit of 2 mericarps united by their faces (*commissure*), each mericarp with 5 primary and sometimes also secondary ribs, usually suspended from a persistent axis (*carpophore*); oil-tubes (*vittae*) usually present in the pericarp.

Ripe fruit characters are important for identification, especially the ornamentation provided by ribs, ridges and spines, and the position of sub-surface vittae. In dorsally compressed fruits the commissure is as wide as the fruit, whereas in laterally compressed fruits it is much narrower than the fruit.

1.	All leaves entire, or ternately lobed to *c.* half-way from base	2
-	At least the lower leaves divided into separate leaflets	3
2.	Glandular-hairy perennial herb; leaves opposite, ternately lobed	**1. Drusa**
-	Annuals or shrubby perennials, glabrous; leaves alternate, entire	**7. Bupleurum**
3.	Fruit with prickles	4
-	Fruit smooth, rugose or winged, but not prickly	6
4.	Bracts several, pinnatifid, equalling the rays	**20. Daucus**
-	Bracts 0-1, entire and shorter than the rays	5
5.	Sepals absent, petals minute; pedicels thicker than the rays in fruit	**2. Anthriscus**
-	Sepals present, petals conspicuous; pedicels not thicker than the rays in fruit	**4. Torilis**
6.	Petioles of basal leaves and lower parts of stem subterranean	**11. Bunium**
-	Petioles and flowering stem entirely aerial	7

7. Beak of fruit at least 4 × as long as the seed-bearing part . . . . **3. Scandix**
- Beak of fruit much shorter than the seed-bearing part 8

8. Fruit winged on the outer edge 9
- Fruit not winged on the outer edge 11

9. Wings of fruit serrated, blackish; bracts 10-12, irregularly cut . . . **12. Melanoselinum**
- Wings of fruit entire; bracts 0-1, entire 10

10. Rays of umbels 2-5; ribs of fruit rugose . . . . **16. Capnophyllum**
- Rays of umbels 30-60; ribs of fruit smooth . . . . **18. Peucedanum**

11. Lateral ribs of fruit swollen and corky; bracts and bracteoles fringed at the margin . . .
. . . . **19. Monizia**
- Lateral ribs of fruit not swollen and corky; bracts and bracteoles entire or absent . . 12

12. Flowers yellow or yellowish 13
- Flowers white, pinkish or greenish 15

13. Plants succulent; bracts half as long as peduncles . . . . **17. Astydamia**
- Plants not succulent; bracts much shorter than the peduncles, or absent 14

14. Bracts and bracteoles absent; fruit 4-10 mm . . . . **15. Foeniculum**
- Bracts 1-3; bracteoles 5-10; fruit 2.5-3 mm . . . . **9. Petroselinum**

15. Fruit globose, mericarps not separating at maturity . . . . **5. Coriandrum**
- Fruit ovoid, mericarps readily separating at maturity 16

16. Fruit not compressed 17
- Fruit laterally compressed 18

17. Styles much elongated in fruit; sepals obvious, erect in fruit (freshwater) . . . .
. . . . **14. Oenanthe**
- Styles not elongated in fruit; sepals minute (maritime rocks) . . . . **13. Crithmum**

18. Ribs of fruit undulate; bracts deflexed . . . . **6. Conium**
- Ribs of fruit smooth; bracts erect or patent, or absent 19

19. Bracts 1- to 2-pinnate, or 3-fid, rarely entire; umbels terminal . . . . **10. Ammi**
- Bracts entire or absent; umbels mostly lateral and leaf-opposed . . . . **8. Apium**

1. Drusa DC.

Glandular-hairy perennial herb. Leaves opposite, ternately lobed. Umbels axillary, short-pedunculate, simple, few-flowered, compact, ebracteate. Petals white. Fruit flattened, with the secondary ridges extended into glochidiate wings.

1. D. glandulosa (Poir.) H. Wolff ex Engl. in Engl., *Pflanzenw. Afr.* **3**(11): 795, *in obs* (1921).
D. oppositifolia DC.
Stems slender, ± prostrate. Hairs simple or forked. Leaves cut to *c.* half-way from base, with lobes dentate, the central lobe 2.5-6 × 4-7.5 cm; petiole and underside of blade sometimes

glochidiate. Umbels 2- to 5-flowered. Fruits 4-8 mm, elliptic.
Fl. ? *Rare plant of the laurisilva and occasionally among shady rocks at low altitudes in Madeira.* ■ M [*Canaries*]

2. Anthriscus Pers.

Annuals or perennials. Leaves 2- to 3-pinnate. Sepals minute or absent; petals white, emarginate. Fruit narrowly oblong with a well-developed beak, commissure constricted, ribs confined to the beak, vitta 1.

1. A. caucalis M. Bieb., *Fl. taur.-caucas.* **1**: 230 (1808).
Slender, wiry, sparsely hairy annual up to 80 cm, often purplish at the base. Leaves 2- to 3-pinnate, lobes 1-10 mm, dentate or pinnatisect, sheaths of upper leaves hairy on the margin. Umbels mostly lateral, pedunculate. Rays 2-6; bracts absent, rarely 1; bracteoles several, linear-lanceolate to ovate, aristate; pedicels elongating and becoming longer than the rays, thickening in fruit. Fruit *c.* 3 mm, ovoid, usually with stiff, spine-like bristles, beak up to 2 mm, glabrous.
Fl. IV - VII. *Recent introduction, first recorded in 1969. Rare, only in the high peaks of Madeira from Pico de Arieiro and Pico Ruivo to Encumeada. Native to Europe and N. Africa.* M

The only specimen seen is probably var. **caucalis** but is too young to determine with any certainty; the bracteoles are much longer than in the typical European form.

A. sylvestris (L.) Hoffm., *Gen. pl. umbell.*: 40 (1814), a robust perennial with shortly-beaked fruits, has been recorded once from São Roque in Madeira. It may have occurred as a casual but no other data are available. M

3. Scandix L.

Annuals. Leaves 3- to 4-pinnate. Sepals absent; petals white, often unequal in the outer flowers, the apex incurved or inflexed. Fruit ± cylindrical, beak much longer than the seed-bearing part, ribs slender, vitta 1, very slender.

1. S. pecten-veneris L., *Sp. pl.* **1**: 256 (1753).
Nearly glabrous annual up to 50 cm, stems hollow with age. Lobes of leaves linear. Umbels pedunculate, with (1-)2(-3) rays; bracts 0-few; bracteoles several, entire, bifid or pinnatifid, margin often membranous and spinose-ciliate. Outer petals often radiate. Fruit 15-80 mm, ± flattened dorsally, beak up to 6 × as long as the seed-bearing part, scabrous on the edge.
Fl. XII - IV. *Infrequent weed of cornfields and waste ground in south-eastern Madeira and in Porto Santo.* MP

All Madeiran plants are subsp. **pecten-veneris**.

4. Torilis Adans.

Annuals or rarely biennials. Leaves 1- to 3-pinnate, lobes toothed. Sepals small and inconspicuous; petals white or pink, the apex inflexed. Fruit linear to ovoid, ribs slender, ciliate, the grooves with spines or tubercles.

1. Plant procumbent; umbels sessile or subsessile, rays short, concealed by flowers or fruits . . . . **1. nodosa**
- Plant erect; umbels pedunculate, rays not concealed by flowers or fruits . **2. arvensis**

1. T. nodosa (L.) Gaertn., *Fruct. sem. pl.* **1**: 82 (1788).
Prostrate or weakly erect annual. Stems up to 50 cm, solid, striate, sparsely hairy with deflexed hairs. Leaves 1- to 3-pinnate, segments ovate, deeply pinnatifid. Umbels sessile or subsessile, leaf-opposed; rays very short, concealed by flowers or fruit; bracts absent; bracteoles exceeding the subsessile flowers. Petals minute. Fruit 2-3 mm, ovoid, outer mericarp with straight, patent, broad-based spines, the inner tuberculate.
Fl. III - V. *Rare in Madeira around Funchal; more common on the peaks and higher slopes of Porto Santo.* **MP**

2. T. arvensis (Huds.) Link, *Enum. hort. berol. alt.* **1**: 265 (1821).
Erect annual up to 40(-100) cm. Leaves very variable, 2-pinnate to 3-foliolate, lobes 2 mm or more wide, coarsely toothed. Umbels mostly terminal, pedunculate, hispid; rays 2-12; bracts 0-1; bracteoles numerous, hispid. Petals sometimes radiate. Fruit 3-6 mm, ovoid, both mericarps spiny or the outer spiny and the inner tuberculate, rarely both tuberculate.
Fl. all year. *Widespread weed of waste ground, paths and roadsides throughout the lower regions of Madeira.* **M**

Four subspecies have been recognized, of which three occur in Madeira:

a) subsp. **arvensis**
Rays 4-12. Outer petals slightly radiate. Styles 2-3 × as long as the stylopodium.
Rather rare; Fajã da Nogueira and along the north coast.

b) subsp. **neglecta** (Schult.) Thell. in Hegi, *Ill. Fl. Mitt.-Eur.* **5**(2): 1055 (1926).
Rays 4-12. Outer petals distinctly radiate. Styles 3-6 × as long as the stylopodium.
The most common subspecies on the island; throughout the range of the species.

c) subsp. **purpurea** (Ten.) Hayek, *Prodr. Fl. Penins. Balcan.* **1**: 1057 (1927).
Upper leaves differing markedly from the lower. Rays (2-)3-4, diverging to an angle of 45°-60°. Petals often purple, the outer scarcely radiate.
Very rare; collected only once from Madeira.

Subsp. *elongata* (Hoffmanns. & Link) Cannon frequently occurs in the Canaries and has been recorded in error from Madeira. It has the upper and lower leaves similar and rays diverging to an angle of 90°. Records of *T. leptophylla* (L.) Rchb. f. from the Madeiran archipelago also appear to be erroneous, with a record from Madeira having been based on a specimen of subsp. *elongata* from Tenerife.

5. **Coriandrum** L.

Annuals. Leaves 1- to 3-pinnate. Sepals persistent, unequal; petals white, the outer large and bifid, the apices inflexed. Fruit globose, ribs not prominent, alternately straight and flexuous, vitta 1.

1. C. sativum L., *Sp. pl.* **1**: 256 (1753).
Glabrous, foetid annual up to 60 cm. Lower leaves 1-pinnate, segments cuneiform, serrate; upper leaves 2- to 3-pinnate, lobes narrowly linear. Umbels mostly terminal, pedunculate; rays 3-5(-8); bracts absent; bracteoles 3, linear, often reflexed. Often only some of the flowers in the umbel hermaphrodite, the remainder functionally male. Fruit 2-6 mm, globose, mericarps not separating readily at maturity.
Fl. XII - I. *Grown in gardens in both Madeira and Porto Santo, often escaping and sometimes becoming established. Native to N. Africa and W. Asia and widely naturalized in Europe.* **M**

Smyrnium olusatrum L., *Sp. pl.* **1**: 262 (1753), a robust, glabrous perennial with basal leaves ternate, the segments 1- to 2-pinnate and flowers yellow has been recorded once from Madeira, in Funchal, probably as a casual. **MP**

6. Conium L.

Annuals or biennials. Leaves 2- to 4-pinnate, lobes pinnatifid or serrate, the ultimate divisions oval or lanceolate. Sepals absent. Petals white, obcordate, the apex inflexed. Fruit laterally compressed, ribs prominent, undulate, vittae absent.

1. C. maculatum L., *Sp. pl.* **1**: 243 (1753).
Nearly glabrous, foetid-smelling annual or biennial up to 2.5 m. Stems hollow, usually purplespotted or -blotched below. Lower leaves broadly triangular in outline, lobes 10-20 mm, ovate to deltate, serrate or crenate. Umbels terminal, pedunculate; rays 10-20, glabrous or puberulent; bracts (0-)5-6, narrowly triangular or lanceolate, deflexed, the margins membranous; bracteoles 3(-6), reflexed, broad at the base and sometimes connate. Fruit 2.5-3.5 mm, ovoid, styles short, divergent.
Fl. V. *A recent introduction, recorded only from Cancela near Funchal. Native to Europe, N. Africa and Asia.* **M**

7. Bupleurum L.

Annuals or shrubby perennials, glabrous. Leaves simple, entire. Sepals absent. Petals yellow, the apex inflexed. Fruit ovoid to oblong, ribs usually conspicuous, vittae 1-several.

1. Plant herbaceous; bracteoles showy and conspicuous; flowers nearly sessile . . . . **1. lancifolium**
-   Plant woody; bracteoles small and inconspicuous; flowers clearly pedicellate . . . . **2. salicifolium**

1. B. lancifolium Hornem, *Hort. bot. hafn.*: 267 (1813).
B. protractum Hoffmanns. & Link; *B. subovatum* Link ex Spreng.
A somewhat variable annual up to 40 cm, often spreading. Lower leaves oblong-lanceolate to ovate, longer than broad, sessile; upper leaves perfoliate, mucronate. Umbels terminal or lateral, pedunculate; rays 2-3(-4), subequal; bracts absent; bracteoles often 4, rarely 5, suborbicular, mucronate, petaloid, yellowish green in flower, whitish in fruit. Flowers very shortly pedicellate. Fruit 3-5 × 2 mm, ovoid, conspicuously tuberculate.
Fl. IV - VI. *An infrequent and probably declining weed of cornfields and gardens in and around Funchal, and in Porto Santo.* **MP**

2. B. salicifolium R. Br. ex Buch, *Phys. Beschr. Canar. Ins.*: 195 (1825).
Shrubby perennial up to 1.5 m, the woody stem with faint leaf-scars. Leaves up to 15 cm, linear to narrowly lanceolate, acute or apiculate, partially clasping the stem. Umbels mostly terminal, pedunculate; rays 5-20, unequal; bracts several, up to c. 1 cm, reflexed; bracteoles 1-2 mm, lanceolate, acute, much shorter than the flowers. Flowers equalling the pedicels. Fruit 7 × 2.5-3 mm, ± ovoid, the ribs prominent, without tubercles. **Plate 27**
Fl. V - XI. *Scattered in dry, rocky places throughout the mountains of Madeira, generally above 1000 m; recorded also from Porto Santo but without further data.* **MP** [*Canaries*]

The above description refers to the Macaronesian endemic (■) subsp. **salicifolium** var. **salicifolium**, which otherwise occurs in the Canaries, and to which plants from the Madeiran archipelago belong. The remaining two infraspecific taxa, subsp. *salicifolium* var. *robustum*

(Burch.) Cauwet & Sunding and subsp. *aciphyllum* (Webb ex Parl.) Sunding & G. Kunkel, are both endemic to the Canaries.

8. Apium L.

Annual, biennial or perennial herbs. Leaves pinnate or ternate. Sepals minute or absent. Petals whitish, entire, acute, the point sometimes inflexed. Fruit broadly ovoid to globose, laterally compressed, ribs stout, vitta 1.

1. Leaves ternate, the segments filiform . . . . **3. leptophyllum**
- Leaves pinnate, lobes broad 2

2. Bracteoles absent; lower segments of lower leaves stalked . . . . **1. graveolens**
- Bracteoles 5-7; segments of lower leaves all sessile . . . . **2. nodiflorum**

1. A. graveolens L., *Sp. pl.* **1**: 264 (1753).
Stout, glabrous biennial up to 1 m, smelling strongly of celery. Stems deeply channeled. Leaves pinnate, segments 0.5-3 cm, deltate to rhombic, serrate or lobed, the lower segments of lower leaves stalked, the upper sessile, cuneiform at base; upper leaves sessile, ternate. Umbels axillary, shortly pedunculate or subsessile, with a long-pedunculate terminal umbel; rays 6-12, unequal; bracts and bracteoles absent. Fruit *c.* 1.5 mm, broadly ovoid to globose. Fl. V - IX. *Scatttered throughout the lower regions of Madeira and Porto Santo, mainly in cultivated areas.* **MP**

Several varieties, including var. **dulce** (Mill.) DC. and var. **rapaceum** (Mill.) DC., are widely cultivated as vegetables.

2. A. nodiflorum (L.) Lag. *Amen. nat. Españ.* **1**: 101 (1821).
Helosciadium nodiflorum (L.) W.D.J. Koch
Glabrous perennial to 1 m. Stems procumbent or ascending, often rooting at the lower nodes. Leaves pinnate, segments (1-)2-6 mm, lanceolate to ovate, serrate, crenate or lobed, sessile, petioles of upper leaves sheathing for *c.* half their length. Umbels axillary, the peduncle shorter than the rays or absent; rays 3-9, unequal; bracts 0(-2); bracteoles 5-7, unequal, ovate to lanceolate, margins pale. Fruit 1.5-2 mm, ovoid, longer than broad; styles 0.2-0.4 mm, recurved.
Fl. V - IX. *In damp places, by paths and levadas, mainly in the mountains and higher regions of Madeira but occasionally also in the lower regions.* **M**

3. A. leptophyllum (Pers.) F. Muell. ex Benth., *Fl. austral.* **3**: 372 (1867).
A. tenuifolium (Moench) Thell.; *Ciclospermum leptophyllum* (Pers.) Sprague; *Helosciadium leptophyllum* DC.
Erect or diffuse, slender, glabrous annual up to 60 cm. Leaves ternately divided, segments filiform, numerous, lower leaves petiolate, upper sessile. Umbels mostly lateral, sessile at the nodes or pedunculate; rays 2-3, slender, each with a partial umbel of many flowers on slender pedicels; bracts and bracteoles absent. Disc broad, convex, scarcely distinct from the very short styles. Fruit 1.5-2 mm, globose to narrowly ovoid, sometimes a little broader than long, ribs very prominent and thick, almost corky, furrows very narrow, vittae 1 in each furrow.
Fl. XII - IX. *Introduced. Common as a weed of roadsides, gardens and waste ground in and around Funchal and other coastal towns in Madeira; recorded also from Porto Santo. Native to S. America but naturalized in many temperate regions.* **MP**

9. Petroselinum Hill

Biennials. Leaves 3-pinnate. Sepals minute. Petals yellowish, emarginate, the apex inflexed. Fruit ovoid, ribs filiform, vitta 1.

1. P. crispum (Mill.) A.W. Hill, *Hand-list herb. pl. Kew*, ed. 3: 122 (1925).
P. sativum Hoffmanns.
Stems up to 75 cm, solid. Lower leaves 3-pinnate, triangular in outline, lobes 10-20 mm, cuneate. Umbels terminal, flat-topped; rays 10-20; bracts 1-3; bracteoles 5-10, linear to ovate. Fruit 2.5-3 mm, broadly ovoid, sometimes with 1 mericarp less well developed than the other.
Fl. VII - VIII. *Introduced and scattered in waste and grassy places around the coast of Madeira; recorded also from Porto Santo. Probably native to S. Europe.* **MP**

10. Ammi L.

Annuals or biennials. Leaves 1- to 3-pinnate or -ternate. Sepals absent or very small. Petals whitish, obcordate, the outermost largest, apex inflexed. Fruit ovoid or oblong-ovoid, slightly laterally compressed, ribs prominent, filiform, vitta 1.

1. Rays erect and thickened in fruit; all leaves with narrow, linear or filiform segments . . . . **1. visnaga**
- Rays patent and slender in fruit; at least the lower leaves with elliptic, obovate or lanceolate lobes . . . . **2. majus**

1. A. visnaga (L.) Lam., *Fl. franç.* 3: 462 (1778).
Stout annual or biennial, up to 1 m, stems leafy. Lower leaves pinnate, upper leaves 2- to 3-pinnate, all with filiform or narrowly linear segments. Umbels terminal or axillary, peduncles slender, up to 30 cm in fruit; rays numerous (up to 100 or more), erect or patent, slender in flower, becoming thickened and erect in fruit; bracts many, 1- to 2-pinnatisect, elongating in fruit; bracteoles linear, subulate, equalling or exceeding the rays in flower. Pedicels erect and thickened in fruit. Fruit 1.5 mm, slightly longer than broad; styles 1 mm, recurved.
Fl. VI - VIII. *An occasional weed of field margins and dry places, scattered throughout lowland Madeira; recorded also from Porto Santo.* **MP**

2. A. majus L., *Sp. pl.* 1: 243 (1753).
A. procerum Lowe
Very variable annual, 30-100 cm. Leaves 2- to 3-pinnate, lobes of lower leaves elliptic or obovate, serrate to dentate; lobes of upper leaves lanceolate-linear, dentate. Umbels terminal, long-pedunculate (up to 25 cm); rays 15-16, slender, erect or patent in flower; bracts 3-fid or pinnatisect, rarely entire; bracteoles lanceolate, acuminate with papery margins. Pedicels slender in fruit. Fruit 3 × 2 mm; styles 0.5 mm or less.
Fl. II - IX. *Weed of dry ground on roadsides, field margins and in waste places in lowland areas, mainly in the southern half of Madeira and in Porto Santo; also in the Desertas and on Selvagem Grande.* **MDPS**

11. Bunium L.

Perennials with tuberous rootstocks and flexuous stems. Leaves 2- to 3-pinnate. Sepals small or absent. Petals white, obcordate, the apex inflexed. Stylopodium abruptly contracted into styles. Fruit laterally compressed, ribs prominent, slender, vittae 1-3.

1. B. brevifolium Lowe in *Trans. Camb. phil. Soc.* **6**: 543 (1837). [*Nozelha, Nozelhinha*]
Erect, unbranched perennial up to 40 cm, with a thin, flexuous, subterranean part of the stem above the tuber. Leaves with narrow, linear-lanceolate segments. Umbels 1-4, terminal, pedunculate; rays 6-12, unequal; bracts and bracteoles absent. Pedicels slightly winged at base. Fruit *c.* 5.2 mm, ovate-oblong; styles 1-1.5 mm, divergent, not reflexed. **Plate 27**
Fl. V - VIII. *Rare endemic of rocky places in the high, central peaks of Madeira.* ● **M**

Very similar to the European *B. bulbocastanum* L., from which it may be distinguished by its lack of bracts and bracteoles, and by the absence of toothing on the upper edge of the pedicels.

12. Melanoselinum Hoffm.

Perennials. Leaves 2- to 3-pinnate, lobes lanceolate or ovate, serrate or dentate, petioles strongly inflated. Sepals small. Petals whitish or purplish, clawed. Fruit compressed dorsally, laterally winged, vittae many.

1. M. decipiens (Schrad. & J.C. Wendl.) Hoffm., *Gen. pl. umbell.*: 177 (1814).
[*Aipo da serra, Aipo do gado*]
Monocarpic. Stems up to 3 m, stout, woody below, with a crown of leaves above. Leaves up to 60 cm with a downy, whitish pubescence, at least when young; petioles pubescent, dilated and winged. Inflorescence 50-90 cm in diameter, terminal above the leaf crown; rays many; bracts 10-20, 20-30 mm, leafy; bracteoles as long as the pedicels. Fruit 12-14 mm, oblong, pubescent, blackish, wings *c*. 1.5 mm wide, strongly dentate. **Plate 29**
Fl. IV - XII. *Endemic to shady rocks and banks in the interior of ravines in northern Madeira; widely cultivated for fodder.* ■ M [*Azores*]

13. Crithmum L.

Succulent perennials. Leaves 1- to 2-pinnate, fleshy, segments lanceolate. Sepals minute; petals yellowish green, obcordate, apex inflexed. Fruit ovoid, not compressed, ribs thick and persistent, vittae several.

1. C. maritimum L., *Sp. pl.* **1**: 246 (1753).
Plant glabrous. Stems up to 50 cm, prostrate or ascending, woody at base. Leaves with acute, linear-lanceolate segments up to 5 cm. Umbels mostly terminal, pedunculate, 1-several; rays 10-20; bracts and bracteoles many, triangular to linear-lanceolate, spreading or deflexed. Fruit 5-6 × 4-5 mm; stylopodium conical, styles very short, erect.
Fl. VI - X. *Common on coastal rocks and sea cliffs in Madeira and Porto Santo; also on Deserta Grande.* MDP

14. Oenanthe L.

Perennials. Leaves 2- to 4-pinnate. Sepals erect in fruit. Petals white, notched, the outer radiate, apex inflexed. Fruit ± cylindrical, lateral ribs grooved or thickened or obscure, vitta 1.

1. O. divaricata (R. Br.) Mabb. in *Botanica Macaronésica* **6**: 63 (1978). [*Aipo preto*]
O. pteridifolia Lowe
Plant tuberous-rooted, the tubers smooth. Stems up to 90 cm, hollow, somewhat succulent. Leaves up to 40 × 20 cm, the segments narrowly ovate to linear. Umbels mostly terminal, pedunculate; rays 7-12, unequal, not thickening in fruit; bracts 1-5; bracteoles *c*. 10, all linear, the margins pellucid. Fruit *c*. 4 × 2 mm, often not forming in outer flowers; styles much elongated in fruit, nearly as long as the mericarps. **Plate 28**
Fl. VI - IX. *Endemic growing by streams and levadas and among wet rocks; common at all altitudes in Madeira; also on Deserta Grande.* ● MD

15. Foeniculum Mill.

Biennials or perennials. Leaves 3- to 4-pinnate. Sepals absent. Petals yellowish, involute at the apex. Fruit scarcely compressed laterally, ribs prominent, stout, vitta 1.

1. F. vulgare Mill., *Gard. dict.*, ed. 8, no. 1 (1768).
F. officinale All.
Glabrous biennial or perennial, the branches dying back annually, foliage smelling strongly of aniseed. Stems up to 2.5 m, becoming hollow with age. Leaves petiolate, glaucous or dark green, triangular in outline, the lobes filiform, not all lying in one plane. Umbels terminal and lateral, pedunculate; rays 4-30; bracts and bracteoles absent. Flowers small. Stigmas sessile, styles developing only in fruit. Fruit 4-10 mm, ovoid-oblong; styles erect or somewhat divergent.
Fl. V - VIII. *On cliffs and rocky places near the sea, mainly around Funchal; also in Porto Santo.* **MP**

Var. *azoricum* (Mill.) Thell. is widely cultivated for its large, tuberous rootstock.

The very similar **Anethum graveolens** L., *Sp. pl.* 1: 263 (1753) is an annual with elliptical, strongly compressed, winged fruits. It is cultivated as a pot-herb in both Madeira and Porto Santo, and may occasionally escape. **MP**

16. Capnophyllum Gaertn.

Usually annuals with slender rootstocks and angular tems. Leaves 3-pinnate. Sepals very small. Petals white, apex involute. Fruit dorsally compressed, ribs very prominent and winged, transversely rugose, vitta 1.

1. C. peregrinum (L.) Lange in Willk. & Lange, *Prodr. fl. hispan.* 3: 33 (1874).
Krubera leptophylla DC.; *K. peregrina* (L.) Hoffm.
Glabrous, short-lived annual. Stems up to 50 cm, wiry, angular, branching regularly in the upper half. Leaf segments bluntly ovate or broadly triangular, lobes entire or with further, linear to lanceolate lobes. Umbels in the angles of the branches, shortly pedunculate; rays 2-5; bracts absent or few; bracteoles 4-6, linear. Pedicels very short and broad. Sepals rigid and persistent in fruit. Fruit 4-6 mm, ovoid-oblong, subsessile; styles *c*. 1 mm, recurved. **Plate 28**
Fl. III - IV. *Mainly a cornfield weed in the Funchal area and in Porto Santo.* ■ **MP** [*Canaries, Cape Verdes*]

17. Astydamia DC.

Succulent perennials with woody stems. Leaves pinnate. Sepals very small, deltate, falling before the flower opens. Petals yellow, the apex somewhat incurved. Fruit laterally compressed, ribs 1-3, weak.

1. A. latifolia (L. f.) Baill., *Hist. pl.* 7: 208 (1879).
Stems up to *c*. 40 cm. Leaves pinnate, very thick and fleshy, the lobes broad, ± obovate, dentate. Umbels 1-several, the lateral umbels sometimes functionally male; rays 10-30; bracts several, ovate, papery, half as long as the peduncles; bracteoles several, ovate, papery, equalling the pedicels. Fruit 12 × 5-6 mm, ovoid; stylopodium swollen, 1.5 × 3 mm, tapering gradually into the styles; stigmas 1 mm, patent.
Fl. II - IV. *A halophyte confined to rocky places on the Salvages.* **S**

18. Peucedanum L.

Perennials, glabrous or nearly so. Leaves 2-ternate. Sepals absent. Petals whitish, apex long, inflexed. Fruit strongly dorsally compressed, lateral ribs winged, dorsal ribs prominent, vittae 1-3.

1. P. lowei (Coss.) Menezes in *Broteria* (Bot.) **23**: 74 (1927).
Imperatoria lowei Coss.; *I. ostruthium* Lowe
Plant shining, glabrous. Stems up to 1.5 m, simple or sparingly branched, terete or striate, hollow. Leaves with lower segments triangular in outline, lobes 5-10 × 4-7 cm, ovate to lanceolate; petioles of cauline leaves strongly inflated. Umbels terminal or lateral; rays numerous, unequal; bracts 0-1; bracteoles 0-few. Fruit 4-5 mm, suborbicular, wings *c*. 1 mm wide; styles persistent, divergent. **Plate 28**
Fl. V - X. *Endemic on wet rock faces; scattered in the central mountains of Madeira.* ● **M**

19. Monizia Lowe

Perennials. Leaves 3- to 4-pinnate; petioles somewhat inflated. Sepals ovate, thickly pubescent, erect. Petals small, dull white, ciliate and fringed at the margin. Fruit dorsally compressed, pubescent, ribs thickened and corky, vittae 6.

An endemic, monospecific genus.

1. M. edulis Lowe in *Hookers J. Bot.* **8**: 57 (1856).   [*Cenoura da rocha, Nozelha*]
Melanoselinum edule (Lowe) Baill.
Long-lived perennial, up to 1 m. Flowering stems annual, stout, woody, single and unbranched. Leaves mostly radical, broadly triangular in outline, bright yellow-green, markedly shining; petioles pubescent. Inflorescence spreading, paniculate; rays 20-25 in each umbel, subequal, ribbed, pubescent; bracts and bracteoles lanceolate or linear, puberulent, fringed at margin. Fruit 10-14 × 5-7 mm, oblong to ellipsoid, pubescent, pale coloured when ripe; styles divergent. **Plate 29**
Fl. III - VI. *Very rare endemic confined to fissures in cliffs above 1500 m in Madeira but much lower, at c. 300 m, on Deserta Grande and Selvagem Grande.* ● **MDS**

20. Daucus L.

Usually perennials. Leaves 2- to 3-pinnate. Umbels subsessile or pedunculate; bracts often pinnatisect. Sepals small or absent. Petals usually white or pinkish, the outer often radiate, apex inflexed. Fruit cylindrical or slightly compressed, the primary ribs ciliate, the secondary ribs spiny.

1. D. carota L., *Sp. pl.* **1**: 242 (1753).
D. neglectus Lowe
Perennial or rarely annual with roughly hairy stems. Segments of lower leaves linear to lanceolate, glabrous to hispid. Umbels mostly terminal, often with 1-few of the central flowers deep purple; rays hispid, very unequal, strongly incurved in fruit; bracts 2 × as long as the rays or shorter, 1- to 2-pinnatisect; bracteoles simple or 2- to 3-pinnatisect. Fruit 2-4 mm, ovoid to ovoid-oblong, spines not longer than the width of the mericarps.
Fl. III - V. *Infrequent weed of cultivated ground and waste places in Madeira.* **M**

Two subspecies occur in Madeira:

a) subsp. carota
Stem and rachis hispid but not densely so. Outer rays of umbel unequal, the umbel becoming strongly contracted in fruit.
Throughout the range of the species.

b) subsp. hispidus (Arcang.) Heywood, *Feddes Reprium* **79**: 68 (1968).
Stem and rachis densely hispid. Outer rays of umbel ± equal, the umbel convex or only slightly contracted in fruit.
Rare.

Hybridization frequently occurs between these subspecies elsewhere, and may also occur in Madeira, making identification difficult.

LXXIX. CLETHRACEAE
J.R. Press

Trees or shrubs. Leaves alternate, simple, exstipulate. Flowers hermaphrodite, actinomorphic. Sepals and petals 5, imbricate, the sepals joined at the base and persistent in fruit. Stamens hypogynous, the anthers inflexed in bud and opening by apical pores. Ovary hairy, superior, 3-locular; style 3-lobed at the apex; ovules numerous. Fruit a dehiscent capsule; seeds numerous.

1. Clethra L.

Evergreen, trees and shrubs. Hairs simple and branched. Leaves serrate, petiolate. Inflorescence a terminal raceme. Capsules septicidal, the septa separating from the central axis.

1. C. arborea Aiton, *Hort. kew.* **2**: 73 (1789). [*Folhado, Folhadeiro*]
Tree or shrub up to 8 m with a bushy crown. Bark brownish or greyish. Twigs with prominent leaf-scars. Young twigs, petioles, inflorescence-axes, pedicels and calyces all densely rusty-villous. Leaves rather crowded at the tips of twigs, pale green and stiff, 9-12(-20) × 4-5(-7) cm, oblanceolate to obovate, acuminate, cuneate at the base, glabrous above, thinly pubescent below especially on the veins; petiole short, reddish. Racemes simple or branched, erect, the flowers secund and somewhat nodding, fragrant; bracts small, caducous. Sepals 3.5-4 × 2-2.5 mm, ovate, obtuse, densely villous on the outside, glabrous within. Petals 6-8 × 4-5 mm, obovate-oblong, emarginate, pure white, with coarse hairs on the inner surface. Stamens 10. Style exserted, straight, persistent. Capsule *c*. 3.5 mm, densely hairy. **Plate 30**
Fl. VIII - X. *A major component of the laurisilva, widespread in the higher regions in Madeira; also planted in parks and along roadsides in regions above 600 m. The foliage is used for fodder.*
■ M [*Azores*]

LXXX. ERICACEAE
J.R. Press

Woody plants ranging from dwarf shrubs to small trees, usually evergreen. Leaves simple, exstipulate. Flowers hermaphrodite, actinomorphic or rarely slightly zygomorphic, usually 4- to 5-merous. Calyx often small, persistent. Corolla usually with petals connate. Stamens usually twice as many as petals and inserted on the fleshy receptacle; anthers opening by pores or slits. Ovary superior, rarely inferior, 4- to 5-celled; style 1. Fruit a capsule or a berry.

The majority of species are calcifuge and mycorrhizal.

1. Leaves alternate, more than 15 mm, oblong to elliptic . . . . **3. Vaccinium**
- Leaves opposite or whorled, less than 15 mm, ± linear or narrowly triangular . . . 2
2. Calyx petaloid, larger than corolla; leaves opposite, with 2 backward-projecting auricles
 . . . . **2. Calluna**
- Calyx much smaller than corolla; leaves in whorls of 3-4, lacking basal auricles
 . . . . **1. Erica**

1. Erica L.
D. McClintock

Stout, evergreen shrubs or small trees. Leaves in whorls of 3-4, glabrous, linear, margins usually revolute. Flowers 4-merous, in terminal or lateral racemes. Pedicels with 2-3 bracteoles. Calyx glabrous, deeply lobed, much shorter than the corolla. Corolla narrowly ovoid or broadly campanulate, shortly lobed, persisting in fruit. Stamens 8; anthers with or without basal appendages, opening by pores. Ovary glabrous; stigma capitate, exserted. Fruit a many-seeded capsule. Seeds minute.

1. Plant much less than 1 m high; flowers 5 mm, appearing in summer . **1. maderensis**
- Plant exceeding 1 m; flowers 2-3 mm, appearing in late spring 2

2. Young shoots pubescent; flowers white; anthers with basal appendages . . **2. arborea**
- Young shoots glabrescent; flowers greenish pink; anthers without basal appendages . . . . **3. scoparia**

1. E. maderensis (Benth.) Bornm. in *Bot. Jb.* **33**: 458 (1904).
E. cinerea auct. mad., non. L. (1753); *E. cinerea* var. *maderensis* Benth.
Stocky shrub up to 80 cm, usually prostrate, in cushions or pendulous but erect in the lower altitudes of the range. Stems of old plants thick, trunk-like. Young shoots pubescent. Leaves 5-9 mm, in whorls of 3, never fasciculate. Bracteoles 3, borne immediately below the calyx. Corolla 5 mm, narrowly ovoid, fawn-pink. Anthers with basal appendages. **Plate 30**
Fl. V - IX. *An infrequent to rare endemic occurring in small numbers in heathy places and on bare rock faces in the higher peaks of Madeira, from 1400-1800 m.* ● **M**

2. E. arborea L., *Sp. pl.* **1**: 353 (1753).   [*Betouro, Urze mollar*]
Tall shrub or small tree, usually up to 5 m but sometimes more, often with a well-defined trunk up to 37 cm diameter. Young shoots pubescent with whitish, simple and branched hairs. Leaves 3(-4) mm, in whorls of 3(-4). Flowers often in a broad panicle. Bracteoles 2-3, in the lower part of the pedicel. Calyx saccate at the base. Corolla 2-2.5 mm, broadly campanulate, white. Anthers with basal appendages. Stigma prominently exserted.
Fl. II - V. *Locally frequent in woods and scrub on Madeira, from sea-level to c. 1300 m.* **M**

3. E. scoparia L., *Sp. pl.* **1**: 353 (1753).   [*Urze durazia*]
E. scoparia subsp. *platycodon* (Webb & Berthel.) A. Hansen & Kunkel
Tall shrub up to 4 m with stems up to 20 cm or more in diameter. Young twigs glabrescent. Leaves 10-12 mm, spreading, in whorls of 3-4. Bracteoles 2, near the middle of the pedicel. Calyx not saccate. Corolla 2-3 mm, broadly campanulate, greenish pink. Anthers without appendages.
Fl. IV - VI. *Common in areas of scrub at most altitudes in Madeira, descending almost to sea-level in the north of the island, but rare and confined to the higher peaks in Porto Santo.* **MP**

Plants from the Maderian archipelago all belong to the endemic (●) subsp. **maderinicola** D.C. McClint., *Heather Soc. Yr Bk* **3**: 35 (1989) **Plate 30**, to which the above description and synonymy applies.

E. vagans L., *Erica*: 10 (1770), a glabrous shrub with leaves in whorls of 4-5, flowers pale lilac and anthers lacking basal appendages, is known from a single locality (Pico do Infante) in Madeira, where it is presumably an escape from cultivation. **M**

2. Calluna Salisb.

Evergreen shrubs. Leaves very small, opposite, sessile. Flowers 4-merous, in loose terminal racemes grouped into panicles. Calyx-segments ± free, petaloid. Corolla deeply lobed, campanulate, slightly shorter than the calyx and persistent in fruit. Stamens 8; anthers with basal appendages and opening by slits. Fruit a septicidal capsule.

1. C. vulgaris (L.) Hull, *Brit. fl.* ed. 2, **1**: 114 (1808).
Much-branched, evergreen shrub. Leaves tiny, with revolute margins and two backward projecting auricles giving the leaf a triangular outline, amplexicaule, those of leading shoots widely spaced, those of short, non-flowering side-shoots densely imbricate. Pedicel with several bracteoles, the upper 4 petaloid and forming an involucre immediately beneath the flower. Calyx and corolla pale pinkish lilac.
Fl. VIII - IX. *Introduced. Known from a single site in Santo da Serra but well-established and abundant there in an open* Eucalyptus *grove. Native in much of Europe and the Azores.* **M**

Various species of **Rhododendron** L. are cultivated in gardens in Madeira. The following two species have been recorded, probably as escapes.

R. ponticum L., *Sp. pl.* ed. 2, **1**: 562 (1762) is a large evergreen shrub with elliptic to oblong, glabrous leaves and dull pinkish purple flowers. Native to the Mediterranean region, it has been recorded from near Paúl da Serra in western Madeira. **M**

R. mucronatum G. Don, *Gen. hist.* **3**: 846 (1834) is a shrub with lanceolate, deciduous spring leaves and oblong, persistent summer leaves, all with dense, soft, grey hairs; flowers are white, fragrant and viscid on the outside. Native to Japan, it has been recorded from near Porto da Cruz in north-eastern Madeira. **M**

3. Vaccinium L.

Deciduous or evergreen shrubs, sometimes small trees. Leaves alternate. Flowers 5-merous, in axillary racemes. Calyx-lobes connate for most or all of their length. Stamens 10. Ovary inferior. Fruit a berry.

1. V. padifolium Sm. in Rees, *Cycl.* **36** n. 22 (1817).   [*Uva da serra*]
V. maderense Link
Semi-evergreen shrub or small tree 2-3(-6) m. Young twigs often reddish, pubescent. Leaves leathery, often flushed dark red in autumn, 2.5-7 × 1-2(-2.5) cm, oblong to elliptic, acute to acuminate, narrowly cuneate at base, pubescent only on the midrib beneath; petiole short, pubescent. Flowers on curved pedicels in erect, axillary, bracteate racemes. Calyx 3-4 mm, with 5 short, broad lobes up to 1.5 mm. Corolla greenish yellow, sometimes tinged reddish, 7-10 mm, globose to campanulate, the lobes very short, recurved. Fruit up to 12 × 10 mm, ± ovoid, ripening blue-black, often with a waxy, white bloom. **Plate 30**
Fl. V - VIII, but often earlier and later. *Very common on open slopes and moorland from c. 800-1700 m in Madeira, also growing in light shade in wooded ravines. The fruits are used in preserves.* ● **M**

LXXXI. MYRSINACEAE
M.J. Short

Trees (in Madeira). Leaves alternate, often clustered at the ends of the branches, simple, coriaceous, glandular-punctate, exstipulate. Flowers regular, hermaphrodite, small, generally in fascicles on short lateral shoots. Sepals 4-6, free or basally connate, persistent. Petals 4-6, ± free or basally connate. Stamens equal in number to and opposite the petals; filaments often fused to the corolla. Ovary superior, 1-locular. Fruit a 1-seeded, indehiscent drupe.

1. Heberdenia Banks ex A.DC.

Leaves petiolate, entire. Flowers pedicellate, 5-merous, in small, axillary fascicles below the leaves. Sepals shortly connate at the base. Petals ± free, spreading stellately. Stamens free, attached to base of petals. Style long and slender; stigma minute. Fruit globose, rather hard, tipped by the slender, persistent style.

1. H. excelsa (Aiton) Banks ex DC. in *Annls Sci. Nat.* **16**: 73 (1841).
Ardisia bahamensis (Gaertn.) DC.; *A. excelsa* Aiton
Glabrous tree, 5-15 m or more high, with a small, ± pyramidal crown. Bark grey, smooth. Leaves all on the present year's growth, (4.5-)6-12 × 2.5-5 cm, obovate to oblong, acute or obtuse, reticulately veined, glossy above, paler below and minutely red glandular-punctate, margins pellucid; petioles 0.5-0.9 mm, brown. Pedicels 0.5-0.8 mm, brown, spreading. Calyx-lobes *c.* 1 mm, ovate, acute. Petals 4-4.5 mm, linear-lanceolate, green without, pale yellowish within and initially with a whitish mealy coating. Fruit 6-9 mm across, purplish red turning to purplish black when ripe. **Plate 32**
Fl. (V-) VII - IX. *Becoming increasingly rare in Madeira, but still found at several sites in the laurisilva of the central mountains, 600-1300 m. Also recorded recently from Ponta do Pedregal on Deserta Grande.* ■ **MD** [*Canaries*]

LXXXII. PRIMULACEAE
M.J. Short

Annual to perennial herbs. Leaves opposite or alternate, the lower sometimes forming basal rosettes, simple, entire, exstipulate. Flowers regular or rarely irregular, hermaphrodite, solitary or in racemes. Calyx 5-lobed, persistent. Petals (3)5, free or fused. Corolla campanulate or rotate. Stamens (3)5, inserted on petal segments, or near base of corolla-tube opposite the lobes, sometimes alternating with staminodes. Ovary superior or semi-inferior, 1-locular; style 1. Fruit a capsule.

1. Flowers in terminal racemes; cauline leaves alternate, lower leaves forming a basal rosette .... **3. Samolus**
- Flowers solitary, axillary; leaves opposite 2
2. Corolla white, of 3 free, unequal petal segments; capsules dehiscing by 3-valves .... **1. Pelletiera**
- Corolla red, pink or blue, rotate, 5-lobed; capsules dehiscing transversely .... **2. Anagallis**

1. Pelletiera A. St.-Hil.

Glabrous, annual herbs. Leaves opposite, sessile. Flowers minute, irregular, solitary, axillary. Calyx divided ± to base. Corolla shorter than the calyx, of 3 free petal segments. Stamens 3(-5), inserted on the petal segments. Capsules dehiscing by 3(4) valves, usually 2-3-seeded.

1. P. wildpretii Valdés in *Candollea* **35**: 645 (1980).
Stems (3.5)6-16 cm, erect, simple or branched from the base. Leaves 4-14 × 1-3.5 mm, linear-elliptic to oblanceolate, acute to subobtuse, sometimes mucronulate, usually thin and clearly reticulately veined. Pedicels up to 4 mm, patent in flower, recurved in fruit. Calyx-lobes 1.3-2.5(-3) mm, linear-lanceolate, acuminate, narrowly scarious. Corolla with 3 unequal, glandular petal segments, *c.* 0.6-0.8 mm; one spathulate and obtuse, one broader and emarginate, the third obovate and bilobed. Stamens 3, inserted towards middle of petal segments. Capsules 1-1.8 mm across, dehiscing by 3 valves. Seeds 0.9-1.2 mm, dark brown, dorsal face with ridges radiating from sunken hilum, ventral face convex, alveolate. **Plate 31**
Fl. III - VI. *Fairly frequent on Selvagem Grande.* ■ **S** [*Canaries*]

Specimens from the Canary Islands have occasionally been observed with four or five stamens per flower and 4-valved capsules.

Records of *Asterolinon stellatum* (L.) Duby from the Salvage Islands are referable to this species.

2. Anagallis L.

Glabrous, annual herbs. Leaves opposite. Calyx divided ± to base. Flowers regular, 5-merous, solitary and axillary, pedicellate. Corolla rotate, red, pink or blue. Ovary superior. Filaments pubescent. Capsules many-seeded, globose, dehiscing transversely, the upper half falling away like a small cup.

1. A. arvensis L., *Sp. pl.* **1**: 148 (1753). [*Murrião*]
A. caerulea L.; *A. latifolia* L.; *A. phoenicea* Scop.
Variable, sprawling or decumbent annual; stems up to 40 cm, usually much-branched, quadrangular. Leaves opposite or rarely some in whorls of 3, sessile, 7-25 × 4-13(-20) mm, ovate, mostly acute, dotted with black glands beneath. Pedicels up to 2.5(-4) cm, slender, erect in flower, curving downwards in fruit. Calyx-lobes 3-5(-8) mm, linear-lanceolate, acuminate, with a keeled midrib and membranous margins. Corolla-lobes (4)5-6 × (3)4-6 mm, obovate to suborbicular, scarlet, salmon pink or bright blue, the apex rounded, subentire to erose, fringed with numerous small, 3-celled hairs, the apical cell globose, glandular. Filaments with long, multicellular, purplish hairs; anthers yellow. Capsules (3-)4-5 mm across, thin-walled, pale brown, usually tipped by the persistent style. Seeds (0.9)1.3 mm, trigonous, dark brown with paler papillae.
Fl. III - XI. *Pastures, waste ground and grassy places, cliffs. Widespread in Madeira and Porto Santo; 0-1000 m. Small-flowered, red-petalled plants are particularly common on Ilhéu de Baixo. Also known from Bugío, Ilhéu Chão, Deserta Grande and Selvagem Grande.* **MDPS**

A. foemina Mill., *Gard. dict.* ed. 8, no. 2 (1768) (*A. arvensis* subsp. *foemina* (Mill.) Schinz & Thell.), like **1**, but distinguished by the always blue corolla and corolla-lobe margins with sparse, 4-celled hairs, the apical cell ellipsoid, has also been recorded from Madeira and the Salvage Islands. However, all blue-flowered specimens seen have been the blue-flowered form of *A. arvensis*.

3. Samolus L.

Glabrous perennial herbs. Leaves alternate, the lower in a basal rosette. Pedicels bracteolate. Flowers regular, 5-merous, in terminal racemes. Calyx divided *c.* ½-way to base; tube fused to ovary. Corolla campanulate, 5-lobed, white. Stamens alternating with staminodes. Ovary semi-inferior. Capsules many-seeded, dehiscing by 5 valves.

1. S. valerandi L., *Sp. pl.* **1**: 171 (1753).
Stems 6-45 cm, erect, leafy. Leaves 10-65 × 6-28 cm, obovate to spathulate, obtuse or the upper sometimes subacute; lower leaves tapering into a winged petiole, forming a lax rosette, the upper sessile. Inflorescence simple or branched, lax, many-flowered. Pedicels 4-13 mm, erect, becoming geniculate at the bracteole. Bracteoles 2-5 mm, linear-elliptic. Calyx 1.5-2 mm in flower; lobes triangular-ovate, acute, slightly accrescent. Corolla 2-4 mm across; lobes obtuse, longer than tube. Capsules 2-3 mm across, enclosed in the persistent calyx. Seeds 0.5-0.6 mm, polyhedral, blackish brown.
Fl. V - X. *Wet rocks and sea cliffs. Fairly common in Madeira and sometimes locally abundant, especially along the north coast, only rarely found inland.* **M**

LXXXIII. PLUMBAGINACEAE
N.J. Turland

Perennial herbs or shrubs. Leaves alternate, simple, without stipules, usually borne in basal rosettes. Inflorescence paniculate, spicate or a capitulum. Flowers regular, 5-merous. Calyx funnel-shaped with membranous limb, or tubular. Corolla with a short tube, or lobes joined only at base, or funnel-shaped with long slender tube and broad limb. Fruit dry, membranous, 1-seeded, surrounded by persistent calyx.

Plumbago auriculata Lam., *Encycl.* **2**: 270 (1786) (*P. capensis* Thunb.) is a glabrous shrub with narrowly elliptic entire leaves; flowers are grouped into terminal spikes and have a tubular calyx with conspicuous stalked glands and funnel-shaped, pale blue corolla *c.* 3 × as long as calyx. Native to S. Africa, it is cultivated for ornamenent in gardens in Funchal and may also be locally established away from cultivation. **M**

1. Leaves narrowly oblanceolate to obovate, entire or sinuately lobed, sometimes withered before flowering; inflorescence a panicle with terminal, 1-sided spikes . **2. Limonium**
- Leaves linear, entire, persistent during flowering; inflorescence a dense, globose, long-scapose capitulum . . . . **1. Armeria**

1. Armeria Willd.

Tufted herb with woody, branched stock covered with overlapping remains of leaf-bases. Leaves linear, entire, acute, persistent during flowering. Inflorescence a dense, globose, many-flowered capitulum, with a basal involucre of overlapping bracts and a tubular, membranous sheath which encloses top of erect, slender, leafless scape.

1. A. maderensis Lowe in *Trans. Camb. phil. Soc.* **6**: 534 (1838).
Leaves 7-12.5 × 0.2-0.7 cm, ascending, with 3-5 parallel veins, glabrous or sparsely hairy. Scapes 25-45 cm, rounded in cross-section, glabrous, or pubescent at base. Capitula 2-3 cm wide; sheath 10-25 mm; involucral bracts 4-10 mm, oblong-ovate, with membranous margin, the outermost smallest and aristate, the inner becoming larger and shorter-awned. Bracts within capitulum 5-12 mm, oblong, membranous. Flowers pink; calyx 6-8 mm; tube *c.* 4 mm, ribbed, densely hirsute along ribs; limb 2-4 mm, erecto-patent with rounded, mucronate lobes. **Plate 31**
Fl. VI - VIII. *A rare endemic confined to dry, open, rocky and sandy places on the summits of the high central peaks of Madeira.* ● **M**

2. Limonium Mill.

Herbs, with or without woody, branched stock. Leaves narrowly oblanceolate to obovate, entire or sinuately lobed, borne in rosettes, sometimes withered before flowering. Inflorescence a panicle with terminal, 1-sided spikes, sometimes also with non-flowering branches (with only a single, tiny scale at apex); inflorescence-branches subtended by a small, membranous, brownish scale. Spikes dense, consisting of 3-bracteate, 1- to 5-flowered spikelets.

1. Leaves sinuately lobed; flowering stems winged . . . . **1. sinuatum**
- Leaves entire; flowering stems not winged 2

2. Leaves oblanceolate, persistent during flowering; flowering stems smooth (Porto Santo) . . . . **2. ovalifolium**
- Leaves obovate, withered before flowering; flowering stems densely papillose (Salvages) . . . . **3. papillatum**

1. L. sinuatum (L.) Mill., *Gard. dict.* ed. 8, no. 6 (1768).
Plant forming groups of acaulescent leaf-rosettes, hispid. Leaves 3-18 × 0.8-3 cm, oblanceolate to narrowly so, gradually narrowed into short petiole, sinuate at margin, with 4-7 rounded lobes on each side; apex with a long filiform arista; petiole dilated and clasping at base. Inflorescence 10-50 cm; main axis and branches with 3-4 wings each 1-3 mm wide, extended at nodes into ascending, linear, leaf-like appendages 10-60 × 1.5-4.5 mm; lowest scale 5-10 mm, ovate, with a long, filiform arista; non-flowering branches absent. Spikes 2-5 cm, with 3-11 spikelets; spikelets 5-flowered. Calyx 10-15 mm; tube 4-6 mm, minutely papillose-hairy; limb 6-9 mm, erecto-patent,

entire, light to dark blue or white. Corolla white, yellow, blue or purple.
Fl. IV - VIII. *Probably recently introduced; naturalized in dry stony pastures, field-margins, roadsides and waste ground in Porto Santo; recorded also from a roadside in Funchal as a probable escape from cultivation. Native to the Mediterranean region.* **MP**

2. L. ovalifolium (Poir.) Kuntze, *Revis. gen. pl.* **2**: 396 (1891).
Statice ovalifolia Poir.
Plant with woody, branched stock, glabrous except for calyx-tube. Leaves 3.5-7 × 1-2 cm (including petiole 1-3.5 cm), oblanceolate, gradually narrowed into petiole, entire, mucronate; petiole dilated and clasping at base, with bases overlapping. Inflorescence 25-35 cm, laxly pyramidal; branches few, smooth, not winged; lowest scale up to 9 mm, lanceolate, acuminate; non-flowering branches absent. Spikes to 1.3 cm, with 1-several spikelets; spikelets 1- to 3-flowered. Calyx 4-5 mm; tube 2-3 mm, white-pubescent; limb *c.* 2 mm, erecto-patent, with ovate lobes. Corolla pale blue.
Fl. V - VII. *Restricted to dry places and maritime cliffs on the northern coast of Porto Santo.* **P**

The above description refers to (●) subsp. **pyramidatum** (Lowe) O.E. Erikss., A. Hansen & Sunding, *Fl. Macaronesia* ed. 2, **1**: 92 (1979) (*Statice ovalifolia* var. *pyramidata* (Lowe) Menezes; *S. pyramidata* Lowe) **Plate 31**, which is endemic to Porto Santo. Other subspecies have been described from the Canaries, W. France and Portugal, while subsp. *ovalifolium* occurs in Morocco.

3. L. papillatum (Webb & Berthel.) Kuntze, *Revis. gen. pl.* **2**: 396 (1891).
L. pectinatum sensu auct. mad., non (Aiton) Kuntze (1891); *Statice pectinata* sensu auct. mad., non Aiton (1789).
Plant with acaulescent leaf-rosettes, glabrous. Leaves 2-5 × 1-3 cm, obovate, entire, slightly retuse at apex with a filiform arista, ± smooth, withered before flowering. Inflorescence 10-35 cm or more high by up to 100 cm across, densely pyramidal; branches many, densely papillose, not winged; lowest scale up to 3 mm, ovate, cuspidate; non-flowering branches present. Spikes to 1.8 cm, with 1-several spikelets; spikelets 1- to 2-flowered. Calyx 4.5-5 mm; tube *c.* 2.5 mm, with long white hairs; limb 2-2.5 mm, erecto-patent, with triangular-ovate lobes.
Fl. II - VII. *Among stones and sand on Selvagem Grande, Selvagem Pequena and Ilhéu de Fora.* ■ **S** [*Canaries*]

The above description refers to (●) var. **callibotryum** Svent. in *Index Seminum., Agron. Investig. Nat. Hispan. Inst.* **1968**: 59 (1969) **Plate 31**, which is endemic to the Salvage Islands. The remaining variety, var. *papillatum*, is endemic to the Canaries.

L. pectinatum (Aiton) Kuntze (*Statice pectinata* Aiton) has been recorded from Selvagem Grande, but in error for **3**; it is otherwise known only from the Canaries.

LXXXIV. SAPOTACEAE
N.J. Turland

Flowers hermaphrodite, regular; calyx divided into segments; corolla falling early, divided into small lobes; stamens alternately fertile and sterile, the latter petaloid; style simple; stigma indistinctly lobed; ovary superior, nearly always hispid, with loculi 1-ovulate. Fruit usually few-seeded.

1. Sideroxylon L.

Calyx and corolla 5-partite or 5-lobed. Stamens 10, inserted into corolla-tube, the 5 sterile petaloid ones alternating with lobes of corolla. Fruit obovoid or globose, containing 1(-3) seeds.

1. S. marmulano Banks ex Lowe in *Trans. Camb. phil. Soc.* **4**: 22 (1831).
[*Barbuzano* (P), *Marmulano* (M)]
Evergreen shrub to 2.5 m; young leafy twigs sparsely to densely and minutely ferruginous-pubescent;

old bare twigs with wrinkled bark and semi-circular leaf-scars, glabrous; sap milky. Leaves alternate, petiolate, simple, 4-15 × 2.5-6.5 cm (including petiole 1-2.5 cm), obovate to oblanceolate or elliptic, rounded or slightly retuse at apex, narrowed into petiole, coriaceous, reticulate-veined, dark green; upper surface glossy, glabrous; underside glabrescent, with prominently raised mid-vein; margin entire, narrowly thickened. Flowers borne in axillary clusters of 1-7; pedicels 4-15 mm, minutely ferruginous-pubescent; calyx-segments 3-5 mm, ovate, overlapping, similarly pubescent; corolla-lobes exceeding calyx, pinkish; petaloid stamens equalling corolla; fertile stamens exserted. Fruit 14-18 × 11-12 mm, obovoid, berry-like, with persistent style, ripening through red to glossy black; seed(s) large, obovoid, hard and woody. **Plate 32**
Fl. XII - I. *Cliffs and rocks near the sea along the northern coast of Madeira from Porto do Moniz to Porto da Cruz, ascending to 750 m in the Ribeira de São Jorge valley. Much rarer on the southern coast, confined to São Gonçalo east of Funchal. Known also in Porto Santo, at Serra de Dentro and Ilhéu de Cima, and on the islet of Doca off the western coast of Deserta Grande and on Bugío.* ■ MDP [*Canaries, Cape Verdes*]

The above description refers to var. **marmulano**, which otherwise occurs in the Canaries and the Cape Verdes, and to which plants from the Madeiran archipelago belong. The remaining two varieties, var. *edulis* Chev. and var. *marginata* (Pierre) Chev. are both endemic to the Cape Verdes.

LXXXV. OLEACEAE
N.J. Turland

Usually glabrous trees or shrubs, sometimes covered with minute scales. Leaves opposite or alternate, without stipules. Flowers usually hermaphrodite, actinomorphic, 4- to 5-merous; calyx campanulate or ± flat or absent; corolla funnel-shaped or ± flat or absent; stamens 2; ovary superior, 2-locular. Fruit a fleshy berry or drupe, or a dry, winged samara.

1. Leaves (1-)3- to 7-foliolate .... **1. Jasminum**
- Leaves always simple 2

2. Young twigs and undersides of leaves covered with minute scales; inflorescence an ebracteate panicle .... **2. Olea**
- Young twigs and undersides of leaves without scales; inflorescence an initially bracteate raceme .... **3. Picconia**

1. Jasminum L.

Mostly glabrous, evergreen or deciduous shrubs with trailing, climbing or erect stems. Leaves opposite or alternate, (1-)3- to 7-foliolate; leaflets entire. Inflorescence terminal, cymose, or a panicle of cymes. Flowers fragrant; calyx campanulate, with 5 teeth; corolla funnel-shaped, with a slender tube and 5 spreading lobes. Fruit a fleshy berry, ripening black.

J. grandiflorum L. [*Jasmineiro*] is a deciduous climbing shrub with opposite, 5- to 7-foliolate leaves and white flowers flushed pink outside. It is native to the mountains of SE Asia and is cultivated for ornament in Madeira.

1. Leaves opposite; flowers white .... **1. azoricum**
- Leaves alternate; flowers yellow .... **2. odoratissimum**

1. J. azoricum L., *Sp. pl.* **1**: 7 (1753). [*Jasmineiro branco*]
Plant evergreen. Stems long and slender, climbing or trailing, scarcely self-supporting. Leaves opposite, trifoliolate; leaflets ovate to ovate-lanceolate, acute to shortly acuminate, subcoriaceous, glossy; terminal leaflet 3-9 × 2-5 cm. Inflorescence terminal, 3-20 × 1.5-4 cm, 5- to 25-flowered,

a lax, leafy, oblong panicle consisting of axillary and terminal cymes. Calyx 3-4 mm including the triangular teeth. Corolla c. 20 mm across, white; tube 10-15 mm; lobes lanceolate. Fruit ± globose. **Plate 33**
Fl. V - X. *Endemic to Madeira, and extremely rare as a wild plant, known only from cliffs in ravines in the upper Ribeira Brava valley, from 900-1200 m, and in the Funchal region. The species has also been cultivated for ornament locally in gardens in and around Funchal.* ● M

J. azoricum has not been recorded from the Azores since the eighteenth century, and must be assumed extinct in that archipelago. It is not clear whether it existed there as a native plant, or purely as an introduction cultivated in gardens.

2. J. odoratissimum L., *Sp. pl.* **1**: 7 (1753). [*Jasmineiro amarello*]
Plant to 1.5 m, evergreen. Stems stout, erect, self-supporting. Leaves alternate, (1-)3(-5)-foliolate; leaflets ovate to lanceolate or elliptic to suborbicular, emarginate, rounded, obtuse or acute at apex, coriaceous, somewhat glossy; terminal leaflet 1-5 × 1-2.5 cm. Inflorescence terminal, 2-10 × 1-10 cm, 5- to 25-flowered, ± densely branched, cymose. Calyx 2-3 mm including the triangular teeth. Corolla 10-15 mm across, yellow; tube 10-14 mm; lobes broadly ovate. Fruit 8-15 × 7-10 mm, ellipsoid, 2-seeded; seeds almost as large as fruit, narrowly elliptic, convex on one side, flat on the other, finely wrinkled longitudinally, black, somewhat glossy. **Plate 33**
Fl. ± all year but mainly II - VI. *Cliffs and rocks, both on the coast and in inland ravines, from São Vicente and Serra de Água eastwards to Portela and Caniço; also on Ilhéu Chão in the Desertas.* ■ MD [*Canaries*]

Fraxinus excelsior L., *Sp. pl.* **2**: 1057 (1753) [*Freixo*] is a deciduous tree to 15 m, with opposite imparipinnate leaves and panicles of flowers lacking both calyx and corolla; fruits are winged samaras. Native throughout most of Europe, it occurs locally in Madeira, probably planted for its timber. M

Ligustrum lucidum W.T. Aiton is an evergreen shrub or small tree to 10 m, with glabrous new growth, opposite simple leaves, white flowers borne in panicles 10-20 cm long, and fleshy blue-black berries. Native to China, Korea and Japan, it is commonly planted for both ornament and shade in the lowlands of Madeira.

L. japonicum Thunb., native to Korea and Japan, is also planted in Madeira. It is very similar to *L. lucidum* but differs in having new growth minutely dark-pubescent, becoming glabrous later.

2. Olea L.

Much-branched, glabrous, evergreen tree or shrub. Young twigs covered with minute scales. Leaves opposite, simple, entire; underside scaly like the twigs. Inflorescence an axillary panicle, many-flowered, ebracteate, densely scaly. Calyx small, campanulate, obscurely 4-toothed. Corolla ± flat, 4-lobed. Fruit a fleshy drupe, ripening black, not pruinose.

1. O. europaea L., *Sp. pl.* **1**: 8 (1753). [*Oliveira, Zambujeiro*]
O. europaea var. *buxifolia* Aiton
Plant to 2.5 m. Twigs straight, slender, grey or whitish. Leaves 1-10 × 0.3-1.4 cm, oblanceolate to narrowly elliptic to linear, often very slender (up to 10 × 0.5 cm), more rarely oblanceolate-oblong or elliptic-oblong to suborbicular (down to 1 × 0.7 cm), subsessile to shortly petiolate, mucronate to cuspidate, revolute at margin, coriaceous, greyish green above, paler beneath. Panicles 2-4.5 cm, narrowly pyramidal. Calyx 1-1.5 mm. Corolla c. 4 mm across, white; lobes ovate. Fruit 12-22 × 9-12 mm, ellipsoid; seeds large, ellipsoid to ovoid, hard and woody, veined.

Fl. III - VI. *Cliffs and rocks up to 500 m along the southern coast of Madeira from Cabo Girão eastwards to Caniço, at Caniçal and on the Ilhéu do Porto da Cruz off the north-eastern coast; on Pico de Ana Ferreira in Porto Santo; Ilhéu Chão and Deserta Grande.* **MDP**

The above description refers to (●) var. **maderensis** Lowe in *Trans. Camb. phil. Soc.* **6**: 537 (1838) **Plate 33**, which is endemic to the Madeiran archipelago. Var. *buxifolia* Aiton has been recorded from near São Gonçalo, on Pico de Ana Ferreira in Porto Santo and on Ilhéu de Chão in the Desertas, but the name merely refers to the shorter to round-leaved growth of young, regenerating or grazed individuals of *O. europaea*, and is not a separate taxon. The cultivated olive is rare in Madeira and also occurs in Porto Santo, having been introduced there from Portugal. It tends to be a larger-growing tree, with narrowly elliptic (not linear) leaves and larger, juicier fruits.

3. Picconia DC.

Like *Olea* but a larger-growing tree, glabrous except for bracts; young twigs and undersides of leaves smooth, not scaly; inflorescence a short axillary raceme, initially bracteate, not scaly; calyx ± flat, with 4 triangular lobes; fruit ripening through red to dark bluish violet, with bluish-pruinose surface.

1. P. excelsa (Aiton) DC., *Prodr.* **8**: 288 (1844). [*Pão branco*]
Notelaea excelsa (Aiton) Webb & Berthel.
Plant to 15 m, but usually smaller; bark pale. Leaves 4.5-13 × 1-5.5 cm, elliptic, shortly petiolate, usually obtuse, rarely acute to shortly mucronate, slightly revolute at margin, coriaceous, dull green above, paler beneath. Racemes 1-5 cm; bracts 4-10 mm, ovate, soon falling, membranous, densely tomentose on inner surface and around margin. Calyx *c.* 2.5 mm across. Corolla *c.* 6 mm across, white; lobes lanceolate. **Plate 32**
Fl. II - VII. *A rare species of laurisilva, thickets, cliffs and rocks, often in ravines, also as isolated trees; in northern and central Madeira; also locally planted in gardens.* ■ M [*Canaries*]

Phillyrea lowei DC., *Prodr.* **8**: 293 (1844) was described from plants in cultivation at Monte, north of Funchal. They were said to have originated on maritime cliffs east of Funchal, but this was probably an error, since no *Phillyrea* has ever been found wild along that stretch of coast or indeed anywhere else in Madeira. The cultivated plants were almost certainly *P. angustifolia* L. introduced from Europe.

LXXXVI. GENTIANCEAE
B.R. Tebbs

Glabrous herbs with a bitter taste. Leaves opposite, rarely alternate, often joined at the base. Flowers 4- to 5-merous, regular. Calyx ± deeply lobed. Corolla-lobes overlapping in bud. Stamens attached to corolla, equal in number to, and alternate with, the lobes. Ovary superior, 1(-2)-locular, with numerous ovules. Fruit a dehiscent capsule. Seeds small, endospermous.

1. Centaurium Hill

Annual or biennial herbs. Flowers (4-)5-merous. Calyx deeply divided, the lobes linear, keeled. Corolla ± funnel-shaped. Anthers linear or linear-oblong, twisting spirally after dehiscence. Style filiform, with 2 caducous stigmas.

1. Flowers yellow, 20-25 mm long . . . . **1. maritimum**
- Flowers pink, 12-14 mm long . . . . **2. tenuiflorum**

1. C. maritimum (L.) Fritsch, *Mitt. Naturw. Ver. Wien* **5**: 97 (1907).
Erythraea maritima (L.) Pers.
Annual or biennial, 10-20 cm, basal leaf-rosette soon withering. Stem solitary, simple or branched above. Leaves 0.5-2 × 0.3-0.6 cm, elliptic-oblong, increasing in length upwards. Flowers 20-25 mm long, yellow, pedicellate, in a loose, few-flowered, corymbose cyme. Calyx ¾ as long to nearly as long as corolla-tube. Corolla-lobes 2-4(-5) mm. Stamens inserted in upper ⅓ or at apex of corolla-tube.
Fl. IV - VI. *Rare plant of high, open, grassy slopes, usually near the sea; scattered mainly in the south-east of Madeira, at Machico, Pico da Silva north-east of Funchal and Estreito de Câmara de Lobos.* **M**

2. C. tenuiflorum (Hoffmans. & Link) Fritsch, *Mitt. Naturw. Ver. Wien* **5**: 97 (1907).
Erythraea pulchella (Sweet) Fries; *E. ramosissima* (Vill.) Pers.; *E. tenuiflora* Hoffmans. & Link
Erect annual, 8-40 cm, basal leaf-rosette weak or absent. Stem slender, solitary, simple or branching above. Leaves 1-2.5 × 0.5-1 cm, ovate to elliptic, increasing in length upwards. Flowers 12-14 mm long, pink, pedicellate, in a few- to many-flowered dichasial cyme. Calyx almost equalling corolla-tube. Corolla-lobes 2-4 mm. Stamens inserted at top of corolla-tube.
Fl. IV - VIII. *Plant of rough grassland, among rocks, along roadsides and paths; scattered and most frequent on the east coast of Madeira from Caniçal to Machico and Praínha, also at Ponta do Clerigo and Achadas da Cruz on the north coast and Paúl do Mar in the south.* **M**

Two subspecies have been recognized in Europe: subsp. **tenuiflorum**, with a basal rosette present, flowers 2-90, with deep pink, entire corolla-lobes, the tube not constricted, and subsp. **acutiflorum** (Schott) Zeltner in *Bull. Soc. neuchâtel. Sci. nat.* **93**: 94 (1970), without a basal rosette, flowers 20-180, with pale pink corolla-lobes notched at apex, the tube constricted below. Specimens from Madeira have been referred to subsp. **tenuiflorum**, although collections seen have some characteristics of both taxa, and it appears that the differences between the two are not as obvious in Madeira as they may be elsewhere. A third subspecies, subsp. *viridense* (Bolle) A. Hansen & Sunding, is endemic to the Cape Verdes.

LXXXVII. APOCYNACEAE
M.J. Short

Perennial herbs, shrubs or trees with milky latex. Leaves opposite or whorled, simple, entire, exstipulate. Flowers hermaphrodite, regular, 5-merous, solitary or cymose. Corolla frequently pubescent within, the lobes often asymmetrical. Stamens 5, inserted on the corolla. Ovary superior, usually of 2 free carpels united above into a single style. Style often enlarged and hairy at the apex. Fruit of 2 follicles (rarely only 1 developing), or a capsule. Seeds often comose.

Several introduced species grown for ornament in the lower regions of Madeira are conspicuous by their large, showy flowers: **Allamanda cathartica** L. [*Alamanda, Flor de manteiga*] is a vigorous climbing shrub with leathery, obovate leaves in whorls of 3-4 and golden-yellow flowers; **Catharanthus roseus** (L.) G. Don (*Vinca rosea* L.), is an erect, perennial herb with oblong leaves and subsessile pink or white flowers up to 3.5 cm across in axillary pairs; **Nerium oleander** L. (*N. odorum* Sol.) [*Aloendro, Cevadilha, Loendro*] is a robust, evergreen shrub up to 4 m high with elliptic leaves with prominent midveins and terminal cymes of pink or white flowers; Frangipani, **Plumeria rubra** L. [*Planta dos dentes*], a small, normally deciduous tree with stout, rather swollen branches and clusters of fragrant, yellow-pink flowers is planted in parks and gardens.

1. Vinca L.

Creeping subshrubs. Leaves opposite, evergreen. Flowers solitary in leaf-axils, long-pedicellate. Corolla tubular, with 5 spreading lobes, blue, the lobes broad, connected by a low ridge. Stamens inserted half-way up the corolla-tube; filaments short, sharply bent at the base; anthers terminating in a flap-like appendage arching over the stigma. Carpels 2, united only by the common style. Fruit of 2 patent follicles. Seeds glabrous.

1. V. major L., *Sp. pl.* **1**: 209 (1753).
Subshrub with numerous arching, then procumbent, vegetative stems up to 1 m long, rooting only at their tips, and shorter erect flowering stems up to 40 cm. Leaves 2-7 × 1.5-4 cm, mostly ovate and acute, base rounded or subcordate, dark green and glossy, margin densely ciliate; petioles 0.5-1.5 cm. Flowers 4-5 cm across, blue, paler towards the centre; pedicels shorter than the subtending leaves. Calyx-lobes 7-12 mm, linear, margins ciliate. Corolla-tube (12-)15-18 mm; lobes obliquely truncate. Follicles 3-5 cm, each with 1-2 dark brown, oblong seeds.
Fl. I-V. *A Mediterranean species well naturalized in Madeira and frequent in shady places, along levadas, tracks and banks, often under* Acacia; *450-c.1100 m.* **M**

LXXXVIII. ASCLEPIADACEAE
M.J. Short

Perennial herbs or shrubs, sometimes climbing, often with milky sap. Leaves usually opposite or whorled, simple, entire; stipules minute or absent. Inflorescence cymose. Flowers 5-merous, hermaphrodite, regular. Calyx deeply lobed. Corolla rotate or campanulate. Corona of 5(10) segments present. Stamens 5; filaments short; anthers fused around the stigma and usually adnate to it; pollen often united in pollinia. Ovary of 2 free carpels; styles 2, united at the stigma. Fruit a pair of follicles (often only one fully developing). Seeds numerous, usually flattened, with a tuft of long silky hairs at one end.

1. Corolla-lobes strongly deflexed at anthesis; herbs or subshrubs with straight stems . . . . **3. Asclepias**
- Corolla-lobes spreading or erect at anthesis; shrubs with young stems twining . . . 2

2. Leaves glabrous . . . . **1. Periploca**
- Leaves densely white-tomentose beneath . . . . **2. Araujia**

1. Periploca L.

Shrubs with erect or climbing stems. Leaves opposite. Flowers fragrant, in terminal or axillary cymes. Corolla-tube very short; lobes linear, spreading at anthesis. Corona segments 5, free, slender, abaxially awned. Anthers free, dehiscing by longitudinal slits; pollen free, in tetrads. Follicles broader towards the base, not spiny.

Literature: K. Browicz, The genus *Periploca* L. A monograph, in *Arboretum korn.* **11**: 5-104 (1966).

1. P. laevigata Aiton, *Hort. kew.* **1**: 301 (1789).
Shrub up to 2 m; stems erect, the young shoots twining, minutely puberulent. Leaves 2-5 × (0.5-)0.8-2.5 cm, elliptic or obovate, obtuse or acute, often mucronulate, glabrous, coriaceous; petioles up to 5 mm. Inflorescence puberulent. Calyx 2.5-3 mm; lobes ovate, obtuse, margin membranous. Corolla-lobes 4-7 mm, linear, obtuse and notched at the apex, glabrous, brown above with a white spot in the centre, greenish yellow below. Corona awns violet, pubescent.

Follicles (5-)7-11 × 0.7-1.1 cm, ± horizontal, finely ribbed when mature.
Fl. I - XII. *Cliffs and dry rocky slopes on Selvagem Grande. Rare.* ■ **S** [*Canaries*]

2. Araujia Brot.

Shrubs with twining stems. Leaves opposite, petiolate. Flowers in few-flowered axillary cymes. Corolla campanulate; tube swollen at base, exceeding the lobes; lobes ± erect at anthesis. Corona-segments 5, free, fleshy. Anthers fused to the stigma, dehiscing apically; pollen in pollinia. Follicles ovoid, sometimes inflated, not spiny.

1. A. sericofera Brot. in *Trans. Linn. Soc. London* **12**: 62 (1818). [*Árvore de seda, Seda*]
Stems up to 10 m, very thin and flexuous when young. Leaves 3.5-8(-10) × 1.3-3.5(-4.5) cm, triangular-oblong, mostly acute, truncate to shallowly cordate at base, sparsely pubescent above, densely white-tomentose beneath; petioles 1-3 cm. Cymes 1-4-flowered. Calyx-lobes 8-14 mm, ovate, acute. Corolla dull whitish pink, puberulous without, darker striped within; tube 11-13 mm; lobes 5-7(-10) mm, ovate, obtuse. Corona-segments hooded, white. Follicles *c*. 10 × 6 cm, patent or subpatent, irregularly ribbed, spongy.
Fl. VII - VIII. *Rare in Madeira; recorded in the nineteenth century as 'quite naturalized' near Machico. Its present distribution and frequency are unknown. Native to S. America.* **M**

3. Asclepias L.
[*Gomphocarpus* R. Br.]

Herbs or subshrubs. Leaves opposite or sometimes whorled. Flowers in axillary, umbelliform cymes. Corolla deeply divided; lobes strongly reflexed at anthesis. Corona usually conspicuous, with 5 free, erect or spreading segments. Anthers fused to the stigma, dehiscing apically; pollen in pollinia. Fruit usually a single follicle, spiny or smooth.

1. Corolla-lobes white; follicles spiny . . . . **1. fruticosa**
- Corolla-lobes crimson; follicles smooth . . . . **2. curassavica**

1. A. fruticosa L., *Sp. pl.* **1**: 216 (1753). [*Árvore de seda, Planta da seda*]
Shrubby perennial up to 1 m or more high; stems erect, slender, simple or sparingly branched, puberulent, leafy. Leaves shortly petiolate, 5-12 × 0.4-1.1 cm, linear-lanceolate, acuminate, margins narrowly revolute, subglabrous. Peduncles, pedicels and calyx white-puberulent. Calyx-lobes 4-7 mm, linear-lanceolate, acuminate. Corolla-lobes 6-7 mm, white, margins tomentulose. Corona-segments laterally compressed, thickened dorsally, pale greenish, the apex with 2 short falcate appendages. Follicles usually solitary, 4-6 × 2-3 cm, inflated, softly spiny, pale green, ± erect on deflexed pedicels. Seeds unwinged.
Fl. all year. *Naturalized along roadsides, on grassy slopes, and in waste places, often growing with* Opuntia; *found up to 1000 m but more often at lower levels. Common in the south of Madeira, particularly along the coast between Funchal and Caniçal; much less frequent in Porto Santo. Native to Africa.* **MP**

2. A. curassavica L., *Sp. pl.* **1**: 215 (1753).
Annual herbs; stems up to 80 cm, erect, woody at the base, sparsely pubescent. Leaves shortly petiolate, 5-12 × 1-3 cm, linear-lanceolate, acuminate, bright green, glabrescent. Peduncles 3-6 cm; pedicels 1-2 cm. Calyx-lobes 2-4 mm, linear-lanceolate, acute. Corolla-lobes 6-7 mm, crimson, sparsely pubescent. Corona-segments boat-shaped, bright orange, with an internal, incurved, horn-like appendage. Follicles usually solitary, 6-10 × 2-3 cm, narrowly ovoid-fusiform, smooth, erect on erect pedicels. Seeds broadly winged.
Fl. XI - III. *Cultivated in gardens as an ornamental and occasionally becoming naturalized in Madeira and Porto Santo. Native to S. America, but now a pantropical weed.* **MP**

LXXXIX. RUBIACEAE
N.J. Turland

Annual or perennial herbs or shrubs, often with small or minute prickles. Leaves opposite, simple, entire; stipules usually leaf-like and usually in whorls of 3-10 (each whorl actually of opposite leaves and a variable number of leaf-like stipules). Inflorescence composed of 1- to many-flowered axillary and terminal dichasia or capitula. Flowers regular; corolla funnel- or cup-shaped or flat, (3-)4- or 5-lobed. Fruit dry, or fleshy and berry-like.

For convenience, leaves and stipules which together form a whorl are all referred to as 'leaves' in this account. Features of the leaves refer to those occurring at the middle part of the stem.

Coffea arabica L. [*Cafeeiro*] is an evergreen shrub with opposite leathery glossy dark green leaves, white fragrant flowers and berry-like fruits which ripen through red almost to black. Native to Arabia and tropical Africa, it has been cultivated for coffee in coastal areas of southern Madeira but is now rarely seen.

1. Evergreen sub-shrub; leaves in whorls of 3, glabrous, not prickly ...... **1. Phyllis**
- Annual or perennial herbs, or scrambling or sub-shrubby perennials; leaves in whorls of 3-10, bearing at least a few prickles or hairs, or both 2

2. Scrambling or sub-shrubby perennials; corolla usually 5-lobed; fruit fleshy, berry-like . ... **4. Rubia**
- Annual or perennial herbs, sometimes scrambling if annual; corolla usually 4-lobed; fruit dry 3

3. Annual herb; flowers lilac to pink, borne in small heads backed by an involucre of connate leaves .... **2. Sherardia**
- Annual or perennial herbs; flowers whitish, greenish or yellowish, in 1- to many-flowered dichasia, not backed by an involucre of connate leaves .... **3. Galium**

1. Phyllis L.

Glabrous evergreen sub-shrubs, not prickly. Stems rounded, smooth. Leaves in whorls of 3, with small ± linear stipules in between, petiolate or with petiole indistinct. Inflorescence of axillary and terminal compound panicle-like dichasia, each many-flowered and bracteate. Calyx ± absent. Corolla (4-)5-lobed, the lobes spreading to strongly recurved, white to green. Fruit dry, of 2 mutually apressed mericarps, glabrous except on inner faces.

1. P. nobla L., *Sp. pl.* 1: 232 (1753). [*Cabreira, Seisim, Seisinho*]
Plant to *c*. 1 m. Leaves 2-14 × 0.5-3 cm (including petiole to 2 cm), narrowly elliptic, rarely almost linear, acute, mid-green, glossy; margin flat or rarely revolute. Inflorescence 3-30 × 2.5-20 cm, pyramidal or ovoid, often narrowly so or shortly cylindric. Corolla 2.5-4.5 mm across, but appearing narrower when lobes are strongly recurved; lobes narrowly elliptic, acute. Fruits 2.5-4 mm, compressed-obovoid; outer face of mericarps convex, longitudinally veined, dark brown when ripe, glabrous; inner face slightly concave, paler, appressed white-hairy. $2n = 22$. **Plate 34**
Fl. X, III - VII. *Cliffs, rocky banks and levada walls from sea-level to 1800 m; widespread in Madeira but rarer in Porto Santo where it occurs on the northern coast near Fonte d'Areia and on Pico do Facho and Ilhéu de Baixo; also on Deserta Grande and the eastern side of Bugío.*
■ **MPD** [*Canaries*]

2. Sherardia L.

Annual herbs. Stems 4-angled, usually scabrid to hispid, sometimes ± smooth. Leaves in whorls of 4-6. Inflorescence of axillary and terminal pedunculate capitula, each backed by an involucre of several connate leaves. Calyx 4- to 6-toothed, persistent in fruit. Corolla funnel-shaped with a long tube, 4-lobed, lilac to pink. Fruit dry, scabrid.

1. S. arvensis L., *Sp. pl.* **1**: 102 (1753).
Stems to 30 cm, slender, procumbent to ascending. Lower leaves obovate, mucronate, withering early; upper leaves 6-15 × 2-4.5 mm, oblanceolate to narrowly elliptic, mucronate, mid-green, sparsely hispid. Heads 4- to 10-flowered. Corolla 4-6 × 3-4 mm. Fruit 2-7 mm.
Fl. III - VII. *Rocky and grassy places, walls, roadsides, paths, fields, vineyards and waste areas in the south-eastern lowlands of Madeira and in the Caldeirão Verde area; also on Pico de Ana Ferreira in Porto Santo; not recorded from Selvagem Grande for over a century.* **MPS**

3. Galium L.

Annual or perennial herbs. Stems 4-angled, often with tiny downward-pointing prickles. Leaves in whorls of 4-10. Inflorescence of axillary and terminal dichasia, each 1- to many-flowered, often bracteate, at least below. Flowers (3-)4(-5)-merous; calyx absent or of short teeth; corolla usually flat, coloured whitish, yellowish, greenish or reddish. Fruit dry, of 2 mericarps, sometimes only 1 developing, glabrous, finely muricate, verrucose or covered with hairs or bristles.

1. Perennials with ± woody persistent stock 2
- Annuals without persistent stock 3

2. Leaves in whorls of 4, ovate or elliptic, with 3 parallel veins from base . **1. scabrum**
- Leaves in whorls of 6-8(-10), oblanceolate to linear, with 1 vein from base. . . . .
. . . . **2. productum**

3. Fruit more than 2 mm; leaves often more than 1.5 mm wide 4
- Fruit less than 2 mm; leaves often less than 1.5 mm wide 6

4. Leaf-margin with forward-pointing prickles or scabrid; fruit prominently verrucose . . .
. . . . **5. verrucosum**
- Leaf-margin with backward-pointing prickles or scabrid; fruit bristly or finely muricate
. . . . 5

5. Peduncles and pedicels convergent and deflexed after flowering; fruit finely muricate; leaves glabrous above . . . . **4. tricornutum**
- Peduncles and pedicels spreading and straight after flowering, sometimes bent just beneath the fruit; fruit densely covered with hooked bristles; leaves sparsely hairy above . . . .
. . . . **3. aparine**

6. Flowers usually in many-flowered dichasia; fruit glabrous . . . . **6. parisiense**
- Flowers 1-2(-4) per whorl, solitary or paired, or on 2-flowered peduncles; fruit bristly
. . . . 7

7. Mericarps cylindrical, curved and separated from each other, sparsely to densely bristly, densely so at apex . . . . **8. murale**
- Mericarps ± obovoid, not curved or separated from eachother, densely and uniformly bristly
. . . . **7. geminiflorum**

1. G. scabrum L., *Sp. pl.* **1**: 108 (1753).
G. ellipticum Willd. ex Hornem., var. *lucidum* Lowe, var. *villosum* (Webb & Berthel.) Lowe
Perennial with ± woody persistent stock. Stems to 50 cm, ascending to erect, patent-hispid. Leaves in whorls of 4, 12-40 × 6-17 mm, ovate or elliptic, with 3 parallel veins from base, hispid along veins. Inflorescence 2.5-9 cm, ± ovoid, many-flowered, much-branched. Corolla 3-4 mm across, yellowish or greenish-yellow; lobes oblong-ovate to oblong-lanceolate, cuspidate. Mericarps 1.5-2 mm, ± reniform, sparsely covered with short hooked bristles.
Fl. VI - VIII. *Cliffs, rocky banks, ledges and moist shady places, often in ravines, from 600-1700 m in the central mountains of Madeira, extending westwards to Rabaçal.* **M**

2. G. productum Lowe in *Trans. Camb. phil. Soc.* **4**: 29 (1831).
G. productum var. *latifolium* Menezes
Usually glabrous perennial with ± woody persistent stock. Stems to 1 m, sprawling or forming dense clumps. Leaves in whorls of 6-8(-10), 3-20 × 0.75-2.5 mm, oblanceolate to linear, with 1 vein from base. Inflorescence 1-55 cm, lax, usually branched and many-flowered; branches often cylindrical and elongate. Corolla 2-4 mm across, white or yellowish; lobes oblong-ovate to oblong-lanceolate, cuspidate. Mericarps 1-1.5 mm, ± ovoid, glabrous. 2n = 22. **Plate 34**
Fl. (I-)IV - XI. *An endemic of rocky places, ravines, walls, field-margins, hedges and roadsides from 150-1850 m in central Madeira, from Pico Ruivo to the southern coast.* ● **M**

Var. *latifolium* differs from the type only in having stems thicker, woody at base, and leaves larger and shorter; it is not separated here.

G. productum is the Madeiran representative of Series *Erecta* Probedimova, which comprises many closely related species and is widespread in Europe.

3. G. aparine L., *Sp. pl.* **1**: 108 (1753). [*Raspa-lingua*]
Annual. Stems to 150 cm, scrambling, with downward-pointing prickles, hairy at nodes. Leaves in whorls of 6-9, 10-45 × 2-8 mm, usually narrowly oblanceolate, cuspidate, sparsely hairy above, with backward-pointing prickles along margin and midrib beneath. Inflorescence to 90 cm, cylindrical, lax; dichasia 1- to 9-flowered, longer than leaves; peduncles and pedicels spreading and straight after flowering, or sometimes bent just under fruit. Corolla 2-3 mm across, whitish; lobes ovate, acute. Mericarps 3-4 mm, subglobose, densely covered with hooked bristles.
Fl. XI, III - VI. *Woodland, hedges, stream-banks, rocky and grassy places, walls, roadsides, waste areas and as a weed of cultivation. Along the southern coast of Madeira, from Ribeira Brava to Caniço, and at São Vicente; also in Porto Santo and Selvagem Grande.* **MPS**

4 G. tricornutum Dandy in *Watsonia* **4**: 47 (1957).
G. tricorne Stokes, pro parte
Like **3** but stems to 30 cm, without hairs at nodes; leaves 8-15 × 1-2 mm, glabrous above; inflorescence narrowly cylindrical; dichasia 2- to 4-flowered, scarcely longer than leaves; pedicels strongly curved downwards after flowering; corolla 1.5-2 mm across; lobes long, acute; mericarps finely muricate.
Fl. II - IV. *A weed of cornfields at the base of Pico do Castelo in Porto Santo but not recorded since the 19th century.* **P**

5. G. verrucosum Huds. in *Phil. Trans. R. Soc.* **56**: 251 (1767). [*Raspa-lingua*]
G. saccharatum All.; *G. valantia* Weber
Annual. Stems to 40 cm, sprawling to erect, with downward-pointing prickles. Leaves in whorls of 5-6(-7), 5-20 × 1.5-3 mm, narrowly oblanceolate, cuspidate, glabrous above, with forward-pointing prickles along margin and midrib beneath. Inflorescence 15-20 cm, narrowly cylindrical, lax; dichasia mostly 3-flowered, shorter than leaves; pedicels curved downwards after flowering. Corolla 2-3 mm across, greenish-white to white; lobes ovate, acute. Mericarps

2.5-4.5 mm, subglobose, prominently verrucose.

Fl. ± all year but mainly XII - VI. *Rocks, walls, stream-banks, levadas, roadsides, cultivated ground and waste areas. Occurring at Paúl do Mar and Seixal in western Madeira, and in the Funchal region in the south-east, but probably more widespread than this in the lowlands; also in Porto Santo.* **MP**

6. G. parisiense L., *Sp. pl.* **1**: 108 (1753).
G. parisiense var. *leiocarpum* Tausch
Annual. Stems 3-35 cm, procumbent to ascending, somewhat sprawling or scrambling, usually much-branched, often rather matted, with minute downward-pointing prickles. Leaves in whorls of 5-8, 2-11 × 0.5-2 mm, linear to narrowly oblanceolate, cuspidate, glabrous, with minute forward-pointing prickles along margin and midrib beneath. Inflorescence 2-30 cm, lax, usually much-branched and then ± narrowly ovoid or pyramidal; dichasia usually much longer than leaves and many-flowered, rarely short and few-flowered (when inflorescence ± cylindrialc). Corolla minute, *c.* 1 mm across, greenish or greenish-white. Mericarps 0.5-1 mm, ± obovoid, glabrous.

Fl. IV - VIII. *Rocky places, mountain pastures, roadsides and margins of terraces from near sea-level to 1600 m; widespread but rather sporadic in Madeira; also in Porto Santo.* **MP**

Plants from Madeira and Porto Santo have glabrous mericarps, whereas some individuals from the Canaries with few-flowered dichasia have fruits sparsely covered with short bristles and may resemble **7**. The latter species differs in being smaller, with more slender stems, and in having no more than 2 flowers per peduncle and mericarps densely covered with rather long bristles.

7. G. geminiflorum Lowe in *Trans. Camb. phil. Soc.* **6**: 541 (1838).
Annual. Stems 2-20 cm, prostrate to erect, very slender, much branched from base, unbranched above, glabrous, sometimes with a few minute downward-pointing prickles. Leaves in whorls of 4-6, 1.5-6 × 0.5-2 mm, narrowly to broadly elliptic or suborbicular, often narrowed into a petiole, mucronate, ± glabrous, with a few minute forward-pointing prickles along margin and midrib beneath. Inflorescence narrowly cylindrical, lax; flowers 1 or 2(-4) per whorl, solitary or paired, or on 2-flowered peduncles. Corolla minute, white. Mericarps *c.* 1 mm, ± obovoid, not curved or separated from each other, densely and uniformly covered with rather long whitish hooked bristles. **Plate 34**

Fl. III - VI. *Rocky places, old walls, mountain pastures and* Cupressus *woodland in Porto Santo, known from Pico do Facho, Pico do Castelo and Ilhéu de Baixo; also recorded from the Desertas and the Salvage islands but without further data.* ■ **DPS** [*Canaries*]

G. geminiflorum appears to be indistinguishable from *G. minutulum* Jord., from SW Europe. If the two taxa were to be treated as conspecific, the name *G. geminiflorum* would have priority.

8. G. murale (L.) All., *Fl. pedem.* **1**: 8 (1785).
Aspera muralis (L.) Lowe
Annual. Stems 3-20 cm, procumbent or ascending from much-branched base, usually unbranched above, very slender, smooth or with minute downward-pointing prickles, patent-hairy above. Leaves in whorls of 4-6, 1-8 × 0.5-1.5 mm, oblanceolate to narrowly so, mucronate, hairy to glabrescent, with minute forward-pointing prickles along margin. Inflorescence narrowly cylindrical, lax; flowers 1-2(-4) per whorl, solitary or paired, or on 2-flowered peduncles. Corolla minute, *c.* 1 mm across, yellowish; lobes ovate, acute. Mericarps 1-1.5 mm, cylindrical, curved and separated from each other, sparsely to densely covered with hooked bristles, densely so at apex.

Fl. IV - VI. *Recorded from a waste place at Praia Formosa west of Funchal in Madeira; also in grassy and open sunny places on the rocky peaks of Porto Santo, and on Deserta Grande and Bugío.* **MDP**

4. Rubia L.

Scrambling or sub-shrubby perennials. Stems 4-angled to rounded, smooth, scabrid or with a few downward-pointing prickles. Leaves in whorls of 3-9, with backward-pointing prickles along margin and midrib beneath. Inflorescence of axillary, often compound and panicle-like dichasia, each few- to many-flowered and bracteate. Calyx minute or absent. Corolla usually 5-lobed, the lobes spreading, green. Fruit fleshy, usually with only 1 berry-like mericarp developing.

Literature: A. Cardona & E. Sierra-Rafols. Contribución al estudio del género *Rubia* I. Taxones mediterraneo-occidentales y macaronésicos. In *An. Inst. bot. A.J. Cavanilles* 37: 557 - 575 (1981).

1. Stems to 3 m, scrambling; leaves in whorls of 7-9, sessile, with margin and midrib beneath closely prickly . . . . **1. agostinhoi**
- Stems to 50 cm, not scrambling; leaves in whorls of 3-6(-7), petiolate, with margin and midrib beneath somewhat remotely prickly . . . . **2. fruticosa**

1. R. agostinhoi Dans. & P. Silva in *Agronomia lusit.* 36: 62 (1974). [*Ruivinho*]
R. angustifolia sensu auct. mad., non L. (1767); *R. peregrina* subsp. *agostinhoi* (Dans. & P. Silva) Valdés Berm. & G. López, var. *angustifolia* sensu Webb & Berthel.
Stems to 3 m, branched, scrambling, 4-angled, glabrous, scabrid or smooth. Leaves in whorls of 7-9, sessile, 10-80 × 2-8 mm, mostly very narrowly elliptic to linear or oblanceolate, acute or mucronate, coriaceous, dark green; margin closely prickly (as is midrib beneath), slightly revolute; leaves on lateral shoots shorter and broader in relation to length than those on main stem. Dichasia 1-6 cm, usually many-flowered. Corolla 3-5 mm across, green; lobes oblong-lanceolate, acuminate. Fruit 5-10 mm, globose, ripening glossy black. 2n=44.
Plate 34
Fl. IV - VIII. *Climbing through shrubs and trees in laurisilva and* Erica *woodland, also on cliffs. Widespread in Madeira, especially in the ravines of the interior and north of the island, from 350-1800 m.* **M**

R. agostinhoi otherwise occurs only in the Azores, Canaries and S. Spain.

2. R. fruticosa Aiton, *Hort. kew.* 1: 147 (1789).
R. gratiosa Menezes
Stems to 50 cm, branched, not scrambling, furrowed, glabrous. Leaves petiolate, ± coriaceous, pale to dark green; margin somewhat remotely prickly (as is midrib beneath), flat to slightly revolute. Corolla 3.5-5 mm across; lobes oblong-lanceolate, obtuse to acute. Fruit not seen.
Plate 34
Fl. II - III. *Cliffs and rocky places near the sea. In Madeira known only from Ponta do Tristão west of Porto do Moniz, and at Caniçal; otherwise occurring at Ilhéu dos Garajaus on the eastern coast of Deserta Grande, and from 20-50 m on the northern sea-exposed cliffs of Pico da Atalaia on Selvagem Grande.* ■ **MDS** [*Canaries*].

A variable species. Madeiran plants appear to be distinct from those on Selvagem Grande and separate descriptions are, therefore, provided here:

Madeira: Stem ± angled, flexuous, smooth or with a few downward-pointing prickles. Leaves in whorls of 4-6(-7), 15-40 × 3-8 mm (including petiole 3-9 mm), narrowly elliptic, obtuse to acute or shortly cuspidate; dichasia up to 7 cm, few- to many-flowered.

Selvagem Grande: Stem rounded, straight and moderately stout, smooth. Leaves in whorls of 3, 20-50 × 10-20 mm (including petiole 5-15 mm), elliptic or broadly so, obtuse; dichasia *c.* 1 cm, few-flowered.

Despite these differences, all plants from the Madeiran archipelago have been referred to subsp. **fruticosa**, which otherwise occurs in the Canaries. The other two subspecies, subsp. *melanocarpa* (Bornm.) Bramwell and subsp. *periclymenon* (Schenck) Sunding are said to be endemic to the Canaries. Var. *pendula* Pit., which is synonymous with subsp. *periclymenon*, was recorded from Selvagem Grande in 1955, but presumably in error, since the record has recently been cited as belonging to subsp. *fruticosa*.

XC. CONVOLVULACEAE
N.J. Turland

Annual or perennial herbs. Stems often twining. Leaves alternate, simple, sometimes minute and scale-like, without stipules. Flowers regular, usually 5-merous, borne in axillary inflorescences, sometimes solitary; sepals usually free; corolla funnel- or bell-shaped, lobed or angled; stamens alternating with corolla-lobes; ovary 1- to 4-chambered, superior; ovules 1 or 2 in each chamber; style terminal. Fruit a capsule.

1. Plant parasitic, lacking green parts; stems thread-like, appearing leafless ... **1. Cuscuta**
- Plant not parasitic, green; stems thicker, obviously leafy 2

2. Stems creeping, rooting at nodes; corolla up to 5 mm, lobed to about halfway .... **2. Dichondra**
- Stems trailing or twining, not rooting at nodes; corolla more than 5 mm, angled but not obviously lobed 3

3. Bracteoles ovate or suborbicular, leaf-like, partly or wholly overlapping sepals .... **3. Calystegia**
- Bracteoles linear to linear-lanceolate, minute or leaf-like, rarely overlapping sepals 4

4. Stigma with 1-3 globose lobes .... **5. Ipomoea**
- Stigma with 2 filiform to cylindric-clavate lobes .... **4. Convolvulus**

1. Cuscuta L.

Herbaceous, usually annual parasites lacking green parts. Stems thread-like, much-branched, twining, smothering other plants and invading their vascular tissues with specialized roots *(haustoria)*. Leaves reduced to minute scales. Inflorescences cymose, globose, dense. Flowers 4- or 5-merous, small, white, yellowish, pink or reddish; corolla bell-shaped with spreading lobes; styles free or united; stigmas elongate.

Flowers are essential for identification.

1. Flowers 3-3.5 mm across, in clusters 7-10 mm across; calyx-lobes membranous, neither swollen nor with a fleshy apical appendage .... **1. epithymum**
- Flowers 2-2.5 mm across, in clusters 5-8 mm across; calyx-lobes swollen, or with a fleshy apical appendage 2

2. Calyx-lobes with a long fleshy apical appendage .... **3. approximata**
- Calyx-lobes without such an appendage, swollen, nearly semi-circular in cross-section . ... **2. planiflora**

1. C. epithymum (L.) L., *Syst. veg.* ed.13: 140 (1774). [*Linheio*]
Stems often reddish or purplish. Flower-clusters 7-10 mm across, each with a bract at its base. Flowers 3-3.5 mm across, 5-merous, mostly sessile, sometimes shortly pedicellate; calyx shorter than corolla-tube; lobes triangular, acute, sometimes lanceolate and acuminate, membranous;

corolla-lobes triangular, acute, rarely acuminate, usually shorter than the bell-shaped tube.
Fl. I, probably also at other times. *Parasitizing* Calendula maderensis *and* Lotus *species on the Ponta de São Lourenço in Madeira, and on a wide range of annual herbs on Selvagem Grande; also recorded as occurring in Porto Santo.* **MPS**

The above description refers to subsp. **epithymum**, which occurs throughout most of Europe, and to which plants from the Madeiran and Salvage archipelagos belong. Subsp. *kotschyi* (Des Moul.) Arcang. occurs in S. Europe, mainly parasitizing dwarf shrubs. It has flowers 2.5 mm across, borne in clusters 5-6 mm across, and has fleshy calyces as long as or slightly shorter than corolla-tubes, with lobes ovate and keeled.

2. C. planiflora Ten., *Fl. napol.* **3**: 250 (1824-1829).
Like **1** but flower-clusters and flowers smaller; calyx as long as corolla-tube; lobes longer than tube, swollen, nearly semi-circular in cross-section, oblong or more rarely ovate; corolla-lobes about as long as tube, often hooded at apex.
Fl. ? *Recorded from Madeira but without further data.* **M**

3. C. approximata Bab. in *Ann. Mag. nat. Hist.* **13**: 253 (1844).
Like **1** but flowers 2-2.5 mm across, in clusters 5-8 mm across; calyx about as long as corolla-tube; lobes short, broader than long, overlapping at base, with rounded sides and a fleshy acute whitish apical appendage longer than the lobe itself; corolla becoming globose or urn-shaped around capsule; lobes triangular-ovate, about as long as tube.
Fl. III - IV. *Parasitizing* Hypericum glandulosum, Micromeria varia *and* Plantago arborescens *on the peaks and higher slopes of Porto Santo.* **P**

The above description refers to subsp. **episonchum** (Webb & Berthel.) Feinbrun in *Israel J. Bot.* **19**: 28 (1970), which otherwise occurs in Spain, Morocco and doubtfully Portugal, and to which plants from Porto Santo belong. Subsp. *approximata* occurs in S. Europe and has flowers 3-4 mm across and calyx-lobes with short obtuse appendages.

The identity of plants recorded from Madeira, at Cabo Girão (as *C. approximata*), and at Serra do Estreito (as *C. calycina* Webb & Berthel.), are uncertain and are omitted from the distributions given here.

2. Dichondra J.R. & G. Forst.

Creeping appressed-pubescent perennial herb, rooting at nodes. Leaves petiolate, entire or lobed at base. Flowers solitary; corolla to 5 mm, divided to about halfway into 5 lobes; styles 2; stigmas capitate.

1. D. micrantha Urb., *Symb. antill.* **9**: 243 (1924).
D. repens sensu auct. mad., non J.R. & G. Forst. (1776).
Stems to 15 cm or more, slender. Leaves 1.5-19 × 2-23 mm, orbicular to reniform; petiole 2-35 mm. Peduncle 1-3 mm. Calyx densely silky-hairy on outer surface, divided to more than halfway; lobes 1.5-1.7 mm, oblong-ovate to ovate. Corolla greenish; lobes 1-1.5 mm, triangular-lanceolate. Capsule 2 mm, globose, usually sparsely hairy, pushed into ground by peduncle.
Fl. III - VI. *Introduced; growing among cobblestones and in cracks of pavements in the lowlands of south-western Madeira, as well as in the Funchal region and at Ribeiro Frio; probably increasing. Native to E. Asia.* **M**

3. Calystegia R. Br.

Glabrous perennial rhizomatous herbs with white latex. Stems trailing or climbing. Leaves petiolate, variously lobed at base. Flowers solitary; bracteoles 2, leaf-like, ovate or suborbicular, partly or wholly overlapping sepals; corolla at least 4 cm, 5-angled, funnel-shaped; stigma with 2 swollen elongate lobes.

1. Stems short, trailing, not or weakly twining; leaves reniform; corolla pink . . . . **1. soldanella**
- Stems long, strongly twining; leaves triangular-ovate, cordate to sagittate at base; corolla white, only rarely pale pink . . . . **2. sepium**

1. C. soldanella (L.) Roem. & Schult., *Syst. veg.* **4**: 184 (1819).   [*Corriola da Praia*]
Stems 2-5 cm (the part above ground-level), trailing, not twining. Leaves 1-4.5 × 1-5 cm, reniform, somewhat fleshy. Bracteoles 11-14 mm, suborbicular. Sepals 12-15 × 6-10 mm, broadly oblong-ovate. Corolla 4-4.5 cm, pink. Stamens *c*. 20 mm; anthers *c*. 4 mm.
Fl. IV - VI. *Frequent on maritime sands along the southern coast of Porto Santo.* **P**

2. C. sepium (L.) R. Br., *Prodr.*: 483 (1810).
Stems long, strongly twining. Leaves 4-8 × 3-5.5 cm, triangular-ovate, cordate to sagittate at base, not fleshy; sinus acute. Bracteoles 18-25 mm, ovate. Sepals *c*. 12 × 2.5 mm, lanceolate, acute. Corolla *c*. 7 cm, white or rarely pale pink. Stamens *c*. 20 mm; anthers *c*. 6 mm.
Fl. I, VI - VII. *Growing among rocks and over walls; very rare, known only from a few widely separated localities in Madeira.* **M**

The above description refers to subsp. **sepium**, which occurs throughout most of Europe, and to which Madeiran plants belong. Subsp. *americana* (Sims) Brummitt occurs in the Azores and along the Atlantic coasts of America, and subsp. *roseata* Brummitt along the coasts of W. Europe; both taxa have bright pink flowers and are usually pubescent.

4. Convolvulus L.

Annual or perennial herbs or shrubs, sometimes with latex. Stems trailing, sprawling, ascending or twining. Leaves petiolate or sessile, entire or lobed. Peduncles with 2, rarely more, minute to leaf-like bracteoles. Corolla funnel-shaped, 5-angled, glabrous except for 5 usually pubescent and often differently coloured lines. Stigma with 2 filiform to cylindric-clavate lobes. Pollen-grains not spinose.

Literature: F. Sa'ad. *The Convolvulus species of the Canary Isles, the Mediteranean region, and the Near and Middle East.* In *Meded. bot. Mus. Rijks -Univ. Utrecht* **281**: 1 - 288 (1967).

1. Leaves sessile, gradually tapered towards base . . . . **1. tricolor**
- Leaves abruptly narrowed into a distinct petiole 2

2. Leaves crenate to deeply cut into several linear lobes . . . . **2. althaeoides**
- Leaves entire, or with hastate to sagittate base 3

3. Plant perennial; stems to several metres, branching throughout, becoming woody; leaves to 11 × 6 cm; peduncles usually 3- to 6-flowered . . . . **3. massonii**
- Plant annual or perennial; stems to 1 m, branched only at ground-level, herbaceous; leaves to 5 × 2.5 cm; peduncles 1(-2)-flowered 4

4. Plant perennial; leaves oblong to linear, hastate to sagittate at base; corolla 12-22 mm, white to pink . . . . **4. arvensis**
- Plant usually annual; leaves ovate to lanceolate, truncate to cordate at base; corolla 8-10 mm, blue . . . . **5. siculus**

1. C. tricolor L., *Sp. pl.* **1**: 158 (1753).
Pubescent annual or short-lived perennial. Stems moderately stout, branching above, sprawling or ascending to 35 cm, not twining, herbaceous. Leaves sessile, 10-28 × 3-7 mm, oblanceolate, gradually tapered towards base, rounded at apex. Peduncles 1-flowered. Bracteoles minute, linear. Sepals with differently shaped distal and proximal portions. Corolla *c.* 2 cm, usually with concentric bands of blue, white and yellow.
Fl. IV - VI. *Waste ground and vineyards in and around Funchal; rare.* **M**

2. C. althaeoides L., *Sp. pl.* **1**: 156 (1753).   [*Corriola* (M), *Corriola brava* (P)]
C. althaeoides var. *virescens* Lowe
Pubescent perennial. Stems slender, branching at or below ground-level, trailing or twining to 70 cm or more, herbaceous. Leaves petiolate, 1-8 × 1-6.5 cm, triangular-ovate, strongly cordate at base, crenate to deeply cut into several linear lobes. Peduncles 1- to 3-flowered. Bracteoles linear, to 13 mm. Sepals 9-13 × 4-7 mm, broadly elliptic to obovate, apiculate to rounded. Corolla 3-4.5 cm, deep pinkish-purple.
Fl. III - XII. *Hedges, banks, terraces, roadsides and rocky ground, to 325 m; frequent, mainly in south-eastern Madeira and in Porto Santo.* **MP**

3. C. massonii F. Dietr., *Nachtr vallst. Lex. Gästn.* **2**: 377 (1816).   [*Corriola*]
C. canariensis var. *massonii* (F. Dietr.) Sa'ad; *C. massonii* var. *uniflorus* Menezes
Glabrous or rarely sparsely pubescent perennial. Stems slender when young, branching throughout, trailing or twining to 4 m or more, becoming woody and thickened with age. Leaves petiolate, 4-11 × 1.5-6 cm, ovate to lanceolate, rounded to cordate at base, acute to acuminate at apex. Peduncles (1-)3- to 6-flowered, usually massed towards tips of stems giving appearance of terminal inflorescence. Bracteoles to 14 × 3 mm, linear-lanceolate. Sepals 9-15 × 4-7 mm, obovate to oblanceolate, apiculate. Corolla 2-2.5 cm, white, narrowly striped with pink. **Plate 35**
Fl. III - VIII. *Growing among rocks, hanging down cliffs or twining through shrubs, mainly in humid gullies and ravines from 400-1000 m in northern and central Madeira; sporadic and rather rare; also on Deserta Grande.* ■ **MD** [*Canaries*]

In the Canaries *C. massonii* is recorded from Tenerife, where it is sympatric with the similar and related *C. canariensis* L., which otherwise occurs on the islands of Gran Canaria, Hiérro, La Gomera and La Palma. The latter species differs in being densely pubescent, with larger leaf-like bracteoles and many-flowered peduncles. The two taxa would perhaps be better treated at subspecific rank.

4. C. arvensis L., *Sp. pl.* **1**: 153 (1753).   [*Corriola* (M), *Corriola mansa* (P)]
Glabrous or pubescent perennial. Stems slender, branched at ground-level, trailing or twining to 1 m, herbaceous. Leaves petiolate, 7-32 × 3-15 mm, oblong or linear, hastate to sagittate at base, rounded or shortly mucronate at apex. Peduncles usually 1-flowered. Bracteles minute, linear, 1-3.5 mm. Sepals 2-3.5 × 1.5-2.5 mm, oblong-ovate to suborbicular, obtuse to emarginate, often apiculate. Corolla 12-22 mm, white to pink.
Fl. III - X. *Fields, gardens, vineyards, waste ground, roadsides and maritime sands; widespread and common both in Madeira and Porto Santo, especially near the sea.* **MP**

Two varieties may be recognized in the Madeiran archipelago:

i) var. arvensis
Leaves oblong; basal lobes spreading or slightly backward-pointing.
Throughout the range of the species.

ii) var. linearifolius Choisy in DC., *Prodr.* **9**: 407 (1845).
Leaves linear; basal lobes strongly backward-pointing.

Gorgulho near Ponta da Cruz west of Funchal, and at Ponta Delgada on the northern coast of Madeira.

5. C. siculus L., *Sp. pl.* **1**: 156 (1753).
Sparsely pubescent annual or short-lived perennial. Stems slender, branched at ground-level, trailing, rarely twining, to 75 cm, herbaceous. Leaves petiolate, 10-50 × 5-25 mm, ovate to lanceolate, truncate to cordate at base, obtuse to acuminate at apex. Peduncles 1(-2)-flowered. Bracteoles to 16 × 3 mm, leaf-like, linear, carried just below and greatly overlapping sepals. Sepals elliptic, acute to acuminate, pubescent in apical half. Corolla 8-10 mm, blue.
Fl. II - V. *Grassy places, fields, vineyards, waste ground, roadsides and sea-cliffs. All along the southern coast of Madeira and on Pico de Ana Ferreira in Porto Santo.* **MP**

The above description refers to subsp. *siculus*, which occurs in the Mediterranean region, and to which plants from the Madeiran archipelago belong. Subsp. *elongatus* Batt. (subsp. *agrestis* auct.) occurs in the Canaries and Africa and has minute bracteoles carried well below the sepals, not or barely overlapping them.

5. Ipomoea L.

Annual or perennial herbs. Stems trailing, sprawling or twining. Leaves petiolate, entire or lobed. Peduncles usually with 2 minute bracteoles. Corolla at least 2.5 cm, funnel-shaped, 5-angled to 5-lobed, ± glabrous. Stigma with 1 to 3 globose lobes. Pollen-grains spinose.

Leaves may vary on the same plant; they are often entire towards the base of the stem but lobed towards apex. The lowest leaves are variable and should be disregarded for identification purposes.

In addition to **1** which is an important food crop, several species are cultivated for ornament in gardens, including **I. quamoclit** L. (*Quamoclit pennata* (Desr.) Bojer), from the tropics, which has pectinate-pinnatifid leaves and crimson flowers.

I. indica (Burm.) Merr., *Interpr. Herb. amboin.*: 445 (1917) (*I. acuminata* (Vahl) Roem. & Schult.; *Pharbitis learii* (Paxton) Lindl.) is usually perennial, with stems twining, becoming woody with age, leaves ovate or broadly so, entire to deeply 3-lobed, sparsely white-pubescent, sometimes densely so beneath, sepals acuminate, gradually tapering from near the densely pubescent base, and flowers variable in colour. It is native to tropical America and may be locally naturalized around Funchal. **M**

I. coccinea L., *Sp. pl.* **1**: 160 (1753) (*Quamoclit coccinea* (L.) Moench) is annual, with stems to 40 cm bearing broadly ovate, entire leaves and orange-red to scarlet flowers. It is native to tropical America and was formerly naturalized at Calheta in south-western Madeira and in batata fields in and around Funchal; it has not been recorded for many years and is probably now extinct on the island. **?M**

1. Plant without tubers; sepals 10-14 mm, not membranous at margin, hispid with swollen-based bristles . . . . **2. purpurea**
- Plant sometimes with large underground tubers; sepals 4.5-11 mm, narrowly membranous at margin, glabrous or sparsely pubescent 2

2. Leaves usually entire; peduncles 1- to 3-flowered; sepals 4.5-6 mm, ovate-lanceolate, shortly apiculate, glabrous; corolla yellow . . . . **3. ochracea**
- Leaves sometimes lobed; peduncles usually several-flowered; sepals 8-11 mm, oblong to elliptic or narrowly lanceolate, aristate, sometimes sparsely hairy; corolla white, or pink to violet . . . . **1. batatas**

1. I. batatas (L.) Poir. in Lam., *Encycl.* **6**: 14 (1804). [*Batateira*]
Batatas edulis Choisy, var. *cordifolia* Lowe, var. *digitata* Lowe
Plant perennating by means of large underground tubers. Stems moderately stout, sprawling or trailing to 1 m, not twining. Leaves 3-10 × 2.5-11 cm, broadly ovate, entire or lobed, truncate or cordate at base, acuminate at apex, glabrous, or very sparsely pubescent on veins; lobes (if present) triangular to lanceolate. Peduncles usually several-flowered. Sepals 8-11 × 2-5 mm, oblong to elliptic or narrowly lanceolate, aristate, glabrous or sparsely hairy; margin narrowly membranous. Corolla 3-4 cm, white, or pink to violet.
Fl. all year but mostly VII - X. *Introduced during the 17th century and at later dates. A major food crop, widely cultivated in fields and gardens in both Madeira and Porto Santo for its edible tubers (batata doce, sweet potato). Native to tropical America.* **MP**

Tubers vary considerably in shape, colour and consistency.

2. I. purpurea (L.) Roth, *Bot. Abh. Beobacht.*: 27 (1787).
Pharbitis purpurea (L.) Voigt
Plant usually annual. Stems slender, twining to 1.5(-3.5) m. Leaves 3-7.5 × 2.5-6.5 cm, broadly ovate, usually entire, cordate at base, acuminate at apex, very sparsely pubescent on both surfaces. Peduncles 1- to several-flowered. Sepals 10-14 mm, narrowly lanceolate, covered with conspicuously swollen-based bristles near base. Corolla 4-5 cm, pink, purple, blue or violet, rarely white, with differently coloured stripes.
Fl. all year. *Introduced; frequently naturalized on waste ground from Campanario eastwards to Caniço in Madeira; also in Porto Santo. Native to tropical America.* **MP**

3. I. ochracea (Lindl.) G. Don, *Gen. hist.* **4**: 270 (1837).
Like **2** but peduncles 1- to 3-flowered; sepals 4.5-6 mm, ovate-lanceolate, shortly apiculate, glabrous, narrowly membranous at margin; corolla 3.5-4 cm, yellow.
Fl. II - VI. *Introduced; locally naturalized in Madeira on waste ground and roadsides from Funchal westwards to Ponta da Cruz. Native to tropical Africa.* **M**

HYDROPHYLLACEAE
M.J. Short

Wigandia caracasana Kunth in Humb., Bonpl. & Kunth, *Nov. gen. sp.* **3**, ed. folio: 110 (1819) [*Vigândia*], an erect, densely pubescent, shrubby perennial up to 4 m high with large, broadly ovate, cordate leaves with coarsely crenate margins, and sessile flowers in scorpioid cymes, the corolla up to 2 cm long with 5 lilac lobes and a paler tube, followed by bilocular, oblong to conical capsules enclosed by the persistent calyx, is cultivated for ornament in Madeira and may occasionally become naturalized. Native to C. and S. America. **M**

XCI. BORAGINACEAE
M.J. Short

Herbs or shrubs, often hispid or scabrid with tubercle-based bristles. Leaves alternate, simple, exstipulate. Flowers regular or rarely somewhat irregular, generally in scorpioid cymes, uncoiling progressively as the flowers open. Calyx 5-lobed, often enlarging in fruit. Corolla often blue or white, 5-lobed, rotate, salverform, tubular, funnelform or campanulate; throat often closed by scales or hairs. Stamens 5, inserted on the corolla and alternate with the lobes. Ovary superior, (2-)4-locular; style simple, gynobasic, rarely terminal; stigma simple or bilobed. Fruit of (2)4, 1-seeded nutlets.

1. Shrubs . . . . **6. Echium**
- Annual to perennial herbs 2

2. Nutlets covered with bristles; corolla with conspicuous darker veins **2. Cynoglossum**
- Nutlets without bristles; corolla without darker veins 3

3. Corolla somewhat irregular, with unequal lobes . . . . **6. Echium**
- Corolla regular, with equal lobes 4

4. Anthers long-exserted, connivent and forming a cone; corolla 20 mm across **3. Borago**
- Anthers included or only very slightly exserted, not connivent; corolla less than 9 mm across 5

5. Flowers white; fruit sessile . . . . **1. Heliotropium**
- Flowers blue (occasionally white but then usually turning blue); fruit pedicellate . . . 6

6. Corolla-tube 7-10 mm long, with a tuft of white hairs closing throat; nutlets reticulate-ribbed . . . . **4. Anchusa**
- Corolla-tube very short, less than 4 mm long, with small, yellow scales closing throat; nutlets smooth, shiny . . . . **5. Myosotis**

1. Heliotropium L.

Annual or perennial herbs, sometimes woody below, or shrubs. Flowers white, yellowish, lilac or purple, mostly sessile, in mainly terminal, ebracteate, branched, scorpioid cymes. Calyx usually divided almost to base, persistent or deciduous. Corolla salverform; tube often pubescent, scales absent; limb shorter than tube, sometimes with a small tooth between each lobe. Stamens included. Style terminal, included; stigma large, entire or bilobed. Nutlets 4, sometimes cohering in pairs.

1. Annual herb with erect stems; leaves elliptic, ovate or obovate, flat, all distinctly petiolate . . . . **1. europaeum**
- Perennial herb with prostrate to ascending stems; leaves linear-lanceolate to oblanceolate, margins revolute, the upper subsessile . . . . **2. ramosissimum**

1. H. europaeum L., *Sp. pl.* 1: 130 (1753).
Annual, green or greyish, with short, appressed hairs; stems 3-55 cm, erect, branched. Leaves up to 5 × 2.5 cm, ovate-elliptic to obovate, subacute to obtuse, cuneate at base, margins flat, entire; petioles up to 3 cm. Inflorescences forked, very dense, elongating greatly and becoming lax in fruit, up to 16 cm long. Flowers creamy-white with a yellow 'eye'. Calyx 2-2.5(-3) mm, persistent, with long, spreading hairs; lobes linear-lanceolate, spreading in fruit. Corolla-tube 2.5-4 mm, pilose without; lobes *c.* 1 mm, ovate; sinuses with a minute tooth. Stigma bilobed. Nutlets usually 4, 1.6-2 mm, slightly tuberculate-rugose, shortly pubescent, brown.
Fl. IV - IX. *Very seldom; cultivated ground, roadsides, and other waste places in Madeira and Porto Santo; also on Ilhéu de Cima, Bugío, and Ilhéu Chão.* **MDP**

2. H. ramosissimum (Lehm.) DC., *Prodr.* 9: 536 (1845).
H. erosum Lehm.
Sprawling, greyish perennial herb, often woody below, densely appressed-pubescent. Stems up to 30 cm, much-branched from the base and above, prostrate to ascending. Leaves up to 3.5 × 0.8 cm, linear-lanceolate to oblanceolate, acute to obtuse, attenuate at base, margins revolute, undulate-crenate to subentire, the lower shortly petiolate, the upper sessile. Inflorescences stiff, often forked, dense in flower, scarcely elongating in fruit, up to 2 cm long. Flowers white with a yellow 'eye'. Calyx 2-2.5 mm, persistent, with ± appressed hairs; lobes linear-lanceolate, spreading in fruit. Corolla-tube 3-3.5 mm, pilose without; lobes *c.* 1 mm,

ovate. Stigma entire. Nutlets usually 4, (1.8-)2-2.5 mm, weakly tuberculate-rugose, glabrous, brown.
Fl. IV - VII. *Locally common on slopes and in dry, stony pastures of Porto Santo.* **P**

H. amplexicaule Vahl, *Symb. bot.* **3**: 21 (1794), a glandular-pubescent, perennial herb from S. America with ascending or decumbent stems, leaves with strongly undulate margins, lilac or purple corollas, and fruit splitting into 2 nutlets, was recorded in 1962 from Madeira as a casual in flower-beds. **M**

H. arborescens L., *Syst. nat.* ed. 10, **2**: 913 (1759) (*H. peruvianum* L.) [*Balsamo de cheiro*], a shrub up to 2 m high with ovate or elliptic-oblong leaves and fragrant, purple flowers, is cultivated as a garden ornamental in Madeira and Porto Santo. **MP**

2. Cynoglossum L.

Pubescent biennial herbs. Flowers blue, purplish or reddish, in mainly terminal, generally ebracteate cymes, usually elongating considerably in fruit. Calyx divided almost to the base; lobes enlarging in fruit. Corolla glabrous, with a short, cylindrical tube, the throat closed by 5 scales; lobes suberect. Stamens included. Style gynobasic, included, persistent in fruit; stigma entire. Nutlets 4, ovoid to orbicular, often dorsally flattened, covered with barbed spines.

1. C. creticum Mill., *Gard. dict.* ed. 8, no. 3 (1768).
Softly pubescent, often rather robust biennial; stem 30-70(-100) cm, erect, usually simple, leafy. Leaves 3-16(-23) × 1-4 cm, greyish; the basal forming a lax rosette, elliptic to oblanceolate, obtuse to acute, long-attenuate at the base, sometimes with a distinct petiole up to 8 cm; cauline leaves oblong, acute, sessile, amplexicaul or auriculate at the base. Inflorescence branched, dense in flower, elongating in fruit up to 40 cm, appressed-pubescent. Pedicels up to 6 mm in flower, up to 14 mm in fruit, recurved. Calyx 4-7 mm in flower, up to 10(-13) mm in fruit; lobes oblong-elliptic, ± obtuse. Corolla mauve to pale blue with conspicuous dark purple reticulate venation, glabrous; tube 3-4 mm, cylindrical; scales oblong, dark blue, pilose; lobes 3-4 mm, obtuse. Nutlets 5-7 mm across, rounded, papillose, densely covered with stout, thick, yellowish, barbed spines.
Fl. III - VII. *Scattered throughout Madeira, along paths, roadsides, and on hillsides, generally in open habitats, up to 800 m; recorded from Ilhota in Porto Santo, but reported to be very rare.* **MP**

3. Borago L.

Hispid annual herbs. Flowers blue, in lax, branched, terminal, bracteate cymes. Calyx divided almost to the base, persistent. Corolla glabrous, with a very short tube; scales short, conspicuous, exserted; lobes widely spreading, acute. Filaments short, thick, with a narrow, horn-like appendage at the apex; anthers large, exserted, connivent and forming a cone. Style gynobasic, included; stigma capitate. Nutlets 4, oblong-ovoid, shortly stipitate, with a thickened, collar-like, basal ring.

1. B. officinalis L., *Sp. pl.* **1**: 137 (1753). [*Borragem*]
Annual, covered with white, bristly hairs; stems 17-70 cm, usually erect, simple or branched, rather stout. Leaves dull green, the basal 4.5-24 × 2-8 cm, ovate, obtuse to acute, margins often undulate; petioles 2.5-14 cm; upper leaves smaller, sessile, amplexicaul. Inflorescence very lax, often large and much-branched, the flowers half-nodding. Bracts linear to lanceolate, acute, the lower leaf-like. Pedicels 10-30 mm, becoming recurved in fruit. Calyx-lobes linear-lanceolate, 8-10 mm in flower, spreading, up to 15 mm in fruit, connivent. Corolla bright blue; tube *c.* 2 mm; scales triangular, glabrous, whitish; lobes 10-14 mm, ovate. Anthers

c. 5 mm, blackish purple. Nutlets enclosed by the persistent calyx, *c.* 5 × 2.5 mm, greyish brown, sharply keeled ventrally, verruculose and longitudinally striate.
Fl. III - IX. *Infrequent; a Mediterranean species naturalized mainly in the lower regions of Madeira on walls, cultivated and waste ground, and in grassy places.* **M**

4. Anchusa L.

Strigose-hispid perennial herbs. Flowers blue, in mainly terminal, bracteate cymes, dense in flower, elongating in fruit. Calyx divided almost to the base. Corolla glabrous, with a straight, cylindrical tube, the throat closed by 5 conspicuous, often papillose scales; lobes spreading. Stamens inserted near or above middle of tube, ± included. Style gynobasic, included; stigma capitate. Nutlets 4, reticulate, with a thickened basal ring.

1. A. azurea Mill., *Gard. dict.* ed. 8, no. 9 (1768).
A. italica Retz.
Hispid, with long, white, spreading, tubercle-based bristly hairs; stem up to 1(-1.5) m, erect, stout, often much-branched. Leaves dull green, the lower (10-)15-30 × 1.5-3.5 cm, elliptic to oblanceolate or oblong, acute, attenuate into a petiole up to 8 cm long, entire to weakly undulate-crenate; upper leaves smaller, sessile. Inflorescence ± lax, paniculate. Bracts linear-lanceolate. Pedicels 2-6.5 mm in flower, lengthening in fruit, erecto-patent. Calyx-lobes linear, acute, 6-10 mm in flower, lengthening in fruit. Corolla deep blue; tube 7-10 mm; scales oblong, densely white papillose-pilose; lobes 4-8 mm, rounded, obtuse. Anthers partially exserted from corolla-tube. Nutlets 6-7 × 2 mm, oblong, yellowish brown, reticulate-ribbed longitudinally.
Fl. IV - VIII. *Infrequent; along roadsides and on cultivated and waste ground in Madeira, up to 750 m.* **M**

5. Myosotis L.

Annual to perennial herbs, ± softly hairy. Flowers usually ultimately blue, in terminal, usually paired, scorpioid cymes, sometimes bracteate below, usually elongating greatly in fruit. Calyx divided from ½-way to nearly to base. Corolla-tube very short, cylindrical; throat closed by 5 short, yellow or white, notched scales; limb patent, the lobes flat or slightly concave, rounded or notched. Stamens included. Style included; stigma capitate. Nutlets 4, ovoid, somewhat compressed, often with a distinct marginal rim, very smooth and shiny, enclosed in the persistent calyx.

Records of *M. azorica* H.C. Watson (*M. maritima* Hochst.) from Madeira are doubtful and lack confirmation.

1. Calyx with at least some spreading, hooked hairs 2
- Calyx with appressed, straight, uniform hairs 4

2. Corolla 7-8.5 mm across, lobes flat; nutlets 1.8-2 × 1.4-1.5 mm . . . . . **3. sylvatica**
- Corolla 3 mm or less across, lobes concave; nutlets 1.1-1.5 × 0.8-1 mm 3

3. Corolla always bright blue, the tube not lengthening; lowermost fruiting pedicels distinctly longer than the calyx . . . . **1. arvensis**
- Corolla opening pale yellow or white, becoming blue later, the tube lengthening with age; fruiting pedicels all shorter than the calyx . . . . **2. discolor**

4. Stem with spreading hairs below . . . . **4. secunda**
- Stem with appressed hairs below . . . . **5. stolonifera**

1. M. arvensis (L.) Hill, *Veg. syst.* **7**: 55 (1764).
M. intermedia Link
Annual to biennial; stem 10-65 cm, erect, simple or often branched from the base, with spreading hairs below. Leaves with spreading hairs, the lower up to 7 × 1.8 cm, oblanceolate, obtuse, attenuate at base and sometimes shortly petiolate; upper leaves oblong, acute to obtuse, sessile. Inflorescence ebracteate, appressed-pubescent. Flowers bright blue with a yellow 'eye'. Calyx up to 5 mm in fruit, divided ½-way or more to base, with appressed, straight or crisped hairs and numerous spreading, hooked hairs at the base; lobes linear-lanceolate, acute, erect. Corolla-limb *c*. 3 mm across, lobes concave. Fruiting pedicels up to 8 mm, spreading, the lowermost exceeding the calyx. Nutlets 1.3-1.5 × 0.9-1 mm, acute, with a marginal rim, blackish brown.
Fl. V - IX. *Rare; along paths, tracks, and grassy places in mountain districts of eastern Madeira; up to 1000 m.* **M**

2. M. discolor Pers., *Syst. veg.* ed. 15: 190 (1797).
M. versicolor Sm.
Annual, 2.5-35 cm; stem erect, very slender, simple or often branched from the base, with spreading hairs below. Leaves with spreading hairs, the lower up to 5 × 1 cm, oblong-spathulate, obtuse, attenuate, sometimes shortly petiolate; upper leaves oblong, acute, sessile. Inflorescence ebracteate, appressed-pubescent. Flowers pale yellow or white with a yellow 'eye', becoming blue. Calyx up to 5 mm in fruit, divided up to *c*. ½-way to base, with straight hairs and numerous hooked hairs at the base; lobes linear, acute, ± erect. Corolla up to 3.5 mm long, tube at length exceeding the calyx; limb 1.5-2.5 mm across, lobes concave. Fruiting pedicels up to 3 mm, ascending, shorter than the calyx. Nutlets *c*. 1.2 × 0.8 mm, acute, with a wide marginal rim, blackish brown.
Fl. III - VII. *Fairly common on walls, banks, paths, cultivated ground, and rocky places in both the lower and montane regions of Madeira, up to 1650 m. Reported to be rare in Porto Santo.* **MP**

Madeiran plants have been referred to subsp. *canariensis* (Pit.) Grau in *Mitt. bot. StSamml. Münch.* **7**: 68 (1968), but do not appear to differ significantly from typical specimens.

M. ramosissima Rochel in Schult., *Oestr. Fl.* ed. 2, **1**: 366 (1814) (*M. hispida* Schltdl.), similar to *M. discolor* but with the corolla always blue and the corolla-tube shorter than the calyx, has been recorded from Madeira. **M**

3. M. sylvatica Hoffm., *Deutschl. Fl.*: 61 (1791).
Biennial to perennial; stems up to 60 cm, erect, often much-branched, leafy, with spreading hairs below. Leaves with spreading hairs, the lower up to 6 × 2 cm, obovate-spathulate, obtuse, attenuate; petioles up to 4 cm; upper leaves oblong-lanceolate, mostly acute, sessile. Inflorescence appressed-pubescent, ebracteate. Flowers bright blue with a yellow 'eye'. Calyx up to 6 mm in fruit, divided up to ¾-way to base, with fine, straight hairs and long, spreading, hooked hairs; lobes triangular-lanceolate, acute, spreading. Corolla 7-8.5 mm across; tube equalling calyx; lobes flat, rounded. Fruiting pedicels up to 10 mm, spreading. Nutlets 1.8-2 × 1.4-1.5 mm, acute, with marginal rim, blackish brown.
Fl. III - IV. *A European species naturalized near the 'Forest Park' and the trout hatchery at Ribeiro Frio, and also along the Levada dos Tornos. First recorded from Madeira in 1971.* **M**

4. M. secunda Al. Murray, *North. Fl.* **1**: 115 (1836).
M. repens D. Don
Annual to biennial up to 40 cm; stems procumbent to erect, with spreading hairs below; rooting stolons produced from axils of the lowermost leaves. Leaves appressed-pubescent, the lower

up to 3(-4) × 1(-1.5) cm, spathulate to obovate, obtuse, attenuate; upper leaves oblong-lanceolate, obtuse to acute, sessile. Inflorescence appressed-pubescent, bracteate below. Flowers pale blue with a yellow 'eye'. Calyx up to 5 mm in fruit, divided more than ½-way to base, appressed-pubescent; lobes linear-lanceolate, acute. Corolla 5-7 mm across. Fruiting pedicels up to 10 mm, becoming recurved. Nutlets 1.3-1.5 × 1-1.2 mm, acute, with a rim, blackish brown. 2n=48.

Fl. IV - X. *Fairly common in Madeira by streams and along levadas in damp ground and wet rocks; 800-1300 m.* **M**

Records of *M. scorpioides* L. and *M. caespitosa* Schultz from Madeira are probably referable to this species.

5. M. stolonifera (DC.) J. Gay ex Leresche & Levier, *Deux excurs. bot.*: 83 (1880).
Perennial up to 50 cm; stems ± erect, slender, and much-branched, appressed-pubescent; rooting stolons produced from axils of the lower nodes. Leaves appressed-pubescent, the lower up to 5 × 1.7 cm, oblong, obtuse, attenuate; upper leaves smaller, obtuse to acute, sessile. Inflorescence usually ebracteate. Calyx up to 4.5 mm in fruit, divided ½-way or more to base, appressed-pubescent; lobes lanceolate, acute. Corolla 6-8 mm across, pale blue or lilac, or occasionally white, with yellow 'eye'; lobes entire or emarginate. Fruiting pedicels up to 12 mm, becoming recurved. Nutlets 1.4-1.5 × 1.1-1.2 mm, obtuse, blackish brown.

Fl. VI. *Rare; in wet places by streams, sometimes forming large colonies; c. 800 m.* **M**

Madeiran material placed here appears to differ from typical *M. stolonifera* by the larger leaves, flowers, and nutlets, and the more pointed calyx-lobes.

6. Echium L.

Annual to biennial herbs or shrubs, hispid or scabrid, hairs often with white, swollen bases. Flowers blue, purple or white, sessile, in bracteate cymes, paniculate, spike-like or thyrsoid. Calyx divided almost to the base, persistent. Corolla somewhat irregular, broadly to narrowly funnel-shaped; tube straight with an open, usually oblique throat, with tufts of hairs or a basal annulus; lobes unequal. Stamens often inserted unequally, exserted or included. Style exserted; stigma bifid. Nutlets 4, ovoid or ovoid-trigonous, tuberculate.

Literature: D. Bramwell, A revision of the genus *Echium* in Macaronesia, in *Lagascalia* 2(1): 37-115 (1972).

1. Annual or biennial herb; corolla (13-)18-32 mm . . . . **1. plantagineum**
- Shrub; corolla 6-11 mm 2

2. Stems with long, spreading hairs; inflorescence (11-)15-35(-47) cm, flowers dark blue .
. . . . **2. candicans**
- Stems with short, appressed hairs; inflorescence 5-16 cm, flowers pale blue . . . . . .
. . . . **3. nervosum**

1. E. plantagineum L., *Mant. pl. alt.*: 202 (1771). [*Invejosa, Vermelhão*]
E. lycopsis auct., non L. (1754)
Variable, softly hispid, annual to biennial herb; stems (10-)20-75(-100) cm, usually erect, occasionally prostrate, simple or often branched from the base. Basal leaves ± rosette-forming, 5-13 × 1-3.5(-8) cm, ovate to oblong-elliptic; petioles up to 10 cm. Cauline leaves sessile, smaller, oblong to lanceolate, cordate-amplexicaul. Cymes usually branched, elongating in fruit. Calyx 7-9 mm in flower, up to 12 mm in fruit; lobes lanceolate. Corolla (13-)18-32 mm, funnel-shaped, deep purplish, becoming pinkish blue with age, occasionally white, sparsely pubescent on veins and lobe margins. Lower 2 stamens exserted, upper 3 included; filaments

with scattered, very long hairs. Style exserted, pubescent. Nutlets 2.5-3 mm, greyish brown. Fl. III - X. *Very common in Madeira and Porto Santo in cultivated and waste ground, grassy banks, rocky slopes, and along paths and roadsides, 0-1100 m; also found on Deserta Grande, Bugío and Ilhéu Chão; collected from Selvagem Grande in 1868, but apparently not recorded since.* **MDPS**

2. E. candicans L.f., *Suppl. pl.*: 131 (1782). [*Massaroco*]
E. candicans var. *noronhae* Menezes
Shrub up to 1.5(-2) m, densely hispid; stems branched; bark papery, greyish white. Leaves greyish green, sessile or subsessile, 7-23 × 1-3(-4) cm, lanceolate to ovate-lanceolate, long acuminate, attenuate to base, veins prominent below, densely hispid, with or without hairs with expanded, white bases. Inflorescence (11-)15-35(-47) cm, dense. Calyx 4-5 mm in flower, up to 7 mm in fruit; lobes lanceolate, persistent, not spreading. Corolla 8-10 mm, narrowly funnel-shaped, deep blue, pubescent without; lobes 2-2.5 mm, obtuse. Stamens exserted; filaments pink. Style exserted, pubescent. Nutlets *c.* 3.5 mm, brown, pubescent. $2n=16$. **Plate 35**
Fl. IV - VIII. *Very rare; in laurisilva and* Erica *scrub on cliffs of high mountain ravines in central Madeira, 800-1400 m. Planted for ornament in gardens and along roadsides, and may occasionally become naturalized.* ● **M**

Records of this species from Porto Santo are believed to be of cultivated specimens.

3. E. nervosum Dryand. in W.T. Aiton, *Hortus kew.* ed. 2, **1**: 300 (1810). [*Massaroco*]
E. fastuosum sensu auct. mad., non Salisb., Dryand. nec Jacq.; *E. nervosum* var. *laxiflorum* Menezes
Shrub up to 1 m, densely strigose; stems branched; bark papery, greyish white. Leaves subsessile, 4.5-12.5 × (0.6-)1.2(-3.4) cm, lanceolate, acute to shortly acuminate, attenuate to base, veins prominent below, densely strigose, some to many hairs with expanded, white bases, especially above. Inflorescence 5-16 cm, dense. Calyx 3-5.5 mm in flower, up to 13 mm in fruit; lobes lanceolate, persistent, spreading in fruit. Corolla 6-11 mm, narrowly funnel-shaped, pale blue with a whitish stripe down the middle of each lobe, or rarely white, pubescent without; lobes 1.5-2.5 mm. Stamens exserted; filaments pink. Style exserted, pubescent below. Nutlets *c.* 3 mm, greyish, glabrous. $2n=16$. **Plate 35**
Fl. I - VIII. *Common in the coastal regions of Madeira, particularly in the south, growing on sea cliffs, dry, rocky slopes and along roadsides, 0-300 m. Reported to be common at Zimbralinho on the south-west coast of Porto Santo, and also found on Pico do Castelo and Pico de Facho. Recorded from the Desertas but without further data.* ● **MDP**

E. simplex DC., *Cat. pl. horti monsp.*: 108 (1813), an endemic monocarpic shrub native to Tenerife (Canary Islands) with white flowers and an unbranched stem, reported to be cultivated in Madeiran gardens, was said to be naturalized for a time on rocks at 'Quinta da Vigia'. **M**

XCII. VERBENACEAE
N.J. Turland

Scabrid shrubs or perennial herbs. Stems 4-angled. Leaves opposite, simple, toothed to laciniate. Inflorescence an axillary, corymbose head, or a terminal panicle of spikes. Flowers 4- to 5-merous, hermaphrodite; calyx usually small, tubular; corolla funnel-shaped, the limb weakly 2-lipped. Fruit a small, fleshy or juicy drupe, or dry and separating at maturity into 4 nutlets.

Species commonly cultivated in parks and gardens in Madeira include **Aloysia triphylla** (L'Hér.) Britton (*Lippia citriodora* (Lam.) Kunth) [*Pecegueiro Inglez*], a white-flowered shrub with lemon-scented leaves from Argentina and Chile, **Clerodendrum speciosissimum** Van Geert ex Morr. (*C. fallax* Lindl.), a red-flowered shrub from Java, and **Duranta erecta** L. (*D. plumieri* Jacq.), a lilac-flowered shrub or tree from tropical America.

1. Shrub; stems often conspicuously prickly; inflorescence an axillary, pedunculate, corymbose head; corolla changing colour with age from white or yellow to pink or red; fruit a small, fleshy or juicy drupe . . . . **1. Lantana**
- Perennial herbs; stems not prickly; inflorescence terminal, a panicle of dense or lax spikes; corolla pink, lilac, purple or violet, not changing colour with age; fruit of 4 nutlets when mature . . . . **2. Verbena**

1. Lantana L.

Shrub. Stems often with small but conspicuous backward-curved prickles. Leaves petiolate, ovate. Inflorescence an axillary, pedunculate, corymbose head. Corolla changing colour with age from white or yellow to pink or red. Fruit a small, fleshy or juicy drupe.

1. L. camara L., *Sp. pl.* **2**: 627 (1753).
Plant to 3.5 m. Leaves 2-6 × 1.5-4 cm (excluding petiole to 1.5 cm), ovate, obtuse to shortly acuminate at apex, serrate-crenate at margins, abruptly narrowed at base, reticulate-rugose, coriaceous. Heads 1.5-3 cm across; peduncles to 7 cm. Corolla with tube 5-10 mm, limb 5-8 mm across.
Fl. almost all year. *Introduced; naturalized in rocky places and along hedges and walls in the Funchal region, up to 550 m. Native to tropical and warm-temperate America.* **M**

2. Verbena L.

Perennial herbs. Stems not prickly. Leaves sessile or petiolate, narrowly elliptic to linear or oblanceolate to narrowly so. Inflorescence a terminal panicle of dense or lax spikes. Corolla pink, lilac, purple or violet, not changing colour with age. Fruit dry, separating at maturity into 4 nutlets.

1. Leaves irregularly deeply crenate to laciniate; spikes lax, slender, elongate . . . .
. . . . **3. officinalis**
- Leaves serrate; spikes dense, ovoid to cylidric or obconic 2

2. Leaves with 5-12 teeth along each side; corolla-tube 6-10 mm, not less than twice as long as calyx, the limb 5-9 mm across . . . . **1. rigida**
- Leaves with 10-25 teeth along each side; corolla-tube 3-4 mm, less than twice as long as calyx, the limb 1.5-2 mm across . . . . **2. bonariensis**

1. V. rigida Spreng., *Syst. veg.* **4**(2): 230 (1827).
V. bonariensis sensu auct. mad., pro parte, non L. (1753); *V. venosa* Gillies & Hook.
Stems creeping at base, rooting at nodes, then erect to 100 cm. Leaves sessile, 3-9 × 1-2 cm, narrowly elliptic to linear, semi-amplexicaul at base, somewhat remotely serrate at margins; teeth 5-12 along each side, cuspidate. Inflorescence a somewhat lax panicle of spikes; spikes 1-6 × 1-2.5 cm, dense, ovoid to cylindric, or appearing obconic when open flowers are grouped towards apex. Corolla purple or violet; tube 6-10 mm, not less than twice as long as calyx; limb 5-9 mm across.
Fl. II - IX. *Introduced; naturalized along hedges, levadas, roadsides and margins of cultivated ground, and on steep banks and in rocky places in eastern Madeira, west to Câmara de Lobos and Santana, up to 500 m. Native to S. Brazil and Argentina.* **M**

2. V. bonariensis L., *Sp. pl.* **1**: 20 (1753). [*Jarvão, Urgebão*]
V. litoralis sensu Menezes (1914), non Kunth (1818).
Like **1** but stems taller, to 200 cm; leaves often larger, to 15 × 2.5 cm, very narrowly elliptic to linear, closely serrate; teeth 10-25 along each side, acute but not cuspidate; inflorescence a somewhat dense panicle, with spikes smaller and more numerous; spikes 1-3 × 0.5-0.7 cm, ovoid to cylindric, not appearing obconic because flowers are much smaller; corolla bluish purple or dark purple; tube shorter, 3-4 mm, less than twice as long as calyx; limb 1.5-2 mm across.
Fl. VI - X. *Introduced; naturalized on roadsides and in wet flushes and waste areas in the lowlands of Madeira, from 200-450 m; known with certainty from two discrete areas: the Funchal region and the area between Prazeres and Calheta in the south-west. Native to S. America.* **M**

Several records of *V. bonariensis* are incorrect and referable to **1**; only those which could be verified have been used for the treatment here.

3. V. officinalis L., *Sp. pl.* **1**: 20 (1753). [*Jarvão, Urgebão*]
Stems erect, to *c.* 80 cm. Leaves 2-10 × 1-4 cm, oblanceolate to narrowly so in outline, irregularly deeply crenate to laciniate, sessile or gradually narrowed into a short petiole. Inflorescence a lax panicle; spikes 10-20 × 0.3-0.4 cm, elongate, slender. Corolla lilac or purple; tube 2.5-3 mm, less than twice as long as calyx; limb 3.5-4 mm across.
Fl. VI - XI. *Eastern Madeira, from Curral das Freiras and Fajã da Nogueira to Funchal and Machico, growing on the borders of laurisilva and in moist places, up to 800 m; recorded also from Porto Santo but without further data.* **MP**

XCIII. CALLITRICHACEAE
M.J. Short

Monoecious, often aquatic herbs. Leaves opposite and decussate, entire, exstipulate. Flowers minute, axillary, often subtended by 2 bracteoles. Sepals and petals absent. Male flowers with 1 stamen. Female flowers with a 4-locular, 4-lobed ovary; styles 2, filiform. Fruit separating at maturity into 4, 1-seeded mericarps.

1. Callitriche L.

Submerged to terrestrial, small, delicate, usually annual herbs with slender stems. Leaves simple; submerged leaves narrow, floating leaves broader, usually in a terminal rosette. Flowers inconspicuous, solitary or a male and female paired in the same leaf-axil. Fruit (not always produced) 4-lobed, the lobes keeled or winged.

1. C. stagnalis Scop., *Fl. carniol.* ed. 2, **2**: 251 (1772). [*Lentilha de água*]
Stems with small, orbicular, scale-like, peltate hairs. Leaves up to 2 cm long, elliptic, obovate-spathulate to suborbicular, obtuse, sometimes shallowly emarginate, rather abruptly narrowed into a short petiole, pale green, the uppermost forming floating rosettes in aquatic plants. Bracteoles membranous. Fruit sessile or very shortly pedicellate, 1.4-1.5 × 1.5-1.6 mm, suborbicular, flattened, pale brown, deeply grooved between the 4, broadly winged mericarps.
Fl. III - IX. *In levadas, pools, water troughs, drainage gutters, and wet mud, in both still and running water. Most common in the north and east of Madeira but generally scattered and rather infrequent, up to c. 800 m; rare in Porto Santo.* **MP**

XCIV. LABIATAE
[LAMIACEAE]
J.R. Press

Herbs or shrubs, the foliage frequently glandular and aromatic. Leaves opposite, usually simple, exstipulate. Flowers usually strongly zygomorphic, sometimes weakly so, usually in contracted and modified cymes in the axils of opposite bracts, each pair of cymes forming a whorl-like verticillaster, the verticillasters forming a simple or compound inflorescence which may be spike-like, cymose, corymbose, paniculate or capitate. Bracts resembling leaves but becoming smaller upwards, or differing from the leaves in size or shape and sometimes brightly coloured. Calyx 4- or 5-toothed, often 2-lipped, the upper lip with 3 teeth, the lower with 2 teeth. Corolla tubular, with a 4- or 5-lobed limb, often 2-lipped, the upper lip usually 2-lobed and the lower lip 3-lobed, rarely upper lip absent with the 5 lobes all forming the lower lip. Stamens usually 4, rarely 2. Ovary with 2 carpels but appearing 4-lobed; style 1. Fruit of 4, 1-seeded nutlets.

Various species are commonly cultivated in Madeira, either for ornament or as pot-herbs, including **Coleus blumei** Benth. [*Côleos*], **Ocimum basilicum** L. [*Mangericão*], **Plectranthus fruticosus** L'Hér., and **Rosmarinus officinalis** L. [*Alecrim*].

Ocimum micranthum Willd., *Enum. pl.*: 630 (1809), an annual native to the New World from Florida to Paraguay, has been recorded once from a roadside in Funchal, where it was presumably a short-lived garden escape. M

Two species have been recorded in error from Madeira: *Glechoma hederacea* L. and *Stachys officinalis* (L.) Trevir. (*Betonica officinalis* L.). The former was an error for *Sibthorpia peregrina* L.

1. Corolla 1-lipped, the upper lip absent 2
 - Corolla 2-lipped or regular 3

2. Annual; leaves lobed; corolla yellow .... **1. Ajuga**
 - Shrub or woody-based perennial; leaves entire, toothed; corolla reddish, purple or cream
   .... **2. Teucrium**

3. Stamens 2 .... **20. Salvia**
 - Stamens always 4, though sometimes reduced in female flowers 4

4. Upper lip of corolla distinctly hooded 5
 - Upper lip of corolla flat or corolla ± regular 13

5. Leaves trifoliolate .... **10. Cedronella**
 - Leaves undivided 6

6. Calyx with broad, leaf-like teeth .... **3. Prasium**
 - Calyx-teeth usually narrow, not leaf-like 7

7. Shrub, densely white-tomentose to lanate .... **5. Sideritis**
 - Herb, glabrous to hirsute but never tomentose 8

8. Calyx-teeth unequal, the upper much broader than the lower 9
 - Calyx-teeth ± equal 10

9. Inflorescence a dense spike, bracts imbricate, purplish, very different from leaves ....
   .... **11. Prunella**
 - Inflorescence of loose, distant verticillasters, bracts similar to leaves .... **12. Melissa**

10. Corolla yellow, with brown markings .... **7. Lamiastrum**
 - Corolla usually purple or white, if yellow then very pale and lacking brown markings
   11

11. Lateral lobes of corolla very small or obscure . . . . **6. Lamium**
-   Lateral lobes of corolla large or obvious 12

12. Calyx-tube trumpet-shaped, the teeth ± as long as wide . . . . **9. Ballota**
-   Calyx-tube cylindrical to obconical, the teeth clearly longer than wide . . . **8. Stachys**

13. Calyx with 10 hooked teeth . . . . **4. Marrubium**
-   Calyx with 5 straight teeth 14

14. Upper tooth of calyx with a broad, apical appendage . . . . **19. Lavandula**
-   Upper tooth of calyx lacking an appendage 15

15. Corolla weakly zygomorphic, 4-lobed, the lobes ± equal 16
-   Corolla strongly zygomorphic, 5-lobed, usually with a 2-lobed upper lip and a 3-lobed lower lip 17

16. Herb; flowers in globular heads or long, ± dense spikes . . . . **18. Mentha**
-   Evergreen shrub; flowers in paniculate heads . . . . **17. Bystropogon**

17. Stamens exserted, longer than upper lip of corolla 18
-   Stamens not exserted, shorter than upper lip of corolla 19

18. Calyx-teeth ± equal; bracts ovate or obovate, imbricate, conspicuous **16. Origanum**
-   Upper calyx-teeth much shorter than the lower; bracts narrowly spathulate, neither imbricate nor conspicuous . . . . **15. Thymus**

19. Leaves ovate to ovate-lanceolate . . . . **13. Clinipodium**
-   Leaves lanceolate, but usually with revolute margins and appearing linear **14. Satureja**

1. Ajuga L.

Perennial herbs. Verticillasters 2- to 4-flowered. Bracteoles small or absent. Calyx tubular-campanulate, with 5 ± equal teeth. Corolla-tube exserted and with a ring of hairs within; upper lip very short; lower lip conspicuous, 3-lobed. Nutlets reticulate.

1. **A. iva** (L.) Schreb., *Pl. verticill. unilab.*: 25 (1773).
Plant villous. Stems up to 20 cm, much-branched, woody towards the base. Leaves up to 20 × 6 mm, linear-oblong, usually with 2-6 short lobes, rarely entire. Calyx-teeth equalling or shorter than tube. Corolla-tube 5.5 mm; lower lip up to 7 mm. Filaments hairy. Nutlets 2.1-2.4 mm, blackish.
Fl. III - VI. *Dry places at the eastern end of Porto Santo, between Pico do Facho and Pico do Castelo, and on Ilhéu de Cima.* **P**

Madeiran plants are var. **pseudiva** (DC.) Benth., *Labiat. gen. spec.*: 699 (1835). They differ from var. *iva* only in having yellow, not pink or purple, corollas.

2. Teucrium L.

Herbs or (in Madeira) shrubs. Calyx campanulate to infundibuliform, usually 2-lipped; teeth 5, the upper tooth larger than the lower 4. Corolla hairy on the outside, with a single, 5-lobed lip (the lower), the 2 uppermost lobes small, the 2 laterals slightly larger, the middle lobe much the largest. Stamens long-exserted.

1. Plant with stellate hairs; bracts similar to leaves . . . . **1. heterophyllum**
- Plant with simple hairs; bracts obviously dissimilar to leaves 2

2. Plant densely puberulent; leaves cuneate; corolla purple . . . . **2. betonicum**
- Plant pubescent to villous; leaves cordate; corolla brownish orange to yellowish . . 3

3. Shrub; leaves ovate-cordate with appressed hairs; corolla brownish orange, with short glandular hairs on the outside . . . . **3. abutiloides**
- Herb, stems woody at base; leaves triangular-ovate, hairs not appressed; corolla greenish yellow, with long hairs on the outside . . . . **4. scorodonia**

1. T. heterophyllum L'Hér., *Stirp. nov.*: 84 (1788).
Shrub up to 1.5 m, very densely and shortly stellate-tomentose. Leaves 3-5 × 0.7-1.5 cm, oblong to oblong-ovate, obtuse, margin entire to bluntly crenate in the upper ⅔, greenish white above, white beneath; petiole less than ⅓ as long as lamina. Verticillasters 2-flowered, distant; bracts similar to leaves. Calyx 6-7 mm, narrowly infundibuliform, the teeth equal, short, triangular. Corolla orange-red, densely hairy on the outside; tube 8-11 mm; lip with lateral lobes suborbicular, the middle lobe 4-5 mm. **Plate 36**
Fl. IV - VI. *Very rare, on rocky sea cliffs at Cabo Girão and Ponta do Garajau in Madeira and on Ilhéu Chão in the Desertas.* ■ MD [*Canaries*]

2. T. betonicum L'Hér., *Stirp. nov.*: 83 (1788).   [*Abrotôna, Erva branca*]
Shrub up to (1-)1.5 m, densely greyish- to white-puberulent, especially on young growth. Leaves 5-12 × 2-4 cm, lanceolate to ovate-lanceolate, acute or obtuse, cuneate at base, bluntly crenate to dentate, greenish white above, white beneath; petiole *c*. ⅓ as long as lamina. Verticillasters usually 2-flowered, in loose terminal and axillary racemes; bracts lanceolate, long-acuminate, entire. Calyx 3.5-4.5 mm, campanulate, often tinged purple, with a ring of hairs in the throat, 2-lipped, the upper tooth ovate, longer and broader than the triangular lower teeth and held erect. Corolla purple, hairy on the outside; tube (8-)10-11.5 mm; lip sharply deflexed, lateral lobes broadly ovate, middle lobe (4.5-)6-7 mm. Nutlets 1.7 mm, with long, white hairs. **Plate 36**
Fl. VI - VIII. *Common in laurisilva in ravines and among rocks in the interior of Madeira and along the north coast, occasionally in lower regions in the south.* ● M

3. T. abutiloides L'Hér., *Stirp. nov.*: 83 (1788).
Straggling shrub to 1.5 m, pubescent to villous, most parts with eglandular hairs, the inflorescence-axis and calyces with both glandular and eglandular hairs. Leaves 8-17 × 6-13 cm, broadly ovate-cordate, acute to acuminate, crenate to serrate, with ± appressed hairs; petiole more than ⅓ as long as lamina. Verticillasters 2- to 4-flowered, in loose to dense racemes; bracts narrowly deltate, entire. Calyx 4-5 mm, campanulate, slightly 2-lipped, the upper tooth ovate and obtuse, the lower teeth narrower and acute. Corolla brownish orange, with dense, short glandular hairs only on the outside; tube (7-)8-11 mm; lip sharply deflexed, lateral lobes narrowly oblong, middle lobe (2.5-)3-4.5 mm. Nutlets 1.5 mm, coarsely reticulate with raised ridges. **Plate 36**
Fl. V - VI. *Rare, in laurisilva in deep ravines in Madeira, in the Ribeiro Frio, Ribeira da Ametade and Ribeira de São Vicente valleys; also on the north coast from São Vicente to Seixal.* ● M

4. T. scorodonia L., *Sp. pl.* 2: 564 (1753).
Rhizomatous perennial with stems woody at base, all parts pubescent to villous, with both glandular and eglandular hairs, the hairs never appressed. Leaves 3.5-8 × 1.5-4 cm, triangular-ovate, obtuse to acute, cordate or rarely subcordate, crenate, rugose; petiole less than ¼ as long as lamina. Verticillasters usually 2-flowered, in loose racemes; bracts ovate-lanceolate, entire. Calyx 4-6.5 mm, campanulate, 2-lipped; teeth aristate, the upper broadly ovate to orbicular, the lower triangular. Corolla greenish yellow, hairy on the outside; tube 6-8.5 mm; lip with lateral lobes

ovate-oblong, the middle lobe 3.5-6 mm. Nutlets 1.4 mm, ± smooth.
Fl. VII - IX. *Rare and rather scattered in the central mountains of Madeira and damp, shady places in deep ravines in the north; also known from near Monte.* **M**
Madeiran plants do not readily fall within any of the currently recognized subspecies.

3. Prasium L.

Shrubby perennials. Verticillasters with 1(-2) flowers, in terminal racemes. Calyx campanulate, 10-nerved, 2-lipped; teeth leaf-like; upper lip 3-toothed; lower lip deeply divided into 2 teeth. Corolla 2-lipped; tube with a ring of scale-like hairs within; upper lip entire, hooded; lower lip 3-lobed. Stamens didynamous, parallel; anther-cells divergent. Nutlets drupe-like.

1. P. majus L., *Sp. pl..* **2**: 601 (1753).
P. majus var. *intermedium* Menezes; *P. medium* Lowe
Plant completely glabrous or softly hairy. Stems erect, branched. Leaves (1-)3.5-7 × (0.5-)1.5-5 cm, ovate to ovate-lanceolate, acute, crenate to crenate-serrate, cordate, truncate or cuneate at base, petiolate. Bracts resembling leaves or entire. Calyx 15 mm, accrescent and reaching 20 mm in fruit; teeth ovate to ovate-lanceolate, aristate. Corolla 13-20 mm, white; upper lip 10-12 mm, oblong, obtuse; lower lip 6-10 mm, with middle lobe largest. Nutlets 2.5-3 mm, ovoid-trigonous, black.
Fl. III - V. *Rare, scattered around the coasts of Madeira and Porto Santo.* **MP**

Glabrous plants occur exclusively in Madeira. Hairy plants also occur in Madeira and are the only form in Porto Santo. They have been variously referred to **P. medium** Lowe in *Trans. Camb. phil. Soc.* **6**: 535 (1838) and to **P. majus** var. **intermedium** Menezes, *Fl. arch. mad.*:144 (1914). However, plants across Europe range from glabrous to glabrescent, hairy or glandular-pubescent and neither taxon is recognized here.

4. Marrubium L.

Perennials. Verticillasters many-flowered. Bracteoles present. Calyx tubular, hairy in the throat, with (in Madeira) up to 10 teeth. Corolla-tube included, glabrous or with an uneven ring of hairs within; upper lip flat, bifid; lower lip 3-lobed. Stamens parallel, the outer longer; anther-cells divergent. Nutlets ovoid with a truncate apex, smooth.

1. M. vulgare L., *Sp. pl.* **1**: 583 (1753). [*Marroios*]
Plant with stellate pubescence. Stems up to 35 cm, much-branched, densely white-lanate at least above, somewhat woody below. Leaves up to 2 cm in diameter, ± orbicular, irregularly crenate, rugose, sparsely white-tomentose above, densely so beneath. Bracteoles filiform, pungent. Calyx-tube 3-5 mm, hairy in the throat; teeth 8-10, spiny and hooked. Corolla white; upper lip *c.* 2 mm, deeply bifid; lower lip 2.5 mm. Nutlets 1.9 mm, black.
Fl. III - VII. *Cliff-tops and roadsides in dry areas of Ponta de São Lourenço, in Porto Santo and on all of the Desertas.* **MDP**

5. Sideritis L.

Small shrubs (in Madeira; annual or perennial herbs elsewhere). Verticillasters with several flowers, forming long, lax to dense spikes. Bracteoles absent. Calyx campanulate, 10-nerved, the 5 teeth equal. Corolla 2-lipped; upper lip ± flat, emarginate; lower lip 3-lobed. Stamens included. Nutlets rounded at apex.

1. S. candicans Aiton, *Hort. kew.* **2**: 289 (1789). [*Erva branca, Selvageira*]
More or less white to greyish, densely tomentose shrub 45-100 cm. Leaves 2.5-12 × 1.5-7.5 cm, the lower ovate-lanceolate to ovate, acute to obtuse, rounded to cordate at base, weakly

crenate to sub-entire, petiolate. Inflorescence up to 30 cm. Calyx 5-6.5 mm; teeth 1-1.5 mm. Corolla creamy yellow; tube 4-5.5 mm, not exserted; upper lip 2-2.5 mm; lower lip 2.2-3.5 mm, the middle lobe shallowly notched. Nutlets 1.5 mm, ovoid.
Fl. III - VII. ● **MDP**

Three varieties have been recognized. The first two in particular are difficult to distinguish from each other and may not merit merit taxonomic separation.

i) var. candicans
S. massoniana var. *longifolia* Lowe
Leaves 4.5-12 × 1.5-7.5 cm, ovate-lanceolate, àcute. Inflorescence usually lax, often much-branched. Corolla glabrescent or somewhat hairy on the outside. **Plate 36**
Widespread in clearings and open, sunny places in Madeira, mostly from c. 600-1700 m.

ii) var. multiflora (Bornm.) Mend.-Heuer in *Vieraea* **3**: 135 (1974).
Leaves 4.5-7 × 2.5-4 cm, ovate-lanceolate, acute. Inflorescence usually dense and little-branched. Corolla usually rather hairy on the outside. **Plate 36**
Dry or exposed places on cliffs at Cabo Girão and Machico in Madeira and on the peaks of Porto Santo.

iii) var. crassifolia Lowe
S. massoniana var. *crassifolia* Lowe in *Trans. Camb. phil. Soc.* **6**: 535 (1838).
Leaves 2.5-4.5 × 2-3.5 cm, ovate to broadly ovate, obtuse or acute. Inflorescence dense, little-branched. Corolla densely hairy on the outside. **Plate 36**
Dry places on the Ponta de São Lourenço in Madeira, and on Bugío in the Desertas.

6. Lamium L.

Annual herbs. Verticillasters densely flowered. Calyx tubular or campanulate, with 5 ± equal teeth. Upper lip of corolla hooded; lower lip 3-lobed, the lateral lobes very small. Anther-cells divaricate, hairy. Nutlets trigonous with a truncate apex.

Literature: J. Mennema, A taxonomic revision of *Lamium*, in *Leiden Bot. Series.* **11**: 1-196 (1989).

1. Bracts usually wider than long, semi-amplexicaul; verticillasters distant . . . .
 . . . . **2. amplexicaule**
- Bracts longer than wide, not amplexicaul; verticillasters crowded . . . **1. purpureum**

1. L. purpureum L., *Sp. pl.* **2**: 579 (1753).
Pubescent annual up to 25 cm, often purple-tinged. Stems branched. Leaves ovate, obtuse, cordate, crenate to deeply incised. Bracts longer than wide, similar to leaves but rounded or truncate at the base, not amplexicaul, the upper ± sessile. Verticillasters crowded. Calyx 5-7 mm; tube somewhat campanulate, equalling or slightly shorter than the teeth. Corolla pinkish purple: tube exserted, straight, usually with a ring of hairs towards the base within; upper lip entire, hairy. Anthers hairy. Nutlets *c.* 2 mm.
Fl. III - VI. **M**

Two varieties occur in Madeira:

i) var. purpureum
Leaves crenate. Calyx-teeth usually as long as tube. Corolla always with a ring of hairs within.
Rare weed of cultivated ground, roadsides and banks around Funchal, Monte, Curral dos Romeiros, Fajã de Nogueira, Porto do Moniz and Chão da Ribeira.

ii) var. **hybridum** (Vill.) Vill., *Hist. pl. dauph.* **2**: 385 (1787).
L. hybridum Vill.
More slender than var. **i**, with often smaller, deeply and irregularly incised leaves. Calyx-teeth mostly shorter than tube. Corolla-tube with only a faint ring of hairs, or hairs absent.
Very rere weed of cultivated ground. First recorded in 1973 at Santo da Serra.

2. L. amplexicaule L., *Sp. pl.* **2**: 579 (1753).
Pubescent annual up to 20 cm. Leaves 1.4-2 × 1.5-2.3, ovate-orbicular to ± orbicular, truncate to cordate, crenate or somewhat lobed. Bracts similar to leaves but usually wider than long, sessile and semi-amplexicaul. Verticillasters distant. Calyx 4-6 mm, teeth slightly shorter than tube. Corolla 15 mm, pinkish purple, straight, long-exserted; upper lip entire, densely hairy. Nutlets *c.* 2 mm.
Fl. XI - V. *Walls and rubbish heaps at Funchal, Monte, Camacha and Porto do Moniz; very rare; also recorded from Porto Santo.* **MP**

Plants from the islands are all var. **amplexicaule**, although plants with abortive flowers have been called var. *clandestina* Rchb.

7. Lamiastrum Heist. ex Fabr.

Like *Lamium* but perennials with stolons. Corolla yellow, the lower lip with 3 ± equal lobes, the middle lobe ± triangular, acute. Anthers glabrous.

1. L. galeobdolon (L.) Ehrend. & Polatschek in *Öst. bot. Z.* **113**: 108 (1966).
Sparsely to densely hairy. Stems erect; stolons leafy and rooting at nodes. Leaves ovate to ovate-orbicular, truncate or cordate at base, usually coarsely toothed. Bracts resembling leaves. Corolla yellow with darker orange or brown markings; tube straight, with a ring of hairs within.
Fl. ? *Recorded from Madeira but without further data.* **M**

8. Stachys L.

Annuals or perennials. Verticillasters forming spike-like inflorescences. Bracteoles present. Calyx tubular or campanulate, 5- to 10-veined; teeth 5, equal. Corolla 2-lipped; upper lip entire or bifid; lower lip 3-lobed. Stamens didynamous. Nutlets obovoid with a rounded apex.

1. Perennial; leaves 50-115 mm; corolla purple with white markings . . . . . **3. sylvatica**
- Annuals; leaves 6-65 mm; corolla pink or purplish, or yellow and white, with dark or purple spots 2

2. Calyx 4-6 mm, with teeth mucronate; corolla 4-6 mm, the lips pink or purplish, the lower with dark spots . . . . **1. arvensis**
- Calyx *c.* 10 mm, with teeth conspicuously white-aristate; corolla 12 mm, the upper lip white, the lower yellow, the thoat with purple spots . . . . **2. ocymastrum**

1. S. arvensis (L.) L., *Sp. pl.* ed. 2, **2**: 814 (1763).
Hirsute to glabrescent annual, sometimes entirely glabrous. Stems up to 35 cm, branched from the base, the branches ascending. Leaves 6-40 × 5-35 mm, ovate-cordate to ovate-oblong, obtuse, bluntly toothed. Verticillasters 2- to 6-flowered. Bracteoles minute. Calyx 4-6 mm, campanulate, with eglandular hairs, often purplish on the dorsal surface; teeth ± equalling or longer than tube, mucronate. Corolla 4-6 mm; tube pale; lips pink or purplish; upper lip 1.5-2 mm; entire, lower lip 1.5-3 mm, with dark spots. Nutlets *c.* 1.5 mm, minutely verrucose.
Fl. all year. *Common on roadsides, banks, walls and in rocky pastures, also as weed of cultivated*

and waste ground. Throughout the lower regions of Madeira and Porto Santo; also on Ilhéu Chão and Deserta Grande. **MDP**

2. S. ocymastrum (L.) Briq., *Lab. Alp. marit.*: 252 (1893).
Hirsute annual. Stems up to 45 cm, branched along their length. Leaves pale green, 15-65 × 10-45 mm, ovate-cordate to ovate-oblong, obtuse, bluntly toothed. Verticillasters 6-flowered. Bracteoles short, linear. Calyx *c.* 10 mm, tubular-campanulate, with glandular and eglandular hairs; teeth ± equalling or longer than tube, conspicuously white-aristate. Corolla 12 mm; upper lip 6-8.5 mm, bifid, white; lower lip 7-10 mm, yellow; throat purple spotted. Nutlets 1.5 mm.
Fl. I - V. *Common on waste ground, rock faces and in grassy places in the lower regions of Madeira and all regions of Porto Santo.* **MP**

3. S. sylvatica L., *Sp. pl.* **2**: 580 (1753).
Hirsute perennial. Leaves 50-115 × 35-65 mm, ovate-cordate, acute, crenate. Verticillasters 3- to 6-flowered. Bracteoles short, linear. Calyx 6-8.5 mm, campanulate, with glandular and eglandular hairs; teeth ± equalling or shorter than tube, mucronate. Corolla 13-15 mm, puberulent, purple with white markings; upper lip 4-5 mm, entire; lower lip 10 mm. Nutlets *c.* 2 mm.
Fl. VI - IX. *Rare plant recorded only at Santo da Serra in eastern Madeira.* **M**

9. Ballota L.

Perennials. Verticillasters many-flowered. Bracteoles present. Calyx funnel-shaped, with 5 awned, ± equal teeth. Corolla-tube not exserted, with a ring of hairs within; upper lip somewhat concave. Stamens parallel, the outer pair longer; anther-cells divergent. Nutlets oblong, apex rounded.

1. B. nigra L., *Sp. pl.* **2**: 582 (1753).
Plant foetid, pubescent. Stems up to *c.* 100 cm, erect, branched. Leaves up to 5.5 × 4 cm, ovate, cuneate to cordate at the base, crenate, with sessile, amber glands. Bracteoles filiform, membranous. Calyx funnel-shaped above; tube 6-7.5 mm; lobes 1.6-2.5 mm, patent, ovate-triangular, abruptly acuminate into an awn. Corolla pinkish purple; tube 6-7 mm; upper lip 4.5-5 mm, entire, densely hairy; lower lip slightly shorter, often with darker markings. Nutlets 2 mm, brownish black.
Fl. V - IX. *A rare weed of roadsides and waste places in northern Madeira from São Vicente eastwards to Machico; also reported from Porto Santo.* **MP**

Madeiran plants have been referred to both subsp. **uncinata** (Fiori & Bég.) Patzak in *Annln naturh. Mus. Wien* **62**: 64 (1958), which has erect or erecto-patent calyx-lobes, and subsp. **foetida** (Vis.) Hayek in *Reprium Spec. nov. Regni veg.* 30(2): 278 (1929) (var. *foetida* Vis.) with curved or patent lobes. They are probably best placed in subsp. *uncinata*, although they differ from that taxon in having amber glands and in some minor details of the calyx-lobes.

10. Cedronella Moench

Aromatic perennial herbs, woody at base. Leaves trifoliolate, petiolate. Inflorescence a terminal spike or head, the lowest verticillasters sometimes remote. Bracts simple, inconspicuous. Calyx tubular to campanulate, with 13-15 veins; teeth 5, equal. Corolla funnel-shaped, 2-lipped, the upper lip 2-lobed, the lower 3-lobed with the middle lobe longest. Stamens curved under the upper lip, equalling the corolla or slightly exserted.

1. C. canariensis (L.) Webb & Berthel., *Hist. nat. Iles Canaries* 3(2, sect. 3): 87 (1845).
[*Hortelã de burro, Hortelã de cabra, Mentastro*]
C. triphylla Moench
Foliage strong and sharp-smelling. Stems up to 150 cm, with scattered, somewhat bulbous-based

hairs below, but pubescent at the nodes and on the axis of the inflorescence. Leaflets 3-8 × 1-3 cm, the terminal larger than the laterals, all lanceolate, acuminate, serrate, sparsely appressed-puberulent to glabrous above, pubescent beneath; petioles pubescent. Calyx up to 8-14 mm, the teeth *c.* half as long as tube, puberulent. Corolla up to 18-20 mm, purplish, pink, lilac or white. Nutlets 1.4-2 mm, broadly oblong, dark brown. **Plate 37**
Fl. VI - VIII. *Very common in shady places, generally above 500 m.* ■ M *[Azores, Canaries]*

11. Prunella L.

Perennial herbs. Verticillasters usually 6-flowered, crowded into dense, terminal, cylindrical spikes subtended by a pair of leaves. Bracts distinct from leaves, sessile, orbicular; bracteoles absent. Calyx tubular-campanulate, 2-lipped, closed in fruit; upper lip 3-toothed; lower lip 2-toothed. Corolla-tube exserted, straight, with a ring of hairs within; upper lip hooded; lower lip minutely toothed. Nutlets oblong.

1. P. vulgaris L., *Sp. pl..* 2: 600 (1753). *[Herva ferrea]*
Pubescent. Stems to 50 cm. Leaves 4-5 × 2 cm, ovate to rhombic-ovate, cuneate at base, entire to minutely toothed; petiole equalling or shorter than lamina. Bracts and calyx with long, white hairs. Calyx 8-9 mm; teeth mucronate, middle one of upper lip broader and longer than laterals, those of lower lip lanceolate, shortly ciliate. Corolla 13-15 mm, violet-blue, rarely white.
Fl. III - IX. *Very common in Madeira up to 1850 m, especially in grassy places.* M

12. Melissa L.

Perennial herbs. Calyx campanulate, 13-nerved, 2-lipped; upper lip 3-toothed; lower lip 2-toothed. Corolla 2-lipped; tube curved and dilated above the middle; upper lip erect or deflexed, sometimes slightly hooded, emarginate; lower lip 3-lobed. Stamens didynamous, included, convergent; anther-cells divergent.

1. M. officinalis L., *Sp. pl.* 2: 592 (1753). *[Herva cidreira]*
Lemon-scented. Stems erect, branched, glabrescent to hairy. Leaves 1-7 × 0.8-5 cm, ovate to rhombic, obtuse or acute, deeply crenate, sparsely hairy above, glandular-puberulent and sparsely hairy beneath. Bracts resembling leaves but smaller. Bracteoles 2.5-8.5 mm, ovate or occasionally linear, entire. Calyx 7-9 mm; teeth of upper lip broadly triangular, those of the lower lip much longer and lanceolate-triangular, all aristate, with long, patent, eglandular hairs and short, glandular hairs. Corolla 8-15 mm, pale yellow, white or pinkish. Nutlets 1.5-2 mm, smooth or minutely reticulate.
Fl. VI - VIII. *Introduced and widely cultivated in Madeira; naturalized along streams in a few places and on rocks at Seixal.* M

Madeiran plants all belong to subsp. **officinalis**.

13. Clinopodium L.

Perennials, sometimes woody at the base. Bracteoles present. Calyx tubular, 13-veined, 2-lipped; upper lip 3-toothed; lower lip 2-toothed and longer than the upper. Corolla-tube straight. Stamens curved, convergent; anther-cells divergent. Nutlets ovoid, smooth.

1. Leaves broadly ovate, cuneate at base, usually obtuse at apex, shallowly crenate at margins, with 3-8 teeth on each side; verticillasters lax . . . . **1. ascendens**
- Leaves ovate to ovate-lanceolate, cuneate to rounded at base, obtuse to acute at apex, shallowly crenate to subentire at margins; verticillasters dense . . . . **2. vulgare**

1. C. ascendens (Jord.) Samp., *Herb. portug.*: 119 (1913). [*Nêveda*]
Calamintha sylvatica subsp. *ascendens* (Jord.) P.W. Ball; *Satureja calamintha* subsp. *sylvatica* Briq., subsp. *sylvatica* var. *calaminthoides* (Rchb.) Briq.
Plant pubescent, stoloniferous. Stems up to 80 cm, little-branched, sometimes straggling, tough and rather woody at the base. Leaves up to 2(-2.5) × 1.5(-2) cm, broadly ovate, cuneate, usually obtuse, shallowly crenate with 3-8 teeth on each side. Verticillasters lax; cymes few-flowered. Bracteoles much shorter than flowers. Calyx-tube 2.5-3.5 mm; upper teeth (1-)1.5-2 mm; lower teeth 2.5(-3.5) mm. Corolla (7-)11-14 mm, pink with purple markings on the lower lip. Nutlets *c*. 1 mm, brown, glandular-punctate.
Fl. III - X. *Among rocks, on open banks and among scrub; common in Madeira, less frequent in Porto Santo.* **MP**

2. C. vulgare L., *Sp. pl.* **2**: 587 (1753).
Satureja clinopodium (L.) Caruel
Plant pubescent to densely villous. Stems 20-70 cm, erect or ascending, sometimes rather woody at base. Leaves 1-4 × 1-2 cm, ovate to ovate-lanceolate, cuneate to rounded at base, obtuse to acute, shallowly crenate to subentire. Verticillasters dense. Bracteoles subulate, about as long as calyx. Calyx-tube 3-5 mm; upper teeth (1-)2.5-3 mm; lower teeth (2-)3.5-4(-5) mm. Corolla 10-17 mm, pink to pinkish purple, rarely white. Nutlets *c*. 1 mm, brown.
Fl. mainly III - VII, but sporadically at most times of year. *Common in Madeira in open areas, among rocks, in scrub and pine woods up to 1800 m.* **M**

Although Madeiran plants have been referred to subsp. **arundanum** (Boiss.) Nyman, *Consp. fl. eur.*: 587 (1881) (subsp. *villosa* (De Noé) Bothmer), which has a longer calyx with longer teeth, they are not readily separable from subsp. **vulgare**.

14. Satureja L.

Perennials. Flowers in loose verticillasters. Calyx tubular, somewhat 2-lipped, 13-veined; tube straight; teeth 5, slightly unequal. Corolla 2-lipped; tube straight. Stamens curved under upper lip.

Literature: P.L. Pérez de Paz, *Revisión del género* Micromeria *Bentham (Lamiaceae-Stachyoideae) en la Region Macaronesica.* (1978) La Laguna; R.H. Willemse, New combinations and names for Macaronesian *Satureja* taxa (Labiatae), in *Willdenowia* **21**: 81-85 (1991).

1. S. varia (Benth.) Webb & Berthel. ex Briq. in Engler & Prantl, *Nat. Pflanzenfam.* **4**(3a): 299 (1896). [*Hysopo, Madeira*]
Micromeria varia Benth.
Plant often flushed with purple. Stems usually much-branched from the base; branches ± prostrate to spreading or erect, glabrous or hairy. Leaves up to 1.5 cm, usually lanceolate but the margins often revolute so the blade appears to be almost linear, hairy to glabrescent, sessile or shortly petiolate. Verticillasters forming a lax or dense, terminal spike. Bracts similar to leaves but smaller. Pedicels very short, up to 1 mm. Calyx often flushed purple, 3-3.5 mm, tubular-cylindrical, narrowed at the base, glabrescent or with ± appressed, white hairs, 13(-15)-nerved, the throat naked within, 2-lipped, the teeth lanceolate, subulate, hairy on the inner face, the lower longer than the upper. Corolla 5-6 mm, white, pink or purplish. Nutlets oblong, rounded at the apex, brown.
Fl. IV - X. *Widespread in Madeira but restricted to the higher parts of Porto Santo; also on Deserta Grande and Bugío.* ■ **MDP** [*Canaries*]

This species is divided into seven subspecies, six of them confined to the Canary Islands, and is represented in Madeira by the endemic (●) subsp. **thymoides** (Sol. ex Lowe) A. Hansen & Sunding in *Sommerfeltia* **17**: 7 (1993) (*Micromeria varia* subsp. *thymoides* (Sol. ex Lowe) P.

Pérez), to which the above description applies. This subspecies is distinguished from the others in having a much-branched, leafy inflorescence, with lax verticillasters, and a corolla 3-3.5 mm long. The following two varieties have been recognized within subsp. *thymoides*.

i) var. **thymoides**
Micromeria varia subsp. *thymoides* (Sol. ex Lowe) P. Pérez var. *thymoides*
Plant up to 35 cm, erect or ascending. Calyx usually not exceeding 3 mm. **Plate 37**
The common form, found throughout the range of the subspecies.

ii) var. **cacuminicolae** (P. Pérez) A. Hansen & Sunding in *Sommerfeltia* **17**: 7 (1993).
Micromeria varia subsp. *thymoides* var. *cacuminicolae* P. Pérez
Plant 5-20 cm, branches pressed agianst the ground. Calix usually 3-3.5 mm, purplish.
Plate 37
Restricted to the high mountains of central Madeira.

15. Thymus L.

Dwarf shrubs or perennial herbs with stems woody at least at the base. Verticillasters 1- to many-flowered, in rather lax to capitate inflorescences. Calyx campanulate to cylindrical, 2-lipped, the teeth of the upper lip differing from the 3 teeth of the lower; tube straight, hairy in the throat. Corolla 2-lipped; tube straight. Stamens exserted.

1. T. caespititius Brot., *Fl. lusit.* **1**: 176 (1804). [*Alecrim da serra*]
T. micans Lowe
Plant caespitose. Stems woody, creeping, with erect flowering branches, 2-7 cm high, and axillary clusters of leaves. Leaves fleshy, *c.* 8 × 2 mm, narrowly spathulate, glabrous except for cilia at the base. Inflorescence lax, the verticillasters pedunculate and 2-flowered. Bracts resembling leaves. Calyx 3-4 mm, campanulate; upper teeth very small; lower teeth flat, somewhat leaf-like, about as long as broad. Corolla 6-14 mm, purplish, pink or white.
Fl. VI - X. *In the higher parts of Madeira, occurring on Paul da Serra and around Pico Ruivo, up to 1850 m; also at Seixal.* **M**

T. vulgaris L. [*Cheiros, Segurelha*] is a small shrub with branches erect to spreading, leaves linear to elliptic, tomentose, the margins revolute, and inflorescences capitate. It is widely cultivated as a pot-herb in gardens in Madeira.

16. Origanum L.

Perennial herbs. Verticillasters few- to many-flowered, grouped into short, terminal or lateral spikes. Bracts distinct from leaves, imbricate, coloured. Calyx tubular, either 2-lipped, with 5 equal teeth, or 1-lipped and deeply slit on one side. Corolla 2-lipped, the upper lip emarginate, the lower 3-lobed. Stamens didynamous, the anther-cells divergent.

Literature: J.H. Ietswaart, A taxonomic revision of the genus *Origanum,* in *Leiden. Bot.* ser. 4 (1980).

1. O. vulgare L., *Sp. pl.* **2**: 590 (1753). [*Oregãos*]
O. virens var. *genuinum* Cout.
Pilose to glabrescent perennial, fragrant. Stems up to 100 cm, erect, woody at base. Leaves 12-30 × 8-15 mm, ovate, acute or obtuse, entire or remotely serrulate. Flowers subsessile in cylindrical or ovoid spikes up to 30 mm long. Bracts imbricate, 5-8 × 3-4 mm, ovate to obovate, membranous, glabrous, yellowish green, sometimes tinged purple. Calyx 3(-3.5) mm, tubular, with conspicuous sessile glands; teeth ± equal. Corolla 7-8 mm, white or rarely tinged pink; upper lip emarginate.

Fl. V - X. *Common throughout Madeira, among rocks and scrub, in open and shady places; also recorded from Porto Santo.* **MP**

Madeiran plants all belong to subsp. **virens** (Hoffmanns. & Link) Ietsw. in *Leiden Bot.* ser. 4: 115 (1980), which is distinguished from other subspecies by its inconspicuously glandular leaves, compact inflorescence, glabrous yellowish green bracts and white flowers.

O. majorana L. (*Majorana hortensis* Moench) [*Mangerona, Oregãos*] is widely cultivated in gardens as a pot-herb. It has bracts small, greyish, leafy-textured and hairy, and calyces 1-lipped.

17. Bystropogon L'Hér.

Small, evergreen shrubs, smelling sweetly of balsam or mint. Young stems usually densely hairy. Leaves dotted with amber oil-glands. Flowers hermaphrodite, sometimes female and then smaller, crowded into loose, capitate glomerules, the whole inflorescence paniculate-racemose. Bracts narrow, inconspicuous. Calyx hairy on the outside, the throat glabrous within, 10-nerved, the nerves prominent; teeth 5, equal, hairy. Corolla sub-bilabiate; tube straight, not exserted, throat very finely hairy within; lobes 4, flat, the upper ± cordate, deeply emarginate, the lower 3-lobed with the middle lobe longer than the laterals, flabelliform and shallowly notched. Stamens didymous, included. Nutlets chestnut or brown, finely tuberculate.

The hairs of the indumentum are either long (0.8-1.5 mm and consisting of usually 5 cells) or short (up to 1 mm and consisting of 1-2 cells) and the indumentum characters are most consistently expressed on the annual growth of the stems and on the petioles. Both species are very variable in leaf size and shape, and in the density and composition of the indumentum, depending on ecological conditions. Some specimens appear intermediate between the two species.

Literature: I.E. La Serna Ramos, Revisión del género *Bystropogon* L'Hér., in *Phanerogamarum Monographiae* **XVIII** (1984).

1. Calyx-teeth subulate; petioles always with at least some long hairs   **1. B. maderensis**
- Calyx-teeth acute; petioles with short hairs only . . . . **2. B. punctatus**

1. B. maderensis Webb & Berthel., *Hist. nat. Iles Canaries* 3(2, sect. 3): 66 (1844).
[*Quebra panella*]
B. maderensis var. *genuinus* Menezes, var. *valdehirsutus* Menezes
Indumentum sparse to dense, containing at least some long hairs or consisting solely of long hairs. Leaves 1.25-7.5 × 0.6-5 cm, variable in shape but usually ovate to lanceolate or elliptic, ± cordate to attenuate at base, crenate to crenate-serrate, ± coriaceous, usually dull, hairy on both surfaces; petiole less than half as long as blade. Calyx (2.5-)2.75-4 mm, infundibuliform-campanulate; teeth triangular-lanceolate, acute, subulate. Corolla white, purplish white or violet, 4.5-6.5 mm (3-3.5(-4) mm in female flowers). Nutlets dark chestnut, 1-1.25(-1.4) × 0.5-0.75(-0.8) mm. **Plate 37**
Fl. VI - VII. *Endemic found mainly in humid parts of the north-east of Madeira, from Ribeiro Frio and Ribeira da Ametade to Portela, Porto da Cruz and Santo da Serra but also occurring on the Paúl da Serra and at Rabaçal.* ● **M**

2. B. punctatus L'Hér., *Sert. angl.*: 20 (1789).
[*Balsamo da rocha, Barbesano, Hortelã da serra*]
B. piperitus Lowe; *B. punctatus* var. *pallidus* Menezes, var. *disjectus* Menezes
Stems, branches and petioles with indumentum of short hairs only, other parts with short hairs or a mixture of short hairs and some long hairs. Leaves 1-6.7 × 0.4-3.6 cm, ovate, lanceolate, rhombate-lanceolate or lanceolate-elliptic, cordate or rounded at base, crenate, crenate-serrate or crenate-dentate, usually shiny, rarely coriaceous, usually shortly hairy on both surfaces, rarely

also with long hairs at base of the midrib; petiole shorter than blade. Calyx 2-2.5(-3.25) mm campanulate to infundibuliform- or tubular-campanulate; tube with short hairs except for the nerves which have long hairs; teeth triangular, short, acute. Corolla white, pinkish violet or purplish violet, (3-)4-5.25(-5.5) mm (2.5-3 mm in female flowers). Nutlets chestnut or dark brown, 1 × 0.5-0.75 mm, tuberculate. **Plate 37**
Fl. VI - VII. *Endemic on cliffs and steep slopes in the wetter parts of central, northern and western Madeira, usually above c. 500 m in central parts but approaching sea-level on the north coast.* ● M

Putative hybrids between species 1 and 2 have been recorded from Ribeiro Frio, as very rare, and have been given the name **B. × schmitzii** Menezes in *Broteria* 4: 179 (1905) (*B. × indiscretus* Menezes). Their exact status is uncertain.

18. Mentha L.

Aromatic perennial herbs with creeping rhizomes. Flowers hermaphrodite or female, borne in dense, many-flowered verticillasters, the verticillasters often crowded into a dense spike or terminal head. Calyx tubular or campanulate, 10- to 13-veined, sometimes weakly 2-lipped, with 5 teeth. Corolla weakly 2-lipped, with 4 ± equal lobes, the upper lobe broadest and emarginate; tube not exserted. Nutlets smooth, reticulate or tuberculate.

1. Calyx hairy in throat, with unequal teeth .... **1. pulegium**
- Calyx glabrous in throat, the teeth ± equal 2

2. Leaves distinctly petiolate 3
- Leaves sessile, or the lower sometimes very shortly petiolate 4

3. Calyx and pedicels with long hairs .... **2. aquatica**
- Calyx and pedicels glabrous or with very short hairs, those of the pedicels deflexed
  .... **3. × piperita**

4. Plant subglabrous .... **4. spicata**
- Plant hairy, at least the undersides of the leaves tomentose to villous 5

5. Leaves with simple hairs beneath; pedicels with numerous long hairs .... **5. longifolia**
- Leaves with branched hairs beneath; pedicels with sessile glands and occasional short hairs
  .... **6. suaveolens**

1. M. pulegium L., *Sp. pl.* **2**: 577 (1753). [*Poejos*]
M. pulegium var. *gibraltarica* (Willd.) Batt. & Trab., var. *tomentella* (Hoffmans. & Link) Cout., var. *vulgaris* Mill.
Plant subglabrous to usually tomentose, strong-smelling. Stems 5-75 cm, ascending to procumbent, sometimes erect. Leaves 5-30 × (3-)4-8(-11) mm, elliptic, entire or with several teeth on each side, hairy at least beneath, shortly petiolate. Verticillasters many-flowered. Bracts leaf-like. Calyx (2-)2.5-2.8(-3.3) mm, tubular, weakly 2-lipped, the throat hairy within; teeth ciliate, unequal, the lower subulate, the upper shorter and broader. Corolla 4-4.5 mm, lilac to mauve; tube gibbous.
Fl. VI - X. *On roadside banks, in grassy places and among rocks from 300-1700 m. Abundant in much of Madeira, also common on the higher peaks of Porto Santo.* **MP**
Somewhat variable in size, habit and indumentum; several varieties have been described.

2. M. aquatica L., *Sp. pl.* **2**: 576 (1753). [*Mentastro, Sandalos*]
M. aquatica var. *hirsuta* (Huds.) Willd.
Plant pubescent to tomentose, strong-smelling. Stems up to 150 cm, erect, often purple-tinged.

Leaves 25-45 × 10-30 mm, ovate to ovate-lanceolate, truncate at base, serrate, distinctly petiolate. Flowers in a terminal head up to 20 × 17 mm, consisting of 2-3(-5) congested verticillasters, sometimes with 1-2 distant verticillasters below. Bracts small and inconspicuous. Pedicels hairy. Calyx 3-3.5 mm, tubular with distinct veins, hairy on the outside, the throat glabrous within; teeth ± equal, subulate to narrowly triangular. Corolla 4.5-5 mm, lilac or purplish.
Fl. VIII - IX. *Infrequent in river beds up to 1000 m, mainly in the Funchal region.* **M**

A very variable species, especially in the indumentum. Madeiran plants are all hairy and have been described as var. **intricata** Menezes, *Fl. Mad.*: 135 (1914).

3. M. × piperita L. , *Sp. pl.* **2**: 576 (1753).
Plant subglabrous or rarely sparsely hairy. Stems up to 40 cm, erect, sometimes thinly hairy on the angles, often purple-tinged. Leaves 30-100 × 15-45 mm, ovate-lanceolate to lanceolate, cuneate to subcordate at base, serrate, distinctly petiolate. Verticillasters forming an oblong spike 30 × 15 mm, often interrupted below, sometimes a subglobose head. Bracts small and inconspicuous. Pedicels usually glabrous, sometimes with very short, deflexed hairs. Calyx 2.5-3(-4) mm, tubular, ± glabrous or shortly and sparsely hairy on the outside, the throat glabrous within; teeth ± equal, ciliate. Corolla 4-5 mm, lilac-pink. Sterile.
Fl. VII - X. *Cultivated in gardens and naturalized by streams in a few, scattered places in Madeira and Porto Santo, usually near villages.* **MP**

A hybrid between **2** and **4**. Hairy plants may arise elsewhere in the wild where both parents are present but all Madeiran plants are ± subglabrous, indicating their origin as garden escapes. Two cultivars are represented. **M. × piperita** nm. **piperita** [*Hortelan pimenta*] has foliage smelling of peppermint. **M. × piperita** nm. **citrata** (Ehrh.) Boivin, *Naturaliste can.* **93**: 1061 (1966) (*M. aquatica* var. *glabrata* Benth.) [*Sandalos*] has foliage smelling of Eau de Cologne.

4. M. spicata L., *Sp. pl.* **2**: 576 (1753). [*Hortelā de leite*]
M. viridis L., var. *hirsuta* Menezes
Plant subglabrous, strong and usually sweet-smelling, sometimes pungent or musty. Leaves 30-53 mm × 10-25 mm, lanceolate or lanceolate-ovate, widest near the base, regularly serrate, the lower surface with a few simple hairs, sessile. Inflorescence a branched spike 11-16 cm × 15-18 mm with numerous verticillasters. Bracts small and inconspicuous. Pedicels glandular. Calyx 1.5-1.8 mm; tube campanulate, glabrous; teeth ± equal, ciliate. Corolla 2-3.2 mm, lilac, pink or white.
Fl. VII - IX. *Cultivated and occasionally naturalized in damp places in Madeira.* **M**

5. M. longifolia (L.) Huds., *Fl. angl.*: 221 (1762). [*Hortelā*]
M. sylvestris L.
Plant musty-smelling. Stems up to 90 cm, whitish-villous. Leaves 25-70 mm × 10-23 mm, oblong elliptic, acute, usually irregularly serrate with spreading teeth, green to grey-tomentose above, grey- or white-villous beneath, the hairs all simple, sessile or the lower leaves very shortly petiolate. Inflorescence usually a branched spike 6-11 cm × 6-10 mm with numerous verticillasters. Bracts small and inconspicuous. Pedicels hairy. Calyx 1.5-2 mm; tube narrowly campanulate, hairy, the throat glabrous within; teeth ± equal, ciliate. Corolla 2.2-3 mm, lilac or white.
Fl. VII - XI. *Sometimes cultivated and naturalized in a few places in Madeira.* **M**

6. M. suaveolens Ehrh. in *Beitr. Naturk. damit. vervandt. Wiss.* **7**: 149 (1792). [*Mentastro*]
M. rotundifolia auct., non (L.) Huds.; *M. rotundifolia* var. *maderensis* Menezes.
Plant sickly-sweet-smelling. Stems sparsely hairy to white-tomentose. Leaves 15-30 mm × 10-20 mm, ovate-oblong to suborbicular, obtuse to acute, widest near the base, strongly rugose, serrate, with the teeth often bent towards the lower leaf surface, hairy above, usually grey- to white-tomentose with branched hairs beneath, sessile. Inflorescence usually a congested, branched

spike 3-9 cm × 5-10 mm, often interrupted below. Bracts small and inconspicuous. Pedicels glandular, with occasional short hairs. Calyx 0.8-1.2 mm, campanulate, sparsely hairy, the throat glabrous within; teeth ± equal. Corolla 2-2.6 mm, whitish or pink.
Fl. VII - VIII. *Scattered in river beds and damp places in Madeira.* **M**

M. × villosa Huds., *Fl. angl.* ed. 2: 250 (1778) (*M. × aromatica* sensu Menezes; *M. rotundifolia* var. *aromatica* Menezes) [*Endros*], is a hybrid between **4** and **6** recorded from Madeira at damp sites at Tabúa and Logar de Baixo, probably as an escape. It is sweet-smelling, with broadly ovate to orbicular, softly hairy leaves with spreading teeth. **M**

19. Lavandula L. [*Rosmaninho*]

Small shrubs. Verticillasters few-flowered, crowded into dense, pedunculate spikes; bracts differing from leaves. Calyx with usually 13-veins; teeth small, ± equal, the uppermost tooth with a broad, cordate to obovate appendage. Corolla 2-lipped; upper lip 2-lobed; lower lip 3-lobed, the lobes equal.

Literature: D.A. Chaytor. A taxonomic study of the genus *Lavandula*, in *J. Linn. Soc., Bot.* **51**: 153-204 (1937); A. Rozeira, A secção *Stoechas* Gingins do género *Lavandula*, in *Broteria* (Sér. Ci. Nat.) **18**: 5-84 (1949).

1. Leaves 1- to 2-pinnatisect . . . . **3. pinnata**
- Leaves entire 2

2. Indumentum of leaves grey; bracts and flowers purplish . . . . **1. stoechas**
- Indumentum of leaves green; bracts greenish, flowers white . . . . **2. viridis**

1. L. stoechas L., *Sp. pl.* **2**: 573 (1753).
L. pedunculata subsp. *ambigua* Menezes, subsp. *maderensis* (Benth.) Menezes
Small, grey-tomentose shrub with stellate hairs. Stems much-branched. Leaves (1-)2-5 cm × (1-)3-5(-6) mm, linear-oblong to linear-lanceolate, entire, margins slightly revolute, sessile. Verticillasters 6- to 10-flowered, crowded into spikes 1.5-4 cm; peduncles (3-)5-15 cm. Bracts orbicular, lilac to purple-tinged, the 3 uppermost erect, up to 2 cm, much longer than the others, oblong-ovate, petaloid, sterile and lacking flowers in their axils. Calyx 13-veined; teeth equal. Corolla-tube purple, not exserted; upper lip emarginate.
Fl. IV - V. *Rather rare, in dry areas in the south-east of Madeira, from Funchal to Santo da Serra and Machico.* **MP**

This polymorphic taxon is widely distributed in the Mediterranean region and is sometimes divided into several, closely related, species. The treatment here follows that of Rozeira in which such divisions are made at subspecific rank. Madeiran plants all belong to the endemic (●) subsp. **maderensis** (Benth.) Rozeira in *Broteria* (Sér. Ci. Nat.) **18**: 5-84 (1949), to which the above description and synonymy apply. It is most easily distinguished from other subspecies by the generally denser indumentum, acute, paler, often more grey fertile bracts.

2. L. viridis L'Hér., *Sert. angl.*: 19 (1789).
Like **1** but indumentum greenish; leaves 2-4.5 cm × 2-4 mm, linear-oblong; peduncles up to 12 cm; bracts green, prominently reticulate-veined, the 3 uppermost ovate, creamy; corolla white.
Fl. III - V (-VI). *Introduced. Grown as a cottage herb in Madeira and naturalized along levadas and near houses in places in the south-east of the island.* **M**

Hybrids between **1** and **2** are known from the Machico area and from between Funchal and Camacha. They are intermediate between the parents in most characters and may have dark or pale sterile bracts and purple or white flowers.

L. dentata L., [*Rosmaninho*] is sparsely grey-tomentose with leaves sessile, 3-4 cm × 3-5 mm, narrowly oblong, pinnatifid, with oblong, blunt lobes; upper bracts narrower than the lower, purplish, somewhat petaloid, and flowers purple. It is cultivated as a cottage herb in Madeira but does not seem to have become naturalized.

3. L. pinnata L. f., *Lavandula*: 11 (1780).
Small, densely grey-tomentose shrub. Leaves 4-8 × 2-4 cm, obovate in outline, 1- to 2-pinnatisect, the lobes oblong to oblong-spathulate, margins inrolled. Verticillasters 2-flowered, in 1-several slender, cylindrical spikes each 4-7 cm and grouped at the tip of a peduncle 11-35 cm long. Bracts all similar, imbricate, 5-6 mm, lanceolate to narrowly ovate, often purple-tinged. Flowers sessile. Calyx 4-4.5 mm, ± campanulate, 15-veined; teeth spreading in fruit, the upper teeth lanceolate and shorter than the ovate lower teeth. Corolla bluish; tube 6 mm; upper lip 2.5-3 mm; lower lip *c*. 1 mm. **Plate 38**
Fl. III (- IX?). *An extremely rare plant of cliffs and rocky places in the Funchal region, mostly near the sea.* ■ M [*Canaries*]

L. angustifolia Mill., (*L. spica* L.) [*Alfazema*], with linear, lanceolate or oblanceolate, entire leaves and 6- to 10-flowered verticillasters in slender spikes is frequently cultivated in gardens in Madeira.

L. rotundifolia Benth. var. *rotundifolia* has been recorded erroneously as occurring in Madeira. The record is based on a specimen from the Cape Verdes, where the species is endemic.

20. Salvia L.

Annual or perennial herbs or shrubs. Verticillasters few- to many-flowered, spikes often interrupted. Bracts different from the leaves, often brightly coloured; bracteoles absent. Calyx tubular or campanulate, 2-lipped, teeth unequal; upper lip 3-toothed or ± entire; lower lip 2-toothed. Corolla 2-lipped, with a straight or invaginated tube; upper lip straight or falcate; lower lip 3-lobed, the central lobe largest. Stamens 2, the connective long and articulated with the filament, the longer arm with a single fertile cell, the shorter with a ± sterile one.

Various species are cultivated as ornamentals in Madeira, including **S. farinacia** Benth., a blue-flowered perennial native to N. and C. America; **S. sessei** Benth., a scarlet-flowered shrub native to Mexico, and **S. splendens** Sellow ex Roem., a scarlet-flowered annual native to Brazil. **S. officinalis** L. [*Salva*], commonly grown in kitchen gardens, is a violet-blue-, pink- or white-flowered shrub, with strongly rugose, undivided leaves and is native to Europe. **S. fruticosa** Mill. (*S. triloba* L. f.) is similar but has 3-lobed leaves and is native to the Mediterranean region.

1. Shrub; inflorescence densely lanate; calyces with purple hairs . . . . . **2. S. leucantha**
- Herb; inflorescence hairy but not lanate; calyces without purple hairs 2

2. Flowers scarlet; upper lip of corolla straight . . . . **3. S. coccinea**
- Flowers blue or lilac; upper lip of corolla falcate . . . . **1. S. verbenaca**

1. S. verbenaca L., *Sp. pl.* **1**: 25 (1753). [*Jarvão* (P)]
S. collina Lowe; *S. verbenaca* subsp. *clandestina* (L.) Briq., var. *dubia* (Lowe) Menezes
Pubescent perennial 6-16(-30) cm. Stems erect, simple or little-branched. Leaves 4-9 × 2-3.5 cm, oblong to ovate, coarsely dentate to pinnatifd with short, broad, crenate lobes, sparsely hairy above, densely hairy beneath. Verticillasters mostly 5- to 6-flowered, in a dense or lax spike. Bracts broadly ovate to wider than long, persistent. Calyx 6-8 mm, accrescent in fruit; upper lip slightly shorter than the lower. Corolla blue or lilac; tube 7-8(-9) mm; upper lip 5-6

mm, falcate; lower lip *c.* half as long as upper.
Fl. I - VI (-VII). *Dry pastures, hillsides and and roadside banks. Common in Porto Santo; rare in Madeira, from Santo Antonio north of Funchal eastwards to Caniço.* **MP**

2. **S. leucantha** Cav., *Icon.* **1**: 16 (1791).
Rounded to spreading shrub up to 120 cm. Stems densely white-tomentose above. Leaves up to 14 × 2 cm, oblong-lanceolate, crenate to serrate, rugose, very shortly pubescent and green above, white-tomentose beneath. Verticillasters 3- to 6-flowered in slender, interrupted spikes. Bracts ovate-elliptic, acuminate, densely lanate, caducous. Calyx 7-9 mm, densely purplish-lanate, often white in the lower part; upper and lower lips ± equal in length. Corolla white; tube 13 mm, invaginated; upper lip 3 mm, straight; lower lip slightly shorter than upper.
Fl. II - VIII. *Introduced. Widely cultivated in gardens in the lower regions of Madeira and naturalized on waste ground, along roadsides and levadas. Native to Mexico.* **M**

3. **S. coccinea** Juss. ex Murray, *Commentat. Soc. Scient. gotting.* **1**: 86 (1778).
S. pseudo-coccinea Jacq.
Annual or perennial up to 70 cm. Stems erect, usually branched, puberulent and with long, spreading, white hairs. Leaves up to 5 × 3 cm, base truncate to cordate, crenate, densely pubescent above and below, usually also with sparse, longer hairs. Verticillasters up to 12(-15)-flowered, in interrupted spikes. Bracts ovate, long-acuminate, caducous. Calyx 8-10 mm, green, densely hairy; lower lip ± equalling or slightly longer than the upper. Corolla scarlet; tube 17 mm; upper lip *c.* 7 mm, straight; lower lip very short.
Fl. III. *Introduced. Widely cultivated in gardens in Madeira, often escaping and becoming naturalized in the lower regions. Native to C. America.* **M**

XCV. SOLANACEAE
M.J. Short

Herbs, shrubs or occasionally small trees. Leaves usually alternate and petiolate, simple to pinnate, exstipulate. Flowers regular or slightly irregular, hermaphrodite, solitary, or in cymes, panicles or spikes. Calyx generally 5-lobed, persistent and often enlarging in fruit. Corolla rotate, funnel-shaped, campanulate or tubular, generally 5-lobed. Stamens usually 5, inserted in the corolla-tube; anthers dehiscing by longitudinal slits or apical pores, sometimes cohering. Ovary superior, 2 (-5)-locular. Style simple; stigma entire to 2-lobed. Fruit a capsule or berry.

1. Corolla with a very short tube and a widely spreading limb, sometimes stellately lobed 2
- Corolla tubular, campanulate or funnel-shaped 4

2. Anthers forming a cone around the style 3
- Anthers not as above .... **7. Normania**

3. Corolla yellow; anthers dehiscing by longitudinal slits .... **8. Lycopersicon**
- Corolla white, blue or violet; anthers dehiscing by terminal pores ..... **6. Solanum**

4. Fruit a capsule 5
- Fruit a berry 8

5. Flowers solitary 6
- Flowers in spikes or panicles 7

6. Flowers erect .... **9. Datura**
- Flowers pendulous .... **10. Brugmansia**

7. Leaves toothed or lobed; flowers in spikes .... **3. Hyoscyamus**
- Leaves simple; flowers in panicles .... **11. Nicotiana**

8. Spiny shrub .... **2. Lycium**
- Annual or perennial herb 9

9. Corolla campanulate; calyx greatly enlarging in fruit and enclosing the berry 10
- Corolla urn-shaped; calyx not enlarging in fruit .... **5. Salpichroa**

10. Glabrous annual; corolla pale blue to lilac .... **1. Nicandra**
- Pubescent perennial; corolla yellow .... **4. Physalis**

1. Nicandra Adans.

Glabrous annual herbs. Leaves alternate, simple. Flowers regular, solitary, axillary. Calyx deeply divided, the lobes sagittate at the base, greatly enlarging in fruit, becoming membranous and bladder-like. Corolla broadly campanulate, shallowly lobed. Stamens included; anthers dehiscing by longitudinal slits. Stigma entire. Ovary 3-5-locular. Fruit a thin-walled, dry berry.

1. N. physalodes (L.) Gaertn., *Fruct. sem. pl.* **2**: 237 (1791).
Stems up to 80 cm or more high, erect, often branched above. Leaves up to 15 × 10 cm, ovate to elliptic, obtuse to acute, the base often unequal, irregularly sinuate-toothed or lobed, attenuate into a narrowly winged petiole. Pedicels 1.5-3 cm. Calyx 12-18 mm in flower, pale green; lobes ovate, cuspidate-acuminate, the margins appressed to form longitudinal wings. Corolla 3-4 cm across, pale blue to lilac with a white throat. Berry 14-18 mm, globose, brownish, nodding, enclosed by the membranous, reticulately-veined, strongly 5-winged calyx. Seeds 1.5-2 mm, brown, reticulate.
Fl. IX - V. *An infrequent casual of waste and cultivated ground, mainly in and around Funchal. Native to Peru.* **M**

2. Lycium L.

Spiny shrubs. Leaves usually in axillary clusters, simple, entire, deciduous. Flowers regular, axillary, solitary or in small clusters. Calyx 5-dentate or weakly 2-lipped with 2-3 teeth, scarcely enlarging in fruit. Corolla narrowly funnel-shaped; lobes spreading. Stamens exserted or included, unequal; anthers dehiscing by longitudinal slits. Stigma weakly 2-lobed. Fruit a small berry.

1. L. europaeum L., *Sp. pl.* **1**: 192 (1753). [*Espinheiro*]
Shrub to 2 m; branches rigid, silvery-white, with stout spines, densely and shortly pubescent when young. Leaves 2-4.5 × 0.3-0.9 cm, mostly oblanceolate, acute to obtuse, attenuate into a short petiole, slightly fleshy, minutely glandular-puberulent below. Flowers solitary or in clusters of 2-3. Calyx 2-2.5 mm, usually 5-dentate. Corolla pale blue or lavender; tube *c.* 10 mm; lobes 3-4 mm, obtuse, shortly ciliate. Stamens usually slightly exserted; filaments glabrous. Berry *c.* 7 mm, ovoid, bright red.
Fl. IV - XII. *Along tracks and roads, sometimes forming large thickets. On Ponta de São Lourenço and scattered along the south coast of Madeira, but uncommon and rarely fruiting. Also rare in Porto Santo, where it has sometimes been planted for hedging.* **MP**

3. Hyoscyamus L.

Annual to perennial herbs, pubescent and viscid. Leaves alternate, simple. Flowers slightly irregular, axillary, forming crowded, bracteate spikes. Calyx campanulate, 5-dentate, enlarging and becoming rigid in fruit. Corolla funnel-shaped, with 5, unequal, obtuse lobes. Stamens included or slightly exserted; anthers dehiscing by longitudinal slits. Stigma capitate. Fruit a capsule, dehiscing by a transverse lid.

1. H. albus L., *Sp. pl.* **1**: 180 (1753). [*Beleno, Meimendro, Meiomento*]
H. major Mill.
Viscid, dull green, usually annual herb; stem (15-)30-100 cm, erect, stout, usually branched, with dense, spreading hairs. Leaves 4-16 × 4.5-12 cm, ovate to orbicular, obtuse, coarsely toothed or lobed, sparsely pubescent; petioles 1.5-9 cm. Flowers sessile, in leafy, 1-sided spikes. Bracts sessile, leaf-like. Calyx 1.3-1.6 mm in flower, densely villous, with short, broadly triangular, acute teeth, 2-3 cm in fruit, becoming swollen below and prominently veined. Corolla 2.3-3.4 cm, pale yellow, the throat usually deep purple within, pubescent without. Filaments purple, pubescent below; anthers cream. Capsules *c*. 1 cm, ovoid, enclosed by the calyx. Seeds *c*. 1.5 mm, oblong, grey, pitted. 2n=68.
Fl. III - X. *Amongst rocks, along roadsides, in fields and on waste ground. Infrequent in Madeira, only at or near sea-level, along the south coast west of Funchal and on the north coast at Porto Moniz. More common in Porto Santo where it mainly grows inland. Also known from Bugío.* **MDP**

4. Physalis L.

Pubescent perennial herbs. Leaves alternate, or 2 per node, unequal in size, simple. Flowers regular, axillary in the leaf axils and forks of the stems, solitary. Calyx campanulate, 5-dentate, greatly enlarging and becoming inflated in fruit. Corolla broadly campanulate; tube short. Stigma capitate; anthers dehiscing by longitudinal slits. Fruit a berry.

1. P. peruviana L., *Sp. pl.* ed. 2: 1670 (1763). [*Tomate, Tomate inglez, Tomateiro inglez*]
Straggling, softly pubescent plant; stem up to 1 m, usually branched. Leaves 5-10 × 4-7 cm, ovate, acuminate, cordate at base, entire or sparingly toothed; petioles 1-4 cm. Pedicels 7-10 mm, erect in flower, drooping in fruit. Calyx 7-9 mm in flower; teeth acuminate. Corolla 1.6-2 cm across, dull yellow with 5 blackish purple spots near the base within. Filaments purple; anthers 3.5-4 mm, blue-purple. Fruiting calyx 3-4 cm, 10-angled, pale green drying to pale brown, pubescent. Berry 11-15 mm, globose, yellow, loosely enclosed by the bladder-like calyx.
Fl. III - IX. *Naturalized on walls and rocks in dry, sunny places. Reported to be a common plant in Madeira in the early part of the nineteenth century, but it has since become scarce and now only occurs very occasionally, mainly in and around Funchal. Also recorded for Porto Santo. Native to S. America, but cultivated elsewhere for its edible fruit.* **MP**

5. Salpichroa Miers

Perennial herbs with leaf-opposed branches. Leaves alternate or 2 per node, unequal in size, simple, entire. Flowers regular, axillary, solitary or in pairs. Calyx divided almost to the base. Corolla urn-shaped, with short, reflexed lobes. Anthers dehiscing by longitudinal slits. Style pubescent below; stigma capitate. Fruit a berry.

1. S. origanifolia (Lam.) Baill., *Hist. pl.* **9**: 288 (1888).
Sprawling herb, becoming somewhat woody below; stems up to 1 m, much-branched, pubescent. Leaves up to 2 × 1.5 cm, ovate-rhombic to suborbicular, subglabrous, ciliate; petioles 5-12 mm. Flowers nodding, in the axils of the upper leaves. Calyx 2-4 mm; lobes narrowly triangular. Corolla 6-10 mm, white or cream; lobes triangular. Berry 1-1.5 cm, ovoid, creamy white, translucent. Seeds 1.5-2 mm, strongly compressed, pubescent.

Fl. ? A S. American species occasionally naturalized along roadsides and on cultivated and waste ground in and around the Funchal area. **M**

Capsicum frutescens L., *Sp. pl.* **1**: 189 (1753) [*Pimenta encarnada, Pimenteira*], a small shrub from tropical America with elliptical leaves, greenish white flowers in groups of 2-3, followed by red berries, is commonly grown in gardens and may occasionally escape. **M**

6. Solanum L.

Herbs, shrubs or small trees. Leaves alternate, simple or pinnate, entire, toothed or lobed. Flowers regular, in pedunculate or sessile cymes, axillary or leaf-opposed. Calyx not or only slightly enlarging in fruit. Corolla stellate and deeply incised to rotate, white, yellow, blue or purplish. Stamens exserted; filaments very short; anthers yellow, forming a cone around the style, dehiscing by terminal pores. Stigma small, capitate. Fruit a berry.

Literature: J.E. Edmonds, Nomenclatural notes on some species of *Solanum* L. found in Europe, in *Bot. J. Linn. Soc.* **78**: 213-233 (1979).

1.	Plant with numerous, stout, yellowish spines	**5. linnaeanum**
-	Plant without spines	2
2.	Shrub or small tree, densely stellate-tomentose	**6. mauritianum**
-	Annual or perennial herb, sometimes shrubby, but then plant glabrous	3
3.	Leaves linear-elliptic to linear-lanceolate; cymes sessile	**4. pseudocapsicum**
-	Leaves rhombic, ovate or ovate-lanceolate	4
4.	Berries black when ripe	**1. nigrum**
-	Berries red or orange when ripe	5
5.	Corolla-lobes white	**2. villosum**
-	Corolla-lobes white with a central purple line	**3. patens**

1. S. nigrum L., *Sp. pl.* **1**: 186 (1753). [*Erva de Santa Maria, Erva moira*]
Annual or short-lived perennial, often bushy; stems 5-50(-100) cm, erect to decumbent, sometimes becoming woody below, often purplish. Leaves 2-7(-11) × 1.5-5(-8) cm, ovate, acute or acuminate, entire to irregularly sinuate-dentate; petioles 0.5-2.5(-7) cm. Cymes (3-)4- to 9-flowered. Peduncles 10-22 mm. Pedicels 7-9 mm in flower, deflexed and up to 12 mm in fruit. Calyx 1-2.5 mm in flower, slightly enlarging in fruit; lobes oblong-ovate. Corolla 10-14 mm across, white, lilac tinged without. Anthers 1.6-2.3 mm, yellow. Berry 0.6-0.9 mm across, globose, dull green turning to black when ripe. Seeds 1.8-2 mm, compressed, finely and shallowly pitted, yellowish brown. **MDPS**

Two subspecies have been recorded:

a) subsp. nigrum.
S. nigrum [var.] α *glabrum* Lowe
Plant subglabrous to sparsely pubescent with short, appressed, eglandular hairs.
Fl. I - XII. *Abundant in Madeira on cultivated and waste ground, walls, in grassy places, and along roadsides, 0-1500 m. Fairly common in Porto Santo and Ilhéu de Cima. Reported to be rare on all the Desertas islands and also recorded from Selvagem Grande.*

b) subsp. **schultesii** (Opiz) Wessely in *Feddes Reprium Spec. nov. veg.* **63**: 311 (1960).
S. nigrum [var.] β *hebecaulon* Lowe
Plant densely pubescent, the hairs fairly long, patent, glandular.
Fl. II - IX. *Roadsides and along levadas, possibly only in the Funchal area of Madeira. Rare.*

S. chenopodioides Lam., *Tabl. encycl.* **2**: 18 (1794) (*S. sublobatum* Willd. ex Roemer & Schultes), a perennial S. American species like **1**, but with ovoid, blackish purple berries, the peduncles sharply deflexed, has been recorded from Funchal but without further data. **M**

2. S. villosum Mill., *Gard. dict.* ed. 8, no. 2 (1768).
S. luteum Mill.
Annual; stems up to 55 cm, decumbent to erect, sparingly branched. Leaves 2-5 × 1-3 cm, ovate to ovate-rhombic, entire to coarsely sinuate-dentate, acute; petioles 0.5-1.5 cm. Cymes 3- to 5-flowered. Peduncles 5-13 mm. Pedicels 5-8 mm in flower, deflexed and up to 10 mm in fruit. Calyx 1-2.5 mm in flower, slightly enlarging in fruit; lobes oblong. Corolla 0.8-1.2 cm across, white. Anthers 1.5-2.5 mm, yellow. Berry 0.5-0.9 cm across, subglose, sometimes slightly longer than wide, yellow- or orange-red when ripe. Seeds 1.6-2 mm, compressed, yellowish brown, finely and shallowly pitted. **M**

Two subspecies have been recorded from Madeira:

a) subsp. **villosum**
S. villosum Lam. [var.] α *velutina* Lowe
Plant villous, with numerous, patent hairs. Stems rounded, smooth.
Fl. VII - IX. *Uncommon on cultivated and waste ground, walls and along roadsides from the Funchal area of Madeira.*

b) subsp. **miniatum** (Bernh. ex Willd.) Edmonds in *Bot. J. Linn. Soc.* **89**(2): 166 (1984).
S. alatum Moench; *S. miniatum* Bernh. ex Willd.; *S. villosum* Lam. [var.] β *laevigata* Lowe
Plant subglabrous to pubescent, the hairs appressed. Stems angled, the ridges narrowly winged and sometimes shallowly denticulate.
Fl. VIII - II. *Very occasional in Madeira.*

3. S. patens Lowe, *Man. fl. Madeira* **2**: 74 (1872).
Like **2b**, but generally larger, with long, spreading branches. Leaves up to 11 × 6 cm, ovate, acute to shortly acuminate, deeply sinuate toothed, at least in the lower half; petioles up to 4 cm. Corolla-lobes white, marked with a central, narrow, purple line. **Plate 38**
Fl. III - XII. *Generally rare in Madeira, but occasionally locally abundant in dry river beds and ravines.* ● **M**

S. patens is not considered by some authors to be distinct from *S. villosum* subsp. *miniatum*.

4. S. pseudocapsicum L., *Sp. pl.* **1**: 184 (1753).
[*Berradura, Erva de Santa Maria, Pimenteira brava*]
Glabrous, shrubby annual or perennial up to 1 m high; stems erect, slender, branched. Leaves alternate or opposite, 2.5-10 × 0.6-2(-2.5) cm, linear-elliptic to -lanceolate, acute, entire to shallowly sinuate; petioles 0.3-2 cm. Cymes 1- to 3(-7)-flowered, sessile. Pedicels 4-10 mm in flower, patent, up to 12 mm in fruit, ± erect. Calyx 3.5-7(-10) mm, enlarging slightly in fuirt; lobes linear-lanceolate. Corolla 1.3-1.5 cm across, stellate, white, the lobes minutely pubescent towards the apex. Anthers 2.5-3.3 mm, orange. Berry 0.9-1.6 cm across, globose, green turning to red when ripe, shiny. Seeds *c.* 4.5 mm, compressed, yellowish brown.
Fl. I - XII. *Common along levadas, roadsides, on rocky banks, and in woodland; 150-700 m.*
M

S. tuberosum L., *Sp. pl.* 1: 185 (1753) [*Semilhas*], a herbaceous perennial with pinnate leaves composed of alternating large and small leaflets, is widely cultivated in Madeira for its edible tubers (potatoes), and to a lesser extent in Porto Santo. Also recorded from Deserta Grande. Native to S. America. **MDP**

5. **S. linnaeanum** Hepper & Jaeger in *Kew Bull.* **41**(2): 435 (1986).
S. sodomeum L.
Shrubby perennial, stellate-pubescent and with numerous stout, yellowish spines up to 14 mm long; stems up to 1 m, erect, branched. Leaves 3.5-12 × 3-9 cm, ± ovate, pinnatifid with rounded, sinuate lobes; petioles 1-3.5 cm. Cymes 2- to 6-flowered, sessile. Pedicels 9-13 mm in flower, erecto-patent, up to 20 mm in fruit, recurved. Calyx 4.5-7 mm in flower, up to 20 mm in fruit; lobes lanceolate. Corolla 2.4-2.8 cm across, pale blue. Anthers 5-6 mm. Berry 1.8-3.2 cm across, globose, marbled green turning to yellow when ripe. Seeds 3.2-3.9 mm, compressed, finely reticulate, light brown. 2n=24.
Fl. VII - XII. *An African species naturalized in the lower regions of Madeira along roadsides, in rough grassland, and waste places, especially near the sea. Fairly frequent on Ponta de São Lourenço.* **M**

6. **S. mauritianum** Scop., *Delic. fl. faun. insubr.* **3**: 16 (1788). [*Tabaqueira*]
S. auriculatum Aiton
Large, foetid, densely stellate-tomentose shrub or small tree up to 4 m with yellowish hairs. Leaves 15-30(-40) × 6-11(-20) cm, ovate to elliptic, acuminate; petioles (2-)4-5 cm, usually with small, leafy pseudo-stipules at the base. Cymes pedunculate, many-flowered, flat-topped. Calyx 5-6 mm; lobes broadly elliptic. Corolla 1.6-2 cm across, stellate to pentagonal, violet-blue. Anthers 2.6-3 mm, pale yellow. Berry *c.* 1.5 cm across, globose, dull yellow. 2n=24.
Fl. I - X. *Introduced to Madeira by the English botanist R.T. Lowe in 1833 and now commonly naturalized along levadas, in ravines, and on cultivated and waste ground, 0-500 m. Native to C. America.* **M**

Records of *S. erianthum* D. Don from Madeira are referable to this species.

7. Normania Lowe

Densely glandular-pubescent herbs or subshrubs. Leaves trifoliolate. Flowers slightly irregular. Calyx campanulate, large and leafy in fruit. Corolla rotate. Filaments very short; anthers unequal in length, with 2 long, 2 medium and 1 very short, dehiscing by subterminal pores and the 4 longest also by longitudinal slits; 4 longest anthers horned in the lower half. Style very long; stigma capitate. Fruit a berry enclosed by the persistent, enlarged calyx.

Literature: J. Francisco-Ortega, J.G. Hawkes, R.N. Lester & J.R. Acebes-Ginovés, *Normania, an endemic Macaronesian genus distinct from Solanum (Solanaceae)*, in *Pl. Syst. Evol.* **185**: 189-205 (1993).

1. **N. triphylla** (Lowe) Lowe, *Man. fl. Madeira* **2**: 87 (1872).
Nycterium triphyllum Lowe; *Solanum trisectum* Dunal
Leaflets with entire to weakly toothed margins; 2 lateral leaflets 3-4 × 1.5-2 cm, oblong, acute, unequal at the base; terminal leaflet larger, 5-7 × 3-4.5 cm, ovate, acute. Calyx-lobes lanceolate, acute. Corolla *c.* 2 cm across, deep violet or blue-purple with a very dark central blotch. Filaments with shortly stalked glandular hairs. Anthers dark purple to blackish, the two longest 8-11 mm, curved, the medium two 6-7 mm, also curved, the shortest 3-4.5 mm, not curved. Style 11-12 mm, white, curved towards the apex. Fruiting calyx with leafy, broadly lanceolate lobes. Berry *c.* 1.5 cm across, globose, orange to crimson. Seeds 4 × 3 mm, margins irregularly crenate. **Plate 38**

Fl. VI - IX. *Extremely rare endemic plant of the Madeiran laurisilva, for a long time known only from early collections made in the last century. However, its continuing presence was confirmed in 1991 by the discovery of two small populations in the north of the island.* ● **M**

8. Lycopersicon Mill.

Glandular-pubescent annual herbs. Leaves alternate, pinnate, with alternating large and small leaflets. Flowers regular, in leaf-opposed cymes. Calyx divided almost to the base, the lobes slightly enlarged and patent to reflexed in fruit. Corolla stellate, deeply 5(-6)-lobed. Filaments very short; anthers coherent, forming a cone around the style, dehiscing by longitudinal slits, produced apically into a slender, sterile appendage. Stigma capitate. Fruit a berry.

1. L. esculentum Mill., *Gard. dict.* ed. 8, no. 2 (1768). [*Tomate, Tomateiro*]
Solanum lycopersicum L.
Strong smelling plant with erect or sprawling stems 30-100 cm long. Leaves 10-40 cm; leaflets ovate to ovate-lanceolate, toothed to lobed. Pedicels articulated near the middle, deflexed in fruit. Corolla 15-25 cm across, yellow; lobes lanceolate, acute. Berry 2-9 cm across, globose, densely glandular-pubescent when young, usually red when ripe. Seeds compressed, suborbicular, yellowish, surrounded by mucilage.
Fl. IV - X. *Cultivated for its edible fruit and found as a casual in the lower regions of Madeira on cultivated ground, in waste places, and among rocks. Also recorded from Porto Santo and Ilhéu de Cima, Ilhéu Chão, and Selvagem Grande. Native to S. & C. America.* **MDPS**

9. Datura L.

Shrubby annual herbs. Leaves alternate, simple. Flowers regular, solitary, axillary or in the forks of the stems, erect. Calyx tubular, 5-toothed, circumscissile, the lower part persistent and enlarging in fruit. Corolla large, funnel-shaped, shallowly 5-lobed, sometimes appearing 10-lobed, the lobes acute to acuminate. Stamens included; anthers dehiscing by longitudinal slits. Stigma 2-lobed. Fruit a spiny capsule, dehiscing by 4 valves or irregularly.

1. Stems glabrous; leaves deeply sinuate-dentate to lobed; capsules erect   **1. stramonium**
- Stems densely pubescent; leaves usually ± entire; capsules nodding . . . . **2. innoxia**

1. D. stramonium L., *Sp. pl.* 1: 179 (1753). [*Bufareira*]
Foetid annual, ± glabrous to puberulent; stems 5-50(-100) cm, erect, simple or branched. Leaves petiolate, 4.5-18 × 2.5-15 cm, ovate to elliptic, acute, deeply sinuate-dentate to lobed. Flowers scentless. Calyx c. 3.5 cm, sharply 5-angled and narrowly winged; teeth 5-9 mm, unequal. Corolla 7-9 cm, white, or blue to purple; lobes terminating in a slender point. Capsules 2.5-4 cms, ovoid, erect. $2n=24$.
Fl. III - XII. *A rather infrequent casual in Madeira and Porto Santo, found along roadsides, in fields, cultivated ground, and waste places. Native to America.* **MP**

Plants with blue to purple corollas are sometimes recognized as var. **tatula** (L.) Torr., *Fl. n. middle United States* **1**: 232 (1824).

2. D. innoxia Mill., *Gard. dict.* ed. 8, no. 5 (1768).
Foetid, softly glandular-pubescent annual; stems up to 1 m, erect, simple or branched above. Leaves petiolate, 5-16 × 4-11 cm, ovate, acute, often unequal at the base, entire or very shallowly sinuate. Flowers fragrant at night. Calyx 6-9 cm; teeth 11-25 mm, unequal. Corolla 11-19 cm, white, appearing 10-lobed, alternate lobes triangular or terminating in a slender

point. Capsules 3-5 cm, ovoid, nodding.
Fl. VI - I. *A rare casual in cultivated ground around Funchal. Native to C. America.* **M**
Records of *D. metel* L. from Madeira are referable to this species.

10. Brugmansia Pers.

Shrubs or small trees. Leaves alternate, simple, entire. Flowers solitary, pendulous. Calyx tubular, sometimes inflated, 5-dentate or separating into 1 or more spathaceous lobes. Corolla large, funnel-shaped; lobes narrowly acute, recurved or spreading. Stamens included; anthers dehiscing by longitudinal slits. Stigma 2-lobed. Fruit an indehiscent berry.

B. sanguinea (Ruiz & Pav.) D. Don (*Datura sanguinea* Ruiz & Pav.), a shrub with orange-red corollas and **B. × candida** Pers. (*Datura candida* (Pers.) Saff.), a shrub or small tree with spathaceous calyces and white corollas with strongly recurved lobes, are cultivated for ornament in Madeiran gardens.

1. B. suaveolens (Humb. & Bonpl. ex Willd.) Bercht. & J. Presl, *Prir. rostlin* 1. Solanaceae 45 (1823). [*Bellas noites, Trombetas, Trombeteira*]
Datura suaveolens Humb. & Bonpl. ex Willd.
Shrub or small tree, 2-4 m high. Leaves petiolate, 13-30 × 6-10 cm, ovate-oblong to elliptic, acute, often very unequal at the base, entire, puberulous on the veins. Flowers fragrant. Calyx 5-toothed, 8-12 cm, glabrous; teeth 11-20 mm, subequal. Corolla 25-30 cm, white; lobes pendent. Anthers coherent into a tube. Fruit narrowly ellipsoid, attenuate at each end.
Fl. IV - XII. *A S. American species commonly planted for ornament in Madeira. Reported by Lowe in the nineteenth century as 'completely choking up with a thick jungle small damp ravines' in the north of Madeira near Ponta Delgada.* **M**

11. Nicotiana L.

Annual to perennial herbs or small shrubs. Leaves alternate, simple and entire. Inflorescences terminal, paniculate. Calyx tubular, 5-dentate, usually much smaller than the corolla. Corolla tubular or funnel-shaped, with 5 shallow lobes. Stamens usually included, one sometimes shorter than the other four; anthers dehiscing by longitudinal slits. Stigma capitate. Fruit a capsule.

1. Shrub; leaves glaucous, glabrous . . . . **1. glauca**
-   Herb; leaves green, viscid-pubescent . . . . **2. tabacum**

1. N. glauca Graham in *Edinb. New. phil. J.* **5**: 175 (1828).
[*Tabaqueira, Tabaqueira azul*]
Slender, loosely branched, glaucous shrub, 1-4 m. Leaves long-petiolate, 3-10(-13) × 1.3-5.5(-7) cm, elliptic to lanceolate or ovate, acute, glabrous. Panicle lax. Flowers pedicellate. Calyx 7-12 mm, tubular; teeth triangular, acute, shortly ciliate. Corolla 28-40 mm, tubular, slightly constricted beneath the limb, yellow, shortly glandular-pubescent; lobes 1.5-3 mm, erect. Stamens subequal. Fruiting pedicels downcurved. Capsules 8-12 mm, ovoid, enclosed by the papery calyx.
Fl. (I-) III - IX. *In open places among rocks, on waste ground, and along roadsides. In the nineteenth century reported to be scarcely established in Madeira, but now a commonly naturalized plant along the south coast, 0-200(-300) m. Frequent in Porto Santo, Ilhéu de Cima, and on Ilhéu de Baixo where it is dominant over much of the island. Also found on cliffs and in sheltered gullies on Selvagem Grande. Native to S. America.* **MPS**

2. N. tabacum L., *Sp. pl.* **1**: 180 (1753). [*Tabaco, Tabaqueira*]
Erect, viscid-pubescent annual or short-lived perennial up to 2 m. Leaves up to 35 × 18 cm, elliptic to oblong-lanceolate, shortly acuminate, decurrent at the base into a very short petiole. Panicle much-branched. Flowers pedicellate. Calyx 14-18 mm, tubular-campanulate; teeth triangular, acuminate, unequal. Corolla 40-55 mm, tubular to funnel-shaped, pale green with a pink throat and limb; lobes short, acuminate, spreading. Stamens 4, unequal, sometimes slightly exserted. Fruiting pedicels erect. Capsules (12-)15-22 mm, ovoid, enclosed by the calyx.
Fl. I - XII. *Reported to be formerly widely cultivated in Madeira and quite commonly naturalized on waste and cultivated ground, but now seldomly seen; also rare in Porto Santo. Recorded from Selvagem Grande in the last century, but no subsequent records known. Believed native to northern S. America.* **MPS**

N. wigandioides K. Koch & Fintelm., *Wschr. gärtn. pflanzenk* **1** (suppl. 2): 5 (1858), a small, slender shrub from Bolivia with conspicuously silvery-white pubescent young stems, many-flowered, pendent panicles, the corollas 10-15 mm long with a greenish white tube and cream, spreading lobes, has been recorded from Porto Santo but without further data. **P**

XCVI. SCROPHULARIACEAE
M.J. Short

Herbs, rarely shrubs or small trees. Leaves exstipulate, alternate, opposite or whorled, simple, entire, toothed or variously lobed. Flowers hermaphrodite, irregular, though sometimes only slightly, in bracteate spikes or racemes, sometimes solitary in the leaf-axils or in cymes. Calyx 4-5(-8)-lobed, occasionally 2-lipped, often persistent. Corolla 4-5(-8)-lobed or clearly 2-lipped, the tube sometimes spurred. Stamens generally 4 or 2, occasionally 5 or more; staminodes sometimes present. Ovary superior, 2-locular. Style 1; stigma mostly capitate. Fruit usually a capsule; seeds usually numerous and small.

Various species belonging to this family are grown in Madeiran gardens as ornamentals and are occasionally observed outside of cultivation, including the C. American **Lophospermum erubescens** D. Don (*Asarina erubescens* (D. Don) Pennell) and **Maurandya scandens** (Cav.) Pers. (*Asarina scandens* (Cav.) Pennell; *Maurandya semperflorens* Ortega) and the S. African **Diascia barberae** Hook.f.

1. Corolla with a very short tube and spreading lobes, not or only weakly 2-lipped ... 2
- Corolla usually with a distinct tube and clearly 2-lipped 4

2. Fertile stamens 2; corolla 4-lobed, blue, pink or white .... **11. Veronica**
- Fertile stamens 5 or more; corolla 5- or more lobed, yellow 3

3. Erect herbs; flowers in large, terminal spikes or panicles; leaves oblong, elliptic or oblanceolate .... **2. Verbascum**
- Creeping herbs with stems rooting at the nodes; flowers axillary; leaves reniform to orbicular .... **12. Sibthorpia**

4. Corolla with a spur or pouch at the base 5
- Corolla without a spur or pouch at the base 8

5. Sprawling or trailing herbs; leaves distinctly petiolate 6
- Erect annuals or perennials; leaves tapering to base 7

6. Flowers lilac with a yellow palate; petioles long, exceeding the lamina **6. Cymbalaria**
- Flowers yellow, cream or white with purple upper lip; petioles short, not exceeding the lamina . . . . **7. Kickxia**

7. Corolla 30-45 mm long; calyx-lobes broadly ovate, ± equal; perennial **4. Antirrhinum**
- Corolla 13-22 mm long; calyx-lobes linear, very unequal; annual . . . . **5. Misopates**

8. Lower corolla-lip entire, saccate, inflated . . . . **8. Calceolaria**
- Lower corolla-lip 3-lobed, neither saccate nor inflated 9

9. Calyx 5-lobed 10
- Calyx 4-lobed 13

10. Leaves opposite 11
- Leaves alternate 12

11. Corolla large, yellow, the tube cylindrical-campanulate . . . . **1. Mimulus**
- Corolla small, green, reddish or white, the tube ± globose . . . . . . **3. Scrophularia**

12. Biennial herb; corolla pinkish purple . . . . **9. Digitalis**
- Large shrub; corolla orange-yellow . . . . **10. Isoplexis**

13. Corolla white, the upper lip flushed with pinkish purple . . . . **14. Bartsia**
- Corolla yellow 14

14. Corolla 10-14 mm; flower spike 1-sided; seeds *c.* 1.5 mm, rather few . **13. Odontites**
- Corolla 16-24 mm; flower spike not 1-sided; seeds *c.* 0.4 mm, numerous . . . .
. . . . **15. Parentucellia**

1. Mimulus L.

Perennial herbs. Leaves opposite, dentate. Flowers solitary in the axils of the upper leaves. Calyx tubular, 5-angled, 5-toothed. Corolla with a cylindrical-campanulate tube, the throat pubescent within; limb 2-lipped, the upper 2-lobed, the lower longer, 3-lobed. Stamens 4, included. Stigma 2-lobed. Capsules oblong or linear, loculicidal, enclosed within the calyx.

1. **M. moschatus** Douglas ex Lindl. in *Bot. Reg.* **13**: t. 1118 (1828).
Viscid, pubescent perennial sometimes smelling of musk; stems up to 40 cm, decumbent to ascending, simple or branched. Leaves all petiolate, 1.5-5 × 0.8-3 cm, ovate or elliptical, distantly and shallowly toothed. Pedicels (4-)10-15 mm. Calyx 8-10 mm; teeth narrowly triangular-lanceolate, subequal. Corolla 1.5-3 cm, pale yellow; mouth of tube open; lobes rounded. Seeds tuberculate. Fl. IV - VI. *Wet or shady places along levadas. A N. American species cultivated for ornament and occasionally naturalized. Recorded from near Camacha and Caniço, south-east Madeira, in the last century and known to be still present at the former locality in 1956, but its current status unknown.* **M**

2. Verbascum L.

Biennial herbs with tall, erect stems, pubescent with simple or branched hairs. Leaves mostly alternate, the basal forming a rosette, entire, toothed or lobed. Inflorescence a long, terminal raceme, spike or panicle. Calyx equally and deeply 5-lobed. Corolla yellow, with a very short tube and a spreading, slightly irregular, 5-lobed limb. Stamens 4 or 5; filaments usually villous. Capsules globose or broadly ellipsoid, septicidal, usually pubescent.

1. Inflorescence usually simple; calyx-lobes 7-10 mm; capsules 6-8 mm . . . **1. virgatum**
-   Inflorescence branched; calyx-lobes 2-4 mm; capsules 2.5-4 mm 2

2. Basal leaves sinuate-pinnatifid; bracts ovate-deltate; filament hairs violet . **2. sinuatum**
-   Basal leaves weakly crenulate to subentire; bracts linear; filament hairs white . . . . .
    . . . . **3. pulverulentum**

1. V. virgatum Stokes in With., *Bot. arr. Brit. pl.* ed. 2, **1**: 227 (1787). [*Verbasco*]
V. blattarioides Lam.
Glandular-pubescent biennial up to 1.5 m. Basal leaves up to 25 × 8 cm, oblong-lanceolate, coarsely crenate, sessile or shortly petiolate, sometimes glabrescent; cauline leaves smaller, sessile. Inflorescence lax, slender, simple or occasionally with axillary racemes. Flowers 1-5 in the axil of each bract. Bracts 9-15 mm, lanceolate to ovate; bracteoles usually present. Calyx 7-10 mm; lobes linear-elliptic. Corolla 3-4 cm across. Stamens 5; filament hairs purple. Capsules 6-8 mm, globose.
Fl. IV - VIII. *Grassy places. Near the top of Pico Branco, Pico do Facho, and Pico do Castelo in Porto Santo; infrequent. Collected from Jardim da Serra in Madeira in the last century and re-found in the same area in 1978.* **MP**

V. creticum (L.) Cav., *Elench. pl. horti matr.*: 39 (1803) (*Celsia cretica* L.), a biennial from the western Mediterranean region with toothed or lobed leaves, serrate bracts and calyx-lobes, and corollas 4-5 cm across with 4, unequal stamens, has been recorded from Madeira, probably only as a garden escape and apparently not persisting. **M**

2. V. sinuatum L., *Sp. pl.* **1**: 178 (1753). [*Verbasco*]
More or less densely grey- to yellowish tomentose biennial up to 1.5 m, sometimes slightly floccose. Basal leaves up to 35 × 15 cm, elliptic to oblanceolate, sinuate-pinnatifid, often undulate, sessile or very shortly petiolate; upper cauline leaves oblong to ovate, acute to acuminate, crenulate to subentire, shortly decurrent. Inflorescence branched; flowers in widely spaced clusters of 2-7. Bracts 3-8(-15) mm, ovate-deltate, shortly cuspidate. Calyx 3-4 mm; lobes triangular-lanceolate, acute. Corolla 1.5-2 cm across. Stamens 5; filament hairs violet. Capsules 3-4 mm, broadly ellipsoid.
Fl. VI - VIII. *Very rare in Madeira, a sporadic ruderal in and around Funchal.* **M**

3. V. pulverulentum Vill., *Prosp. Hist. pl. Dauphiné*: 22 (1779). [*Verbasco*]
V. floccosum Waldst. & Kit., *V. haemorrhoidale* Aiton
Biennial up to 1.5 m, densely white-floccose, especially when young. Basal leaves up to 40 × 15 cm, oblong-obovate to oblanceolate, weakly crenulate to entire, sessile or shortly petiolate; cauline leaves sessile, smaller, the upper cordate, acuminate. Inflorescence branched, fairly dense; flowers 2-10 in the axil of each bract. Bracts 2.5-5 mm, linear, acute. Calyx 2-3 mm; lobes linear to linear-lanceolate, acute. Corolla *c.* 2 cm across. Stamens 5; filament-hairs white. Capsules 2.5-4 mm, subglobose.
Fl. II - IX. *A rare casual from Europe, scattered and local in Madeira, its current status unknown.* **M**

V. densiflorum Bertol., *Rar. Ital. pl.* **3**: 52 (1810) (*V. thapsiforme* Schrad.) [*Verbasco*] and **V. thapsus** L. subsp. **thapsus**, *Sp. pl.* **1**: 177 (1753), both greyish, whitish or yellowish pubescent biennials with the upper cauline leaves long-decurrent, a dense, woolly, usually simple, spike-like raceme, and the three upper filaments with white or yellowish hairs, the two lower glabrous, have been reported from Madeira. The former, with a spathulate stigma, was recorded in the last century as a very rare casual in and around Funchal, and a single plant of the latter, with a capitate stigma, has been recorded more recently from Chão da Ribeira above Seixal. **M**

3. Scrophularia L.

Annual or perennial herbs with erect, often square stems. Leaves usually opposite, serrate or crenate. Flowers in axillary cymes (sometimes reduced to one flower), forming terminal, lax panicles or racemes. Calyx deeply 5-lobed. Corolla dull red, yellowish green or rarely white, urn-shaped to almost globose, usually ± 2-lipped, shortly 5-lobed, the upper 2 lobes joined at the base. Fertile stamens 4; scale-like staminode usually present, inserted under upper lip of corolla. Capsules septicidal. Seeds globose to ovoid.

Literature: V. Dalgaard, Biosystematics of the Macaronesian species of *Scrophularia*, in *Op. bot. Soc. bot. Lund* **51**: 1-64 (1979).

1. Annual; corolla 3-5 mm; capsules conical, beaked 2
- Perennial; corolla 5.5-10 mm; capsules ovoid to subglobose, not beaked 3

2. Corolla brownish red; stem usually densely glandular-pubescent .... **4. arguta**
- Corolla white with purple markings; stem usually with sparse, sessile glands **5. lowei**

3. Calyx-lobes eglandular-pubescent; bracts all leaf-like .... **1. scorodonia**
- Calyx-lobes glabrous or with a few glands; at least the upper bracts very different from the leaves 4

4. Calyx-lobe margins broadly scarious, brown and often dentate; anthers yellow with brownish apex; leaves single serrate to crenate, the base often with a pair of small lobes beneath
   .... **3. racemosa**
- Calyx-lobe margins not or only narrowly scarious and then white and entire; anthers usually dark red; leaves usually double serrate to crenate, the base without small lobes beneath
   .... **2. hirta**

1. S. scorodonia L., *Sp. pl.* **1**: 620 (1753).
Robust perennial herb up to 1.5 m; stem erect, simple or branched above, pilose, sometimes glabrous below. Leaves 3.5-16 × 1.8-9.5 cm, triangular-ovate, margins double serrate to double serrate-crenate, pilose below, glabrescent above, margins pubescent; petioles 0.4-4(-9) cm. Inflorescence with scattered eglandular hairs and numerous long-stalked glands. Bracts usually all leaf-like. Bracteoles narrowly linear-lanceolate. Calyx-lobes 3-4.5 mm, oblong-ovate, ± obtuse, eglandular-pubescent, margin scarious, pale brown. Corolla 7-10 mm, yellowish green tinged with brownish purple. Stamens slightly exserted; anthers yellow. Capsules 5-8 × 4-6 mm, ovoid, acute. 2n=58.
Fl. IV - X. *Common in eastern Madeira along levadas and roadsides, on walls, steep slopes and rock faces, and at the margins of laurisilva; 100-900 m.* **M**

2. S. hirta Lowe in *Trans. Camb. phil. Soc.* **4**: 21 (1831).
S. confusa Menezes; *S. hirta* subsp. *ambigua* Menezes; *S. longifolia* Benth.; *S. pallescens* Lowe ex Menezes
Variable perennial herb up to 1.5(-2) m; stem erect or hanging from cliffs, usually branched near base, glabrous or shortly villous. Leaves 3.5-25 × 2-10 cm, oblong-lanceolate to ovate, margins double- or less commonly single-serrate to -crenate, glabrous or pilose-villous, margins glabrous; petioles 0.8-9 cm. Inflorescence sparsely to densely glandular with sessile or shortly stalked glands. Lowermost bracts often leaf-like; upper bracts oblong-elliptic to oblong-obovate. Bracteoles similar to upper bracts, but smaller. Calyx-lobes 2.5-4 mm, oblong-ovate to suborbicular, obtuse, glabrous, margin not or narrowly scarious, white and entire. Corolla 5.5-8.5 mm, dark red. Stamens exserted; anthers red, rarely yellow. Capsules 4-6 × 4-7 mm,

ovoid to subglobose, acute to obtuse. 2n=58. **Plate 39**
Fl. V - IX. *Fairly common on cliffs, banks, rocks, walls, along levadas and roadsides, usually in damp places, mainly on the north coast of Madeira and in the mountains of the central ridge, but also occurring near Prazeres in the south-west; 10-1800 m.* ● M

S. hirta × scorodonia
S. × *spuria* Menezes
Stems and leaves glabrous to densely pilose-villous. Lower bracts leaf-like, uppermost reduced. Calyx-lobes with broad scarious margin, eglandular hairs absent, sometimes with a few stalked glands. Capsules abortive.
Hybrid specimens have been collected from Ribeira Frio, Ribeiro d'Ametade, and between Espigão Amarelo and Porto da Cruz. Plants with ± developed capsules have been interpreted as backcrosses by Dalgaard (1979: 22). ● M

Records of *S. smithii* Hornem. and *S. langeana* Bolle from Madeira are referable either to *S. hirta* or *S. hirta* × *scorodonia*.

3. S. racemosa Lowe in *Trans. Camb. phil. Soc.* **4**: 20 (1831).
Perennial herb up to 1.5 m; stem erect or hanging from cliffs, branched above, glabrous, usually dark purple. Leaves 3.5-20 × 1.5-7.5 cm, oblong-lanceolate to narrowly ovate, the base sometimes markedly oblique and often with a pair of small lobes beneath, glabrous or with sparse sessile glands, margin usually single serrate to crenate-serrate; petioles 0.5-14 cm. Inflorescence with few, more or less sessile glands. Bracts narrowly linear-lanceolate to -elliptical. Bracteoles similar but smaller. Calyx-lobes 2.5-4 mm, oblong-ovate, glabrous, margin broadly scarious, brown and often dentate. Corolla 5.5-8.5 mm, dark red to violet-brown. Stamens exserted; anthers yellow. Capsules 4-6 × 3.5-5 mm, ovoid to subglobose, acute to obtuse. 2n=84. **Plate 39**
Fl. V - IX. *Locally abundant in Madeira on wet rocks, steep slopes, along levadas and in other damp places. Recorded from Ribeira de S. Jorge in the north of the island and to the east and north of Funchal in the south; 100-600 m.* ● M

Records of *S. auriculata* L. from Madeira are referable to this species.

S. racemosa × scorodonia
S. moniziana Menezes
Leaves double crenate-serrate, sparsely pilose beneath. Inflorescence with numerous stalked glands. Bracts small, not leaf-like. Calyx with glands; eglandular hairs absent. Capsules undeveloped or very small.
Fl. ? *Recorded in Madeira where the parent species grow together.* ● M

Records of *S. laxiflora* Lange and *S. laevigata* Vahl from Madeira are referable to this hybrid.

4. S. arguta Sol. ex Aiton, *Hort. kew.* **2**: 342 (1789).
Annual, 2-75 cm; stem erect, simple or branched, densely glandular villous, often reddish or purplish. Leaves opposite or alternate, 1.5-6(-9) × 1.5-5(-6) cm, ovate, acute, sparsely to densely glandular-pubescent, margins double-serrate; petioles 0.3-3.5(-4.5) cm. Inflorescence densely glandular-pubescent. Bracts leaf-like. Bracteoles narrowly triangular. Pedicels generally 3-5 mm long, c. 0.5 mm wide. Calyx-lobes 2-3.5 mm. oblong-ovate, keeled. Corolla 3.5-5 mm, brownish red. Stamens included; anthers brownish red. Cleistogamous flowers sometimes produced near the base of the plant. Capsules 4-8 × 2.5-5 mm, conical, beaked.
Fl. III. *Rocky slopes and cliffs on Selvagem Grande.* S

Records for *S. arguta* from Madeira and Porto Santo are probably referable to *S. lowei* Dalgaard, but see comments under that species.

5. S. lowei Dalgaard in *Op. bot. Soc. bot. Lund* **51**: 37 (1979). [*Focinho de burro*]
Annual, 4-90 cm; stem erect, simple or branched, sparsely covered with sessile or stalked glands, sometimes villous at base, often purplish. Leaves opposite or alternate, 1.5-7.5(-12) × 1.2-5(-12) cm, ovate, acute, glabrous or with scattered sessile glands or less commonly glandular hairs, pale green or sometimes purplish, margin double serrate; petioles 0.4-3(-6) cm. Inflorescence with shortly stalked glands. Bracts leaf-like. Bracteoles narrowly linear-elliptic. Pedicels generally 4-9 mm long, *c.* 0.3 mm wide. Calyx-lobes 1.7-4 mm, oblong-ovate, keeled. Corolla 3-4.5 mm, white with purplish markings within and without. Stamens included; anthers purplish red. Capsules 5-9 × 2.5-5 mm, conical, beaked. 2n=40. **Plate 39**
Fl. III - V. *Rare; on rocky slopes, walls, cliffs, and waste ground. Occurring in the south of Madeira in ribeiras near Funchal, also recorded from Porto Moniz in the north; 0-200 m. In Porto Santo known from the peak of Ana Ferreira, Sitio do Lombo, and Enguias, as well as on Ilhéu de Baixo and is very common on Ilhéu de Cima. Also recorded from Ilhéu Chão and Bugío.* ● **MDP**

Dalgaard distinguished *S. lowei* from *S. arguta* (from the Salvage Islands) on the basis of an absence of cleistogamous shoots, longer and more slender pedicels, longer and broader bracteoles, a sparse indumentum of mostly sessile glands, and a white corolla. However, specimens of *S. lowei* with glandular hairs on the stems and leaves are not uncommon and two collections, one from Selvagem Grande and one from Ilhéu de Cima, Porto Santo, are white-flowered, but otherwise more closely resemble *S. arguta* with their short, thick pedicels and numerous stalked glands on the stem.

4. Antirrhinum L.

Perennial herbs. Leaves usually opposite below and alternate above, subsessile, entire. Flowers in terminal, bracteate racemes. Calyx deeply 5-lobed, the lobes shorter than the corolla-tube, ± equal. Corolla with a broad, ± cylindrical tube, gibbous at the base; limb 2-lipped; upper lip 2-lobed; lower lip 3-lobed with a prominent basal palate closing the tube mouth. Stamens 4, included. Capsules asymmetrical, ovoid, dehiscing by 3 apical pores. Seeds numerous.

Literature: D.A. Sutton (1988), *A revision of the tribe Antirrhineae*: 67-93. London.

1. A. majus L., *Sp. pl.* **1**: 617 (1753). [*Bocca de peixa*]
Short-lived perennial herb, often becoming woody at the base, usually glandular-pubescent above and glabrous below; stems 30-80 cm, erect or ascending, simple or sparingly branched. Leaves 30-70 × 4-20 mm, linear to lanceolate, attenuate. Inflorescence dense in flower, becoming lax in fruit. Bracts lanceolate to ovate. Pedicels 1-7 mm in flower, becoming stout in fruit. Calyx 5-8 mm; lobes broadly ovate, obtuse. Corolla 30-45 mm, glandular-pubescent without, usually purplish pink with yellow palate. Capsules 8-12 mm, the locules unequal, usually glandular-pubescent. Seeds 0.8-1 mm, reticulate, blackish brown.
Fl. IV - VII. *Cultivated for ornament and occasionally naturalized on old walls, buildings, and dry, waste places in and around Funchal.* **M**

5. Misopates Raf.

Annual herbs. Leaves opposite below, alternate above, subsessile, entire. Flowers in leafy or bracteate racemes, very shortly pedicellate. Calyx deeply and very unequally 5-lobed. Corolla with a ± cylindrical tube, gibbous at the base; limb 2-lipped, the upper lip 2-lobed, the lower lip with a prominent basal palate closing the tube mouth. Stamens 4, included. Capsules asymmetrical, ovoid, dehiscing by 3 apical pores.

Literature: D.A. Sutton (1988), *A revision of the tribe Antirrhineae*: 145-154. London.

1. Capsules ovoid, 5-6.5 mm wide; bracts 15-30(-45) mm (Madeira and Porto Santo) . 2
- Capsules narrowly oblong-ovoid, 3.5-4 mm wide; bracts 3-5 mm (Salvage Is.) . . . . **3. salvagense**

2. Inflorescence lax in flower; corolla 13-16 mm, usually pink or white . . . **1. orontium**
- Inflorescence dense in flower; corolla 20-22 mm, white, usually with purple markings on upper lip . . . . **2. calycinum**

1. M. orontium (L.) Raf., *Autik. bot.*: 158 (1840).
[*Boca de peixe, Bocca de peixa, Focinho de burro*]
Antirrhinum orontium L.
Stems 15-70 cm, erect, solitary, simple or branched above, glabrous or pubescent. Leaves subsessile, 20-60(-75) × 2-10(-12) mm, linear to linear-lanceolate, acute. Inflorescence lax, usually glandular-pubescent. Bracts 15-30(-45) mm, linear, leaf-like. Pedicels *c.* 1 mm in flower, 3-4 mm in fruit. Calyx-lobes 6-16(-18) mm, linear, with scattered long hairs. Corolla 13-16 mm, usually pink or lilac with purple veins, palate usually pale; upper lip sinus 1.3-1.7(-2.5) mm; lower lip sinus 1-1.5(-1.7) mm. Capsules 6-9 × 5-6.5 mm, ovoid, usually densely glandular-pubescent. Seeds 0.8-1.2 mm, the dorsal face (bearing hilum) with low median longitudinal ridge, otherwise smooth, the ventral face with a discrete, thick, sinuate but unbranched marginal ridge and a low median longitudinal ridge. n=8.
Fl. III - X. *Very common in the lower regions of Madeira on waste and cultivated ground, walls, and grassy places, up to c. 600 m; frequent in Porto Santo and also known on Deserta Grande.* **MDP**

Represented in the Madeiran archipelago by subsp. **orontium**, to which the above description refers.

A plant resembling *M. orontium* in most respects was collected from Selvagem Grande in 1985. However, with seeds differing in some morphological features and only about half the size of those of typical *M. orontium*, it may represent a distinct taxon.

2. M. calycinum (Vent.) Rothm. in *Feddes Reprium Spec. nov. veg.* **136**: 112 (1956).
Antirrhinum orontium var. *calycinum* (Vent.) Lange
Like **1**, but inflorescence dense in flower, the flowers overlapping; corolla 20-22 mm, white, usually with purple veins on the upper lip, the tube sometimes flushed with pink; upper lip sinus 2.5-4 mm; lower lip sinus 2-3.5 mm.
Fl. III - IX. *Fairly rare, growing in similar habitats to* M. orontium. **M**

Specimens with characters intermediate between *M. calycinum* and *M. orontium* are found and it has been suggested that these two species may hybridize where they occur together.

3. M. salvagense D.A. Sutton, *Rev. Antirrhineae*: 151 (1988).
Like **1**, but bracts 3-5 mm; capsules 6-8 × 3.5-4 mm, narrowly oblong-ovoid; seeds 0.65-0.85 mm, the dorsal face (bearing hilum) with several low longitudinal ridges anastomosed with a marginal ridge, the ventral face with a thick, branched marginal ridge anastomosed with the prominent, longitudinal median ridge, giving an irregularly honeycomb-like appearance.
Plate 40
Fl. ? *Selvagem Grande.* ● **S**

This recently described species with its distinctive seed morphology was based on a single fruiting specimen lacking both leaves and flowers, collected from Selvagem Grande between 1860 and 1868. No further specimens have been found.

6. Cymbalaria Hill

Slender, trailing, short-lived perennial herbs or sometimes annuals. Leaves mostly alternate, long-petiolate, reniform to suborbicular, lobed. Flowers long-pedicellate, solitary in the leaf-axils. Calyx deeply and ± equally 5-lobed. Corolla with a cylindrical tube, spurred at the base; limb 2-lipped, the upper lip 2-lobed, the lower 3-lobed with a prominent basal palate closing the tube mouth. Stamens 4, included. Capsules ± globose, each locule dehiscing by a pore with 3 or more valves.

Literature: D.A.Sutton, *A revision of the tribe Antirrhineae*: 158-160 (1988). London.

1. C. muralis P. Gaertn., B. Mey. & Scherb., *Oekon. Fl. Wetterau* 2: 397 (1800).
Plant glabrous. Stems up to 60 cm or more long, creeping or trailing. Leaves 5-30 × 6-30 mm, reniform to suborbicular, 5-7-lobed with rounded, often mucronulate lobes, rather thick, sometimes purplish beneath; petioles exceeding the lamina. Pedicels up to 25 mm in flower, elongating up to 65 mm or more in fruit, reflexed. Calyx-lobes 1.5-2.5 mm, linear-lanceolate. Corolla 8.5-11 mm (including spur), lilac to violet with a yellow palate; spur 2-2.8 mm. Capsules 3-4.5 mm. Seeds 0.8-1 mm, conspicuously rugose.
Fl. III - X. *Common in Madeira on walls and sea cliffs, up to 1000 m.* **M**

Madeiran specimens belong to subspecies **muralis**, to which the above description refers.

7. Kickxia Dumort.

Annual herbs with prostrate or climbing stems. Leaves mostly alternate, petiolate or subsessile, entire, dentate or lobed. Flowers usually solitary in leaf-axils, pedicellate. Calyx deeply and subequally 5-lobed. Corolla with a cylindrical tube, spurred at the base; limb 2-lipped, the upper 2-lobed, erect, the lower 3-lobed with a basal palate partly closing the mouth of the tube. Stamens 4, included. Capsules ± globose, each locule dehiscing by a single pore.

Literature: D.A. Sutton (1988), *A revision of the tribe Antirrhineae*: 169-179. London.

1. Upper leaves hastate; pedicels pubescent just below the calyx but otherwise glabrous; calyx-lobes scarious towards the base . . . . **1. elatine**
- Upper leaves not hastate; pedicels pubescent throughout; calyx-lobes not scarious . . 2

2. Pedicels 8-20 mm; calyx-lobes 1.5-2.5 mm broad, ovate; corolla yellow . . . **2. spuria**
- Pedicels 0.5-2 mm; calyx-lobes 0.4-0.5 mm broad, linear-lanceolate; corolla white or cream . . . . **3. lanigera**

1. K. elatine (L.) Dumort., *Fl. belg.*: 35 (1827).
Elatinoides elatine (L.) Wettst.
Sparsely villous annual; stems up to 50(-100) cm, usually prostrate, much-branched from base, simple or sparingly branched above, slender. Leaves shortly petiolate; the lower 15-30 × 14-20 mm, oblong-ovate to suborbicular, obtuse, entire or somewhat toothed; upper leaves smaller, acute, hastate, entire, the petioles sometimes twining. Pedicels 6-29 mm, pubescent just below the calyx, otherwise glabrous. Calyx-lobes 3-4(-5.5) mm, lanceolate, acute, margins scarious towards the base. Corolla 7.5-9.5 mm (including spur), subglabrous to pubescent, yellow with a purple upper lip; spur 4-7 mm, straight. Capsules 3-4.5 mm, globose, thin-walled, glandular-puberulent at the apex. Seeds 1-1.1 mm. 2n=36.
Fl. V - XI. *Rather infrequent, on cultivated and waste ground in Madeira, occurring mainly in the north between Santa and Faial, but with occasional records from central and southern areas; 0-450 m.* **M**

Madeiran plants all belong to subsp. **elatine**. Records for subsp. *crinita* (Mabille) Greuter from Madeira are in error.

2. K. spuria (L.) Dumort., *Fl. belg.*: 35 (1827). [*Orelha de rato*]
Elatinoides spuria (L.) Wettst.; *Linaria spuria* (L.) Mill.
Glandular-villous annual; stems up to 55 cm, prostrate, usually much-branched from the base and often above. Leaves shortly petiolate, (6-)10-35 × (4-)8-25 mm, ovate to suborbicular, subacute to obtuse, sometimes mucronate, mostly entire, the lower rarely sparingly denticulate. Pedicels 8-20 mm, villous. Calyx-lobes 3-4 × 1.5-2.5 mm in flower, up to 6(-8) mm in fruit, ovate, acute, cordate at base. Corolla 10-15 mm, pubescent, yellow with a purple upper lip; spur 6-7 mm, curved. Capsules 3-5 mm, subglobose, sparsely glandular-pubescent. Seeds 1-1.2 mm.
Fl. V - X. *Occasional on cultivated and waste ground in the south of Madeira between Funchal and Ponta do Sol; 0-600 m. Recorded from Selvagem Grande in the last century.* **MS**

The above description applies to subsp. **integrifolia** (Brot.) R. Fern. in Heywood, in *Bot. J. Linn. Soc.* **64**(2): 74 (1971), to which all Madeiran specimens are referable. Records of subsp. *spuria* from the Madeiran archipelago are in error.

3. K. lanigera (Desf.) Hand.-Mazz. in *Annln naturh. Mus. Wien* **27**(4): 403 (1913).
Elatinoides lanigera (Desf.) Cout.
Glandular-villous annual; stems up to 50(-90) cm, procumbent, branched from the base and with many branches above. Leaves subsessile or shortly petiolate, up to 20 × 13 mm, ovate to broadly ovate, acute to obtuse, often apiculate, entire or dentate. Pedicels 0.5-2 mm, villous. Calyx-lobes 2.5-3 × 0.4-0.5 mm, linear-lanceolate, acute. Corolla 7-10 mm, glandular-pubescent, white or cream with violet upper lip, violet spotted palate, and pale purple spur; spur 3-5 mm, somewhat curved. Capsules 1.5-2 mm, subglobose, thin-walled, glandular-pubescent at apex. Seeds 0.9-1 mm.
Fl. V - XI. *Rare; on cultivated and disturbed ground in Madeira. Only a few records are localized, but apparently occurring mainly around Funchal. No recent records known.* **M**

Linaria maroccana Hook.f. in *Curtis's bot. Mag.* **98**: t. 5983 (1872), a Moroccan endemic annual with erect stems, linear leaves, and a dense inflorescence of reddish purple, long-spurred flowers, has been recorded from waste ground near Funchal Stadium, Madeira, presumably as a garden escape. **M**

8. Calceolaria L.

Annual herbs. Leaves opposite and usually decussate, often pinnately divided. Flowers in lateral cymes. Calyx 4-lobed. Corolla 2-lipped, yellow; tube very short; upper lip hooded, lower lip entire, saccate, inflated, much larger than the upper. Stamens 2. Capsules conical-ovoid, septicidal.

1. C. tripartita Ruiz & Pav., *Fl. peruv.* **1**: 14 (1798).
Plant shortly glandular-pubescent; stem 10-75(-100) cm high, erect, procumbent or ascending, rather succulent and brittle, often rooting at the base. Leaves 2.5-8 × 1.4-8 cm, lobed or more usually pinnatisect with 1-2 pairs of elliptic or lanceolate, lateral lobes and a larger terminal lobe, all serrate-margined; petioles 1-3.5 cm. Pedicels 1.4-3 cm. Calyx-lobes 4.5-7 mm, ovate, acuminate, margins usually shallowly dentate. Corolla glabrous; upper lip 3-4 mm; lower lip 11-16 mm. Anther connective elongated; lower cell sterile. Capsules 7-10 mm. Seeds 0.7-0.9 mm, ellipsoid, longitudinally ridged, light brown. 2n= c. 64.
Fl. XI - V. *Cultivated in Madeira for ornament and recorded as a garden escape around Funchal; also occasionally naturalized in damp places along levadas and roadsides, on banks*

and at the base of rock-faces, mainly in the north-east, 400-900 m. Native from S. Mexico to Bolivia, but a widespread pantropical weed. **M**

Specimens from Madeira have often been incorrectly identified as *C. chelidonioides* Kunth or *C. pinnata* L. in the past.

9. Digitalis L.

Biennial or perennial herbs. Leaves alternate, the lower forming a basal rosette. Flowers often nodding, in terminal, bracteate, one-sided racemes. Calyx deeply 5-lobed, shorter than corolla-tube. Corolla with cylindrical-campanulate tube, constricted near the base; limb very shallowly 5-lobed, weakly 2-lipped; lips erecto-patent, the lower longer than the upper. Stamens 4, included. Stigma 2-lobed. Capsules ovoid, septicidal.

1. D. purpurea L., *Sp. pl.* **2**: 621 (1753). [*Dedaleira, Teijeira*]
Usually biennial but sometimes a short-lived perennial or flowering in the first year; stem 50-150 cm, erect, simple, greyish tomentose or sometimes glabrous below. Leaves up to 20(-30) cm, ovate to lanceolate, acute, crenate-dentate, green and softly pubescent above, grey-tomentose beneath; the lower attenuate into a long, narrowly winged petiole, the upper becoming sessile. Bracts sessile, linear-lanceolate, entire. Raceme simple, few to many-flowered, rather lax. Pedicels 4-10 mm, tomentose. Calyx-lobes 10-12 mm, oblong-ovate or lanceolate, acute and mucronate, densely glandular-pubescent. Corolla 3-5 cm, pinkish purple, spotted with dark purple within on the white, lower throat; lobes pubescent along the margin and with scattered hairs on the lower lip within. Capsules *c.* 1 cm, pubescent.
Fl. III - IX. *Grassy places and woodland; very common in Madeira, except for the coastal region; also recorded from the Desertas but with no further data.* **MD**

10. Isoplexis (Lindl.) Benth.

Shrubs. Leaves alternate, simple, serrate-margined. Flowers in terminal, bracteate racemes. Calyx deeply 5-lobed, shorter than the corolla-tube. Corolla with cylindrical-campanulate tube; limb 2-lipped, the upper entire, the lower 3-lobed, $\pm$ equalling the upper. Stamens 4, slightly exserted. Stigma capitate. Capsules ovoid, septicidal.

1. I. sceptrum (L.f.) Loudon, *Encycl. pl.*: 528 (1829). [*Isoplexis*]
Callianassa sceptrum (L.f.) Webb
Shrub 0.5-2(-4) m high. Leaves congregated at the ends of the branches, 10-40 × 4-14 cm, obovate-oblong, mostly acute, serrulate or serrate, attenuate at the base, glabrescent above, sparsely pubescent, gland-dotted, and paler beneath. Peduncles up to 18 cm, erect. Bracts linear. Flowers subsessile, in a dense, glandular-pubescent inflorescence. Calyx-lobes 10-16 mm, linear, acute. Corolla 2-3.5 cm, orange-yellow with purplish veins; upper lip rounded, the lower with 3 rounded lobes; lobes glandular-pubescent. Capsules 9-11 mm, glandular-pubescent. Seeds 0.9-1.2 mm. 2n=56. **Plate 40**
Fl. VI - VIII. *A characteristic shrub of the Madeiran laurisilva, very rare in the wild and usually found growing on steep cliffs and ravines. Many individuals are reported to have been planted along levadas and roadsides by the Madeiran Forestry Service. 600-1000(-1200) m.*
● **M**

11. Veronica L.

Annual or perennial herbs. Leaves opposite; bracts (the lower sometimes leaf-like) usually alternate. Flowers blue or less commonly white or pink, in terminal or axillary racemes or spikes, or solitary in the leaf-axils. Calyx deeply and often unequally 4-lobed. Corolla with a very short tube and 4 spreading lobes, the upper larger and the lower smaller than the lateral lobes. Stamens 2, exserted. Capsules often laterally compressed, apex emarginate or bilobed. Seeds flat or boat-shaped, smooth or rugose.

1. Flowers in bracteate racemes or spikes 2
- Flowers solitary in the leaf-axils 4

2. Racemes axillary .... **1. anagallis-aquatica**
- Racemes or spikes terminal 3

3. Leaves glabrous, entire or weakly crenulate; fruiting pedicels ± equalling calyx .... **2. serpyllifolia**
- Leaves pubescent, crenate-serrate; fruiting pedicels shorter than calyx ... **3. arvensis**

4. Leaves 3-5(-7)-lobed; capsules glabrous; calyx-lobes cordate at base .. **4. hederifolia**
- Leaves crenate-serrate; capsules pubescent; calyx-lobes not cordate at base 5

5. Corolla 8.5-12.5 mm across; capsules compressed, the lobes sharply keeled and divergent, sparsely pubescent .... **5. persica**
- Corolla 5-8 mm across; capsules not or scarcely compressed, the lobes not keeled, erect, densely pubescent 6

6. Calyx-lobes ovate, ± acute; corolla uniformly bright blue .... **6. polita**
- Calyx-lobes oblong to oblong-ovate, ± obtuse; corolla usually pale blue with the lower lobe(s) whitish .... **7. agrestis**

1. V. anagallis-aquatica L., *Sp. pl.* **1**: 12 (1753).
V. anagallis auct.; *V. anagallis-aquatica* var. *elata* Hoffmanns.; *V. transiens* Rouy
Annual or perennial herb, glabrous below, shortly glandular-pubescent above. Stem 17-100(-150) cm, erect, often with decumbent branches at the base, thick, hollow, and rather fleshy. Leaves 3-8 × 1-3.2 cm, ovate-elliptic to lanceolate, subentire to serrulate, the lowermost obtuse, shortly petiolate, the upper ± acute, sessile, amplexicaul. Flowers in opposite, axillary, rather lax racemes. Bracts linear to linear-lanceolate, acute. Calyx-lobes lanceolate, acute. Corolla 4-5 mm across, pale blue to lilac. Fruiting pedicels 3.5-7 mm, ± patent. Capsules 2.5-3.5 mm wide, orbicular, glabrous or very sparsely glandular-pubescent. Seeds 0.4-0.5 mm, ovoid-ellipsoid, plano-convex. $2n=36$.
Fl. IV - X. *Fairly common in Madeira along roadsides and tracks, in wet flushes, pools, streamsides, and other damp places; 0-1000 m.* **M**

V. officinalis L., *Sp. pl.* **1**: 11 (1753), a perennial herb with branched, creeping and often mat-forming stems, ascending above, oblong to broadly elliptic, serrate leaves, alternate, dense, spike-like racemes of lilac flowers, followed by triangular-obcordate capsules, has been recorded in Madeira from Pico Jorge and near Pico Canário. No specimens have been seen but the Madeiran plants are reported to be glabrous except for a few hairs on the petioles, differing from typical specimens which are softly hairy throughout. It has been suggested that they may represent a new, as yet undescribed, subspecies. **M**

2. V. serpyllifolia L., *Sp. pl.* **1**: 12 (1753).
Perennial herb; stems 7-20 cm, branched from base, procumbent, rooting, puberulent, the flowering stems mostly ascending to erect. Leaves sessile or very shortly petiolate, 0.7-1.6 cm, oblong to ovate, entire or weakly crenulate, glabrous. Flowers in terminal racemes. Upper bracts oblong-lanceolate, sparsely and shortly ciliate; lower bracts becoming leaf-like. Calyx-lobes oblong. Corolla *c.* 6 mm across, pale blue or whitish with darker veins. Fruiting

pedicels 2-5 mm, puberulent, ± equalling calyx. Capsules 3.5-4.7 mm wide, broader than long, 2-lobed, ciliate. Seeds 0.7-0.8 mm, oblong-ellipsoid, flat, smooth.
Fl. IV - X. *Infrequent in damp, shady places along levada paths in Madeira;* c. *800-900 m.* **M**

V. peregrina L., *Sp. pl.* **1**: 14 (1753), native to N. and S. America, a glabrous, erect annual with entire or remotely serrate-crenate, ovate-oblong leaves, small whitish corolla, 2-3 mm across, and glabrous, scarcely emarginate capsules, has been recorded as an introduction in Madeira, but its status on the island is unknown. **M**

3. **V. arvensis** L., *Sp. pl.* **1**: 13 (1753).
Pubescent annual; stems 5-30 cm, ascending to erect, simple or branched. Leaves 0.6-1.6 cm, ovate, crenate-serrate, the lower shortly petiolate, the upper sessile. Flowers in terminal racemes, elongating and becoming very lax in fruit. Upper bracts lanceolate; lower bracts becoming leaf-like. Calyx-lobes lanceolate. Corolla *c.* 3 mm across, deep blue. Fruiting pedicels 1-2 mm, shorter than calyx. Capsules 3-3.5 mm wide, about as long as broad, obcordate, ciliate. Seeds *c.* 0.8 mm, oblong-ellipsoid, flat, smooth.
Fl. II - VII. *Infrequent in Madeira and Porto Santo, along roadsides, on walls, and in cultivated ground, up to* c. *800 m.* **MP**

4. **V. hederifolia** L., *Sp. pl.* **1**: 13 (1753).
Pubescent annual; stems 5-30 cm, usually branched from the base, decumbent. Leaves petiolate, up to 1.6 cm, suborbicular, 3-5(-7)-lobed, ciliate, pale green. Flowers solitary in the leaf-axils. Pedicels up to 1.5 cm. Calyx-lobes ovate-cordate, accrescent in fruit, ± glabrous, long-ciliate. Corolla 5-9 mm across, pale blue. Capsules *c.* 4 mm wide, subglobose, scarcely emarginate, glabrous. Seeds *c.* 2.2 mm, boat-shaped, weakly rugose.
Fl. I - VII. *A rare weed of cultivated ground in Madeira, occasionally becoming locally common.* **M**

5. **V. persica** Poir. in Lam., *Encycl.* **8**: 542 (1808).
Pubescent annual; stems 6-35(-50) cm, usually branched from the base, decumbent. Leaves shortly petiolate, up to 1.5(-2) cm, ovate, crenate-serrate, minutely ciliate. Flowers solitary in the leaf-axils. Pedicels up to 3 cm, the distal end recurved in fruit. Calyx-lobes lanceolate to ovate, accrescent and prominently veined in fruit, ciliate. Corolla 8.5-12.5 mm across, bright blue with darker veins and very pale lower lobe. Capsules 6-8 mm wide, about twice as wide as long, compressed, deeply 2-lobed, the lobes sharply keeled, divergent, sparsely pubescent, ciliate. Seeds 1.7-2 mm, boat-shaped, rugose, yellowish. $2n=28$.
Fl. I - VI. *Formerly a rare plant in Madeira, first collected in 1922, but now fairly widespread and becoming increasingly common in grassy areas along paths and levadas, along roadside banks and on waste ground; up to 1000 m.* **M**

6. **V. polita** Fr., *Novit. fl. svec.*: 63 (1819).
Pubescent annual; stems 4-25 cm, branched from the base, decumbent. Leaves shortly petiolate, 5-12 mm long, ovate, crenate-serrate, ciliate, dark green. Flowers solitary in the leaf-axils. Pedicels 4-10 mm, ± curving downwards in fruit. Calyx-lobes ovate, acute or subacute, accrescent and conspicuously veined in fruit, ciliate. Corolla 5-8 mm across, bright blue. Capsules *c.* 4 mm wide, wider than long, scarcely compressed, 2-lobed, the lobes erect, not keeled, densely pubescent with short, eglandular hairs and scattered, longer, glandular hairs. Seeds *c.* 1.1 mm, boat-shaped, rugose.
Fl. III - VI. *Very rare weed of cultivated ground and walls in Madeira and Porto Santo.* **MP**

7. V. agrestis L., *Sp. pl.* **1**: 13 (1753).
Like **6**, but leaves irregularly crenate-serrate and lighter green; sepals oblong to oblong-ovate, obtuse to subobtuse, faintly veined; corolla pale blue or rarely pinkish with the lower lobe or lobes whitish; and capsules obscurely keeled, with long glandular hairs.
Fl. III - IV. *A very rare garden weed in Madeira and Porto Santo.* **MP**

12. Sibthorpia L.

Creeping perennial herbs with stems rooting at the nodes. Leaves alternate, petiolate, reniform to orbicular, crenate. Flowers axillary, long-pedicellate. Calyx usually 6-lobed. Corolla with a very short tube and (5-)6(-8) entire, spreading, subequal lobes. Stamens equal in number to the corolla-lobes. Capsules globose, loculicidal.

1. S. peregrina L., *Sp. pl.* **2**: 631 (1753). [*Hera terrestre, Herva redonda, Herva terrestre*]
Plant hirsute; stems up to 1 m, prostrate and far-creeping. Leaves 15-60 mm wide, reniform to orbicular, crenate; petioles 10-70 mm. Flowers 1-6 in the axils of the leaves; pedicels 15-80 mm, slender, somewhat coiled or curved in fruit. Calyx 3.5-5 mm; lobes lanceolate, acute. Corolla 9-12 mm across, pale yellow, with rounded lobes. Capsules 2.5-3.5 mm wide, shorter than the calyx, pubescent near the apex. Seeds 1-1.2 mm, orbicular, reddish brown. 2n=20.
Fl. IV - X. *Common in Madeira in woodland, along levadas, and on banks, in damp, shady places; (150-)500-1400 m. Rare in Porto Santo, recorded from Pico do Facho.* ● **MP**

13. Odontites Ludw.

Hemiparasitic annual herbs. Leaves opposite or sometimes alternate above, sessile, toothed. Flowers in bracteate, spike-like, one-sided racemes. Calyx tubular to campanulate, subequally 4-toothed. Corolla with a cylindrical tube; limb 2-lipped, the upper hooded, entire, the lower 3-lobed. Stamens 4; anthers pubescent, mucronate. Capsules loculicidal. Seeds rather few, longitudinally ridged.

1. O. holliana (Lowe) Benth. in DC., *Prodr.* **10**: 550 (1846).
Plant appressed strigulose-pubescent; stem 15-55 cm, erect, branched from base and sometimes above, branches patent to ascending. Leaves 1.5-5 × 0.5-1.5 cm, lanceolate, deeply and bluntly serrate. Inflorescence lax. Bracts becoming leaf-like below. Pedicels 1-2(-6) mm. Calyx 5.5-9.5 mm in flower, slightly shorter than corolla-tube; teeth oblong, obtuse. Corolla 10-14 mm, pale yellow, pubescent without; upper lip entire, lobes of lower lip emarginate. Anthers with tuft of hairs, included. Capsules 6-10 mm, oblong to oblong-obovate, pubescent in upper half. Seeds *c.* 1.5 mm, ellipsoid. **Plate 40**
Fl. VI - XI. *Very rare, along levadas, on rock-ledges, and in grassy places in the mountains of central Madeira, 800-1650 m.* ● **M**

The calyx of one unlocalized Madeiran specimen is unusually long (up to 12 mm), with lobes exceeding the corolla-tube.

14. Bartsia L.

Annual hemiparasitic herbs. Leaves opposite and decussate, sessile. Inflorescence a terminal raceme or spike. Calyx tubular to campanulate, unequally 4-lobed. Corolla tubular, 2-lipped; upper lip hooded, entire; lower lip 3-lobed, deflexed. Stamens 4. Capsules ovoid to subglobose, loculicidal. Seeds numerous, minute, longitudinally ribbed.

1. B. trixago L., *Sp. pl.* **2**: 602 (1753).
Bartsia versicolor (Willd.) Pers.; *Bellardia trixago* (L.) All.
Plant often flushed with purple above; stem 14-50 cm, erect, usually simple, with downwardly

pointing hairs below. Leaves 24-60 × 4-12 mm, linear to linear-lanceolate, distantly and obtusely serrate. Inflorescence short, dense, glandular-pubescent. Flowers subsessile, *c.* 2 cm long. Calyx 8-10 mm, the lobes triangular. Corolla pubescent without, white, the upper lip suffused with pinkish purple; lower lip with two conspicuous swellings. Anthers densely brown-villous. Capsules 7-10 mm, pubescent. Seeds 0.6-0.8 mm, white.
Fl. III - VI. *On grassy slopes and pastures of Porto Santo, common and in places abundant. Also reported from Santana in Madeira, but no recent records known.* **MP**

15. Parentucellia Viv.

Annual, hemiparasitic, glandular-viscid herbs. Leaves opposite or the uppermost sometimes alternate, sessile, toothed. Flowers in terminal, bracteate, spike-like racemes. Calyx tubular, 4-lobed, membranous between the veins. Corolla with a cylindrical tube, 2-lipped, the upper lip forming a hood, entire or emarginate, the lower lip 3-lobed, longer than the upper. Stamens 4, included in the upper lip; anthers pilose, mucronate. Capsules oblong, loculicidal. Seeds numerous, minute.

1. P. viscosa (L.) Caruel in Parl., *Fl. ital.* **6**: 482 (1885).
Glandular-pubescent. Stem 20-50 cm, erect, simple or occasionally branched from near base. Leaves 15-45 × 5-15 mm, oblong-lanceolate, acute to obtuse, coarsely serrate. Bracts leaf-like. Pedicels *c.* 1 mm. Calyx 10-16 mm; lobes linear-lanceolate, enlarging and persistent in fruit. Corolla 16-24 mm, yellow, caducous. Anthers pilose. Capsules 8-10 × 3-4 mm, oblong-fusiform, pubescent. Seeds *c.* 0.4 mm.
Fl. V - VII. *Pastures, cultivated fields, and grassy places. Very rare in Madeira, only known from a few collections of the last century from the north and south-east of the island and one recent record (1986) from the levada between Portela and Machico.* **M**

XCVII. GLOBULARIACEAE
M.J. Short

Small shrubs or perennial herbs. Leaves usually alternate, exstipulate, simple. Flowers irregular, hermaphrodite, 5-merous. Corolla with a narrow tube, usually 2-lipped, the upper lip short or almost absent, the lower with three long lobes. Stamens 4, inserted on the corolla-tube. Ovary superior, 1-locular; style simple; stigma bilobed or capitate. Fruit dry, enclosed in the persistent calyx.

1. Globularia L.

Small shrubs. Leaves entire, evergreen. Flowers numerous, in dense, globose heads, surrounded by involucral bracts. Calyx deeply lobed. Corolla blue or whitish. Stamens exserted. Style exserted; stigma bilobed. Fruit small.

1. G. salicina Lam., *Encycl.* **2**: 732 (1788). [*Globulária, Malfurada*]
Lytanthus salicinus (Lam.) Wettst.
Erect shrub up to 1.5 m with slender branches. Leaves 3.5-7 × 0.5-3 cm, narrowly elliptic, acute, attenuate at the base, entire, glabrous. Inflorescences up to *c.* 1 cm across, axillary, often crowded towards the tops of the stems. Peduncles 1-3 cm, tomentulose. Bracts ovate, tomentulose to glabrescent, margins densely long ciliate. Calyx-lobes narrowly linear-lanceolate, margins very long ciliate. Corolla *c.* 4 mm long, pale powder blue or whitish. Fruit *c.* 1 mm, oblong, dark brown. **Plate 41**
Fl. III - XI. *Common in Madeira on hillsides, cliffs, slopes, among rocks, and rough grassland; mostly below 300 m, but up to 500(-700) m in the south, seldom above 400 m in the north. Rare in Porto Santo, recorded recently from Sitio do Lombo. Also known from Deserta Grande.* ■ **MDP** [*Canaries*]

XCVIII. BIGNONIACEAE
M.J. Short

Trees, shrubs or vines. Leaves usually opposite, mainly pinnately compound, the terminal leaflet sometimes a tendril; stipules absent. Flowers hermaphrodite, irregular, in panicles or racemes, or solitary. Calyx 5-lobed. Corolla with a conspicuous tube and 5 lobes, often 2-lipped. Stamens usually 4, inserted on the corolla-tube, the fifth staminodal. Ovary superior, 2-locular. Stigma 2-lobed. Fruit mostly a capsule. Seeds usually flattened and broadly winged.

Many members of this family are grown as ornamentals in parks and gardens of Madeira, including: **Catalpa bignonioides** Walter, a tree with large, simple, cordate-ovate leaves and white flowers marked with purple spots and yellow stripes within; **Jacaranda mimosifolia** D. Don [*Jacarandá*], a conspicuous street-tree in parts of Funchal with bipinnate leaves and panicles of purplish blue, densely pubescent flowers followed by red-brown, orbicular capsules; **Kigelia africana** (Lam.) Benth. [*Árvore das salsichas, Quigélia*], a tree with large, yellow flowers turning to red, followed by large, persistent, pendulous, oblong-cylindrical, woody and indehiscent fruits; **Pandorea jasminoides** (Lindl.) K. Schum. (*Tecoma jasminoides* Lindl.) [*Bignónia*], a vine with pinnate leaves and white, shortly pubescent corollas with a red to deep purple throat; **Podranea ricasoliana** (Tanfani) Sprague [*Ricasoliana*], a scrambling shrub with pink corollas up to 6 cm long; **Pyrostegia venusta** (Ker Gawl.) Miers (*Bignonia venusta* Ker Gawl.) [*Bignónia, Gaitas, Gaitinhas*], a vine with bifoliolate leaves and orange flowers with a curved, tubular, glabrous corolla-tube and linear, puberulent lobes; **Spathodea campanulata** P. Beauv. [*Chama da floresta, Espatódea*], a tree with brilliant scarlet-orange, broadly campanulate corollas, the lobes with crispate margins; **Tecoma stans** (L.) Juss. ex Humb., Bonpl. & Kunth [*Bois de pissenlit, Fausse bignone*], a shrub with serrate leaflets and yellow corollas; and **Tecomaria capensis** (Thunb.) Spach (*Tecoma capensis* (Thunb.) Lindl.) [*Camarões*], a small shrub with flexuous branches, crenate-serrate leaflets, and orange, tubular, slightly curved corollas.

1. Macfadyena A.DC.

Vines. Leaves bifoliolate, sometimes with a terminal, trifid, claw-like tendril; leaflets entire. Inflorescences axillary, often reduced to 1(-3) flowers. Calyx cupuliform, often irregularly lobed, membranous. Corolla yellow, tubular-campanulate. Capsules linear, flattened. Seeds winged.

1. M. unguis-cati (L.) A.H. Gentry in *Brittonia* 25: 236 (1973).   [*Bignónia unha-de-gato*]
Bignonia unguis-cati L.; *Doxantha unguis-cati* (L.) Rehder
Climber, rooting frequently from the nodes. Pseudostipules ovate. Leaves 2-foliolate, sometimes terminating in a tendril; leaflets 5-10(-15) × 1-7 cm, lanceolate to ovate, acute to acuminate, membranous; petioles 1-4.5 cm; petiolules 0.5-2.5 cm. Flowers mostly solitary or paired. Pedicels 1.5-5 cm. Calyx *c.* 1 cm long, margins sinuate. Corolla yellow, striped with orange in the throat; tube 4-7 cm long, glabrous; lobes 1.2-3 cm, spreading. Capsules up to 40(-70) × 1-1.8 wide. Seeds (2.3-)3-4 × *c.* 1 cm; wings brown, membranous.
Fl. IV - V. *A tropical American species cultivated as an ornamental in the lower regions of Madeira, especially in and around Funchal, and sometimes becoming naturalized; 0-450 m.*
M

XCIX. ACANTHACEAE
M.J. Short

Herbs, sometimes climbers, or shrubs. Leaves simple to pinnately-lobed, exstipulate. Flowers hermaphrodite, irregular, often subtended by conspicuous coloured bracts and bracteoles. Calyx deeply 4- or 5-lobed. Corolla 2-lipped, or 1-lipped with the upper lip absent. Stamens 2 or 4, staminodes sometimes present. Ovary superior, 2-locular; style simple, persistent, stigma usually bifid. Fruit a capsule.

1. Acanthus L.

Herbaceous perennial herbs with simple, erect stems. Leaves mostly in a basal rosette, pinnatifid to pinnatisect. Flowers large, in dense terminal spikes. Bracts large, spiny-toothed. Bracteoles linear to lanceolae, entire, aristate. Calyx 4-lobed, the lateral lobes much smaller than upper and lower lobes. Corolla-tube very short; lower lip 3-lobed, upper lip absent. Stamens 4, included. Capsules oblong-conical, 4-seeded. Seeds ± orbicular, flattened.

1. A. mollis L., *Sp. pl.* 2: 639 (1753).   [*Acanto, Erva gigante*]
Plant up to 1 m, robust and often forming large clumps, ± puberulent. Basal leaves long-petiolate, dark glossy green, 20-60 × 10-20 cm, ovate, deeply pinnately-lobed, the lobes oblong to triangular, dentate with soft but not spiny teeth, sparsely strigose; upper cauline leaves 1-5 cm, ovate, spinose-dentate, ± sessile. Flowers sessile. Bracts c. 4 cm, oblong-ovate, 5-7-veined, often tinged with purple. Upper calyx-lobe 4.5-5.5 cm, oblong, greenish or violet, arching over the corolla; lower lobe oblong, bifid. Corolla 4.5-5.5 cm, white, sometimes flushed with purple; lobes rounded.
Fl. V - VII. *A Mediterranean species introduced to the lower regions of Madeira as an ornamental, sometimes escaping and becoming naturalized along paths and roadside banks.*
M

Asystasia gangetica (L.) T. Anderson in Thwaites, *Enum. pl. zeyl.*: 235 (1860), a straggling, slightly woody, branched herb up to c. 60 cm with ovate, acuminate leaves and terminal, 1-sided racemes of flowers with white, yellow or purple, funnel-shaped corollas up to 4 cm long, followed by club-shaped capsules, is cultivated in Madeiran gardens and may occasionally escape. Native to the Old World tropics. M

Thunbergia gregorii S. Moore in *J. Bot.* 32: 130 (1894) [*Tumbérgia*], a hirsute trailing or climbing herb with brownish hairs, triangular-ovate leaves with winged petioles, long-pedunculate flowers, solitary in the axils of the leaves, the bright orange, 5-lobed corollas up to 5 cm long, is cultivated as an ornamental in Madeira and may occasionally escape, scrambling over walls or cultivated ground. Native to tropical Africa. M

Records of *T. alata* Bojer ex Sims may be referable to the above species.

C. OROBANCHACEAE
M.J. Short

Herbaceous root-parasites lacking chlorophyll. Stems erect, fleshy, usually swollen at the base. Leaves alternate, simple and scale-like, exstipulate. Flowers hermaphrodite, irregular or almost regular, in terminal, bracteate spikes or racemes. Calyx 4-5-lobed, sometimes 2-lipped. Corolla tubular, often curved, subequally 5-lobed or 2-lipped. Stamens 4, in two pairs of unequal length; filaments inserted towards base of corolla-tube. Ovary superior, 1-locular; style 1; stigma capitate or bilobed. Fruit a 2-valved, loculicidal capsule, usually remaining enclosed by the persistent calyx and corolla. Seeds very numerous, minute.

1. Corolla with 5 nearly equal lobes; calyx-lobes obtuse . . . . **1. Cistanche**
- Corolla distinctly 2-lipped; calyx-lobes acuminate . . . . **2. Orobanche**

1. Cistanche Hoffmanns. & Link

Perennial herbs. Stems simple, stout, with numerous, imbricate, scale-like leaves. Flowers sessile, showy, in a very dense spike. Bracteoles 2. Calyx campanulate, 5-lobed, the lobes imbricate, rounded. Corolla broadly funnel-shaped; limb 5-lobed, spreading, scarcely bilabiate, the lobes broad, nearly equal. Stamens included; anthers woolly. Style curved at tip, persistent in fruit; stigma large, subentire.

1. C. phelypaea (L.) Cout., *Fl. Portugal*: 571 (1913).
Phelypaea lutea Desf.
Glabrous. Stem 15-70 cm. Leaves 6-22 mm, oblong to oblong-ovate, obtuse, brown, with scarious margins. Spike 8-20 cm long. Bracts 15-23 mm, similar in shape to leaves, minutely denticulate; bracteoles oblong-obovate, minutely denticulate, slightly shorter than calyx. Calyx 12-17 mm, with short, suborbicular, scarious-margined, denticulate lobes. Corolla 3-4.5 cm, bright shining yellow; tube usually curved; lobes ovate-orbicular, obtuse. Filaments pubescent at the base. Capsules 12-20 mm, ovoid. Seeds *c*. 1 mm, black, alveolate.
Fl. II - ? *Recorded from Selvagem Grande, Selvagem Pequena, and Ilhéu de Fora. Reported to be parasitizing* Suaeda *on Selvagem Pequena.* **S**

2. Orobanche L.

Erect perennial or annual herbs, often glandular-pubescent. Stems simple or branched, slender or stout. Flowers sessile or very shortly pedicellate, in usually dense spikes or racemes. Bracteoles 2, adnate to calyx, or absent. Calyx campanulate or tubular, 4-dentate, or divided almost to the base into 2 lateral, bifid segments. Corolla distinctly 2-lipped; tube cylindrical to funnel-shaped, straight or curved; upper lip erect, entire or shortly bilobed, lower lip spreading, 3-lobed. Stamens included; anthers glabrous. Style nearly erect or curved at tip; stigma bilobed.

1. Flowers subtended by 2 bracteoles as well as by a bract; corolla conspicuously contricted above ovary .... **1. ramosa**
- Bracteoles absent; corolla not constricted above ovary 2

2. Stem with eglandular hairs; corolla 20-25(-30) mm, white, lilac-veined, the tube straight or indistinctly curved .... **2. crenata**
- Stem with glandular hairs; corolla 12-20 mm, pale yellow, the tube distinctly curved 3

3. Corolla pale yellow tinged with dull violet .... **3. minor**
- Corolla pale yellow, often tinged with brownish red .... **4. calendulae**

1. O. ramosa L., *Sp. pl.* **2**: 633 (1753).
Stems up to 15 cm, branched or sometimes simple, slender, brownish, subglabrous to sparsely glandular-pubescent. Leaves ovate to ovate-lanceolate. Inflorescence 2-8 cm, rather lax, glandular-pubescent. Bracts 5-8 mm, ovate-lanceolate; bracteoles 2, linear-lanceolate, adnate to calyx. Calyx 5-7 mm; tube campanulate; teeth 4, subequal, triangular, long-acuminate. Corolla 10-15 mm, constricted above the ovary and inflated below, pale yellowish at the base, violet distally, glandular-pubescent; lobes subentire or denticulate, ciliate. Filaments glabrous or very sparsely pubescent below. Stigma shallowly bilobed, white or yellow. Capsules 5-6 mm.
Fl. IV - V. *Very rare; recorded from the coastal region of southern Madeira, but no recent records known. Reported to be parasitizing* Scorpiurus muricatus *var.* sulcatus *(L.) Fiori.* **M**

Plants in Madeira belong to subsp. **nana** (Reut.) Cout., *Fl. Portugal*: 566 (1913) (*O. nana* (Reut.) Beck), to which the above description applies.

2. O. crenata Forssk., *Fl. aegypt.-arab.*: 113 (1775).
Stem up to 80 cm, simple, stout, yellowish, sparsely villous. Leaves linear-lanceolate. Spike up to 35(-50) cm, many-flowered, dense above, often lax below, glandular-pubescent. Bracts 14-20 mm, lanceolate, long-acuminate; bracteoles absent. Calyx 10-15(-20) mm, divided into two lateral halves, the segments deeply bifid with slender, unequal teeth. Corolla 20-25(-30)

mm, funnel-shaped, straight or slightly curved, white, lilac-veined, very sparsely glandular-puberulent; lips with broad, crenulate-dentate lobes. Filaments pubescent below. Stigma white, yellow or orange. Capsules *c.* 10 mm.
Fl. IV - V. *Very rare in Madeira, recorded from the garden of Quinta S. Roque where it was reported to be parasitizing the garden 'sweet pea', and also from Quinta do Bom Successo, Funchal.* **M**

3. O. minor Sm. in Sowerby, *Engl. bot.*: t. 422 (1797).
O. barbata sensu auct. mad.
Stem 13-50 cm, simple, creamy yellow or pinkish purple, glandular-pubescent. Leaves linear-lanceolate. Spike 4.5-34 cm, ± dense above, lax below, glandular-pubescent. Flowers sessile. Bracts 10-15 mm, lanceolate, acuminate; bracteoles absent. Calyx 7-13 mm, divided into two lateral halves, the segments free, deeply bifid or bidentate, the teeth very slender, usually unequal. Corolla 12-19 mm, tubular, curved, subglabrous to glandular-pubescent, pale creamy yellow, often pinkish purple distally; lower lip with subequal lobes or the middle one larger; lobes crenulate. Filaments sparsely pubescent below. Stigma purple or yellow. Capsules *c.* 8 mm.
Fl. III - VI. *Scattered along roadsides, on cliffs and walls, in pastures and on grassy slopes; mainly in the coastal zone at low altitudes on Madeira; in low areas but also on peaks of Porto Santo. Recorded parasitizing several different species, including* Andryala glandulosa *subsp.* glandulosa, Crepis, Plantago coronopus, Digitalis purpurea, *and various Leguminosae. This is the most frequent species of* Orobanche *encountered in Madeira and Porto Santo. Also recently recorded from Deserta Grande.* **MDP**

4. O. calendulae Pomel in *Bull. Soc. Sci. Alger.* **11**: 110 (1874).
O. mauretanica Beck
Like **3**, but leaves oblong-lanceolate; bracts 15-20 mm; calyx 10-15 mm, the segments equally bifid; corolla glandular-pubescent, pale yellow, often veined and tinged with brownish red; lower lip with equal lobes or the middle one larger; stigma purple or red.
Fl. ? *Very rare. Recorded from Porto Santo on dunes at Ponta da Calheta and also from Pico do Castelo. Host plants on Porto Santo are unknown, but elsewhere this species has been reported parasitizing species of Compositae, among others.* **P**

MYOPORACEAE
M.J. Short

Myoporum tenuifolium G. Forst. (*M. acuminatum* R. Br.) [*Mioporo*], a small evergreen tree native to Australia with a rather dense, rounded crown, glossy, elliptical leaves, and axillary clusters of flowers, the white corolla with a short tube and five spreading, bearded, purple-spotted, blunt lobes, followed by small, globular, blackish purple drupes is planted as an ornamental street and shade tree in Madeira and Porto Santo, and is particularly frequent on the latter island. **MP**

CI. PLANTAGINACEAE
J.R. Press

Annual or perennial, usually scapose herbs, or dwarf shrubs. Leaves usually in basal rosettes and spirally arranged, sometimes cauline and opposite. Flowers usually 4-merous, actinomorphic, hermaphrodite or rarely unisexual, in bracteate, racemose heads or spikes. Sepals connate at the base, persistent. Corolla gamopetalous, scarious. Stamens with long filaments and large anthers. Ovary superior. Fruit a capsule. Seeds often mucilaginous when wet.

1. Plantago L.

Herbs or dwarf shrubs. Flowers 4-merous, mostly hermaphrodite, numerous in heads or cylindrical spikes. Ovary 2- to 4-locular; ovules 2-many. Capsule circumscissile.

1. Leaves opposite 2
- Leaves in basal rosettes 3

2. Annual .... **10. afra**
- Small shrub .... **9. arborescens**

3. Corolla-tube hairy; leaves often toothed to 1- to 2-pinnatifid, sometimes ± entire .... **1. coronopus**
- Corolla-tube glabrous; leaves entire, remotely denticulate or with very shallow teeth 4

4. Anterior sepals connate for more than half their length 5
- Anterior sepals free for more than half their length 7

5. Bracts and sepals densely villous .... **3. lagopus**
- Bracts and sepals glabrous or very shortly hairy 6

6. Plant caulescent, stem up to 6 cm .... **4. leiopetala**
- Plant acaulescent .... **2. lanceolata**

7. Corolla-lobes erect .... **8. myosurus**
- Corolla-lobes spreading 8

8. Annual; bracts broader than long; sepals with a keel extending to not more than half-way .... **6. loeflingii**
- Perennial; bracts longer than broad; anterior sepals keeled ± to the apex 9

9. Leaves lanceolate or elliptic to suborbicular, usually at least 1 cm wide ... **7. major**
- Leaves linear or linear-lanceolate, usually less than 1 cm wide .... **5. ovata**

1. P. coronopus L., *Sp. pl.* **1**: 115 (1753).   [*Diabelha*]
P. coronopus var. *latifolia* DC., var. *pseudo-macrorrhiza* sensu Menezes, var. *vulgaris* Gren. & Godr.
Annual, biennial or perennial, occasionally with a short stem. Rosettes solitary, rarely several. Leaves fleshy, up to 18 cm, linear-oblong to oblanceolate in outline, ± entire or with few teeth to 1- to 2-pinnatifid, glabrous or pubescent, usually ciliate on the margins, 1- to 3-veined. Scapes up to 28 cm, ± prostrate to ascending, terete, hairs upwardly appressed above, patent below. Spikes 2-20 cm, dense, cylindrical. Bracts 1.5-2.5 mm, triangular-ovate, acuminate, with broad scarious, ciliate margins. Sepals free, 1.7-2.4 mm, ovate, the anterior ciliate on the narrow scarious margins, the posterior ciliate on the margins and keel. Corolla-tube hairy; lobes 1.1-1.3 mm, ovate, acute. Seeds (3-)4(-5), 1.1-1.2 mm, ellipsoid.
Fl. III - V. *Very common throughout the archipelagos, mainly near the coast but occasionally also inland.* **MDPS**

Specimens with proliferating spikes sometimes occur among normal populations. This polymorphic species is treated here in a broad sense to include small, slender plants from the Salvage Islands which are always annual, with leaves ± entire or with few teeth, scapes ± prostrate, geniculate beneath the inflorescence, and capsules always with 3 seeds. These plants probably represent the Macaronesian endemic (■) **P. aschersonii** Bolle in *Bot. Jb.* **14**: 251 (1891), a species otherwise endemic to the Canary Islands but which is considered by some authors to be conspecific with *P. coronopus*.

2. P. lanceolata L., *Sp. pl.* **1**: 113 (1753). [*Orelha de cabra, Prados, Tanchagem*]
P. lanceolata var. *capitata* Presl, var. *contigua* Menezes, var. *eriophora* (Hoffmanns. & Link) Cout., var. *timbali* (Jord.) Gaut.
Perennial with 1-several rosettes. Leaves up to 33 × 3.5 cm, usually lanceolate, entire or remotely denticulate, narrowing into a distinct petiole, ± glabrous to pubescent with long, white hairs, veins 3-5(-7). Scapes up to 80 cm, dense, cylindrical. Bracts 3-7.5 mm, rhombic-ovate, long-acuminate, glabrous or shortly hairy. Sepals 2.5-3.3 mm, with broad scarious margins, the anterior fused for almost their whole length but the midribs free, the posterior usually shortly hairy above. Corolla-tube 2-2.5 mm, glabrous; lobes 2-2.2 × 1-1.3 mm, ovate, acute, with a brown central nerve. Seeds 2, 3 mm long, ellipsoid, the hylum in a long, shallow, groove.
Fl. IV - VII. *Along tracks, on walls and in open areas; common in Madeira, much rarer in Porto Santo, Deserta Grande and Bugío.* **MDP**

Plants with proliferating spikes sometimes occur.

3. P. lagopus L., *Sp. pl.* **1**: 114 (1753). [*Prados*]
Like **2** but sometimes annual and often with a short stem to 6 cm. Leaves 5-17 cm, lanceolate to elliptic, usually 5-veined and remotely denticulate. Scapes up to 40 cm. Spikes 1-5 cm, the bracts and sepals giving the whole a villous appearance. Bracts 3-4 mm, densely white-villous in the upper half. Sepals 2.5 mm and, like the bracts, villous in the upper half. Corolla-lobes to 1.5 mm broad, acute to acuminate. Seeds 1.5-1.9 mm.
Fl. III - VII. *Common on open slopes and grassy places in most parts of Madeira and Porto Santo.* **MP**

4. P. leiopetala Lowe in *Trans. Camb. phil. Soc.* **4**: 17 (1853).
Like **2** but caulescent with a thick, woody stem up to 6 cm, white-hairy, bearing the bases of old petioles and scapes and surmounted by the leaf-rosette; leaves generally shorter, up to 21 cm; spikes 1-3 cm; sepals usually glabrous, sometimes shortly villous above; corolla-lobes (2.1-)2.5-3 mm. **Plate 41**
Fl. III - VII. *Along cliffs of the north coast of Madeira, from São Jorge west to Porto do Moniz and on the higher peaks and rocky areas of Porto Santo.* ● **MP**

P. malato-belizii Lawalrée in *Bolm Soc. broteriana*, II, **33**: 183 (1959) differs from **4** only in its larger leaves with 11-15 veins. Known from a single locality at Caldeirão do Inferno, it is probably not specifically distinct.

5. P. ovata Forssk., *Fl. aegypt.-arab.*: 31 (1775).
Villous perennial with 1-several rosettes. Leaves up to 10 × 0.5 cm, linear to linear-lanceolate, entire or remotely denticulate. Scapes slightly longer than the leaves, terete. Spikes (0.5-)0.7-2.5 cm, dense and cylindrical. Bracts 2.5-3(-3.5) mm, broadly ovate, glabrous with very broad, scarious margins. Sepals 2.5-3 mm, free, hairy, the scarious margins broad, the keel extending to tip of sepal. Corolla-tube glabrous, 2.4 mm; lobes 2-2.5mm, ± orbicular, spreading. Seeds 2.
Fl. III - IV. *Dry banks, verges and grassy areas in Porto Santo only.* **P**

6. P. loeflingii L., *Sp. pl.* **1**: 115 (1753).
Like **5** but a rather stiffly hairy annual smaller in all its parts. leaves sometimes remotely denticulate; bracts broader than long; sepals keeled only below the middle; corolla-lobes ovate-lanceolate.
Fl. ? *Recent introduction to Porto Santo where it is a rare plant found mainly on sandy soils. Native to Iberia.* **P**

7. P. major L., *Sp. pl.* **1**: 112 (1753). [*Azinhaga do forte* (P), *Tanchagem*]
Sparsely hairy to glabrous perennial. Leaves erect, 7-10.5 × 5-7 cm, ovate, entire or sometimes very shallowly toothed, 5-veined, ± abruptly narrowed into a petiole equalling or shorter than the blade. Scapes erect, up to 30 cm, striate at least above, with upwardly appressed hairs. Spikes 7-11 cm, narrowly cylindrical. Bracts 1.6-2 mm, ovate. Sepals free, 1.6-1.8 mm, ovate, with scarious margins, the mid-vein extending to the apex, keeled. Corolla-tube glabrous; lobes 1 mm, lanceolate, spreading. Seeds numerous, *c.* 1 mm, ellipsoid-trigonous.
Fl. V - IX. *Common throughout the lower regions of Madeira, recorded as very rare in Porto Santo.* **MP**

Plants from the islands are all subsp. **major**.

8. P. myosurus Lam., *Tabl. encycl.* **1**: 342 (1792).
Sparsely white-villous annual. Leaves 8-12 cm, lanceolate to elliptic, acute, 5-veined, margins ciliate, narrowed at the base into a petiole. Scapes to 25 cm, erect, slightly sulcate, with patent hairs. Spikes *c.* 11 cm. Bracts ovate to triangular, keeled and with long, white hairs on the back. Sepals free, up to 2 mm, ovate, with a green, keeled mid-vein and broad, scarious margins. Corolla white; tube 2 mm, glabrous; lobes erect, 2.5 mm, ovate, acute. Seeds 3.
Fl. V - VI. *A recent introduction, found along paths and roadsides in and around Funchal. First recorded in 1969, it has rapidly become established and is spreading beyond the town. Native to S. America.* **M**

Madeiran plants are all subsp. **myosurus**.

9. P. arborescens Poir. in Lam., *Encycl.* **5**: 389 (1804).
P. maderensis Decne.
Glandular puberulent shrub. Stems 15-50(-100) cm, much-branched. Leaves opposite, usually somewhat crowded towards the tips of branches, 1.5-7 cm × 1.5-2 mm, linear, minutely glandular-puberulent. Spikes capitulate, usually 4- to 6-flowered, pedunculate. Sepals free, 2.5-4 mm, ovate to broadly ovate, margins scarious, sometimes with short setae. Corolla glabrous; tube 4-6 mm; lobes 2-2.7 mm, ovate-lanceolate, shortly acuminate, with a brown central vein. Seeds 2, to 4 mm, black.
Fl. III - VIII. *Common among rocks, on banks, roadsides and in open areas, particularly in coastal regions of Madeira but reaching 1100 m in inland parts; common throughout Porto Santo; also present on Deserta Grande.* ■ **MDP** [*Canaries*]

Plants from the Madeiran islands all belong to the endemic (●) subsp. **maderensis** (Decne.) A. Hansen & G. Kunkel in *Monographiae biol. canar.* **3**: 69 (1972) **Plate 41**, to which the above description and synonymy apply. The species otherwise occurs only in the Canaries, as subsp. *arborescens*, which differs in having at least some long hairs on the leaves and often also on stems, bracts and sepals.

10. P. afra L., *Sp. pl.* ed 2, **1** : 168 (1762).
P. afra var. *obtusata* (Svent.) A. Hansen & Sunding
Glandular-pubescent annual. Stem up to 30 cm, erect, branched, the branches straight and ascending. Leaves opposite, up to 6 cm × 4 mm, linear to very narrowly linear-lanceolate, entire or occasionally with a few, slender teeth. Spikes up to 10 mm, ovoid. Bracts 3-6 mm, lanceolate to ovate-lanceolate, the margins broadly scarious below. Sepals free, 2.5-3.5 mm, ovate, the margins broadly scarious. Corolla glabrous; lobes 1.5-1.6 mm, with a dark patch at base. Seeds (1-)2, 1.6-2 mm.
Fl. II - IV. *Rare plant of dry and sandy places; recorded from Praia Formosa and Funchal in Madeira and from Selvagem Pequena.* **MS**

Plants from the Salvages have been separated as an endemic taxon, var. *obtusata* (Svent.) A. Hansen & Sunding, on the basis of being more densely branched above, with longer leaves overtopping the very dense inflorescence and with lower floral bracts linear-lanceolate, not acuminate, shorter or scarcely overtopping the flowers. However, from the little material available they do not appear to differ from plants from the rest of Macaronesia.

CII. CAPRIFOLIACEAE
J.R. Press

Woody or rarely herbaceous perennials. Leaves opposite, usually lacking stipules. Flowers 5-merous, hermaphrodite, actinomorphic or zygomorphic. Calyx small. Corolla-tube variously developed. Stamens 5, inserted on the corolla, alternating with corolla-lobes; anthers longitudinally dehiscent. Ovary inferior, 2- to 5-locular; ovules 1-many, pendent. Fruit a drupe or berry.

1	Leaves pinnate; flowers actinomorphic	**1. Sambucus**
-	Leaves simple; flowers zygomorphic	**2. Lonicera**

1. Sambucus L.

Small trees, shrubs or herbs. Stems with large pith. Leaves pinnate, deciduous, with or without stipules. Flowers small, actinomorphic, in corymbs or panicles. Calyx 5-lobed. Corolla rotate. Stigma sessile, 3- to 5-lobed. Ovary 3- to 5-locular. Fruit a drupe with 3-5 compressed seeds.

1	Herb; stipules present, conspicuous	**1. ebulus**
-	Shrub or small tree; stipules absent	2
2.	Leaflets sparsely pubescent beneath; ripe fruits black	**3. nigra**
-	Leaflets glabrous; ripe fruits yellowish grey	**2. lanceolata**

1. S. ebulus L., *Sp. pl.* **1**: 269 (1753). [*Engos*]
Perennial herb to *c*. 60 cm, with extensive creeping rhizome. Stems stout, erect, usually simple, shallowly grooved. Leaflets 3-11, 4-11 × 1.5-4.5 cm, oblong to oblong-lanceolate, acuminate, the base often asymmetric, sometimes with the lamina extending onto the rachis, acutely serrate. Stipules leaf-like, up to 3.5 × 1.5 cm, ovate. Inflorescence with (2-)3 primary rays. Flowers *c*. 8 mm across, white; corolla-lobes acute; anthers purple. Fruit globose, ripening black, sour.
Fl. VII. *Introduced. Shady places by roadsides and houses, especially in the north of Madeira. Formerly cultivated for medicinal use. Native to Europe and Asia.* M

2. S. lanceolata R. Br. in Buch, *Phys. Beschr. Canar. Ins.*: 195, n. 284 (1828)[*Sabugueiro*]
S. maderensis Lowe; *S. nigra* var. *lanceolata* (R. Br.) Lowe
Glabrous shrub or small tree to *c*. 7 m; bark with narrow wings or ridges and scattered raised lenticels. Leaflets 5-7(-11), 2-17.5 × 1.5-7 cm, oblong to oblong-lanceolate, occasionally asymmetric at the base, acutely serrate. Stipules absent. Inflorescence with 5 stout primary rays. Flowers *c*. 6 mm across, white or creamy; corolla-lobes rounded; anthers yellow. Fruit globose, ripening yellowish grey, rarely black, sweet. **Plate 42**
Fl. V - VI. *Endemic in damp woods and thickets in ravines in Madeira; sometimes also near houses.* ● M

3. S. nigra L., *Sp. pl.* **1**: 269 (1753).
Like **2** but leaflets sparsely pubescent beneath; flowers fragrant; fruits ripening black.
Fl. IV - VI. *Occasionally grown in gardens. Naturalized on roadsides at Santo Antonio, and*

recorded from Monte as a garden escape; possibly overlooked elsewhere. Native to Europe and W. Asia. **M**

2. Lonicera L.

Deciduous or evergreen, woody climbers. Leaves entire, lacking stipules. Flowers in axillary pairs or terminal heads or whorls. Bracts usually present. Bracteoles free or connate, rarely absent. Calyx 5-lobed. Corolla 2-lipped (in Madeira), the upper lip 4-lobed; tube sometimes gibbous. Stamens 5. Stigma capitate. Ovary 2- to 3(-5)-locular, the ovaries of paired flowers sometimes with united walls. Fruit a few-seeded berry.

Several species are cultivated as ornamentals.

1. Flowers in axillary pairs; berries ripening black . . . . **1. japonica**
- Flowers in heads or whorls; berries ripening red . . . . **2. etrusca**

1. L. japonica Thunb., *Fl. jap.*: 89 (1784).
Semi-evergreen, woody climber. Twigs hirsute. Leaves up to 8 × 4 cm, ovate to oblong-ovate, acute, rouded or subcordate at base, pubescent when young, later glabrous, ciliate. Flowers in axillary pairs. Bracts leaf-like. Corolla 1.5-5 cm, white tinged with purple; tube narrow, almost as long as limb, glandular-pubescent. Berries ripening black.
Fl. V - IX. *Introduced: cultivated for ornament and locally naturalized close to habitation throughout Madeira. Native to E. Asia.* **M**

2. L. etrusca Santi, *Viagg. Montamiata*: 113 (1795). [*Madresilva*]
Deciduous or semi-evergreen, woody climber. Leaves up to 6.5 × 4 cm, broadly elliptical or ovate, obtuse or subacute at apex, glaucous or whitish green beneath, the upper pair connate, the next sessile or shortly petiolate. Inflorescences terminal, the whorls solitary, or 2-3 clustered at the tips of the branches. Corolla *c.* 2.5 cm, yellowish white, often tinged with purple; tube *c.* 1½ × as long as limb, narrow. Berries fused towards the base, ripening red.
Fl. IV - V. *Possibly introduced, naturalized on banks and in hedges throughout Madeira but always close to cultivated ground. Native to S. Europe.* **M**

Several varieties of *L. etrusca* have been distinguished; Madeiran plants, which are wholly glabrous, have been called var. **glabra** Lowe, *Man. fl. Madeira* **1**: 382 (1868).

CIII. VALERIANACEAE
N.J. Turland

Annual or perennial herbs. Leaves opposite, without stipules. Inflorescence cymose, bracteate, often dense. Flowers hermaphrodite or unisexual; calyx variously developed, usually toothed; corolla funnel-shaped, sometimes with a long tube with a swelling or spur near base, the limb with (3-)5 ± unequal lobes; ovary inferior, with 3 cells, one of them with a single ovule, the other two sterile. Fruit dry, indehiscent, usually with a persistent, often enlarged calyx.

1. Annuals; stems dichotomously branched; corolla-tube with a slight swelling, but not spurred; stamens 3 . . . . **1. Valerianella**
- Annual or perennial; stems not dichotomously branched; corolla-tube with a swelling or spur near base; stamen 1 . . . . **2. Centranthus**

1. Valerianella Mill.

Dichotomously branched annuals. Leaves and bracts strap-shaped. Inforescence of terminal clusters of flowers, and sometimes also single flowers borne in the forks of the stems. Flowers very small, hermaphrodite; calyx variously developed; corolla very small, pinkish or bluish, the limb with 5 ± unequal lobes, the tube not more than twice as long as limb and with a slight swelling (but not spurred); stamens 3. Sterile chamber of fruit variously developed.

Fruits are usually required for accurate identification.

1. Calyx well-developed . . . . **3. dentata**
- Calyx reduced to minute teeth above fruit 2

2. Fertile chamber of fruit with a thickened and spongy outer wall as thick as cavity of chamber
. . . . **1. locusta**
- Fertile chamber of fruit without a thickened and spongy outer wall 3

3. Sterile chambers of fruit reduced to slender ribs . . . . **4. microcarpa**
- Sterile chambers of fruit moderately well-developed, though usually smaller than the fertile . . . . **2. carinata**

1. V. locusta (L.) Laterr., *Fl. bordel.* ed. 2: 93 (1821).   [*Alface da terra, Saboia*]
V. olitoria (L.) Pollich
Plant to 30 cm, sparsely hairy. Leaves 2-10 × 1.5-2 cm, narrowly oblanceolate, gradually tapered into petiole, ± entire to weakly toothed. Bracts like leaves but smaller, linear-oblong. Corolla pale lilac or pale blue. Fruits 2-2.5 mm, borne in dense, hemispherical clusters and also singly in forks of stems, lenticular, glabrous to minutely puberulent; fertile chamber with a thickened and spongy outer wall as thick as cavity of chamber; sterile chambers as large as fertile, separated by a shallow, longitudinal groove. Calyx reduced to a minute tooth above each chamber.
Fl. III - VIII. *Banks, grassy places, walls and cornfields, from 200-800 m; scattered in Madeira, in the area from Seixal and Ribeira Brava to Porto da Cruz and Machico.* **M**

2. V. carinata Loisel., *Not. fl. France*: 149 (1810).
Like **1** but fruits narrowly oblong-ovoid, obtusely 4-angled, the fertile chamber without a thickened and spongy outer wall, the sterile chambers usually smaller than the fertile; calyx reduced to a single, minute tooth above fertile chamber.
Fl. III - IV. *Recently discovered in Madeira, occurring in the Funchal region and on the northern coast at Faial, growing on roadsides.* **M**

3. V. dentata (L.) Pollich, *Hist. pl. palat.* **1**: 30 (1776).
V. dentata var. *dasycarpa* Rchb., var. *leiocarpa* (DC.) W.D.J. Koch; *V. morisonii* (Spreng.) DC, var. *leiocarpa* DC., var. *lasiocarpa* (W.D.J. Koch) Lowe
Like **1** but bracts narrowly lanceolate to linear, with a few distinct teeth near base; corolla pale pink; fruits 1.5-2 mm, inverted-pear-shaped, the side with sterile chambers flattened, ± glabrous to densely pubescent, the sterile chambers reduced to ribs and separated by an ovate, flat area; calyx well-developed, with 5-6 unequal teeth, 1 of them very long.
Fl. IV - VII. *Maritime cliffs, walls and cornfields in Madeira; rare, restricted to the area between Porto do Moniz, São Vicente and the Encumeada pass in northern Madeira.* **M**

4. V. microcarpa Loisel., *Not. fl. France*: 151 (1810).
V. bracteata Lowe; *V. microcarpa* var. *puberula* sensu auct. mad., non (Bertol. ex Guss.) Gaut.; *V. puberula* sensu auct. mad., non (Bertol ex Guss.) DC. (1830)
Like **1** but bracts linear, with 2 spreading, subulate lobes near base; fruits *c*. 1.5 mm, ovoid, the side with sterile chambers flattened, ± glabrous to densely hairy, the fertile chamber without a thickened and spongy outer wall, the sterile chambers reduced to slender ribs, separated by an ovate, flat area; calyx reduced to a minutely toothed, narrow rim above fruit.

Fl. IV - VII. *A weed of cornfields and sugar-cane plantations; occurring along the northern coast of Madeira, from Seixal to Porto da Cruz, and in the south-east, from the Funchal region to Machico.* **M**

Species **1** and **4** sometimes grow together.

V. bracteata Lowe, described from cultivated ground and yam-beds in the Porto da Cruz area of north-eastern Madeira, has been considered probably referable to **4**, but in fact appears closer to species **1** and **2**. Its exact status remains uncertain.

2. Centranthus DC.

Glabrous, annual or perennial herbs. Stems not dichotomously branched. Leaves entire to deeply pinnatifid. Inflorescence a dichasium or panicle of dichasia often compressed into dense clusters or heads, at least when young. Flowers hermaphrodite or unisexual; calyx with 5-25 teeth forming a plumose pappus in fruit; corolla red to pink to white, the tube distinct, cylindric, with a spur or conical swelling near base, the limb with 5 usually unequal lobes; stamen 1. Sterile chambers of fruit minute.

1. Plant perennial; all leaves entire (the upper sometimes slightly dentate); corolla-tube with a distinct spur 4-6 mm long .... **1. ruber**
- Plant annual; upper leaves pinnatifid or ± so; corolla-tube with a small, somewhat indistinct spur or swelling .... **2. calcitrapae**

1. C. ruber (L.) DC. in Lam. & DC., *Fl. franç.* ed. 3, **4**: 239 (1805).
Plant perennial, with ± shrubby habit, glaucous. Stems to 90 cm. Leaves 5-15 × 1.5-5 cm (including petiole 0-3.5 cm), lanceolate to narrowly so, usually shortly acuminate at apex, usually ± entire at margins, the lower gradually narrowed into petiole, the upper usually sessile and sometimes slightly and irregularly dentate. Inflorescence usually a panicle of ± dense dichasia, 2-14 cm, ovoid, pyramidal or ± globose. Flowers scented; corolla deep red, pink or white; tube 6-9 mm, slender, with a distinct slender spur 4-6 mm; limb 4-5 mm across. Fruits *c.* 4.5 mm, lanceolate.
Fl. most of the year. *Introduced; cultivated for ornament in Madeira, and naturalized as an escape on walls, roadsides, rocky places and in cultivated and waste areas, mainly in the Funchal region. Native to the Mediterranean region.* **M**

The above description refers to subsp. **ruber**, to which Madeiran plants belong.

2. C. calcitrapae (L.) Dufr., *Hist. nat. Valér.*: 39 (1811).
Plant annual. Stem 15-80 cm, simple or branched. Leaves 2-10 × 1-5 cm (including petiole 0-4 cm), glaucous, often purplish; lower leaves suborbicular to ovate, crenate-dentate, abruptly contracted into petiole; median and upper leaves progressively becoming more deeply pinnatifid and more shortly petiolate, so that uppermost leaves are sessile and cut almost to base into 7 oblanceolate to linear lobes, the terminal lobe usually larger and broader than the laterals and more prominently toothed. Inflorescence a terminal dichasium, dense and capitate in flower, becoming lax and spreading in fruit. Corolla pink or lilac to white; tube short and not very distinct, with a short spur or swelling near base; limb 2-2.5 mm across. Fruits 2.5-3 mm, lanceolate, glabrous.
Fl. III - VI. *Scrub, cliffs, banks, grassy and rocky places, roadsides and disturbed ground, from sea-level all along the northern coast of Madeira up to Paúl da Serra and high peaks of the east; apparently absent from the southern coast.* **M**

The above description refers to subsp. **calcitrapae**, characterized by its glabrous fruits, and to which Madeiran plants belong.

CIV. DIPSACACEAE
M.J. Short

Annual to perennial herbs. Leaves opposite or whorled, simple or pinnately lobed, exstipulate. Flowers borne in dense capitula subtended by receptacular bracts, usually irregular, hermaphrodite or plant gynodioecious. Ovary enclosed by an *involucel* of united bracteoles, often subtended by a receptacular scale. Calyx cupuliform, often divided into 4-8 bristles or teeth. Corolla sometimes 2-lipped, 4-5-lobed. Stamens 2 or 4 exserted. Ovary inferior, 1-locular. Style 1; stigma simple or 2-lobed. Fruit an achene enclosed by the persistent involucel.

1. Stem with prickles; receptacular bracts stiff, spiny-tipped . . . . **1. Dipsacus**
- Stem without prickles; receptacular bracts herbaceous, not spiny-tipped 2

2. Leaves all simple; corolla 4-lobed, 6-7.5 mm long . . . . **2. Succisa**
- At least the middle and upper leaves lyrate or pinnatifid; corolla 5-lobed, 12-18 mm long
. . . . **3. Scabiosa**

1. Dipsacus L.

Biennials with prickly stems. Leaves simple or pinnately-lobed, often connate at the base. Capitula ovoid-globose. Receptacular bracts in 1-2 rows, stiff, spine-tipped, erect or patent. Receptacular scales spiny-tipped. Involucel 4-angled, terminating in a short 4-lobed cup. Calyx cupuliform, ciliate, often falling before the fruit. Corolla unequally 4-lobed. Stigma simple. Achenes appressed-pubescent.

1. D. ferox Loisel., *Fl. gall.*: 719 (1807).
Plant 30-100 cm high; stem simple or branched above. Leaves mostly basal, 8-15 × 2.5-5 cm, oblong-lanceolate, irregularly and coarsely crenate, sometimes lobed or sinuate, densely pubescent and with sparse to numerous, short, stout, yellowish prickles mainly on the margins and the veins below. Pedicels stout, with numerous prickles. Receptacular bracts lanceolate-subulate, strong and rigid, curving upwards and exceeding the capitula. Receptacular scales subulate, densely pubescent. Corolla densely appressed-pubescent, pinkish white. Anthers pink to purple.
Fl. V - VI. *Occasional in Porto Santo on peaks and in lowland areas.* **P**

2. Succisa Haller

Gynodioecious perennials. Leaves simple. Capitula hemispherical, long-pedunculate. Receptacular bracts herbaceous, in 2-3 rows. Involucel 4-angled, with 4 triangular, herbaceous lobes. Calyx shallowly cupuliform, prolonged above into (4-)5 bristle-tipped teeth, persistent in fruit. Marginal and central florets subequal. Corolla subequally 4-lobed.

1. S. pratensis Moench, *Methodus*: 489 (1794).
Scabiosa succisa L.; *Succisa praemorsa* Asch.
Adpressed-pubescent. Stems up to 100 cm, erect or ascending, usually branched. Basal leaves in a rosette, petiolate, 5-15(-30) cm, elliptic to broadly lanceolate, entire or weakly serrate, glabrescent; cauline leaves lanceolate or bract-like. Capitula up to 2.5 cm across. Receptacular bracts linear-lanceolate. Calyx bristles 1.5-2 mm, reddish black. Corolla 6-7.5 mm, lilac or sometimes nearly white. Achenes *c.* 5 mm.
Fl. VII - IX (XII). *A rare plant of shady, rocky places, mainly in the mountains of central and northern Madeira.* **M**

3. Scabiosa L.

Annuals or biennials. Leaves simple to pinnately-lobed. Capitula long-pedunculate, hemispherical, the outermost florets radiant. Receptacular bracts herbaceous, in 1-3 rows. Involucel 8-ribbed, expanded above into a scarious, funnel-shaped corona inrolled at the apex. Calyx cupuliform, prolonged above into 5 long bristles, persistent in fruit. Marginal florets usually longer than central. Corolla unequally 5-lobed.

1. S. atropurpurea L., *Sp. pl.* **1**: 100 (1753). [*Saudades*]
Scabiosa maritima L.
Subglabrous to adpressed-pubescent annual or biennial. Stems 30-100 cm, erect, branched. Lower leaves petiolate, 2.5-8 cm, elliptic to oblong-spathulate, simple or lyrate, crenate or serrate; upper leaves pinnatifid, the lobes linear to oblanceolate, becoming entire. Capitula 2-3 cm across. Receptacular bracts linear-lanceolate. Florets 12-18 mm, lilac or white to dark purple, fragrant. Involucel-tube conspicuously grooved throughout. Calyx setae 3-7 mm, yellowish red, stipitate. Fruiting capitula ovoid-oblong; achenes *c*. 3.5 mm.
Fl. IV - VIII. *Cultivated and waste ground in Madeira, mainly in and around Funchal; occasional in Porto Santo.* **MP**

CV. CAMPANULACEAE
B.R. Tebbs

Annuals to herbaceous perennials, usually with latex present. Leaves alternate, simple. Flowers regular, or strongly zygomorphic (*Lobelia*). Calyx tube 3- to 5-lobed. Corolla ± deeply lobed, the edges of lobes touching but not overlapping. Stamens free or joined, as many as corolla lobes and alternating with them, attached to the base of the corolla or to the disc surrounding the ovary. Style 1, with 3-5 stigmas. Ovary inferior, 2- to 5-chambered. Fruit a capsule, dehiscing irregularly or by valves or pores. Seeds numerous.

1. Flowers zygomorphic; stamens connate around style . . . . **7. Lobelia**
- Flowers actinomorphic; stamens free 2

2. Capsule cylindrical, 6-9 × as long as broad . . . . **3. Legousia**
- Capsule less than 3 × as long as broad 3

3. Flowers in capitula or corymbs 4
- Flowers in panicles or solitary 5

4. Inflorescence a capitulum; leaves linear-oblong, entire . . . . **2. Jasione**
- Inflorescence a corymb; leaves bi-serrate . . . . **5. Trachelium**

5. Perennial; flowers bright yellow or reddish brown . . . . **4. Musschia**
- Annual; flowers blue or violet 6

6. Capsule dehiscing by lateral pores; pedicels short . . . . **1. Campanula**
- Capsules dehiscing by apical valves; pedicels long . . . . **6. Wahlenbergia**

1. Campanula L.

Annual herbs. Flowers 5-merous, blue, purple or white, in racemes, cymes or panicles. Calyx-teeth often exceeding ovary, lobes flat or folded at sinus. Corolla campanulate or infundibuliform. Style without basal disc. Capsule pendent or erect, ovate or turbinate, 3- to 5-celled, dehiscing by lateral pores.

1. Plant hispid; corolla 3-5 mm . . . . **1. erinus**
- Plant glabrous or pubescent; corolla 10-20 mm . . . . **2. lusitanica**

1. C. erinus L., *Sp. pl.* **1**: 169 (1753).
Hispid annual, 3-30 cm, bushy-branched, stems striate, weakly angled. Leaves 0.8-2 cm, opposite or alternate, sessile, bristly-hairy, ovate to obovate, crenate-dentate or slightly lobed. Flowers sessile, terminal or axillary. Calyx-teeth acute, spreading after pollen is released, shorter than the corolla. Corolla 3-5 mm, pale blue. Capsule flask-shaped, pendent.
Fl. IV - VIII. *Common on rock ledges and in dry places at all altitudes in Madeira, Porto Santo and on Deserta Grande.* **MDP**

2. C. lusitanica Loefl., *Iter Hispan.*: 111, 302 (1758).
Glabrous or pubescent annual, 10-35 cm, diffusely branched. Leaves ovate to ovate-lanceolate, crenate or the upper slightly serrate. Inflorescence simple or widely branched, pedicels filiform. Calyx-teeth linear, much longer than the tube. Corolla 10-20 mm, infundibuliform to campanulate, lobed ± half-way to the base. Capsule erect.
Fl. IV - VII. *Roadsides near the power station at Fajã da Nogueira. This may represent a very recent introduction, possibly connected with the building of the power station. Native to the Iberian Peninsula.* **M**

2. Jasione L.

Annual, biennial or perennial herbs. Leaves linear, oblong, entire. Flowers in a capitulum surrounded by 1 or more rows of involucral bracts, small, numerous, sessile or subsessile. Calyx 5-toothed. Corolla divided to the base into 5 lobes. Stamens 5; anthers connate at base. Stigmas 2, short, stout. Capsule dehiscing by 2 short valves.

1. J. montana L., *Sp. pl.* **2**: 928 (1753)
Pubescent herb, normally biennial, but sometimes annual or perennial. Stems 5-40 cm, usually decumbent at base, simple or branched, upper parts leafless. Leaves up to 5 cm, linear-oblong to linear-lanceolate, blunt, margins crenate or undulate, petioles short. Inflorescence 5-35 mm, globose. Bracts shorter than inflorescence, triangular to ovate-cuspidate, numerous. Calyx-teeth subulate, as long as the un-opened corolla. Corolla *c.* 5 mm, blue, persistent.
Fl. V - VIII. *Weed in the Serviços Florestais plant nursery at Santo de Serra. Most probably introduced with foreign plants or seeds. Native to Europe and the Mediterranean region.* **M**

3. Legousia Durande

Annual herbs with paniculate or racemose inflorescences. Calyx 5-lobed. Corolla 5-lobed, circular-spreading or broadly campanulate. Stamens 5; anthers free. Ovary 6-9 × as long as broad, cylindrical. Stigmas 3. Capsule dehiscing by 3 apical pores.

1.	Flowers axillary; calyx-lobes spreading or recurved	**1. falcata**
-	Flowers terminal; calyx-lobes erect	**2. hybrida**

1. L. falcata (Ten.) Fritsch in *Mitt. naturw. Ver. Univ. Wien* **5**: 100 (1907).
Specularia falcata (Ten.) A. DC.
Pubescent annual to 50 cm. Leaves obovate, margins undulate. Flowers axillary, solitary or several, in a lax spike at least half the total length of the stem. Calyx lobes 10-12 mm long, narrow, spreading or recurved. Corolla *c.* ⅓ as long as calyx lobes, violet. Capsule 15-20 mm, gently curving, not narrowed at apex.
Fl. III - VI. *Rare plant of cornfields and among rocks up to 500 m; at scattered localities in Madeira from São Vincente and Serra de Água to Santa Cruz.* **M**

2. L. hybrida (L.) Delarbre, *Fl. Auvergne* ed. 2: 47 (1800).
Specularia hybrida (L.) DC.
Shortly hispid annual 10-35 cm. Leaves oblong or oblong-obovate, margins undulate. Flowers

3-5, mostly in terminal clusters. Calyx lobes 5-7 mm long, erect. Corolla *c.* ½ as long as calyx lobes, purple to lilac. Capsule 15-30 mm, straight, distinctly narrowed at apex.
Fl. IV - VII. *On disturbed ground around Funchal, Pico dos Barcellos and São Gonçalo.* **M**

4. Musschia Dumort.
N.J. Turland

Robust, rosette-forming perennials with a woody stock. Leaves bi-serrate. Inflorescence paniculate, much-branched, many-flowered, pyramidal, with branches held perpendicular to the main axis. Bracts leaf-like. Sepals curving inwards, flushed with same colour as corolla. Corolla-lobes curving outwards. Stigma with large, conspicuous lobes. Capsule dehiscing by means of numerous lateral slits in the sides of the chambers. Seeds abundant, very small, pale brown.

An endemic genus.

1. Plant glabrous, with shiny, coriaceous leaves; corolla-lobes bright yellow . . **1. aurea**
- Plant pubescent, with soft, dull leaves; corolla-lobes reddish brown . . . **2. wollastonii**

1. M. aurea (L.f.) Dumort., *Comment. Bot.*: 28 (1822). [*Músquia*]
Campanula aurea L. f.; *Musschia aurea* var. *angustifolia* (Ker. Gawl.) DC.
Plant glabrous, to 50 cm; not monocarpic. Leaf-rosettes arising from short, congested stock. Leaves 10-35 × 2-5.5 cm, ovate to narrowly elliptic, coriaceous, shiny, not clasping stem; petiole winged, distinct from lamina. Inflorescence to 40 cm. Sepals 12-18 × 4-10 mm, triangular-ovate, cuspidate. Corolla-lobes 11-16 × 3-4 mm, narrowly triangular-lanceolate, acuminate, bright yellow. Anthers 5-6 mm. Stigma-lobes 7-11 mm. Ripe capsule 13-20 × 6.5-10 mm, ovoid-obconical. **Plate 42**
Fl. VII - XII. *Rare plant of rock crevices on cliffs, either facing the sea or in moist inland valleys in Madeira, also on Deserta Grande and the adjacent islet of Doca.* ● **MD**

2. M. wollastonii Lowe, *Hooker's J. Bot.* **8**: 298 (1856). [*Tanjeiro brava*]
Plant pubescent, usually very robust, up to 2 m; usually monocarpic. Leaf-rosette usually solitary, becoming elevated on a stout, ascending, woody stem. Leaves 14-19 × 3.5-17 cm, narrowly oblanceolate, soft, dull, glabrescent except on the veins, clasping the stem; petiole usually indistinct from lamina. Inflorescence to 1 m. Sepals 20-25 × 6 mm, narrowly triangular-lanceolate, cuspidate. Corolla-lobes 21-28 × 3-5 mm, narrowly triangular-lanceolate, acuminate. Anthers 8-10 mm. Stigma-lobes 14-15 mm. Ripe capsule 11-14 × 11-13 mm, globose-obconical. **Plate 42**
Fl. VIII - XI. *Very rare, in crevices of rock faces in very humid, shaded valleys in the laurisilva in the interior of Madeira from 400-900 m.* ● **M**

5. Trachelium L.

Perennial herbs with corymbose inflorescences. Calyx 5-lobed. Corolla 5-lobed, tubular. Stamens 5; filaments glabrous, anthers free. Style extending beyond corolla, thickened at apex; stigmas 2-3. Capsule dehiscing by 2-3 basal pores.

1. T. caeruleum L., *Sp. pl.* **1**: 171 (1753).
Perennial up to 100 cm, stems mostly glabrous. Leaves ovate-lanceolate, bi-serrate, usually ciliolate, the upper sessile. Flowers fragrant. Corolla tube very slender, about 3 × as long as lobes, blue. Style noticeably extended beyond corolla. Capsule pear-shaped.
Fl. VI - VIII. *Common in damp shady places and on ledges above levadas in the Funchal region.* **M**

Madeiran plants all belong to subsp. **caeruleum**.

6. Wahlenbergia Schrad. ex Roth

Annuals. Flowers in panicles, pedicels long. Calyx 3- to 5-lobed. Corolla campanulate or infundibuliform. Stamens 3-5, broader at base; anthers free. Style enclosed in corolla, with up to 5 short stigmas. Capsule dehiscing by apical valves.

1. W. lobelioides (L. f.) Schrad. ex Link, *Handbuch* **1**: 632 (1829).
Erect, slightly fleshy annual, 20-50 cm. Leaves lanceolate to ovate-lanceolate, shallowly to deeply dentate, sessile, clustered towards base. Inflorescence a few-flowered panicle, flowers with long pedicels, drooping. Calyx tubular, with 5 narrow lobes, reddish in fruit. Corolla with 3-4 but mostly 5 spreading lobes, pinkish lilac to purple or rarely white. Capsule ovoid. **Plate 42**
Fl. III - V. *Found in rocky places and on waste ground in moist valleys throughout the Madeiran archipelago and on Selvagem Grande.* **MDPS**

The above description and synonymy refer to the Macaronesian endemic (■) subsp. **lobeliodes**, which otherwise occurs in the Canaries and Cape Verdes, and to which plants from the Madeiran and Salvage islands belong. Subsp. *nutabunda* (Guss.) Murb., from the Mediterranean region, and subsp. *riparia* (A. DC.) Thulin, from tropical Africa, both have ± hirsute stems.

7. Lobelia L.

Annuals or herbaceous to shrubby perennials. Flowers solitary in leaf axils, or in racemes, strongly zygomorphic. Calyx-tube 5-lobed, joined to ovary. Corolla-tube curved, deeply split dorsally, 2-lipped, the upper 2 lobes smaller than the lower 3. Stamens 5; filaments connate in a tube around the style; anthers 2, bearded. Stigma capitate, 2-lobed. Capsule dehiscing by 2 apical valves.

Besides *L. erinus* and *L. laxiflora*, several other species are cultivated in gardens.

1.	Shrub; flowers scarlet-yellow, up to 3 cm long	**1. laxiflora**
-	Herb; flowers blue, up to 2 cm long	2
2.	Inflorescence few-flowered; flowers up to 15 mm across	**2. erinus**
-	Inflorescence 10- to 20-flowered; flowers up to 5 mm across	**3. urens**

1. L. laxiflora Kunth in Humb., Bonpl. & Kunth, *Nov. gen. sp.* **3**: 311 (1820).
Shrub up to 175 cm. Stems stout, pubescent on younger shoots. Leaves obovate to ovate-lanceolate, entire, serrate. Flowers *c.* 3 cm long. Calyx with reflexed lobes. Corolla scarlet-yellow, with exserted, bearded stamens.
Fl. I - III. *Grown as a garden plant in Madeira, and occasionally naturalized. Native to Mexico.*
M

2. L. erinus L., *Sp. pl.* **1**: 932 (1753).
Lax, semi-decumbent annual to 30 cm. Leaves dentate, lower ovate to ovate-lanceolate, upper narrowly lanceolate. Calyx-lobes long, narrow, erect at first, later spreading. Corolla up to 15 mm wide, usually blue but varying from white to purple in cultivated forms. Capsule within persistent calyx-teeth.
Fl. IV - X. *Widely cultivated as an ornamental and occasionally naturalized. Native to S. Africa.*
M

3. L. urens L., *Sp. pl.* **2**: 931 (1753) [Cabreira]
Erect perennial to 60 cm. Lower stems and leaves tinged with purple. Leaves dentate, the lower elliptic-lanceolate, the upper linear-lanceolate. Calyx-lobes short. Corolla up to 5 mm wide,

blue to purple. Capsule erect.

Fl. VIII - IX. *Frequent in Madeira, growing at the margins of woods and on the banks of levadas from, 500-1000 m.* **M**

CVI. COMPOSITAE
[ASTERACEAE]
J.R. Press

Herbs or shrubs, often with latex. Leaves alternate, opposite or in rosettes, exstipulate. Flowers small (*florets*), functionally male, female or sterile, aggregated into heads (*capitula*) resembling a single flower with a calyx-like involucre of 1 or more rows of bracts. Capitula usually pedunculate, solitary or in corymbose or racemose inflorescences, with or without receptacular scales each subtending a floret, florets all similar (*homogamous*) or the inner and outer florets differing (*heterogamous*), when the inner are usually hermaphrodite and the outer usually female. Calyx (*pappus*) represented by a corona, auricle, scales, setae or hairs, or absent. Corolla of 3 main types: *(a)* tubular with usually (4-)5 short, ± equal teeth; *(b)* tubular, 2-lipped; *(c)* ligulate, with the tube extended on one side into a long, usually 3- to 5-toothed ligule; corolla of female florets rarely filiform or absent. Homogamous capitula may be composed of any of the 3 floret types; heterogamous capitula usually have central (*disc*) florets of type *(a)* and outer or marginal (*ray*) florets of type *(b)* or type *(c)*. Stamens 5, epipetalous, the anthers often sagittate or tailed at the base and usually connate into a tube around the style. Ovary inferior, 1-locular. Fruit an achene crowned with the pappus and sometimes with a slender beak.

1.	Plant with latex; florets all ligulate	2
-	Plant usually without latex; at least the inner florets not ligulate	19
2.	Leaves spiny	3
-	Leaves not spiny	6
3.	Receptacle with scales which enfold achenes	**51. Scolymus**
-	Receptacle without scales	4
4.	Achenes not compressed; spines on leaves bulbous-based	**58. Helminthotheca**
-	Achenes compressed; spines on leaves not bulbous-based	5
5.	Achenes abruptly narrowed at apex into a beak	**62. Lactuca**
-	Achenes not beaked	**61. Sonchus**
6.	Florets pale blue	**52. Cichorium**
-	Florets variously coloured but never blue	7
7.	Receptacle with scales	**56. Hypochoeris**
-	Receptacle without scales	8
8.	At least some achenes without pappus	**64. Lapsana**
-	Achenes with pappus of scales or hairs	9
9.	Achenes strongly compressed	10
-	Achenes not compressed	11

10. Pappus of uniform, deciduous or persistent hairs; at least the outer involucral bracts with a scarious margin . . . . **60. Launaea**
- Pappus of a few scabrid, deciduous hairs and ± persistent, softer, inner hairs in fascicles; involucral bracts without a scarious margin . . . . **61. Sonchus**

11. At least some achenes with a pappus of scales or of scales and hairs 12
- All achenes with a pappus entirely of hairs 14

12. At least inner achenes with some plumose pappus hairs . . . . **57. Leontodon**
- All achenes with a pappus entirely of scales or of scales and simple hairs 13

13. Achenes 1-2 mm . . . . **53. Tolpis**
- Achenes 5-6 mm . . . . **54. Hedypnois**

14. Plants scapose, the scapes unbranched . . . . **63. Taraxacum**
- Plants with leafy flowering stems, not scapose 15

15. At least some pappus hairs plumose 16
- All pappus hairs simple 18

16. Involucral bracts in several rows . . . . **58. Helminthotheca**
- Involucral bracts in 1 row 17

17. Leaves entire . . . . **59. Tragopogon**
- Leaves lobed . . . . **55. Urospermum**

18. Plants stellately hairy and usually viscid, at least above . . . . **66. Andryala**
- Plants without stellate hairs, sometimes with glandular hairs but not viscid **65. Crepis**

19. Involucral bracts spiny or with a usually fimbriate, lacerate or dentate (rarely entire) apical appendage 20
- Involucral bracts not spiny, lacking an apical appendage 33

20. Leaves marbled with white above, at least on the veins 21
- Leaves uniformly coloured above 23

21. Leaves decurrent on stem, forming a narrow wing; outer florets much large than the inner, sterile . . . . **43. Galactites**
- Leaves not decurrent on stem; outer florets not enlarged, fertile 22

22. At least some pappus hairs plumose . . . . **42. Notobasis**
- Pappus hairs not plumose . . . . **45. Silybum**

23. At least some involucral bracts with a fimbriate, lacerate or dentate (rarely entire) apical appendage 24
- Involucral bracts often spine-tipped but lacking a well-defined apical appendage . . 28

24. Outer involucral bracts leafy 25
- Outer involucral bracts not leafy 26

25. Florets yellow to orange-red . . . . **49. Carthamus**
- Florets blue . . . . **50. Carduncellus**

26. Capitula with ligulate ray florets and tubular disc florets . . . . **37. Arctotis**
 - Capitula with tubular florets only, the outer sometimes larger 27

27. Small shrub . . . . **46. Cheirolophus**
 - Herb . . . . **48. Centaurea**

28. Involucral bracts shiny, spreading when dry; florets creamy . . . . **38. Carlina**
 - Involucral bracts dull, erect when dry; florets purplish or rarely white 29

29. Outer involucral bracts hooked at apex . . . . **39. Arctium**
 - Outer involucral bracts with straight apices 30

30. Stem with a spiny wing; pappus of simple, scabrid hairs . . . . **40. Carduus**
 - Stem not winged; pappus of plumose hairs 31

31. Receptacle fleshy; involucral bracts with apical spine more than 10 mm; leaves with stiff, stout spines . . . . **44. Cynara**
 - Receptacle not fleshy; involucral bracts unarmed or with apical spine less than 10 mm; leaves with soft, slender spines 32

32. Pappus of plumose hairs . . . . **41. Cirsium**
 - Pappus of slender scales and setae . . . . **47. Mantisalca**

33. Capitula unisexual 34
 - Capitula hermaphrodite 35

34. Leaves palmately lobed; female capitula with hooked spines . . . . . . **22. Xanthium**
 - Leaves 2-pinnatisect; female capitula with 5-7, straight, spinose teeth . **21. Ambrosia**

35. At least some leaves opposite 36
 - All leaves alternate or basal 42

36. Upper leaves alternate; capitulum very large, up to 30 cm, nodding . **18. Helianthus**
 - All leaves opposite; capitula usually small, erect 37

37. Stems prostrate; leaves 2-pinnatifid . . . . **33. Cotula**
 - Stems erect to ascending; leaves entire or pinnate 38

38. Ligules absent; disc florets white, lilac or blue; plant usually strong-smelling 39
 - Ligules usually present; disc florets yellow; plant not strong-smelling 40

39. Herbaceous to shrubby perennial . . . . **1. Ageratina**
 - Annual . . . . **2. Ageratum**

40. Pappus of 2-4 bristles . . . . **19. Bidens**
 - Pappus of scales, or a corona 41

41. Plant glandular-hairy; pappus of scales . . . . **20. Galinsoga**
 - Plant lacking glandular hairs; pappus a rim-like corona . . . . **17. Eclipta**

42. Pappus of hairs 43
 - Pappus of scales, or a corona, or absent 57

43. Involucral bracts in 1 row, sometimes with small, supplementary bracts at the base of the capitulum 44
- Involucral bracts in (2-)3 or more rows 45

44. Shrub; florets purple .... **35. Pericallis**
- Herb or climber, sometimes woody at base; florets yellow .... **34. Senecio**

45. Involucral bracts subtending the outer florets 46
- Involucral bracts not subtending the outer florets 47

46. Pappus-hairs apically plumose .... **10. Ifloga**
- Pappus-hairs not apically plumose .... **7. Logfia**

47. Involucral bracts mostly dull brown or green 48
- Involucral bracts entirely white, yellow or shining golden brown (rarely pinkish) . 56

48. Disc florets yellow 49
- Disc florets purple 55

49. Ligules absent 50
- Ligules present but often very short 52

50. Annual .... **6. Conyza**
- Shrubby perennial 51

51. Leaves dentate .... **13. Phagnalon**
- Leaves entire .... **15. Schizogyne**

52. Ligules white, pinkish or purple 53
- Ligules yellowish 54

53. Annual or biennial, erect to ascending; leaves entire, sessile .... **4. Aster**
- Perennial, usually trailing; leaves with 3 or more lobes, petiolate ..... **5. Erigeron**

54. Annual .... **6. Conyza**
- Viscid perennial .... **14. Dittrichia**

55. Pappus-hairs connate at base .... **9. Gamochaeta**
- Pappus-hairs free .... **8. Filago**

56. Shrub, woody-based perennial, or robust and strong-smelling biennial **12. Helichrysum**
- Annual, not strong smelling .... **11. Pseudognaphalium**

57. Plant scapose, leaves all in a basal rosette .... **3. Bellis**
- Plant not scapose, flowering stems leafy 58

58. Pappus of free scales; ligules shorter than outer involucral bracts .... **16. Nauplius**
- Pappus a corona or absent; ligules longer than outer involucral bracts or absent .. 59

59. Inner achenes strongly curved to annular, muricate or rugose ..... **36. Calendula**
- Inner achenes at most slightly curved, not muricate 60

60. Receptacular scales present, at least among the inner florets 61
- Receptacular scales absent 63

61. Capitula numerous, corymbose; involucre less than 5 mm in diameter . . **25. Achillea**
- Capitula solitary; involucre more than 5 mm in diameter 62

62. Disc florets saccate, obscuring apex of achene . . . . **26. Chamaemelum**
- Disc florets swollen at base in fruit but not obscuring apex of achene . . **27. Anthemis**

63. Ligules present 64
- Ligules absent 68

64. Ligules yellow or yellow and white 65
- Ligules wholly white, cream or pink 66

65. Leaves entire . . . . **31. Coleostephus**
- Leaves deeply divided . . . . **28. Chrysanthemum**

66. Ligules less than 10 mm . . . . **23. Tanacetum**
- Ligules more than 10 mm 67

67. Achenes cylindrical; leaves entire to shallowly lobed . . . . **30. Leucanthemum**
- At least outer achenes 3-angled and usually with 1-4 wings; leaves usually deeply divided, up to 2-pinnatifid . . . . **29. Argyranthemum**

68. Small shrub, densely white- or grey-tomentose . . . . **24. Artemisia**
- Prostrate annual, pubescent to villous 69

69. Capitula sessile, surrounded by leaves . . . . **32. Soliva**
- Capitula with long, slender peduncles . . . . **33. Cotula**

Subfam. ASTEROIDEAE
Tribe EUPATORIEAE
1. Ageratina Spach

Perennial herbs or shrubs, occasionally scandent, with usually opposite leaves. Capitula in corymbose inflorescences. Involucral bracts c. 30, in 2-3 rows, ± equal and spreading when mature. Receptacle slightly convex, glabrous or with minute hairs. Florets eligulate, hermaphrodite, white or pinkish, with a slender basal tube and campanulate limb; corolla-lobes distinctly longer than wide, densely papillose on the inner surface. Achenes cylindrical to prismatic, usually 5-ribbed; pappus a single row of very slender, often easily detachable bristles.

1. Leaves broadly cuneate, the margins toothed towards the base; main pair of lateral veins arising from the base . . . . **1. adenophora**
- Leaves narrowly cuneate, the margins entire towards the base; main pair of lateral veins arising above the base . . . . **2. riparia**

1. A. adenophora (Spreng.) R. King & H. Rob. in *Phytologia* 19: 211 (1970).

[*Abundancia, Inça muito*]

Eupatorium adenophorum Spreng.
Strong-smelling, glandular-hairy herb or undershrub up to 150 cm. Leaves with lamina up to

9 × 7 cm, rhombic-ovate, broadly cuneate at base, acuminate, crenate-serrate, three-veined with the two lateral veins arising from the base. Capitula 6 mm across; involucral bracts up to 4.5 mm long, with two pronounced central veins; florets white; styles pale lilac. Achenes 1.5-2 mm, black, glabrous.
Fl. all year. *Introduced. An invasive weed, now becoming common throughout the lowlands of Madeira up to 1100 m. Usually found in damp places on banks, slopes and cliffs, along levadas and in woods, occasionally in drier situations. Rare in Porto Santo, with a few records from around Vila Baleira. Native to Mexico, the W. Indies and Pacific Islands.* **MP**

2. A. riparia (Regel) R. King & H. Rob. in *Phytologia* **19**: 216 (1970).
Eupatorium riparium Regel
Like **1** but with eglandular hairs; leaves up to 11 × 3 cm, lanceolate, narrowly cuneate and with margins entire towards the base; capitula 4-5 mm across; achenes thinly puberulous; pappus pale pink.
Fl. III - VIII. *Introduced; growing in similar habitats to 1 in the Funchal region, and at Santo da Serra and Portela; apparently spreading. Native to C. America.* **M**

Only a few, scattered records are known for this species but it is easily confused with **1** and may be under-recorded.

2. Ageratum L.

Like *Ageratina* but sometimes annual; leaves sometimes alternate; involucral bracts 30-40, hardened and often with scarious margins; receptacle conical, glabrous or with scales; florets white, blue or lavender, with or without a distinct basal tube; corolla-lobes about as long as wide; achenes prismatic, with 4-5 ribs, sometimes shortly setulose on the ribs; pappus of 5-6 flattened, awn-like scales or a short corona, or absent.

Literature: M.F. Johnson, A monograph of the genus *Ageratum* L. (Compositae-Eupatorieae), in *Ann. Mo. bot. Gdn* **58**: 7-88 (1971).

1. Involucral bracts ± glabrous, fimbriate at the apex, abruptly contracted into a short tip
   .... **1. conyzoides**
- Involucral bracts pilose, entire and gradually tapering into an acuminate tip ....
   .... **2. houstonianum**

1. A. conyzoides L., *Sp. pl.* **2**: 839 (1753).
Foetid, pubescent annual with erect stems up to 60 cm. Leaves with lamina 3.5-5.5 × 2-3 cm, ovate to elliptic-oblong, cuneate, acute, crenate, sometimes dotted with minute, yellow glands. Involucral bracts in 2 rows, ± glabrous, 3.5-4 mm, oblong-lanceolate, erose to shortly fimbriate at the apex, abruptly contracted into a short, scabrid tip. Florets white with blue corolla-lobes. Achenes 1.5 mm, 5-angled, scabrid on angles, black; pappus 1.5-2 mm long, the 5 scales extended at the tip into a long, scabrid awn.
Fl. XI - V. *Introduced. Found occasionally by ditches and roads, mainly around Funchal. A pantropical weed.* **M**

Madeiran plants belong to subsp. **conyzoides**.

2. A. houstonianum Mill., *Gard. Dict.* ed. 8, no. 2 (1768).
A. mexicanum Sims
Like **1** but the involucral bracts 4-4.5 mm, conspicuously pilose, narrowly lanceolate, entire

and uniformly tapering into an acuminate apex bearing short, stipitate glands.
Fl. I - IV. *Eastern Madeira, usually in the vicinity of gardens, and possibly only occurring as an escape.* **M**

Tribe ASTEREAE

3. Bellis L.

Annual or perennial herbs. Leaves alternate, often in a basal rosette, entire or toothed. Capitulum solitary. Involucral bracts in (1-)2 rows, leafy. Receptacle conical or flat, pitted; scales absent. Outer florets female, ligulate; ligules entire, white or pink; inner florets hermaphrodite, the corolla 4- to 5-lobed, yellow. Achenes obovate, compressed, the margin thickened; pappus absent.

1. B. perennis L., *Sp. pl.* **2**: 886 (1753).
Scapose perennial herb with fleshy roots and short stolons. Leaves in basal rosettes, 4-7 × 1-1.7 cm, spathulate to oblanceolate, usually shallowly toothed, appressed-hairy, often slightly fleshy and narrowing into a distinct petiole about as long as the blade. Capitula 17-27 mm in diameter on hairy peduncles up to 17 cm long. Ligules 5-8 mm, white, often flushed red or purplish beneath. Achenes hairy.
Fl. III - VI. *Uncommon plant of grassy places, mainly in eastern Madeira.* **M**

4. Aster L.

Perennial herbs, sometimes annual or biennial. Leaves alternate, simple. Capitula solitary or in branched inflorescences. Involucral bracts imbricate, usually in 3 or more rows, the outermost shortest. Receptacle flat, pitted. Outer florets ligulate, female or sterile, in a single row, blue to purplish, pink or white. Inner florets tubular, hermaphrodite, yellow, sometimes becoming purplish. Achenes oblong, compressed, hairy; pappus of 1-several rows of hairs.

1. A. squamatus (Spreng.) Hieron. in *Bot. Jb.* **29**: 19 (1900).
Glandular-punctulate annual or biennial with erect or ascending stems. Leaves 2-3 cm, linear to linear-lanceolate, entire, sessile. Capitula numerous, obconical. Involucral bracts oblong, acute, tinged purple, appressed in 3 rows, unequal, the outer *c.* 2 mm, the innermost 5-6 mm. Ligules *c.* 1.5 mm, purplish blue. Achenes 2.5-3 mm.
Fl. VII. *Recently introduced; first recorded in 1969 at Ponta do Sol and now spreading along the south coast of Madeira; also recorded from Porto Santo in 1981. Native to C. and S. America.* **MP**

5. Erigeron L.

Annuals, biennials or perennials. Leaves entire or lobed. Capitula 1-several forming a loose panicle. Outer florets ligulate, female; inner florets tubular, hermaphrodite, the outermost sometimes filiform and female. Achenes hairy; pappus of hairs, scales or occasionally both.

1. E. karvinskianus DC., *Prodr.* **5**: 285 (1836).
E. mucronatus DC.
Sparsely hairy perennial. Stems long, much-branched, trailing or ascending, woody at the base. Leaves 1-3.5 cm, the blade narrowing into the petiole; lower leaves rhombic to obovate in outline with 3, rarely more, mucronate lobes, often with short, leafy axillary shoots; upper leaves narrowly elliptic to oblanceolate, entire. Capitula near tips of stems, pedunculate. Involucre 7-10 mm in diameter; bracts in several rows, linear-lanceolate with somewhat scarious margins. Ligules of outer florets narrow, acute, entire, white or pink above, purplish below; inner florets all

hermaphrodite, yellow. Achenes *c.* 1 mm, oblong-ovoid, laterally compressed with thickened margins; pappus a single row of white hairs.

Fl. III - IX. *Originally introduced as a garden ornamental but now completely naturalized, widespread and common on moist rocks and walls almost everywhere in Madeira. Native to Mexico.* **M**

6. Conyza Less.

Mostly herbs, rarely shrubs. Leaves alternate, entire to shallowly lobed. Involucral bracts in several rows. Receptacle flat, scales absent. Outer florets numerous, in several rows, female, the corolla tubular-filiform, with minute ligules present or absent; inner florets few, hermaphrodite, the corolla tubular, 4- or 5-lobed. Achenes laterally compressed; pappus of hairs.

1. Involucre campanulate, bracts ± glabrous; inner florets 4-lobed . . . . . **1. canadensis**
- Involucre broadly cylindrical, bracts evenly hairy; inner florets 5-lobed 2

2. Inflorescence columnar, not glandular; ligulate outer florets present; pappus pale grey . . . . **3. sumatrensis**
- Inflorescence pyramidal, glandular, sticky; ligulate florets absent; pappus yellowish brown . . . . **2. bonariensis**

1. C. canadensis (L.) Cronquist in *Bull. Torrey bot. Club* **70**: 632 (1943). [*Avoadeira*]
Erigeron canadensis L.
Pubescent annual with spreading hairs. Stems 30-150 cm. Leaves 4-9 cm, entire to remotely toothed, relatively thin-textured, margins usually with long, white cilia, the lower leaves oblong-lanceolate, cuneate, petiolate; upper leaves narrower, sessile. Inflorescence long, ± columnar, panicle-like, with numerous, small capitula. Capitula 3-5 mm in diameter, campanulate; involucral bracts linear-lanceolate to subulate, pale green with scarious margins, glabrous or subglabrous. Female florets numerous; ligules 0.5-1 mm, cream or greenish white. Hermaphrodite florets few; corolla 4-lobed, yellow. Achenes *c.* 1 mm, oblanceolate, hairy; pappus of long, grey or pale brown hairs.
Fl. VI - XII. *Common weed, especially of cultivated and disturbed ground; occurring throughout Madeira up to 1000 m.* **M**

2. C. bonariensis (L.)Cronquist in *Bull. Torrey bot. Club* **70**: 632 (1943).
C. ambigua DC.; *C. crispa* Pourr.
Densely pubescent annual. Stems up to 100 cm, erect, sometimes decumbent. Leaves variable, mostly (5-)7-10 × 0.3-2 cm, linear to obovate in outline, entire to toothed or shallowly pinnately lobed, narrowing gradually into a distinct petiole, somewhat thick in texture; both surfaces with short, appressed hairs; midrib and the margins towards the base with a few long cilia. Inflorescence ± pyramidal with lateral branches frequently overtopping the main axis, glandular and sticky. Capitula (4-)6-9(-12) mm in diameter, almost cylindrical; bracts purple or crimson towards the tips, evenly white-hairy. Female florets numerous; ligules absent. Hermaphrodite florets 5-lobed, creamy. Achenes *c.* 1 mm, oblanceolate, hairy; pappus yellowish brown.
Fl. VI - IX. *A common weed of roadsides, cultivated and waste ground throughout the lower regions of Madeira and Porto Santo.* **MP**

3. C. sumatrensis (Retz.)E. Walker in *J. Jap. Bot.* **46**: 72 (1971).
C. floribunda Humb., Bonpl. & Kunth
Like **2** but often taller; inflorescence with numerous branches, the lateral branches not overtopping the main axis, not glandular; capitula 5-7 mm in diameter; at least some female florets with minute ligules; pappus pale grey.

Fl. ? *A recent introduction, occurring on dry roadsides near Santa Cruz, in Fajã da Nogueira and probably elsewhere. Native to S. America but a cosmopolitan weed.* **M**

Tribe INULEAE

Literature: A. Anderberg, Taxonomy and phylogeny of the tribe *Gnaphalinieae* (Asteraceae), in *Op. bot. Soc. bot. Lund.* **104**: 1-195 (1991).

7. Logfia Cass.

Annual herbs. Leaves alternate, sessile, tomentose to villous, margins flat or slightly concave, entire. Capitula heterogamous, disciform, few together, surrounded by a ray of leaves. Involucral bracts in a few rows, cartilaginous or apically minutely papery, brownish, transparent, folded around the female florets. Receptacle conical, scales absent. Florets purple, the outer female, pistillate and filiform, the inner hermaphrodite and fewer than the female florets. Achenes small, oblong, those of the outer florets somewhat laterally compressed, all sparsely hairy with short, clavate, myxogenic twin-hairs; pappus of 1 row of slender, scabrid bristles, ciliate towards the base.

1. Clusters of capitula overtopped by subtending leaves; leaves linear-subulate; middle involucral bracts saccate, completely enclosing outer florets . . . . **1. gallica**
- Clusters of capitula not overtopped by subtending leaves; leaves oblong to linear-lanceolate; middle involucral bracts weakly saccate, not completely enclosing outer florets . . . . **2. minima**

1. L. gallica (L.) Coss. & Germ. in *Annals Sci. nat.* II, **20**: 291 (1843).
Filago gallica L.; *Xerotium gallicum* (L.) Bluff & Fingerh.
Silky-hairy, grey-white annual. Stem up to *c.* 20 cm, erect or somewhat ascending, much-branched, at least above. Leaves 5-20 mm, linear-subulate, held somewhat erect. Clusters of capitula overtopped by subtending leaves. Outer involucral bracts small, ovate; middle bracts lanceolate triangular with a saccate base completely enclosing the outer female florets; inner bracts lanceolate, subtending but not enclosing some central florets; all bracts stellately spreading in fruit. Outer achenes 0.7 mm, smooth, lacking a pappus, the inner 0.4 mm, with transparent papillae.
Fl. VII - IX. *Dry soils, from Monte to Machico, Santana and São Vicente.* **M**

2. L. minima (Sm.) Dumort., *Fl. belg.*: 68 (1827).
Filago minima (Sm.) Pers.; *F. montana* auct. mad.
Silky-hairy, greyish annual. Stem 2-15 cm, erect or ascending, usually branched above and below. Leaves up to 7 mm, narrowly oblong to linear-lanceolate, acute and mucronate, erect and somewhat appressed. Capitula not overtopped by subtending leaves. Outer involucral bracts short, narrowly lanceolate; middle bracts broadly lanceolate, saccate at the base but not completely enclosing the outer florets; inner bracts not saccate. Outer achenes 0.7 mm, smooth; pappus absent; inner achenes 0.5 mm, with translucent papillae.
Fl. VI - VII. *Common in high pastures in central Madeira and the high parts of Deserta Grande.* **MD**

8. Filago L.

Tomentose annuals. Leaves alternate, sessile, entire. Capitula heterogamous, disciform, few together, surrounded by a ray of leaves. Involucral bracts in few rows, cartilaginous or apically minutely papery, brownish, transparent, not enclosing the florets. Receptacle conical; scales absent. Florets all tubular, purple, the outer female, filiform and pistillate, the inner hermaphrodite or functionally male, fewer than the outer florets. Achenes small, oblong, those of the outer florets somewhat laterally compressed, sparsely hairy; pappus of 1 row of slender, scabrid, free hairs with patent cilia near the base.

1. Involucre weakly 5-angled; bracts in 4-6 rows, divergent in fruit . . . **2. pyramidata**
- Involucre strongly 5-angled; bracts in 3 rows, not divergent in fruit . . . **1. lutescens**

1. F. lutescens Jord., *Observ. pl. nouv.* **3**: 201 (1846).
Tomentose annual 10-30 cm. Stems ascending, rarely erect, usually much-branched. Leaves up to 25 × 5 mm, oblong to oblanceolate, apiculate, semi-amplexicaul. Capitula half-submerged in the tomentum, overtopped by 1-2 subtending leaves. Involucre weakly 5-angled; bracts in vertical rows of 3(-4), erect, lanceolate, with a straight arista.
Fl. V - VII. *Scattered in Madeira, rare in Selvagem Grande; records from Bugío require confirmation.* **M ?D S**

Only subsp. **atlantica** Wagenitz in *Willdenowia* **5**: 56 (1968) (*Filago micropodioides* auct. mad., non Lange (1861)) is present in the islands. It is distinguished by its grey-white tomentum, capitula up to 9 mm across, and involucral bracts entirely yellow, or purplish at the base only.

2. F. pyramidata L., *Sp. pl.* **2**: 1199 (1753).
Like **1** but stems decumbent or ascending, always branched; leaves oblong to broadly spathulate; involucre strongly 5-angled, the bracts in 4-6 rows and divergent in fruit with a recurved arista.
Fl. V - VIII. *Recorded from Monte and between Santo Antonio and Pico de Arieiro.* **M**

Madeiran plants are var. **pyramidata**.

9. Gamochaeta Wedd.

Annual or perennial herbs. Leaves alternate, sessile, entire. Capitula small and inconspicuous, heterogamous, disciform, in head-like clusters or ± elongate spikes. Involucral bracts papery, brownish or sometimes tinged with purple, transparent. Receptacle flat; scales absent. Florets all tubular, purple, the outer female and filiform, the inner fewer and hermaphrodite. Achenes small, oblong, with globose, myxogenic twin-hairs; pappus of 1 row of slender, barbellate hairs connate at the base and deciduous as a single unit.

1. Stem branched along its whole length; upper and lower leaf surfaces similar in colour and hairiness . . . . **1. calviceps**
- Stem simple or branched only at base; upper and lower leaf surfaces different in colour and hairiness . . . . **2. pensylvanica**

1. G. calviceps (Fern.) Cabrera in *Boln Soc. Argent. Bot.* **9**: 368 (1961).
Gnaphalium calviceps Fern.
Silky-hairy annual up to 20 cm. Stem branching along its whole length. Basal leaves not persisting at flowering; caluine leaves 15-25 × 2-4 mm, linear to linear-lanceolate, dull grey-green and silky-hairy on both surfaces, the uppermost with blades becoming folded and curved upwards at the inflorescence. Capitula in loose, terminal and stalked, axillary clusters forming a spike or paniculate inflorescence. Lower inflorescence-leaves as long as the subtended cluster, the upper obscure or absent. Capitula tomentose at the base only.
Fl. VI. *A recently introduced, and as yet rare, ruderal weed at Serra de Água, Ribeira Brava and Terreira de Luta.* **M**

2. G. pensylvanica (Willd.) Cabrera in *Boln Soc. Argent. Bot.* **9**: 368 (1961).
Gnaphalium pensylvanicum Willd.; *G. purpureum* auct. mad., non (Willd.) Cabrera (1961); *G. spathulatum* Lam.
Annual up to 40 cm. Stem solitary, erect, simple or branched only from near the base. Basal leaves not persisting at flowering; cauline leaves 20-70 × 6-15 mm, oblanceolate to spathulate, thinly tomentose and greenish above, densely grey-green-tomentose below. Capitula in dense, terminal and sessile, axillary clusters forming a leafy spike. Lower inflorescence-leaves plane, spreading and distinctly longer than the subtended cluster; upper smaller but evident. Capitula densely tomentose in the lower ⅔.
Fl. IV - VI. *Introduced weed of waste ground and streets, scattered through many of the lowland areas of Madeira.* M

G. filaginea (DC.) Cabrera in *Boln Soc. Argent. Bot.* **9**: 371 (1961), a perennial with spathulate, ± obviously curved leaves tomentose on both surfaces. It has been recorded from Madeira but without further data; the record requires confirmation. M

10. Ifloga Cass.

Ericoid annuals. Leaves alternate, sessile, tomentose on the upper surface, entire. Capitula few together in the leaf-axils along the stem and branches. Involucral bracts papery, brownish, transparent. Receptacle flat; scales absent. Florets all tubular, the outer female, filiform, purple and often subtended by the outer involucral bracts, the inner usually fewer, hermaphrodite or functionally male, yellow. Achenes small, oblong, with globose, myxogenic twin-hairs; pappus of 1 row of slender, apically plumose bristles with patent cilia towards the base.

Literature: O.M. Hilliard, A revision of *Ifloga* in Southern Africa, with comments on the northern hemisphere, in *Bot. J. Linn. Soc.* **82**: 293-312 (1981).

1. I. spicata (Forssk.) Sch. Bip. in Webb & Berthel., *Hist. nat. Iles Canaries* 3(2, sect. 2): 310 (1845).
Small tufted annual with erect or decumbent stems. Leaves linear, tomentose, sometimes papery below. Female florets solitary in axils of bracts outside the involucre of hermaphrodite florets, and usually twice as many as hermaphrodite florets. Pappus of female florets absent.
Fl. ? *Occurring on Selvagem Grande only.* S

Plants from Selvagem Grande are recorded as belonging to susbp. **spicata**.

11. Pseudognaphalium Kirp.

Annual, biennial or perennial herbs. Leaves alternate, sessile, straight, decurrent or not, hairy to velutinous, entire. Capitula heterogamous, disciform, in small, terminal corymbs. Involucral bracts in a few rows, papery, yellowish or white to pink, transparent, entire. Receptacle flat; scales absent. Florets all tubular, yellow, the outer female filiform, the inner hermaphrodite and fewer than the female florets. Achenes small, oblong, glabrous or sparsely hairy with short, clavate, myxogenic twin-hairs; pappus of 1 row of slender, barbellate hairs with patent cilia near the base.

1. P. luteo-album (L.) Hilliard & B.L. Burtt in *Bot. J. Linn. Soc.* **82**: 206 (1981).
Gnaphalium luteo-album L.
Densely white-tomentose annual to 50 cm. Stem erect or ascending, simple or branched. Leaves 15-80 × 3-10 mm, the basal spathulate to oblanceolate, the cauline oblong and semi-amplexicaul, not decurrent, all tomentose on both surfaces. Corymb dense; capitula sessile, ± glabrous, 4-12 in semi-globose clusters lacking subtending leaves. Involucral bracts yellowish, shining, ovate

to oblong, obtuse. Florets tinged reddish above. Achenes 0.3 mm, tuberculate.
Fl. all year. *Damp places, cliffs and roadsides. Common in the lower regions of Madeira, but in Porto Santo known only from Serra de Dentra; very rare on Deserta Grande.* **MDP**

12. Helichrysum Mill.

Perennial or annual herbs, in Madeira mainly subshrubs. Leaves alternate, straight, often pilose to velutinous, eglandular; margin usually flat, entire. Capitula heterogamous, disciform, borne in loose, flat-topped, terminal corymbs or solitary. Involucral bracts in few to many rows, entire, papery, monomorphic, monochromous; lamina brown, yellow, pink or white, transparent. Receptacle flat; scales usually absent. Female florets yellow, pistillate, filiform, in 1-several rows, not outnumbering the disc florets. Disc florets perfect. Corolla yellow. Achenes small, oblong, sparsely hairy with short, myxogenic hairs or with long, non-myxogenic hairs, or glabrous. Pappus monomorphic, of basally free or somewhat connate, barbellate or subplumose, capillary bristles in 1 row.

1. Biennial; capitula more than 10 mm in diameter, involucral bracts bright yellow . . . . 5. **foetidum**
- Small shrub or woody-based perennial; capitula less than 8 mm in diameter, involucral bracts white or brownish yellow, sometimes tinged pinkish or purplish 2

2. Leaves abruptly contracted towards the base before narrowing into the petiole . . . . 6. **petiolare**
- Leaves not abruptly contracted into the petiole, or leaves sessile 3

3. Involucral bracts white, rarely flushed pinkish; leaves lanceolate; flowers scented . 4
- Involucral bracts yellowish brown, rarely flushed violet; leaves obovate to elliptic; flowers not scented 5

4. Leaves shortly petiolate, 1-nerved . . . . 1. **melaleucum**
- Leaves subsessile, 3-nerved . . . . 2. **devium**

5. Capitula 5-7 mm, obconical, involucral bracts remaining erect . . . . . . 3. **obconicum**
- Capitula 3 mm, hemispherical, involucral bracts becoming patent to spreading . . . . 4. **monizii**

1. H. melaleucum Rchb. ex Holl, *Flora* **13**: 382 (1830).   [*Perpetua, Propéta*]
H. melanophthalmum (Lowe) Lowe
Small, densely white- or grey-tomentose shrub to 100 cm. Leaves 30-80 × 4-13(-15) mm, lanceolate, shortly petiolate, 1-nerved, tomentose on both surfaces. Capitula 5-8 × 4-6 mm, hemispherical. Involucral bracts white, sometimes flushed pinkish towards the base, erect at first, finally wide-spreading, the outer ovate and obtuse to hemispherical at apex, the inner obovate to oblong and obtuse. Florets blackish to dark purplish brown, sweet-scented. Pappus-hairs echinulate along their whole length. **Plate 43**
Fl. III - VI. *Sunny cliffs and steep rocks up to 1700 m; common along the north coast and in central Madeira south to Funchal, though apparently absent east of a line from Caniçal to Santo da Serra; rarer on the peaks of Porto Santo and on Deserta Grande.* ● **MDP**

Plants with pinkish-tinged involucral bracts have been referred to as **H. melanophthalmum** var. **rosea** Lowe, *Man. fl. Madeira* **1**: 483 (1868).

2. H. devium J.Y. Johnson, *Gdnrs. Chron.* **2**: 62 (1888).
Like **1** but leaves 50-70 × 7-20 mm, only gradually narrowed at base, subsessile and

semi-amplexicaul, 3-nerved, less thickly tomentose; involucral bracts sometimes flushed pink towards the base. **Plate 43**
Fl. III - IV. *Sea cliffs on Ponto de São Lourenço and west to Espigão Amarelo and Caniçal.*
● **M**

Occurring further east and in drier places than **1**, but very similar to it and possibly not specifically distinct.

3. H. obconicum DC., *Prodr.* **6**: 181 (1837). [*Morrão, Murrão, Murrião*]
Small, densely white-tomentose shrub up to 60 cm. Leaves 25-50 × 8-20 mm, obovate to elliptic, petiolate, tomentose on both surfaces. Capitula 5-7 × 6-8 mm, obconical. Involucral bracts yellowish to golden brown, erect, not spreading, the outer very small, the inner lanceolate to narrowly oblong. Florets yellowish, not scented. Pappus-hairs scarcely echinulate below, densely so at the tip. **Plate 43**
Fl. VIII - X. *Frequent but rather scattered on bare, coastal rocks and sea cliffs.* ● **M**

4. H. monizii Lowe, *Man. fl. Madeira* **1**: 481 (1868).
Small, densely tomentose shrub to 100 cm. Leaves 20-50 × 6-11 mm, elliptic, petiolate, greyish above, white beneath. Capitula 3 × 2-3 mm, ± hemispherical. Involucral bracts yellowish brown, sometimes flushed violet, erect at first, becoming patent and ultimately stellately spreading, the outer ovate, the inner oblong to oblong-lanceolate. Florets not scented. Pappus-hairs evenly echinulate along their whole length. **Plate 43**
Fl. (VI-)XI - II. *Very rare plant of gorges and sea cliffs at Cabo Girão, Cabo Garajau and Praia Formosa.* ● **M**

5. H. foetidum (L.) Cass., *Dict. Sci. Nat.* **25**: 469 (1822). [*Perpetua*]
Strong-smelling, foetid biennial up to 150 cm. Stems erect, simple, branching only in the inflorescence. Leaves rather viscid, up to 100 × 30 mm, the lower oblong to obovate and petiolate, the middle and upper lanceolate-cordate and amplexicaul, all tipped with a short mucro, shortly hairy and dull green above, densely white-tomentose beneath. Capitula up to 20 mm in diameter. Involucral bracts ovate, acute, membranous, very shiny, bright yellow, spreading. Florets yellow. Pappus-hairs evenly echinulate along their whole length.
Fl. V - VIII (-X). *On banks, along paths, tracks and forest margins above 500 m; originally introduced at Santo da Serra, now widespread, mainly in eastern Madeira; increasing.* **M**

Madeiran plants all belong to var. **citreum** Less., *Syn. gen. Compos.*: 285 (1832), which is characterized by its yellow bracts and pappus-hairs, and is more widespread than var. *foetidum*, which has white involucral bracts and pappus-hairs.

6. H. petiolare Hilliard and B.L. Burtt in *Notes R. bot. Gdn Edinb.* **32**: 357 (1973).
H. petiolatum auct. mad., non (L.) DC.
Diffuse, white-tomentose, woody perennial up to 1 m, often hanging or trailing down steep banks. Leaves up to 35 × 20 mm, ovate to subcordate, blunt, abruptly narrowed towards the base into the petiole, tomentose on both surfaces. Capitula numerous, in terminal corymbs, 3-5 mm in diameter, hemispherical. Involucral bracts ovate-oblong, rounded to hemispherical, creamy white.
Fl. VIII - X. *Introduced to gardens and now naturalized along paths and roadsides at Queimadas and near Camacha.* **M**

13. **Phagnalon** Cass.

Perennial herbs or dwarf shrubs. Leaves alternate, usually sessile, sparsely hairy, the margins dentate and revolute. Capitula usually solitary at the ends of branches, heterogamous, disciform. Involucral bracts generally cartilaginous but often with a papery, apical portion. Receptacle flat;

scales absent. Florets all tubular, yellow, the outer female and filiform, the inner fewer and perfect. Achenes sparsely hairy with elongated, non-myxogenic twin-hairs. Pappus of 1 row of slender, barbellate hairs connate at the base.

1. Involucral bracts all acute to acuminate, with narrow, membranous margins **1. saxatile**
    At least outer involucral bracts with a broad, obovate apex and membranous margins .
    . . . . **2. bennettii**

1. P. saxatile (L.) Cass. in *Bull. Soc. philomath. Paris* **1819**: 174 (1819). [*Isca*]
Spreading and rather straggling shrub, the whole plant lanate to floccose. Leaves 20-35(-40) × 2-4(-5) mm, linear-oblong to narrowly oblanceolate, the margins often revolute and making the leaves appear narrower still, remotely denticulate to dentate, rarely entire. Capitula 10-15 mm in diameter, hemispherical to ovate-globose, solitary and terminal on stems and branches; peduncles 8-10 cm. Involucral bracts oblong-lanceolate, all acute to acuminate, greenish, the tips brown, the margins membranous and erose. Florets pale yellow, the acute to obtuse lobes with short papillae at the tips. Achenes with white bristles.
Fl. III - XII. *Mostly on cliffs and slopes near the sea, but ascending to 1000 m. In Madeira along the south coast from Prainha west to Câmara de Lobos; on Ilhéu Chão in the Desertas and on Selvagem Grande; also recorded from Porto Santo but without further data.* **MDPS**

2. P. bennettii Lowe, *nomen nudum*.
P. rupestre auct. mad., non (L.) DC. (1836).
Shrub to 0.5 m, the whole plant white-lanate. Leaves 45-50(-60) × 3-5 mm, linear-lanceolate to narrowly oblanceolate, sessile but narrowing gradually at the base, entire to remotely denticulate, greyish above, white below. Capitula 10-13(-15) mm in diameter, hemispherical to campanulate, solitary and terminal at tips of stems and branches; peduncles 10-14 cm. Involucral bracts scarious, at least the outer and usually the middle oblong-spathulate with a broad, obtuse and often erose apex. **Plate 43**
Fl. II - VI. *Endemic of dry slopes; apparently uncommon in south-east Madeira, from the valleys of the Ribeira dos Soccoridos and Ribeira de João Gomes and the Ponta de São Lourenço; rare in Porto Santo, on Pico do Castelo and Pico de Ana Ferreira; also recorded from the Desertas but without further data.* ● **MDP**

The name *P. bennettii* was never validly published and cannot correctly be used for this plant. A new name, **P. hanseni** Quaiser & Lack, has been suggested but it has yet to be published. For want of any alternative, the name *P. bennettii* is used here.

14. Dittrichia Greuter

Usually perennial herbs, rarely biennials or small shrubs. Leaves alternate. Capitula medium-sized to small, solitary or in corymbose or paniculate inflorescences. Involucral bracts imbricate, in many rows. Receptacle ± flat; scales absent. Florets yellow, the outer female and ligulate (the ligule often very short), the inner tubular and hermaphrodite. Achenes cylindrical, abruptly contracted below the pappus; pappus of simple hairs connate towards the base.

1. D. viscosa (L.) Greuter in *Exsicc. genav. Conserv. Bot. disrib.* **4**: 71 (1973).
Inula viscosa (L.) Aiton [*Alfabaca, Alfavaca*]
Stout, unpleasant-smelling perennial, glandular-hairy and viscid. Stems 30-60 cm, woody at the base with erect or ascending branches. Leaves 50-70 × 8-12 cm, oblong-lanceolate to linear or rarely ovate, denticulate, the lower gradually narrowed into a petiole, the upper sessile and semi-amplexicaul. Capitula 4-10(-12) mm in diameter. Involucral bracts narrowly lanceolate. Florets yellow, the inner sometimes with purple-tinged corolla lobes; ligules exceeding the involucre.

Achenes hairy; pappus-hairs echinulate.

Fl. III - X. *Weed of waste ground along the south coast of Madeira from Funchal to Ponta do Sol; spreading slowly.* **M**

Madeiran plants are subsp. **viscosa**.

15. Schizogyne Cass.

Like *Dittrichia* but always shrubby; capitula in clusters; ray florets absent; pappus a row of scabrid hairs.

1. S. sericea (L. f.) Schultz Bip. in DC., *Prodr.* **5**: 285 (1832).
S. obtusifolia var. *sericea* DC.
Densely and shortly grey- to white-tomentose shrub up to 1 m. Leaves *c.* 40 × 1-1.5 mm, linear, entire, obtuse, flat. Capitula 3-6 mm, broadly conical. Outer involucral bracts 1.5 mm, ovate-triangular; inner bracts 2-3.5 mm, lanceolate. Ray florets absent; disc florets yellow. Achenes 1.3-1.5 mm, pubescent; pappus 2-2.5 mm.
Fl. II. *Maritime rocks on Selvagem Grande and Selvagem Pequena.* ■ **S** [*Canaries*]

16. Nauplius (Cass.) Cass.

Annuals or subshrubs. Leaves alternate, sessile. Capitula solitary. Involucre campanulate; bracts in several rows, the outer leafy with a coriaceous base, the inner coriaceous with scarious margins. Receptacle with scales. Outer florets female, the ligule 3-lobed; inner florets hermaphrodite with a terete corolla-tube. Outer achenes triangular in cross-section, often slightly compressed, the inner terete, all strigose-hispid, with 3-6 ribs; pappus of free or connate scales.

The genus consists of mainly Macaronesian subshrubs but only the single, annual species is present in Madeira.

Literature: A. Wiklund, The genus *Nauplius* (Asteraceae-Inuleae), in *Nord. J. Bot.* **7**: 1-23 (1987).

1. Nauplius aquaticus (L.) Cass., in *Dict. sci. nat.* **34**: 273 (1825).
Asteriscus aquaticus (L.) Less.
Sericeous annual with erect stems 10-40 cm high, pseudo-dichotomously branched above, the branches erecto-patent. Leaves oblanceolate, entire, obtuse, semi-amplexicaul, densely covered with sessile glands. Involucral bracts with coriaceous bases, the outer 1-2 cm long, exceeding the ligules, with a long, leafy, rounded apex, the inner with or without a short, green apex. Receptacular scales slightly boat-shaped. Florets yellow, glandular, the inner with 2 large and 3 smaller corolla-lobes. Achenes obovate to oblong, densely strigose-hispid; pappus of *c.* 5 free scales 1-1.5 mm long.
Fl. IV - VI. *Coastal species of dry, stony pastures, rocky areas and disturbed ground in Porto Santo and on Deserta Grande.* **PD**

Tribe HELIANTHEAE

Guizotia abyssinica (L. f.) Cass. in *Dict. Sci. Nat.* **59**: 248 (1829) is a tall annual reaching 2 m high with sessile, amplexicaul leaves opposite below, alternate above and large yellow flowers. Native to E. Africa, it is a common constituent of bird-seed and sometimes occurs as a casual; it has been recorded as such from Funchal. **M**

17. Eclipta L.

Annuals. Leaves opposite. Capitula pedunculate, several in axils of leaves. Involucral bracts in 2 rows. Receptacular scales present. Outer florets female, ligulate; inner florets hermaphrodite, tubular, the corolla usually 4-lobed. Outer achenes triangular in cross-section; inner ± cylindrical; pappus a corona.

1. E. prostrata (L.) L., *Mant. pl. alt.*: 286 (1771).
E. alba (L.) Hassk.; *E. erecta* L., nom. illegit. superfl.
Strigose annual to 45 cm or more. Stems erect or ascending, usually much-branched. Leaves up to 8 × 2 cm, lanceolate to ovate-lanceolate, shallowly to remotely serrate, shortly petiolate. Involucre 6-10 mm in diameter, hemispherical; bracts 4-5.5 mm, broadly ovate, herbaceous; receptacle flat; scales setaceous. Ligules *c.* 6 mm, white; tubular florets white. Achenes *c.* 2 mm, sparsely strigose at the apex; pappus a very narrow, rim-like, toothed corona.
Fl. IX - XII. *Introduced; an occasional to rare weed of water-courses and wet places in Madeira. Native to tropical America.* M

Eleutheranthera ruderalis (Sw.) Sch. Bip. in *Bot. Zeit.* **24**: 239 (1866), a pantropical annual with glandular leaves and discoid capitula, has been recorded from Madeira without further data. It is often mistaken for *Eclipta*. M

Montanoa bipinnatifida K. Koch in *Wochenschr.* **7**: 407 (1864), native to Mexico, is planted in Madeira and may perhaps occur as a rare escape. It is a stiffly and shortly white-hairy shrub with deeply pinnatifid leaves up to *c.* 30 × 13 cm and numerous capitula with yellow ligules 3 cm long. M

18. Helianthus L.

Annuals. Lower leaves opposite. Involucral bracts herbaceous, in 2-several rows. Receptacle ± flat; scales partially enclosing achenes. Outer florets sterile, ligulate; inner florets hermaphrodite, the corolla tubular, 5-lobed. Achenes compressed; pappus of 1-4 deciduous setae or scales.

1. H. annuus L., *Sp. pl.* **2**: 904 (1753) .   [*Girasol*]
Stout, unbranched annual. Leaves large, broadly ovate, the lower cordate at the base, 3-veined, toothed, hispid with appressed hairs, petiolate. Capitula very large, up to 30 cm in diameter, usually solitary, nodding; ligules 2 cm or more long, yellow; inner florets brownish. Achenes obovoid, pubescent, usually white with black streaks.
Fl. VI - X. *Cultivated in gardens and on terraces, often appearing on waste ground in Madeira but possibly not truly naturalized. Native to N. America.* M

19. Bidens L.

Annual or perennial herbs. Leaves opposite, entire to 2-pinnatisect. Capitula solitary. Involucre campanulate to hemispherical, bracts in 2 rows, the outer usually herbaceous and leafy, the inner membranous and often with scarious margins. Receptacle flat or slightly convex; scales present. Ray florets, if present, usually sterile; disc florets hermaphrodite. Achenes tipped with 1-8 retrosely barbed bristles.

Literature: E.E. Sherff, The Genus *Bidens,* in *Field Mus. Pub. Bot. ser.* **16**: 1-709 (1937).

1. Leaves 1- to 2-pinnate; outermost involucral bracts gradually narrowed towards the apex; mature achenes greatly exceeding inner involucral bracts . . . . **2. biternata**
- Leaves pinnate; outermost involucral bracts not gradually narrowed towards apex; mature achenes not or barely exceeding inner involucral bracts . . . . **1. pilosa**

1. B. pilosa L., *Sp. pl.* **2**: 832 (1753).   [*Armores de burro, Malpica, Setas*]
Erect, ± glabrous annual. Stems up to 45 cm, branched, ridged, sometimes purplish. Leaves pinnate, with 3-5 petiolulate leaflets, the terminal longer than the laterals; leaflets ovate, acute to acuminate, serrate. Outer involucral bracts green, linear-oblong to linear-spathulate; inner

bracts blackish with scarious margins. Ray florets, if present, white; disc florets yellow. Achenes up to 10 mm, equalling or slightly longer than inner involucral bracts, fusiform, ribbed, papillose and with a few erect setae, the 2(-3) terminal bristles slightly spreading.
Fl. all year. *Introduced. Very common weed on roadsides, waste and disturbed ground throughout the lowlands of Madeira; also occurring on Selvagem Grande and Selvagem Pequena. Native to S. America.* **MS**

There appear to be three varieties in Madeira:

i) var. **pilosa**
Kerneria pilosa var. *discoidea* Lowe
Ray florets absent.

ii) var. **minor** (Blume) Sherff in *Bot. Gaz.* **80**: 387 (1925).
Ray florets white; ligules 5-8 mm, shorter than the capitulum.

iii) var. **radiata** Sch. Bip. in Webb & Berthel., *Hist. nat. Iles Canaries* 3(2, sect. 2): 242 (1844).
Kerneria pilosa var. *radiata* (Schultz Bip.) Lowe
Ray florets white; ligules 7-15 mm, longer than the capitulum.

2. B. biternata (Lour.) Merr. & Sherff in *Bot. Gaz.* **88**: 293 (1929).
Like **1** but leaves 1- to 2-pinnate; outer involucral bracts gradually narrowed towards the apex; ray florets, if present, *c.* 4 mm long; achenes up to 20 mm, exceeding inner involucral bracts, with 2-4(-5) terminal bristles.
Fl. all year. *Introduced but data on the distribution in Madeira are lacking. Native to Old World tropics and subtropics.* **M**

Cosmos bipinnatus Cav. resembles *Bidens biternata* but has 2-pinnate leaves with filiform segments, prominent purple, pink, white or yellow ligules and achenes produced into a beak. It is commonly cultivated in Madeira.

20. Galinsoga Ruiz & Pav.

Annuals. Leaves opposite. Capitula in few-flowered dichasia. Involucral bracts ovate, in 1-2 rows with several short outer bracts. Receptacle conical; scales present. Outer florets female, ligulate; inner florets hermaphrodite, tubular, the corolla 5-lobed. Achenes obovoid-prismatic; pappus of scales.

1. Stems with ± appressed eglandular and a few glandular hairs less than 0.5 mm; involucral bracts persistent; receptacular scales trifid; pappus-scales not aristate . **1. parviflora**
- Stems with dense, spreading eglandular and numerous glandular hairs more than 0.5 mm; involucral bracts deciduous; receptacular scales entire; pappus-scales aristate . . . .
. . . . **2. quadriradiata**

1. G. parviflora Cav., *Icon.* **3**: 41 (1795).
Erect, much-branched annual to 45 cm. Stem ± glabrous below, pubescent above with ± appressed eglandular hairs and a few short glandular hairs less than 0.5 mm long. Leaves up to 4 cm, ovate, acute to acuminate, shallowly toothed, sparsely pubescent, shortly petiolate. Involucre 5 mm in diameter, with 2-4 short, persistent outer bracts; inner bracts 2.5-3 mm, broadly ovate, ± glabrous, herbaceous, green, often tinged purple at the tips, persistent. Receptacular scales trifid. Outer florets usually 5, the ligule *c.* 1 mm, 3-toothed, white; inner florets 1-1.5 mm, yellow. Achenes 1.3-1.5 mm, black, shortly setose, those of the outer florets slightly flattened, 3-sided,

those of inner florets 5-sided; pappus-scales 0.8 mm, fimbriate.
Fl. IV. *Recently introduced weed of cultivated and waste ground, roadsides and streets; spreading. Native to S. America but a cosmopolitan weed.* **M**

2. G. quadriradiata Ruiz & Pav., *Syst. veg. fl. peruv. chil.* **1**: 198 (1798).
G. ciliata (Raffin.) S.F. Blake
Like **1** but stems densely pubescent with numerous, spreading, glandular hairs more than 0.5 mm; leaves up to 6 cm, more deeply toothed; involucre with 1-2 short outer bracts, all bracts deciduous; receptacular scales entire; pappus-scales aristate.
Fl. III - IX. *Very recent introduction. Weed of waste ground, roadsides and similar habitats; spreading rapidly.* **M**

21. Ambrosia L.

Monoecious annuals. Leaves opposite or alternate. Capitula unisexual, the males drooping in terminal, bracteate cymes, the females of 1-3 florets in axils of the upper leaves. Achenes enclosed in a nut-like, spiny involucre; pappus absent.

1. A. artemisifolia L., *Sp. pl.* **2**: 988 (1753).
A. elatior L.
Erect annual, pubescent to shortly hispid. Leaves rhombic-ovate to oblong-ovate in outline, 2-pinnatisect with mostly lanceolate lobes, dull green, at least some leaves alternate. Male capitula 3-4 mm in diameter, hemispherical; bracts connate, broadly and shallowly toothed, pubescent; male florets 10-20, 1.3-1.5 mm, the corolla tubular with 5 obtuse lobes. Female capitula 3-3.5 mm in fruit, obovoid, with a beak, glabrescent but with 5-7 spinose teeth; corolla absent. Achene 3-3.5 mm, obovoid, smooth.
Fl. VII - IX. *Very rare plant of cultivated ground near Porto do Moniz; possibly recently re-introduced from Europe.* **M**

22. Xanthium L.

Monoecious annuals. Leaves alternate, petiolate. Capitula in axillary clusters with the males above the females. Male capitula subglobose, with a single row of free involucral bracts; receptacular scales present; florets numerous, with a tubular 5-lobed corolla. Female capitula ovoid to ellipsoid, the outer row of involucral bratcs small and free, the inner larger, spiny, connate and leathery, each capitulum ending in 2 beaks; florets 2, consisting only of ovary and style, with corolla and pappus absent. Fruit a tough burr containing 2 ovoid achenes.

1. X. strumarium L., *Sp. pl.* **2**: 987 (1753).
Leaves ovate to triangular, blade often broader than long, base cordate or sometimes cuneate, ± palmately lobed, both surfaces with short, stiff, appressed hairs; petiole equalling or longer than the blade. Fruiting capitulum pubescent and covered with hooked spines, with 2 ± straight beaks.
Fl. VII - IX. *Infrequent, scattered in the lower regions of Madeira along roadsides, field margins and on waste ground.* **M**

A very variable species. Plants from Madeira have been placed in subsp. **italicum** (Monetti) D. Löve in *Bot. J. Linn. Soc.* **71**(4): 271 (1975) [1976], an aromatic plant with ripe fruits 15-35 mm, yellow or brown and covered with stout spines, but all material seen has ripe fruits 12-17 mm, greyish green and covered with slender spines and belongs to subsp. **strumarium**.

Tribe HELENIEAE

Two species of the tribe Helenieae have been recorded from the islands as casuals; neither appears to be fully naturalized.

Tagetes minuta L., *Sp. pl.* **2**: 887 (1753), a S. American species naturalized in S. Europe, has been recorded from Funchal. It has strong-smelling, pinnatifid leaves, numerous yellowish green capitula with cylindrical involucres of 3-4 bracts which are connate almost to their tips. **M**

Gaillardia pulchella Foug. in *Mem. Acad. Sc. Par.* **1786**: 5 (1788), from N. America, has slightly fleshy leaves, large, solitary capitula and reddish purple ligules, yellow towards the tips. It is cultivated for ornament and sometimes escapes onto roadsides and waste ground in Porto Santo. **P**

Tribe ANTHEMIDEAE

23. Tanacetum L.

Perennial herbs. Leaves usually alternate, pinnatifid to pinnatisect. Involucral bracts in (2-)3 or more rows. Receptacle flat or convex; scales absent. Ray florets white. Achenes usually oblong, 5- to 12-ribbed, often with sessile glands; pappus a corona.

1. T. parthenium (L.) Sch. Bip., *Tanaceteen*: 55 (1844). [*Alfinetes de Senhora, Artemija*]
Chrysanthemum parthenium (L.) Bernh.; *Pyrethrum parthenium* (L.) Sm.
Perennial with gland-dotted, pungently aromatic and bitter-tasting foliage. Stems ridged, erect, branching above. Leaves up to 6.5 × 3 cm, ovate-oblong in outline, pinnatifid to pinnatisect, the segments oblong to ovate-elliptic, pinnatifid and toothed, sparsely to densely hairy. Capitula numerous. Involucre 6 mm in diameter; bracts lanceolate, ciliate with a narrow, scarious border laciniate towards the apex. Ligules 3-4 mm, white, 3-toothed; disc florets tubular, 5-toothed, yellow. Achenes narrowly obconical with 5-8 pale ribs and dotted with shining, yellow glands; pappus a short, membranous corona.
Fl. IV - X. *Introduced and naturalized in various parts of Madeira, usually near habitation. Cultivated as a cottage garden and medicinal herb; a form with double flowers is also sometimes grown. Native to S. Europe, N. Africa and SW Asia.* **M**

24. Artemisia L.

Small shrubs (in Madeira). Leaves usually alternate, lobed or dissected. Inflorescence a long panicle of pendent capitula. Involucral bracts in (2-)3 or more rows. Receptacle flat or conical, hairy; scales absent. Ray florets absent. Disc florets with corolla yellow or purplish. Achenes obovoid, usually glabrous, sometimes hairy; pappus absent.

1. A arborescens L., *Sp. pl.* ed. 2, **2**: 1188 (1763). [*Losna*]
A. argentea L'Hér.
Small, freely branching shrub to 75 cm or more, shortly and densely white- or grey-tomentose, the old leaves persistent. Leaves pungently aromatic when bruised, triangular in outline, 1- to 2-pinnatisect, 3-6.5(-8) × 3-5.5(-8) cm, the lobes 1.5-5 mm wide. Capitula unilateral, often nodding, in dense panicles. Involucre 3-5 mm in diameter, hemispherical; bracts ovate, tomentose. Receptacle hairy. Florets yellow, the outermost female and narrower than the hermaphrodite inner florets. **Plate 45**
Fl. IV-VIII. *Maritime plant of rocks and pastures near the sea, occasionally cultivated in gardens; rare on the south-west coast of Madeira, Porto Santo and Deserta Grande; not recorded for*

Bugío but common and conspicuous on Ilhéu Chão and some of the smaller islets including Ilhéu de Cima off Porto Santo where it has recovered much of its former abundance after being reduced by fuel cutters during the 19th century. **MDP**

Plants with conspicuous, purple corolla-lobes occur among the normal, entirely yellow-flowered populations on Ilhéu de Cima (and possibly elsewhere); they appear to be merely a colour form.

25. Achillea L.

Perennial herbs. Leaves usually alternate, entire to pinnatisect. Capitula usually small and corymbose. Involucral bracts in (2-)3 or more rows. Receptacle flat to conical, elongated in fruit; scales present. Ray florets white, with a short, wide limb and flattened tube; disc florets flattened, slightly saccate around the apex of the achene, white. Achenes dorsiventrally flattened, with 2(-3) ribs; pappus absent.

1. A. millefolium L., *Sp. pl.* **2**: 899 (1753). [*Feiteirinha macellão*]
Pubescent, stoloniferous perennial. Stem to 20 cm or more, erect. Leaves lanceolate to oblong in outline, terete, 2- to 3-pinnatisect, the middle cauline 4-5 × *c*. 1 cm; rachis up to 1 mm wide. Capitula numerous, in a terminal corymb. Involucre 3.5-2.5 mm in diameter, ovoid; bracts with long, white hairs and dark brown margins. Florets white, the ligule 1-1.2 mm.
Fl. VII. *Introduced to Madeira, allegedly from Britain. Infrequent; sometimes cultivated in gardens and naturalized on roadsides, waste ground and in pastures. Native to Europe and Asia.* **M**

A. ageratum L. [*Maçela*], a native of the W. Mediterranean region, is cultivated in gardens in Madeira. It has yellow florets.

Anacyclus radiatus Loisel., *Fl. gall.*: 582 (1807) subsp. **radiatus** is a widespread Mediterranean annual with finely 3-pinnatisect leaves, inner involucral bracts with an expanded, membranous, hyaline apex, yellow ray florets and broadly winged achenes. It has been recorded once as a ruderal from Funchal. **M**

26. Chamaemelum Mill.

Annuals or perennials. Leaves usually alternate, pinnatisect. Involucral bracts in (2-)3 or more rows. Receptacle conical; scales present. Ray florets white, sometimes absent; disc florets saccate around the apex of the achene, yellow. Achenes obovoid, with 3 very thin ribs and longitudinal rows of myxogenic cells; pappus absent.

1. Annual; stems not creeping and rooting; leaves 1- to 2-pinnatisect . . . . . **1. mixtum**
- Perennial; stems creeping and rooting; leaves 2- to 3-pinnatisect . . . . **2. nobile**

1. C. mixtum All., *Fl. pedem.* **1**: 185 (1785). [*Margaça*]
Anthemis mixta L.; *Ormenis mixta* (L.) Dum.
Hairy annual. Stems to 30 cm, decumbent to erect, branches spreading. Leaves to 5.5 cm or more, oblong in outline, 1- to 2-pinnatisect, the uppermost usually 1-pinnatisect, the segments narrow with a bulbous, sharply pointed tip. Involucre (6-)7-8 mm in diameter; bracts obovate, obtuse, with a broad scarious margin somewhat fimbriate towards the apex, pubescent. Ray florets compressed, the tube 2-winged, the ligules 7-11 mm and white; disc florets yellow, the tube abaxially oblique and saccate at base, enclosing apex of achene. Receptacle conical; scales elliptic, acute, carinate and enclosing corolla-tube and achene. Achenes *c*. 1 mm, ovate, compressed, brown with weak, white striae.
Fl. V - X. *Introduced. On waste ground, roadsides, among crops and along borders of fields*

and levadas. Infrequent and scattered in the lower regions of Madeira, it has recently appeared in Porto Santo. Native to the Mediterranean region. **MP**

2. C. nobile (L.) All., *Fl. pedem.* **1**: 185 (1785). [*Maçela*]
Anthemis nobilis L.
Like **1** but a creeping and rooting perennial with decumbent to ascending flowering stems; leaves sharply aromatic, dotted with pale glands, all 2- to 3-pinnatisect, the segments very narrow; ligules sometimes absent; receptacular bracts not or only weakly carinate.
Fl. VI - VIII. *Introduced. Native to the Mediterranean region.* **M**

Two varieties are present in Madeira.

i) var. **nobile**
Ligules present.
Apparently very rare, recorded only from Camacha, and not at all in recent years.

ii) var. **discoideum** (Boiss. ex Willk.) A. Fern. in *An. Soc. broteriana* **45**: 38 (1979).
[*Maçela de botão*]
Anthemis aurea Webb ex Nyman; *A. nobilis* var. *aurea* (L.) Cout.; *Ormenis aureus* Coss. & Durieu
Ligules absent.
The more common form in Madeira, occurring in dry, sunny mountain pastures in the south-east of the island.

27. Anthemis L.

Annuals. Leaves usually alternate, pinnatisect. Capitula usually solitary. Involucral bracts in (2-)3 or more rows. Receptacle convex to narrowly conical; scales usually present, at least in the upper part. Ray florets white; disc florets with tube conspicuously swollen at the base in fruit. Achenes top-shaped, sometimes compressed, with up to 11 ribs; pappus absent.

1. A. cotula L., *Sp. pl.* **2**: 894 (1753). [*Margaça*]
Glabrous to sparsely pubescent, somewhat foetid annual. Stem to 60 cm, ascending or erect, corymbosely branched. Leaves to 4 × 3 cm, ovate-oblong in outline, 2- to 3-pinnatisect, glandular, lobes narrow, mucronate. Capitula on long peduncles. Involucre 6-8 mm in diameter, bracts with scarious, brown margins. Receptacle hemispherical in flower, becoming conical in fruit; scales present in the upper ⅔, absent in the lower ⅓, subulate, shorter than or equalling the disc florets. Ray florets sterile, the ligule 7-9 mm, white; disc florets fertile, yellow. Achenes 1.3-12.5 mm, with 9-11 verrucose ribs; pappus absent.
Fl. IV - IX. *Introduced. Frequent and widespread weed of roadsides, disturbed, waste, and cultivated ground in the lower regions of Madeira and Porto Santo. Native to Europe, N. Africa and W. Asia.* **MP**

28. Chrysanthemum L.

Annuals. Leaves usually alternate, deeply divided. Capitula solitary or few. Involucral bracts in (2-)3 or more rows. Receptacle convex; scales absent. Ray florets wholly yellow, or white towards the tips. Achenes of ray florets 3-angled, with broad lateral wings and a narrow wing or rib on the back; achenes of disc florets prism-shaped with a narrow wing, or cylindrical and ribbed; pappus absent.

1. Leaves 2-pinnatisect, not glaucous . . . . **1. coronarium**
 - Leaves ± entire to 1- to 2-pinnatifid, glaucous . . . . **2. segetum**

1. C. coronarium L., *Sp. pl.* **2**: 890 (1753).   [*Pajito, Pajeita*(P), *Sejamos amigos*]
Pinardia coronaria (L.) Less.
Glabrescent perennial. Stems erect. Leaves up to 9 × 6 cm, 2-pinnatisect, not glaucous, the segments oblong, sharply dentate. Involucre 12-20 mm in diameter, hemispherical; outer bracts with brown margin and scarious border; inner bracts with scarious border and apical appendage. Ligules 8-16 mm, yellow, or cream with yellow base; disc florets yellow. Achenes to 3 mm, the outer strongly 3-angled, the angles winged, the abaxial faces with 3 ribs; inner achenes similar but laterally compressed, with narrower wings and with ribs on the lateral faces.
Fl. IV - V. *Introduced, probably as a cornfield weed, now common on waste ground and waysides in Porto Santo, and to a lesser extent in Madeira; also occurring on the Salvage Islands. Native to Mediterranean Europe.* **MPS**

2. C. segetum L., *Sp. pl.* **2**: 889 (1753).   [*Malmequer, Pampilho*]
Glabrous annual. Stem 20-50 cm, erect with spreading branches. Leaves oblong to obovate in outline, slightly fleshy, glaucous; lower and middle cauline leaves coarsely toothed to 1- to 2-pinnatifid, the lobes toothed; upper leaves ± entire, amplexicaul. Peduncles clavate. Involucre *c.* 15 mm in diameter, hemispherical; bracts ovate, yellowish green with a brown, scarious margin, the inner also with a scarious appendage. Florets all yellow, the ligules 9-11 mm. Achenes *c.* 2.75 mm, those of ray florets laterally compressed, 3-sided with 2 lateral wings, the lateral faces 2-ribbed, the abaxial faces 3-ribbed; achenes of disc florets ± cylindrical, 10-ribbed.
Fl. (II-) III - VII. *Probably introduced. Weed of waste ground, the fringes of arable land and edges of levadas. Common in the lower regions of Madeira, rarer in Porto Santo and on the Salvage Islands. Native to SW Asia and the E. Mediterranean region.* **MPS**

Records of *Ismelia carinata* Sch. Bip. (*Chrysanthemum carinatum* Schousb.), from Selvagem Pequena, are erroneous and based on a misidentification of *Argyranthemum thalassophilum*.

29. Argyranthemum Webb ex Sch. Bip.

Suffruticose perennials. Leaves usually alternate, dissected. Capitula solitary or few in a lax, corymbose inflorescence. Involucral bracts in (2-)3 or more rows. Receptacle convex to conical; scales absent. Ray florets white, rarely yellow or pink; disc florets usually yellow. Achenes of ray florets 3-angled, strongly winged, sometimes coalesced into groups; achenes of disc florets usually laterally flattened and winged, sometimes cylindrical and wingless; true pappus absent but achenes often apically produced to give a pappus-like corona. 2n=18.

The entire genus is endemic to Macaronesia.

Literature: C.J. Humphries, A revision of the Macaronesian genus *Argyranthemum* Webb ex Schultz Bip. (Compositae-Anthemideae), in *Bull. Br. Mus. nat. Hist.* (Bot.) **5**(4): 147-240 (1976); Ø.H. Rustan, Infraspecific variation in *Argyranthemum pinnatifidum* (Asteraceae-Anthemideae) in Madeira, in *Bocagiana* **55**: 1-18 (1981).

1.	Leaves 1- to 2-pinnatisect with 2-6 primary segments	2
-	Leaves ± entire to pinnatifid, or if pinnatisect, then with 10 or more segments ..	3
2.	Disc florets purplish red, the ligules pink to white; outer involucral bracts not or only slightly carinate towards the base	**3. haemotomma**
-	Disc florets yellow, the ligules white; outer involucral bracts distinctly carinate towards the apex	**2. thalassophilum**

3. Achenes of disc florets with a single wing; leaves pinnatisect-pectinate, the segments narrow
   .... **1. dissectum**
- Achenes of disc florets irregularly ribbed but lacking wings; leaves ± entire to pinnatifid, rarely shallowly pinnatisect with broad segments .... **4. pinnatifidum**

1. A. dissectum (Lowe) Lowe, *Man. fl. Madeira* 1: 464 (1868).
Chrysanthemum dissectum Lowe
Stems 60-120 cm, lax, branched throughout. Leaves 2-9 × 0.2-3.5 cm, ovate-oblong to lanceolate in outline, 1- or usually 2-pinnatisect, pectinate, petiolate, glabrous; primary lobes 10-16, linear-ligulate to narrowly lanceolate, parallel; secondary lobes dentate. Inflorescence with 1-5 capitula; peduncles 10-25 cm, slender. Involucre 14-20 mm in diameter; bracts triangular to spathulate, the outer fleshy with a narrow scarious margin, the inner scarious and hyaline, expanded at the apex. Ray florets 15-22 mm, white, emarginate to 3-toothed at the apex; disc florets c. 3 mm, the corolla-lobes yellow. Outermost achenes 4-5 × 2-3 mm, turbinate, trigonous, with 2 large, coriaceous, laciniate lateral wings and a diminutive, carinate ventral wing; inner achenes obconical, slightly compressed to terete or ± quadrangular, 1-winged; pappus coroniform. **Plate 44**
Fl. III - VII. *An endemic plant of sea-cliffs, wet rocks, banks and hedges, mainly in the north-west of Madeira, from Seixal to Ponta do Pargo, but also recorded from scattered localities in the central mountains around Pico Grande, Ribeiro Frio, and from Cabo Girão.* ● M

2. A. thalassophilum (Svent.) Humphries in *Bull. Br. Mus. nat. Hist.* (Bot.) 5(4): 209 (1976).
Stems up to 90 cm, ascending, branched throughout. Leaves 2.4-6 × 0.7-3.5 cm, obovate-rhombate, 1- to 2-pinnatisect, petiolate, glabrous; primary lobes 2-6, 3-25 × 2-8 mm; secondary lobes obtuse, sometimes absent. Inflorescence with 2-6 capitula. peduncles 5-12 cm. Involucre 12-18 mm in diameter; bracts triangular to obovate, the outer fleshy and carinate towards the apex, the inner scarious with a rounded, crenulate, hyaline apex and divergent veins. Ray florets 16-20 mm, creamy white, entire or emarginate; disc florets c. 4 mm, the corolla-lobes yellow, the tube paler or whitish. Outermost achenes 4.5-6 × 4.5-8 mm, trigonous, 3-winged, the wings coriaceous; inner achenes narrower, obconical, laterally compressed to terete, 1-winged; pappus coroniform, irregularly dentate. **Plate 44**
Fl. III - IV. *Endemic on coastal rocks on Selvagem Grande and Selvagem Pequena.* ● S

3. A. haemotomma (Lowe) Lowe, *Man. fl. Madeira* 1: 463 (1868).
Chrysanthemum barretti Costa; *C. haemotomma* Lowe
Stems 60-120 cm, somewhat decumbent to ascending, ± branched from the base. Leaves 2-7 × 1-3 cm, obovate or ovate, 1- to 2-pinnatisect, petiolate, glabrous, rigid, fleshy; primary lobes 2-6, 5-25 × 2-8 mm; secondary lobes dentate, alternate, acute. Inflorescence with 1 or 3-4 capitula; peduncles up to 20 cm. Involucre 15-20 mm in diameter; outer bracts triangular, scarious, with fleshy midribs, sometimes slightly carinate towards the base; inner bracts obovate with an expanded, scarious, hyaline apex. Ray florets 12-15 × c. 3.5 mm, pink, rose or white, the apex obtuse to emarginate; disc florets 2-3 mm, the corolla-lobes deep purple, the tubes pale pink or white. Outermost achenes 4-6 × 3-6 mm, unequally triquetrous, arcuate, 3- to 4-winged, the lateral wings carinate and diminutive, the ventral wings 1-2 and wider than laterals, with sinuate margins; inner achenes smaller, obconical, laterally compressed to ± quadrangular, 2-winged, those near centre of disc 1-winged; pappus coroniform, sometimes absent completely in the innermost achenes. **Plate 44**
Fl. V - VII. *An extremely rare endemic known only from wet, coastal cliffs from 50-450 m near Porto do Moniz and on Bugío.* ● MD

4. A. pinnatifidum (L. f.) Lowe, *Man. fl. Madeira* 1: 460 (1868). [*Malmequer, Pampilhos*]
Chrysanthemum pinnatifidum L. f.; *Pyrethrum grandiflorum* sensu Holl et auct mad., non Willd. (1809).

Stems 30-150 cm, decumbent or ascending, with several stems of equal length to give a candelabra-like habit, leafy on upper parts only. Leaves 4-20 × 0.5-7.5 cm, obovate or oblong-lanceolate, gradually cuneate at the base, ± entire to pinnatifid, or pinnatisect, shortly petiolate to sessile, glabrous; lobes 3-9, dentate. Inflorescence with 2-30 capitula; peduncles stout, up to 30 cm. Involucre 6-15 mm in diameter; outer bracts triangular, with scarious margins and fleshy midribs; inner bracts scarious with a hyaline, expanded, laciniate apex. Ray florets 10-20 × 3-6 mm, white, 1- to 3-toothed at apex; disc florets 3-4 mm, the corolla-lobes yellow, the tube white. Outermost achenes 3-4 × 3-4 mm, ± obconical, unequally trigonous, arcuate, occasionally with 1-3 diminutive wings; inner achenes smaller, obconical, laterally compressed to quadrangular, irregularly ribbed, wingless; pappus a coriaceous ridge, sometimes ± coroniform, or absent.

Fl. III - VII. *Endemic, common in open areas on wet rocks, cliffs and in ravines at all altitudes and in most parts of Madeira, though absent in much of the south-west.* ● M

1. Leaves not succulent; capitula usually more than 10 . . . . . . a) subsp. **pinnatifidum**
- Leaves succulent; capitula usually less than 10 2

2. Lobes of leaves ovate to elliptic; involucre 10-14 mm in diameter b) subsp. **montanum**
- Lobes of leaves triangular; involucre 14-17 mm in diameter . c) subsp. **succulentum**

a) subsp. **pinnatifidum**
Leaves thin-textured, pinnatifid, sometimes pinnatisect; lobes 13-35(-45) mm. Inflorescence with (2-)10-30 capitula. **Plate 44**
More or less throughout the range of the species from c. 300-1500 m; widely cultivated in gardens.

b) subsp. **montanum** Rustan in *Bocagiana* **55**: 15 (1981).
A. pinnatifidum var. *flaccida* Lowe, pro parte.
Leaves succulent, pinnatifid; lobes 6-12 mm, ovate to elliptic. Inflorescence with 1-2(-5) capitula. Involucre 10-14 mm in diameter. **Plate 44**
Rare, in the high peaks above 1500 m between Pico de Arieiro and Pico Ruivo.

c) subsp. **succulentum** (Lowe) Humphries in *Bull. Br. Mus. nat. Hist.* (Bot.) **5**(4): 224 (1976).
A. pinnatifidum var. *succulentum* Lowe; *Chrysanthemum mandonianum* Coss.
Leaves succulent, shallowly pinnatifid to ± entire; lobes up to 10(-15) mm, triangular. Inflorescence with 1-2(-5) capitula. Involucre 14-17 mm in diameter. **Plate 44**
Sea cliffs along the north coast, from Seixal to the Ponta de São Lourenço and its islets.

30. Leucanthemum Mill.

Perennial herbs. Leaves usually alternate, entire to pinnatifid. Capitula solitary or few. Involucral bracts in (2-)3 or more rows. Receptacle convex or sometimes conical; scales absent. Ray florets white; disc florets swollen and spongy at the base in fruit. Achenes with *c.* 10 ribs; pappus usually absent, sometimes a corona.

1. L. vulgare Lam., *Fl. franç.* **2**: 137 (1779).
Chrysanthemum leucanthemum L.
Stems sparsely hairy, creeping, rooting at the base with erect flowering stems. Leaves up to 7 × 1 cm; lower leaves obovate-spathulate to oblanceolate, petiolate, toothed or lobed; upper leaves oblong, sessile, toothed and semi-amplexicaul. Involucre 13-15 mm in diameter, hemispherical; bracts ovate-lanceolate with a dark, marginal band and a scarious border broad at the apex. Ligules 15 mm, white; disc florets yellow. Achenes *c.* 1.7 mm, ± cylindrical, black with 8-10 pale ribs; pappus usually absent, in ray florets sometimes forming a corona.
Fl. I - VI. *Introduced and naturalized at Santo da Serra but apparently rare elsewhere. Native to Europe.* M

L. lacustre (Brot.) Samp., *Herb portug.*: 132 (1913) (*Chrysanthemum lacustre* Brot.), native to coastal regions of C. Portugal, has been recorded once from Madeira, at Balcões, near Ribeiro Frio, probably as an escape from cultivation. It is like **1** but has leaves ovate to lanceolate, serrate, and achenes of ray florets with auricles. **M**

31. Coleostephus Cass.

Annuals. Leaves usually alternate, spathulate, toothed. Capitula solitary or few. Involucral bracts in (2-)3 or more rows. Receptacle convex to conical; scales absent. Ray florets golden yellow; corolla-lobes of disc florets with appendages, the tube very swollen and spongy at the base in fruit. Achenes curved, with 8-10 white ribs fused towards the base into a bulbous callus; pappus an oblique, scarious corona.

1. C. myconis (L.) Rchb. f., *Icon fl. Germ. helv.* **16**: 49 (1853). [*Pampilho*]
Glabrous. Stems to 40 cm, erect, branching at or near the base. Leaves 2.5-5.5 × 1.3-1.5 cm, the lower spathulate to obovate, obtuse and petiolate, the upper obovate to oblong, serrate, sessile and semi-amplexicaul. Involucre 15 mm in diameter; bracts obovate to spathulate with a narrow scarious border. Ligules 6-12 mm; disc florets yellow. Achenes with 8-10 pale ribs, those of ray florets compressed and with an oblique, papery pappus about equalling length of achene and sheathing the corolla-tube; those of disc florets cylindrical, slightly curved, the pappus shorter, sheathing lower half of corolla-tube only.
Fl. IV - IX. *Introduced; in cultivated fields and open pastures in the lower regions of Madeira. Formerly common as a cornfield weed but declining. Native to S. Europe and N. Africa.* **M**

32. Soliva Juss.

Small herbs. Leaves alternate, pinnatisect. Capitula sessile and surrounded by leaves. Involucral bracts in 2 rows, the margins scarious. Receptacle flat; scales absent. Outer florets female, lacking corollas; inner florets functionally male with tubular, usually 4-lobed corollas. Achenes compressed, with transversely sulcate wings and a persistent, spinescent style; pappus absent.

1. S. stolonifera (Brot.) R. Br. ex G. Don in Loudon, *Hort. brit.*: 364 (1830).
Gymnostyles stolonifera (Brot.) Tutin
Pubescent to villous annual, stems creeping and rooting at nodes. Leaves 0.5-2 cm, with 3-4 pairs of lanceolate, obtuse and occasionally toothed segments. Capitula 4-6 mm in diameter. Achenes 2 × 1.5 mm including the narrow wing, the upper ends of which form divergent points; persistent style about as long as achene, recurved; both style and upper part of achene densely villous.
Fl. III - V. *Introduced. A weed among the stones of paths and streets. Formerly described as common, especially in the Funchal region, now apparently rare though possibly overlooked. Native to S. America.* **M**

Eriocephalus africanus L., *Sp. pl.* **2**: 926 (1753), native to S. Africa, is sometimes planted in Madeira and may perhaps be becoming naturalized. It is a much-branched shrub to *c.* 1 m with leaves linear or lobed, silvery-hairy, capitula small, borne in clusters, with white ray florets, purple disc florets and very woolly inner involucral bracts which become conspicuous in fruit. **M**

33. Cotula L.

Annuals. Leaves sometimes opposite, pinnatifid. Capitula solitary, long-pedunculate. Involucral bracts in (2-)3 or more rows. Receptacle flat to conical; scales absent. Disc florets with (3-)4 lobes, tube sometimes saccate, short or absent in the outer florets. Achenes flattened, often winged, usually with blunt hairs; pappus absent.

1. C. australis (Sieber ex Spreng.) Hook. f., *Fl. nov.-zel.* 1(2): 128 (1852).
Hairy annual with prostrate, mat-forming stems. Leaves 3.5 × 1.7 cm, ± oblong in outline, 2-pinnatifid, the segments narrow and acute. Capitula on long, slender peduncles. Involucre 4-5 mm in diameter; bracts ovate to obovate, green, with pale, scarious margins. Outer florets lacking corollas, the pedicels long, *c.* 5 mm; inner florets with tubular, pale or greenish yellow corollas, the pedicels very short. Achenes 1.3 mm; outer achenes obovate, strongly laterally compressed, with short glandular hairs and a scarious wing; inner achenes plano-convex, not compressed, lacking a wing; pappus absent.
Fl. III - XII. *Introduced within the last 100 years, spreading and becoming naturalized in many parts of Madeira, mainly on roadsides and in streets, on banks, among cobbles and rocks; it has also spread to Porto Santo and Deserta Grande but is not yet common on these islands. Native to Australia.* **MDP**

C. coronopifolia L., *Sp. pl.* 2: 892 (1753) is like **1** but glabrous, with entire or only slightly divided leaves. It has been recorded from Madeira by several authors as a very rare plant known from a single locality at Ponto do Sol but it is probably no longer present on the island. **?M**

C. leptalea DC. has been recorded once from Madeira, apparently in error for **1**.

Tribe SENECIONEAE

34. Senecio L.

Annual or perennial herbs, occasionally climbers. Leaves spirally arranged, often pinnately divided. Inflorescence usually corymbose. Involucral bracts in 1 row but often with shorter supplementary bracts present at the base and sometimes forming a calyx-like structure (*calycle*). Receptacle flat; scales absent. Florets yellow, the outer female and ligulate, sometimes absent, the inner hermaphrodite and tubular. Achenes ± cylindrical, ribbed; pappus of simple or toothed hairs, occasionally absent.

1.	Leaves triangular to orbicular in outline, palmately veined	2
-	Leaves oblong to oblanceolate in outline, pinnately veined	3
2.	Glabrous climber; ligulate florets absent	**1. mikanoides**
-	Pubescent, shrubby perennial; ligulate florets present	**2. petasitis**
3.	Plant usually densely hairy, at least some hairs glandular; supplementary bracts 0-3, entirely green	**4. sylvaticus**
-	Plant glabrous or at most thinly hairy, all hairs eglandular; supplementary bracts 5 or more, conspicuously black-tipped	4
4.	Ligules more than 2 mm long; leaf-margins revolute	**3. incrassatus**
-	Ligules less than 2 mm or absent; leaf-margins flat	**5. vulgaris**

1. S. mikanoides Otto ex Walp., *Allg. Gartenztg* **13**: 41 (1845).
Glabrous, climbing perennial. Stems slender, twining up to 6 m, much-branched, fleshy above, woody towards the base. Leaves up to 8(-11) cm, triangular-ovate to ± orbicular, cordate, with 5-11 short, triangular lobes, slightly fleshy, bright green. Capitula 4-6 mm in diameter, numerous in dense panicles. Involucre obconical; bracts 3-4 mm, the margins scarious; supplementary

bracts 1-4, up to half as long as involucre, entirely green. Florets yellow, eligulate. Achenes 1.4-1.6 mm, cylindrical, glabrous.
Fl. XII - III. *Introduced; frequent on cliffs, roadsides and waste ground in many parts of the lower regions of Madeira. Native to S. Africa.* **M**

2. S. petasitis (Simms) DC., *Prodr.* **6**: 431 (1838).
Pubescent, suffruticose perennial up to 2 m. Leaves up to 17 cm, triangular to almost orbicular, truncate to sub-cordate at the base, shallowly palmately lobed, the margins with small callose teeth, palmately veined, dull green above, paler and more densely hairy beneath; petiole equalling or exceeding lamina. Panicle with numerous capitula, each 2-3 cm in diameter and cylindrical to narrowly obconical. Involucral bracts 8-10 mm, the margins scarious, pale and sometimes tinged purplish; supplementary bracts absent. Ligulate florets 4-5, the ligules 7-10 mm, bright yellow. Achenes 3 mm, glabrous, ribbed.
Fl. II - IV. *Introduced. Cultivated as an ornamental and naturalized along levadas and in damp places from 400-800 m, mainly in eastern Madeira. Native to Mexico.* **M**

3. S. incrassatus Lowe in *Trans. Camb. phil. Soc.* **6**: 538 (1838). [*Doiradinha*]
Glabrous to thinly arachnoid-hairy annual. Stems 6-25 cm, succulent, angled, usually much-branched. Leaves crowded, up to 7 cm, oblong to oblanceolate, shallowly pinnatifid to 1- to 2-pinnatisect, fleshy; lobes broad, rounded and entire to narrow, acute and irregularly toothed, the margins revolute; lower and middle cauline leaves narrowed into a petiole; upper leaves amplexicaul. Capitula numerous, in dense cymes, the peduncles slightly clavate. Involucre 4-5 mm in diameter, cylindrical to obconical; bracts 3.8-5 mm, with pale scarious margins and a tuft of hairs at the apex; supplementary bracts 5-10(-16), gibbous at base, black at apex, scarcely forming a calycle and intergrading with peduncular bracts. Ligulate florets 5-8, the ligules 2.5-5.5 × 1.2-3.5 mm, bright yellow, spreading, later reflexed or revolute. Achenes 1.6-2.3 mm, brown, ribbed, with stiff, white hairs. **Plate 45**
Fl. XII - IV. *Usually found on light, disturbed soils but also on cliffs and rocky outcrops, always near the sea. Throughout the archipelagos, common in Porto Santo and on the Salvages; in Madeira only from Cabo Girão to Cabo Garajau and along Ponto de São Lourenço; rather rare on the Desertas.* ● **MDPS**

Some specimens are very similar to *Senecio glaucus* subsp. *coronopifolius* (Maire) Alexander, a widespread N. African species which usually has leaves with narrower lobes, and much larger capitula with fewer supplementary and peduncular bracts.

4. S. sylvaticus L., *Sp. pl.* **2**: 868 (1753).
Thinly arachnoid-hairy to floccose annual with both glandular and eglandular hairs, at least in the inflorescence. Stems up to 65 cm, erect, simple or sparsely branched above. Leaves 2-5 cm, oblong to oblanceolate, pinnatifid with irregularly toothed lobes, revolute at margins, slightly fleshy; lower cauline leaves narrowed into a petiole; middle and upper leaves amplexicaul. Capitula in loose cymes. Involucre cylindrical, 2.5-3(-4) mm in diameter, bracts 6-7.5 mm, glandular-pubescent with a narrow, scarious margin; supplementary bracts 0-3, green. Ligulate florets 8-10, the ligules 1.5 mm, yellow and strongly revolute. Achenes 2.2-2.4 mm, black or grey, shortly and stiffly hairy between the ribs.
Fl. V - VII. *By paths, in clearings, on banks and among rocks, especially on disturbed or recently burnt ground. Common throughout Madeira, especially above 500 m, rare in Porto Santo and on the Desertas.* **MDP**

S. lividus L., *Sp. pl.* **2**: 867 (1753), similar to **4** but with usually unbranched stems, capitula 6-10 mm in diameter, dark-tipped supplementary bracts and achenes 3-4 mm, may also be present in Madeira but available material is poor and the record of this species requires confirmation. **?M**

5. S. vulgaris L., *Sp. pl.* **2**: 867 (1753).
Slightly fleshy annual, ± glabrous to pubescent or sparsely arachnoid-hairy. Stems to 50 cm, usually branching from or near the base. Leaves up to 5(-6) cm, oblong to oblanceolate, pinnatifid, lobes irregularly toothed, margins flat; lower cauline laves narrowed into a petiole; middle and upper leaves amplexicaul-auriculate. Capitula subsessile at first; peduncles later elongating to give a loose inflorescence. Involucre ± cylindrical, 4-5 mm in diameter; bracts 5.5-7.5 mm, glabrous with a narrow, scarious margin, sometimes black-tipped; supplementary bracts 11-20, up to ⅓ as long as the involucre but usually less and forming a calycle, conspicuously black-tipped. Ligulate florets, if present, with revolute ligules less than 2 mm. Achenes 2.3-2.6 mm, brown, shortly and stiffly hairy between the ribs.
Fl. III - VI. *Mainly on loose or disturbed soils, in cultivated and waste ground, along tracks and roads. Frequent but scattered throughout the lower regions of Madeira, rare in Porto Santo and on Deserta Grande.* **MDP**

The common form on the islands has discoid, eligulate capitula.

Emilia javanica (Burm. f.) C.B. Rob. in *Philipp. J. Sci.* **3**: 217 (1908) (*E. coccinea* (Sims) G.Don; *E. sagittata* (Vahl) DC.), an annual with capitula 10-12 mm long and red florets, has been recorded as an escape near Água da Pena on the east coast of Madeira. It is a widespread weed in the tropics and subtropics. **M**

35. Pericallis D. Don

Like *Senecio* but never annual; leaves palmately veined and lobed; supplementary bracts absent; florets white, pink or purple; achenes often heteromorphic.

1. P. aurita (L'Hér.) B. Nord. in *Op. bot. Soc. bot. Lund* **44**: 20 (1978).

[*Erva de coelho*]
Senecio auritus (L'Hér.) Lowe; *S. maderensis* DC; *S. populifolia* auct mad.
Slender, open shrub to 1.5 m. Branches spreading or trailing, flexuous, thinly floccose. Leaves 5-12 cm, often as broad as long, ovate-cordate to ovate-triangular, obtuse, usually with very shallow, triangular lobes, minutely callose-toothed at margins, thinly floccose becoming glabrescent above, densely white tomentose beneath; petioles floccose, those of upper leaves with two large, entire, semicircular auricles at the base, those of lower leaves with small auricles or auricles absent. Capitula numerous, in broad corymbs. Involucre 5-6 mm in diameter, campanulate, glabrous; bracts 2.5-4 mm, oblong, acute, strongly 3-veined, with a very narrow scarious margins and dark purple tip. Peduncular bracts numerous. Florets bright purple; outer (ligulate) florets 5(-6), *c*. 9 mm, with a spreading, oblong 3-toothed limb 6 mm long and paler than the inner florets; inner florets tubular, *c*. 4 mm, 5-lobed. Achenes 2 mm, with 8 ribs alternating with rows of short, white hairs; pappus a single row of toothed hairs. **Plate 45**
Fl. V - VII. *Common in ravines and on rocky slopes in the higher parts of Madeira above 1000 m, in Porto Santo only on the summit of Pico Branco.* ● **MP**

Tribe CALENDULAE

36. Calendula L.

Annuals or perennials, sometimes suffrutescent and often glandular. Leaves alternate, simple. Involucral bracts in 1-2 rows, linear, acuminate, subequal. Receptacular scales absent. Outer florets female, ligulate, yellow or orange; inner florets tubular, functionally male, usually concolorous with the outer florets, sometimes brownish. Outer achenes usually with a narrow beak, sometimes 3-winged, sometimes boat-shaped; inner achenes falcate or annular, muricate or rugose on the back; pappus absent.

1. Ligules at least 2 × as long as involucral bracts; stems always erect; leaves crowded ± towards the apex . . . . **3. officinalis**
- Ligules less than 2 × as long as involucral bracts; stems spreading, rarely erect; leaves remote or crowded below 2

2. Perennial, stems woody towards the base; inner achenes falcate; leaf-margins ± entire but thickly fringed with hairs . . . . **1. maderensis**
- Annual, not woody towards the base; at least some inner achenes annulate; leaf-margins denticulate . . . . **2. arvensis**

1. C. maderensis DC., *Prodr.* **6**: 454 (1837).   [*Vacoa, Vaqueira*]
Viscid perennial, densely arachnoid-hairy, especially the younger parts, and densely pubescent with long eglandular and shorter glandular hairs. Stems to 30 cm or more, prostrate or ascending, much-branched, woody at the base. Leaves thick and rather fleshy, crowded on the lower parts of the stem, obovate-spathulate, obtuse or acute, entire, often with a thick fringe of eglandular hairs at margins, sessile. Involucre *c*. 1 cm in diameter. Ligules 10-15(-18) mm; florets all yellow. Outer achenes straight or slightly incurved, usually beaked, mostly glandular-hispid and muricate with very broad, toothed wings; inner achenes falcate, never annular, muricate, with or without a beak. 2n=32. **Plate 45**
Fl. XII - VI. *A littoral endemic, rather rare in Madeira occurring mainly on the north coast and on the Ponta do São Lourenço; common on all three Desertas.* ● **MD**

Very similar to the Mediterranean species *C. suffruticosa* Vahl and perhaps only a subspecies of it.

2. C. arvensis L., *Sp. pl.* ed. 2, **2**: 1303 (1763).   [*Vaqueira*]
C. aegyptiaca Desf.
Annual, thinly arachnoid-hairy, at least when young, and with short, sometimes sparse glandular and eglandular hairs. Stems 10-30 cm, rarely up to 1 m, slender, erect or spreading, loosely branched. Leaves rather remote, up to 6.5 cm, lanceolate to oblong-obovate, acute or obtuse, usually denticulate with a few small but conspicuous, callose teeth, sessile, the upper sometimes amplexicaul. Involucre 1-1.8 cm in diameter. Ligules mostly less than 10 mm, less than 2 × as long as the involucral bracts; florets all yellow. Outer achenes incurved or patent, long-beaked, with 2 spiny wings and dorsal rows of long spines; middle achenes incurved, boat-shaped, sometimes with toothed wings and dorsal spines; inner achenes annular, muricate.
Fl. III - VI. *Very common weed of waste and cultivated ground, margins of pastures and roadsides throughout the lower and middle regions of Madeira and the upper regions of Porto Santo; also occurring on Bugío and the Salvage Islands.* **MDPS**

3. C. officinalis L., *Sp. pl.* **2** :921 (1753).   [*Cuidados*]
Usually annual, sometimes perennial, aromatic, rather viscid and glandular-pubescent to thinly arachnoid-hairy. Stems to 30 cm, erect, stout and much-branched. Leaves crowded along whole length of stems, up to *c*. 7 cm, oblanceolate, oblong or spathulate, entire or with small teeth, amplexicaul. Capitula large, up to 7 cm in diameter with the ligules 2 × as long as involucral bracts; florets usually all orange or yellow, or disc florets brownish. Achenes all incurved, mostly boat-shaped, muricate on the back, winged or not, sometimes alternating with larger and narrower achenes with or without beaks; inner achenes annular.
Fl. III - VII. *Widely cultivated in gardens everywhere and naturalized in various places in Madeira and Porto Santo.* **MP**

Tribe ARCTOTIDEAE

37. Arctotis L.

Perennials, often woody at the base. Leaves alternate, entire to pinnatisect. Capitula solitary, axillary, on a long peduncle. Involucral bracts in 4-6 rows, apices sometimes with an appendage, margins often membranous. Receptacle flat, pitted; scales absent. Outer florets with a narrow, 3-toothed ligule; inner florets tubular, the corolla 5-lobed, abruptly narrowed below. Achenes obovoid with 3 narrow wings or ridges on one side and a basal tuft of hairs; pappus a corona of hyaline scales.

1. A. venusta Norl. in *Bot. Notiser* **118**: 406 (1965).
A. stoechadifolia auct. mad., non Bergius (1767).
Whole plant densely tomentose with long, thin arachnoid hairs, scattered thicker hairs and sessile glands. Stems decumbent, woody at the very base. Leaves 8-15 × 1.5-3 cm, ± entire to deeply pinnatifid, the sinuses broad, the lobes sharply toothed. Capitula up to 7 cm in diameter. Involucral bracts membranous, in 4(-5) rows, those in the outer 2 rows triangular-ovate with a long, cylindrical, tomentose appendage, those in the inner 2 rows oblong-ovate, without an appendage. Outer florets female with a ligule up to 30 mm, white or cream above, red to maroon below; middle florets hermaphrodite and inner florets male, both with dark purplish corolla-lobes. Pappus-scales in 2 rows.
Fl. I - V. *Introduced. Widely cultivated and planted as a roadside ornamental in or near towns, sometimes escaping and becoming naturalized. Native to S. Africa.* **M**

Tribe CYNAREAE

38. Carlina L.

Shrubby perennials (in Madeira). Leaves alternate, entire, usually with spiny margins. Capitula solitary or in corymbs. Outer involucral bracts leafy, the inner scarious, shiny, stiff and spreading when dry. Receptacle flat; scales divided into linear segments; bristles also sometimes present. Florets hermaphrodite, all with a tubular, 5-lobed corolla. Achenes oblong, hairy; pappus of 1 row of plumose hairs which are united at the base into clusters.

1. C. salicifolia (L. f.) Cav., in *An. Cienc. nat. Madrid* **4**: 81 (1801).
Low shrub to 1 m. Stems branched, white-tomentose in the upper parts and with prominent leaf-scars. Leaves deciduous but long persistent after withering, crowded towards the ends of the branches, 6-10 cm × 6-15 mm, lanceolate, coriaceous, green and glabrescent above, densely white-tomentose beneath, subsessile and with a few ciliate spines at the base. Involucre 15-30 mm in diameter (excluding outer bracts), discoid to hemispherical. Outer involucral bracts foliaceous, of varying lengths, lanceolate to ovate, spreading; inner involucral bracts shorter than the outer, scarious, recurved, blackish or purplish brown. Receptacular scales persistent, often tipped with red. Florets creamy yellow. Achenes 3 mm, with dense, appressed, shiny, brown hairs; pappus-hairs 2- to 3-branched, plumose.
Fl. V - VIII. *Cliffs and rocky slopes throughout much of Madeira; also in Porto Santo and on the Desertas.* ■ **MDP** [*Canaries*]

Two varieties occur in the islands:

i) var. salicifolia
Margins of leaves and involucral bracts with slender, ciliate spines. **Plate 46**
Throughout the range of the species. ■ [*Canaries*]

ii) var. **inermis** Lowe, *Man. fl. Madeira* 1: 515 (1868).
Margins of leaves and involucral bracts unarmed, ± entire. **Plate 46**
Rarer than var. i, growing in dry places at low altitudes on the south coast of Madeira. ■ [*Canaries*]

39. Arctium L.

Robust biennial herbs with erect stem and long, stout tap-root. Leaves alternate, entire or remotely toothed. Capitula solitary or in racemes. Involucral bracts numerous, imbricate and appressed below, narrowing above into a long, rigid, subulate point, the apex hooked in at least the outer bracts. Receptacle flat; scales numerous, rigid, subulate. Florets hermaphrodite, all with a tubular, reddish purple or white corolla. Achenes oblong, compressed, rugose; pappus of several rows of scabrid, golden-yellow hairs.

1. A. minus Bernh., *Syst. Verz.*: 154 (1800). [*Bardana, Murruca, Teijeira, Tinjeira*]
Lappa minor (Schk.) DC.
Stems pubescent. Basal leaves up to 35 × 30 cm, ovate, cordate, entire or repand with rounded, very shallow, mucronulate lobes; cauline leaves much smaller, ovate, truncate to cuneate, irregularly and shallowly crenate-denticulate; petioles hollow. Racemes terminating in a solitary capitulum. Fruiting capitula 20-30 mm in diameter, ± globose and closed at the top, sessile or with peduncles up to 1.5 cm. Involucral bracts green or purple-tinged, with spiny, hooked apices. Florets purple, exceeding the involucral bracts. Achenes 5 mm, brown.
Fl. VII - IX. *Woods and shady waste places in ravines. Scattered throughout many parts of Madeira.*
M

40. Carduus L.

Annuals, biennials or perennials with spinose stems and leaves. Stem winged. Leaves alternate, entire to pinnatisect. Capitula solitary or in terminal clusters. Involucral bracts in many rows, imbricate, spine-tipped. Receptacle with dense bristles. Florets hermaphrodite, usually purple; corolla tubular, the limb with 1 lobe longer than the others. Achenes smooth or ribbed, glabrous; pappus deciduous, of many rows of scabrid hairs, the inner longer, all united below to form a ring.

1. Leaves undivided, 2-crenate, the margins spinulose; florets white . . . . **3. squarrosus**
- Leaves pinnatifid with triangular lobes, the margins with strong spines; florets purple 2

2. Stem with wing up to 10 mm; involucral bracts ± glabrous, the middle bracts with scarious margins . . . . **1. tenuiflorus**
- Stem with wing up to 5 mm; involucral bracts arachnoid-hairy, the middle bracts not scarious at margins . . . . **2. pycnocephalus**

1. C. tenuiflorus Curtis, *Fl. lond.* 2(6): t. 55 (1793).
Annual or biennial to 40 cm or more. Stems parsely branched above, arachnoid-hairy and with a spiny wing up to 10 mm wide. Leaves pinnatifid, segments broadly triangular, margin spiny, glabrous above, arachnoid-hairy beneath, the lower petiolate, the upper sessile. Capitula sessile, in clusters of (1-)2-3. Involucre 7-10 mm in diameter, cylindrical; bracts lanceolate with an apical spine, purple-flushed, glabrous, the middle bracts with scarious margins. Florets purple. Achenes 4 mm, swollen, black-flecked.
Fl. III - VIII. *Roadsides, field margins and waste ground; common ± throughout Madeira at all elevations; also present in Porto Santo.* **MP**

2. C. pycnocephalus L., *Sp. pl.* ed. 2, **2**: 1151 (1763).
Like 2 but stem with wing up to 5 mm; capitula broader, subsessile or pedunculate, solitary or in clusters of 2-3; bracts arachnoid-hairy, the middle bracts not scarious at the margins, faintly 3-veined; achenes compressed.
Fl. ? *Waste ground; known only from Funchal and apparently very rare.* M

3. C. squarrosus (DC.) Lowe in *Trans. Camb. phil. Soc.* **6**: 540 (1838).
Annual to 75 cm or more. Stem simple or branched above, arachnoid-hairy and with a spinulose wing up to 5 mm wide. Leaves up to 14 × 4 cm, oblanceolate to obovate, shallowly bi-crenate, the margins spinulose, glabrous above, arachnoid-hairy beneath. Capitula in groups of 3-5, sessile. Involucre 7-13 mm in diameter, campanulate to ovoid; bracts lanceolate, glabrous, the outer patent or recurved, acuminate with a pungent tip, green centre and broad, scarious, straw-coloured margins. Florets white, shorter than the involucral bracts. Achenes 3 mm, obovoid, truncate at apex, flecked with black; some pappus-hairs clavate or broad and flattened at the tip. **Plate 46**
Fl. V - VIII. *A rare endemic of coastal rocks and laurisilva at low altitudes in Madeira.* ● M

41. Cirsium Mill.

Spinose, perennial herbs. Leaves alternate, entire. Capitula solitary. Involucral bracts in many rows, imbricate, spine-tipped. Receptacle with dense bristles. Florets hermaphrodite; corolla tubular. Achenes oblong, gibbous, compressed; pappus of many rows of plumose hairs, the outer shorter and sometimes simple, all united at the base.

1. C. latifolium Lowe in *Trans. Camb. phil. Soc.* **4**: 28 (1831).   [*Tangerão manso*]
Perennial with thick, fleshy roots. Stem to 60 cm or more, erect, hollow, arachnoid-hairy and branched above. Leaves up to 36 × 14 cm, ovate to obovate, obtuse, the basal coarsely bi-crenate and narrowing to form a broad petiole, the cauline ± entire, sessile and somewhat auriculate-amplexicaul, all glabrescent and shiny above, arachnoid-lanate beneath, the margins spinulose with slender, soft spines. Capitula solitary, pedunculate. Involucre 20-30 mm in diameter, ± globose; bracts lanceolate, ciliate at margins, with a soft apical spine less than 1 mm long. Florets purple. Achenes 5 mm. **Plate 46**
Fl. VI - VIII. *A locally frequent endemic of cliffs and steep rocky slopes in damp ravines in Madeira.* ● M

42. Notobasis Cass.

Spinose annuals. Leaves alternate, white-veined above, thinly arachnoid-hairy beneath. Capitula solitary or in racemose clusters. Involucral bracts imbricate, spine-tipped. Florets hermaphrodite; corolla tubular. Achenes compressed, smooth; pappus of numerous stout, plumose outer hairs and a ring of short inner hairs united at the base.

1. N. syriaca (L.) Cass., *Dict. Sci. Nat.* **35**: 171 (1825).   [*Cardo*]
Cirsium syriacum (L.) Gaertn.
Robust plant with stem usually branched above. Basal leaves herbaceous, pinnatifid, the lobes sometimes very shallow and spinose-dentate; cauline leaves increasingly coriaceous and deeply divided, the uppermost rigid and pinnatisect with the lobes reduced almost to stout spines. Involucre 18-25 mm in diameter, campanulate. Florets *c.* 2 cm. Achenes 6.5 mm, pale brown; outer pappus-hairs 15 mm, the inner *c.* 1 mm.
Fl. V - VI. *Rare plant of dry, sunny soils on the Ponta de São Lourenço and between Funchal and Caniço.* M

43. Galactites Moench.

White-tomentose annuals. Leaves alternate, margins spiny. Capitula solitary, in corymbs or cymes. Involucral bracts imbricate, all but the outer tipped with a rigid, somewhat spreading spine. Receptacle densely hairy. Florets purple or white, the outer large, funnel-shaped and sterile, the inner small, tubular and hermaphrodite. Achenes ± cylindrical, striate, glabrous; pappus of white, plumose hairs.

1. G. tomentosa Moench, *Methodus*: 558 (1794). [*Cardo*]
Stems 30-80 cm, erect, usually unbranched. Leaves glabrous, marbled with white on the veins above, white-tomentose beneath; basal leaves oblanceolate, serrate but soon withering; cauline leaves pinnatifid, the margins with spines up to 8 mm, decurrent and forming a narrow spiny wing on the stem. Involucre ovoid-conical; bracts arachnoid-hairy, with a long, greenish apical spine with a dark blotch at its base. Florets *c.* 2.5 cm, usually purple, rarely white, fragrant. Achenes 4 mm, somewhat compressed, silvery.
Fl. (II-)IV - VI(-IX). *Weed of pastures, waste and cultivated ground, ± common everywhere in Madeira and Porto Santo and in the higher parts of Deserta Grande.* **MDP**

Plants from the Ponta de São Lourenço in which the upper leaves are linear, with at most 1 or 2 short spines, have been called var. **crinita** Lowe, *Man. fl. Madeira* **1**: 496 (1868).

44. Cynara L.

Perennials with erect stems, sometimes acaulous. Leaves alternate or basal, dentate to pinnatifid, usually spinose. Capitula solitary or in corymbose cymes. Involucral bracts in many rows, imbricate, glabrous, all but the inner tipped with a stout spine or ovate to triangular appendage. Receptacle fleshy and bristly. Florets hermaphrodite, purple, blue or white; corolla tubular, 5-lobed. Style long-exserted. Achenes obovoid, glabrous; pappus of many rows of plumose hairs united at the base.

Literature: A. Wiklund, The genus *Cynara* L. (Asteraceae-Cardueae), in *Bot. J. Linn. Soc.* **109**: 75-123 (1992).

1. C. cardunculus L., *Sp. pl.* **2**: 827 (1753). [*Cardo, Cardo da gente, Pencas*]
C. horrida Aiton
Stem up to 70 cm, stout, erect, simple or branched above. Leaves up to 40 cm, 1- to 2-pinnatifid, thinly tomentose above, densely white-tomentose beneath; spines 15-30 mm, slender, stiff, yellow, borne singly at the tip of the segments and in clusters at the base of the segments and at the base of the leaf itself. Involucre 40-70 mm in diameter, ovoid-globose; bracts with the apex rather abruptly narrowed into a long, stout, patent spine. Florets lilac to bluish, rarely white. Achenes 7 mm, speckled black.
Fl. VII - X. *Dry, sandy soils and pastures. Rare in Madeira, known only from Caniçal and the Ponta de São Lourenço, more common in Porto Santo.* **MP**

Plants from Madeira and Porto Santo, all of which are spiny, have been called var. **ferocissima** Lowe, *Man. fl. Madeira* **1**: 498 (1868).

45. Silybum Adans.

Spinose annuals or biennials. Leaves alternate, white-veined or sometimes variegated. Capitula solitary. Involucral bracts in many rows, imbricate, the outer and middle bracts with a leafy spinose-dentate appendage which usually ends in a long spine. Receptacle densely hairy. Florets hermaphrodite, reddish purple; corolla tubular, deeply 5-lobed. Achenes obovoid, compressed, glabrous, with a membranous apical border; pappus of many rows of white, scabrid hairs united at the base to form a ring.

1. S. marianum (L.) Gaertn., *Fruct. sem. pl.* **2**: 378 (1791).
[*Cardo, Cardo de Santa Maria*]
Stem erect, grooved, green, glabrous to somewhat arachnoid-hairy. Leaves pale shiny green with conspicuous white variegation along the veins above; basal leaves pinnatifid, glabrescent; cauline leaves smaller and less deeply divided, sessile and auriculate-amplexicaul, the whitish marginal spines up to 8 mm. Capitula large, erect or ±nodding. Involucral bracts glabrous; outer and middle bracts with an appendage ending in a recurved, yellowish, canaliculate spine; innermost bracts with a shorter, erect spine. Achenes black, flecked with grey; pappus pure white.
Fl. V - VI. *Waste places and the margins of cultivated land. Recorded as rare in the lower regions of Madeira, in Porto Santo only from Pico do Concelho, but common on Deserta Grande and Bugío.* **MDP**

46. Cheirolophus Cass.

Perennial herbs or small shrubs. Leaves alternate, entire or pinnatifid, unarmed. Capitula solitary on stems and branches, peduncles clavate. Involucral bracts in several rows, imbricate, coriaceous, the apical appendages palmate with 7-9 subequal fimbriae. Florets purplish pink to yellow or white, the outer female, the inner hermaphrodite; corolla tubular. Achenes compressed, the attachment-scar transverse; pappus of inner achenes of several rows of very slender, barbellate hairs, absent on outer achenes.

1. C. massonianus Lowe in *Hooker's. J. Bot.* **8**: 297 (1856).
Centaurea massoniana Lowe
Small, profusely branched shrub, the branches white-tomentose in the upper parts and with prominent leaf-scars. Leaves crowded at the ends of the branches, lanceolate to elliptic, entire, thinly hairy, shortly ciliate on the margins and dotted on both surfaces with numerous glands. Peduncles long, purple. Involucre 16-25 mm in diameter, ovoid-globose; appendages of bracts semi-circular, deeply lacerate. Florets pinkish purple. Achenes 6 mm, pale brown. **Plate 47**
Fl. V - VI. *Very rare endemic of steep cliffs and ledges, in Madeira found only near Cabo Girão and at Alegría above Funchal; in Porto Santo on Pico Branco and Pico do Concelho.* ● **MP**

47. Mantisalca Cass.

Biennials or perennials. Capitula solitary at tips of branches. Involucral bracts in several rows, imbricate and appressed, with a short, erect to deflexed, deciduous apical spine. Achenes ± compressed, transversely rugose, with 10-15 ribs anastomosing at base and apex; pappus of an outer row of setae equalling or longer than the achene and an inner row of long, slender scales.

1. M. salmantica (L.) Briq. & Cavill. in *Archs Sci. phys. nat.* V, **12**: 111 (1930).
Centaurea salmantica L.; *Microlonchus salmantica* (L.) DC.
Perennial up to 40 cm. Stems slender, stiff, sparsely branched. Basal leaves lyrate-pinnatisect, crenate, pubescent, petiolate; cauline leaves linear to linear-lanceolate, dentate to ± entire, glabrous, sessile. Involucre 10-20 mm in diameter, ovoid; bracts ovate, dark at the apex, apical spine 0.5-0.7 mm and readily deciduous. Florets c. 25 mm, purple, the marginal florets spreading. Achenes 3.5-3.8 mm, brown with black spots.
Fl. V - VIII. *Rare on dry, sunny soils in south-east Madeira from Caniço to Machico, and on*

Pico de Ana Ferreira in Porto Santo; *also recorded from the Salvage Islands but not in recent times.* **MPS**

48. Centaurea L.

Annuals to perennial herbs, rarely suffruticose. Leaves alternate, entire to pinnatisect, usually without spines. Capitula solitary or 2-3 at tips of branches. Involucral bracts in several rows, imbricate, usually with a variously dentate to fimbriate or spiny, apical appendage. Florets all tubular and hermaphrodite or the outer sometimes larger, radiate and sterile. Achenes ± compressed; pappus of 2-many rows of scabrid to plumose hairs, or of scales, the innermost row of hairs or scales shortest, usually differing in shape and texture from the outer and sometimes connate at the base, rarely pappus absent.

1. Appendage of bracts decurrent, fimbriate, with a single slender spine .... **5. diluta**
- Appendage of bracts neither decurrent nor fimbriate, spines several 2

2. Appendage with pinnately arranged spines; annual or biennial 3
- Appendage with palmately arranged spines; perennial 4

3. Upper leaves decurrent; florets yellow; pappus present .... **4. melitensis**
- Upper leaves not decurrent; florets purple; pappus absent .... **1. calcitrapa**

4. Upper leaves decurrent; pappus present .... **2. sonchifolia**
- Upper leaves not decurrent; outer achenes without a pappus .... **3. sphaerocephala**

1. C. calcitrapa L., *Sp. pl.* 2:917 (1753).
Thinly pubescent and glandular annual or biennial. Stem divaricately branched. Leaves grey-lanate at first, the basal pinnatifid, remotely serrate; cauline leaves pinnatifid, not decurrent, the uppermost narrowly lanceolate, spinulose on the margins. Capitula solitary, shortly pedunculate. Involucre 6-8 mm in diameter, ovoid-conical; appendages of bracts with a pale,stout, spreading central spine up to 25 mm and several much shorter, slender lateral spines near the base. Florets pinkish purple. Pappus absent.
Fl. V - VI. *Rare, along roadsides and beaches in south-east Madeira, from Funchal to Machico.* **M**

2. C. sonchifolia L., *Sp. pl.* 2: 915 (1753)
Scabrid to very shortly hispid, glandular perennial up to 30 cm, with creeping roots. Stems widely branched, scabrid. Basal leaves lyrate-pinnatifid, petiolate; upper cauline leaves oblong to oblong-lanceolate, spinulose-dentate, sessile and decurrent on the stem but not forming a continuous wing. Capitula solitary. Involucre 7-15 mm in diameter, ovoid, slightly arachnoid-hairy; appendages of bracts deflexed, with 5-10 palmate spines, the central spine 4-5.5 mm and slightly longer than the others. Florets purple or the inner whitish. Achenes *c*. 4.5 mm; pappus very short.
Fl. V - VII. *Naturalized in fields at a single locality between Funchal and Caniço.* **M**

3. C. sphaerocephala L., *Sp. pl.* 2: 916 (1753).
Like **2** but sometimes arachnoid-tomentose, viscid; cauline leaves ovate-lanceolate, not decurrent on the stem.; outer achenes without a pappus.
Fl. VII. *A recent introduction, naturalized on roadsides at a single locality at Camacha.* **M**

4. C. melitensis L., *Sp. pl.* 2: 917 (1753). [*Beija mão*]
C. melitensis var. *conferta* Webb & Berthel., var. *vulgaris* Webb & Berthel.
Glandular annual 25-45 cm. Stems erect, branched, arachnoid-hairy when young. Leaves puberulent to pubescent, the basal lyrate-pinnatifid to sinuately lobed, the upper cauline lanceolate to oblong,

entire to denticulate, long decurrent on the stem giving a ± continuous wing. Capitula solitary or in groups of 2-3. Involucre 6-10 mm in diameter, ovoid, arachnoid-hairy to glabrous; appendages of bracts with a pinnate spine 4-6.5 mm which is purplish when young. Florets yellow, glandular. Achenes 2.3-2.5 mm, grey-brown with pale, whitish lines; pappus half as long to as long as achene.
Fl. III - IX. *Dry, sunny soils, usually near the sea. In Madeira mostly on the south coast from Funchal to Ponta de São Lourenço but also elsewhere; widespread in Porto Santo and surrounding islets, rather rare on all the Desertas and Selvagem Grande.* **MDPS**

5. C. diluta Aiton, *Hort. kew.* **3**: 261 (1789).
Scabrid to shortly hispid, glandular perennial with erect, branched stems. Lower leaves lyrate to incise-dentate; upper leaves entire. Capitula solitary. Involucre ovoid; appendages of bracts decurrent on the bract, orbicular-ovate, the margins membranous and irregularly fimbriate, the apex with a short, slender spine. Florets purple, the outer patent. Pappus of outer achenes very short, of inner achenes *c.* as long as the achene.
Fl. ? *Recorded from near Funchal. Native to Spain and N. Africa, common as a casual elsewhere; possibly not fully naturalized in Madeira.* **M**

49. Carthamus L.

Usually spinose annuals or biennials, rarely woody-based perennials. Indumentum usually glandular and arachnoid hairy. Leaves entire to pinnate but usually pinnatifid. Capitulum solitary. Involucral bracts in many rows, imbricate, spiny, the outer leafy, the inner sometimes with an apical appendage. Florets hermaphrodite; corolla tubular. Achenes oblong to obpyramidal, glabrous, the outer usually rugose and lacking a pappus, the inner smooth and with a pappus of many rows of narrow scales.

1. Plant glabrous; leaves undivided . . . . **2. tinctorius**
- Plant arachnoid-hairy; leaves pinnatifid to pinnatisect . . . . **1. lanatus**

1. C. lanatus L., *Sp. pl.* **2**: 830 (1753). [*Cardo*]
Kentrophyllum lanatum (L.) DC. & Duby
Branched annual to 40 cm with thinly arachnoid-lanate indumentum and numerous short-stalked, amber glands. Leaves up to 7 cm, deeply pinnatifid to pinnatisect, spinose-dentate. Involucre 20 × 20 mm, ovoid; outer bracts leaf-like; middle bracts with a scarious basal portion and leaf-like appendage; inner bracts ± entirely scarious, the appendage dentate to entire. Florets 3 cm, yellow. Achenes 4.5 mm; pappus-scales setaceous, scabrid, persistent.
Fl. V - VIII. *In dry, sunny places along margins of fields and roadsides in the lower regions of Madeira, mainly in the south; in Porto Santo known only from Serra de Dentro.* **MP**

2. C. tinctorius L., *Sp. pl.* **2**: 830 (1753). [*Açafrão, Açafroa*]
Branched, ± glabrous annual. Leaves somewhat coriaceous, up to 6.5 cm, ovate, usually undivided, spinose-dentate. Involucre 30 × 30 mm, ± globose; outer bracts leaf-like; middle bracts with a scarious basal portion and leaf-like appendage; inner bracts ± entirely scarious, the appendage dentate to entire. Florets 2.5 cm, orange-red. Pappus absent or the inner achenes sometimes with short scales.
Fl. IV - VI. *Sometimes cultivated for culinary use (the dried florets), especially in Porto Santo, and occasionally naturalized in waste and cultivated ground.* **MP**

50. Carduncellus Adans.

Perennials with simple or branched stems, sometimes acaulous, usually spiny and lanate to arachnoid-hairy. Basal leaves pinnate to lyrate; cauline leaves sinuate to dentate with often spinose teeth. Involucral bracts in many rows, imbricate, spiny, the outer leaf-like, the inner with a lacerate to fimbriate appendage. Florets hermaphrodite; corolla tubular, usually blue to purple. Achenes usually obpyramidal and rugose or grooved towards the apex, glabrous; pappus of many rows of narrow scales or ± plumose setae, free or connate at the base, deciduous or persistent.

1. C. caeruleus (L.) C. Presl, *Fl. Sicul.*: xxx (1826).
Stem up to 65 cm, thinly arachnoid-lanate to glabrescent. Cauline leaves 4-12 cm, oblong to lanceolate, spinose-dentate, the veins very prominent beneath. Involucre 1.5-2.5 × 12-2.5 cm, ± cylindrical; outer bracts leafy, about as long as the inner; inner bracts with a brown, scarious, laciniate appendage. Florets blue. Achenes 3.5 mm; pappus-scales not more than 2 × as long as achene, very narrow, scabrid, persistent.
Fl. V - VII. *Rare plant of dry, rocky soils in fields and along roads to the west of Funchal, at Caniço and near the airport at Santa Cruz; also in Porto Santo.* **MP**

Subfam. CICHORIOIDEAE

51. Scolymus L.

Spiny annuals. Leaves alternate, pinnatifid. Receptacular bracts in several rows. Florets ligulate, yellow. Receptacular scales present, narrowly winged and tightly sheathing the compressed achenes. Pappus of a few hairs or absent.

1. S. maculatus L., *Sp. pl.* **2**: 813 (1753).   [*Tigarro* (P)]
Glabrous, rigid annual, the wings of the stem and the leaves with conspicuous pale, thickened margins and veins. Stem up to 1 m, erect with continuous, spiny wings along the whole length. Leaves up to 20 cm; basal leaves obovate, pinnatifid, less spiny and rigid than the cauline; cauline leaves oblong, sinuate-pinnatifid with triangular lobes, the uppermost with pectinate spines. Capitula *c.* 15 × 10 mm, in sub-corymbose clusters, surrounded by several erect, leafy bracts. Involucral bracts membranous, entire. Florets few, erect, yellow with black, woolly hairs mainly towards the base. Receptacle conical, the scales truncate at the apex. Achenes 2-3 mm; pappus absent.
Fl. V - VII. *Localized or infrequent plant of dry, rocky ground and poor soils along the south coast of Madeira, mainly the towards Ponta de São Lourenço; more widespread in Porto Santo.* **MP**

52. Cichorium L.

Perennials, rarely annuals. Leaves alternate, runcinate or dentate. Capitula terminal or axillary. Cylindrical involucre with 2 rows of bracts, the inner longer. Receptacle flat; scales absent. Florets usually blue. Achenes obovoid, obscurely angled, truncate; pappus of 1-2 rows of blunt scales.

1. C. endivia L., *Sp. pl.* **2**: 813 (1753).   [*Almeirante, Almirante* (P), *Almeirão*]
Annual or biennial. Stems to 70 cm, stiff, widely branched from the base, often flexuous. Basal leaves up to 20 cm, oblanceolate, runcinate-pinnatifid, dentate, the teeth with spinose tips, shortly petiolate; cauline leaves similar but with fewer teeth or entire, sessile, amplexicaul-auriculate. Capitula *c.* 10 mm long, cylindrical, the terminal with peduncles thickened at the top. Outer involucral bracts oblong to ovate, obtuse, purplish-tinged; inner bracts linear-lanceolate, acuminate, green. Florets pale blue. Achenes 2 mm, brown; pappus-scales 0.3-0.5 mm.
Fl. III - IX. *Dry soils on waste ground, roadsides and fields; in the lower regions of Madeira; widespread in Porto Santo.* **MP**

Madeiran plants all belong to subsp. **divaricatum** (Schousb.) P.D. Sell in *Bot. J. Linn. Soc.* **71**: 240 (1976), which is distinguished from the cultivated salad plant subsp. *endivia* by its pubescent basal leaves.

53. Tolpis Adans.

Annual, biennial or perennial herbs or shrubs. Stems simple or branched. Leaves alternate, entire to pinnatisect, sometimes amplexicaul, herbaceous to subcoriaceous or succulent. Inflorescence cymose; capitula 1-many. Involucral bracts in 2-3 rows. Receptacle flat, pitted; scales absent. Florets yellow or yellow and purplish brown. Achenes sub-cylindrical to obconic, 5-ribbed, sometimes of 2 kinds; pappus usually of long hairs interspersed with numerous short, scale-like hairs.

1. Annual; achenes dimorphic .... **1. barbata**
- Woody perennial or small shrub; achenes monomorphic 2

2. Perennial herb with woody, rhizomatous stems; mature achenes light brown .... **3. macrorhiza**
- Small shrub; mature achenes dark brown to black .... **2. succulenta**

1. T. barbata (L.) Gaertn., *Fruct. sem. pl.* **2**: 372 (1791).
Crepis crinita Sol. ex Lowe; *Tolpis crinita* (Sol. ex Lowe) Lowe; *T. umbellata* Bertol.
Pubescent annual. Stems 3-100 cm, simple or branched. Leaves 1.5-20 × 0.3-4.2 cm, narrowly lanceolate or elliptic to oblanceolate, entire to dentate or sub-pinnatifid with 0-10 teeth or lobes, glabrous or the petiole white-floccose. Involucre 4-20 mm across; inner bracts 4-9 mm, shorter than the outer, lanceolate, erect; outer bracts linear, dentate, incurved but not appressed. Outer florets 5-12 mm, lemon yellow with an olive band on the outer face; inner florets smaller, yellow or purplish brown. Achenes 1.1-2.5 mm, dark brown to black, pubescent, dimorphic, the outer incurved and stouter than the inner; long pappus-hairs few, whitish.
Fl. IV - VIII. *Scattered throughout the lowlands of Madeira up to 900 m, on sunny banks and among rocks.* **M**

A very variable species. Madeiran plants all belong to subsp. **barbata**. A small form with completely yellow ligules is sometimes separated as *T. umbellata* Bertol. It occurs sporadically throughout the range of the species, including Madeira, but is treated here as a part of the variation within **1**.

2. T. succulenta (Dryand. in Aiton) Lowe, *Man. fl. Madeira* **1**: 525 (1868).
Crepis pectinata Lowe; *C. tenuifolia* Banks & Sol.; *Tolpis fruticosa* Schrank; *T. succulenta* var. *ligulata* Lowe, var. *linearifolia* Lowe, var. *multifida* Lowe, var. *oblongifolia* Lowe
Perennial, becoming shrubby with age. Stems 10-100 cm, brownish, glabrous, becoming woody with a smooth, slightly glossy bark, rugose with age; branches erect, ascending or decumbent, bearing leaf-rosettes with short axes in the leaf-axils. Leaves 2-15 × 0.5-6 cm, narrowly to broadly elliptic or oblanceolate, entire or simply dentate to pinnatifid or bipinnatisect, with 2-7 teeth or lobes, herbaceous to somewhat succulent, glabrous, or occasionally somewhat woolly-floccose; petiole glabrous to somewhat white-farinose at the base. Involucre 10-18 mm across, white-tomentose at base; inner bracts 5-12 mm, longer than the outer, white-tomentose along the scarious margins. Florets 8-14 mm, yellow, with an olive band on the outer face. Achenes 1.5-1.9 mm, all subcylindrical, glabrous, dark brown to black; long pappus-hairs 9-15(-17), whitish. **Plate 47**
Fl. IV - X. *Common in Madeira, Porto Santo and on all of the Desertas, growing on sea cliffs, rocks and rocky banks up to 1000 m; in Madeira also occasionally up to 1500 m in open habitats.*
■ **MDP** [*Azores*]

3. T. macrorhiza (Lowe ex Hook.) DC., *Prodr.* 7: 87 (1838).
Perennial up to 10 cm with woody, rhizomatous stems, rosette-forming, producing annual leafy, flowering axes. Stems simple or branched with age, glabrous, rugose, dark brown or black. Leaves 6-16 × 1.4-5.2 cm, elliptic to lanceolate or oblanceolate, dentate, with 3-11 teeth, slightly succulent, sub-coriaceous, glabrous at maturity but grey-farinose when young; basal leaves with a glabrous petiole; upper cauline leaves ± sessile. Involucre 0.7-13 cm across, somewhat white-farinulose at the base; inner bracts 5-8 mm, longer than the outer, white-tomentose towards the scarious margins; outer bracts without scarious margins. Florets 6-8 mm, the outer with an olive band on the outer face. Achenes 1.7-2 mm, glabrous, light brown; long pappus-hairs 20-30, unequal, yellowish. 2n=18. **Plate 47**
Fl. VI - VIII. *Frequent but rather scattered at altitudes above c. 700 m in shady, moist, rocky areas of the steeper valleys of central Madeira; also on rocky outcrops in exposed areas above 1400 m.* ● **M**

54. Hedypnois Mill.

Annuals with several stems. Leaves basal and cauline, entire to lobed. Capitula 1-many. Involucral bracts in 2 rows, the inner much longer than the outer. Receptacle flat; scales absent. Ligules yellow. Achenes cylindrical; pappus of scales, those of the inner achenes long-aristate, sometimes also mixed with hairs.

1. H. cretica (L.) Dum. Cours., *Bot. cult.* 2: 339 (1802).   [*Leitua* (P)]
H. cretica var. *rhagadioloides* (L.) Cout.; *H. rhagadioloides* (L.) F.W. Schmidt
Stems 4-20 cm, sparsely hairy. Leaves up to 12 cm, oblanceolate, ± entire to deeply lobed, sparsely hairy with more dense hairs on margins and veins beneath, the basal narrowed into a short petiole, the cauline sessile. Peduncles thickened below the capitula. Involucral bracts in 2 rows, the outer 2-3 mm, the inner 7-9 mm and partially enclosing the incurved outer achenes. Achenes 5-6 mm, tapering towards the apex, covered with rigid, tooth-like hairs; pappus-scales of outer achenes forming a corona up to 1 mm, those of inner achenes up to 4 mm, long-aristate.
Fl. I - VI. *Common on dry soils of waste and cultivated ground in Madeira and Porto Santo, usually near the sea.* **MP**

Rhagadiolus stellatus (L.) Gaertn., *Fruct. sem. pl.* 2: 354 (1791), from S. Europe, is somewhat similar to **1**. It has leaves subentire to pinnatifid, inner involucral bracts usually 8, incurved, hispidulous to scabridulous and enveloping the long, persistent outer achenes, pappus absent. It has been recorded from both Madeira and Porto Santo but without further data; the records require confirmation. **?M ?P**

55. Urospermum Scop.

Annuals to perennials with solitary stems. Leaves dentate to pinnatifid. Capitula few. Involucral bracts in 1 row, connate at the base. Receptacular scales absent. Ligules yellow, sometimes with red stripes. Achenes beaked; pappus of 2 rows of plumose hairs.

1. U. picroides (L.) Scop. ex F.W. Schmidt, *Samml. phys.-ökon. Aufs.*: 275 (1795).
[*Leituga de burro* (P)]
U. picroides var. *asperum* DC.
Pale green, hispid and spinulose annual. Stem 10-60 cm, solitary, erect, branched. Leaves up to 22 × 10 cm; basal and lower cauline leaves obovate, dentate to runcinate-pinnatifid with triangular to oblong lateral lobes and a large ovate-triangular terminal lobe, sparsely hispid and spinulose at least on the veins beneath, densely and minutely so on the margins, the petioles winged; upper cauline leaves similar but ovate to linear, with narrower lobes, shortly petiolate

to sessile and auriculate-amplexicaul. Peduncles thickened towards the apex. Involucre 12-20 × 6-25 mm; bracts ovate-lanceolate, long-acuminate, some with dark, narrow margins, the remainder with broad, pale, scarious margins. Ligules *c.* 17 mm, pale yellow. Achenes 3-3.5 mm, laterally compressed, the broad faces with truncate flanges; beak 7-10 mm, hollow, bulbous and swollen at the base, the upper part slender and cylindrical with short, bristly hairs; pappus white.
Fl. III - VI. *Weed of waste places, roadsides and the fringes of cultivated ground, scattered but frequent in the lower regions of Madeira and the higher parts of Porto Santo and on Deserta Grande; rare on Bugío.* **MDP**

56. Hypochoeris L.

Annuals or perennials with usually branched scapes often thickened below the capitula. Leaves usually all basal, entire to pinnatifid. Involucral bracts lanceolate, imbricate, in several rows. Receptacle flat; scales numerous, scarious. Ligules yellow. Achenes cylindrical to fusiform, at least the inner usually with a long, slender beak; pappus of 1-2 rows of hairs, the inner long, plumose, the outer short, plumose or scabrid, rarely of fimbriate scales.

1. Perennial; florets 2 × as long as involucral bracts; capitula opening even in dull weather
   . . . . **1. radicata**
- Annual; florets ± equalling the involucral bracts; capitula opening only in full sun . .
   . . . . **2. glabra**

1. H. radicata L., *Sp. pl.* **2**: 811 (1753).
H. radicata var. *rostrata* Moris
Perennial up to 90 cm. Scapes several, erect, branched, with a few small scales in the upper part. Leaves all basal, 6-25(-30) × 1.5-3.5 cm, oblanceolate to narrowly oblong, sinuate dentate to pinnatifid, usually hispid but sometimes almost glabrous. Involucral bracts 6-16(-20) mm, glabrous or bristly on the midribs, tips dark purple. Florets up to 2 × as long as the involucral bracts, deep bright yellow, the outer often with an olive-green stripe on the back. Achenes dark reddish brown, 3-4.5 mm (excluding beak), muricate, all beaked, the inner achenes with the beak longer than the body, the outer with the beak shorter than or equalling the body; pappus 2 rows of hairs.
Fl. V - VII. *Common in fields, rocky pastures, on cliffs, banks and by paths in most parts of Madeira up to c. 1500 m.* **M**

2. H. glabra L., *Sp. pl.* **2**: 811 (1753).
H. glabra var. *glabra*, var. *loiseleuriana* Godr. in Gren. & Godr.
Annual 5-30 cm. Scapes sparsely branched, with a few small bracts in the upper part. Leaves mostly basal, rarely 1 or 2 cauline, 2-6 × 0.5-1.5 cm, oblanceolate, sinuate-dentate, glabrous to hispid. Capitula opening only in full sun. Involucal bracts 5.5-8.5(-15) mm. Florets pale yellow, about equalling the involucral bracts. Achenes dark reddish brown, 3 mm (excluding beak), muricate, beak of inner achenes longer than the body, that of outer achenes short or more usually absent; pappus of 2 rows of hairs.
Fl. I - VII. *Common in short-grazed pastures, on banks and roadsides, especially on dry soils in the higher regions of Madeira, Porto Santo and Deserta Grande.* **MDP**

This species has been divided into several varieties based both on hairiness of the leaves and on the presence or absence of erostrate outer achenes (the common Madeiran form has outer achenes lacking beaks). At least some of the variation appears to be environmentally induced and none of the varieties is recognized here.

Plants apparently representing the hybrid **1** × **2** occasionally occur where the two species grow together. They are annual, with the habit and leaves of *H. glabra* and the long florets of *H. radicata*.

57. Leontodon L.

Usually scapose perennials, sometimes annuals or biennials, often with branched hairs. Leaves entire to pinnatifid, all in a basal rosette. Scapes solitary or numerous, branched or unbranched. Involucral bracts in several imbricate rows. Receptacle pitted, the pits often with toothed or hairy margins; scales absent. Ligules usually yellow, often with a dull or reddish stripe on the back. Achenes muricately ribbed, sometimes beaked; pappus usually of 2 rows of hairs with dilated bases, the inner plumose, the outer usually simple; pappus of marginal achenes of hairs or scales, or sometimes absent.

1. L. taraxacoides (Vill.) Merat, *Annals Sci. nat.* **22**: 108 (1831). [*Leituga*]
Leontodon saxatilis subsp. *rothii* (Ball) Maire; *L. nudicaulis* sensu auct. mad., non (L.) Schinz & R. Keller
Perennial, rarely annual or biennial, with eglandular, simple or long-stalked 2- to 3-fid hairs. Leaves (2-)6-15(-25) × (0.5-)1-2(-3) cm, oblanceolate, almost entire, shallowly toothed or pinnatifid below, hispid; hairs all 2- to 3-fid; petioles red. Scapes 7-30(-50) cm, sparsely hispid below, glabrous above. Capitula solitary, nodding in bud; outer involucral bracts very small, often blackish or purplish; inner bracts 8-11 mm, oblong to lanceolate, glabrous to hispid, especially on the midrib. Florets 8-15 mm, yellow, the outer with a broad greyish red or olive-green stripe on the back. Outer achenes 4.5-5 mm, cylindrical, with a pappus of fimbriate scales and sheathed by the inner involucral bracts; inner achenes 4-6.5 mm, cylindrical to fusiform, dark brown, muricate, narrowing into a slender beak with a pappus of 2 rows of hairs, the outer simple, the inner plumose.
Fl. all year. *Common and widespread in rough pastures and in grassy places on cliffs, by levadas, roadsides and uncultivated ground in Madeira; less frequent in Porto Santo and rare on both Deserta Grande and Bugío.* **MDP**

Plants from the islands all have inner achenes with beaks 2-3 mm long and appear to belong to subsp. **longirostris** Finch & P.D. Sell in *Bot. J. Linn. Soc.* **71**: 247 (1976), although they are nearly always perennial, not annual. There are two distinct forms in the islands, one with glabrous involucres (=*Thrincia hispida* var. or forma *gymnocephala* Lowe), the other and more common with ± densely hispid involucres (=*Thrincia hispida* var. or forma *chaetocephala* Lowe).

Specimens of *L. rigens* (Dryand. in Aiton) Paiva & Ormonde, including the holotype, have been wrongly localized in Madeira. This species is endemic to the Azores.

58. Helminthotheca Vaill. ex Zinn

Annuals or biennials, hispid, the hairs mostly with 2-4 hooked branches. Stems usually solitary and branched. Leaves basal and cauline, alternate, sinuate-dentate to pinnatisect. Involucral bracts in several imbricate rows, the outermost sometimes broad and resembling an epicalyx. Receptacle flat, pitted; scales absent. Florets yellow. Achenes curved, transversely wrinkled or muricate, sometimes beaked; pappus usually of 2 rows of deciduous hairs, the inner plumose, the outer simple or plumose.

1. H. echioides (L.) Holub in *Folia geobot. phytotax.* **8**(2): 176 (1973).
[*Lingua de vaca*]
Helmintia echioides (L.) Gaertn.; *Picris echioides* L.
Bristly greyish green annual or biennial with slender spines and numerous stiff, long-stalked,

mostly 4-fid hairs often with swollen white bases. Stems ribbed, slightly succulent. Leaves oblanceolate, sinuate-dentate, thickly fringed with spinules, the basal and lower cauline oblanceolate and narrowing into a winged petiole, the middle and upper cauline lanceolate to ovate, sessile and amplexicaul. Capitula numerous, on short peduncles. Outer involucral bracts 7-17 mm, broadly ovate-cordate, leafy; inner bracts 10-20 mm, lanceolate, acuminate, with a slender, long-toothed awn near the tip. Florets c. 10 mm, yellow, the outer sometimes reddish purple on the back. Achenes 2.5-4 mm (excluding beak), the outer curved and whitish, the inner straight, oblong and golden brown, all shallowly transversely-wrinkled and with a slender beak about as long as or shorter than the body of the achene; pappus pure white, deciduous together with the upper part of the beak.
Fl. III - IX. *Common in fields, rough pastures and waste places, on banks and roadsides throughout the islands except for the Salvages.* **MDP**

59. Tragopogon L.

Annuals to perennials. Stems usually solitary, simple to sparsely branched. Leaves long, narrow, sheathing at the base. Capitula few. Involucral bracts in 1 row. Receptacular scales absent. Florets yellow, pink or purplish. Achenes fusiform, ribbed, often narrowed into a long, slender beak with an annulus immediately below the pappus; pappus usually of 1 row of plumose hairs, in the outer achenes sometimes of scabrid hairs.

1. T. hybridus L., *Sp. pl.* **2**: 789 (1753). [*Cravo de seara*]
Geropogon glaber L.
Glabrous annual. Stem to 30 cm, sparsely branched. Leaves up to 25 cm, linear. Capitula few, peduncles slightly thickened towards the top. Involucral bracts 3.5-5 cm, linear, slightly keeled. Florets half as long as involucral bracts, pinkish. Achenes very slender and extending into a long beak but lacking an annulus, the outer 4-4.5 cm with a pappus of 5 unequal, rigid, scabrid hairs, the inner 2.5-3 cm with a pappus of numerous, plumose hairs.
Fl. III - V. *A plant of drier soils, common especially in cornfields in Porto Santo, rare in Madeira and mainly in the south-east extending to the Ponta de São Lourenço.* **MP**

60. Launaea Cass.

Shrubs (in Madeira) with dichotomously branched stems, shrubby and spiny. Involucral bracts in several rows, imbricate. Receptacular scales absent. Florets hermaphrodite, yellow, the ligules striped olive beneath. Achenes cylindrical or compressed, ribbed; pappus of hairs in several rows.

1. L. arborescens (Batt.)Murb. in *Acta Univ. Lund* nov. ser., **19**: 65 (1923).
Glabrous, spiny dwarf shrub less than 40 cm high. Leaves mostly basal, up to 5.5 cm, pinnatisect with linear lobes. Capitula terminal, very shortly pedunculate. Involucral bracts with broad, scarious margins. Achenes strongly narrowed at base; ribs 4.
Fl. ? *Confined to rocks above the sea at Baia de Abra, on the Ponta de São Lourenço.* **M**

61. Sonchus L.

Annuals, biennials or herbaceous perennials (subgenus *Sonchus*), or perennials woody at the base, sometimes shrubs or small trees (subgenus *Dendrosonchus*). Leaves entire to pinnatisect, denticulate, sometimes spiny, the cauline amplexicaul. Capitula few to many. Involucral bracts in 3 rows. Receptacular scales absent. Ligules yellow. Achenes compressed, narrowed at both ends, lacking a beak, with 1-4 ribs on each face; pappus of 2 types of hairs: coarse, solitary, deciduous bristles, and finer, flexuous, persistent hairs in fascicles.

Literature: A. Aldridge, A critical reappraisal of the Macaronesian *Sonchus* subgenus *Dendrosonchus* s.l. (Compositae-Lactuceae), in *Bot. Macaronésica* 2: 25-58 (1977).

1. Annual or biennial (rarely perennial) herbs, never woody 2
- Shrubs, small trees, or perennials with a woody caudex 4

2. Achenes winged, smooth between ribs; cauline leaves with clearly rounded auricles . . . . **2. asper**
- Achenes not winged, rugose between ribs; cauline leaves with acute or sometimes slightly rounded auricles 3

3. Ligule about as long as corolla-tube; leaf-lobes oblong to oblong-deltoid, terminal lobe usually larger than the upper pair of lateral lobes . . . . **1. oleraceus**

- Ligule longer than corolla-tube; leaf-lobes narrowly oblong to linear, terminal lobe about as large as upper pair of lateral lobes . . . . **3. tenerrimus**

4. Leaves petiolate . . . . **5. pinnatus**
- Leaves sessile, the bases sheathing the stem 5

5. Shrub, sometimes a small tree, 1-4 m high, with short, thick branches . **4. fruticosus**
- Perennial with short, woody caudex . . . . **6. ustulatus**

Subgen. SONCHUS
Annuals, biennials or herbaceous perennials. Leaves distributed along the whole length of the stem.

1. S. oleraceus L., *Sp. pl.* 2: 794 (1753). [*Serralha*]
Sonchus oleraceus var. *integrifolius* Wallr., var. *lacerus* Wallr., var. *laciniatus* Lowe, var. *rotundifolius* Hoff. & Link, var. *triangularis* Wallr.
Annual or biennial. Stem up to 50 cm, often glandular-hairy in the upper part and white-tomentose at the base of and just beneath the capitula. Leaves 20 × 8 cm, variable; lower cauline leaves pinnatisect to pinnatifid, lyrate or runcinate, rarely entire, the lobes oblong-deltoid, denticulate but not spiny, the terminal lobe usually larger than the uppermost pair of lateral lobes, the petiole winged with acute, spreading auricles; upper cauline leaves similar but with the petiole shorter and more broadly winged. Florets yellow, ligule about as long as the corolla-tube. Achenes compressed, oblanceolate with 3 ribs on each face, rugose between the ribs.
Fl. all year. *A very variable and widespread weed of cultivated and waste ground, common everywhere but the highest regions.* **MDPS**

2. S. asper (L.) Hill, *Herb. brit.* 1: 47 (1769). [*Serralha*]
Like **1**, but stems and bases of capitula never white-tomentose; leaves with the terminal lobes usually smaller than the uppermost pair of lateral lobes, the margins sinuate, dentate to spinose, the auricles rounded and appressed; ligules shorter than the corolla-tube; achenes elliptic to oblanceolate, strongly compressed, winged, smooth between the ribs.
Fl. (IV)V - X. *Mainly a weed of cultivated and waste ground but also occuring in open areas in woods and scrub. Scattered throughout the archipelagos, occurring on all islands except Selvagem Pequena, but nowhere common.* **MDPS**

Two subspecies are present in the islands:

a) subsp. **asper**
S. asper var. *vulgaris* Coss. & Germ., var. *integrifolius* Lowe
Annual. Leaves thin, green, mostly cauline. Achenes with wings and ribs ± smooth.
The common subspecies in Madeira and the Desertas.

b) subsp. **glaucescens** (Jord.) Ball in *J. Linn. Soc.* (Bot.) **16**: 548 (1878).
Biennial. Leaves coriaceous, glaucous, often forming a rosette. Wings and ribs of achenes with numerous recurved spinules.
Less frequent than subsp. a, occurring in Madeira and the Salvage Islands.

3. S. tenerrimus L., *Sp. pl.* **2**: 794 (1753).
Like **1** but stems branched; leaves with sub-rounded or acute auricles, the lower with few lobes, the upper with many; lobes variable, narrowly oblong, lanceolate or linear, the terminal lobe about as large as the upper pair of lateral lobes, the margins entire or denticulate; ligules longer than corolla-tube; achenes narrowly oblanceolate, abruptly narrowed at base.
Fl. III - ? *Introduced; a ruderal recorded from Madeira at Funchal and Boaventura, and from Selvagem Grande; probably present elsewhere. Native to the Mediterranean region.* **MS**

Often difficult to separate from **1**, at least in Madeira, and some records of that species may in fact belong here.

Subgen. DENDROSONCHUS Webb ex Sch. Bip.
Shrubs, small trees or perennials with a woody caudex. Leaves mostly crowded at the tips of the stems and branches.

4. S. fruticosus L. f., *Suppl. pl.*: 346 (1781). [*Lingua de vaca, Serralha da rocha*]
Shrub, sometimes a small tree 1-4 m high, the branches short and stout. Leaves 28-67 × 7-18 cm, pinnatifid, sinuate, divided more than half-way to the midrid, the lobes rounded or triangular, with a much larger, ± triangular terminal lobe, floccose when young, becoming glabrous, denticulate at margins, sessile with a sheathing base. Inflorescence an umbel-like cyme up to 30 cm in diameter with numerous capitula, the peduncles lengthening as the capitula mature. Capitula 15-30 × 6-15 mm. Florets yellow. 2n=18. **Plate 48**
Fl. IV - VIII. *An endemic on steep rocks, especially in wet ravines from 800-1200 m. Common in the central and higher parts of Madeira where it is also cultivated; rare in Porto Santo.* ● **MP**

5. S. pinnatus Aiton, *Hort. kew.* **3**: 116 (1789). [*Leituga*]
S. pinnatus var. *angustilobus* Lowe, var. *latilobus* Lowe
Shrubby perennial up to 2 m. Leaves glabrous, 13-35 × 4-10 cm, pinnatisect, the lobes lanceolate and irregularly toothed to subentire, petiolate. Inflorescence a branched, spreading corymbose cyme with the main axis of the inflorescence extending up to 30 cm or more beyond the terminal rosette of leaves. Capitula numerous, 7-10 × 4 mm, hemispherical to obconical. Florets bright yellow. 2n=18. **Plate 48**
Fl. IV - VIII. *An endemic frequent on cliffs and steep rocks in ravines from 1000-1400 m in Madeira.* ● **M**

6. S. ustulatus Lowe in *Trans. Camb. phil. Soc.* **4**: 22 (1831).
Perennial with woody caudex up to 25 cm high, but usually much shorter, producing a single flowering stem 15-50 cm long. Leaves glaucous, especially beneath, 14-40 × 5-12 cm, pinnatifid to pinnatisect, the lobes linear-lanceolate to triangular or ovate, sometimes overlapping, toothed or entire, sessile with a sheathing base. Inflorescence a regularly branched, corymbose cyme.

Capitula few, (4-)8-20, 7-12 × 4-5 mm, hemispherical. Florets bright yellow.
Fl:X - I. *Endemic, on cliffs and rocks near the sea, common in Madeira, rare in Porto Santo and the Desertas.* ● **MDP**

a) subsp. **ustulatus**
S. ustulatus var. *angustifolia* Lowe
Leaves fleshy, lobes linear-lanceolate, more than 2 × as long as wide, toothed. **Plate 48**
Dry, sunny rocks along the southern coast of Madeira.

b) subsp. **maderensis** Aldridge, in *Bot. Macaronésica* 2: 91 (1976).
S. ustulatus var. *imbricata* Lowe, var. *latifolia* Lowe
Leaves not, or only slightly fleshy, lobes ovate, rarely rhombic, not more than 2 × as long as wide, ± entire. **Plate 48**
Damp and shady places on the north coast of Madeira; also in Porto Santo and on the Desertas.

62. Lactuca L.

Annuals, biennials or perennials, usually with overwintering leaf-rosettes. Stems solitary, branched, leafy. Leaves entire to pinnatifid, often prickly. Capitula small, solitary to numerous. Involucre cylindrical; bracts unequal, in several rows. Receptacle pitted; scales absent. Florets (in Madeiran taxa) yellow, exceeding the involucre. Achenes compressed, ribbed, narrowed abruptly into a beak; pappus of 2 equal rows of soft hairs.

1. Bracts of inflorescence with rounded, appressed auricles; achenes dark reddish black with a white beak, the apex with a few palmate bristles . . . . **1. virosa**
- Bracts of inflorescence sagittate, with spreading auricles; achenes pale olive-grey with a white beak, the apex with numerous simple bristles . . . . **2. serriola**

1. L. virosa L., *Sp. pl.* **2**: 795 (1753).
L. patersonii Menezes; *L. scariola sensu* Lowe, non L. (1763).
Annual or biennial. Stems erect, usually purplish red, sometimes prickly below. Leaves obovate-oblong, dentate to deeply pinnatifid with broad lobes, the basal narrowed into a petiole, the cauline sessile and amplexicaul, with appressed, cordate auricles, all glaucous, stiff and spinulose on the margins and main veins beneath. Inflorescence a long, pyramidal panicle, the bracts with rounded, appressed auricles. Involucre 8-13 mm; tips of bracts purplish red. Achenes 4.4-5 mm, elliptic, narrowly winged, each face with 5(-7) ribs, the apex with a few palmate bristles, dark reddish black with a white beak.
Fl. VI - IX. *Scattered in wooded ravines and rocky slopes, generally above c. 800 m but sometimes at lower altitudes in Madeira, rare on the higher parts of Deserta Grande.* **MD**

2. L. serriola L., *Cent. pl.* II: 29 (1756).
Annual, more rarely biennial. Stems erect, whitish or occasionally reddish, sometimes prickly below. Leaves obovate-oblong, simple or the lower cauline runcinate-pinnatifid with curved lateral lobes, the uppermost simple and sagittate, all glaucous, stiff and spinulose on the margins and main veins beneath. In full sun the cauline leaves are held vertically in the north-south plane. Inflorescence a long pyramidal or spike-like panicle, the bracts sagittate with spreading auricles. Involucre 8-10 mm. Achenes 3.7-4 mm, elliptic, each face 5-7-ribbed, the apex with numerous simple bristles, pale olive-grey with a white beak.
Fl. VI - IX. *In disturbed ground and waste places in Madeira.* **M**

L. sativa L., *Sp. pl.* **2**: 795 (1753) [*Alfaca*], is widely cultivated in Madeira and Porto Santo and has been recorded as a casual, from Madeira. It has entire ovate-oblong to orbicular cauline leaves completely lacking spinules and grey achenes with 0-few simple bristles at the apex. **M**

63. Taraxacum F.H. Wigg.

Tap-rooted perennials. Leaves in basal rosettes, entire or toothed to deeply pinnately lobed. Capitula solitary, on unbranched scapes. Involucral bracts in 2 rows, the outer often spreading and shorter than the erect inner bracts. Receptacular scales absent. Ligules usually yellow, often with a darker stripe beneath. Achenes narrow, usually with a slender beak separated from the achene by a conical swelling; pappus of numerous rows of hairs.

Literature: J.L. van Soest, *Taraxaca* from Madeira collected by Johannes Lid 1968, in *Nytt Mag. Bot.* **17**: 99-100 (1970); C.I. Sahlin & J.L. van Soest, Two new *Taraxaca* from the Macaronesian islands, in *Agronomia lusit.* **35**: 313-316 (1974); A.J. Richards, Sectional nomenclature in *Taraxacum* (Asteraceae), in *Taxon* **34**: 633-644 (1985); J. Kirschner, & J. Štěpánek, Again on the sections in *Taraxacum* (Cichoriaceae) (Studies in *Taraxacum* 6), in *Taxon* **36**: 608-617 (1987).

Apomictic reproduction is widespread in *Taraxacum* and many species have been named and described. Several species of this taxonomically difficult genus have been recorded from Madeira. However, they are extremely difficult to identify accurately, especially autumn-flowering plants or those without ripe fruits. There is also considerable dispute about the circumscription, rank and correct nomenclature of many of the taxa. A broad, simplistic treatment has been adopted here. Only sections are described, and the species within them are merely listed, together with their known localities and the source of the records.

Sect. VULGARIA Dahlst.
Robust plants. Leaves entire to strongly triangular-lobed, completely green, unspotted; petioles often winged. Scapes usually hairy. Outer involucral bracts erect to recurved, lanceolate to linear, dark with a paler inner surface, the border (if present) inconspicuous. Ligules usually brownish-striped beneath. Achenes (excluding beak and cone) 2.5-3.5 mm, spinulose or tuberculate. *Mainly plants of grasslands and waste places.*

T. cacuminatum G.E. Haglund, in *Acta. Horti gothoburg.* **11**: 23 (1936).
Recorded from São Antonio, near Funchal (det. van Soest, 1970).

T. duplidentifrons Dahlst. in *Rep. botl Exch. Club Br. Isl.* **8**: 624 (1929)
?*T. raunkieri* Wiinst.
Recorded from Camacha, around Poiso and Queimadas (det. van Soest, 1969).

T. hamatum Raunk., *Dansk Exkurs.-Fl.* ed. 2: 255 (1906).
Recorded from Paúl da Serra (det. Doll, 1986).

T. lidianum Soest in *Nytt Mag. Bot.* **17**: 99 (1970).
Recorded from roadsides in and around Funchal (det. van Soest, 1970).

T. maderense Sahlin & Soest in *Agronomia lusit.* **35**: 313 (1974).
Recorded near Monte (det. Sahlin & van Soest, 1974).

Sect. OBOVATUM Soest
Leaves entire or shallowly lobed, dark, glabrous, held horizontally; petiole short, wide, green. Scapes lanate. Capitula 25-30 mm across. Outer involucral bracts erect, ovate to ovate-lanceolate with a pale border and a small appendage near the tip. Ligules often with a grey stripe beneath. Achenes (excluding beak and cone) 3.5-4 mm, strongly rugose.

T. obovatum DC., *Rapp. voy.* **2**: 83 (1813).
Recorded as a common ruderal, especially in streets in and around Funchal, and along roadsides between Camacha and Choupana (det. Lundevall, 1969). Many, if not most, of the early records from Madeira probably also belong here.

T. officinale agg. *Scattered in many parts of Madeira up to 1300 m, often as a ruderal* (det. Doll, 1986).

Sect. SPECTABILIA Dahlst. emend. A.J. Richards
Medium-sized plants with entire or shallowly lobed, often dark, spotted and hairy leaves; petioles narrow. Mid-ribs, petioles and scapes often red or purple. Outer involucral bracts spreading to appressed, ovate-lanceolate to lanceolate, with a narrow, pale border. Ligules often striped reddish purple or greyish beneath. Achenes (excluding beak and cone) 4-4.5 mm, smooth or slightly tuberculate.
Generally plants of wet hilly grasslands.

T. adamii Claire in *Bull. Soc. bot. Rochelaise* **12**: 49 (1891).
Said to be rather frequent in Madeira, in damp, grassy places and along footpaths and levadas, mainly in the central mountains above c. *900 m.* (det. Malato-Beliz, 1958).

T. lainzii Soest, *Trab. Jard. bot. Univ. Santiago* **7**: 5 (1954).
Recorded from Monte (det. van Soest 1970).

T. praestans H. Lindb. in *Acta Soc. Fauna Flora fenn.* **29**: 24 (1907).
Recorded from Monte (det. van Soest 1970).

64. Lapsana L.

Annuals to perennials. Stems solitary, branched, leafy. Capitula small, numerous, in loose corymbose panicles. Involucral bracts in 1 row with additional scale-like bracts at the base. Receptacular scales absent. Florets yellow. Achenes slightly compressed with *c.* 20 ribs, the outer curved and longer than the inner; pappus absent.

1. L. communis L., *Sp. pl.* **2**: 811 (1753).
Stems 40-80 cm, erect, glandular or eglandular hairy below, glabrous above. Leaves sparsely eglandular hairy, the lower ovate to lyrate-pinnatifid, with small, oblong lateral lobes and a large terminal lobe, the upper ovate to lanceolate, all shortly toothed or the upper rarely entire. Involucral bracts 4-6 mm, linear-oblong, slightly keeled, glabrous; additional bracts at the base of involucre few, less than 1 mm, dark. Ligules up to twice as long as involucre. Achenes straw-coloured, the outer 3.2-4 mm.
Fl. I - VI. *Common weed scattered along roadsides and in waste places throughout the lower regions of Madeira.* **M**

Madeiran plants are all subsp. **communis**.

65. Crepis L.

Annual to perennial herbs with 1-many usually branched, stems. Leaves ± toothed to pinnatisect. Capitula few to numerous. Involucral bracts in 2 rows, the outer much shorter than the inner. Receptacle with ciliate pits. Ligules yellow, usually reddish on the outer face. Achenes 10-ribbed, beaked or not; pappus of 1-many rows of soft, white hairs.

1. Achenes muricate, sometimes slightly narrowed above but lacking a distinct beak. . . . . **4. capillaris**
- Achenes smooth, with a distinct beak 2

2. Plant not succulent . . . . **1. vesicaria**
- Plant succulent 3

3. At least lower branches ascending at a narrow angle to the stem; florets completely yellow . . . . **2. divaricata**
- Branches decumbent or at 90° to the stem; outer florets red on the outer face . . . . **3. noronhaea**

1. C. vesicaria L., *Sp. pl.* **2**: 805 (1753)   [*Almeirante, Almeirão, Letubra mansa*]
Perennial or biennial, sometimes annual, up to 60 cm. Leaves very variable, the basal oblanceolate, denticulate or dentate to runcinate-pinnatifid, or pinnatisect to 2-pinnate, glabrous or pubescent, the petiole often purplish; cauline leaves similar, becoming narrower, sessile, the uppermost lanceolate and amplexicaul with rounded auricles. Capitual numerous, ± cylindrical. Involucres 10-15 mm in fruit, usually tomentose, especially towards the base, sometimes with glandular hairs; outer bracts erect to ascending, ¼-½ the length of the inner. Receptacle with glandular pits. Florets yellow. Achenes 7-8 mm, fusiform, beaked, muricate.
Fl. III - VI. **MD**

a) subsp. **haenseleri** (Boiss. ex DC.) P.D. Sell in *Bot. J. Linn. Soc.* **71**: 254 (1976).
Barkhausia laciniata Lowe; *Crepis laciniata* (Lowe) F.W. Schultz, var. *integrifolia* Lowe, var. *pinnatifida* Lowe
Biennial or annual, glabrous or pubescent but not or rarely setose. Outermost ligules reddish purple on the back. **Plate 49**
Waste or cultivated ground, waysides; common, mainly in the south and east of Madeira but reaching the north coast; also occurring on Deserta Grande.

b) subsp. **andryaloides** (Lowe) Babc. in *Univ. Calif. Publs Agric. Sci.* **6**: 369 (1939).
Barkhausia comata Lowe; *B. dubia* Lowe; *B. hieracioides* Lowe; *Crepis andryaloides* Lowe; *C. comata* Banks & Sol. ex Lowe; *C. dubia* (Lowe) F.W. Schultz; *C. hieracioides* (Lowe) F.W. Schultz, var. *laevigata* Lowe, var. *nigricans* Lowe
Perennial or biennial with numerous, often black, glandular or eglandular setae, especially on the upper parts of stems and on involucres. Florets completely lemon-yellow. **Plate 49**
An endemic of rocks and cliffs on the north coast and in the higher central regions of Madeira; apparently formerly more widespread. ●

Both subspecies are variable, each showing a number of extreme forms all linked by intermediates. Additionally, intermediates between the subspecies occur, especially along the north coast of Madeira.

2. C. divaricata (Lowe) F.W. Schultz in *Flora, Jena* **23**: 719 (1840).   [*Almeirante*]
Barkhausia divaricata var. *robusta* Lowe; *Crepis divaricata* var. *robusta* (Lowe) Lowe
Succulent biennial or perennial. Stems up to 45 cm, pubescent especially in the upper parts, and with black or brown, glandular setae, branches ascending, the lower making a narrow angle with the stem. Basal leaves in a dense rosette, elliptic to oblanceolate, glabrous, denticulate to shallowly pinnate, undulate, the petiole ¼-⅓ the length of the blade; cauline leaves narrower, sessile, the upper entire with rounded auricles. Capitula cylindrical to turbinate. Involucre 10-12 mm in fruit, pubescent at the base and with a few glandular hairs; outer bracts becoming spreading, mostly ⅓ as long as the inner. Florets yellow. Achenes 5-6(-7) mm, fusiform, beaked, muricate.

Fl. III - VII. *A rare endemic of dry, gravelly soils. In Madeira found only on the islands of the Ponta de São Lourenço; also rare in Porto Santo but more common on the Desertas; recorded once from Selvagem Grande, over a century ago, but never seen since.* ● **MDPS**

Plants from the Desertas are more densely pubescent and glandular-setose than those from the other islands. *C. divaricata* appears scarcely to differ from *C. vesicaria* subsp. *andryaloides* in herbarium specimens. The only distinguishing feature appears to be succulence.

3. **C. noronhaea** Babc. in *Univ. Calif. Publs Agric. Sci.* **6**: 369 (1939).
Barkhausia divaricata var. *pumila* Lowe; *Crepis divaricata* var. *pumila* (Lowe) Lowe
Like **2** but somewhat smaller in all its parts; lower branches decumbent or spreading at 90° to the stem; petiole ⅓-½ as long as the blade; involucral bracts more densely pubescent; outer ligules red on the back. **Plate 49**
Fl. II - V. *Plant of dry soils on cultivated and waste ground and on roadsides in Porto Santo, once quite common but rarer now. Records for the Desertas are erroneous.* ● **M**

Very similar to **2**, and perhaps only a variant of it, as originally described.

4. **C. capillaris** (L.) Wallr. in *Linnaea* **14**: 657 (1841)
Annual or biennial to 25 cm. Stems 1 to many, branched from the base or above. Basal leaves numerous, lanceolate to oblanceolate, obtuse to acute, minutely toothed to lyrate, narrowed at the base to form a petiole, glabrous or sparsely eglandular-hairy; cauline leaves similar but smaller, sessile, amplexicaul with a sagittate base. Capitula numerous, borne on slender peduncles. Involucres *c.* 6 mm, glabrous to tomentose or with sparse, glandular hairs; outer bracts appressed, ⅓-½ the length of the inner. Ligules yellow, usually reddish on the outer face. Achenes smooth, not beaked.
Fl. V. *Introduced. A scarce plant, recorded for the first time from Madeira in 1973, at Santo da Serra. Native to Europe, naturalized elsewhere.* **M**

Madeiran plants appear to belong to var. **capillaris**.

66. Andryala L.

Annual, biennial or perennial herbs, stellate-hairy. Stems solitary or numerous, branched. Leaves entire to pinnatisect, the cauline often sessile, rounded to amplexicaul. Capitula few to numerous, rarely solitary. Involucral bracts in 1-2 rows. Receptacle pitted; pits with laciniate-dentate margins. Florets sulphur-yellow, gold or orange, sometimes with a red stripe on the back of the ligule. Achenes oblong to obconical with 8-10 ribs extending into teeth at the truncate apex; pappus a ring of greyish hairs, falling entire.

1. Leaves entire to shallowly pinnatifid with lobes 2 mm or more wide . . **1. glandulosa**
- Leaves deeply pinnatisect, the segments ± filiform, less than 2 mm wide . . . .
 . . . . **2. crithmifolia**

1. **A. glandulosa** Lam., *Encycl.* **1**: 154 (1783).
A. cheiranthifolia L'Hér.; *A. varia* Lowe ex DC.
Villous to tomentose and viscid perennial, with a tawny to whitish indumentum. Stems up to 100 cm, from a stout, woody stock. Leaves ovate to lanceolate, entire to pinnatisect. Florets golden yellow. Achenes 1-1.5 mm, dark brown with 10 paler ribs.
Fl. (III-)IV - X(-XII). **MDP**

A very variable species within which two subspecies are distinguishable. Several varieties and formas have also been described but none is recognized here.

a) subsp. **glandulosa** [*Alfavaca, Boffe de burro*]
A. cheiranthifolia var. *congesta* Lowe; *A. varia* subsp. *congesta* (Lowe) Menezes
Plant up to 50 cm; stems glandular-hairy and viscid. Leaves entire to sinuate-dentate, rarely with short, blunt lobes; basal leaves 5-25 cm, often withered at anthesis; cauline leaves rather crowded, ovate to lanceolate. Inflorescence rather dense; peduncles usually stout. Involucre 11-20 × 10-15 mm. **Plate 49**
Common in Madeira, mainly on the north coast eastwards to the islands of the Ponta de São Lourenço; also in Porto Santo and on all of the Desertas; always on coastal rocks and sea-cliffs except in Porto Santo where it also occurs inland on Pico do Facho.

b) subsp. **varia** (Lowe ex DC.) R. Fern. in *Anu. Soc. broteriana* **25**: 28 (1959).
[*Flor de coelho*]
A. cheiranthifolia var. *sparsiflora* Lowe; *A. varia* subsp. *sparsiflora* (Lowe) Menezes
Plant up to 100 cm, the stems glandular-hairy and viscid above, not or sparsely so below. Leaves ± entire or obscurely toothed to pinnatifid with lobes 2 mm or more wide; basal leaves (3-)5-45 cm; cauline leaves usually sparse, lanceolate, the uppermost entire. Inflorescence rather lax; peduncles slender. Involucre 8-10(-12) × 6-8 mm. **Plate 49**
Everywhere in Madeira up to 1500 m, almost always at inland sites, usually on dry rocks and banks. ■ [*Canaries*]

2. A. crithmifolia Aiton, *Hort. kew.* **3**: 129 (1789).
A. cheiranthifolia var. *sparsiflora* subvar. *coronopifolia* Lowe; *A. glandulosa* subsp. *varia* var. *varia* forma *coronopifolia* (Lowe) R. Fern.; *A. varia* subsp. *sparsiflora* forma *coronopifolia* (Lowe) Menezes; *A. varia* var. *crithmifolia* DC.
Perennial or biennial with a stout, woody stock. Stems annual, up to 200 cm, erect, fistular, tomentose and glandular-hairy and viscid above, thinly tomentose to pubescent below. Leaves crowded towards the base of the stems, sparse above, mealy-pubescent, glaucous; basal leaves up to 15 cm, deeply and finely 1- to 2-pinnatisect, the segments less than 2 mm wide; uppermost leaves entire. Inflorescence an open, flat-topped or domed corymb. Capitula 5-10 mm across; florets yellow. Achenes 1-1.3 mm, dark brown with 10 pale ribs. **Plate 49**
Fl. V - VII. *A rare maritime endemic found only at a few sites along the south coast of Madeira, from Funchal west to Madalena do Mar.* ● M

Lowe (1868) stated that an extreme maritime form of *A. glandulosa* (as *A. cheiranthifolia* var. *sparsiflora* subvar. *coronopifolia*) very closely resembled *A. crithmifolia*. However, herbarium specimens show no discernable differences between the taxa. Moreover, both *A. crithmifolia* and *A. cheiranthifolia* var. *sparsiflora* subvar. *coronopifolia* occur only on the south coast of Madeira while *A. glandulosa* is almost exclusively found in inland and northern, coastal sites. *A. cheiranthifolia* var. *sparsiflora* subvar. *coronopifolia* is here included within *A. crithmifolia*.

MONOCOTYLEDONES

CVII. ALISMATACEAE
J.R. Press

Perennial aquatics or plants of wet places, sometimes with floating or submerged leaves. Leaves alternate or basal, petioles sheathing at base. Flowers actinomorphic, usually hermaphrodite, bracteate. Sepals 3, persistent. Petals 3, usually fugacious. Stamens (3-)6(-many). Ovary superior. Carpels 3-many; ovules 1. Fruit a whorl of achenes.

1. Alisma L.

Glabrous, scapose perennials. Inflorescence a much-branched panicle, the branches whorled. Carpels numerous, free, in a single whorl, each with 1 ovule; styles lateral. Ripe achenes laterally compressed.

1. A. lanceolatum With., *Arr. Brit. pl.* ed.3, **2**: 326 (1796).
A. plantago-aquatica L. var. *lanceolatum* (With.) Lej.
Leaves aerial, up to 18 × 3.5 cm, lanceolate to elliptic, acuminate, base cuneate; petioles up to 22 cm. Petals *c.* 3.5 mm, purplish-pink. Fruiting heads 4-5 mm in diameter. Achenes 2-2.5 mm; style ± straight, arising near the top of the fruit.
Fl. V - VI. *A rare plant of pools and the margins of streams and rivers, found in Madeira only at Santana and Funchal.* **M**

CVIII. POTAMOGETONACEAE
J.R. Press

Aquatic herbs. Leaves all submerged or both submerged and floating, alternate or opposite, with stipular sheaths at the base. Flowers hermaphrodite, 4-merous, in ebracteate, pedunculate spikes. Perianth sepaloid. Stamens sessile at base of perianth-segments. Carpels (1-)4, free or fused at the base; styles short. Fruits achenes or drupes.

1. Potamogeton L.

Mostly perennials, often rhizomatous or perennating by winter buds. Leaves all or mostly alternate; submerged leaves thin and translucent, with or without a distinct petiole, floating leaves (if present) usually thick and opaque, always clearly petiolate. Stipules free or fused to the leaf base below, sometimes tubular in the lower part. Perianth-segments 4, valvate and narrowed into a short claw. Fruit of several small drupes.

1. Floating leaves present 2
- Floating leaves absent . . . . **1. pusillus**

2. Floating leaves elliptical, more than 3 × as long as wide, cuneate to obtuse at base . . . . **2. nodosus**
- Floating leaves ovate to lanceolate, 2-3 × as long as wide, cuneate to subcordate at base . . . . **3. polygonifolius**

1. P. pusillus L., *Sp. pl.* **1**: 127 (1753).
P. panormitanus Biv.
Rhizome absent, perennating by winter buds. Leaves all submerged, linear and grass-like, 0.5-2 mm wide. Stipules tubular towards the base. Fruits 1.7 mm.
Fl. VII. *Very rare plant in Madeira, found only in the Ribeira do Faial.* **M**

2. P. nodosus Poir. in Lam., *Encycl. Suppl.* **4**: 535 (1816).
P. fluitans Roth pro parte; *P. leschenaultii* Cham. & Schldl.; *P. machicanus* Lowe (nom. inval.)
Rhizome present. Floating leaves with blades 6-14 × 1.5-2.5 cm, narrowly elliptic, cuneate to rarely obtuse at the base; submerged leaves minutely denticulate, often disappearing early. Stipules not tubular. Fruits 3-4 mm.
Fl. VI - VIII. *Very rare plant in rivers around Machico and Santa Cruz.* **M**

Records of *P. gramineus* L. refer to this species.

3. P. polygonifolius Pourr. in *Hist. Mém. Acad. Toulouse* **3**: 325 (1788).
P. cyprifolius Lowe (nom. inval.)
Rhizome present. Floating leaves with blades (3.5-)4-7 × (1.5-)2-3.5 cm, ovate to lanceolate, cuneate to truncate or subcordate at the base; submerged leaves entire, often disappearing early. Stipules not tubular. Fruits 2-2.2 mm.
Fl. V - VI. *Plant of streams, ponds and marshy ground, mainly in the hills above Santana and Porta da Cruz but also above Funchal, Ponta de Pargo and Porto do Moniz.* **M**

Records of *P. natans* L. refer to this species.

CIX. RUPPIACEAE
J.R. Press

Submerged aquatic perennials of saline water. Leaves linear to filiform, with sheathing bases. Flowers hermaphrodite, small, in pedunculate spikes. Perianth absent. Stamens 2. Carpels usually 4, long-stipitate in fruit. Fruit indehiscent.

1. Ruppia L.

Leaves mostly alternate but the 2 involucral leaves subtending each spike subopposite and with inflated sheaths. Spikes with 2 flowers, one above the other on opposite sides of the axis.

1. R. maritima L., *Sp. pl.* **1**: 127 (1753).
R. rostellata W.D.J. Koch
Leaves at most 1 mm wide. Sheaths of involucral leaves only slightly inflated. Fruiting pedicels up to 2 cm. Fruits 1.5-2 mm, ovoid and curved.
Fl. IV - V. *A very rare plant occurring in salt pans and brackish pools and marshes at Paúl do Mar on the south coast of Madeira.* **M**

CX. LILIACEAE
A.R. Vickery

Annuals or perennials, usually herbs with bulbs, corms or rhizomes; more rarely climbers or small shrubs. Inflorescence a raceme, spike, panicle, umbel or cyme. Flowers regular or slightly zygomorphic, usually perfect; perianth-segments 6, petaloid, ovary 3-locular, superior, styles 1 or 3. Fruit a capsule or berry.

Lilium candidum L. [*Açucena*] is cultivated in gardens and is recorded as appearing on banks and roadsides close to gardens in the Funchal area; it is probably not fully naturalized.

1. Stems woody; scrambling climbers or erect shrubs 2
- Stems herbaceous; usually scapose herbs 5

2. Climbers or scramblers 3
- Erect subshrubs .... **11. Ruscus**

3. Leaves petiolate with tendrils at the base of the petioles .... **12. Smilax**
- Leaves minute, scarious, functionally replaced by broad or narrow cladodes which lack tendrils and well defined stems 4

4. Flowers in small cymes at margins of cladodes .... **10. Semele**
- Flowers in umbel-like clusters or racemes, or solitary, not at margins of cladodes ....
  .... **9. Asparagus**

5. Leaves coriaceous, margins spinose-dentate .... **3. Aloe**
- Leaves not coriaceous, margins entire 6

6. Inflorescence an umbel 7
- Inflorescence a raceme or panicle, or corymbiform 9

7. Rootstock tuberous; perianth exceeding 3 cm, usually bright blue (rarely white) .... **6. Agapanthus**
- Rootstock a bulb, perianth less than 2 cm; white, pink, greenish, red or purple .. 8

8. Plants smelling of onion or garlic when crushed; umbels sometimes containing bulbils; style gynobasic .... **7. Allium**
- Plants not smelling of onion or garlic; umbels never containing bulbils; style terminal . ... **8. Nothoscordum**

9. Plants with bulbs 10
- Plants lacking bulbs 11

10. Perianth lilac or blue .... **5. Scilla**
- Perianth white or creamy white .... **4. Ornithogalum**

11. Inflorescence arching, often with rosettes of leaves arising from some of its nodes ... . **2. Chlorophytum**
- Inflorescence erect, without rosettes of leaves at nodes .... **1. Asphodelus**

1. Asphodelus L.

Scapose annuals or perennials with short rhizomes. Leaves linear. Inflorescence a raceme or panicle. Fruit a locucidical capsule.

1. A. fistulosus L., *Sp. pl.* **1**: 309 (1753). [*Cebolinho de burro* (P)]
A. madeirensis E. Simon
Annual or short-lived perennial. Leaves 3-30 cm. Scape to 65 cm, erect, usually branched. Perianth-segments *c.* 5 mm, oblong, white to pale pink with dark central vein on outside. Capsule *c.* 5 × 5 mm.
Fl. III - IV. *Roadsides, cultivated ground, and dry sandy places; in Madeira only on the Ponta de São Lourenço and its islets; common in Porto Santo; also on Ilhéu Chão. Native to the Mediterranean region.* **MDP**

2. Chlorophytum Ker Gawl.

Herbaceous perennials with rhizomes and fibrous or fleshy roots. Leaves mostly basal, linear. Inflorescence a simple or branched raceme. Flowers white. Fruit a trigonous capsule.

1. C. comosum (Thunb.) Jacques in *Soc. imp. & centr. d'Hortic.* **8**: 345 (1862).
Anthericum comosum Thunb.; *Chlorophytum capense* O. Kunze; *C. elatum* R. Br.
Leaves 20-45 cm × 5-20 mm, sometimes with pale yellow stripes. Inflorescence 30-80 cm, branched or simple, lax, arching, sometimes with rosettes of leaves often with roots arising from some of the nodes. Flowers 1-3 per node, pedicels *c.* 8 mm, perianth-segments *c.* 1 cm, white. Capsule *c.* 5 mm.
Fl. I - XII. *Cultivated as an ornamental in Madeira and occasionally naturalized. Native to S. Africa.* **M**

Hemerocallis lilioasphodelus L. is native to Europe but is widely cultivated for ornament in Madeira and possibly naturalized in places. It is a rhizomatous, ± scapose, herb with an inflorescence consisting of 2 subequal cymose branches with fragrant yellow or reddish flowers 6-10 cm. **M**

3. Aloe L.

Perennial, scapose herbs. Leaves usually in a dense rosette, linear, coriaceous, succulent, spinose-dentate. Inflorescence an axillary (often subterminal), raceme or panicle. Flowers red, orange or yellow. Fruit a loculicidal capsule.

1. A. vera (L.) Burm.f., *Fl. indica*: 83 (1768).   [*Babosa*]
A. perfoliata Buch. var. *vera* L.; *A. barbadense* Mill.
Short-stemmed, stoloniferous herb. Leaves 30-6 × 5-8 cm. Scape simple or sparingly branched. Perianth 2.5-3 cm, yellow, outer segments connate for ⅓-½ of their length.
Fl. III - IV. *Cultivated for ornament and naturalized in a few places in Madeira. Native to Arabia and NE Africa.* **M**

4. Ornithogalum L.

Bulbous, scapose perennials. Leaves linear. Inflorescence a corymbiform or elongate raceme. Flowers showy, usually white. Fruit a loculicidal capsule.

1. O. arabicum L., *Sp. pl.* **1**: 307 (1753).
Leaves 25-60 × 2-3.5 cm. Scape 30-80 cm, unbranched. Inflorescence ± corymbiform. Bracts prominent, acute, shorter than pedicels. Perianth-segments 18-25 × *c*. 8 mm, white or creamy white.
Fl. V. *Only two collections from Madeira, both made in the mid-nineteenth century, have been seen and no recent records are known. Probably only an escape from cultivation. Native to the Mediterranean region.* **M**

5. Scilla L.
A.R. Vickery & N.J. Turland

Bulbous scapose perennials. Leaves lanceolate to oblong-lanceolate or linear. Flowers in a raceme. Perianth-segments free, usually blue or lilac, rarely pale to white. Fruit a subglobose 3-lobed capsule.

1. S. madeirensis Menezes in *Broteria* (Bot.) **22**: 24 (1926).
S. hyacinthoides Aiton (1789), non L. (1767); *S. madeirensis* var. *melliodora* Svent., var. *melliodora* forma *pallida* Svent.
Bulb large, deep purple. Leaves 4-11, 18-35(-40) × 2.5-7 cm, somewhat canaliculate, glabrous. Scape 24-50 cm (including inflorescence 12-20 cm). Inflorescence conical-cylindric, many-flowered, somewhat lax. Bracts 2 per pedicel, 1 located immediately below base of pedicel, small, linear, membranous, white, the other similar but much smaller and located at side of base of pedicel. Pedicels 7-20 mm, elongating to 25-30 mm in fruit, patent. Perianth-segments 3-5 × 1-2.5 mm, obtuse, mucronate, lilac or pale blue, yellowish in bud. Filaments and ovary more deeply coloured bluish. Capsule 10-12 × 8-10 mm, greenish yellow; seeds black. **Plate 50**
Fl. IX - XII. *A rare endemic of vertical cliffs and other rocky places from near sea-level to 800 m. In Madeira occurring in a western and northern coastal band extending from Prazeres to São Vicente, and in the Funchal region; recently discovered on Deserta Grande and the adjacent islet of Doca; recorded as occurring in Porto Santo but without further data; also*

present in sandy and gravelly soils between rocks on the Salvage Islands, (very rare on Selvagem Grande but common on both Selvagem Pequena and Ilhéu de Fora). ● **MDPS**

Plants from the Salvage islands have been treated as a separate, endemic variety, var. *melliodora* Svent. in *Ind. Sem. Agron. Investig. Nat. Hispan. Inst.* **1968**: 57 (1969). This plant appears to differ only in having sweetly scented flowers and scarcely merits formal taxonomic recognition. Another species, *S. latifolia* Willd., otherwise known only from the Canaries and Morocco, has also been recorded from the Salvages. This plant is very similar and closely related to *S. madeirensis* and may even be conspecific (in which case, *S. latifolia*, being the earlier name, would have priority). Further study is required to ascertain the relative taxonomic position of the Salvage Island plants. It seems highly unlikely that these two apparently vicariant species should co-exist in such a small geographic area. These Macaronesian species are related to *S. hyacinthoides* L., which occurs from S. Europe eastwards to Iraq.

6. Agapanthus L'Hér.

Scapose herbs forming dense clumps from tuberous rootstocks. Leaves linear. Infloresence an umbel. Flowers slightly zygomorphic, showy, blue (rarely white). Fruit a loculicidal capsule.

1. A. praecox Wiild., *Enum. pl.*: 353 (1809).
A. orientalis F.M. Leight.
Leaves 25-35 × 2-2.8 cm. Scape 45-60 cm, unbranched. Umbels with many (more than 50) flowers. Pedicels 2.5-6.5 cm. Perianth-segments 3.2-5 cm, connate at base.
Fl. VI - VIII. *Widely planted in gardens, parks and along roadsides in Madeira; abundantly naturalized along roads and terrace margins. Native to S. Africa.* **M**

Madeiran plants are subsp. **orientalis** (F.M. Leight.) F.M. Leight. in *Jl S. Afr. Bot.*, suppl. **4**: 21 (1965).

7. Allium L.

Bulbous, scapose herbs, usually with a distinctive smell of onion or garlic. Leaves always linear (in Madeiran species). Inflorescence an umbel, initially enclosed within a membranous spathe, sometimes containing bulbils which occasionally replace the flowers. Fruit a membranous cpasule.

The following species are widely cultivated and may occasionally become naturalized:

A. cepa L., *Sp. pl.* **1**: 301 (1753) [*Cebolo*] is of unknown origin but is a widely cultivated vegetable of which there are numerous cultivars. Leaves fistular, sheathing lower part of scape. Scape 20-55 (-100) cm, lower part usually inflated. Spathe usually 3-valved, persistent, shorter than umbel. Umbel 4-9 cm, in diameter, dense, many-flowered. Perianth stellate, segments 3-4.5 × 2-2.5 mm, white with green stripe. **M**

A. ampeloprasum L., *Sp. pl.* **1**: 294 (1753) [*Alho porro*] is native to S. and W. Europe. Leaves linear, sheathing lower ½ of scape. Scape 25-100 cm, terete. Spathe 1-valved, caducous. Umbel 5-9 cm in diameter, dense, many-flowered, with or without bulbils. Perianth campanulate, segments 4-5.5 × 1.3-2.4 mm, pale pink to dark red. **M**

A. sativum L., *Sp. pl.* **1**: 296 (1753) [*Alho*] is widely cultivated in Europe. Leaves linear, flat, sheathing lower ½ of scape. Scape 25-100 cm, terete. Spathe 1-valved with a long beak, caducous. Umbel 2.5-5 cm in diameter, with few flowers and many bulbils. Perianth cup-shaped, segments 3-5 mm, usually greenish white or pink. **M**

1. Umbel with bulbils only, lacking flowers . . . . **4. A. vineale**
- Umbel always with some flowers, sometimes also with bulbils 2

2. Leaves all basal, or almost so 3
- Leaves sheathing at least the lower ¼ of the scape . . . . **3. A. paniculatum**

3. Stigma 3-lobed . . . . **2. A. triquetrum**
- Stigma entire . . . . **1. A. neapolitanum**

1. A. neapolitanum Cirillo, *Pl. rar. neapol.* **1**: 13(1788).
Bulbs 1-2 cm in diameter. Leaves linear essentially basal, or sheathing lowest part of scape. Scape 20-55 cm, triquetrous. Spathe 1-valved, entire, persistent, shorter than pedicels. Umbel 5-8 cm in diameter, lax, many-flowered, without bulbils. Perianth cup-shaped; segments 7-12 × 4-6 mm, white. Capsule *c.* 5 mm.
Fl. III - VI. *On roadsides in the Funchal region.* **M**

A. roseum L., *Sp. pl.* **1**:296 (1753) is like **1** but with the spathe 3- to 4- lobed and perianth campanulate, pink or white. It has been recorded for Maedeira but without further data. **M**

2. A. triquetrum L., *Sp. pl.* **1**: 300 (1753). [*Alho bravo*]
Bulbs *c.* 1.5 cm in diameter. Leaves linear, ± basal. Scape 10-45 cm, triquetrous. Spathe 2-valved, usually persistent, ± equalling the pedicels in length. Umbel 4-7 cm in diameter, lax 3- to 15-flowered, often drooping to one side, without bulbils. Perianth campanulate, segments 10-18 × 2-5 mm, white with longitudinal green stripe. Capsule *c.* 7 mm.
Fl. III - V. *Along levadas, roadsides and open woodland everywhere in the lower regions of Madeira, often in great numbers.* **M**

3. A. paniculatum L., *Syst. nat.*, ed. 10, **2**: 978 (1759). [*Cebolhino*]
Bulbs 1-2.5 cm in diameter. Leaves linear, sheathing the lower ⅓-½ of the scape. Scape 30-70 cm, terete. Spathe 2-valved, the valves unequal, exceeding the pedicels in length. Umbel 3.5-7 cm in diameter, diffuse, many-flowered, very rarely with bulbils. Pedicels of equal length. Perianth campanulate, segments 4.5-7 × 1.5-2.5 mm, varying in colour from greenish grey to brownish purple, pink or lilac. Capsule *c.* 5 mm.
Fl. VIII - IX. *By walls, on roadsides and dry, rocky slopes, apparently mainly in south-east Madeira.* **M**

4. A. vineale L., *Sp. pl.* **1**: 299 (1753).
Bulbs 1-2 cm in diameter. Leaves fistular, sheathing lower ⅓-⅔ of scape. Scape 30-120 cm, terete. Spathe 1-valved, caducous. Umbel (in Madeiran plants) lacking flowers, composed entirely of densely packed bulbils.
Fl. VI. *Margins of cultivated ground in north-east and north-west Madeira, probably also elsewhere.* **M**

8. Nothoscordum Kunth

Scapose bulbous herbs, not smelling of onion or garlic. Leaves linear. Inflorescence an umbel initially contained within a spathe, bulbils absent. Fruit a membranous capsule.

1. N. gracile (Aiton) Stearn in *Taxon* **35**: 338 (1986). [*Alho americano, Alho bravo*]
Allium gracile Aiton; *A. fragrans* Vent.; *Nothoscordum fragrans* (Vent.) Kunth; *N. inodorum* (Aiton) Nicholson
Bulb *c.* 15 cm in diameter. Leaves 20-40 cm × 4-10 mm. Scape 30-60 cm. Umbel with 10-15

fragrant flowers; pedicels 2-4 cm, unequal. Perianth infundibuliform, segments 10-15 × 4 mm, white with pink midrib. Capsule *c.* 6.5 mm.
Fl. VII - XI. *Naturalized in waste places, very common in most lowland areas of Madeira. Native to subtropical S. America.* **M**

9. Asparagus L.

Herbs, shrubs or climbers. Leaves insignificant, functionally replaced by cladodes. Flowers small, in umbel-like clusters or racemes, unisexual, or functionally unisexual. Fruit a berry.

Literature: B. Valdes, Revision del genero *Asparagus* (Liliaceae) en Macaronesia, in *Lagascalia* **9**(1): 65-107 (1979).

Several species are cultivated in Madeira for ornament, including **A. densiflorus** (Kunth) Jessop (*A. sprengeri* Regel) and **A. setaceus** (Kunth) Jessop; none are known to be naturalized there.

1. Cladodes more than 5 mm broad, leaf-like . . . . **1. asparagoides**
- Cladodes less than 2 mm broad, needle-like 2

2. Young stems with papillose longitudinal ribs . . . . **4. umbellatus**
- Young stems smooth, without papillose ribs 3

3. Cladodes ± erect; pedicels of fruit shorter than cladodes . . . . **2. scoparius**
- Cladodes spreading; pedicels of fruit equalling or longer than cladodes . . . **3. nesiotes**

1. A. asparagoides (L.) Druce in *Rep. Bot. Exch. Club Brit. Isles* **3**: 414 (1914).
[*Alegra campo de folho miuda*]
A. medeoloides (L.f.) Thunb.; *Medeola asparagoides* L.; *Myrsiphyllum asparagoides* (L.) Willd.
Stems twining, to 1.5 m. Cladodes 20-35(-40) × 8-15 mm, ovate to ovate-lanceolate, solitary. Flowers solitary or in pairs. Perianth-segments 4.5-5.5 mm. Berry 6-8 mm.
Fl. III - V. *Grown for ornament in Madeira and naturalized in forest, along levadas and on rocks and walls in various places. Native to S. Africa.* **M**

2. A. scoparius Lowe in *Trans Camb. phil. Soc.* **4**: 11 (1831). [*Esparto*]
Stems to 3 m, smooth. Cladodes 5-15 mm, erect, needle-like, in fascicles of 2-20. Flowers in fascicles of 3-12. Perianth-segments 3-4 mm. Fruiting pedicels shorter than cladodes. **Plate 50**
Fl. XII - VI. *Dry, rocky places. Rare in Madeira, only known from Gorgulho and Caniço; also recorded from Selvagem Grande but probably no longer present there.* ■ **MS** [*Canaries, Cape Verdes*]

3. A. nesiotes Svent., *Index Sem. Aurantapae*: 56 (1969).
A. acutifolius auct., non L. (1753)
Stems to 4 m, smooth. Cladodes 5-12 mm, spreading, needle-like, in fascicles of 7-25. Flowers in fascicles of 1-6. Perianth-segments 3-4 mm. Fruiting pedicels equalling or longer than the cladodes. 2n=60. **Plate 50**
Fl. VIII - IX. *Maritime plant confined to coastal rocks on the Salvages.* ■ **S** [*Canaries*]

Salvage Island plants belong to (●) subsp. **nesiotes**. Subsp. purpuriensis Marrero & Ramos is confined to the Canaries.

4. A. umbellatus Link in Buch, *Phys. Beschr. Canar. Inseln*: 140 (1825).
A. lowei Kunth; *A. scaber* Lowe, non Brignoli (1810); *A. umbellatus* var. *scaber* (Lowe) Baker
Stems to 5 m, with papillose longitudinal ribs when young. Cladodes (6-)20-30 mm, needle-like, in fascicles of 4-30. Flowers in fascicles of 4-15. Perianth-segments 5-7 mm. **Plate 50**
Fl. XI - I. *Very rare plant of cliffs and roadsides near the sea in Madeira, along the north coast eastwards to Santana; also cultivated in gardens.* ■ M

Madeiran plants are the (●) subsp. **lowei** (Kunth) Valdés in *Lagascalia* **9**: 86 (1979). Subsp. *umbellatus*, from the Canaries, differs in the generally shorter cladodes (5-20 mm) and the markedly scabrid pedicels.

10. Semele Kunth

Climbers with short rhizomes. Leaves inconspicuous, scarious, functionally replaced by persistent, leaf-like cladodes. Flowers in small cymes, usually arising at the margins of the cladodes, unisexual, cream. Fruit an orange-red berry.

1. S. androgyna (L.) Kunth, *Enum. pl.* **5**: 277 (1850). [*Alegra campo*]
Ruscus androgynus L.; *Semele maderensis* G.C. da Costa; *S. menezesi* G.C. da Costa; *S. pterygophora* G.C. da Costa; *S. tristonis* G.C. da Costa
Climber to 7 m. Leaves 3-8 mm. Cladodes 2.5-14 × 1-7.8 cm, subsessile, broadly ovate to ovate-lanceolate, rarely compound with several small, axillary cladodes arising from the margins. Inflorescences arising at margins of cladodes or rarely at the apex of a ± flattened stem. Perianth-segments *c.* 3 × 1.5 mm. Berry *c.* 7 mm in diameter, spherical. **Plate 51**
Fl. IV - VI. *Rare in the wild in Madeira, mainly in rocky, wooded ravines of the interior, occasionally in damp places on the north coast; widely cultivated in gardens.* ■ M [*Canaries*]

In addition to the 4 species listed here in synonymy, da Costa (*Broteria* **19**: 57-78 (1950), described a large number of subspecific taxa from Madeira. These all appear to be slightly abnormal individuals, and are not considered to be of taxonomic significance.

11. Ruscus L.

Small, rhizomatous, erect shrubs. Leaves inconspicuous, functionally replaced by persistent, leaf-like cladodes. Flowers in condensed racemes arising from the centre of the cladodes, unisexual, dull green. Fruit a scarlet berry.

1. R. streptophyllus P. F. Yeo in *Feddes Repert.* **73**: 117 (1966).
R. hypophyllum auct. mad., var. *lancifolius* Lowe, var. *latifolius* Lowe
Stems unbranched, to 60 cm. Leaves 7-14 × 3.5-8 mm. Cladodes 5-18 × 1.7-8 cm, broadly elliptic to lanceolate, subsessile, twisted at base. Perianth-segments 3.5 × 1.7-2.5 mm. Berry 10-15 mm in diameter, spherical. **Plate 51**
Fl. X - I. *Rare endemic of laurisilva, shady banks and rock ledges in central parts of Madeira.*
● M

12. Smilax L.

Climbing or scrambling, sometimes prickly, herbs or shrubs. Leaves alternate, simple; petiole with 2 tendrils near the base. Inflorescence an umbel or fascicle. Flowers dioecious. Fruit a 1- to 3-seeded berry.

1. Flowers in fascicles; leaves cordate, hastate or sagittate at base . . . . **1. aspera**
- Flowers in solitary, long-stalked umbels; leaves rounded or subcordate at base . . . . . .
 . . . . **2. canariensis**

1. **S. aspera** L., *Sp. pl.* **2**: 1028 (1753).
S. aspera var. *altissima* Morris & de Not.; *S. mauritanica* Poir.; *S. pendulina* Lowe
Climbing, creeping or scrambling shrub. Stems usually unarmed in Madeiran plants. Leaves usually heart-shaped, coriaceous. Flowers 5-30, in fascicles or subsessile umbels on axillary or terminal branches. Perianth-segments 2-4 mm, greenish white. Berries *c.* 6 mm in diameter, black or red.
Fl. VIII - X. *Frequent on sea cliffs, in thickets and rocky places in Madeira.* M

The species, as presented here, includes plants with leaves ranging from densely prickly (the typical form) to unarmed.

2. **S. canariensis** Brouss. ex Willd., *Sp. pl.* **2**: 784 (1806).
S. pseudochina Buch
Scrambling or climbing shrub. Stems unarmed. Leaves elliptical, lanceolate or ovate, coriaceous. Flowers 4-12, in axillary umbels; peduncles 7-20 cm. Perianth-segments 4-6 mm.
Plate 52
Fl. ? *Apparently a rare plant occurring in laurisilva and on rocky slopes along the north coast of Madeira.* ■ M [*Canaries*]

Madeiran records of this plant are based on a few collections of mostly sterile material and require confirmation. It is otherwise confined to the Canaries.

CXI. AGAVACEAE
N.J. Turland

Glabrous trees or very large suckering herbs. Leaves borne in rosettes, strap-shaped, pointed, coriaceous, sometimes succulent. Inflorescence a large, terminal panicle. Flowers hermaphrodite, actinomorphic; perianth-segments 6, united at least at the base; stamens 6. Ovary superior or inferior. Fruit a fleshy capsule.

1. Dichotomously branched tree with terminal leaf-rosettes; leaves with smooth margins, not succulent . . . . **1. Dracaena**
- Very large stoloniferous herb forming groups of acaulescent leaf-rosettes; leaves with spinose margins, succulent . . . . **2. Agave**

1. Dracaena Vand. ex L.

Umbrella-shaped tree. Trunk, branches and young stems all cylindrical, thick, very stout, branching dichotomously from tips when terminal inflorescence is produced. Leaves many, in terminal rosettes, strap-shaped, flared and clasping at base, pointed but not spinose at apex, coriaceous but not succulent; margin smooth. Inflorescence with many spreading, spike-like branches, each bearing flowers clustered into groups. Perianth-segments free almost to base. Ovary superior. Fruit ± globose, 3-lobed, containing 1-3 seeds.

1. **D. draco** (L.) L., *Syst. nat.* ed.12, **2**: 246 (1767). [*Dragoeiro*]
Plant up to 15 m or more; trunk to 5 m or more in circumference; bark smooth, scaly, greyish-brown. Leaves 30-60 × 1-4 cm, glaucous, brownish at base. Inflorescence to *c.* 75 cm; flowers fragrant. Perianth-segments 7-9 mm, narrowly lanceolate, obtuse, creamy white. Stamens not exceeding perianth; filaments tapered towards apex; anthers small, *c.* 1.6 mm, versatile. Style slender, *c.* 3 mm; stigma minute, *c.* 0.3 mm, with 3 rounded lobes. Fruit *c.* 1 cm wide, ripening orange. **Plate 51**
Fl. VIII - X. *Extinct in Porto Santo and extremely rare in Madeira as a wild plant; known only from the Ribeira Brava area, where two individuals survive on a cliff. A solitary old tree at*

Ponta do Garajau, east of Funchal, fell into the sea during a storm in autumn 1982. The species was a common component of the original vegetation of the lower, more arid zones of both Madeira and Porto Santo (up to 200 m). It is still planted for ornament on both islands, mainly in parks and gardens. ■ MP [*Azores, Canaries, Cape Verdes*]
Records from the Azores probably refer to planted, rather than native, individuals.

2. Agave L.

Very large suckering herb forming groups of acaulescent, monocarpic leaf-rosettes. Leaves strap-shaped, narrowed above broad sheathing base and towards stoutly spinose apex, succulent, coriaceous; margin remotely spinose-dentate. Inflorescence erect, very stout, pyramidally branched in upper part; branches horizontal, with erect flowers grouped into dense clusters at the tips. Perianth-segments united for most of their length. Ovary inferior. Fruit oblong, trigonous, containing many seeds.

1. A. americana L., *Sp. pl.* 1: 323 (1753). [*Piteira*]
Leaves to 200 × 30 cm, glaucous. Inflorescence 4-10 m. Flowers 7-9 cm (including ovary); perianth-segments linear-oblong, obtuse, greenish yellow; stamens and style exserted; filaments 7-8 cm, tapered towards apex; anthers large 3-3.5 cm, versatile; stigma small, 3-lobed.
Fl. III - V. *Introduced; naturalized on rocks on the southern coast of Madeira; also cultivated. Native to Mexico.* M

A. attenuata Salm-Dyck, *Hort. dyck.*: 303 (1834) has 1-several, ascending stems to 1.5 m, with leaves to 70 × 16 cm, not prickly at margins, and a cylindric inflorescence to 3.5 m, narrowed towards apex, usually recurved, densely crowded with yellowish-green flowers. Native to Mexico, it is commonly cultivated in Madeira and appears to be becoming naturalized in a few places in the south-east, from Funchal to the airport at Santa Cruz. M

CXII. AMARYLLIDACEAE
J.R. Press

Bulbous, usually scapose, perennials. Leaves all basal, strap-shaped. Flowers in an umbel (rarely solitary), enclosed in bud in a 1- or 2-valved, scarious spathe. Flowers hermaphrodite, actinomorphic or slightly zygomorphic, 3-merous. Perianth-segments 6, petaloid; corona sometimes present. Stamens 6. Ovary inferior, 3-locular. Fruit a capsule, rarely a berry.

A number of species commonly cultivated in gardens occasionally escape and may become locally established, usually along roadsides and levadas or in open fields. The following have been recorded in Madeira:

Alstroemeria pulchella L.f., *Suppl. pl.*: 206 (1782). Leaves lanceolate, up to 7.5 cm. Flowers *c.* 4.5 cm long, funnel-shaped, the upper perianth-segments longer than the lower, dark red, spotted with green and brown. M

Crinum bulbispermum (Burm.) Milne-Redh. & Schweick. in *Bot. J. Linn. Soc.* **52**: 161 (1939). Robust plant with lanceolate leaves. Umbels with up to 10 large, white or pinkish flowers on a solid peduncle. Perianth with a long, curved, slender tube and spreading, lanceolate segments. M

Nerine sarniensis (L.) Herb. *Curtis's bot. Mag.*: sub. t. 2124 (1820). Leaves up to 30 cm, appearing after the flowers. Flowers deep rose to red (in Madeira), perianth-segments recurved, with crisped margins. M

1. Flowers with a trumpet- or rim-like corona . . . . **2. Narcissus**
- Flowers lacking a corona . . . . **1. Amaryllis**

1. Amaryllis L.

Leaves distichous, becoming spiral. Umbel with 3-8(-12) slightly nodding flowers. Spathe 2-valved. Perianth campanulate, slightly zygomorphic, the tube short, the lobes slightly spreading; corona absent. Stamens inserted on throat of perianth-tube, declinate. Fruit subglobose, slightly fleshy.

1. A. belladonna L., *Sp. pl.* **1**: 293 (1753). [*Bella donas, Beladona*]
Brunsvigia rosea (Lam.) Hannibal
Bulbs large, up to 9 cm diameter, with a tough tunic, adhering in clumps, the tips often visible just above soil level. Leaves up to 35 cm, appearing after the flowers. Scape 30-60 cm, naked, purplish. Flowers fragrant. Perianth *c*. 9 cm, lobes pinkish, paler or white towards the base. Fl. IX - XI. *Cultivated and widely naturalized in open areas and through woodland, throughout Madeira, mostly above 300 m.* **M**

2. Narcissus L.

Flowers solitary or in umbels of 2-15. Spathe 1-valved, scarious. Perianth with a short tube, the segments all similar and usually spreading, yellow or white; corona large and trumpet-shaped or reduced to a small ring. Capsule ± globose to ellipsoid.

1. Narcissus jonquilla L., *Sp. pl.* **1**: 290 (1753). Up to 30 cm. Leaves sub-cylindrical, usually dark green. Flowers in umbels of 2-5, fragrant, yellow, the corona *c*. 2.5 mm long. Double forms are common.
Fl. X - II. *Widely cultivated and naturalized in fields and grassy places above c. 300m in Madeira. Native to the Iberian Peninsula.* **M**

Three other species are also cultivated and may be naturalized in places.

N. × medioluteus Mill., *Gard. dict.* ed. 8, no. 4 (1768). Leaves flat, glaucous. Flowers in umbels of 2, fragrant, perianth white, corona 3-5 mm, bright yellow. Native to S. France. **M**

N. × odorus L., *Cent. pl.* **II**: 14 (1756). Up to 35 cm. Leaves keeled. Spathe 5-6 cm. Flowers in umbels of 1-2(-3), perianth yellow, segments 16-20 mm, corona *c*. 10 mm, yellow. Of garden origin. **M**

N. pseudonarcissus L., *Sp. pl.* **1**: 289 (1753). Leaves usually glaucous. Flowers usually solitary, large, perianth white or yellow, the corona 15-45 mm, also white or yellow and concolorous with, or darker than, the perianth. Native to W. Europe. **M**

CXIII. DIOSCOREACEAE
J.R. Press

Dioecious perennial herbs. Stems arising from underground tubers and often twining. Leaves alternate, rarely opposite, usually entire. Flowers in spikes, racemes or panicles. Perianth-segments 6, fused at the base into a short tube. Stamens 6. Ovary inferior, 3-locular; styles 3. Fruit usually a capsule, rarely a berry.

1. Tamus L.

Tuber large. Stems twining, annual. Leaves alternate. Inflorescence racemose. Perianth campanulate. Fruit a berry, incompletely 3-locular. Seeds globose.

1. T. edulis Lowe in *Trans. Camb. phil. Soc.* **4**: 12 (1833). [*Norça*]
Glabrous climber. Leaves up to 25 × 20 cm, very broadly ovate to broadly triangular, deeply cordate at base, acuminate, entire, with 9(-11) prominent, curved veins. Flowers numerous, in branched, axillary panicles. Perianth-segments a lurid purplish colour, 1-1.5(-2) mm, narrowly oblong. Ovary 4-5 mm; styles stout, strongly recurved. Fruit 12-15 × 10 mm, ovoid, reddish. **Plate 52**
Fl. XII - II. *A very rare plant recorded in the wild only from sea-cliffs in eastern Madeira at Porto da Cruz and Garajau, but formerly cultivated in the Ribeira da Janela to Achadas da Cruz region for the edible tubers.* ■ M [*Canaries*]

CXIV. IRIDACEAE
J.R. Press

Perennial, usually scapose herbs with rhizomes, corms or bulbs. Leaves long, narrow, not differentiated into lamina and petiole, often distichous, folded longitudinally, overlapping and sheathing at the base. Inflorescence terminal. Flowers hermaphrodite, 3-merous, actinomorphic or zygomorphic; spathe of 1 or 2 bracts usually present. Perianth-segments all petaloid, often persistent after withering, connate below to form a straight or curved tube. Stamens 3. Ovary inferior, usually 3-locular; style 3-branched, the branches entire or divided, sometimes petaloid. Fruit a loculicidal capsule.

Literature: C. Innes. (1985). *The World of Iridaceae*. Sussex.

Various species and their hybrids are cultivated in parks and gardens and may be encountered as escapes. The following have been recorded from Madeira:

Dietes iridioides (L.) Sweet, *Hort. brit.* ed. 2,: 497 (1830) (*D. vegeta* (L.) N.E. Br.). Very similar to *Iris* but leaves in a flat, evergreen fan; flowers white to pale blue, all 6 perianth-segments spreading. Native to S. Africa. M

Ferraria crispa Burm., *Nova Acta Leopolina* **2**: 199 (1761) (*F. undulata* L.). Old corms persistent. Leaves numerous, fleshy and glaucous, completely sheathing scape. Flowes large, perianth-lobes spreading, dark brown to maroon or blackish, often with yellow markings, foetid. Native to S. Africa. M

Ixia maculata L., *Sp. pl.* ed. 2: 1664 (1763). Resembles *Sparaxis* but more slender, with reddish-brown, concolorous bracts *c.* 8 mm long, the outer 3-toothed at the apex, the inner with 2 teeth. Flowers orange to yellow with reddish markings on the outside and brownish-purple centres. Native to S. Africa. M

Tigridia pavonia (L.f.) DC. in Redouté, *Liliac.* **1**: t. 6 (1802). Bulbs ovoid. Leaves pleated, keeled. Flowers range in colour from white and yellow to orange, pink and red, often with coloured blotches in the centre. Outer perianth-lobes *c.* 10 cm, the lower portions forming a bowl, the upper portions spreading and reflexed, the inner lobes much smaller. Stamens and style forming an erect tube in the centre of the flower. Native to C. America. M

1.	Style branches broad and petaloid	**1. Iris**
-	Style branches slender, not petaloid	2
2.	Style branches deeply divided	3
-	Style branches entire or shallowly bifid at the apex	6

3. Inflorescence with 1-3 flowers, not spicate; scape subterranean at flowering, up to 10 cm above ground in fruit . . . . **9. Romulea**
- Inflorescence with more than 3 flowers, spicate; scape more than 10 cm above ground 4

4. Inflorescence not secund . . . . **5. Watsonia**
- Inflorescence secund 5

5. Leaves distichous; perianth-tube slender and cylindrical for ± its full length . . . . **8. Anomatheca**
- Leaves not distichous; perianth-tube funnel- shaped, gradually expanding towards the mouth . . . . **2. Freesia**

6. Flowers actinomorphic and erect . . . . **3. Sparaxis**
- Flowers zygomorphic, curved or held horizontally 7

7. Perianth-tube much shorter than lobes; flowers usually purplish . . . . . **4. Gladiolus**
- Perianth-tube slightly shorter to longer than lobes; flowers reddish to orange 8

8. Plant with stolons; perianth-tube expanding gradually towards the mouth; lobes of ± equal length . . . . **7. Crocosmia**
- Plant without stolons; perianth-tube narrow at base, expanding abruptly to a broader cylinder above, upper lobe much longer than the others . . . . **6. Chasmanthe**

1. Iris L.

Plants with bulbs or rhizomes, usually caulescent. Leaves often distichous, folded longitudinally and overlapping towards the sheathing bases. Flowers actinomorphic, large, showy. Perianth-tube short; outer perianth-segments spreading or deflexed, inner usually smaller and erect, with a clearly defined limb and claw. Style branches broad, bifid and petaloid, each lying closely over a stamen and the claw of 1 outer segment; stigma an apical flap on the underside of the style branch. Ovary 3-locular. Capsule trigonous.

Various species and hybrids with a range of flower colours are grown in gardens in the islands.

I. japonica Thunb. in *Trans. Linn. Soc. Lond.* **2**: 327 (1794), with slender stolons, white to lavender flowers, the perianth-segments with frilly margins and an orange crest, style lobes fimbriate, has been recorded from Porto da Cruz but is doubtfully naturalized. Native to C. China and Japan. **M**

I. xiphium L., *Sp. pl.* **1**: 40 (1753) is widely cultivated and may perhaps be naturalized in Madeira. It is a bulbous perennial with fibrous roots and narrow, glaucous leaves less than 8 mm wide present throughout winter. The flowers show a range of colours but are typically purple and yellow. Seeds yellowish brown. Native to W. and C. Mediterranean. **M**

1. Scape angled on one side; flowers livid violet-yellow; seeds bright orange . . . . **1. foetidissima**
- Scape not angled; flowers yellow; seeds brown . . . . **2. pseudacorus**

1. I. foetidissima L., *Sp. pl.* **1**: 39 (1753).
Plant dark green, foetid. Stem simple, slightly compressed and angled on one side. Leaves evergreen, the basal about as long as the scape. Flowers 1-4, *c.* 7 cm in diameter, livid violet

and yellow; pedicels up to 8 cm. Capsule c. 4.5 cm, oblong-ellipsoid. Seeds bright orange-red, persisting in the capsule after dehiscence.
Fl. V. *Naturalized in woods, probably in several sites between Santo Antonio and Eira do Serrado. Native to W. Europe and N. Africa.* **M**

2. I. pseudacorus L., *Sp. pl.* **1**: 38 (1753).
Plant somewhat glaucous. Stem slightly compressed but not angled on one side. Leaves deciduous, with a conspicuous midrib. Flowers 4-12, 8-10 cm in diameter, yellow; pedicels up to 5 cm. Capsule ellipsoid, shortly beaked. Seeds brown.
Fl. IV - VI. *Found in wet places and apparently rare but the precise distribution in Madeira is unknown.* **M**

2. Freesia Eckl. ex Klatt

Corms small, globose-conical, with finely fibrous tunics. Scape leafy, erect, simple or branched. Inflorescence a secund spike held horizontally. Flowers fragrant, zygomorphic. Perianth-tube ± straight, narrowly funnel-shaped; lobes short, subequal or the lower ones smaller. Style branches filiform, deeply divided with recurved lobes.

1. F. refracta (Jacq.) Eckl. ex Klatt in *Linnaea* **34**: 673 (1866).
Up to 45 cm, the narrow leaves equalling or exceeding the zig-zag scape. Bulbils present in axils of lower leaves. Bracts emarginate. Flowers fragrant, 4-5 cm, creamy with yellow markings in the throat, 2-lipped. Perianth-lobes 1.2-1.6 cm.
Fl. III - V. *Commonly cultivated in Madeira and naturalized on grassy banks and roadsides around Funchal and other towns.* **M**

3. Sparaxis Ker Gawl.

Corms small, globose with fine, pale tunics. Leaves mostly basal, distichous. Scape simple or branched. Inflorescence a spike with few to several flowers; bracts scarious with numerous brown streaks. Flowers actinomorphic. Perianth-tube short, expanded above; lobes ± equal, spreading. Stamens erect or sometimes spreading and asymmetrically arranged. Style slender, erect, branches entire. Capsule globose, membranous. Seeds smooth.

1. S. tricolor (Schneev.) Ker Gawl. in *Ann. Bot. Lond.* **1**: 225 (1804).
Corm 1-2 cm in diameter. Leaves up to 10, arranged in a fan. Scape erect, up to 20 (-40) cm. Bracts entire or slightly lacerate. Flowers variously coloured but perianth usually yellow towards the base with reddish or purple markings on the outside. Perianth-lobes 2.5 cm, oblong-lanceolate.
Fl. II - VI. *Cultivated in Madeira, escaping and becoming naturalized on old terraces, waste ground and paths.* **M**

Several other species, including **S. grandiflora** (D. Delaroche) Ker Gawl., are cultivated as garden ornamentals and may occur close to houses.

4. Gladiolus L.

Corms small to medium, globose to ovoid, tunics usually coarsely fibrous. Leaves distichous, the cauline resembling bracts. Scape simple or branched. Inflorescence a secund or distichous spike; bracts usually large, green, the inner bifid. Flowers usually weakly zygomorphic. Perianth-tube usually short, curved, funnel-shaped, the lobes subequal and spreading or unequal and bi-labiate, the upper lobe largest and hooded. Style filiform, the short branches expanding into shortly bilobed apices with fimbriate margins. Capsule ± globose to cylindrical. Seeds winged or not.

1. G. italicus Mill., *Gard. dict.* ed. 8, no. 2 (1768).
G. segetum Ker Gawl.
Corm *c*. 2 cm in diameter. Basal leaves up to 60 × 1.5 cm. Scape up to 80 cm. Inflorescence lax, somewhat distichous, flowers 4-12, usually reddish purple, sometimes paler, 4-5(-6) cm; tube very short; lobes narrowed towards the base into a claw, the upper lobe slightly longer and *c*. twice as broad as the laterals. Anthers longer than the filaments. Capsule 10-12 mm. Seeds top-shaped, not winged.
Fl. III - VI. *Cultivated in gardens and widely naturalized, usually in cultivated ground, in Madeira and Porto Santo. Native to the Mediterranean region.* **MP**

Garden hybrids of the type sometimes known as **G. × hortulanum** Bailey and its derivatives are widely grown in gardens and parks and sometimes escape. They are robust herbs over 1 m high with large flowers in various colours but often orange-red. The perianth-lobes overlap, the upper hooded, the remainder reflexed or revolute at the tips.

5. Watsonia Mill.

Corms large. Leaves distichous, the upper often entirely sheathing and resembling bracts, erect to curved, the margins and midrib often thickened and hyaline. Scape simple or branched. Inflorescence with few-many flowers; bracts usually reddish or brown at least towards the apex, the inner sometimes emarginate or forked. Flowers zygomorphic. Perianth-tube curved and in 2 parts, the lower slender, erect, cylindrical, becoming horizontal and abruptly expanded above, the upper cylindrical or funnel-shaped; inner lobes slightly larger than the outer. Style slender, branches divided to halfway, recurved. Capsule oblong, leathery. Seeds winged.

Literature: P. Goldblatt, The genus *Watsonia*, in *Ann. Kirstenbosch bot. Gdns* **19**: 1-148 (1989).

1. W. meriana (L.) Mill., *Gard. dict.* ed 8, no. 1 (1768).
Corm up to 4.5 cm in diameter with a tough, grey-brown tunic. Leaves 4-6, the lower 3-4 basal. Scape 50-200 cm, simple or with few, short branches. Flowers red. Perianth-tube *c*. 5 cm long, strongly curved, the upper part cylindrical; lobes *c*. 2.4 cm, ovate and spreading at 90° to the tube, the upper segment slightly hooded. Anthers violet. Capsules narrowly oblong.
Fl. IV - VI. *Cultivated in gardens in Madeira, naturalized in waste and fallow ground near the sea, especially around Santa Cruz. Native to S. Africa.* **M**

W. borbonica (Pourr.) Goldblatt subsp. **ardernei** (Sanders) Goldblatt in *Ann. Kirstenbosch. bot. Gdns* **19**: 35 (1989) (*W. ardernei* Sanders), also from S. Africa, is more widely cultivated in gardens and along roadsides above 200 m, especially around Santana in northern Madeira, and may occasionally escape. It has flowers with the upper perianth-tube flared or funnel-shaped, horizontal to slightly pendulous, the lobes 25-30 mm, obovate, spreading. Both white- and pink-flowered forms are grown. **M**

6. Chasmanthe N.E. Br.

Corms depressed-globose, with papery to slightly fibrous tunics. Leaves distichous. Scape simple or branched. Inflorescence a many-flowered spike; bracts sometimes membranous, the inner bifid at the apex. Flowers zygomorphic. Perianth with a long tube very slender at the base, abruptly expanded and tubular above; lobes unequal, the upper largest and slightly hooded, the others reflexed. Style slender, the branches recurved. Capsule depressed-globose. Seeds few, large, often brightly coloured.

1. C. aethiopica N.E. Br. in *Trans. R. Soc. S. Afr.* **20**: 272 (1932).
Antholyza aethiopica L.
Leaves up to 100 × 4.5 cm, firm textured. Scape up to 160 cm. Bracts membranous. Perianth

bright scarlet on the upper side, yellowish-green on the lower; tube curved, 300 mm long, the lower ¼ 1 mm in diameter, the upper ¾ 6 mm in diameter; upper lobe 20-30 mm, 2-3 × as long as the other lobes.

Fl. IX - IV. *Widely cultivated and naturalized along roads and levadas, mainly in lowland regions of Madeira. Native to S. Africa.* **M**

7. **Crocosmia** Planch.

Corms small, globose, with fibrous tunics. Leaves distichous, erect. Inflorescence a lax spike with few-many flowers; bracts herbaceous. Flowers zygomorphic. Perianth-tube widening gradually from the base, lobes subequal or the upper lobe hooded and larger than the lower 3 which form a lip. Style slender, the branches short and sometimes bifid at the tip. Capsule depressed-globose, membranous.

1. C. × crocosmiflora (G. Nicholson) N.E. Br. in *Trans. R. Soc. S. Afr.* **20**: 264 (1932).
Tritonia × crocosmiflora (Lemoine) G. Nicholson
Plant with stolons; corms *c*. 2 cm in diameter. Leaves 4-8, up to 60 × 2.5 cm, strongly ribbed. Scape up to 120 cm, simple or with a few short branches; bracts 6-8 mm, brown, the inner bifid at the apex. Perianth 2 cm, tube slightly curved above. Anthers yellow.

Fl. VI - VIII. *Cultivated and frequently naturalized on the edges of woods and in shady places in various parts of Madeira. Of garden origin.* **M**

8. **Anomatheca** Ker Gawl.

Corms small, conical, with fibrous tunics. Basal leaves distichous, longitudinally folded, forming an erect fan. Cauline leaves small. Scape simple or branched. Inflorescence a secund spike of up to 10 flowers; bracts herbaceous or membranous, the inner bifid. Flowers zygomorphic with a long perianth-tube and subequal lobes. Style branches deeply divided and recurved. Capsule smooth or rough.

Literature: P. Goldblatt, A revision of the genera *Lapeirousia* Pourret and *Anomatheca* Ker Gawl. in the winter rainfall region of South Africa, in *Contr. Bolus Herb.* **4**: 1-111 (1972).

1. A. laxa (Thunb.) Goldblatt in *Jl S. Afr. Bot.* **37**: 442 (1971).
A. cruenta Lindl.; *Lapeirousia laxa* (Thunb.) N.E. Br.; *L. cruenta* (Lindl.) Bak.
Corms up to 1.5 cm in diameter, rounded at the base. Basal leaves up to 25 cm. Scape up to 45 cm, with up to 6 flowers. Flowers deep or salmon pink. Perianth-tube 2-3 cm, erect, slender and cylindrical; lobes *c*. 1 cm, ovate to oblong, at 90° to the tube, the lower 3 each with a dark red patch at the base. Seeds *c*. 1.5 mm, dark, glossy red.

Fl. III - VI. *Cultivated and naturalized in cultivated and waste ground, on roadsides and cliffs in many parts of Madeira but usually near the coast Native to S. Africa.* **M**

Unlike native plants described from S. Africa, those from Madeira have leaves shorter than the scapes.

9. **Romulea** Maratti

Corms small, asymmetrical with a crescent-shaped ridge at the base of one side, tunics hard, smooth and entire. Leaves linear to terete, grooved and sometimes coiled. Scapes usually subterranean, at least at flowering. Flowers usually solitary, with floral bracts at base of ovary; pedicels recurved after flowering. Perianth actinomorphic, the tube short, the lobes cupped or spreading. Styles filiform, with deeply divided branches. Capsule globose.

Flowers usually open late morning and close in late afternoon.

1. R. columnae Sebast. & Mauri, *Fl. rom. an. prodr.*: 18 (1818).
Corm 5-10 mm in diameter, asymmetrical with a basal ridge on one side. Basal leaves 2, cauline leaves erect or coiled. Scape subterranean at flowering, up to 10 cm above ground in fruit. Flowers 1-3, lilac to pale mauve or violet, the veins purple, the throat yellow. Bracts herbaceous, purplish or mottled red-brown; bracteole entirely membranous, densely mottled. Stigmas held below the top of the anthers. Capsule subglobose, 3-lobed.
Fl. III - VI. *In short turf in mountain meadows and grassy places, especially in well-trampled ground such as footpaths.* **MP**

There is some doubt about the precise identities and relationships of subspecific taxa within *R. columnae*. Typical Madeiran plants are very small with pale flowers and correlate well with descriptions of subsp. *columnae*. Larger, more robust plants with dark flowers are here referred to subsp. *grandiscapa*.

a) subsp. columnae
Basal leaves 12 cm × 1 mm. Perianth lilac to pale mauve, the tube 3.5-4 mm, the lobes 6-7 mm. Fruiting scape 5(-8) cm above ground. Capsule 4-6 mm.
Present in Madeira above 500 m at Paúl da Serra, Balcões and between Camacha and Portela.

b) subsp. grandiscapa (Webb) G. Kunkel in *Monographiae biol. canar.* **3**: 25 (1972).
R. columnae var. *grandiscapa* (Webb) Pit.; *R. grandiscapa* (Webb) Gay
Generally larger in all its parts than **a**. Basal leaves up to 30(-45) cm × 2 mm. Perianth purple, the tube 4-5(-9) mm, the lobes 12-18 mm. Fruiting scape up to 10 cm above ground. Capsule up to 12 mm.
Much less common than a, only recorded from Pico da Silva (east of Funchal) in Madeira and from Pico de Ana Ferreira in Porto Santo. ■ [*Canaries*]

CXV. JUNCACEAE
M.J. Cannon

Annual or perennial herbs, often tufted or rhizomatous. Stems cylindrical or flattened, sometimes nodular. Leaves cylindrical or flattened and grass-like, with a sheathing base, sometimes reduced to scales. Inflorescence usually cymose or paniculate; flowers in clusters or spikes, or solitary, hermaphrodite, 3-merous, the perianth in 2 whorls, sometimes cleistogamous. Fruit a capsule; seeds with a membranous tegument often prolonged into an appendage.

1. Leaves flat, grass-like, ciliate at least when young, the sheath entire; capsule 3-seeded . . . . **1. Luzula**
- Leaves usually cylindrical, rarely flat, glabrous, the sheath usually split; capsule many-seeded . . . . **2. Juncus**

1. Luzula DC.

Perennial or rarely annual herbs. Leaves grass-like, the margin often ciliate, at least when young, the base sheathing, without auricles. Inflorescence spreading or condensed into heads, cymose, or apparently paniculate. Flowers in clusters or solitary. Stamens 6. Ovary 1-celled. Fruit a 3-seeded capsule; seeds often with conspicuous appendages.

Stamens are best observed in fresh material as the ratio of anther length to filament length is difficult to determine in dried specimens.

1. Flowers solitary; inflorescence widely spreading, paniculate . . . . **1. elegans**
- Flowers in clusters of 2 or more; inflorescence capitate, ± cymose or corymbose . 2

2. Leaves up to 10 mm broad, densely ciliate; seeds without a basal appendage . . . .
   . . . . **2. seubertii**
- Leaves up to 6 mm broad, sparsely ciliate; seeds with a basal appendage half as long as the seed 3
3. Plants loosely caespitose, stolons numerous; anthers 2-6 × as long as the filaments . .
   . . . . **3. campestris**
- Plants densely caespitose, stolons absent or few; anthers equalling or slightly longer than the filaments . . . . **4. multiflora**

1. **L. elegans** Lowe in *Trans. Camb. phil. Soc.* **6**: 532 (1838).
L. purpurea (Masson ex Buchenau) Link; *Ebingeria elegans* (Lowe) Chrtek & Křísa
Annual, rarely biennial. Stems up to 40(-50) cm, erect, rather wiry. Basal leaves up to 12 cm × 3(-4) mm, sparsely ciliate. Inflorescence widely spreading, ± paniculate; bract shorter than the inflorescence; bracteoles glabrous; flowers solitary, some sterile. Perianth-segments 2.5-3 mm, purplish, strongly aristate, the arista of outer segments 0.5 mm, those of the inner slightly less. Anthers about ½ as long as the filaments. Styles shorter than the ovary. Capsule globose, shorter than the perianth-segments; seeds 0.7-0.9 mm, reddish brown, without appendages. **Plate 53**
Fl. III - VIII. *Along levadas, on rocks and walls. A very scattered and localized endemic in Madeira; more frequent in higher parts of the central eastern region of the island but nowhere common.*
● M

This species has been placed in the monotypic genus *Ebingeria* by Chrtek & Křísa. It differs from *Luzula* sensu stricto only in minor anatomical characters and the predominantly annual habit and is here included in *Luzula*.

2. **L. seubertii** Lowe *Hooker's J. Bot.* **8**: 300 (1856).
Caespitose, stoloniferous perennial up to 90 cm, the tufts bearing the persistent veins of old leaf-bases. Basal leaves up to 35 cm × 10 mm; all leaves ciliate. Inflorescence sub-umbellate; bracts much shorter than the inflorescence; bracteoles long-ciliate. Perianth-segments 5-6 mm, papery, narrowly aristate-acuminate, pale brownish, all ± equal in length; anthers about ½ as long as the filaments. Styles ⅔ as long as the ovary. Capsule ovoid-globose, about ½ as long as the perianth-segments. Seeds 1.2-1.5 mm, brown, without appendages. **Plate 53**
Fl. V - VIII. *A rare endemic, scattered throughout the central mountains of Madeira from Serra do Seixal to Poiso, generally above 1000 m.* ● M

3. **L. campestris** (L.) DC. in Lam. & DC., *Fl. franç.* ed. 3, **3**: 161 (1805).
Loosely caespitose perennial 10-30(-40) cm, with numerous short stolons. Basal leaves up to 10 cm × 2.4 mm, sparsely ciliate. Inflorescence paniculate, of 1 sessile and 3-6 pedunculate globose heads each with 3-12 flowers; bracts sparsely ciliate, often longer than the inflorescence at anthesis; bracteoles almost glabrous. Perianth-segments 3-4 mm, all similar, narrowly lanceolate, brownish. Anthers 2-6 times as long as the filaments. Styles longer than the ovary. Capsule obovoid, shorter than the perianth-segments. Seeds 1.1-1.3 mm, brown, with a basal appendage up to ½ as long as the seed.
Fl. III - VII. *Common in damp, grassy areas above* c. *500 m in Madeira; doubtfully recorded from Porto Santo.* **M ?P**

4. **L. multiflora** (Retz.) Lej., *Fl. Spa* **1**: 169 (1811).
L. campestris subsp. *multiflora* (Retz.) Buchenau
Densely caespitose perennial up to 30(-50) cm, usually without stolons. Basal leaves up to 15 cm × 3-4(-6) mm, sparsely ciliate. Inflorescence sub-umbellate of up to 10 slender branches

with elongated heads of 8-16 flowers, or short-branched with rounded or lobed heads; bracts very sparsely ciliate, as long as or longer than the inflorescence at anthesis; bracteoles barely ciliate. Perianth-segments 2.5-3.5 mm, lanceolate, ± equal, brownish. Anthers equalling or slightly longer than the filaments. Styles as long as the ovary. Capsule globose, equalling or shorter than the perianth-segments. Seeds 1.1-1.3 mm, brown, with a basal appendage ½ as long as the seed.

Fl. IV - VII. *Scattered in damp, shady places in the higher regions of the eastern half of Madeira, from Monte and Queimadas eastwards to Portela.* **M**

Two subspecies occur in Madeira:

a) subsp. **multiflora**
Flower-clusters stalked; lower bract not longer than the inflorescence; perianth-segments equalling the capsule.
Throughout the range of the species.

a) subsp. **congesta** (Thuill.) Hyl. in *Uppsala Univ. Årsskr.* **1945**(7): 110 (1945).
L. congesta (Thuill.) Lej.
Flower-clusters subsessile; lower bract longer than the inflorescence; perianth-segments longer than the capsule.
Throughout the range of the species, but less common than subsp. a.

L. purpureo-splendens Seub. (*L. elegans* Guthnick, non Lowe), endemic to the Azores, has been recorded from Madeira but probably as a mis-identification.

2. Juncus L.

Annual or perennial herbs, often caespitose and rhizomatous. Leaves glabrous, flattened or cylindrical, sometimes reduced to scales; sheaths split, sometimes auriculate; pith, if present, in 1 tube (*unitubulose*) or 2 (*bi-tubulose*). Inflorescence cymose, sometimes condensed into terminal or apparently lateral heads, the subtending bract appearing to be a continuation of the stem; bracteoles 3 or more forming an involucre or absent. Flowers clustered or solitary; outer perianth-segments often longer than the inner; stamens 6, rarely 3; ovary 1- to 3-locular with numerous ovules. Fruit a many-seeded capsule; seeds with or without appendages.

1. Leaves cylindrical, resembling the stem 2
 - Leaves flat or cylindrical, not resembling the stem 5

2. Leaves sharply and stiffly pointed . . . . **1. acutus**
 - Leaves not sharply and stiffly pointed 3

3. Stems smooth, or very weakly striate . . . . **3. effusus**
 - Stems clearly ridged 4

4. Inflorescence compact; stems green, the pith continuous . . . . **4. conglomeratus**
 - Inflorescence spreading; stems bluish, the pith interrupted . . . . **2. inflexus**

5. Leaves ± cylindrical but not resembling the stem, transversely septate within . . . 6
 - Leaves grass-like, flat to sub-cylindrical, not septate within 7

6. Leaves bulbous at the base, bi-tubulose; proliferating clusters of leaves often replacing flowers . . . . **11. bulbosus**
 - Leaves not bulbous at base, unitubulose; without clusters of leaves replacing flowers .
   . . . . **12. articulatus**

7. Involucral bracteoles absent . . . . **10. capitatus**
- Involucral bracteoles present 8

8. Plants perennial; basal leaves with auricles . . . . **5. tenuis**
- Plants annual; basal leaves without auricles . . . . **6-9. bufonius group**

1. J. acutus L., *Sp. pl.* **1**: 325 (1753). [*Junco*]
Stout, densely caespitose perennial. Stems up to 150 cm, bearing 2-5 sharply and stiffly pointed leaves. Inflorescence compact or somewhat spreading; one bract sometimes longer than the inflorescence; involucral bracteoles absent. Perianth-segments 2.5-4 mm, subequal in length, reddish brown, with wide scarious margins; inner segments broader and blunter than the outer. Anthers several times as long as the filaments. Capsule *c.* 2 × as long as the perianth, obovoid to ovoid, the tip obtuse or acute; seeds 1.5 - 2.5 mm with subequal appendages.
Fl. II - V. *Seashores and coastal areas; common in Madeira; rare in Porto Santo, known only from the north-east of the island.* **MP**

Two subspecies occur in the Madeiran archipelago. They are known to hybridize, and plants exhibiting intermediate features do occur in the islands.

a) subsp. **acutus**
Capsule 5-6 mm, ovoid, conical at the apex, colour variable.

b) subsp. **leopoldii** (Parl.) Snogerup in *Bot. Notiser* **131**: 187 (1978).
J. acutus var. *multibraceatus* (Tineo) Cout.
Capsule 4-5 mm, usually obovoid, obtuse at the apex, dark or reddish brown.

2. J. inflexus L., *Sp. pl.* **1**: 326 (1753). [*Junco*]
J. glaucus Sibth.
Caespitose perennial up to 60 cm, greyish or bluish green. Stems dull, with 12-18 ridges most prominent just below the inflorescence; pith interrupted. Leaf-sheaths glossy, dark brown or blackish. Inflorescence usually lax and open, many-flowered; lowest bract longer than the inflorescence, appearing as a continuation of the stem; involucral bracteoles present. Perianth-segments 2.5-4 mm, unequal, lanceolate, acuminate to subulate. Anthers 1-1½ times as long as the filaments. Capsule about equalling the inner perianth-segments, ovoid, somewhat trigonous, dark brown, the tip acute to obtuse, mucronate; seeds *c.* 0.5 mm, obliquely ovoid, reticulate, without appendages.
Fl. VI - X. *Common in damp places throughout Madeira, up to 1800 m; also present in Porto Santo.* **MP**

3. J. effusus L., *Sp. pl.* **1**: 326 (1753). [*Junco*]
Caespitose perennial forming large tufts up to 150 cm, bright or yellowish green. Stems smooth and glossy with many faint striae; pith continuous. Leaf-sheaths reddish to dark brown, dull. Inflorescence usually lax, many-flowered; lowest bract longer than the inflorescence, appearing as a continuation of the stem; involucral bracteoles present. Perianth-segments 1.5-3 mm, subequal, lanceolate, finely pointed. Anthers as long as the filaments. Capsule shorter than the perianth, broadly ovoid, blunt at the tip, not mucronate, yellowish brown; seeds 0.5 mm, obliquely ovoid, reticulate, without appendages.
Fl. III - X. *Infrequent in damp places in the eastern half of Madeira.* **M**

4. J. conglomeratus L., *Sp. pl.* **1**: 326 (1753).
Caespitose perennial forming small tufts up to 100 cm, bright to greyish green. Stems dull, with up to 30 conspicuous ridges prominent just below the inflorescence; pith continuous. Leaf-sheaths broad, brownish, dull. Inflorescence usually a dense, compact head; lowest bract longer than

the inflorescence, appearing as a continuation of the stem; involucral bracteoles present. Perianth-segments *c.* 2.4 mm, subequal, lanceolate. Anthers usually shorter than the filaments. Capsule about as long as the perianth, ovoid, brownish, with the remains of the style elevated in a depression at the tip; seeds *c.* 0.3 mm, ± ovoid, without appendages.
Fl. X. *Very rare in Madeira, growing in grassland near Poiso; perhaps also elsewhere.* **M**

5. J. tenuis Willd., *Sp. pl.* **2**: 214 (1799). [*Junco*]
Weakly caespitose perennial up to 80 cm. Stems with a few basal leaves and leafless sheaths; pith absent. Leaves flat, curved, channelled, the bases expanded above into conspicuous, whitish auricles. Inflorescence lax; lowest bract leafy, much exceeding the inflorescence; involucral bracteoles present. Perianth-segments 3-4 mm, subequal, narrowly lanceolate or ovate, very sharply pointed, greenish to straw-coloured. Anthers ⅓-½ as long as the filaments. Capsule shorter than the perianth, broadly ovoid, obtuse, mucronate, greenish or pale-coloured; seeds 0.3-0.4 mm, obliquely ovoid with short appendages, becoming very mucilaginous.
Fl. VI - IX. *Along paths and in open ground ± throughout Madeira.* **M**

6-9. J. bufonius group
Tufted annuals up to 50 cm, often shorter. Stems erect, ascending or recurved, bearing several or very few flat leaves 1-12 cm × 0.5-5 mm. Inflorescence lax to densely capitate, fan-shaped; lower bracts leaf-like; involucral bracteoles present. Perianth-segments unequal, the outer longer than the inner, the margins scarious. Capsule ovoid to ellipsoid or trigonous; seeds ovoid to barrel-shaped, smooth or striate.

The *J. bufonius* group is a complex of closely related species, of which four are present in Madeira. Many of the specimens seen are probably of hybrid origin; several appear to be hybrids between species **6** and **7**.

1. Inflorescence lax, with the flowers widely spaced on the branches 2
 - Inflorescence densely or loosely capitate 3

2. Leaves 1.5-5 mm wide; seeds with prominent striae . . . . **6. foliosus**
 - Leaves not more than 1.5 mm wide; seeds smooth . . . . **8. bufonius**

3. Inner perianth-segments long-acuminate, longer than the caspule . . . . . **7. sorrentinii**
 - Inner perianth-segments obtuse to subacute, equalling or shorter than the capsule . . . .
 . . . . **9. hybridus**

6. J. foliosus Desf., *Fl. atlant.* **1**: 315 (1798).
Leaves broad, up to 5 mm wide. Inflorescence lax, with the flowers widely spaced on the branches. Outer perianth-segments 4-7 mm, with a dark stripe between the margin and central, green portion, the inner shorter and broader, all long-acuminate to apiculate. Capsule about as long as the inner perianth-segments, broad, obtusely angled; seeds 0.5-0.65 mm, ovoid, with 20-30 prominent striae.
Fl. ? *Recently introduced; a rare plant of ditches and marshy ground; in Madeira known from Achadas da Cruz, Achadas do Marques and Terreiro da Luta; also on Pico do Castelo in Porto Santo.* **MP**

7. J. sorrentinii Parl., *Fl. ital.* **2**: 356 (1857).
Leaves narrow, up to 1 mm wide, mostly borne towards the base. Inflorescence densely or loosely capitate, of 1-3 fan-shaped heads. Outer perianth-segments 6-8 mm, long-acuminate, not dark-lined. Capsule shorter than the inner perianth-segments, trigonous, ± sharply angled, abruptly contracted

into a conical apex; seeds 0.4-0.5 mm, broadly ellipsoid, smooth.
Fl. ? *Recently introduced; established near Santa Cruz and Caniçal; rare.* **M**

8. J. bufonius L., *Sp. pl.* **1**: 328 (1753).
Leaves narrow, up to 1.5 mm. Inflorescence lax, with the flowers widely spaced on the branches. Perianth-segments all sharply pointed, the outer 4-7 mm, longer and not much wider than the inner, not dark lined. Capsule shorter than the inner perianth-segments, ovoid to ellipsoid, mucronate; seeds 0.4-0.55 mm, ellipsoid, smooth.
Fl. ? *Frequent in the eastern half of Madeira; rare in the western half and in Porto Santo.* **MP**

9. J. hybridus Brot., *Fl. lusit.* **1**: 513 (1804).
Stems generally stout. Leaves 0.5-1 mm. Inflorescence capitate, of 1-10 dense heads. Outer perianth-segments 3.5-8.5 mm, acute, mostly scarious; inner segments ± obtuse, not dark-lined. Capsule 3.5-4.5 mm, ± equalling inner perianth-segments; seeds 0.3-0.4 mm, ellipsoid to ovoid, usually obtuse, smooth.
Fl. ? *Recently introduced; recorded from south-east Madeira from Machico to Caniçal.* **M**

10. J. capitatus Weigel, *Observ. bot.*: 28 (1772).
Small and caespitose or single-stemmed annual, up to 20 cm, often much less. Stems leafless; pith absent. Leaves narrow, flat, channelled; sheaths short, not auricled. Inflorescence of 1 or several clusters of (1-)2-8 flowers; bracts 2, overtopping the inflorescence; involucral bracteoles absent. Perianth-segments unequal, greenish, often becoming reddish, the outer 3-4 mm and ovate, with long recurved tips, the inner shorter and scarious. Anthers ½ as long as the filaments. Capsule shorter than the perianth, ovoid, blunt, scarcely mucronate; seeds *c*. 3 mm, ovoid, strongly reticulate.
Fl. V - IX. *A rare plant of open scrub, banks and walls in a few scattered localities in south-east Madeira, on Paúl da Serra and at Santa in the north-west.* **M**

Plants may be either wind-pollinated, with a long style, or self-pollinating, with a short style.

11. J. bulbosus L., *Sp. pl.* **1**: 327 (1753).
J. supinus Moench
Caespitose or much-branched, creeping perennial, sometimes aquatic. Creeping stems up to 100 cm; erect stems 1-30 cm, with a bulbous swelling at the base. Leaves mostly basal; pith bi-tubulose, imperfectly septate, the septa often indistinct. Inflorescence simple or irregularly branched with up to 20 heads, often with small shoots in the heads, or proliferating without flowers; involucral bracteoles absent. Perianth-segments 2-3 mm, the inner equalling or slightly longer than the outer, all with a scarious margin, light reddish brown. Anthers *c*. ⅓ as long as the filaments. Capsule 2.2-3.5 mm, equalling or slightly shorter than the perianth, trigonous, blunt or truncate at the tip; seeds 0.5 - 0.6 mm, top-shaped, reticulate.
Fl. V - VII. *A very rare plant in Madeira; its precise distribution is unknown.* **M**

12. J. articulatus L., *Sp. Pl.* **1**: 327 (1753).
J. lamprocarpus Ehrh. ex Hoffm.
Caespitose perennial up to 60(-100) cm, or with a creeping rhizome, or stems submerged and rooting and branching at the nodes. Stems with 0-2 basal sheaths and a few cauline leaves. Leaves curved; pith unitubulose, septate. Inflorescence branched, sometimes sparingly so, with few to many heads of (4-)5-8(-15) flowers; involucral bracteoles absent. Perianth-segments *c*. 3 mm, ± equal, lanceolate, acute to mucronate, dark brown or black, the inner with broad, scarious margins. Anthers equalling or longer than the filaments. Capsule longer than the perianth, long-ovoid, trigonous, abruptly acuminate; seeds 0.5-0.6 mm, ovoid, reticulate, without appendages.
Fl. V - IX. *Scattered and infrequent along streams and levadas in Madeira.* **M**

CXVI. COMMELINACEAE
M.J. Cannon

Annual or perennial herbs. Stems often fleshy and often swollen at the nodes. Leaves alternate, entire with a basal, tubular sheath. Flowers hermaphrodite or rarely unisexual (with male and female flowers on the same plant), borne in terminal or axillary cymes. Sepals 3. Petals 3, rarely 1 much reduced or absent; stamens 6, or 3 and 3 staminodes. Ovary (2-)3-celled, ovules 1-several per cell; stigma capitate or 3-lobed. Fruit a capsule, seeds with a thickening indicating the position of the embryo.

1.	Petals all similar	**3. Tradescantia**
-	Petals dissimilar	2
2.	Stamens 6, all fertile	**2. Tinantia**
-	Stamens 3; staminodes 2-3	**1. Commelina**

1. Commelina L.

Perennial or rarely annual herbs, often tuberous-rooted. Stems often creeping and rooting at the nodes. Inflorescence subtended by a bract (spathe), sometimes with one exserted cyme and one enclosed cyme. Sepals 3, one free, the others ± connate. Petals 3, free, one often smaller than the others, zygomorphic. Fertile stamens 3, often appearing together on one side of the flower, one with a very short filament; staminodes 2-3. Ovary 2- to 3-celled, with 1-2 ovules per cell. Seeds often truncate at the apex, smooth or rugose.

1.	Spathe closed, staminodes 2	**1. benghalensis**
-	Spathe open, staminodes 3	**2. diffusa**

1. C. benghalensis L., *Sp. pl.* 1: 60 (1753).
Sprawling or erect annuals, ± pubescent. Stolons often bearing subterranean, cleistogamous flowers. Leaves up to 3(-7) × 2(-4) cm, ovate or elliptic, pubescent; petiole sheathing at the base, the sheath membranous with long weak hairs at the mouth. Spathe funnel-shaped, fused at the anterior edges. Inflorescence of 2 cymes, one included within the spathe, one exserted. Petals blue, the external hyaline and boat-shaped, the inner clawed with limb broader than long. Median anther large, the filament short, the 2 lateral anthers often only partially complete. Staminodes 2. Capsule 3-valved, the dorsal valve slow in dehiscing; seeds 5, 2.2 × 1.5 mm, obovoid, rugose-pitted.
Fl. ? *Native to the tropical Asia and Africa, grown in gardens and naturalized on rocks and walls in Funchal and possibly elsewhere.* **M**

2. C. diffusa Burm.f., *Fl. Indica*: 18 (1768). [Herva menina]
C. agraria Kunth
Sprawling or ± climbing annuals, rooting at the nodes, ± glabrous. Leaves up to 1.5(-6) × 1-2 cm, oblong-lanceolate, shortly stalked above the sheath; sheath membranous, glabrous, barely ciliate at the mouth. Spathe open, sometimes folded, 1 cyme with 1-3 sterile flowers, the other with 2-3 fertile flowers. External petal white, very short; internal petals blue, clawed, the limb suborbicular. Median anther large, curved, the 2 lateral anthers straight, fertile. Staminodes 3. Capsule 3-valved; seeds 5, 2.5-3 × 1.5 mm, cylindrical, reticulate.
Fl. VI - IV. *Common in damp places in south-east Madeira. Native to the tropics.* **M**

2. Tinantia Scheidw.

Erect annuals or perennials. Inflorescence pedunculate, terminal or rarely also axillary, 3-branched, simple or umbellate, the flowers pedicellate on the branches. Sepals 3, subequal. Petals 3, obovate, subequal. Stamens 6, all perfect; filaments bearded, free, or the upper 3 united at the base. Ovary 3-locular; locules with 2-5 ovules.

1. T. erecta (Jacq.) Schltdl. in *Linnaea* **25**:185 (1852).
T. fugax Jacq.
Erect annual up to 1 m. Leaves up to 5 × 3 cm, elliptic to oblanceolate; sheaths short, membranaceous, scarcely ciliate at the mouth. Bracts unequal, leaf-like. Inflorescence glandular-hairy; cymes in groups of 1-4; peduncles up to 5 cm; pedicels up to 2 cm. Petals 1.5 × 1.5 cm, blue or purplish. Stamens 6; filaments long-bearded. Capsule 7-11 × 4.5 mm, oblong. Seeds 3-3.5 mm, rugose, 2-3 per locule.
Fl. IX - X. *Cultivated in gardens, sometimes escaping and becoming naturalized in the Funchal region. Native to C. and S. America.* M

3. Tradescantia L.

Perennials, creeping and rooting at the nodes or erect. Inflorescence usually terminal; cymes 2, subtended by paired, boat-shaped bracts. Sepals 3, usually free. Petals 3, free or connate at the base. Stamens 6, all fertile and ± similar; filaments usually hairy. Ovary 3-celled, with 2 or rarely 1 ovule per cell. Seeds reticulate-rugose.

1. Petals united at the base .... **1. zebrina**
- Petals free 2

2. Leaves ovate to ovate-oblong, up to 3 or 4 cm, flowers 1-1.5 cm in diameter ......
  .... **2. fluminensis**
- Leaves linear, up to 35 cm, flowers 2.5-3.5 cm in diameter .... **3. virginiana**

1. T. zebrina Hort. ex Bosse Vollst., *Handb. Bl.-gärtn.* **4**: 655 (1846).
Zebrina pendula Schenizl.
Perennials, creeping and rooting at the nodes or pendent. Leaves up to 5 × 2.5 cm, ovate-oblong to broadly ovate, bluish green, often striped with silver above, purplish beneath. Inflorescence terminal, stalked; bracts unequal, leaf-like. Flowers 1-1.5 cm in diameter; petals pink or violet-blue, united at the base into a tube up to 1 cm; stamens borne on the petals.
Fl. all year. *Widely cultivated in gardens in and around Funchal, escaping and becoming naturalized. Native to Mexico.* M

2. T. fluminensis Vell., *Fl. flumin.*: 140 (1825).
Perennials, creeping and rooting at the nodes. Leaves up to 5 × 2 cm, ovate to ovate-oblong, asymmetric, entirely green or resembling those of **1**. Inflorescence terminal, short-stalked; bracts unequal, leaf-like; pedicels recurved after flowering. Flowers 1-1.5 cm in diameter; petals white, free; stamens not borne on the petals, filaments long-hairy.
Fl. III - VI. *Common along levadas, on banks terraces and walls in many parts of lowland Madeira. Native to N. and S. America.* M

Hybrids between *Tradescantia virginiana* L. and related species are widely cultivated as ornamentals. They are easily distinguished from **1** or **2** by their erect habit, long, linear leaves (sometimes up to 35 cm), and showy flowers 2.5-3.5 cm in diameter.

CXVII. GRAMINEAE
[POACEAE]
T.A. Cope

Annual or perennial herbs, rarely woody shrubs; stems hollow in the internodes, closed at the nodes. Leaves solitary at the nodes, alternate and distichous, comprising sheath, ligule and blade; sheaths encircling the stem, with the margins free or connate; ligule adaxial, at junction of sheath and blade, membranous or reduced to a fringe of hairs, rarely absent; blades usually long and narrow, flat or inrolled. Inflorescence made up of spikelets arranged in a panicle or in racemes, the latter solitary, digitate or disposed along a central axis, rarely gathered into leafy false panicles. Spikelets comprising distichous bracts arranged along a slender axis (*rhachilla*); the 2 lower bracts (*glumes*) empty; the succeeding 1 to many bracts (*lemmas*) each enclosing a flower and opposed by a hyaline scale (*palea*), the whole (lemma, palea and *flower*) termed a floret; base of spikelet or floret sometimes with a horny prolongation downwards (*callus*); glumes and lemmas often bearing 1 or more stiff bristles (*awns*). Flowers usually bisexual, sometimes unisexual, small and inconspicuous; perianth represented by 2 minute scales (*lodicules*); stamens 1-6, usually 3; ovary 1-locular; fruit a caryopsis with thin pericarp adnate to the seed.

The following cereals and fodder grasses are grown and may occur as crop remnants:

Oryza sativa L., *Sp. pl.* **1**: 333 (1753) [*Aroz*] is cultivated in Madeira. The inflorescence is a panicle with simple, raceme-like primary branches; spikelet has 1 fertile floret with 2 sterile lemmas at its base; glumes are absent but the sterile lemmas are easily mistaken for them; the coriaceous lemma enfolds the 1-keeled palea and may be awnless or awned; stamens 6; ligule of lower leaves up to 45 mm long.

Secale cereale L., *Sp. pl.* **1**: 84 (1753) is, or was, grown in Madeira and possibly also in Porto Santo. It resembles *Elymus* but is annual and has linear-subulate glumes.

Triticum L. resembles *Secale* but has broad, truncate, asymmetrical glumes. Three species are cultivated in the islands:

T. aestivum L., *Sp. pl.* **1**: 85 (1753) (*T. vulgare* Vill.) has a mealy, gluten-rich endosperm ideal for making bread. The glumes are keeled in the upper half only, rounded below, and the lax raceme is more than 3 × as long as wide. 2n=42. Madeira and Porto Santo.

T. compactum Host, *Icon. descr. gram. austriac.* **4**: 4 (1809) has a mealy, low-gluten, protein-poor endosperm better used as a flour for making biscuits and pastry. It resembles *T. aestivum* but has a dense raceme less than 3 × as long as wide. 2n=42. Madeira.

T. turgidum L., *Sp. pl.* **1**: 88 (1753) has a mealy endosperm and is used mainly for livestock feed. The glumes are keeled throughout. 2n=28. Madeira.

Saccharum officinarum L., *Sp. pl.* **1**: 54 (1753) (including var. *violaceum* Pers., var. *litteratum* Hassk. and var. *subobscurum* Menezes) [*Cana de asugar*] was formerly widely cultivated in Madeira and Porto Santo and is still occasionally grown on a small scale. It is a tall perennial reaching a height of 4-5 m; it has a plumose panicle up to 1 m, and broad, flat leaves up to 4 cm wide; the ligule is a scarious rim or line of fine hairs.

Coix lacryma-jobi L., *Sp. pl.*: **2**: 972 (1753) is sometimes grown in Madeira as an ornamental but it can also be utilized as cattle fodder. It is a monoecious annual with the female raceme enclosed in a bony utricle. The utricles, of which many shape and colour varieties exist, can be threaded on a cord to make necklaces.

Zea mays L., *Sp. pl.* **2**: 971 (1753) was formerly grown in Madeira either for its grain or as a fodder crop, but is now more likely to be found in gardens or as an occasional relic in ruderal sites. It is a robust, monoecious annual; the female inflorescence is axillary, wrapped in several sheaths, with protruding stigmas ('silks'), the numerous spikelets arranged in rows on a thickened, woody 'cob'; the male inflorescence ('tassel') is a terminal panicle.

1. Bamboos with woody stems .... **1. Pseudosasa**
- Herbs, sometimes cane-like or reed-like 2

2. Spikelets proliferating .... **12. Poa**
- Spikelets not proliferating 3

3. Spikelets grossly dimorphic, mostly with 1 fertile spikelet suurounded by 1-6 male or sterile spikelets surrounding it 4
- Spikelets sometimes slightly different in size but not grossly dimorphic, usually solitary 6

4. Sterile spikelets persistent .... **9. Cynosurus**
- Sterile and fertile spikelets falling in groups 5

5. Sterile spikelets many-flowered; lemma of fertile spikelet awned ... **10. Lamarckia**
- Sterile spikelets reduced to 2 papery glumes; lemma of fertile spikelet awnless .... **30. Phalaris**

6. Tall reed-like plants 2-5 m high, the leaves sometimes confined to a basal tussock and with harsh, saw-like edges; panicle large and feathery, (15-)30-100 cm 7
- Smaller plants, not reed-like; if panicle a feathery head then plant scarcely more than 1 m high 9

7. Leaves confined to a basal tussock .... **43. Cortaderia**
- Leaves all cauline 8

8. Ligule a membrane .... **44. Arundo**
- Ligule a line of hairs .... **45. Phragmites**

9. Inflorescence a dense, ovoid or cylindrical panicle with the spikelets clothed and ± concealed in long silky or fluffy white hairs 10
- Inflorescence not as above 11

10. Plant annual; hairs of panicle fluffy .... **34. Lagurus**
- Plant perennial; hairs of panicle silky .... **65. Imperata**

11. Inflorescence a false panicle of 1-10 raceme-pairs, each enclosed in a leafy spatheole .. .. **69. Hyparrhenia**
- Inflorescence structure variable, but never with leafy spatheoles 12

12. Spikelets in pairs (threes), 1 sessile and bisexual, 1(-2) pedicelled and male or sterile; inflorescence never a solitary, bilateral raceme 13
- Spikelets not in pairs (threes) with 1 sessile and pedicelled, occasionally spikelets in threes but then inflorescence a solitary, bilateral raceme 15

13. Inflorescence an open panicle .... **66. Sorghum**
- Inflorescence comprising subdigitate racemes 14

14. Pedicels and raceme internodes solid, terete .... **67. Dichanthium**
- Pedicels and raceme internodes flattened, hyaline and translucent between the thickened margins .... **68. Bothriochloa**

15. Inflorescence a single, terminal, bilateral raceme or false raceme	16
– Inflorescence of several racemes or an open to contracted panicle, sometimes the panicle reduced to a unilateral raceme	24
16. Spikelets embedded in hollows in the axis and covered by the glume(s)	17
– Spikelets broadside to the axis and not embedded	18
17. Both glumes present, placed side by side	**17. Parapholis**
– Lower glume suppressed, except in the terminal spikelet where they are opposite, not side by side	**18. Hainardia**
18. Lower (inner) glume absent except in the terminal spikelet or reduced to a small scale; spikelets edgeways to the axis	**6. Lolium**
– Both glumes present, well developed or sometimes awn-like	19
19. Spikelets in threes at each node of the axis; glumes awn-like	**39. Hordeum**
– Spikelets solitary at each node of the axis	20
20. Lemma with a dorsal, geniculate awn	**24. Gaudinia**
– Lemma awnless or with a straight awn from near the tip	21
21. Plants perennial	22
– Plants annual	23
22. Spikelets shortly pedicelled	**37. Brachypodium**
– Spikelets quite sessile	**38. Elymus**
23. Lemma 5-nerved, 2.7-5 mm	**7. Micropyrum**
– Lemma 7- to 9-nerved, 7.5-10 mm	**37. Brachypodium**
24. Inflorescence of 2 or more racemes	25
– Inflorescence an open to contracted panicle, sometimes reduced to a solitary, unilateral raceme (if spikelets enclosed in the uppermost sheath with only the filaments and stigmas protruding see *Pennisetum clandestinum*)	36
25. Racemes conjugate or digitate or, subdigitate with the axis shorter than the raceme	26
– Racemes scattered along an elongated axis which is always longer than the racemes	31
26. Spikelets dorsally compressed, falling entire	27
– Spikelets laterally compressed, breaking up above the persistent glumes	28
27. Upper lemma with inrolled margins clasping only the edges of the palea	**57. Paspalum**
– Upper lemma with thin, flat margins ± concealing the palea	**62. Digitaria**
28. Spikelets 1-flowered	**52. Cynodon**
– Spikelets with 3 or more flowers	29
29. Spikelet with 1 fertile floret and 2-3 sterile rudiments; fertile lemma hairy	**51. Chloris**
– Spikelet with several fertile florets; lemmas glabrous	30

30.	Racemes terminating in a spikelet	**48. Eleusine**
-	Racemes terminating in a rigid naked point	**49. Dactyloctenium**

31. Racemes very short, of 1-3 spikelets deeply embedded in the thickened, corky axis .... **60. Stenotaphrum**
- Racemes free, appressed or divergent 32

32. Lower glume awned .... **53. Oplismenus**
- Lower glume awnless, sometimes absent 33

33. Lower glume absent .... **57. Paspalum**
- Lower glume present 34

34. Racemes with spikelets mostly in 4 rows; spikelets gibbously plano-convex, cuspidate to awned .... **55. Echinochloa**
- Racemes 1- to 2-rowed 35

35. Spikelets borne singly, ovate; upper lemma granulose .... **59. Paspalidium**
- At least some spikelets borne in pairs; upper lemma rugulose ..... **56. Brachiaria**

36. Spikelets subtended by an involucre of bristles 37
- Spikelets not subtended by an involucre of bristles 39

37. Bristles persisting on the axis after the spikelets have fallen .... **58. Setaria**
- Bristles falling with the spikelets 38

38. Bristles free to the base .... **63. Pennisetum**
- Bristles connate below into a disc or shallow cup .... **64. Cenchrus**

39. Spikelets strictly 1-flowered, without sterile or male florets above or below the fertile 40
- Spikelets 2-flowered or more, often with only 1 fertile floret, but this accompanied by 1 or more male or sterile florets 48

40. Lemma with 3 equal, straight awns from the tip .... **46. Aristida**
- Lemma awnless or with a single awn, if with 3 awns then the laterals straight and the central geniculate, the latter arising from the back 41

41. Lemma toughened at maturity; glumes membranous throughout 42
- Lemma hyaline or membranous at maturity, rarely cartilaginaous tand with glumes indurated and gibbously swollen below, membranous above 44

42. Floret dorsally compressed .... **4. Oryzopsis**
- Floret terete or laterally compressed 43

43. Floret terete, the awn central .... **2. Stipa**
- Floret gibbously ellipsoid, the awn eccentric .... **3. Nassella**

44. Glumes and lemma similar in texture, the glumes shorter than the lemma .... **50. Sporobolus**
- Glumes longer and usually firmer than the hyaline lemma 45

45.	Lemma 3-awned; panicle densely ovoid-cylindrical	**32. Triplachne**
-	Lemma awnless or 1-awned	46
46.	Spikelets falling entire, attached to the pedicel	**35. Polypogon**
-	Spikelets breaking up above the persistent glumes	47
47.	Panicle dense, spike-like; glumes indurated and gibbously swollen below, membranous above	**33. Gastridium**
-	Panicle lax or contracted, but not spike-like; glumes membranous throughout	**31. Agrostis**
48.	Ligule a line of hairs or a narrow ciliate rim	49
-	Ligule a membrane, sometimes very short and obscure, but never ciliate	54
49.	Spikelets 2-flowered, falling entire or sometimes the upper lemma falling before the rest of the spikelet; florets always 2, not laterally compressed	50
-	Spikelets usually breaking up, the florets usually falling separately from the ± persistent glumes; florets rarely falling entire and then spikelets more than 2-flowered and strongly laterally compressed	51
50.	Upper lemma crustaceous, the margins inrolled and clasping only the edges of the palea; lower lemma awnless	**54. Panicum**
-	Upper lemma thinly cartilaginous, the margins flat and ± enfolding the palea; lower lemma usually awned	**61. Melinis**
51.	Glumes shorter than the lowest lemma	**47. Eragrostis**
-	Glumes longer than the lowest lemma	52
52.	Plants annual	**42. Schismus**
-	Plants perennial	53
53.	Lemma hairy on the margins, mucronate	**40. Danthonia**
-	Lemma with tufts of hair across the back, with a geniculate awn	**41. Rytidosperma**
54.	Spikelets always falling entire at maturity and 2-flowered	**27. Holcus**
-	Spikelets usually breaking up above the glumes at maturity and often also between the florets, rarely falling entire and then with 1 fertile floret below and a clavate mass of sterile lemmas above	55
55.	Spikelets with only 1 fertile floret, this accompanied by 1 or 2 male or barren and often much reduced florets below, or by a clavate mass of sterile lemmas above	56
-	Spikelets with 2 or more fertile florets, sometimes the uppermost progressively reduced	59
56.	Lowest floret fertile, subsequent florets reduced to a clavate mass of sterile lemmas	**20. Melica**
-	Uppermost floret fertile, the 1 or 2 lower florets male or barren	57
57.	Lower lemmas reduced to small chaffy or fleshy scales at the base of the fertile lemma, sometimes 1 of them absent	**30. Phalaris**
-	Lower lemma(s) well developed	58

58. Spikelet 2-flowered, the lower floret male, shorter than the fertile . . . . **22. Arrhenatherum**
- Spikelet 3-flowered, the lower florets sterile, longer than the fertile . . . . **29. Anthoxanthum**

59. Lemma orbicular to oblate, with broad membranous margins, appressed to the lemma above . . . . **11. Briza**
- Lemma narrower, the margins less distinct and often inrolled 60

60. Glumes clearly longer than the lowest lemma, often as long as the whole spikelet and enclosing it; awn not terminal 61
- Glumes shorter than the lowest lemma with all florets well exserted, rarely slightly exceeding the lowest lemma and then awn, if present, clearly terminal 65

61. Plants annual 62
- Plants perennial 64

62. Spikelets pendulous, 10-50 mm . . . . **23. Avena**
- Spikelets not pendulous, 2.5-7.5 mm 63

63. Lemma with a straight awn from just below the tip . . . . **25. Rostraria**
- Lemma with a geniculate awn arising from below the middle of the back . . **28. Aira**

64. Lemma 2-toothed at the tip, the teeth aristulate; central awn arising from near the middle, geniculate, very conspicuous . . . . **21. Helictotrichon**
- Lemma 4-toothed at the tip; awn arising from below the middle, inconspicuous . . . . **26. Deschampsia**

65. Lemma with a distinctly subapical awn 66
- Lemma awnless, mucronate, apiculate or with a terminal awn 67

66. Spikelets 3-7.5 mm; awn 1-3 mm; plant annual . . . . **25. Rostraria**
- Spikelets (12-)18-70 mm; awn (6-)12-60 mm, sometimes less than 3 mm but then plant perennial . . . . **36. Bromus**

67. Plants perennial 68
- Plants annual 72

68. Lemmas rounded on the back, at least in the lower part 69
- Lemmas keeled throughout 70

69. Lemma hyaline at the tip, 7-nerved . . . . **19. Glyceria**
- Lemma firm at the tip, 5-nerved . . . . **5. Festuca**

70. Lemma 3-nerved; rhachilla hairy . . . . **14. Parafestuca**
- Lemma 5-nerved; rhachilla glabrous but floret-callus often with long fine woolly hairs 71

71. Lemma acute, at most shortly hairy on the keel; spikelets evenly distributed in the panicle . . . . **12. Poa**
- Lemma acuminate, spinulose on the keel; spikelets gathered into 1-sided clumps at the ends of the panicle-branches which may be appressed to the main axis . . **13. Dactylis**

72. Lemma awned .... **8. Vulpia**
- Lemma awnless, at most shortly apiculate 73

73. Lemma 3-nerved .... **15. Sphenopus**
- Lemma 5-nerved 74

74. Lemma rounded on the back or keeled only towards the tip ...... **16. Catapodium**
- Lemma keeled throughout .... **12. Poa**

1. Pseudosasa Nakai

Shrubby bamboo with slender rhizomes and bearing branches singly; culm-sheaths persistent. Inflorescence a small panicle, the branches naked or subtended by minute bracts. Spikelets 3- to 8-flowered; glumes 2; stamens 3, rarely 4; stigmas 3.

1. P. japonica (Siebold & Zucc. ex Steud.) Makino in *J. Jap. Bot.* **2**(4): 15 (1920).
Stems reaching to 6 m, arching, hollow, woody, clothed below by broad sheaths with rudimentary blades, freely branching above.
Fl. ? *Native to Japan and Korea. The only bamboo to have naturalized in the archipelago, it is grown as an ornamental in Funchal and is found as an escape on roadsides above the town.* **M**

2. Stipa L.

Annuals or perennials; ligule a membrane. Inflorescence a panicle. Spikelets 1-flowered without rhachilla-extension, the floret fusiform, terete, with pungent callus; glumes longer than the floret; lemma firmly membranous to coriaceous, the margins overlapping, the tip entire or shortly bilobed, awned.

1. Annual; glumes pallid .... **1. capensis**
- Perennial; glumes deeply suffused with purple .... **2. neesiana**

1. S. capensis Thunb., *Prodr. pl. cap.* **1**: 19 (1794).
S. tortilis Desf.
Annual up to 40 cm. Panicle partially enclosed in the uppermost sheath. Glumes narrowly lanceolate-acuminate, hyaline, pallid, the lower 16-24 mm, the upper 14-22 mm; lemma, including the callus, 5.5-8 mm; awns 6.5-10 cm, bigeniculate, with hairy column and scabrid limb, without a crown of hairs around the base, eventually twisted together to form a tail at the summit of the panicle.
Fl. III - V. *Dry rocky places, usually near the sea; southeast Madeira, Porto Santo, Bugío and Selvagem Grande.* **MDPS**

2. S. neesiana Trin. & Rupr., *Sp. gram. stipac.*: 27 (1842).
S. eminens Nees, non Cav. (1799).
Tufted perennial up to 100 cm. Panicle exserted from or scarcely enclosed at the base by the uppermost sheath. Glumes lanceolate-subulate, hyaline, deeply suffused with purple, the lower 11-22 mm, the upper 10-19 mm; lemma, including the callus, 7.5-14.5 mm; awns 5-12 cm, bigeniculate, with pubescent column and scabrid limb, with a crown of hairs 0.6-1.6 mm around the base, not twisted together into a prominent tail though the uppermost may be entangled.
Fl. IV - VII. *Native of S. America cultivated as an ornamental, but often becoming a weed and naturalized on roadsides and in waste places around Funchal.* **M**

3. Nassella Desv.

Perennials; ligule a membrane. Inflorescence a panicle. Spikelets 1-flowered without rhachilla-extension, the floret gibbously ellipsoid, laterally compressed, the callus obtuse to bluntly acute; glumes longer than the floret; lemma coriaceous, with overlapping margins, awned, the awn eccentric.

1. N. trichotoma (Nees) Hack. ex Arechav. in *An. Mus. nac. Montevideo* **1**: 336, t.19 (1895).
Stems up to 50 cm. Panicle effuse, the branches bare at the base and the spikelets borne on long filiform pedicels, the whole deciduous at maturity. Glumes narrowly ovate-subulate, hyaline, green or purple below, pallid above, 4-8.5 mm; lemma 1.5-2.5 mm; awn 1.5-3 cm, scabrid throughout.
Fl. ? *An introduced weed found on hillsides at c. 550 m. Native of S. America.* **M**

4. Oryzopsis Mich.

Perennial; ligule a membrane. Inflorescence a panicle. Spikelets 1-flowered without rhachilla-extension, the floret narrowly lanceolate to ovate, dorsally compressed, the callus very short and obtuse; glumes longer than the floret; lemma coriaceous, dark in colour, the margins seldom overlapping, the tip entire or minutely 2-toothed, awned.

1. O. miliacea (L.) Asch. & Schweinf. in *Mem. Inst. égypt.* **2**:169 (1887).
Piptatherum miliaceum (L.) Coss.
Stems wiry, up to 150 cm. Panicle lax, the branches in whorls. Glumes ovate to lanceolate, 3-4 mm; lemma obovate, 1.5-2 mm, glabrous or sometimes with 2 tufts of short hair on the callus, obtuse, minutely 2-toothed; awn deciduous, 3-4.5 mm, subterminal.
Fl. (I-)IV - VIII. *Rocky hillsides in southeast Madeira and in Porto Santo from sea-level to 1050 m.* **MP**

5. Festuca L.

Perennials; ligule a membrane; leaves sometimes flat, but mostly inrolled to filiform. Inflorescence an open to contracted panicle. Spikelets 2- to several-flowered, breaking up above the glumes and between the florets; lemmas membranous to coriaceous, rounded on the back, 5-nerved, acute to shortly awned; floret-callus and rhachilla glabrous.

1.	Widest leaves more than 5 mm wide, flat	2
-	All leaves less than 5 mm wide, folded or setaceous	3
2.	Spikelets 10-12 mm; lemma more than 6 mm, with an awn up to 4 mm	**1. arundinacea**
-	Spikelets 5.5-6.5 mm; lemma up to 5.2 mm, awnless	**2. donax**
3.	Rhachilla-internodes clearly visible between the florets at anthesis	**3. jubata**
-	Rhachilla-internodes hidden by the florets at anthesis	4
4.	Ligule without auricles; leaves of the flowering stems usually flat, 0.5-3 mm wide	**4. rubra**
-	Ligule with short but conspicuous auricles, all leaves folded-setaceous and less than 1 mm across	**5. ovina**

1. F. arundinacea Schreb., *Spic. fl. lips.*: 57 (1771).
F. elatior L.
Densely tufted, up to 200 cm, sometimes forming large tussocks; leaves flat, 3-12 mm wide;

ligule up to 2 mm. Panicle 10-50 cm, nodding. Spikelets 3- to 5-flowered, 10-12 mm; glumes lanceolate, the lower 3-6 mm, the upper 4.5-7 mm; lemmas 6-9 mm, scabrid, with an awn up to 4 mm; anthers 3-4 mm.
Fl. VI. *Native to Europe, temperate Asia and NW Africa. Recorded around Queimadas, probably introduced for fodder.* M

The only specimen known is atypical of most of *F. arundinacea*, having a remarkably long awn (4 mm) and relatively long, conspicuously scabrid glumes and lemmas. The species is almost certainly introduced in Madeira, not having been collected before 1985, but its exact position in the poorly understood *F. elatior* complex is not at all certain.

2. F. donax Lowe in *Trans. Camb. phil. Soc.* **4**: 9 (1831).
Densely tufted, up to 120 cm; leaves flat, up to 10 mm wide, scabrid on the margins; ligule up to 5 mm. Panicle (12-)18-30 cm, nodding. Spikelets 2- to 3(-4)-flowered, 5.5-6.5 mm; glumes narrowly lanceolate-acuminate, the lower 4.3-5.3 mm, the upper 5-6.2 mm; lemmas 4.4-5.2 mm, faintly scaberulous, awnless; anthers *c*.3 mm. **Plate 53**
Fl. VI - VIII(-IX). *Endemic occurring on rock ledges in central Madeira from 900-1800 m.* M

3. F. jubata Lowe, in *Trans. Camb. phil. Soc.* **6**: 538 (1838).
F. agustinii subsp. *mandonii* (St.-Yves) A. Hansen
Tufted, up to 60 cm; leaves folded, 0.3-0.6 mm across, somewhat recurved, smooth. Panicle 3-10 cm, slender. Spikelets 4- to 5-flowered, 7-8(-10) mm; glumes narrowly lanceolate, acute, the lower 5-5.8 mm, the upper 7-9.2 mm; lemma 6-7.5 mm, glabrous to puberulous, long-acuminate, with an awn 3-5(-8) mm; anthers *c*.4 mm. **Plate 53**
Fl. V - IX. *Rocky pastures, in Madeira both inland, in the Curral das Freiras, and on the north coast around Boca do Risco; also from Porto Santo.* ■ MP [*Azores*]

4. F. rubra L., *Sp. pl.* **1**: 74 (1753).
Tufted or stoloniferous, up to 70 cm; leaves of sterile shoots inrolled or setaceous, those of flowering shoots more or less flat and 0.5-3 mm wide; ligule very short, without auricles.
Fl. VI. *Native to Europe. Introduced around Porto do Moniz.* M

F. rubra sensu lato has been reported from the archipelago since 1969, but no attempt has yet been made to determine the precise identity of the plants. The taxonomy of the aggregate in Europe is still far from completely understood and until it is it will not be possible to decide exactly which species occur(s) in Madeira.

5. F. ovina L., *Sp. pl.* **1**: 73 (1753).
Tufted, up to 50 cm; all leaves folded-setaceous, up to 1 mm across; ligule very short, with small but conspicuous rounded auricles.
Fl. VI. *Native throughout North temperate regions. Introduced to Madeira where it has been reported both as common in the mountains and as of rare and accidental occurrence. Definite records are available only from Monte and Santana, and possibly from Funchal and Porto Moniz.* M

Although collections from the Madeiran archipelago which date back to the first half of the last century are available, records of *F. ovina* (as one or more of its many segregates) have been published only since 1906. Several names are used in this literature:

F. ovina subsp. *duriuscula* (L.) Kozlovski, (*F. duriuscula* L.). Portuguese and Madeiran authors have consistently misapplied this name to what is now regarded as either *F. lemanii* Bastard or *F. indigesta* Boiss.

F. longifolia Thuill. This name has been misapplied to both *F. lemanii* and *F. brevipila* Tracey.

F. ovina sensu lato. The taxonomy of *F. ovina* has not yet been fully resolved and it is not possible to determine exactly what the Madeiran taxa within the complex may be. Therefore the species is treated here in the broad sense.

The hybrid ×**Festulolium loliaceum** (Huds.) P. Fourn. (*Glyceria loliacea* (Huds.) Godr.), i.e. *Festuca pratensis* × *Lolium perenne*, has been reported from Madeira. It is not native since only one of its parents (itself an introduction) occurs in the islands. It resembles a robust *Lolium perenne* but with a tendency towards branched racemes, and a small lower (inner) glume is usually present.

6. Lolium L.

Annuals, biennials or perennials; ligule a membrane. Inflorescence a bilateral raceme with spikelets in 2 opposite rows, edgeways to the rhachis. Spikelets several- to many-flowered; lower (inner) glume absent (except in terminal spikelet); upper (outer) glume shorter than the adjacent lemma to as long as the spikelet (and exceeding the uppermost floret), coriaceous; lemma membranous to coriaceous, with or without a subterminal awn.

1. Lemmas turgid at maturity; mature fruit not more than 3 times as long as wide . . . . **5. temulentum**
-   Lemmas not turgid at maturity; mature fruit more than 3 times as long as wide . . 2

2. Perennial with non-flowering shoots at anthesis; leaves flat or folded in the young shoot . . . . **1. perenne**
-   Annual or biennial without non-flowering shoots at anthesis, or biennial; leaves rolled in the young shoot 3

3. Spikelets usually 11- to 22-flowered; glume up to ½ as long as spikelet **2. multiflorum**
-   Spikelets usually 4- to 9(-11)-flowered; glume more than ½ as long as spikelet . . 4

4. Lemma awnless or with an awn up to 10 mm; rhachis usually stout . . . . **3. rigidum**
-   Lemma awned, the awn up to 15 mm; rhachis always slender . . . . . . **4. canariense**

1. L. perenne L., *Sp. pl.* **1**: 83 (1753).
Perennial up to 90 cm; leaves flat or folded in the young shoot. Raceme with slender rhachis. Spikelets 5-23 mm, 2- to 10-flowered; glume lanceolate or narrowly oblong, one third as long to as long as the spikelet, rarely exceeding the uppermost floret; lemmas oblong or lanceolate-oblong, 3.5-9 mm, usually awnless, not turgid at maturity. Mature fruit more than 3 times as long as wide.
Fl. (I-)III - V. *Introduced; native to temperate Eurasia and N. America. A common weed in southeast Madeira; also on Pico do Castelo in Porto Santo.* **MP**

Hybrids between **1** and **2** occur where the species grow together in Madeira, and possibly also occur in Porto Santo.

2. L. multiflorum Lam., *Fl. franç.* **3**: 621 (1778).
Annual or biennial (rarely a short-lived perennial) up to 130 cm; leaves rolled in the young shoot. Raceme with slender flexuous rhachis. Spikelets 8-30 mm, (5-)11- to 22-flowered; glume lanceolate to narrowly oblong, ¼ to ½(-¾) as long as the spikelet; lemmas oblong to lanceolate-oblong, 4-8 mm, usually awned, not turgid at maturity; awn up to 15 mm. Mature fruit more than 3 times as long as wide.

Fl. IV - VIII(-XII). *Introduced weed; native to C. & S. Europe and the Mediterranean region. Occurs in southeast Madeira; rare in Porto Santo.* **MP**

3. L. rigidum Gaudin, *Agrost. helv.* **1**: 334 (1811).
Arthrochortus loliaceus Lowe, non *Rottboellia loliacea* Bory & Chaub.; *Lolium lowei* Menezes; *L. parabolicae* Sennen & Samp.
Annual up to 45(-70) cm; leaves rolled in the young shoot. Raceme with slender to very stout rhachis, in the latter case the spikelets more or less sunk in concavities and concealed by the glume. Spikelets 5-18 mm, usually 5- to 8(-11)-flowered; glume oblong-lanceolate, (½-)¾ as long to as long as the spikelet, slightly exceeding the uppermost floret; lemmas oblong to oblong-lanceolate, usually awnless but sometimes with a slender awn, not turgid at maturity; awn up to 10 mm. Mature fruit more than 3 times as long as wide.
Fl. III - VII. *Scree slopes, dunes, cultivated areas, roadsides and waste places up to 60 m. Common in southeast Madeira from Funchal to Ponta de São Lourenço, in Porto Santo and Ilhéu Chão; also from Selvagem Grande.* **MDPS**

Two segregates are often recognized: subsp. *rigidum* (sometimes as var. *rigidum*) with slender rhachis (including *L. parabolicae*); and subsp. *lepturoides* (Boiss.) Sennen & Mauricio (sometimes as var. *rottbollioides* Heldr. ex Boiss.) with stout rhachis (including *L. lowei*). The latter is the more common in Madeira. The true value and status of these segregates has not been thoroughly investigated.

L. subulatum Vis. is also recorded from Madeira. It is so like *L. rigidum* subsp. *lepturoides* that it is hard to make a distinction between them. *L. subulatum* is said to have longer glumes and larger florets. With only sparse material available from Madeira, no attempt has been made to separate these two taxa.

4. L. canariense Steud., *Syn. pl. glumac.* **1**: 340 (1854).
Annual up to 65 cm; leaves rolled in the young shoot. Raceme with slender rhachis. Spikelets 6-25(-35) mm, 4- to 9(-11)-flowered; glume lanceolate to narrowly lanceolate, half as long to as long as the spikelet and in extreme cases twice as long as the distance between the base of the spikelet and the tip of the uppermost floret; lemmas ovate-lanceolate, awned, not turgid at maturity; awn up to 15 mm. Mature fruit more than 3 times as long as wide.
Fl. IV. *Confined to Pico de Castelo in Porto Santo.* ■ **P** [*Canaries*].

5. L. temulentum L., *Sp. pl.* **1**: 83 (1753).
L. temulentum var. *arvense* (With.) Junge
Annual up to 120 cm; leaves rolled in the young shoot. Raceme stiff and rather stout. Spikelets 8-28 mm, 2- to 15-flowered; glume lanceolate, three quarters as long to as long as the spikelet, up to 1.5 times the distance from the base of the spikelet to the tip of the uppermost floret; lemmas elliptical to ovate, usually awned, very turgid at maturity; awn up to 23 mm. Mature fruit 2-3 times as long as wide.
Fl. III - IV. *Native to the Mediterranean region and SW. Asia. An introduced weed of roadsides and waste places around Funchal harbour; also recorded from Porto Santo and the Salvages but without further data.* **MPS**

7. Micropyrum (Gaudin) Link

Annuals; ligule a membrane. Inflorescence a bilateral raceme, the spikelets broadside to the rhachis in 2 opposite rows, rarely branched below. Spikelets several- to many-flowered, breaking up above the glumes and between the florets; glumes subequal, coriaceous, rounded on the back; lemmas coriaceous, rounded on the back, 5-nerved, obtuse to emarginate, sometimes awned.

1. M. tenellum (L.) Link in *Linnaea* **17**: 398 (1843).
Catapodium tenellum (L.) Trab.; *Nardurus lachenalii* var. *festucoides* (Bertol.) Cout.
Stems up to 60 cm, slender, rather stiff. Raceme 1-20 cm, rarely branched below, the spikelets subsessile. Spikelets 4-9(-14) mm, appressed at anthesis; glumes 2-6 mm, the upper slightly the longer; lemmas 2.7-5 mm, awnless or with an awn as long as the body; anthers 0.5-1.3 mm.
Fl. VII. *Confined to the high, central peaks of Madeira. Records from the Desertas are thought to be erroneous.* **M**

8. Vulpia C.C. Gmel.

Annuals; ligule a membrane. Inflorescence a scanty, more or less secund panicle or reduced to a unilateral raceme. Spikelets several-flowered, the uppermost florets often progressively reduced and male or barren, breaking up above the glumes and between the fertile florets; glumes very unequal, the lower sometimes minute; lemmas thinly coriaceous, rounded on the back, 5-nerved, narrow, tapering into a long straight awn; stamens 1-3, often small and cleistogamous.

1. Florets chasmogamous, the anthers usually 3, 2-5 mm . . . . **1. geniculata**
- Florets cleistogamous, the anthers 1(-3), 0.4-0.8(-1.8) mm 2

2. Lemma 1.3-1.9 mm wide; lower glume 2.5-5 mm, ½-¾ as long as the upper . . . . . .
. . . . **2. bromoides**
   Lemma 0.8-1.3 mm wide; lower glume 0.5-3 mm, usually less than ½ as long as the upper 3

3. Inflorescence usually well exserted from the uppermost sheath; lower glume ¼-½ as long as the upper . . . . **3. muralis**
   Inflorescence usually not fully exserted from the uppermost sheath; lower glume ¹/₁₀-²/₅ as long as the upper . . . . **4. myuros**

1. V. geniculata (L.) Link, *Hort. berol.* **1**: 148 (1827).
Stems up to 60 cm. Panicle diffuse, 5-15 cm, the branches erect or somewhat spreading. Spikelets 6.5-9.5 mm (excluding the awns), the florets chasmogamous, most of them fertile; lower glume 2.5-5.5 mm, ²/₅-³/₅ as long as the upper; lemma 5-7.5 mm, with an awn as long as or a little longer than the body; anthers usually 3, 2-5 mm, mostly extruded at anthesis.
Fl. IV - VI. *Native to the W. Mediterranean region. An introduced weed, mostly of waste places in southeastern Madeira; rare in Porto Santo.* **MP**

V. ligustica (All.) Link is very like **1** but with the lower glume 0.2-0.8 mm and less than ¹/₆ as long as the upper. It has been recorded for Madeira but almost certainly in error for **1**.

2. V. bromoides (L.) Gray, *Nat. arr. Brit. pl.* **2**: 124 (1821).
V. sciuroides (Roth) C.C. Gmel.
Stems up to 50 cm. Panicle sparingly branched or rarely reduced to a unilateral raceme, erect, 1-11 cm, usually well exserted from the uppermost sheath. Spikelets 6.5-11.5 mm (excluding the awns), the florets cleistogamous, most of them fertile; lower glume 2.5-5 mm, ½-¾ as long as the upper; lemmas 4.5-7.5 × 1.3-1.9 mm, with an awn about as long as the body; anthers usually 1, 0.4-0.7(-1.8) mm, usually included at anthesis.
Fl. IV - VIII. *Common in Madeira and the higher parts of Porto Santo; also present on Selvagem Grande; records from the Desertas probably refer to* V. muralis. **MPS**

3. V. muralis (Kunth) Nees in *Linnaea* **19**: 694 (1847).
Stems up to 60 cm. Panicle sparingly branched, rarely reduced to a unilateral raceme, erect, 3-15 cm, usually well exserted from the uppermost sheath. Spikelets 5-10 mm (excluding the

awns), the florets cleistogamous, most of them fertile; lower glume 1-3(-6) mm, ¼-½ as long as the upper; lemma 4-7(-10) × 0.8-1.3 mm, with an awn 2-3 times as long as the body; anthers usually 1, 0.3-0.7 mm, mostly included at anthesis.
Fl. III - VIII. *Roadsides and waste places; up to 1800 m. Common throughout Madeira; rare in Porto Santo and Bugío.* **MDP**

4. V. myuros (L.) C.C. Gmel., *Fl. bad.* **1**: 8 (1805).
Stems up to 65 cm. Panicle sparingly branched or reduced to a unilateral raceme, erect or nodding, usually not fully exserted from the uppermost sheath. Spikelets 6-10.5 mm (excluding the awns), the florets cleistogamous, most of them fertile; lower glume 0.4-2.5 mm, ¹⁄₁₀-⅖ as long as the upper; lemma 4.5-7.5 × 0.8-1.3 mm, with an awn up to twice the length of the body; anthers usually 1, 0.4-0.8(-1.3) mm, usually included at anthesis.
Fl. (I-)III - VII. *Pastures and ruderal sites, mainly in east central Madeira; recorded also from Porto Santo. Records for the Desertas are thought to be erroneous.* **MP**

9. Cynosurus L.

Annuals or perennials; ligule a membrane. Inflorescence a spike-like or capitate, ± unilateral panicle, bearing paired spikelets, the outer of each pair sterile and covering a fertile spikelet. Fertile spikelet (1-)2- to 5-flowered, breaking up above the glumes and between the florets; glumes narrow; lemmas coriaceous, rounded on the back, 5-nerved, acute, obtuse or shortly 2-toothed, awned. Sterile spikelet reduced to a pectinate cluster of sterile lemmas, persistent on the panicle.

1. Perennial; lemma of fertile spikelet with an awn shorter than the body . . **1. cristatus**
- Annual; lemma of fertile spikelet with an awn longer than the body 2

2. Leaves 3-9 mm wide; panicle dense; upper lemmas of sterile spikelet not much shorter or wider than the lower . . . . **2. echinatus**
- Leaves 1-3 mm wide; panicle rather lax and lobed; upper lemmas of sterile spikelet much shorter and wider than the lower . . . . **3. effusus**

1. C. cristatus L., *Sp. pl.* **1**: 72 (1753).
Perennial up to 75(-90) cm; leaves 0.5-2 mm wide. Panicle up to 7(-8.5) × 0.7 cm, narrowly oblong. Spikelets 3-5 mm, clearly distichous; glumes and lemmas of sterile spikelet similar, linear-lanceolate, shortly awned, the upper lemmas shorter than the lower; lemmas of fertile spikelets lanceolate, shortly awned.
Fl. VI - VIII. *Recorded from Madeira but without further data.* **M**

2. C. echinatus L., *Sp. pl.* **1**: 72 (1753).
Annual up to 60 cm; leaves 3-9 mm wide. Panicle 1-4 × 0.7-1.5 cm (excluding the awns), ovoid to oblong, unilateral, dense. Spikelets 8-10 mm, not obviously distichous; glumes and lemmas of sterile spikelets similar, narrow, the upper lemmas not much shorter or wider than the lower, all with awns 6-15 mm; lemmas of fertile spikelets lanceolate, 2-toothed at the tip and with a long awn from the sinus.
Fl. IV - VII. *Common throughout Madeira from sea-level to 1850 m, less so in Porto Santo; also present on Selvagem Grande.* **MPS**

3. C. effusus Link in Schrad., *J. Bot.* **1799**: 315 (1800).
C. elegans auct., non Desf. (1798), var. *brizoides* (Lowe) Menezes
Similar to **2** but smaller, with leaves 1-3 mm wide; panicle lax and lobed; and with upper lemmas of sterile spikelets much shorter and wider than the lower.
Fl. V - VIII. *Scattered in ravines throughout Madeira.* **M**

10. Lamarckia Moench

Annuals; ligule a membrane. Inflorescence a condensed, secund panicle, bearing deciduous clusters of 3 sterile spikelets concealing 2 smaller spikelets, one of them fertile and the other reduced. Fertile spikelet 1-flowered (rarely more) plus an awned rudiment; glumes narrow, longer than the lemma; floret raised on a long rhachilla-internode; lemma membranous, rounded on the back, 5-nerved, 2-toothed at the tip and awned from the sinus. Sterile spikelets many-flowered; lemmas membranous, empty, obtuse.

1. L. aurea (L.) Moench, *Methodus*: 201 (1794).
Stems up to 20 cm; uppermost sheaths inflated. Panicle oblong, up to 6 × 2.5 cm, golden yellow or tinged with purple; pedicels villous. Sterile spikelets 5-8 mm; glumes linear-lanceolate; lemmas ovate, distichous and imbricate.
Fl. II - V. *Fields and roadsides, frequently near the sea; up to 500 m. Widespread in southern and eastern Madeira, less frequent in Porto Santo and all three islands of the Desertas.* **MDP**

11. Briza L.

Annuals; ligule a membrane. Inflorescence an open panicle. Spikelets several- to many-flowered, ovate to rotund, laterally compressed, breaking up above the glumes and between the florets; glumes cordate, suborbicular; lemmas orbicular, chartaceous, appressed to the florets above, gibbous on the back and lightly keeled, 5- to 11-nerved, obtuse or bilobed.

1. Spikelets up to 12 in the panicle, (10-)15-25 mm . . . . **1. maxima**
- Spikelets numerous in the panicle, 3-5 mm . . . . **2. minor**

1. B. maxima L., *Sp. pl.* **1**: 70 (1753).
Stems up to 60 cm. Panicle 2-10 cm, secund, bearing 1-12 pendulous spikelets. Spikelets (10-)15-25 mm, 7- to 20-flowered; glumes 5-7 mm, concave, suborbicular, subcoriaceous; lemmas 6-8 mm, similar to the glumes.
Fl. (II-)III - VII. *Fields and vineyards. Common throughout Madeira, up to 1500 m; rare in Porto Santo and Deserta Grande.* **MDP**

2. B. minor L., *Sp. pl.* **1**: 70 (1753).
Stems up to 60 cm. Panicle 3-20 cm, erect, loose and much-branched, bearing numerous pendulous spikelets. Spikelets 3-5 mm, 4- to 8-flowered; glumes 2-3.5 mm, concave, hooded, suborbicular, subcoriaceous; lemmas 1.5-2 mm, similar to the glumes.
Fl. (III-)VI - VII. *Stony fields, rough grassland and cliff-tops. Common throughout Madeira, up to 1500 m; rare in Porto Santo.* **MP**

12. Poa L.

Annuals or perennials, the basal sheaths sometimes persistent and thickened into a bulb-like structure; ligule a membrane. Inflorescence an open or contracted panicle. Spikelets 2- to several-flowered, breaking up above the glumes and between the florets, rarely proliferating; lemmas herbaceous with hyaline margins, keeled throughout, the keel glabrous or ciliate, 5-nerved, awnless; floret-callus often with a web of fine woolly hairs, the rhachilla glabrous; palea-keels smooth, scaberulous or stiffly ciliolate.

1. Plant bulbous at the base; spikelets usually proliferating . . . . **5. bulbosa**
- Plant not bulbous at the base; spikelets not proliferating 2

2. Plant usually annual, rarely a perennial with compressed and procumbent stems . . . . **1. annua**
   Plants always perennial, with erect stems 3

3. Stems strongly compressed . . . . **4. compressa**
   Stems terete 4

4. Plant with short leafy stolons; ligule pointed, 4-10 mm . . . . **2. trivialis**
   Plant with creeping rhizomes; ligule blunt, usually less than 1 mm but up to 3 mm on upper leaves . . . . **3. pratensis**

1. P. annua L., *Sp. pl.* **1**: 68 (1753).
Annual, up to 30 cm, the stems often compressed, rarely perennial and then the stems markedly compressed and usually procumbent; ligule 2-5 mm. Panicle 3-8 cm, pyramidal, the lower branches spreading or deflexed after anthesis. Spikelets 3-10 mm, 3- to 5(10)-flowered; lemmas thinly hairy on the keel and nerves, without wool on the callus; palea-keels ciliate or rarely glabrous; anthers 0.6-0.8(-1) mm.
Fl. all year. *Very common in a wide variety of habitats throughout Madeira; recorded from Porto Santo but without further data.* M

2. P. trivialis L., *Sp. pl.* **1**: 67 (1753).
Tufted perennial with short leafy stolons, up to 90 cm, the stems erect, terete; ligule pointed, 4-10 mm. Panicle (9-)15-25 cm, ovoid-pyramidal, the branches densely scabrid. Spikelets 3-4 mm, 3- to 4-flowered; lemmas hairy on the keel, glabrous on the nerves, with a tuft of wool on the callus; palea-keels scaberulous; anthers 1.5-2 mm.
Fl. (IV-)V - VII. *Very common throughout Madeira.* M

3. P. pratensis L., *Sp. pl.* **1**: 67 (1753).
Tufted perennial with creeping rhizomes, up to 50(-70) cm, the stems erect, terete; ligule blunt, mostly less than 1 mm but up to 3 mm on upper leaves. Panicle 6-15 cm, ovoid to pyramidal, the branches sparsely scabrid. Spikelets 4-6 mm, 2- to 5-flowered; lemmas hairy on the keel and nerves and with a tuft of wool on the callus; palea-keels scaberulous; anthers 1.5-2 mm.
Fl. V - VI. *Grassy hillsides from 1000-1850 m. Scattered in Madeira.* M

4. P. compressa L., *Sp. pl.* **1**: 69 (1753).
Loosely tufted perennial with creeping rhizomes, up to 40 cm, the stems erect and strongly compressed; ligule blunt, 0.5-3 mm. Panicle 2-10 cm, narrowly oblong to ovate, the branches sparsely scabrid. Spikelets 3-8 mm, (2)4- to 8-flowered; lemmas thinly hairy on the keel and nerves, with a few strands of wool on the callus, or quite glabrous; palea-keels scaberulous; anthers 1-1.2 mm.
Fl. ? *Native to Europe and SW. Asia. Known only from Funchal and probably introduced.* M

5. P. bulbosa L., *Sp. pl.* **1**: 70 (1753).
Densely tufted perennial, up to 40 cm, the basal sheaths persistent and thickened into a bulb-like structure, the stems terete; ligule acute, up to 3 mm. Panicle 2-6 cm, ovoid, compact, the branches scabrid; spikelets purple-tinged, usually proliferating, the glumes and lemmas leaf-like; non-proliferating spikelets 3-5 mm, 2- to 6-flowered; lemmas hairy on the keel and nerves and with a tuft of wool on the callus; palea-keels scabrid; anthers 1-1.5 mm.
Fl. V. *Rocky ground; confined to Pico de Arieiro and Pico Ruivo at from 1500-1800 m.* M

Madeiran plants all seem to be of the proliferating kind usually called var. **vivipara** Koeler, but which is perhaps better accommodated at an even lower rank than variety.

13. Dactylis L.

Tufted perennials, the vegetative shoots strongly compressed; ligule a membrane. Inflorescence a contracted, lobed, 1-sided panicle, the spikelets crowded in compact clusters at the ends of the short main branches. Spikelets 2- to 5-flowered, strongly laterally compressed, breaking up above the glumes and between the florets; glumes keeled; lemmas thinly coriaceous, keeled, 5-nerved, spinously ciliate on the keel, entire or 2-toothed, mucronate to shortly awned.

Literature: P.F. Parker, Studies in *Dactylis* II. Natural variation, distribution and systematics of the *Dactylis smithii* Link. complex in Madeira and other Atlantic islands, in *New Phytol.* **71**(2): 371-378 (1972).

1. D. glomerata L., *Sp. pl.* **1**: 71 (1753).
D. glomerata subsp. *hispanica* (Roth) Nyman, subsp. *smithii* (Link) Stebbins & Zohary; *D. smithii* Link, subsp. *marina* (Borrill) Parker, subsp. *hylodes* Parker
Tufted or sub-shrubby, up to 60 cm, but usually much less. Panicle 2-18 cm, oblong and contracted or ovoid and rather lax. Spikelets 5-9 mm; lemmas 3-7 mm.
Fl. IV - IX(-XII). *Sea cliffs on the north coast of Madeira and on inland hills in the centre and southeast of the island; isolated populations occur on the south coast of Madeira, and on Pico Branco in Porto Santo; recorded from the Desertas but without further data.* **MP**

Two quite distinct morphological forms of *D. glomerata* occur in Madeira. One of these is a tall tussock-forming plant typical of the species over much of Europe and probably introduced for fodder. The other is a small sub-shrubby plant of sea-cliffs and inland hills and is also known from the Canaries and the Mediterranean region. Some authors have accorded this local form specific rank, as *D. smithii*, and it has been further subdivided into three subspecies. As a whole, *D. glomerata* is a complex comprising a number of widespread tetraploids with enclaves, mostly in the Mediterranean region, of diploids of narrow geographical range. Individual races can be separated according to such features as stomatal dimensions, pollen size, chromosome number and average population characteristics, but since they overlap so much it is not yet clear at what rank they should be treated. Taxonomy over the whole range of the species is very uneven and no wholly satisfactory account is available. While the two forms in Madeira are readily distinguishable, the matter is further complicated by the fact that two subspecies of *D. smithii* are said to occur in Madeira, subsp. *marina* on the coast and subsp. *hylodes* inland. They are both tetraploid. With the current state of our understanding of *Dactylis*, it seems best for now to acknowledge that these different populations exist without formally assigning species or subspecies rank to them.

14. Parafestuca Alexeev

Perennials; ligule a membrane. Inflorescence a panicle. Spikelets several-flowered, breaking up above the glumes and between the florets; lemmas coriaceous, keeled throughout, 3-nerved; floret-callus and rhachilla shortly hairy.

An endemic, monospecific genus.

1. P. albida (Lowe) Alexeev in *Byull. Mosk. Obshch. Ispyt. Prir.*, (Biol.) **90**: 107 (1985).
Festuca albida Lowe
Densely tufted, up to 100 cm, the stems densely pubescent; lower sheaths disintegrating into fibres; leaves 3-10(-12) mm wide. Panicle narrowly ovoid, (8-)15-35 cm, dense. Spikelets 6-8(-10) mm, 2- to 3(4)-flowered; glumes lanceolate, the lower 5.5-8 mm, the upper 6.5-9 mm; lemmas lanceolate, 6.5-9 mm, scabrid on the keel, acute, awnless; anthers 2.5-3.5 mm. **Plate 54**
Fl. V - VIII. *Endemic on rocky hillsides between 1100 and 1200 m, mainly in eastern and central Madeira,* ● **M**

15. Sphenopus Trin.

Annuals; ligule a membrane. Inflorescence an effuse panicle. Spikelets several-flowered, on long pedicels that gradually widen upwards, breaking up above the glumes and between the florets; glumes unequal, shorter than the first lemma; lemma membranous, keeled, 3-nerved, obtuse to subacute, awnless.

1. **S. divaricatus** (Gouan) Rchb., *Fl. germ. excurs.*: 45 (1830).
Stems up to 30 cm. Panicle up to 10 cm, the branches mostly paired and naked for more than half their length. Spikelets 2-3 mm, 2- to 5-flowered; lower glume 0.1-0.4 mm; upper glume 0.6-1 mm; lemma 1.5-2 mm; anthers c.0.4 mm.
Fl. VI. *Shady places in the mountains of Madeira; apparently rare.* **M**

16. Catapodium Link

Annuals; ligule a membrane. Inflorescence a panicle with short stiff branches or reduced to a raceme. Spikelets several- to many-flowered on stout pedicels, breaking up above the glumes and between the florets; glumes subequal, coriaceous; lemmas coriaceous, rounded on the back or keeled only towards the tip, 5-nerved, glabrous, awnless or at most shortly apiculate.

1. **C. rigidum** (L.) C.E. Hubb. in Dony, *Fl. Bedfordshire*: 437 (1953).
Scleropoa rigida (L.) Griseb.
Stems up to 35 cm. Inflorescence 1-12(-18) cm, rigid, usually branched below, the branches often bare of spikelets in the lower part. Spikelets 4-20 mm, 5- to 12-flowered, readily breaking up; glumes subequal, the lower 1.3-2 mm, the upper 1.5-2.3 mm; lemmas 2-2.6(-3) mm, acute to obtuse, often shortly apiculate; rhachilla-joints 0.5-1 mm.
Fl. (I-)III - V. *Wall-tops, hillsides, roadsides and as a weed of gardens; in and around Funchal, on the Ponto de São Lourenço, Porto Santo and Ilhéu Chão.* **MDP**

All records of the similar *C. marinum* (L.) C.E. Hubb. in *Bull. misc. Inf. R. bot. Gdns, Kew* **1954**: 375 (1954) from Madeira and the Desertas are almost certainly referable to 1. *C. marinum* is distinguished by having the spikelets in 2 rows, the lower glume 2-3 mm and the upper 2.3-3.3 mm.

17. Parapholis C.E. Hubb.

Annuals; ligule a membrane. Inflorescence a single cylindrical bilateral raceme with fragile axis, the spikelets alternate in 2 opposite rows, sessile and more or less sunk in the rhachis-joint. Spikelets 1-flowered; glumes appressed to the rhachis, subequal, placed side by side, exceeding and covering the floret, coriaceous; lemma hyaline, entire, awnless. Grain with liquid endosperm.

1. Keel of glumes distinctly winged; anthers over 2 mm; plants elongate and slender . . . . **1. filiformis**
- Keel of glumes not winged; anthers under 1 mm; plants usually strongly curved . . . . **2. incurva**

1. **P. filiformis** (Roth) C.E. Hubb. in *Blumea*, suppl. **3**: 14 (1946).
Stems up to 25 cm, slender, erect or decumbent; uppermost sheath usually not or only slightly inflated. Raceme 3-15 cm, straight or almost so; spikelets 4-6 mm; glumes linear-lanceolate, acuminate, with distinctly winged keel; anthers 2-3.5 mm.
Fl. ? *Recorded for Madeira and Porto Santo but without further data.* **MP**

2. **P. incurva** (L.) C.E. Hubb. in *Blumea*, suppl. **3**: 14 (1946).
Stems up to 25 cm, erect or decumbent, often strongly curved; uppermost sheath inflated. Raceme

1-10 cm, usually strongly curved; spikelets 4.5-6 mm; glumes lanceolate, acute, the keel not winged; anthers 0.5-0.9 mm.
Fl. IV - V. *Infrequent on sandy soils in Porto Santo; also recorded for Madeira but without further data.* **MP**

18. Hainardia Greuter

Annuals; ligule a membrane. Inflorescence a single cylindrical bilateral raceme with fragile axis, the spikelets alternate in 2 opposite rows, sessile and more or less sunk in the rhachis-joint. Spikelets 1-flowered; lower glume suppressed except in the terminal spikelet, the upper appressed to the rhachis, exceeding and covering the floret, coriaceous; lemma hyaline, entire, awnless. Grain with solid endosperm.

1. H. cylindrica (Willd.) Greuter in *Boissiera* **13**: 177 (1967).
Monerma cylindrica (Willd.) Coss. & Durieu
Stems up to 35 cm, erect or ascending. Raceme up to 25 cm, straight or curved, rigid; spikelets 5-8 mm; glume ovate-lanceolate, acuminate; anthers 2.5-3.5 mm.
Fl. IV - VIII. *Waste ground near the sea in southeast Madeira; also recorded from Porto Santo and the Desertas.* **MDP**

19. Glyceria R. Br.

Perennials; ligule a membrane; sheath-margins connate. Inflorescence a panicle. Spikelets several-flowered, the uppermost floret reduced, breaking up above the glumes and between the florets; glumes persistent, shorter than the adjacent lemma; lemmas rounded on the back, 7-nerved, the nerves conspicuously parallel, hyaline at the tip, awnless.

1. G. declinata Bréb., *Fl. Normandie*, ed. 3: 354 (1859).
G. fluitans var. *pumila* Fr. ex Andersson, non Wimm. & Grab. (1827)
Stems up to 50 cm, often ascending. Panicle 3-25 cm, narrow, secund, with few spikelets, contracted in fruit; branches 1-3 at the central nodes, the longer with 1-6, the shorter with 1(-2) spikelets. Spikelets 13-25 mm; lower glume ovate, 1.5-2.5 mm, subacute to acute; upper glume elliptic, 2.5-4 mm, subobtuse; lemma oblong-ovate, 3.5-4.5 mm, prominently 7-nerved, scabrid on the back, with 3-5 distinct obtuse or acute teeth at the tip; palea deeply divided into 2 aristate teeth distinctly exceeding the lemma; anthers 0.6-1.1 mm.
Fl. III - IX. *In rivers, ditches and other moist places, scattered through most of Madeira above 600 m.* **M**

G. spicata Biv. ex Guss. (*G. fluitans* var. *spicata* (Biv. ex Guss.) Trabut) is like **1** but with spikelets mostly subsessile; lemma 5-6 mm, weakly nerved, rounded to truncate and shallowly crenulate; anthers 1.1-1.6 mm. It has been recorded for Madeira but probably in error for **1**.

20. Melica L.

Perennials; ligule a membrane; sheath-margins connate. Inflorescence a panicle. Spikelets several-flowered, the uppermost florets reduced and contracted to a clavate mass of sterile lemmas, breaking up below the glumes or below the lowest floret, reluctantly between the florets; glumes shorter than the adjacent lemma, papery; lemmas rounded on the back, 5- to 9-nerved, the nerves not parallel, hyaline towards the tip, usually awnless.

1.	Spikelets 7-9 mm; lemma 5.5-6.5 mm, with hairs 4-5 mm	**1. ciliata**
-	Spikelets 4.5-6(-6.5) mm; lemma 5-6 mm, with hairs 3-4 mm	**2. canariensis**

1. M. ciliata L., *Sp. pl.* **1**: 66 (1753).
Stems sometimes creeping at the base, up to 60(-120) cm. Panicle 12-25 cm, usually lobed. Spikelets with 1 fertile floret, 7-9 mm; lower glume 4.5-5 mm; upper glume as long as the spikelet; fertile lemma 5.5-6.5 mm, ciliate on the margins and lateral nerves with hairs 4-5 mm long.
Fl. IV - VIII. *Rocky hillsides and roadsides from 300-750 m in southern Madeira; also on Salvagem Grande.* **MS**

All Madeiran material seems to be referable to subsp. **magnolii** (Gren. & Godr.) Husn., *Graminées*: 56 (1898), to which the above description applies. Subsp. *ciliata*, from Europe, has a panicle usually less than 8 cm and spikelets 5-7 mm.

2. M. canariensis W. Hempel in *Feddes Reprium* **75**: 109 (1967).
Like **1** but with spikelets 4.5-6(-6.5) mm and lemma 5-6 mm with hairs 3-4 mm long. **Plate 54**
Fl. VI-IX. *Apparently infrequent in southern Madeira; also recorded from the Salvages.* ■ **MS** [*Canaries*]

It is probable that *M. canariensis* does not warrant rank above subspecies of *M. ciliata*, but there is not enough material available for its true status to be properly determined. It has the spikelets of *M. ciliata* subsp. *ciliata*, but the panicle of subsp. *magnolii*.

21. Helictotrichon Schult.

Perennials; ligule a membrane. Inflorescence a panicle. Spikelets 2- to several-flowered, the uppermost 1 or 2 florets reduced, breaking up above the glumes and between the florets; glumes unequal, 1- to 5-nerved, rounded on the back, shorter than the spikelet; lemmas rounded on the back, firmly membranous to coriaceous, with a geniculate awn from the back.

1. H. marginatum (Lowe) Röser in *Dissnes bot.* **145**: 136 (1989).
Avena marginata Lowe; *A. sulcata* Gay ex Boiss.; *Avenula marginata* (Lowe) Holub; *Helictotrichon sulcatum* (Gay ex Boiss.) Henrard
Stems up to 100 cm. Panicle 10-20 cm, narrow. Spikelets 10-20 mm; lower glume 9-14 mm; upper glume 12-17 mm; lemma 10-16 mm, longitudinally grooved in the lower half, yellowish green or reddish brown below, bifid at the tip, the teeth with bristles 0.8-2 mm long.
Fl. VI - VIII. *Infrequent in rocky fields and on slopes up to c. 1300 m, mainly in the central mountains of Madeira around Pico de Arieiro.* **M**

22. Arrhenatherum P. Beauv.

Perennials; basal internodes often swollen into globose corm-like structures; ligule a membrane. Inflorescence a narrow panicle. Spikelets 2-flowered, with or without an additional rudiment, the lower floret male with a geniculate awn, the upper bisexual and weakly awned or awnless, breaking up above the glumes, the florets falling together; glumes unequal, the upper as long as the spikelet; lemmas firmly membranous to subcoriaceous, rounded on the back, 2-toothed at the tip.

1. A. elatius (L.) P. Beauv. ex J. & C. Presl, *Fl. čech.*: 17 (1819).
Stems up to 150 cm; basal internodes swollen into 1-6 corm-like structures 6-10 mm in diameter. Panicle oblong or lanceolate, 10-25(-30) cm, loose or rather dense. Spikelets 7-10 mm; florets separated by a rhachilla-internode up to 0.6 mm; lower glume 4-6 mm; upper glume as long as the spikelet; first lemma 7-10 mm, glabrous or sparsely hairy, the hairs up to 1 mm; awn 10-20 mm, arising in the lower third of the lemma.
Fl. V - IX. *Throughout most of Madeira.* **M**

Madeiran plants belong to subsp. **bulbosum** (Willd.) Schübl. & M. Martens, *Fl. Würtemberg*: 70 (1834), to which the above description applies. Records of subsp. *elatius* (L.) P. Beauv. ex Presl, which lacks swollen basal internodes, should probably be refered to subsp. *bulbosum*.

A. album (Vahl) Clayton in *Kew Bull.* **16**: 250 (1962) (*A. elatius* subsp. *erianthum* (Boiss. & Reut.) Trab.) is similar to **1** but with the first lemma *c.* 8 mm, villous in the lower half, the hairs 2-3 mm; awn 15-25 mm, arising from near the base of the lemma. It was reported in 1914 as being a rare plant in Madeira but has not been recorded since. In the absence of further information its presence on the island requires confirmation. **M**

23. Avena L.

Annuals; ligule a membrane. Inflorescence an open or contracted panicle with large pendulous spikelets. Spikelets 2- to 6-flowered, the lower 1-3 florets fertile, the remainder progressively reduced, breaking up above the glumes, sometimes also between the florets (all florets persistent in cultivated species); glumes lanceolate to ovate or elliptic, equal or nearly so, 7- to 11-nerved, rounded on the back, as long as the spikelet or almost so; lemmas lanceolate to ovate or oblong, rounded on the back, herbaceous but hardening in fruit, with a geniculate or rarely straight awn from the back, or awnless in some cultivated species.

The mode of disarticulation of the rhachilla is an important character in *Avena*, particularly as to whether it breaks up between the florets. If it does so, then the articulation can be seen as an oblique line from the side and as a bearded horseshoe-shaped join in face-view. In non-disarticulating spikelets there is no such join and the rhachilla-internodes are generally shorter and stouter.

Two species of oat are cultivated in Madeira:

A. strigosa Schreb. is cultivated on a small scale but may persist as a weed of other cereals. The lemma-teeth bear awns 1-9 mm long.

A. brevis Roth, is similarly grown on a small scale and may also persist. The lemma-teeth are awnless or sometimes have a short awn-point up to 1 mm.

1. Rhachilla continuous between the florets, not breaking up at maturity ... **1. sterilis**
   - Rhachilla articulated between the florets, these falling separately at maturity 2

2. Tip of lemma with 2 awned teeth, the awns up to 12 mm .... **2. barbata**
   - Tip of lemma with 2 acute teeth .... **3. fatua**

1. A. sterilis L., *Sp. pl.*, ed. 2, **1**: 118 (1762).
Stems up to 150 cm. Panicle up to 40 cm, usually patent. Spikelets 30-50 cm, 3- to 5-flowered; rhachilla disarticulating at maturity above the glumes but not between the florets; lemma narrowly lanceolate, 25-40 mm, stiffly hairy in the lower ⅔, 2-toothed at the tip, the teeth acute; awn 30-90 mm.
Fl. III - VIII. *Weed of fields and roadsides in southeast Madeira and on Ilhéu Chão.* **MD**

2. A. barbata Pott ex Link in Schrad., *J. Bot.* **2**: 315 (1799).
A. lusitanica (Morais) Baum
Stems up to 100 cm. Panicle up to 30(-50) cm, subsecund. Spikelets 20-30 mm, 2- to 3-flowered; rhachilla disarticulating at maturity above the glumes and between the florets; lemma narrowly lanceolate, 16-20 mm, villous in the lower half, with 2 awned teeth at the tip, the awns 3-12 mm; central awn 30-60 mm.
Fl. IV - VII. *Common on roadsides and in waste places, often near the sea; southern and eastern Madeira; Porto Santo; also reported for the Desertas but without further data.* **MDP**

A. lusitanica is sometimes treated as a separate species on account of having minor differences in lemma-tooth morphology and lodicule structure, but it is doubtful whether either of these features has any taxonomic significance.

3. A. fatua L., *Sp. pl.* **1**: 80 (1753).
A. occidentalis Durieu
Stems up to 150 cm. Panicle up to 40 cm, nodding, sometimes subsecund. Spikelets 18-25 mm, 2- to 3-flowered; rhachilla disarticulating at maturity above the glumes and between the florets; lemma ovate-lanceolate, 12-23 mm, glabrous or hairy on the back below, 2-toothed at the tip, the teeth acute; awn 25-40 mm.
Fl. V - VII. *Weed of wheatfields in Madeira and Porto Santo; records from the Desertas appear to be erroneous.* **MP**

A. occidentalis has been treated as a separate species but differs only in the shape of the scar at the base of the upper florets. It is a difficult character to see (slightly reniform instead of elliptic) and is of very doubtful taxonomic value. Other characters used concern lodicule morphology and epiblast form, but these are almost inaccessible and are likewise of doubtful value.

The hybrid *A. fatua* × *A. sativa* has been reported from Funchal. Since *A. sativa* is not cultivated in Madeira it was probably introduced accidentally. It is not known whether it has persisted since its discovery in 1974.

24. Gaudinia P. Beauv.

Annuals or biennials; ligule a membrane. Inflorescence a fragile bilateral raceme, fracturing at the base of each internode, the spikelets sessile in opposite rows and broadside to the axis. Spikelets several-flowered, falling entire; glumes equal or unequal, shorter than the spikelet; lemmas thinly coriaceous, weakly keeled, acute, with a dorsal geniculate awn.

1. G. fragilis (L.) P. Beauv., *Ess. Agrostogr.*: 95, 164 (1812).
Annual up to 120 cm. Inflorescence up to 35 cm. Spikelets 10-18 mm; lower glume 3-6 mm; upper glume 8-10 mm; lemma c.7 mm, glabrous or hairy, the awn c.10 mm.
Fl. IV - VII. *Introduced weed of roadsides, gardens and waste places in southeast Madeira. Native to the Mediterranean region.* **M**

25. Rostraria Trin.

Annuals; ligule a membrane. Inflorescence a loose or dense spike-like panicle. Spikelets 2- to 5-flowered, breaking up above the glumes and between the florets; glumes subequal or unequal, keeled; lemmas usually a little longer than the glumes, acute or obtuse, with a straight or slightly curved awn from just below the bifid tip.

1. Glumes unequal, the lower 1-nerved, shorter and narrower than the 3-nerved upper, thinly hairy or glabrous, rarely woolly; rhachilla not produced, subglabrous or with hairs up to 0.5 mm long . . . . **1. cristata**
- Glumes subequal, 3-nerved, the lower often shortly and densely woolly, the upper rarely so; rhachilla produced, densely villous with hairs *c.*1 mm long . . . . **2. pumila**

1. R. cristata (L.) Tzvelev in *Nov. Sist. Vysshikh Rast.* **7**: 47 (1971).
Lophochloa cristata (L.) Hyl.; *Koeleria phleoides* (Vill.) Pers.
Stems up to 60 cm. Panicle cylindrical and dense or pyramidal, lax and more or less lobed, 1-12 cm. Spikelets 3- to 6-flowered, 3-7.5 mm; rhachilla not produced, subglabrous or with hairs not more than 0.5 mm; glumes glabrous or hairy (rarely woolly), acute, the lower narrowly

elliptic, 1-nerved, 2-3 mm, the upper elliptic, 3-nerved, 2.5-3.5 mm; lemmas elliptic, 2.5-3.5 mm, glabrous or thinly hairy, smooth or minutely papillose, acute; awn 1-3 mm, usually straight.
Fl. III - VI. *Weed of gardens and waste places; scattered in southeast Madeira and throughout Porto Santo; also recorded from the Desertas without further data.* **MP**

2. R. pumila (Desf.) Tzvelev in *Nov. Sist. Vysshikh Rast.* 7: 48 (1971).
Lophochloa pumila (Desf.) Bor
Stems up to 35 cm. Panicle ovate-oblong, lax and sometimes more or less lobed, 1-6 cm. Spikelets usually 4-flowered, 3.5-5 mm; rhachilla produced, densely villous with hairs c.1 mm long; glumes subequal, elliptic, 3-nerved, 2.5-3 mm, acute, the lower often shortly and densely woolly, the upper less so or glabrous; lemma elliptic, 3-3.5 mm, glabrous, acute; awn 1.5-2.5 mm, straight or slightly curved.
Fl. V. *Rare; known only from Campo de Baixo in Porto Santo.* **P**

26. Deschampsia P. Beauv.

Perennials; ligule a membrane. Inflorescence usually an open panicle. Spikelets 2-flowered with well developed hairy rhachilla-extension, breaking up above the glumes and between the florets; glumes subequal, as long as the spikelet, or almost so; lemmas hyaline to polished-cartilaginous, rounded on the back, 5-nerved, 4-toothed at the tip, with an inconspicuous awn from the base or lower half.

1. Anthers 2.6-3.5 mm; lemma 5.4-6.3 mm . . . . **1. maderensis**
- Anthers 1.3-1.8 mm; lemma 2.9-4.4 mm . . . . **2. argentea**

1. D. maderensis (Hack. & Bornm.) Buschm. in *Phyton, Horn* 2(4): 276 (1950).
D. foliosa auct. mad., non Hack.
Stems up to 30 cm; leaves folded-setaceous, c.0.35 mm across. Panicle 5-6.5 cm, narrow, the branches ascending, not whorled. Lower glume 5.3-7.3 mm; upper glume 5.8-7.5 mm; lemmas 5.4-6.3 mm, the upper slightly exceeding the glumes, the outer apical teeth longer than the inner, awned from near the base; anthers 2.6-3.5 mm. **Plate 54**
Fl. VI - VIII. *Endemic to Madeira on rocky slopes from c. 1500-1600 m around Pico de Arieiro, Pico Ruivo and Ribeiro Frio.* ● **M**

Similar in overall appearance to *D. foliosa* Hack., with which it is often confused, but the leaves are marginally finer; the most obvious differences are in lemma and anther length. *D. foliosa* occurs in the Azores; Madeiran plants so-named all belong to *D. maderensis*.

2. D. argentea (Lowe) Lowe in *Trans. Camb. phil. Soc.* 6: 529 (1838).
D. argentea var. *gomesiana* Menezes; *D. argentea* var. *prorepens* Hack. & Bornm.
Stems up to 85 cm; leaves folded-setaceous, 0.6-1.5 mm across. Panicle (7-)12-35 cm, lax, often nodding, the branches whorled, filiform. Lower glume 3.5-4.4(-5.7) mm; upper glume 3.7-5.1(-6.4) mm; lemmas 2.9-4.4 mm, the upper slightly exceeding the glumes, the outer apical teeth longer than the inner, awned from near the base; anthers 1.3-1.8 mm. **Plate 54**
Fl. VI - IX. *Endemic to damp woodland on rocky slopes from 150-1500 m; mostly in eastern central Madeira.* ● **M**

This species is also very similar to *D. foliosa*, differing only in the size of the plant as follows:
D. argentea: height 30-85 cm; leaf 0.6-1.5 mm wide; ligule 5.5-13 mm; panicle (7-)12-35 cm
D. foliosa: height 11-32 cm; leaf 0.4-0.7 mm wide; ligule 2.5- 6.3 mm; panicle 3.5-9 cm.
Although the taxa are isolated on different islands, in all these features variation is quite continuous and in length of glumes, lemmas and anthers the two are identical. It is possible that the rank of subspecies would be more appropriate.

27. Holcus L.

Perennials; ligule a membrane. Inflorescence a moderately dense panicle. Spikelets 2-flowered, falling entire, the lower floret bisexual and the upper male; glumes subequal, enclosing the florets, papery; lemmas cartilaginous, polished, rounded on the back, the lower awnless, the upper with a dorsal awn from the upper third.

1. Awn hooked, not protruding beyond the glume-tips, but sometimes protruding from the side; plant softly hairy . . . . **1. lanatus**
- Awn faintly geniculate, protruding beyond the glume-tips; plant glabrous except for the conspicuously bearded nodes . . . . **2. mollis**

1. H. lanatus L., *Sp. pl.* **2**: 1048 (1753).
Softly hairy, tufted and often short-lived, up to 100 cm. Panicle 3-20 × 1-8 cm, somewhat loose to very dense, whitish or tinged with pale green, pink or purple. Spikelets 4-6 mm; glumes lanceolate, the upper a little longer and wider than the lower; upper lemma with a strongly hooked awn *c*.2 mm.
Fl. V - VIII. *Introduced weed, native to the temperate Old World; roadsides, pastures and woodland margins. Common in most of Madeira; present in Porto Santo, Deserta Grande and Selvagem Grande.* **MDPS**

2. H. mollis L., *Syst. nat.* ed. 10, **2**: 1305 (1759).
Rhizomatous, forming compact tufts or loose mats, glabrous except for the conspicuously bearded nodes, up to 100 cm. Panicle 4-12 × 1-5 cm, compact or somewhat loose, whitish or tinged with grey or purple. Spikelets 4-6 mm; glumes slightly unequal, the lower lanceolate, the upper ovate; upper lemma with a faintly geniculate awn 3.5-5 mm.
Fl. VII. *Infrequent to rare; north-west Madeira.* **M**

28. Aira L.

Annuals; ligule a membrane. Inflorescence an open or contracted panicle. Spikelets 2-flowered, without rhachilla-extension, breaking up above the glumes, the florets arising at about the same level; glumes equal, as long as the spikelet; lemmas lanceolate, thinly coriaceous, rounded on the back, acuminately 2-toothed at the tip, with a geniculate awn arising from the lower half.

1. Panicle contracted, spike-like, the pedicels shorter than the spikelets; sheaths smooth . . . . **1. praecox**
- Panicle lax with long spreading branches, the pedicels at least as long as the spikelets; sheaths retrorsely scaberulous . . . . **2. caryophyllea**

1. A. praecox L., *Sp. pl.* **1**: 65 (1753).
Delicate plant up to 20 cm; sheaths smooth. Panicle contracted, spike-like, 0.5-5 cm, silvery, purplish or pale green, the branches and pedicels short, the latter shorter than the spikelets. Spikelets 2.5-3.5 mm.
Fl. V - VIII. *Frequent in rocky places in the central mountains of Madeira from 1100-1800 m.* **M**

2. A. caryophyllea L., *Sp. pl.* **1**: 66 (1753).
Delicate plant up to 35(-50) cm; sheaths retrorsely scaberulous. Panicle very loose with widely spreading branches, 1-12 cm long and wide, silvery or tinged with purple, the branches bare of spikelets below and the pedicels at least as long as the spikelets. Spikelets (2-)2.5-3.5 mm.
Fl. IV - VII. *Frequent throughout most of Madeira, less common in Porto Santo.* **MP**

Two subspecies are usually recognized in Europe and both are said to occur in Madeira. Subsp. *caryophyllea* is the more widespread, forming loose tufts up to 35 cm high, with spikelets 2.5-3.5 mm on pedicels up to 5 mm; subsp. *multiculmis* (Dumort.) Bonnier & Layens is restricted to SW Europe, forms denser tufts up to 50 cm high and has smaller spikelets (2-2.5 mm) on pedicels usually less than 5 mm. However, the morphological differences are so ill-defined that it is doubtful whether these variants are worth formal recognition.

29. Anthoxanthum L.

Perennials; ligule a membrane. Inflorescence a contracted spike-like panicle. Spikelets 3-flowered, breaking up above the glumes, the 2 lower florets reduced to sterile lemmas; glumes unequal, the upper exceeding and enclosing the florets; sterile lemmas exceeding the fertile, membranous, the lower with a short straight awn from above the middle, the upper with a geniculate awn from below the middle; fertile lemma cartilaginous, the flower strongly protogynous.

1. A. odoratum L., *Sp. pl.* **1**: 28 (1753).
A. odoratum var. *villosum* Loisel. ex Rchb.
Tufted, sweetly aromatic plant up to 75(-100) cm. Panicle 2-9(-12) cm, often interrupted below. Spikelets 7-9.5 mm; lower glume 3.7-5.2 mm; sterile lemmas 2.8-3.6 mm, hairy; fertile lemma 1.7-2.4 mm, smooth and glossy; anthers 2, 3.5-5 mm.
Fl. III - VI. *Hillsides in damp woodland in the higher parts of Madeira, from 1000-1800 m.*
M

30. Phalaris L.

Annuals or perennials; ligule a membrane. Inflorescence a contracted or spike-like panicle. Spikelets all alike or rarely in deciduous clusters of 6 or 7, of which only 1 or 2 are fertile, 3-flowered, with the 2 lower florets reduced or rarely one of them sometimes completely suppressed, breaking up above the glumes; glumes subequal, papery, boat-shaped, keeled, the keel often broadened into a wing, enclosing the florets; sterile florets seldom more than half as long as the fertile, chaffy or reduced to a fleshy scale; fertile lemma glabrous or pubescent, often glossy, laterally compressed, hardened.

1.	Fertile lemma ± glabrous; fertile spikelets 1(-2) surrounded by an involucre of male or sterile spikelets and usually falling as a whole cluster at maturity	2
-	Fertile lemma sparsely to densely pubescent; spikelets all fertile and falling singly at maturity	3
2.	Annual	**1. paradoxa**
-	Perennial with tuberous base	**2. coerulescens**
3.	Perennial	**3. aquatica**
-	Annual	4
4.	Sterile floret 1	5
-	Sterile florets 2	6
5.	Glumes acute in outline, the wing narrowed above	**4. minor**
-	Glumes obtuse in outline, the wing truncate above	**5. maderensis**
6.	Sterile florets reduced to minute fleshy scales less than 1 mm	**6. brachystachys**
-	Sterile florets broad and chaffy, at least half as long as the fertile floret	**7. canariensis**

1. P. paradoxa L., *Sp. pl.* ed. 2, **2**: 1665 (1763).
Annual up to 100 cm. Panicle oblong, up to 9 × 2 cm, dense, spike-like; spikelets falling in groups of 6 or 7 with 5 or 6 sterile spikelets clustered about a single fertile spikelet. Glumes of fertile spikelets 5.5-8 mm, acuminate to subulate, winged above, the wing-margin with a tooth-like projection near the middle; sterile florets of fertile spikelet 2, obsolete, represented only by knob-like projections; fertile floret 2.5-3.5 mm, glabrous or with a few short hairs near the tip.
Fl. III - V(-VIII). *An introduced weed of arable fields and waste places; scattered along the southern coast of Madeira; around Vila Baleira in Porto Santo; also Ilhéu Chão. Native to the Mediterranean region and SW Asia.* **MDP**

2. P. coerulescens Desf., *Fl. atlant.* **1**: 56 (1798).
Tufted perennial up to 150 cm, the stems bulbous at the base. Panicle cylindrical or ovate-oblong, 3-11 × 1-2.5 cm; spikelets falling singly or in groups of 6 or 7, the latter with 1 or 2 fertile surrounded by a cluster of male spikelets. Glumes of fertile spikelet 5.5-9 mm, acute or acuminate, winged, the wing-margin markedly toothed or erose; sterile lemmas of fertile spikelet 2, obsolete, represented only by knob-like projections; fertile floret 2.7-4.4 mm, glabrous or with a few short hairs at the base.
Fl. III - VI(-X). *Native to the Mediterranean region and S. Europe. An introduced weed of arable fields, screes and waste places; frequent from Funchal to the Ponta de São Lourenço; also in Porto Santo and Deserta Grande.* **MDP**

3. P. aquatica L., *Amoen. acad.* **4**: 264 (1759).
P. tuberosa L.; *P. nodosa* Menezes, non Murray (1774); *P. altissima* Menezes
Tufted or shortly rhizomatous perennial up to 150 cm, the base often tuberous. Panicle cylindrical and spike-like or occasionally lobed at the base, 1.5-12 × 1-2.5 cm; spikelets falling singly. Glumes 4.4-7.5 mm, acute, broadly winged, the wing-margin entire; sterile floret 1, subulate, 0.2-2.2 mm (sometimes with a short second floret up to 0.5 mm); fertile floret 3.1-4.6 mm, densely pubescent.
Fl. IV - VIII. *Native to the Mediterranean region. An introduced weed of gardens and waste places in Madeira and Porto Santo; also reported from the Desertas but without further data.* **MDP**

P. arundinacea L. is similar to **3** but has long, scaly rhizomes and a dense, lobed panicle spreading at anthesis. The variegated var. **picta** L. is cultivated in gardens in Madeira.

4. P. minor Retz., *Observ. bot.* **3**: 8 (1783).
Annual up to 100 cm. Panicle ovate-oblong, 1-6 × 1-2 cm; spikelets falling singly. Glumes 4-6.5 mm, acute, broadly winged, the wing narrowed above, its margin usually toothed or erose, rarely entire; sterile floret 1, either obscure and 0.2-0.3 mm or well developed and 1.1-1.8 mm; fertile floret 2.7-4 mm, pubescent.
Fl. IV - VII. *Native to the Mediterranean region eastwards to the NW Himalaya. An introduced weed of fields, gardens and waste places in and around Funchal; around Vila Baleira and Serra de Dentro in Porto Santo.* **MP**

In the form of this species in which the sterile floret is obscure it closely resembles the callus of the fertile lemma; care must be taken not to mistake the callus for a second sterile lemma. Only *P. brachystachys* among the annuals has two obscure sterile florets, but the wing of the glume is entire.

5. P. maderensis (Menezes) Menezes, *Gram. Madeira*: 23 (1906).
Annual up to 50 cm. Panicle oblong to cylindrical, 2.5-5 × 1 cm, spike-like; spikelets falling singly. Glumes 4.5-4.8 mm, acute, broadly winged above, the wing truncate at the tip, its margin

entire or shallowly toothed; sterile floret 1, subulate, 1.3-1.4 mm; fertile floret 3.2-3.4 mm, densely pubescent. **Plate 55**
Fl. IV - V. *Known for certain only from the Ilhéus de Ponta de São Lourenço also reported from Porto Santo but without further data.* ■ M [*Canaries*]

Differs from **4** only in the shape of the glume-wing and in the more cylindrical, rather than ovate-oblong, panicle.

6. **P. brachystachys** Link in *Neues J. Bot.* **1**: 134 (1806).
P. canariensis sensu Brot., non L. (1753)
Annual up to 90 cm. Panicle ovoid, 1.5-4 × 0.8-1.8 cm, spike-like; spikelets falling singly. Glumes 6.3-8.5 mm, acute, broadly winged, the wing-margin entire; sterile florets 2, swollen and fleshy, 0.6-1.2 mm; fertile floret 4.4-5.5 mm, densely pubescent.
Fl. III - VII. *Native to the Mediterranean region and the Canaries. An introduced weed of arable fields; from sea-level to 600 m; apparently scattered in southeast Madeira; also on Pico do Facho in Porto Santo.* **MP**

7. **P. canariensis** L., *Sp. pl.* **1**: 54 (1753).
Annual up to 100 cm. Panicle ovate to oblong-ovate, 1.5-4 × 1.5-2 cm; spikelets falling singly. Glumes 7-10 mm, acute, broadly winged, the wing-margin entire; sterile florets 2, broad and chaffy, 2.5-4.5 mm; fertile floret 4.8-6.8 mm, densely pubescent.
Fl. IV - VIII. *A weed of gardens and waste places in and around Funchal.* **M**

31. Agrostis L.

Annuals or perennials; ligule a membrane. Inflorescence an open or contracted panicle. Spikelets 1-flowered without rhachilla-extension, breaking up above the glumes; glumes at least as long as the lemma, membranous, 1-nerved; lemma hyaline, thinner than the glumes, rounded on the back, (3-)5-nerved, awned or awnless, the awn dorsal and usually geniculate; palea shorter than the lemma, often minute.

1. Palea inconspicuous, less than ⅙ as long as the lemma; plant annual . . **4. pourrettii**
 - Palea conspicuous, more than ⅙ as long as the lemma; plant perennial 2

2. Pedicels at least twice as long as the spikelets, capillary, the panicle very lax . . . . **3. obtusissima**
 - Pedicels about as long as the spikelets, not capillary 3

3. Glumes acuminate; lemma often hairy below, 3- to 5-nerved, the nerves often raised above and excurrent for 0.2-0.5 mm; panicle-branches branched 2-3 times, bare of spikelets below . . . . **1. castellana**
 - Glumes acute; lemma glabrous except on the callus, 5-nerved, the nerves neither raised nor excurrent; panicle-branches branched 1-2 times, often spiculate from the base . . . . **2. stolonifera**

1. **A. castellana** Boiss. & Reut., *Diagn. pl. nov. hisp.*: 26 (1842).
A. hispanica Boiss. & Reut.; *A. olivetorum* Godr.
Perennial with short rhizomes, sometimes stoloniferous, up to 80 cm. Panicle 6-16 cm, the branches spreading at anthesis but usually contracted afterwards, bare of spikelets below, branched 2-3 times. Spikelets 2.5-4 mm, greenish yellow to purplish; glumes lanceolate, acuminate, scabrid on the keel especially above, sometimes thinly hairy; lemma *c.* ⅔ as long as the glumes, 3- to 5-nerved, the lateral nerves often raised in the upper part and excurrent for 0.2-0.5 mm but

in awnless forms often neither raised nor excurrent, the body usually hairy below, awned from the back below, the awn up to 5 mm; palea ½-⅘ as long as the lemma.
Fl. V - VIII. *Common in fields and woods, especially in eastern Madeira, from 150-1350 m.*
M

An extremely variable species widespread in Macaronesia but in the past seldom recognized for what it was. The three main variants have been called var. *communis* Menezes (awned), var. *mutica* Hack. (awnless) and var. *mixta* Hack. (with a mixture of both types in a single inflorescence).

2. A. stolonifera L., *Sp. pl.* **1**: 62 (1753).
A. alba auct., non L. (1753)
Decumbent, creeping perennial, the stems rooting and forming stolons; flowering stems erect, up to 100 cm. Panicle 1-30 cm, the branches spreading at anthesis but usually contracted afterwards, often spiculate from the base, branched only once or twice. Spikelets 2-3 mm, greenish or tinged with purple; glumes lanceolate, acute, scabrid on the keel, otherwise glabrous; lemma ⅔-¾ as long as the glumes, 5-nerved, the nerves not excurrent, the body glabrous, usually awnless, rarely with a short awn from near the tip; palea ½-⅔ as long as the lemma.
Fl. ? *Previously reported as common in grassy areas in Madeira but known for certain only from Faial; records for Porto Santo are probably erroneous.* **M**

3. A. obtusissima Hack., *Öst. bot. Z.* **52**: 107 (1902).
A. reuteri auct. macar., non Boiss.
Tufted perennial with creeping woody base, up to 45 cm; leaves mostly in a basal tuft. Panicle 10-12 cm, very lax, the filiform pedicels mostly at least twice as long as the spikelets. Spikelets 2.2-2.6 mm, tinged with purple; glumes lanceolate, obtuse, the tip truncate and often irregularly toothed or erose, smooth; lemma ⅔-¾ as long as the glumes, 5-nerved, denticulate at the tip with the nerves excurrent, the body glabrous, awnless or with a short awn; palea *c.* ½ as long as the lemma. **Plate 55**
Fl. VI - VIII(-X). *Endemic confined to the high, central peaks of Madeira.* ● **M**

4. A. pourrettii Willd. in *Neue Schr. Ges. naturf. Fr. Berl.* **2**: 290, t. 8/4 (1808).
A. salmantica (Lag.) Kunth
Annual up to 50 cm. Panicle up to 13 cm, lax, the branches spreading both during and after anthesis. Spikelets 2-2.5 mm, greenish; glumes lanceolate, the lower longer than the upper, scabrid on the keel; lemma less than ½ as long as the glumes, 5-nerved, the lateral nerves long-excurrent, awned above the middle, the awn *c.* 3 mm; palea extremely small.
Fl. V - VI. *Weed of gardens and waste places, mostly around Funchal, occasionally elsewhere.*
M

32. Triplachne Link

Annuals; ligule a membrane. Inflorescence a spike-like panicle. Spikelets 1-flowered with rhachilla-extension, breaking up above the glumes. Glumes equal, longer than the floret, membranous, 1-nerved, acute; lemma hyaline, gibbously ovate in profile, rounded on the back, truncate at the tip, with 2 straight lateral awns and a geniculate dorsal awn; palea almost as long as the lemma.

1. T. nitens (Guss.) Link, *Hort. berol.* **2**: 241 (1833).
Stems up to 20 cm, usually geniculately ascending. Panicle densely ovoid-cylindrical, 1-5 cm. Spikelets 3.5-4 mm; glumes lanceolate, shining, scabrid on the keel; lemma 1.3-1.5 mm, denticulate, pilose with long appressed hairs; lateral awns about as long as the lemma; dorsal awn c.4 mm,

dark below, pale above.
Fl. IV - VI. *Coastal foreshores and dunes, coastal rocks and other dry open places in Porto Santo, Ilhéu Chão and Bugío; in Madeira only from the Ponta de São Lourenço.* **MDP**

33. Gastridium P. Beauv.

Annuals; ligule a membrane. Inflorescence a spike-like panicle. Spikelets 1-flowered, with or without a rhachilla-extension, breaking up above the glumes. Glumes unequal, longer than the floret, membranous above, tough and gibbously swollen below, 1-nerved, acuminate; lemma coriaceous, rounded on the back, truncate and denticulate, awnless or with a geniculate dorsal awn, awned and awnless lemmas mixed in the same panicle.

1. Awned lemma densely hairy all over; rhachilla-extension well developed, hairy . . . . **1. phleoides**
- Awned lemma thinly hairy on sides; rhachilla-extension scarcely developed, glabrous . . . . **2. ventricosum**

1. G. phleoides (Nees & Meyen) C.E. Hubb. in *Kew Bull.* **9**: 375 (1954).
Stems up to 35 cm. Panicle 3.5-11 cm, dense. Spikelets (5-)6-8 mm; lower glume as long as the spikelet, the upper *c.* ¾ as long; awned lemma 1-1.3 mm, densely hairy all over, with well developed hairy rhachilla-extension; awn 4-6 mm.
Fl. III - VII(-IX). *Native to the Mediterranean region. An introduced weed, frequent in lowland Madeira; Pico do Castelo in Porto Santo; Ilhéu Chão; Selvagem Grande.* **MDPS**

2. G. ventricosum (Gouan) Schinz & Thell. in *Vjschr. naturf. Ges. Zürich* **58**: 39 (1913).
G. lendigerum (L.) Desv.
Stems up to 45 cm. Panicle 2-8 cm, dense, sometimes lobed. Spikelets 3-5 mm; lower glume as long as the spikelet, the upper *c.* ¾ as long; awned lemma 0.8-1 mm, thinly hairy only on the sides, with scarcely developed glabrous rhachilla-extension; awn 3-4 mm.
Fl. V - VII. *Occasional, mostly in the Funchal region.* **M**

Probably all records of this species from Porto Santo and the Desertas are referable to **1**.

34. Lagurus L.

Annuals; ligule a membrane. Inflorescence an ovoid, spike-like panicle. Spikelets 1-flowered with rhachilla-extension, breaking up above the glumes. Glumes equal, as long as the spikelet, narrowly lanceolate, membranous, 1-nerved, acuminate to a slender awn; lemma membranous, rounded on the back, 2-awned at the tip and with a geniculate dorsal awn from the upper third.

1. L. ovatus L., *Sp. pl.* **1**: 81 (1753).
Greyish-pubescent plant up to 50 cm, with inflated sheaths and velvety-pubescent leaves. Panicle 0.5-6 cm, very softly and densely hairy. Spikelets 7-9 mm (excluding awns), those at the base of the panicle mostly sterile; glumes villous on the back, the setaceous tip with hairs c.2 mm; lemma 3-5 mm with apical awns 2-6 mm and dorsal awn 8-20 mm.
Fl. (I-)III - IV(-VIII). *Common in the lower regions of Madeira and Porto Santo; also on Deserta Grande and Bugío.* **MDP**

35. Polypogon Desf.

Annuals or perennials; ligule a membrane. Inflorescence a contracted to spike-like panicle. Spikelets 1-flowered without rhachilla-extension, falling entire together with the pedicel or part of it. Glumes equal, as long as the spikelet, chartaceous, scabrid, 1-nerved, entire to bilobed, often with a slender awn; lemma hyaline, rounded on the back, the nerves sometimes excurrent, awned or awnless.

1. Perennial; glumes awnless . . . . **1. viridis**
- Annual; glumes awned 2

2. Glumes firm and prickly below, hyaline and ciliate above, 2-lobed . . . **2. maritimus**
- Glumes evenly chartaceous, scabrid on the back, thinly hairy on the margins, slightly notched at the tip 3

3. Awn of glumes 2-3 × as long as the body of the glume . . . . **3. monspeliensis**
- Awn of glumes scarcely exceeding the body of the glume, usually shorter . . **4. fugax**

1. P. viridis (Gouan) Breistr. in *Bull. Soc. bot. Fr.* **110** (Sess. Extr.): 56 (1966).
P. semiverticillatus (Forssk.) Hyl.
Stoloniferous perennial up to 100 cm. Panicle pyramidal, lobed, 2-15 cm. Spikelets 1.5-2.5 mm; glumes obtuse, scabrid on the back; lemma about half as long as the glumes, denticulate, awnless.
Fl. III - IX. *Streams and wet ditches. Common in southern Madeira from Machico to Ponta do Sol, the Ribeira Brava valley, and São Vicente and Faial in the north; rather rare in Porto Santo.* **MP**

2. P. maritimus Willd. in *Neue Schr. Ges. naturf. Fr. Berl.* **3**: 442 (1801).
Annual up to 25 cm. Panicle spike-like, cylindrical or sometimes lobed, 1-6 cm. Spikelets 1.5-3 mm; glumes shortly connate, firm and prickly below, hyaline and ciliate above, 2-lobed, awned from between the lobes, the awn up to 7 mm; lemma about half as long as the glumes, denticulate, awnless.
Fl. IV - VI. *Infrequent; in Madeira only from Machico; also from Porto Santo and Ilhéu de Cima; Ilhéu Chão.* **MDP**

3. P. monspeliensis (L.) Desf., *Fl. atlant.* **1**: 67 (1798).
Annual up to 80 cm. Panicle ovate to narrowly oblong, dense, cylindrical or lobed, 1.5-16 cm. Spikelets 2-3 mm; glumes slightly notched at the tip, scabrid on the back especially below, shortly hairy on the margins, with an awn 2-3 times as long as the body; lemma about half as long as the glumes, awnless or with a short awn.
Fl. IV - VI. *Damp ground along the south coast of Madeira from Tabua to the Ponta de São Lourenço; in the lower regions of Porto Santo and on Ilhéu de Baixo; recorded for the Desertas but without further data.* **MDP**

4. P. fugax Nees ex Steud., *Syn. pl. glumac.* **1**: 184 (1854).
Annual up to 60 cm. Panicle narrowly ovate to oblong, dense, cylindrical or lobed, 3-15 cm. Spikelets 1.8-2.4 mm; glumes slightly notched at the tip, scabrid on the back especially below, minutely hairy on the margins, with an awn shorter than to as long as, or rarely exceeding, the body; lemma about half as long as the glumes, awnless or with a short awn.
Fl. V - VIII. *Apparently rare.* **M**

36. Bromus L.

Annuals or perennials; ligule a membrane; sheaths with margins connate for most of their length. Inflorescence a panicle, the spikelets all alike. Spikelets several-flowered, laterally compressed, breaking up above the glumes and between the florets; glumes herbaceous, persistent; lemmas herbaceous to subcoriaceous, often with membranous margins, entire or 2-toothed, with subapical awn; ovary capped by a hairy lobed appendage.

Whilst generally not a difficult genus, depauperate plants can be troublesome to place correctly as the inflorescence may be reduced to just a single spikelet.

1. Lemmas keeled; plant perennial . . . . **1. catharticus**
- Lemmas rounded on the back; plants annual 2

2. Spikelets elliptic to ovate at maturity; the lower glume 3- to 7-nerved, upper glume 5- to 9-nerved 3
- Spikelets cuneate at maturity; the lower glume 1-nerved, upper glume 3-nerved . . 4

3. Panicle dense; spikelets 12-22 mm; lemma 8-11 mm . . . . **2. hordeaceus**
- Panicle lax; spikelets 25-40 mm; lemma 11-14 mm . . . . **3. lanceolatus**

4. Lemma more than 20 mm (excluding the awn) . . . . **4. diandrus**
- Lemma less than 20 mm (excluding the awn) 5

5. Panicle drooping, very lax, the branches at least as long as the spikelets . . **5. sterilis**
- Panicle erect, most branches shorter than the spikelets 6

6. Panicle lax, the spikelets not densely crowded; panicle-branches 10 mm or more; lemma at least 3 mm wide (when flattened) . . . . **6. madritensis**
- Panicle very dense, the spikelets crowded on branches 1-10 mm; lemma less than 3 mm wide (when flattened) . . . . **7. rubens**

1. B. catharticus Vahl, *Symb. bot.* **2**: 22 (1791).
B. unioloides Kunth, non *Festuca unioloides* Willd.; *B. willdenowii* Kunth; *F. unioloides* Willd.
Short-lived perennial tussock-grass up to 100 cm. Panicle 10-40 cm, loose. Spikelets oblong-ovate, 16-40 mm, strongly laterally compressed; glumes narrowly lanceolate in profile, acuminate, the lower 10-15 mm, the upper a little longer; lemmas narrowly lanceolate in profile, 15-20 mm, laterally flattened and sharply keeled, 2-toothed and with a short awn up to 3 mm.
Fl. I - IV(-VII). *Probably an escape from former cultivation for fodder and now naturalized in waste places, especially around Funchal. Native of S. America.* **M**

2. B. hordeaceus L., *Sp. pl.* **1**: 77 (1753).
B. mollis L.
Annual up to 60 cm. Panicle up to 10 cm, very dense. Spikelets ovate-lanceolate, 12-22 mm, softly hairy; glumes subequal, elliptic, 5-9 mm, the lower 3- to 7-nerved, the upper 5- to 9-nerved; lemmas obovate to oblanceolate, curved or bluntly angled on the margins, 8-11 mm; awn 6-10 mm.
Fl. III - VI. *Cultivated fields and waste places from 750-1050 m. In Madeira mainly in eastern and central parts; throughout Porto Santo; also recorded for the Desertas but without further data.* **MDP**

The following subspecies occur in the region:

a) subsp. **hordeaceus**
Awns slender, terete, straight at maturity; lemma with bluntly angled margins.
Throughout the range of the species.

b) subsp. **molliformis** (Lloyd) Maire & Weiller, in Maire, *Fl. Afr. Nord.* **3**: 255 (1955).
Awns stout, flattened and often twisted below, patent or recurved at maturity; lemma with smoothly curved margins.
Only known from around Funchal and Monte.

3. B. lanceolatus Roth, *Catal. bot.* **1**: 18 (1797).
B. lanceolatus var. *lanuginosus* (Poir.) Dinsm.
Annual up to 65 cm. Panicle 6-16 cm, narrow with short erect branches. Spikelets narrowly elliptic, 25-40 mm, hairy or glabrous; glumes lanceolate to narrowly ovate, the lower 3- to 5-nerved, 7-9 mm, the upper 5- to 9-nerved, 9-11 mm; lemmas narrowly elliptic, curved along the margins, 11-14 mm; awn flattened at the base, 15-25 mm, becoming recurved and twisted at maturity.
Fl. IV - VI. *A rare weed along roadsides at the harbour in Funchal and at Terra de Maria in Porto Santo.* **MP**

4. B. diandrus Roth, *Bot. Abh. Beobacht.*: 44 (1787).
B. rigens var. *gussonii* (Parl.) Cout., subsp. *maximus* (Desf.) Menezes; *B. rigidus* Roth
Annual up to 90 cm. Panicle 15-25 cm, lax with spreading branches to dense with erect branches. Spikelets cuneate at maturity, 25-70 mm, glabrous; glumes narrowly lanceolate, the lower 1-nerved, 15-25 mm, the upper 3-nerved, 20-35 mm; lemmas linear-lanceolate to lanceolate, 20-35 mm; awn 30-60 mm, stout, flattened below, straight at maturity.
Fl. IV - VIII. *Very common weed of roadsides and waste places throughout Madeira and Porto Santo.* **MP**

Until recently two taxa were recognized within *B. diandrus*: *B. diandrus* sensu stricto (2n=56) and *B. rigidus* (2n=42). Their separation has always been problematical but recent analysis has clearly demonstrated that a morphological separation is untenable, despite the different chromosome numbers.

5. B. sterilis L., *Sp. pl.* **1**: 77 (1753).
Annual up to 100 cm. Panicle 10-20 cm, very lax, the branches mostly much longer than the spikelets, these 1(-3) to a branch. Spikelets cuneate at maturity, 20-35 mm, hairy or glabrous; glumes unequal, the lower subulate, 1-nerved, 6-14 mm, the upper linear-lanceolate, 3-nerved, 10-20 mm; lemmas narrowly lanceolate, 14-20 mm; awn 15-30 mm, slender straight.
Fl. ? *From the Funchal region, but apparently not recorded in recent years.* **M**

6. B. madritensis L., *Cent. pl.* **I**: 5 (1755).
B. madritensis var. *ciliatus* Guss.
Annual up to 60 cm. Panicle 3-15 cm, erect, lax, the branches 1-3 cm. Spikelets cuneate at maturity, 30-50 mm, hairy or glabrous; glumes linear-lanceolate, the lower 1-nerved, 5-10 mm, the upper 3-nerved, 10-15 mm; lemmas narrowly oblong-lanceolate, 12-20 × 3-3.5 mm when flattened; awn 12-20 mm, slender, straight or becoming weakly recurved.
Fl. III - VII. *Weed of roadsides and waste places. Common in the Funchal region and on Ponta de São Lourenço in Madeira; scattered in Porto Santo, Ilhéu Chão and Bugío.* **MDP**

7. B. rubens L., *Cent. pl.* **I**: 5 (1755).
Annual up to 40 cm. Panicle 2-10 cm, very dense, cuneate below, the branches and pedicels up to 1 cm, much shorter than the spikelets. Spikelets cuneate at maturity, 18-25 mm, hairy or glabrous; glumes unequal, the lower subulate, 1-nerved, 5-7 mm, the upper lanceolate, 3-nerved, 8-10 mm; lemmas narrowly oblong-elliptic, 10-13 × 2-3 mm when flattened; awn 8-12 mm, slender, straight.
Fl. ? *Recorded only once in Madeira and possibly only a casual.* **M**

37. Brachypodium P. Beauv.

Annuals or perennials; ligule a membrane. Inflorescence a tough bilateral raceme, the pedicels 1-3 mm. Spikelets solitary at each node, several-flowered, breaking up above the glumes and between the florets; glumes persistent, 3- to 9-nerved, shorter than the adjacent lemmas; lemmas herbaceous, sometimes becoming coriaceous at maturity, 7- to 9-nerved, rounded on the back, obtuse or awned from the tip; ovary tipped by a small fleshy hairy appendage.

1. Annual .... **1. distachyum**
- Perennial 2

2. Awn 7-14 mm, at least as long as the lemma; plant tufted, without rhizomes ....
  .... **2. sylvaticum**
- Awn up to 2.5 mm, much shorter than the lemma; plant with creeping rhizomes ....
  .... **3. phoenicoides**

1. B. distachyum (L.) P. Beauv., *Ess. Agrostogr.*: 101, 155 (1812).
Trachynia distachya (L.) Link
Annual up to 40 cm; leaves up to 12 cm × 4 mm, flat, rather stiff, glaucous. Raceme with 1-3 or rarely more crowded spikelets. Spikelets oblong-lanceolate, 20-30 mm, slightly compressed; glumes unequal, hairy or glabrous, the lower 5-6 mm, the upper 7-8 mm; lemmas lanceolate-oblong, 7.5-10 mm, inrolled, glabrous, sparsely bristly or pubescent; awn up to 15 mm.
Fl. III - VIII. *Dry rocky ground and seashores up to 60 m. Common throughout the Madeiran archipelago.* **MDP**

2. B. sylvaticum (Huds.) P. Beauv., *Ess. Agrostogr.*: 101, 155 (1812).
Tufted perennial without rhizomes, up to 90 cm; leaves up to 35 cm × 4-12 mm, flat, flaccid and drooping, rough, loosely hairy above. Raceme with 4-12 distant spikelets. Spikelets narrowly oblong or lanceolate, 20-40 mm, terete; glumes unequal, usually hairy, the lower 6-8 mm, the upper 8-11 mm; lemmas lanceolate-oblong, 7-11 mm, shortly and stiffly hairy; awn 7-14 mm.
Fl. VI - IX. *Common in Madeira in open, grassy places, scrubland and along roadsides from 150-1100 m; also reported from Porto Santo but without further data.* **MP**

3. B. phoenicoides (L.) Roem. & Schult., *Syst. veg.* 2: 740 (1817).
Rhizomatous perennial up to 100 cm; leaves 10-40 cm × 3-5 mm, flat and flaccid or inrolled and stiff, smooth or rough, hairless. Raceme with 6-9 distant spikelets. Spikelets linear, often falcate, 30-60 mm, terete; glumes unequal, glabrous, the lower 4-6 mm, the upper 5-7.5 mm; lemmas ovate-lanceolate, 8-10 mm, glabrous, mucronate or shortly awned; awn, when present, not more than 2.5 mm.
Fl. IV - VI. *Reported from Madeira but without further data.* **M**

38. Elymus L.

Perennials; ligule a membrane. Inflorescence a bilateral raceme, the spikelets sessile and appressed to a tough or fragile rhachis. Spikelets solitary at each node, several-flowered, laterally compressed; glumes lanceolate to narrowly oblong, firmly membranous to coriaceous, the lower shorter than the adjacent lemma, 3- to 9-nerved; lemma coriaceous, 5-nerved, rounded on the back, obtuse, acute or 2-toothed, muticous or awned; ovary tipped by a small fleshy hairy appendage.

1. Rhachis fragile, breaking up above each spikelet at maturity, glabrous on the edges ...
  .... **1. farctus**
- Rhachis tough, not breaking up, spinose-ciliate on the edges .... **2. repens**

1. E. farctus (Viv.) Runemark ex Melderis in *Bot. J. Linn. Soc.* **76**: 382 (1978).
Agropyron junceiforme (Á. & D. Löve) Á. & D. Löve; *Elytrigia junceiformis* Á. & D. Löve
Rhizomatous perennial up to 80 cm, stout; leaves inrolled, up to 5 mm wide. Raceme 15-35 cm, the rhachis breaking up at maturity just above each spikelet, glabrous on the edges. Spikelets

10-25 mm, glabrous; glumes 10-18 mm, asymmetrically keeled, obtuse, awnless; lemmas 10-18 mm, keeled towards the tip, obtuse, awnless.
Fl. VII. *On sandy coasts of Selvagem Pequena & Ilhéu de Fora.* **S**

2. E. repens (L.) Gould in *Madroño* **9**: 127 (1947).
Agropyron repens (L.) P. Beauv.; *Elytrigia repens* (L.) Nevski
Rhizomatous perennial up to 120 cm, slender; leaves usually flat, 3-10 mm wide. Raceme 5-15(-20) cm, the rhachis tough, spinose-ciliate along the edges. Spikelets 8-17 mm, glabrous; glumes 5-15 mm, acute, mucronate or shortly awned; lemmas 6-11 mm, keeled towards the tip, awnless or with a subulate tip.
Fl. VII - IX. *Weed of roadsides, gardens and waste places in Madeira.* **M**

39. Hordeum L.

Annuals or perennials; ligule a membrane. Inflorescence an oblong false raceme bearing spikelets in triads at each node, the triads comprising 1 central bisexual spikelet and 2 male or barren laterals, the rhachis fragile (tough in cultivated species). Central spikelet 1-flowered with bristle-like rhachilla-extension, dorsally compressed; glumes placed side by side, usually awn-like or expanded below; lemma rounded on the back, acuminate to a conspicuous awn; ovary tipped by a small fleshy hairy appendage. Lateral spikelets usually smaller than the central.

One species is cultivated in Madeira:

H. vulgare L. (*H. distichon* L., *H. hexastichon* L.) is a cultivated cereal recognized by the tough rhachis. Two forms are grown in Madeira: 2-rowed barley in which only the central spikelet of a triad is fertile; and 6-rowed barley in which all three are fertile.

1. Perennial . . . . **1. secalinum**
- Annual 2

2. Glumes of central spikelet long-ciliate below . . . . **2. murinum**
- Glumes of central spikelet scabrid . . . . **3. marinum**

1. H. secalinum Schreb., *Spic. fl. lips.*: 148 (1771).
Perennial up to 70 cm; upper sheaths not inflated. Raceme 2-5 cm, linear. Central spikelet sessile; glumes setaceous, up to 14 mm, scabrid; lemma 6-9 mm, with an awn 6-12 mm. Lateral spikelets pedicellate; glumes similar to those of central spikelet; lemma 4-6 mm, with an awn up to 3 mm.
Fl. ? *Reported for both Madeira and Porto Santo but with no other data.* **MP**

2. H. murinum L., *Sp. pl.* **1**:85 (1753).
Annual up to 50 cm; upper sheaths inflated. Raceme 2-7 cm, oblong. Central spikelet sessile or pedicellate; glumes lanceolate, up to 26 mm (including the awn), ciliate on the margins below; lemma 7-12 mm, with an awn 18-50 mm. Lateral spikelets pedicellate; glumes 16-30 mm including the awn, the inner lanceolate, ciliate below, the outer setaceous, scabrid; lemma 7-11 mm, with an awn 10-40 mm.
Fl. III - VI. *Common on dry rocky ground and in waste places in the lower regions of Madeira and Porto Santo; also on Ilhéu Chão and Selvagem Grande.* **MDPS**

Two subspecies occur in the region:

a) subsp. **glaucum** (Steud.) Tzvelev in *Nov. Sist. Vysshikh Rast.* **1971**: 67 (1971).
H. glaucum Steud.; *H. murinum* auct. mad., pro parte
Leaves glaucous; anthers of central spikelet 0.2-0.5 mm; rhachilla-extension of lateral spikelets

stout, orange-brown.
Present only in the Funchal region.

b) subsp. **leporinum** (Link) Arcang., *Comp. fl. ital.*: 805 (1882).
H. leporinum Link; *H. murinum* auct. mad., pro parte
Leaves green; anthers of central spikelet 0.7-1.4 mm; rhachilla-extension of lateral spikelets slender, green.
Throughout the range of the species.

3. **H. marinum** Huds., *Fl. angl.* ed. 2, **1**: 57 (1778).
Annual up to 60 cm; upper sheaths inflated. Raceme 1.5-5 cm, oblong. Central spikelet sessile; glumes setaceous, up to 26 mm, scabrid; lemma 6-8 mm, with an awn up to 24 mm. Lateral spikelets pedicellate; glumes subulate, up to 26 mm including the awn, scabrid; lemma 3-5 mm, with an awn 3-5 mm.
Fl. IV - V. *Rather rare in grassy fields and by streamsides from Machico to Caniçal in Madeira and on Pico do Castelo in Porto Santo.* **MP**

Two subspecies occur in the region:

a) subsp. **marinum**
One glume of the lateral spikelet with a well developed wing on one side.
Porto Santo only; rare.

b) subsp. **gussoneanum** (Parl.) Thell. in *Vjschr. naturf. Ges. Zürich* **52**: 441 (1908).
Both glumes of the lateral spikelet subulate, or one of them slightly swollen but not winged.
Throughout the range of the species.

40. **Danthonia** DC.

Perennials; ligule a line of hairs. Inflorescence a sparse panicle but the lower sheaths also often bearing cleistogamous spikelets in their axils. Spikelets several-flowered, breaking up above the glumes and between the florets; glumes almost as long as the spikelet, papery, 3- to 9-nerved; lemmas firmly membranous, 7- to 9-nerved, hairy on the margins.

1. **D. decumbens** (L.) DC., *Fl. franç.* ed. 3, **3**: 33 (1805).
Festuca decumbens L.; *Sieglingia decumbens* (L.) Bernhardt; *Triodia decumbens* (L.) P. Beauv.
Densely tufted, up to 50 cm; leaves stiff, rolled, 5-25 cm × 2-4 mm. Panicle narrow, 2-7 cm, with 3-8 spikelets. Glumes 6-12 mm; lemmas 5-7 mm, 2-toothed and with a short mucro from between the teeth.
Fl. IV- VII(-IX). *Grassy places in eastern Madeira; 300-900 m.* **M**

41. **Rytidosperma** Steud.

Perennials; ligule a line of hairs. Inflorescence a contracted panicle. Spikelets several-flowered, the upper florets imperfect, breaking up above the glumes and between the florets; glumes almost as long as the spikelet, narrow, papery, 3- to 7-nerved; lemmas 5- to 9-nerved, bearing tufts of hair arranged in 2 transverse series, bilobed, with an awn in the sinus.

1. **R. tenuis** (Steud.) A. Hansen & Sunding, *Fl. Macar.*, ed. 2, **1**: 93 (1979).
Notodanthonia tenuior (Steud.) S.T. Blake; *Plinthanthesis tenuior* Steud.
Tufted, up to 80 cm. Glumes subequal, 8-11.5 mm; lemma membranous, 7-8.5 mm including the oblong callus and the 4.5-6 mm awned lobes; central awn geniculate, twisted below, 11-12 mm.
Fl. VI. *Native to Australia. Introduced, but apparently not cultivated, and now thoroughly naturalized at Quinta do Palheiro Ferreiro, northeast of Funchal.* **M**

42. Schismus P. Beauv.

Annuals; ligule a line of hairs. Inflorescence a contracted or spike-like panicle. Spikelets several-flowered, falling entire, or the upper florets falling singly and the lower florets, glumes and pedicel falling together later; glumes as long as the spikelet or almost so, membranous with hyaline margins, prominently 5- to 7-nerved; lemmas membranous, 7- to 9-nerved, hairy on the back and margins, bilobed or merely notched, more or less mucronate in the sinus.

1. S. barbatus (L.) Thell. in *Bull. Herb. Boissier* II, **7**: 391 (1907).
Festuca barbata L.; *S. marginatus* P. Beauv.
Stems up to 20 cm. Glumes 3-4.5(-6) mm; lowest lemma 1.5-2 mm including the 0.3-0.4(-0.7) mm broadly triangular lobes; palea 1.6-2.1 mm, usually reaching almost to the tips of the lemma-lobes, sometimes exceeding them.
Fl. IV - V. *Infrequent in Porto Santo and Ilhéu de Cima.* **P**

43. Cortaderia Stapf

Gynodioecious tussock-forming perennials with basal leaves; ligule a line of hairs. Inflorescence a large plumose panicle. Spikelets several-flowered, breaking up above the glumes and between the florets; glumes longer than the lowest lemma, ⅔ as long to as long as the spikelet, narrow, hyaline, 1-nerved; lemmas 3- to 5(7)-nerved, those of the bisexual spikelets glabrous or almost so, those of the female spikelets hairy on the back, entire, drawn out into a long, acute or filiform tip.

1. C. selloana (Schult.) Asch. & P. Graebn., *Syn. mitteleur. Fl.* **2**(1): 325 (1900).
Arundo dioeca Spreng., non Lour. (1790); *A. selloana* Schult.; *Gynerium argenteum* Nees
Leaves 1-2m × 6-10 mm, with rough saw-edged margins, confined to a basal tussock. Panicle 30-60 × 10-15 cm, on a stem 2-3m high, white or tinged with silver or pink.
Fl. VIII - X. *Native of S. America. Cultivated in Madeira and occasionally naturalized.* **M**

44. Arundo L.

Perennial reeds; ligule a membrane with a minutely ciliolate margin. Leaves cauline. Inflorescence a large plumose panicle, the spikelets all alike. Spikelets several-flowered, the upper florets imperfect, breaking up above the glumes and between the florets. Glumes as long as the spikelet, 3- to 5-nerved; lemmas membranous, 3- to 7-nerved, plumose below the middle, entire or 2-toothed, with a short straight awn from between the teeth.

1. A. donax L., *Sp. pl.* **1**: 81 (1753).
Stems up to 5m. Leaves conspicuously distichous, rounded or cordate at the base, up to 60 × 5 cm. Panicle 30-60 cm. Spikelets 10-15 mm; lemmas (6-)8.5-13 mm, with an awn c.1.5 mm.
Fl. VIII - X. *Widely planted for poles etc., but often escaping and naturalizing in Madeira and Porto Santo; also reported for the Desertas but with no further data.* **MDP**

45. Phragmites Adans.

Perennial reeds; ligule a very short membrane with a long-ciliate margin. Leaves cauline, the blades deciduous. Inflorescence a large plumose panicle, the spikelets all alike. Spikelets several-flowered with the lowest floret male or empty, breaking up above the glumes and between the florets. Glumes shorter than the lowest lemma, 3- to 5-nerved; fertile lemmas hyaline, 1- to 3-nerved, glabrous, but with a linear plumose callus, long-caudate (though very fragile and the long narrow tip often breaking off), entire.

1. P. australis (Cav.) Trin. ex Steud., *Nomencl. bot.* ed. 2, **2**: 324 (1841).
Arundo australis Cav.; *P. communis* var. *congesta* (Lowe) Menezes; *P. congesta* Lowe
Stems up to 4 m; leaves 20-60 cm × 8-32 mm, flat, smooth beneath, the tip filiform and flexuous. Panicle 15-20 cm. Spikelets with rhachilla-hairs 8-12.5 mm; lower glume ovate, 2.7-5 mm; upper glume lanceolate, 5.5-9 mm; first lemma 9.5-17 mm; fertile lemmas 10-14 mm.
Fl. VIII - IX. *Lagoons and saltmarshes along south coast of Madeira; also recorded from Porto Santo but with no further data.* **MP**

46. Aristida L.

Annuals or short-lived perennials; ligule a line of hairs. Inflorescence a panicle. Spikelets 1-flowered without rhachilla-extension, breaking up above the persistent glumes. Glumes 1-nerved; lemma coriaceous, wrapped around and concealing the palea, 3-awned, the awns more or less connate at the base.

1. A. adscensionis L., *Sp. pl.* **1**: 82 (1753).
A. coerulescens Desf.
Stems up to 60 cm. Panicle usually contracted about the primary branches, these either spreading or appressed to the main axis. Glumes unequal, the lower 6-7 mm, the upper 7.2-8.5 mm; lemma, including the callus, 7-9 mm, laterally compressed, scabrid on the keel and sometimes also on the flanks; awns terete, the central 1.7-2.1 cm, the laterals similar or a little shorter.
Fl. X - III. *Open rocky places in Madeira, mostly along the south coast.* **M**

47. Eragrostis Wolf

Annuals or perennials; ligule a line of hairs. Inflorescence an open or contracted panicle. Spikelets several-flowered, variously breaking up at maturity; glumes 1(-3)-nerved; lemmas 3-nerved, membranous or papery, glabrous, entire, awnless.

1.	Densely tufted perennial	**1. curvula**
-	Annual	2
2.	Caryopsis subrotund, 0.4-0.6 mm in diameter	**2. cilianensis**
-	Caryopsis broadly oblong to elliptic-oblong, 0.6-1 mm long	3
3.	Leaves with crateriform glands along the margins; caryopsis broadly oblong, (0.5-)0.6-0.8 mm	**3. minor**
-	Leaves without crateriform glands; caryopsis elliptic-oblong, 0.6-1 mm	**4. barrelieri**

1. E. curvula (Schrad.) Nees, *Fl. Afr. austral. Ill.* **1**: 397 (1841).
Densely tufted perennial up to 120 cm; basal sheaths silky-hairy. Panicle 6-30 cm, open or contracted. Spikelets dark leaden-grey, 4-10 mm; glumes unequal, ovate, the lower 1.2-1.8 mm, the upper 1.8-2.3 mm; lemmas oblong-lanceolate in profile, firmly membranous, 1.8-2.6 mm, scaberulous at least above, obtuse to subacute.
Fl. I - VI. *Native to S. Africa. Introduced, possibly for fodder, and persisting in waste places and sometimes as a garden weed.* **M**

2. E. cilianensis (All.) Vign. ex Janch. in *Mitt. naturw. Ver. Univ. Wien* **5**(9): 110 (1907).
Annual up to 75 cm; leaves with or without crateriform glands along the margins. Panicle 5-15 cm, usually contracted with short pedicels, these and the branches often bearing crateriform glands. Spikelets green to leaden-grey, 4.5-15 mm; glumes subequal, lanceolate, 1.5-2.7 mm;

lemmas elliptic-ovate in profile, papery, 2-2.8 mm, scaberulous, subacute to obtuse. Caryopsis usually subrotund, 0.4-0.6 mm in diameter.
Fl. III - VII(-XI). *Introduced; a weed around Funchal. Native to tropical and warm temperate Old World.* **M**

3. E. minor Host, *Icon descr. gram. austriac.* **4**: 15 (1809).
Poa eragrostis L.
Annual up to 60 cm; leaves with crateriform glands along the margins. Panicle 3-20 cm, rather open but with short pedicels, these and the branches also bearing crateriform glands. Spikelets green to olive, leaden-grey or reddish, 4.8-9 mm; glumes subequal, lanceolate, 1.3-1.8 mm; lemmas ovate in profile, papery, 1.4-1.8 mm, scaberulous, obtuse. Caryopsis broadly oblong, (0.5-)0.6-0.8 mm long.
Fl. XII. *Known only from cliff-tops at Paúl do Mar in Madeira.* **M**

4. E. barrelieri Daveau in *J. Bot. Paris* **8**: 289 (1894).
Annual up to 60 cm; leaves never with crateriform glands. Panicle 3-15 cm, the spikelets evenly spaced or gathered into fascicles, the branches and pedicels often bearing crateriform glands. Spikelets pallid, greyish or reddish, 6-15 mm; glumes unequal, lanceolate, acute, the lower 1-1.5 mm, the upper 1.3-2.1 mm; lemmas oblong-lanceolate in profile, papery, 1.7-2.3 mm, scaberulous above, obtuse. Caryopsis elliptic-oblong, 0.6-1 mm long.
Fl. nearly all year. *Introduced weed, native to the Mediterranean region, tropical Africa and SW Asia. Common on roadsides and in waste places in Madeira; rare in Porto Santo.* **MP**

48. Eleusine Gaertn.

Annuals or perennials; ligule a membrane, usually with a ciliate fringe. Inflorescence composed of digitate or subdigitate racemes, these with imbricate spikelets and terminating in a fertile spikelet. Spikelets laterally compressed, several-flowered, breaking up above the glumes and between the florets; lemmas keeled, membranous, glabrous, obtuse or acute.

1. Annual; racemes slender, 3.5-15.5 cm × 3-5.5 mm . . . . **1. indica**
- Perennial; racemes broadly oblong, 1-3.5 cm × 7-15 mm . . . . **2. tristachya**

1. E. indica (L.) Gaertn., *Fruct. sem. pl.* **1**: 8 (1788).
Annual up to 85 cm. Inflorescence comprising (1-)5-10 slender digitate racemes 3.5-15.5 cm × 3-5.5 mm (a few often set below the rest). Spikelets elliptic, 4.6-7.8 mm; glumes unequal, acute, the lower 1.1-2.3 mm, the upper 1.8-2.9 mm; lemmas lanceolate in profile, 2.4-4 mm long, acute to subacute, with 1 or 2 additional nerves on either side of the keel. Caryopsis elliptic or lanceolate-elliptic, obliquely striate with very fine close perpendicular lines running between the striae.
Fl. IV - IX. *From sea-level to 300 m. Introduced weed, native throughout the tropics; common and widespread in Madeira; also found around Vila Baleira in Porto Santo.* **MP**

2. E. tristachya (Lam.) Lam., *Tabl. encycl.* **1**: 203 (1791).
Densely tufted perennial up to 45 cm. Inflorescence comprising 1-3(-4) broadly oblong digitate racemes 1-3.5 cm × 7-15 mm. Spikelets elliptic-ovate, 5.5-8 mm; glumes unequal, subacute, the lower 1.6-2.6 mm, the upper 2.6-3.4 mm; lemmas oblong-lanceolate in profile, 3.2-3.8 mm, obtuse or subacute, with 2 additional nerves on either side of the keel. Caryopsis broadly trigonous, obliquely striate and coarsely granular.
Fl. VI - IX. *Introduced weed, mostly in streets in and around Funchal, Monte and Camacha. Native of S. America.* **M**

49. Dactyloctenium Willd.

Annuals or perennials; ligule a membrane. Inflorescence composed of digitate racemes, these bearing imbricate spikelets and terminating in a naked point, eventually falling from the tip of the stem. Spikelets several-flowered, laterally compressed, breaking up above the glumes but not between the florets; upper glume with an oblique awn from just below the tip; lemmas strongly keeled, membranous, glabrous, acute to shortly awned, the tip often recurved.

1. Annual .... **1. aegyptium**
- Perennial .... **2. australe**

1. D. aegyptium (L.) Willd., *Enum. pl.*: 1029 (1809).
Annual, often stoloniferous and mat-forming, up to 70 cm; leaves papillose-hispid especially on the margins. Inflorescence comprising 3-9 linear to narrowly oblong racemes 1.2-6.5 cm. Spikelets broadly ovate, 3.5-4.5 mm; glumes subequal, 1.5-2.2 mm, the upper with an awn half to twice the length of the body; lemmas narrowly ovate in profile, 2.6-4 mm, acute and with a stout cusp or mucro up to 1 mm.
Fl. IX - XII. *An introduction reported for Madeira but without further data.* M

2. D. australe Steud., *Syn. pl. glumac.* 1: 212 (1854).
Stoloniferous perennial up to 80 cm; leaves softly pilose with spreading tubercle-based hairs especially on the lower surface. Inflorescence comprising (3-)4-5 linear-oblong racemes 3.2-5 cm. Spikelets elliptic-oblong, 5-5.5 mm; glumes subequal, 1.7-1.9 mm, the upper with a stout awn almost twice as long as the body; lemmas lanceolate to narrowly ovate in profile, 3.5-3.7 mm, the tip flexuous, acuminate-aristulate, the awn-point 0.5-0.7 mm.
Fl. IX - X. *Native of S. Africa. Introduced as a lawn grass and now naturalized in a few places.* M

50. Sporobolus L.

Perennials; ligule a line of hairs. Inflorescence an open or contracted panicle. Spikelets 1-flowered without rhachilla-extension, fusiform, breaking up below the floret; glumes unequal, deciduous, awnless; lemma thinly membranous, 1-nerved, entire, awnless. Fruit with free pericarp, this commonly swelling when wet and expelling the seed to the tip of the spikelet.

1. S. africanus (Poir.) Robyns & Tournay in *Bull. Jard. bot. État Brux.* **25**: 242 (1955).
Agrostis spicata Thunb., non Vahl (1790); *A. africana* Poir.; *S. indicus* auct., non (L.) R. Br.
Stems up to 110 cm. Panicle linear with short branches, 10-35 cm, the branches appressed and densely spiculate from the base. Spikelets (1.5-)2.1-2.8 mm; lower glume 0.4-0.9 mm; upper glume about half as long as the lemma; lemma as long as the spikelet.
Fl. I - VII. *Roadsides in Madeira from Santa Cruz to Machico.* M

51. Chloris Sw.

Annuals or perennials; ligule a short membrane with ciliate margin. Inflorescence comprising tough, unilateral, digitate racemes. Spikelets laterally compressed, with 1 fertile floret and 2-3 reduced lemmas, breaking up above the glumes but not between the florets; glumes acute; fertile lemma 3-nerved, keeled, entire or bilobed, conspicuously awned from or just below the tip.

1. Fertile lemma obliquely obovate in profile, the keel slightly gibbous, with a crown of spreading hairs 1.5-4 mm at the tip; annual .... **1. virgata**
- Fertile lemma lanceolate in profile, ciliate along the margins and without a crown of spreading hairs at the tip; stoloniferous perennial .... **2. gayana**

1. C. virgata Sw., *Fl. Ind. occid.* **1**: 203 (1797).
Annual up to 70 cm. Inflorescence a head of 4-8 racemes each 2.5-5 cm. Spikelets 3-flowered, 2-awned; fertile lemma obliquely obovate in profile, the keel slightly gibbous, 2-3.6 mm, with a crown of spreading hairs 1.5-4 mm at the tip, the awn 2.5-8.5 mm; second lemma represented by an oblong glabrous awned scale and third lemma by a clavate glabrous awnless scale.
Fl. IV. *Only known from Praia Formosa in Madeira. Native throughout the tropics and probably introduced.* **M**

2. C. gayana Kunth, *Révis. gramin.* **1**: 293, t.58 (1830).
Stoloniferous perennial up to 100 cm. Inflorescence a head of 9-12 racemes each 3-12 cm. Spikelets 3- to 4-flowered, 2-awned; fertile lemma lanceolate in profile, 2.9-3.2 mm, ciliate on the margins and keel, the awn 4-5 mm; second floret with a palea and often a male flower, the lemma lanceolate, ciliate on the margins, awned; third lemma represented by a scabrid oblong or clavate awnless scale, and fourth lemma, if present, by a glabrous clavate awnless scale.
Fl. VIII. *A rare introduced weed in Madeira; also reported for Porto Santo but without further data. Native to tropical and southern Africa, and Arabia.* **MP**

52. Cynodon Rich.

Perennials; ligule a short membrane with a ciliate margin. Inflorescence composed of tough, unilateral, digitate racemes. Spikelets strongly laterally compressed, 1-flowered, with or without a rhachilla-extension; lemma keeled, firmly cartilaginous, entire, awnless.

1. C. dactylon (L.) Pers., *Syn. pl.* **1**: 85 (1805).
Stoloniferous sward-forming plant up to 40 cm, also with slender underground rhizomes. Racemes 3-6, 1.5-6 cm. Glumes shorter than the floret; lemma 2-3 mm, subglabrous to silky-pubescent on the keel.
Fl. IV - VII(-XI). *From sea-level to 600 m. Open rocky ground, grassland and seashores in southern and eastern Madeira, and in Porto Santo.* **MP**

53. Oplismenus P. Beauv.

Trailing perennials; ligule a membrane with a ciliate fringe; leaves lanceolate to ovate. Inflorescence comprising unilateral racemes along a central axis, bearing paired spikelets (the lower often reduced). Spikelets 2-flowered, falling entire, laterally compressed; glumes ½-¾ the length of the spikelet, the lower, or both, tipped by a viscid awn; lower lemma acute to shortly awned; upper lemma subcoriaceous, smooth and shining, dorsally compressed, acute, its margins inrolled and broadly clasping the palea.

1. O. hirtellus (L.) P. Beauv., *Ess. Agrostogr.*: 54, 169 (1812).
Stems up to 100 cm or more; leaves 1-13 cm × 4-20 mm. Inflorescence 3-15 cm, comprising 3-9 racemes each 0.5-3 cm, the spikelets contiguous or the upper racemes reduced to fascicles. Spikelets 2-4 mm with awns 3-14 mm.
Fl. ? *Reported for Madeira but wihtout further data.* **M**

54. Panicum L.

Annuals or perennials; ligule a short ciliate membranous rim. Inflorescence a panicle. Spikelets 2-flowered, falling entire; glumes hyaline to membranous, the lower shorter than the spikelet, the upper usually as long; lower lemma resembling the upper glume; upper lemma crustaceous, the margins inrolled and clasping only the edges of the palea.

1. Perennial 2
- Annual 3

2. Upper lemma rugose .... **1. maximum**
- Upper lemma smooth .... **2. repens**

3. Spikelets 4-5.5 mm .... **3. miliaceum**
- Spikelets 1.8-3.3 mm .... **4. capillare**

1. P. maximum Jacq., *Icon. pl. rar.* **1**: 2, t.13 (1781).
Densely tufted perennial up to 3m; leaves 10-60(-80) cm × 4-20(-40) mm, flat, glabrous. Panicle ovate, 10-45 cm, open or contracted, the branches bare below, the lowermost conspicuously whorled. Spikelets oblong, 2.5-3.6(-4) mm, glabrous or pubescent, acute or subobtuse; lower glume orbicular, 1/4-1/3 the length of the spikelet; upper lemma pallid, rugose.
Fl. V - X. *Native to Tropical Africa. Introduced; formerly cultivated for fodder and persisting in waste places, roadsides and banana plantations in Madeira; reported for Porto Santo but without further data.* **MP**

2. P. repens L., *Sp. pl.* ed. 2, **1**: 87 (1762).
Subglabrous perennial up to 1m, with long rhizomes and often also surface stolons; leaves 7-25 cm × 2-8 mm, often stiff and pungent, distichous. Panicle narrowly oblong, 5-20 cm, the branches bare below. Spikelets ovate 2.5-3 mm, acute; lower glume broadly ovate, ⅓ the length of the spikelet and clasping its base; upper lemma pallid, smooth and glossy.
Fl. VII - X. *Common on waste ground and rocky slopes in lowland Madeira, often near the coast.* **M**

3. P. miliaceum L., *Sp. pl.* **1**: 58 (1753).
Robust, sparsely to densely hispid annual up to 1.5m; leaves 15-40 cm × 8-24 mm, cordate to amplexicaul. Panicle narrowly oblong to pyramidal, 15-35 cm, often lax and rather 1-sided, the branches bare below. Spikelets ovate to ovate-oblong, (4-)4.5-5.5 mm, glabrous, acute to shortly acuminate; lower glume ovate, ½-¾ the length of the spikelet, separated from the upper by a short internode; upper lemma orange or yellowish, smooth and glossy, usually persistent.
Fl. VII. *This species is reported for Madeira where it was doubtless introduced and at one time cultivated as a cereal (Broomcorn millet), but there are appear to be no recent records. It originated in India, but is widely naturalized in warm temperate regions.* **M**

4. P. capillare L., *Sp. pl.* **1**: 58 (1753).
Hispid annual up to 80 cm; leaves 7-30 cm × 5-14 mm. Panicle broadly ovate or broadly oblong, 15-50 cm, the branches filiform with long capillary pedicels, the whole inflorescence eventually breaking off. Spikelets elliptic or oblong, 1.8-2.5(-3.3) mm, glabrous, acute or acuminate; lower glume broadly ovate, ⅓-½ the length of the spikelet; upper lemma pallid or yellowish to olive-brown, smooth and glossy.
Fl. ? *A rare introduced weed. Native to N. America.* **M**

55. Echinochloa P. Beauv.

Annuals; ligule often absent. Inflorescence comprising racemes along a central axis, the spikelets paired or in short secondary racemelets, typically densely packed in 4 rows. Spikelets 2-flowered, falling entire, flat on the front, convex or gibbous on the back, often hispidulous, cuspidate or awned at the tip; lower glume ⅓ the length of the spikelet, the upper as long; lower lemma often stiffly awned; upper lemma crustaceous, smooth and glossy, the margins inrolled, clasping only the edges of the palea; upper palea with the tip briefly reflexed and more or less protuberant between the lemma-margins.

1. Spikelets acuminate to awned, 3-4 mm, hispid; racemes untidily 2- to several-rowed, the longest 2-10 cm and usually with short secondary racemelets at the base . **1. crus-galli**
- Spikelets acute to cuspidate, 1.5-3 mm, pubescent; racemes neatly 4-rowed, seldom over 3 cm, simple .... **2. colona**

1. E. crus-galli (L.) P. Beauv., *Ess. Agrostogr.*: 53, 161 (1812).
Panicum crus-galli var. *hostii* (M. Bieb.) K. Richt.
Coarse plant up to 100 cm; ligule absent. Inflorescence linear to ovate, 6-22 cm, the racemes untidily 2- to several-rowed, the longest 2-10 cm, usually with short secondary racemelets at the base. Spikelets ovate-elliptic, 3-4 mm, hispidulous; lower lemma acuminate or with an awn up to 5 mm.
Fl. VII - IX. *Native to warm temperate and subtropical regions. An introduced weed of waste places and banana plantations, scattered all around the coast of Madeira.* M

2. E. colona (L.) Link, *Hort. berol.* **2**: 209 (1833).
Stems up to 100 cm; ligule absent. Inflorescence typically linear, 1-15 cm, the racemes neatly 4-rowed, seldom over 3 cm, simple, commonly ½ their length apart and appressed to the axis. Spikelets ovate-elliptic to subglobose, 1.5-3 mm, pubescent; lower lemma acute to cuspidate, rarely with a subulate point up to 1 mm.
Fl. VII - X. *Native throughout the tropics and subtropics. An introduced weed, scattered along the south coast of Madeira.* M

56. Brachiaria (Trin.) Griseb.

Perennials; ligule a line of hairs. Inflorescence composed of racemes along a central axis; rhachis narrowly winged, bearing paired spikelets, their lower glume adaxial. Spikelets 2-flowered, falling entire; lower glume shorter than the spikelet, the upper as long; lower lemma resembling the upper glume; upper lemma crustaceous, its margins inrolled and clasping only the edges of the palea; upper palea with its tip tucked within the lemma.

1. B. mutica (Forssk.) Stapf in Prain, *Fl. trop. Africa* **9**: 526 (1919).
Panicum barbinode Trin.
Sprawling plant up to 125 cm, prostrate and rooting from the lower nodes. Inflorescence of 5-20 racemes on an axis 7-20 cm; racemes 2-10 cm, bearing paired spikelets or these sometimes on short secondary racemelets below or borne singly above, the rhachis 0.5-1 mm wide. Spikelets elliptic, 2.5-3.5 mm, glabrous, acute; lower glume ¼-⅓ the length of the spikelet; upper lemma rugulose.
Fl. ? *Native throughout the tropics. Introduced and formerly cultivated for forage in Madeira; persisting in wet places.* M

57. Paspalum L.

Perennials; ligule a short membrane. Inflorescence composed of single, digitate or scattered racemes; rhachis flat, sometimes winged, bearing single or paired spikelets. Spikelets 2-flowered, falling entire, plano-convex; lower glume absent; upper glume and lower lemma similar, as long as the spikelet; upper lemma coriaceous to crustaceous, its margins inrolled and clasping only the edges of the palea; upper palea obtuse, if acute then the tip not reflexed.

1. Spikelets with a ciliate fringe from the margins of the upper glume . . . **1. dilatatum**
- Spikelets glabrous or minutely pubescent, without a ciliate fringe 2

2. Spikelets ovate, plumply plano-convex; upper glume and lower lemma somewhat leathery; upper glume minutely pubescent . . . . **2. distichum**
- Spikelets narrowly ovate-elliptic, markedly flattened; upper glume and lower lemma thinly papery; upper glume glabrous . . . . **3. vaginatum**

1. P. dilatatum Poir. in Lam., *Encycl.* **5**: 35 (1804).
Tufted plant up to 180 cm. Inflorescence composed of (2-)3-5(-11) racemes borne along an axis 2-20 cm; racemes 4-11 cm, the spikelets paired, in 4 rows. Spikelets ovate, 2.8-3.8 mm, yellowish green; upper glume sparsely pilose on the sides, ciliate on the margins; lower lemma similar, but not ciliate.
Fl. III - IX. *Introduced weed of wet places, common and widespread in lowland Madeira. Native of S. America, but now widely introduced.* **M**

2. P. distichum L., *Syst. nat.* ed. 10, **2**: 855 (1959).
P. paspalodes (Mich.) Scribn.
Creeping stoloniferous plant up to 50 cm. Inflorescence composed of a conjugate pair of racemes, rarely with 1 or 2 additional racemes below; racemes 1.5-7 cm, the spikelets single, in 2 rows. Spikelets ovate, 2.5-3.5 mm, pale green; upper glume somewhat leathery, appressed-pubescent; lower lemma similar, but glabrous.
Fl. VII - XI. *Introduced weed, common in wet places in lowland Madeira. Native to tropics and subtropics throughout the world.* **M**

3. P. vaginatum Sw., *Prodr.* : 21 (1788).
Creeping stoloniferous plant up to 60 cm. Inflorescence composed of a conjugate pair of racemes, rarely with 1-3 additional racemes below; racemes 1.5-7.5 cm, the spikelets single, in 2 rows. Spikelets narrowly ovate-elliptic, 3-4.5 mm, pale brownish green; upper glume papery, glabrous; lower lemma similar.
Fl. ? *Introduced weed of wet places at Funchal and Machico. Native to the tropics.* **M**

58. Setaria P. Beauv.

Annuals or perennials; ligule a short ciliate rim. Infloresence an open or spike-like panicle, the spikelets subtended by 1 or more scabrid bristles which persist on the axis. Spikelets 2-flowered, falling entire, more or less gibbous; glumes shorter than the spikelet or the upper equalling it; lower lemma as long as the spikelet; upper lemma crustaceous, its margins inrolled and clasping only the edges of the palea.

S. megaphylla (Steud.) T. Durand & Schinz, *Consp. fl. afric.* **5**: 773 (1894) is cultivated in Madeira and may persist near gardens. It is a tall perennial 1-3 m high with broad, conspicuously pleated leaves. The panicle is rather narrow, 20-60 cm, with short stiff ascending branches. It is a native of tropical Africa and tropical America. **M**

Records of the Indian species *S. palmifolia* (König) Stapf which is similar to *S. megaphylla* but with the long, flexuous panicle-branches are probably errors for *S. megaphylla*.

1. Bristles retrorsely barbed, tenaciously clinging . . . . **4. verticillata**
- Bristles antrorsely barbed, not clinging 3

2. Upper glume as long as the spikelet, concealing the upper lemma . . . . . . **3. viridis**
- Upper glume shorter than the spikelet, exposing part of the upper lemma 3

3. Plant annual . . . . **1. pumila**
- Plant perennial with short knotty rhizomes . . . . **2. parviflora**

1. S. pumila (Poir.) Roem. & Schult., *Syst. veg.* **2**: 891 (1817).
S. glauca auct., non (L.) P. Beauv.
Annual up to 130 cm; leaves flat. Panicle spike-like, cylindrical, 1-10(-20) cm; bristles 3-12 mm, antrorsely scabrid, commonly tawny brown. Spikelets ovate, 1.5-3.5 mm; glumes ⅓-½ the length of the spikelet; upper lemma rugose to corrugate, rarely almost smooth.
Fl. III - IX. *Introduced weed of gardens, arable fields and waste places, especially near the coast. Native to tropical and warm temperate regions of the Old World.* **M**

2. S. parviflora (Poir.) Kerguélen in *Lejeunia* **120**: 161 (1987).
S. geniculata auct., non (Willd.) P. Beauv.; *S. gracilis* Kunth
Tufted, short-lived perennial up to 90 cm, with a short knotty rhizome; leaves flat. Panicle spike-like, cylindrical, 2-9 cm; bristles 3-15 mm, antrorsely scabrid, commonly tawny brown. Spikelets ovate, 2-2.5 mm; glumes ⅓-½ the length of the spikelet; upper lemma rugulose.
Fl. IV - VIII. *Introduced weed of roadsides and waste places. Native to tropical and subtropical America, Australia and Asia.* **M**

Specimens in which the base is missing are virtually indistinguishable from **2** which is annual.

3. S. viridis (L.) P. Beauv., *Ess. Agrostogr.*: 51, 171, 178 (1812).
Annual up to 50 cm; leaves flat. Panicle spike-like, cylindrical or lobed below, 2-12 cm; bristles 3-12 mm, antrorsely scabrid, commonly green, sometimes tinged with purple. Spikelets ellipsoid, 2-2.5(-3) mm; lower glume ¼-⅓ the length of the spikelet; upper glume as long as the spikelet; upper lemma finely rugose.
Fl. VIII - IX. *Garden weed in Funchal.* **M**

Probably the only native species of *Setaria* in Madeira, being found in cooler regions than the others.

4. S. verticillata (L.) P. Beauv., *Ess. Agrostogr.*: 51, 178 (1812).
S. adhaerens (Forssk.) Chiov.
Annual up to 100 cm; leaves flat. Panicle spike-like, cylindrical or untidily lobed, 2-15 cm, often entangled; bristles 3-8 mm, retrorsely barbed and tenaciously clinging, commonly green. Spikelets ellipsoid, 1.5-2.5 mm; lower glume ⅓-½ the length of the spikelet; upper glume as long as the spikelet; upper lemma finely rugose.
Fl. III - VIII. *An introduced weed of gardens and waste places in many places in Madeira and Porto Santo; also reported for Porto Santo but without further data. Tropical and warm temperate regions throughout the world.* **MDP**

59. Paspalidium Stapf

Perennials; ligule a ciliate rim. Inflorescence comprising short racemes, these overlapping along a central axis, the spikelets borne singly in 2 neat rows, their lower glume abaxial. Spikelets 2-flowered, falling entire; lower glume shorter than the spikelet; upper glume and lower lemma similar, as long as the spikelet or the glume a little shorter; upper lemma crustaceous, its margins inrolled and clasping only the edges of the palea.

1. P. geminatum (Forssk.) Stapf in Prain, *Fl. trop. Africa* **9**: 583 (1920).
Plant with creeping or floating spongy rhizomes; stems up to 60 cm, prostrate and rooting from the lower nodes. Inflorescence 5-30 cm; racemes 0.5-4 cm, less than their own length apart, their rhachis narrowly winged. Spikelets ovate, 1.6-2.6 mm; lower glume ¼-⅓ upper glume ⅔-⅘ the length of the spikelet; upper lemma granulose.
Fl. ? *Reported for Madeira but without further data.* **M**

60. Stenotaphrum Trin.

Creeping perennials; ligule a line of hairs. Inflorescence comprising very short racemes more or less embedded in a thickened central axis, the spikelets borne singly and with abaxial lower glume. Spikelets 2-flowered, falling entire; glumes both short or the upper as long as the spikelet; lower lemma coriaceous; upper lemma chartaceous with flat margins.

1. S. secundatum (Walter) Kuntze, *Revis. gen. pl.* **2**: 794 (1891).
S. americanum Schrank
Robust perennial up to 30 cm; leaves 5-15 cm × 5-10 mm, obtuse. Inflorescence 5-10 cm, straight or curved; racemes 5-7 mm, deeply embedded in a corky axis and each bearing 1-2(-3) spikelets; main axis breaking up into segments at maturity, each segment carrying one raceme. Spikelets 4-5 mm.
Fl. III - XI. *Introduced as a lawn grass and now naturalized in a number of places in Madeira and Porto Santo. Native to the Atlantic shores of Africa and America.* **MP**

61. Melinis P. Beauv.

Perennials; ligule a ciliate rim. Inflorescence a panicle with capillary branches. Spikelets 2-flowered, falling entire, slightly laterally compressed; lower glume small or suppressed; upper glume as long as the spikelet, sometimes awned; lower lemma narrower than the upper glume, usually awned; upper lemma often deciduous before the rest of the spikelet, thinly cartilaginous, laterally compressed, its margins flat and more or less enfolding the palea.

1. M. minutiflora P. Beauv., *Ess. Agrostogr.*: 54, t. 11/4 (1812).
Stems often matted, up to 100 cm; leaves and sheaths with hairs that are often sticky and sometimes smelling strongly of linseed oil. Panicle lanceolate to narrowly ovate, 10-30 cm, dense, often purplish. Spikelets narrowly oblong, 1.5-2(-2.4) mm; upper glume obtusely bilobed, with or without a short mucro; lower lemma awnless or with an awn up to 15 mm.
Fl. VII. *Native of tropical Africa. Introduced, probably for fodder (Molasses grass), and persisting around Ponta Delgada.* **M**

62. Digitaria Haller

Annuals; ligule a short scarious or membranous rim. Inflorescence composed of digitate or subdigitate racemes, the spikelets in pairs. Spikelets 2-flowered, falling entire, flattened on the front but convex on the back; lower glume small or suppressed; upper glume as long as the spikelet or much shorter; lower lemma usually as long as the spikelet; upper lemma chartaceous to cartilaginous, its flat hyaline margins enfolding and concealing the palea.

1. Nerves of the lower lemma scaberulous . . . . **1. sanguinalis**
- Nerves of the lower lemma smooth . . . . **2. ciliaris**

1. D. sanguinalis (L.) Scop., *Fl. carniol.* ed. 2, **1**: 52 (1772).
Stems decumbent at the base, geniculately ascending, up to 60 cm. Inflorescence comprising 2-16 racemes each 3-20 cm. Spikelets narrowly elliptic to ovate-elliptic, 2.3-3.5 mm; upper glume ⅓-½ the length of the spikelet; lower lemma scaberulous on the nerves.
Fl. VII - X. *Native to warm temperate regions throughout the world, sometimes penetrating into the tropics. An introduced weed.* **M**

2. D. ciliaris (Retz.) Koeler, *Descr. Gram.*: 27 (1802).
Like **1** but lower lemma quite smooth on the nerves.
Fl. V - XI. *Native throughout much of the tropics. An introduced weed of roadsides and waste places.* **M**

63. Pennisetum Rich.

Perennials; ligule a line of hairs. Inflorescence a spike-like panicle, usually cylindrical, bearing rosette-like clusters comprising 1-4 spikelets surrounded by a deciduous involucre of bristles which are free right to the base. Spikelets 2-flowered, falling with the involucre; lower glume up to ½ the length of the spikelet, sometimes suppressed; upper glume shorter than to as long as the spikelet; lower lemma equally variable in length; upper lemma membranous to coriaceous, its thin flat margins covering about half the palea.

1. Inflorescence reduced to a cluster of 2-4 subsessile spikelets enclosed in the uppermost sheath, with long protruding filaments and stigmas . . . . **1. clandestinum**
- Inflorescence a spike-like panicle, conspicuously exserted 2

2. Dwarf plant up to 20(-40) cm, with dense ovoid panicle 2-4 × 1-2 cm, excluding the bristles; bristles softly villous, the longest 35-65 mm . . . . **2. villosum**
- Tall plant up to 6 m, with linear panicle 7-30 × 1.5-2 cm, excluding the bristles; bristles glabrous or thinly ciliate below, the longest 10-40 mm . . . . **3. purpureum**

1. P. clandestinum Hochst. & Chiov. in Pirotta, *Fl. Eritrea*: 41 (1903).
Low sward-forming plant with slender rhizomes and stout, much-branched stolons; stems up to 20 cm; leaves up to 15 cm × 2-5 mm. Inflorescence reduced to a cluster of 2-4 spikelets enclosed within the uppermost sheath, only their tips, the filaments (up to 5 cm) and stigmas (up to 3 cm) protruding; involucral bristles delicate, ½-¾ the length of the spikelets. Spikelets 13-20 mm.
Fl. ? *Reported from Madeira but without further data. Native to the E. African highlands and probably introduced as a pasture grass (Kikuyu grass).* **M**

2. P. villosum R. Br. ex Fresen. in Salt, *Voy. Abyss., app.*: 62 (1814).
Low, mat-forming rhizomatous plant; stems up to 20(-40) cm; leaves 5-20 cm × 3-5 mm. Inflorescence ovoid to subspherical, 2-4 × 1-2 cm (excluding the bristles), conspicuously exserted; involucral bristles softly villous below, the longest 35-65 mm. Spikelets 9-12 mm.
Fl. ? *Native to E. Africa and Arabia.; cultivated for ornament in Madeira, often escaping and persisting outside gardens.* **M**

3. P. purpureum Schum., *Beskr. Guin. pl.*: 44 (1827).
Robust plant froming large clumps 2-6 m high; leaves up to 120 × 5 cm. Inflorescence linear, 7-30 × 1.5-2 cm (excluding the bristles), conspicuously exserted; involucral bristles thinly ciliate, sometimes glabrous, the longest 10-40 mm. Spikelets 5-7 mm.
Fl. ± all year. *Native of tropical Africa. Introduced for its value as a fodder grass (Elephant grass, Napier grass), but now persisting only as a weed of waste ground, gardens and banana plantations in southeast Madeira.* **M**

64. Cenchrus L.

Perennials; ligule a line of hairs. Inflorescence spike-like, cylindrical, bearing rosette- or burr-like clusters comprising 1-8 spikelets surrounded by a deciduous involucre or 1 or more whorls of bristles, at least the inner whorl flattened and united below into a disc or cupule. Spikelets 2-flowered, falling with the involucre; lower glume up to ½ the length of the spikelet, sometimes suppressed; upper glume a little shorter than the spikelet; upper lemma firmly membranous to coriaceous, the thin flat margins covering up to ⅔ of the palea.

1. C. ciliaris L., *Mant.* **2**: 302 (1771).
Pennisetum cenchroides Rich.
Stems up to 150 cm, ascending, wiry or sometimes almost woody. Inflorescence cylindrical to ovoid, 2-12 × 1-2.5 cm; involucral bristles 6-16 mm, the inner greatly exceeding the spikelets, one of them longer and stouter than the rest, flattened below, connate only at the base to form a disc 0.5-1.5 mm across or sometimes connate for up to 0.5 mm above the rim of the disc. Spikelets 2-4 per burr, 2-5.5 mm.
Fl. ? III - X. *Introduced for fodder (African foxtail-grass, Buffel grass) but now thoroughly naturalized in waste places in many coastal parts of Madeira and Porto Santo. Native throughout Africa and eastwards to India.* **MP**

65. Imperata Cirillo

Perennials; ligule a short scarious membrane. Inflorescence a narrow, often spike-like panicle with tough branches bearing paired similar spikelets, each spikelet of a pair borne on a slender pedicel. Spikelets 2-flowered, terete, enveloped in long silky hairs arising from the callus and the glumes; glumes as long as the spikelet, membranous; lower floret reduced to a hyaline lemma; upper floret bisexual, with an awnless hyaline lemma.

1. I. cylindrica (L.) Reusch., *Nomencl. bot.* ed. 3 : 10 (1797).
Aggressively rhizomatous, up to 120 cm; leaves up to 100 cm × 2-20 mm, stiffly erect. Panicle 3-22 cm. Spikelets 2.2-6 mm.
Fl. V - VI. *Native from the Mediterranean region to SW Asia, tropical Africa, tropical Asia and Australia. Reported from Madeira but without further data.* **M**

66. Sorghum Moench

Annuals or perennials, usually robust; ligule a scarious membrane. Inflorescence a panicle with persistent branches bearing short fragile racemes (tough in cultivated races) of paired dissimilar spikelets, one of a pair sessile, the other pedicelled. Sessile spikelet 2-flowered, dorsally compressed; lower floret reduced to a hyaline lemma; upper floret bisexual, the lemma 2-toothed and with a glabrous geniculate awn. Pedicelled spikelet male or sterile, usually smaller than the sessile, awnless.

S. bicolor (L.) Moench, (*S. vulgare* Pers.), Sorghum or Guinea-corn is cultivated in Madeira on a small, non-commercial basis. It is an annual and its large persistent grains are exposed at maturity by the gaping glumes.

1. S. halepense (L.) Pers., *Syn. pl.* **1**: 101 (1805).
Extensively rhizomatous perennial up to 3m; leaves 20-90 cm × 5-40 mm. Panicle 10-55 × 3-25 cm, the primary branches bare below, divided above, ultimately bearing racemes composed of 1-5 spikelet-pairs and terminating in a triad of 1 sessile and 2 pedicelled spikelets. Sessile spikelet 4.5-5 mm, cream varying through mahogany to brown or blackish. Pedicelled spikelet 4.5-6.5 mm, male, eventually deciduous.
Fl. IV - VIII. *Common weed of vineyards, cultivated fields, roadsides and waste ground.* **M**

67. Dichanthium Willemet

Perennials; ligule a membrane. Inflorescence comprising subdigitate fragile racemes bearing paired, dissimilar spikelets, one of a pair sessile, the other pedicelled; internodes and pedicels solid, terete. Sessile spikelet 2-flowered, dorsally compressed; lower floret reduced to a hyaline lemma; upper floret bisexual, the lemma entire and passing into a glabrous geniculate awn. Pedicelled spikelet male or sterile, awnless.

1. D. annulatum (Forssk.) Stapf in Prain, *Fl. trop. Africa* **9**: 178 (1917).
Stems up to 100 cm; nodes conspicuously bearded. Inflorescence comprising 2-15 racemes each 3-7 cm, the lowest 0-6 spikelet-pairs on each homogamous. Sessile spikelet narrowly oblong, 2-6 mm; lower glume slightly concave, pubescent to villous below the middle and with long bulbous-based hairs on the margins above, obtuse to subacute; awn of upper lemma 8-25 mm.
Fl. III - IX. *Native from tropical Africa to SE Asia. An introduced weed, found on roadsides and rocky areas near the sea in and around Funchal.* **M**

68. Bothriochloa Kuntze

Perennials; ligule a membrane. Inflorescence composed of subdigitate fragile racemes bearing paired dissimilar spikelets, one of a pair sessile, the other pedicelled; internodes and pedicels hyaline and translucent between thickened margins. Sessile spikelet 2-flowered, dorsally compressed; lower floret reduced to a hyaline lemma; upper floret bisexual, the lemma entire and passing into a glabrous geniculate awn. Pedicelled spikelet male or sterile, awnless.

1. B. ischaemum (L.) Keng in *Contr. biol. Lab. Sci. Soc. China* (Bot.) **10**: 201 (1936).
Stems tufted, up to 80 cm; nodes glabrous or shortly bearded. Inflorescence comprising 5-15 racemes each 4-6 cm (the lowest being longer than the axis), without homogamous spikelet-pairs below. Sessile spikelet oblong-lanceolate, 3.5-5 mm; lower glume hairy below the middle, acute; awn of upper lemma 12-15 mm.
Fl. VI - IX. *Reported from Madeira without further data but possibly in error for* Dichranthium annulatum. **M**

69. Hyparrhenia E. Fourn.

Perennials; ligule scarious. Inflorescence comprising pairs of fragile racemes bearing paired dissimilar spikelets, one of a pair sessile, the other pedicelled; raceme-pairs enclosed in a linear to ovate, bract-like spatheole and crowded together into a false panicle. Sessile spikelet 2-flowered, dorsally compressed; lower floret reduced to a hyaline lemma; upper floret bisexual, the lemma 2-toothed and bearing a hairy geniculate awn. Pedicelled spikelet male or sterile, awned or awnless.

1. H. hirta (L.) Stapf in Prain, *Fl. trop. Africa* **9**: 315 (1919).
Tufted rhizomatous plant up to 60 cm. False panicle typically scanty, up to 30 cm, bearing 2-10 raceme-pairs; spatheoles linear-lanceolate, 3-8 cm, eventually turning reddish, about as long as the enclosed peduncle; racemes 2-4 cm, villous. Sessile spikelet 4-6.5 mm, subacute to acute; awn of upper lemma 10-35 mm.
Fl. III - VII. *From 150-750 m. Common in fields, along roadsides and in waste places in many parts of Madeira and Porto Santo.* **MP**

CXVIII. ARACEAE
N.J. Turland & M.J. Cannon

Perennial, monoecious glabrous herbs with subterranean, tuberous stems or rhizomes. Leaves basal, petiolate; petiole usually long, sheathing at base. Inflorescence consisting of an often conspicuous petal- or leaf-like bract (*spathe*), which subtends and sometimes envelops a flower-bearing spike (*spadix*), which may be prolonged above the flowers into a sterile fleshy *appendix*. Flowers unisexual, the male borne higher on spadix than the female, perianth absent; club- or hair-like sterile flowers (*pistillodes* or *staminodes*) sometimes also present. Fruit a berry containing 1-many seeds.

1. Leaves peltate . . . . **2. Colocasia**
- Leaves not peltate 2

2. Leaves pedately divided into segments . . . . **5. Dracunculus**
- Leaves not pedately divided 3

3. Lower margins of spathe fused to form a tube . . . . **4. Arisarum**
- Lower margins of spathe free to base, though overlapping 4

4. Spadix without an appendix, covered with flowers to apex . . . . . . **1. Zantedeschia**
- Spadix with a conspicuous sterile appendix borne above the floral zones . . . **3. Arum**

1. Zantedeschia Spreng.

Plant with short rhizomes. Leaves usually sagittate with broad, obtuse basal lobes; petiole shortly sheathing at base. Inflorescence longer than leaves; spathe persistent in fruit, eventually withering but not falling, rhombic-ovate, the lower margins overlapping and closely enfolding base of spadix, the limb wide and open, cuspidate and recurved at apex; spadix shorter than spathe, with male and female floral zones adjacent, the male zone much longer; appendix absent. Staminodes present at base of each female flower.

1. **Z. aethiopica** (L.) Spreng., *Syst. veg.* 3: 765 (1826). [*Jarro*]
Richardia aethiopica (L.) Spreng. [sphalm.]; *R. africana* Kunth
Plant 50-100 cm. Leaves with lamina 15-40 × 6-20 cm; petiole to 30 cm or more. Spathe 10-20 cm, green at base, the limb white. Spadix about half as long as spathe, the upper part densely covered with yellow anthers; staminodes conspicuous, yellow. Fruits ripening yellow. Fl. ± all year but mainly XI - VI. *Introduced; widely cultivated in Madeira both for ornament and the cut flower trade; commonly planted and naturalized on shady banks, roadsides and along water-courses up to c. 800 m. Native to S. Africa.* **M**

2. Colocasia Schott

Plant robust, with often large tubers and slender, subterranean stolons. Leaves large, coarse, cordate-sagittate, peltate, the basal lobes rounded or angular, the apex acute; petiole moderately sheathing at base. Inflorescence much shorter than leaves, 1-several in each leaf-axil; lower part of spathe persistent in fruit, constricted between the overlapping basal margins and the hooded or flattened limb; spadix shorter than spathe, with male floral zone much longer than the female and separated from it by a zone of sterile flowers; appendix present, short or long. Male flowers with 3-6 stamens fused to form a column.

1. **C. esculenta** (L.) Schott in Schott & Endl., *Melet. bot.*: 18 (1832).
[*Inhame, Inhame do Enxerto*]
C. antiquorum Schott
Leaves with lamina to 90 cm, with thick raised veins beneath; petiole inserted much nearer to sinus than to apex of lamina.
Fl. said never to flower in Madeira. *Introduced; widely cultivated on a small field scale in Madeira for its edible tubers; naturalized along streams and in wet places. Native habitat unknown, possibly India.* **M**

3. Arum L.

Plant with tubers, dormant during summer. Leaves narrowly hastate; petiole shortly sheathing at base. Inflorescence shorter than to ± equalling leaves; spathe not persistent in fruit, the lower margins overlapping each other and enfolding lower part of spadix, the limb usually concave, the apex drooping; spadix shorter than spathe, with a zone of pistillodes below and staminodes above the male floral zone; appendix present, stout, with a distinctly stalked base.

1. A. italicum Mill., *Gard. Dict.* ed. 8, no. 2 (1768). [*Bigalhó*]
Plant to 40 cm or more. Leaves appearing in autumn; lamina 6-15 × 4-11 cm, deep green; petiole to 22 cm, dull purple. Peduncle dull purple. Spathe 14-23 cm; limb greenish white, sometimes slightly purplish at base; tube purple inside. Spadix ⅓-½ as long as spathe; appendix yellow. Fruits ripening red. **Plate 56**
Fl. III - VI. *Woodland, including laurisilva, other shady places and stream-sides; infrequent but scattered throughout Madeira.* **M**

The above description refers to the (■) subsp. **canariense** (Webb & Berthel.) P.C. Boyce, *Genus Arum*: 77 (1993) (*A. canariense* Webb & Berthel.; *A. italicum* var. *canariense* (Webb & Berthel.) Engler), which otherwise occurs in the Azores and Canaries and to which Madeiran plants belong.

4. Arisarum Mill.

Plant with irregularly shaped tubers or rhizomes, dormant during summer. Leaves ovate to hastate-sagittate; petiole slightly sheathing at base. Inflorescence shorter than to exceeding leaves; spathe not persistent in fruit, the lower margins fused to form a tube, the upper part open and usually curved at apex; spadix ± exserted from or almost completely concealed by spathe-tube, with a short, basal, floral zone lacking sterile flowers; appendix present, long.

1. A. vulgare Targ.-Tozz. in *Annali Mus. Sto. nat. Firenze* **2**: 67 (1810).
Plant to *c*. 30 cm. Leaves with lamina 5-12 × 3-10 cm; petiole to 30 cm, purple-spotted. Peduncle 7-30 cm, slender. Spathe 3.5-6 × 1-1.5 cm, greenish or brownish-purple; appendix greenish. Female flowers 4-5. Fruits ripening greenish, containing 2-6 seeds.
Fl. XII, III - IV(-V). *Water-courses and fields in the Funchal region.* **M**

5. Dracunculus Mill.

Plant with tubers, dormant during summer. Leaves pedately divided into 5-7 ± lanceolate segments; petiole sheathing at base, the sheaths strongly convoloute into a weak to moderately robust pseudostem. Inflorescence longer than leaves; spathe persistent in withered state in fruit, the lower margins overlapping each other and enclosing base of spadix, the limb ± flat; spadix shorter than spathe, the floral zone without sterile flowers and much shorter than appendix.

1. D. canariensis Kunth, *Enum. pl.* **3**: 30 (1841).
D. vulgaris sensu auct. mad., non Schott (1832).
Plant to 1.5 m, usually glaucous green and unspotted, rarely with very pale irregular blotches on pseudostem. Leaves with segments of lamina to 15 × 4 cm; petiole to 40 cm, about twice as long as sheath. Peduncle to 50 cm. Spathe-tube greenish; limb 20-30 × 5-6 cm, cream or greenish white. Appendix narrow, robust, pale yellow. Fruits ripening orange-red.
Fl. III - V. *Laurisilva; always rare in Madeira; recorded from São Martinho and São Gonçalo, either side of Funchal, and from the Santana area on the northern coast, but not seen for at least eighty years and possibly now extinct.* ■ **M** [*Canaries*]

CXIX. LEMNACEAE
M.J. Cannon

Small aquatic, free-floating or submerged herbs. Stems and leaves usually undifferentiated, fronds with or without roots. Plants monoecious, rarely dioecious, inflorescence of 1 female and 1-2 male flowers in pouches or sheaths. Fruit a utricle. Seeds smooth or ribbed.

1. Lemna L.

Fronds more than 1 mm, free-floating or lying on wet mud; veins 1-5; roots 1 per frond or 0. Budding pouches 2, one vegetative and giving rise to new fronds, the other to flowers and fruits, both opening by transverse slits coinciding with the margin of the frond. Inflorescence of 1 female and 2 male flowers surrounded by a spathe. Perianth 0. Male flowers with 1 stamen. Female flowers with a sessile, globular ovary. Fruit ribbed, rarly smooth, ± compressed, slightly or distinctly winged.

1. Fronds frequently swollen and gibbous beneath, if flat then cell mesh with a few large distinct cells in the centre; fruit winged, with 2 seeds . . . . **1. gibba**
- Fronds always flat, cell mesh with numerous ± uniform cells in the centre; fruit not winged, 1-seeded . . . . **2. minor**

1. L. gibba L., *Sp. pl.* **2**: 970 (1753).   [*Patinha de agua*]
Fronds usually floating, up to 6 mm, ovate to sub-orbicular, usually asymmetric, rounded at the apex and base, ventral surface usually gibbous, with cell mesh divisions larger in the central portion; veins 3-5, fused at the base. Inflorescence in a ± triangular pouch. Ovules 2-6. Fruit winged, 2-seeded. Plants rarely produce flowers or fruit.
Fl. III - V. *Common in Madeira in still and slow-moving water in ditches and ponds, occasionally also on wet rocks.* M

Plants are sometimes found without the gibbous ventral surface and are extremely difficult to assign to the correct species, especially where this and the following species grow together.

2. L. minor L., *Sp. pl.* **2**: 970 (1753).
Like **1** but usually smaller, up to 5 mm and almost symmetrical with flat ventral surface and cell mesh of nearly uniformly sized small cells; veins usually 3, if 4-5 then the lateral vein connected to the inner some way above the base. Ovules 1. Fruit scarcely winged, 1-seeded.
Fl. ? *Common in Madeira, mainly in still water and on wet rocks.* M

CXX. CYPERACEAE
N.J. Turland

Annual or perennial herbs, often rhizomatous, usually glabrous. Stems usually erect, simple, cylindric or 3-angled. Leaves grass-like, sometimes reduced to sheaths. Bracts leaf-like to filiform, or absent. Flowers insignificant, wind-pollinated, hermaphrodite or unisexual, each subtended by a scale-like bracteole (*glume*) and arranged in 1- to many-flowered spikelets, the axis of which is termed the *rhachilla*. Spikelets solitary or aggregated into compound, usually bracteate inflorescences. Perianth absent, or composed of minute bristles surrounding ovary. Stamens 1-3. Stigmas 2-3. Ovary 1-celled, with a single ovule. Fruit a small nut, usually with 2-3 angles.

Most of the species grow in wet places.

1. All leaves reduced to sheaths; inflorescence a single terminal spikelet; bracts absent . . . . **4. Eleocharis**
- Usually at least some leaves with obvious blades; inflorescence of 1-many spikelets; bracts often conspicuous, sometimes erect and resembling a prolongation of stem 2

2. Inflorescence a raceme or spike-like head of 1 or more spike-like panicles, each panicle a dense aggregation of 1-flowered spikelets; flowers unisexual, the males and females often in separate panicles . . . . **9. Carex**
- Inflorescence of 1 or more sessile clusters of 1-many spikelets, or an umbel with rays terminating in clusters of spikelets; flowers hermaphrodite 3

3. Glumes emarginate at apex, conspicuously aristate 4
- Glumes rounded to acuminate at apex, sometimes very shortly mucronate but never aristate 5

4. Spikelets 7-11 × 3.5-6 mm . . . . **2. Schoenoplectus**
- Spikelets 20-25 × 5-7 mm . . . . **1. Bolboschoenus**

5. Bracts absent, or 1-2 with 1 bract erect and resembling a prolongation of stem . . . 6
- Bracts 2-many, forming an involucre surrounding inflorescence, not resembling a prolongation of stem 8

6. Rhizomatous perennials; stems solitary or tufted, with leaves reduced to sheaths, the uppermost sheath sometimes with a short blade; stamens 3; stigmas 2; nut plano-convex . . . . **6. Juncellus**
- Annual or perennial, without rhizomes; stems tufted, leafy; stamens 1-2(-3); stigmas 2-3; nut 3-angled 7

7. Inflorescence of dense globose heads of spikelets, with at least 1 sessile head and 2-4 short rays each terminating in a similar head; bracts 2, 1 of them much longer than inflorescence . . . . **5. Cyperus**
- Inflorescence a sessile cluster of 1-2 ovoid to lanceolate-oblong spikelets; bracts 1(-2), shorter than or up to 5 × as long as inflorescence, or absent . . . . **3. Isolepis**

8. Inflorescence of 1(-3) sessile dense heads of 1-flowered spikelets; glumes persistent . . . . **8. Kyllinga**
- Inflorescence umbellate, or of sessile clusters of spikelets; spikelets 4- to many-flowered; glumes deciduous 9

9. Stigmas (2-)3; nut 3-angled, rarely plano-convex, with a face against rhachilla . . . . **5. Cyperus**
- Stigmas 2; nut lenticular, with an angle against rhachilla . . . . **7. Pycreus**

1. Bolboschoenus Asch. ex Palla

Rhizomatous perennial. Stems 3-angled, leafy. Inflorescence terminal, usually with an involucre of a few leaf-like bracts, consisting of a sessile cluster of 1-several spikelets or an umbel with each ray terminating in a cluster of 1-6 spikelets. Glumes emarginate and conspicuously aristate at apex (giving spikelets a bristly appearance). Flowers hermaphrodite; perianth-bristles 1-6, rarely absent; stamens 3; stigmas 2-3. Nut plano-convex or 3-angled.

1. B. maritimus (L.) Palla in W.D.J. Koch, *Syn. deut. schweiz. Fl.* ed. 3, **3**: 2532 (1905). *Scirpus maritimus* L.; *S. maritimus* var. *genuinus* Gren. & Godr.
Stems to *c*. 50 cm. Spikelets 20-25 × 5-7 mm, lanceolate; glumes dark brown. Perianth-bristles shorter than to longer than nut, with backward-pointing barbs. Nut shiny brown to black, smooth. Fl. V. *Restricted to Porto Santo; known only from a wet place at Serra de Dentro.* **P**

An old record of *Scirpus maritimus* var. *monostachyus* Webb & Berthel., from Madeira, is erroneous and referable to *Schoenoplectus triqueter* (L.) Palla.

2. Schoenoplectus (Rchb.) Palla

Rhizomatous perennial. Stems 3-angled, with leafless sheaths below, the uppermost sheath usually with a short blade. Inflorescence appearing lateral, a sessile congested cluster of 1-3 spikelets subtended by an erect bract resembling a prolongation of stem. Glumes emarginate and conspicuously aristate at apex (giving spikelets a bristly appearance). Flowers hermaphrodite; perianth-bristles 3-4; stamens 3; stigmas 2. Nut plano-convex.

1. S. triqueter (L.) Palla in *Verh. zool.-bot. Ges. Wien* **38**: 49 (1888) et in *Bot. Jb.* **10**: 299 (1888).
Scirpus maritimus var. *monostachyus* sensu auct. mad., non Webb & Berthel.; *S. pungens* var. *sarmentoi* Menezes; *S. triqueter* L.
Stems 40-80 cm. Bract much longer than inflorescence. Spikelets 7-11 × 3.5-6 mm, ovoid to ovoid-lanceolate; glumes mid-brown. Perianth-bristles shorter than to equalling nut, with backward-pointing barbs. Nut brownish black, smooth.
Fl. VII - VIII. *Very rare in Madeira, known only from moist places near the Ribeira de São João in the Funchal region; recorded also from Porto Santo but without further data.* **MP**

3. Isolepis R. Br.

Annual or perennial. Rhizomes absent. Stems tufted, cylindric, slender, usually leafy below. Inflorescence a sessile cluster of 1-2 spikelets, appearing terminal and ebracteate, or lateral, when subtended by an erect bract resembling a prolongation of stem. Glumes rounded to obtuse at apex, very shortly mucronate but not aristate. Flowers hermaphrodite; perianth-bristles absent; stamens 1-2; stigmas 2-3. Nut 3-angled.

1. Plant annual or perennial; stems and leaves slender but not filiform; bract 2-5 times as long as inflorescence; nut usually glossy, with prominent longitudinal ribs . . . . **1. setacea**
- Plant always annual; stems and leaves filiform; bracts absent or very short, only very rarely longer than inflorescence; nut matt, finely papillose or almost smooth . . . **2. cernua**

1. I. setacea (L.) R. Br., *Prodr. fl.* **1**: 222 (1810).
Scirpus setaceus L.
Annual or perennial. Stems 2-14 cm, densely tufted and forming a small tussock; stems and leaves slender but not filiform. Bract 1 (rarely with a second much smaller bract), 2-5 × as long as inflorescence. Spikelets 2.5-3.5 mm, ovoid; glumes purplish brown with a green mid-vein. Nut usually glossy, brown or black, with prominent longitudinal ribs.
Fl. VI. *A rare plant of moist trodden ground and other wet places; occurring between the high peaks of eastern Madeira and Faial on the northern coast, from 900-1700 m.* **M**

2. I. cernua (Vahl) Roem. & Schult., *Syst. veg.* **2**: 106 (1817).
Scirpus cernuus Vahl; *S. savii* Sebast. & Mauri
Like **1** but always annual; stems 7-20 cm; stems and leaves filiform; bract absent or very short, only rarely longer than inflorescence (when up to 5 × as long); spikelets 2.5-5 × 2-2.5 mm, usually ovoid, sometimes lanceolate-oblong; glumes white or greenish; nut matt, finely papillose or almost smooth.
Fl. IV - IX. *Streams, levadas, wet crevices and ledges of cliffs and other wet places, from sea-level to 1000 m; widespread and frequent in Madeira east to Faial and Funchal; very rare in Porto Santo, known only from the Fonte das Pombas.* **MP**

4. Eleocharis R. Br.

Rhizomatous perennial. Leaves reduced to sheaths. Inflorescence a single terminal spikelet; bracts absent. Glumes obtuse at apex, with a ± broad membranous margin. Flowers hermaphrodite; perianth-bristles usually 4, rarely absent; stamens 3; stigmas 2. Nut biconvex, with a persistent and usually swollen base to the style (*stylopodium*).

1. E. palustris (L.) Roem. & Schult., *Syst. veg.* **2**: 151 (1817).
Stems 25-75 cm. Spikelets 7-20 × 2.5-6 mm, narrowly ellipsoid; glumes dark brown, the margin and usually the centre pale; lowest 2 glumes empty, each half-encircling base of spikelet. Perianth-bristles variably developed. Nut pale yellow or dark brown, minutely dotted.
Fl. ? *Very rare in Madeira, known from a single collection from Porto do Moniz.* M

5. Cyperus L.

Annuals or perennials. Rhizomes present or absent. Stems solitary or tufted, usually leafy. Inflorescence terminal, umbellate or of sessile clusters of spikelets surrounded by an involucre of leaf-like bracts, rarely apparently lateral (when bracts 2, 1 of them erect and resembling a prolongation of stem). Spikelets 4- to many-flowered; glumes rounded to acuminate at apex, sometimes very shortly mucronate but never aristate, deciduous. Flowers hermaphrodite; perianth-bristles absent; stamens 1-3; stigmas (2-)3. Nut 3-angled, rarely plano-convex, with a face against rhachilla.

1. Annuals; rhizomes absent; stems tufted 2
- Usually perennials; rhizomes present; stems solitary or tufted 3

2. Inflorescence appearing lateral, subtended by 2 bracts, 1 of them erect and resembling a prolongation of stem; spikelets 2-3 × c. 1 mm, the glumes rounded at apex, not mucronate
. . . . **7. difformis**
- Inflorescence terminal, surrounded by an involucre of usually 3 bracts; spikelets 4-10 × 1.5-2 mm, the glumes obtuse to acute at apex, often very shortly mucronate **6. fuscus**

3. Plant leafless; bracts 16-21, all ± equal and much longer than inflorescence . . . .
. . . . **5. involucratus**
- Plant leafy; bracts 2-9, unequal, not all longer than inflorescence 4

4. Spikelets oblong-lanceolate, the glumes not overlapping . . . . **4. eragrostis**
- Spikelets very narrowly lanceolate to linear, the glumes overlapping 5

5. Each ray of umbel terminating in a somewhat lax panicle of spikelets, with spikelets separated from each other by a gap of approximately their own width; glumes yellowish brown or greenish brown . . . . **3. esculentus**
- Each ray of umbel terminating in a dense panicle or ± sessile cluster of spikelets; glumes chestnut-brown or darker 6

6. Stems to c. 60 cm; most bracts longer than inflorescence; rays of umbel 4-10 cm, each terminating in a dense panicle of many (at least 15) spikelets; spikelets 10-15 mm . . . .
. . . . **1. longus**
- Stems shorter, 13-30 cm; bracts sometimes all small and shorter than inflorescence; rays of umbel shorter, 1.5-6 cm, each terminating in a ± sessile cluster of few (up to 8) spikelets; spikelets 10-30 mm . . . . **2. rotundus**

1. C. longus L., *Sp. pl.* **1**: 45 (1753). [*Junça*]
C. longus subsp. *badius* (Desf.) Murb.; *C. longus* subsp. *genuinus* Cout.
Perennial. Rhizomes stout, 3-10 mm wide, covered with broad scales; tubers absent. Stems to 60 cm, leafy. Bracts 2-4, unequal, 2-3 of them longer than inflorescence. Inflorescence a terminal umbel with 2-5 rays, each ray 4-10 cm, terminating in a dense panicle of many (at least 15) spikelets; a few ± sessile clusters of spikelets also present at base of rays. Spikelets 10-15 × 1.5-2 mm, linear; rhachilla broadly winged; glumes closely overlapping, obtuse at apex,

chestnut-brown, with a greenish keel and somewhat indistinct veining. Stamens 3. Stigmas 3. Nut 3-angled, brownish to black.
Fl. V - X. *By water-courses and in other wet places; occasional in Madeira; rare Porto Santo, near Serra de Fora.* **MP**

2. C. rotundus L., *Sp. pl.* **1**: 45 (1753).   [*Coquinho*]
Like **1** but rhizomes slender, *c.* 1 mm wide, with somewhat remote narrow scales; tubers sometimes present; stems often shorter, 13-30 cm; bracts sometimes all small and shorter than inflorescence; rays often shorter, 1.5-6 cm, terminating in sessile clusters of few (up to 8) spikelets; spikelets 10-30 × *c.* 2.5 mm; glumes moderately closely overlapping, dark brown.
Fl. throughout the year. *In Madeira mainly in the Funchal region, growing as a weed in cultivated and waste areas from sea-level to 150 m; rare in Porto Santo, occurring at Vila Baleira.* **MP**

3. C. esculentus L., *Sp. pl.* **1**: 45 (1753).
Like **1** but annual or perennial; rhizomes long, slender, scaly; tubers sometimes present; stems 13-40 cm, leafy; bracts sometimes all shorter than inflorescence; rays 3-7 cm, terminating in somewhat lax panicles of spikelets, with spikelets separated from each other by a gap of approximately their own width; spikelets 8-14 × 2-2.5 mm; glumes loosely overlapping, yellowish brown to greenish brown, with prominently veining; nut dark grey or reddish.
Fl. IV - X. *Scattered around the coast of Madeira, growing in marshy places, cultivated areas and on roadsides, from near sea-level to 250 m.* **M**

4. C. eragrostis Lam., *Tabl. encycl.* **1**: 146 (1791).
C. vegetus Willd.
Perennial. Rhizomes short and thick. Stems 25-60 cm, leafy. Bracts 6-9, unequal, 3-6 of them much longer than inflorescence. Inflorescence a terminal umbel with 4-7 rays, each ray 1.5-5 cm, terminating in a dense rounded head of spikelets; a few ± sessile clusters of spikelets also present at base of rays. Spikelets 7-10 × 3-4 mm, oblong-lanceolate; rhachilla not winged; glumes not overlapping, acuminate at apex, green to yellowish brown. Stamen 1. Stigmas 3. Nut 3-angled, grey.
Fl. I - X. *Introduced; naturalized on stream-banks, by levadas and in other wet and cultivated places in Madeira, from sea-level to 600 m; occurring mainly on the southern coast, from Canhas to the Funchal region; rarer on the northern coast, known from the Santana area. Native to tropical America.* **M**

5. C. involucratus Rottb., *Descr. icon. rar. pl.*: 22 (1773).
C. alternifolius sensu auct. mad., non L. (1767)
Similar to **4** but stems taller, to *c.* 90 cm, leafless; bracts 16-21, all ± equal and much longer than inflorescence; rays numerous; spikelets smaller, 4.5-6 × 2-3 mm, lanceolate.
Fl. ? *Introduced; apparently a well-established escape from cultivation in south-eastern Madeira, from the Funchal region to Machico, growing in waste places. Native to Africa.* **M**

6. C. fuscus L., *Sp. pl.* **1**: 46 (1753).
Annual. Rhizomes absent. Stems 5-20 cm, tufted, leafy. Bracts usually 3, unequal, all longer than inflorescence. Inflorescence a terminal umbel with up to 4 rays, each ray to 2.5 cm, terminating in a ± lax rounded cluster of spikelets; at least 1 sessile cluster of spikelets also present at base of rays; sometimes rays absent and only 1 or more sessile clusters present. Spikelets 4-10 × 1.5-2 mm, oblong-lanceolate to oblong-linear; rhachilla not winged; glumes obtuse to acute at apex, often very shortly mucronate, dark brown, often greenish centrally. Stamens 2. Stigmas (2-)3. Nut 3-angled, rarely plano-convex, white.
Fl. ? *Very rare in Madeira, growing by water-courses; known only from two records, both from the northern coast, one between Porto do Moniz and Seixal, the other from Santana.* **M**

7. C. difformis L., *Cent. pl.* **2**: 6 (1756).
Annual. Rhizomes absent. Stems 13-30 cm, tufted, leafy. Inflorescence appearing lateral, of dense globose heads of spikelets, with at least 1 sessile head and 2-4 short rays, each ray 5-12 mm and terminating in a similar head; bracts 2, 1 of them short and ± patent, the other erect and much longer than inflorescence, resembling a prolongation of stem. Spikelets 2-3 × *c*. 1 mm, narrowly oblong; rhachilla not winged; glumes very loosely overlapping, rounded at apex, not mucronate, dark brown. Stamens 1(-3). Stigmas 3. Nut 3-angled, pale yellow to yellowish green.
Fl. IX - X. *Introduced; probably naturalized in the Funchal region, growing in waste places. Native to tropical Africa and Asia.* **M**

6. Juncellus C.B. Clarke

Rhizomatous perennial. Rhizomes usually long, slender. Stems solitary or tufted. Leaves reduced to sheaths, the uppermost sheath sometimes with a short blade. Inflorescence appearing lateral, a sessile cluster of 1-40 spikelets subtended by 2 bracts, 1 of them short and spreading, often very short or absent, the other erect and at least twice as long as inflorescence, resembling a prolongation of stem. Spikelets many-flowered; glumes obtuse at apex, deciduous. Flowers hermaphrodite; perianth-bristles absent; stamens 3; stigmas 2. Nut plano-convex, with a face against rhachilla.

1. J. laevigatus (L.) C.B. Clarke in Hook. f., *Fl. Brit. India* **6**: 596 (1893). [*Junquilho* (P)]
Cyperus laevigatus L.
Stems 8-60 cm. Spikelets 5-15 × *c*. 2.5 mm, oblong-lanceolate to linear, acute; glumes closely overlapping, pale yellowish brown. Nut greyish brown or yellowish brown.
Fl. IV - VIII. *Very rare in Madeira, growing in saturated crevices of maritime cliffs and other wet places between Ribeira Brava and Ponta do Sol on the south coast (known here since 1861 at least); commoner in Porto Santo, occurring near Serra de Fóra and at Vila Baleira.* **MP**

The above description refers to subsp. **laevigatus**, to which plants from Madeira and Porto Santo belong.

7. Pycreus P. Beauv.

Annual. Rhizomes absent. Stems tufted, leafy. Inflorescence a terminal umbel with up to 4 rays, surrounded by an involucre of leaf-like bracts, each ray terminating in a lax rounded cluster of spikelets; at least 1 sessile cluster of spikelets also present at base of rays; sometimes rays absent and only 1 or more sessile clusters present. Spikelets 8- to many-flowered; glumes obtuse at apex, deciduous. Flowers hermaphrodite; perianth-bristles absent; stamens 2-3; stigmas 2. Nut lenticular, with an angle against rhachilla.

1. P. flavescens (L.) Rchb., *Fl. germ. excurs.* **1**: 72 (1830).
Cyperus flavescens L.
Stems 7-30 cm. Bracts 2-4, unequal, at least 1 bract (often 2) longer than inflorescence. Rays to 4 cm. Spikelets 7-10 × 2-3 mm, oblong-lanceolate; glumes somewhat loosely overlapping, yellowish green to yellowish brown. Nut brownish black.
Fl. VII - X. *Mainly on the southern coast of Madeira, from Ponta do Sol to Caniçal; rarer on the northern coast, occurring around Santana; growing on wet maritime cliffs, in marshy places and along water-courses and roadsides, from near sea-level to 200 m.* **M**

8. Kyllinga Rottb.

Rhizomatous perennial. Rhizomes slender. Stems solitary or tufted, leafy. Inflorescence terminal, of 1(-3) sessile dense heads of 1-flowered spikelets, surrounded by an involucre of a few leaf-like bracts. Glumes acuminate at apex, persistent. Flowers hermaphrodite; perianth-bristles absent; stamens 1-3; stigmas 2. Nut lenticular, with an angle against rhachilla.

1. K. brevifolia Rottb., *Descr. icon. rar. pl.*: 13 (1773).
Cyperus brevifolius (Rottb.) Hassk.
Stems 20-30 cm. Bracts 3 (sometimes with a small extra bract), much longer than inflorescence; 2 bracts opposite each other, the other perpendicular to them. Heads 4-8 mm, broadly ovoid. Glumes whitish with a greenish keel. Nut yellowish to brown.
Fl. IX. *Introduced; locally established in springs and other marshy places in Madeira; known from Ribeira Brava in the south, São Jorge in the north, and Água de Pena and Caniçal in the east. Native to tropical America and Asia.* M

9. Carex L.
J.R. Press

Rhizomatous perennials. Stems usually leafy. Leaves usually sheathing at base, flat or keeled and involute, with a membranous *ligule* at the junction of blade and sheath. Inflorescence a raceme or spike-like head of 1 or more spike-like panicles, each panicle (termed a *spike* here) a dense aggregation of 1-flowered spikelets and subtended by a leaf-like to filiform bract. Flowers unisexual, the males and females often in separate panicles; perianth-bristles absent; male flowers with 2-3 stamens; female flowers enclosed in a sac-like perigynium (*utricle*) with a beak from which project the 2-3 stigmas. Nut 3-angled or biconvex.

Inflorescence and utricle characters, in particular sizes and colours, refer to the time when the fruit is ripe.

1. Spike solitary . . . . **10. peregrina**
- Spikes 2 or more 2

2. Spikes all similar in appearance 3
- Spikes dissimilar in appearance 5

3. Plant with long, creeping rhizomes, not caespitose . . . . **3. divisa**
- Plant without creeping rhizomes, densely caespitose 4

4. Lower spikes separated from each other by a gap of at least their own length . . . . . .
  . . . . **2. divulsa**
- Lower spikes overlapping or separated by a gap of less than their own length . . . . . .
  . . . . **1. muricata**

5. Utricles puberulent . . . . **9. pilulifera**
- Utricles glabrous 6

6. Terminal male spike less than 3 cm, erect 7
- Terminal male spike more than 3 cm, ± pendulous 9

7. Utricles ascending; leaves canaliculate, greyish . . . . **7. extensa**
- Utricles patent or deflexed, at least in the lower part of the spike; leaves flat or folded, green 8

8. Utricles shining; largest spikes more than 2 × as long as wide . . . . . . **6. punctata**
- Utricles not shining; largest spikes not more than 2 × as long as wide . . **8. viridula**

9. Spike dense; utricles 3-3.5 mm, the beak less than 1 mm . . . . **4. pendula**
- Spike lax; utricles 4-5.5 mm, the beak more than 1 mm . . . . **5. lowei**

Subgen. VIGNEA
Monoecious, with 2 or more spikes all similar in appearance, hermaphrodite or unisexual, in a simple or paniculately branched inflorescence. Lateral spikes without a scale between the bract and the lowest glume. Stigmas usually 2.

1. C. muricata L., *Sp. pl.* **2**: 974 (1953).
C. echinata auct. mad., non Murray (1770); *C. muricata* var. *tenuior* Kük.
Caespitose. Roots blackish-brown. Stem trigonous, obtusely angled, usually weakly scabrid. Basal sheaths on stems pale brown, becoming blackish. Leaves 1.5-3 mm wide. Inflorescence dense, (1-) 1.5-3 cm; spikes 4-7, crowded and overlapping, usually male above, female below; lowest bract glumaceous. Female glumes pale brown. Utricles 3-3.5 mm, ovoid, yellowish to dark shiny brown, evenly narrowed into the 2-fid beak, erecto-patent. Stigmas 2.
Fl. V - VIII. *In woods and scrub, and along the margins of fields and paths. Scattered in central and southern Madeira; recorded as very rare in Porto Santo.* **MP**

Madeiran plants all belong to subsp. **lamprocarpa** Čelak. in *Anal. Kvét. Ceská*: 88 (1879) (*C. pairaei* F.W. Schultz), to which the above description and synonymy apply. Subsp. *muricata*, from northern and eastern Europe, differs principally in its more robust habit, strongly scabrid stems, globose spikes, much darker female glumes and larger and darker, strongly patent utricles.

2. C. divulsa Stokes in With., *Bot. arr. Brit. pl.* ed. 2, **2**: 1035 (1787).
Caespitose, rhizomes short, not creeping. Roots brown or blackish-brown. Stems trigonous, striate. Basal sheaths on stems pale brown, becoming dark to blackish-brown. Leaves 2-3 mm wide. Inflorescence 4-15 cm; spikes 5-8 (-10), the upper contiguous, female, the lowest usually well-separated from each other and often on 1-3 short, lateral branches, male above, female below; lowest bract setaceous. Female glumes whitish or pale greenish brown. Utricles (3-) 3.5-4 mm, ovoid to diamond-shaped, pale brown, faintly veined at base, evenly narrowed into the deeply 2-fid beak, appressed to erecto-patent. Stigmas 2.
Fl. IV - VIII. *Woods, scrub and field margins. Very common in Madeira; present in Porto Santo and recorded from Deserta Grande.* **MDP**

Only subsp. **divulsa** is present in Madeira. The more widespread but strongly calcicole subsp. **leersii** (Kneuck.) W. Koch differs in its rigid habit, yellowish foliage, shorter and stouter inflorescence and larger, shiny red-brown utricles.

3. C. divisa Huds., *Fl. angl.*: 348 (1762).
C. divisa var. *moniziana* (Lowe) Menezes; *C. moniziana* Lowe ex Boott
Rhizomes stout, far-creeping. Stems 20-50 cm, erect, trigonous, ± weakly scabrid towards the apex. Basal sheaths pale to dark brown. Leaves 1-3(-5) mm wide. Spikes 3-8(-12), the upper often entirely male, in a dense ovoid to ± lax, lobed, oblong inflorescence (1-)1.5-3 cm; lowest bract glumaceous to leaf-like and greatly exceeding the inflorescence. Female glumes equalling or slightly exceeding utricles, acute and cuspidate to aristate, dark to reddish brown, usually with a narrow scarious margin. Utricles 3.5-4 mm, dark reddish brown with prominent, slender veins, ± gradually narrowed into a short somewhat scabrid, 2-fid beak. Stigmas 2.
Fl. IV - VII. *Rare, in damp places in a few sites in south-east Madeira east of Funchal and at Caniço and Machico.* **M**

C. spicata Huds. has been recorded from Madeira but probably in error for **1** or **2**. It differs by the long, acute ligules and acuminate female glumes (versus obtuse ligules and ovate glumes), and the bulbous-based utricles with narrow beaks. The leaf-sheaths, bracts and sometimes glumes are strongly flushed red.

Subgen. CAREX
Monoecious. Spikes 2 or more, dissimilar in appearance and usually unisexual, at least the terminal male and the basal female, in a usually simple inflorescence, usually with 3 stigmas Lateral spikes usually with a scale between the bract and the lowest glume.

4. C. pendula Huds., *Fl. angl.*: 352 (1762).   [*Palha de amarrar vinha*]
C. pendula var. *myosuroides* (Lowe) Boott, var. γ Boott
Densely caespitose with short rhizomes. Stems 70-80 cm, sharply trigonous, smooth. Basal sheaths reddish or purplish brown. Leaves 9-16 mm wide. Spikes cylindrical; males 1(-2), 10-19 cm; females (3-)5-7, up to 25 cm, pendent, the lower often with peduncles half this length or more. Female glumes shorter than or exceeding utricles, reddish-brown. Utricles 2.5 mm, pale to brownish green, glabrous, faintly veined, slightly outwardly curved and abruptly narrowed into a short beak. Stigmas 3.
Fl. IV - VI. *Rare, in damp places. Scattered mainly along the north coast of Madeira from Ribeiro do Inferno to Santana, occasionally elsewhere.* **M**

5. C. lowei Bech. in *Candollea* **8**: 15 (1939).
C. elata Lowe, non All. (1785); *C. lowei* forma *simplex* Bornm.
Caespitose with short, creeping rhizomes. Roots dull brown. Stems 80-150 cm, erect, trigonous, smooth. Basal sheaths pale to dark reddish-brown. Leaves flat, 7-8 mm wide. Spikes cylindrical; males 1-2, 50-110 mm; females usually 5, remote, branched from the base, 70-110 mm; lowest bract leaf-like, sheathing. Female glumes as equalling but usually much longer than the utricles, greenish to reddish brown, long-aristate, the margins scarious. Utricles 4.5-5.5m m, ellipsoid, glabous, membranous, pale to reddish brown, slender veined, ± abruptly narrowed into a long, 2-fid beak. Stigmas 3. **Plate 55**
Fl. V - VI. *Endemic, among rocks and along watercourses in damp, wooded valleys of central and northern Madeira.* ● **M**

C. malato-belizii Raymond in *Contr. Inst. bot. Univ. Montréal* **70**: 73 (1957), described from a single locality in the north of Madeira and supposedly differing in size and branching of the spikes and the size of the achenes, does not appear to be specifically distinct from **5**.

6. C. punctata Gaudin, *Agrost. helv.* **2**: 152 (1811).
Somewhat caespitose, with short rhizomes. Stems up to 30 cm, obtusely trigonous, smooth. Basal sheaths pale to dark brown, entire. Leaves *c*. half as long as stems, 2-3 mm wide, bright green. Male spike 1(-2), 10-15 × 2-3 mm. Female spikes 2-3, 5-25 × 5 mm, distant, erect, cylindrical or oblong; lowest bract about as long as or longer than the inflorescence, erect. Female glumes mostly scarious except for green mid-rib. Utricles 3-3.5(-4) mm, slightly inflated, ± patent, greenish and speckled with reddish brown dots, shining, veins slender, rather abruptly narrowed into the beak. Stigmas 3.
Fl. VI - VIII. *Very rare. Known only from São Jorge, and Rocha do Navio at Santana.* **M**

Madeiran and Azorian plants, which differ from European plants in the longer (1 mm or more) and slightly scabrid beaks of the utricles, have been called (■) var. **laevicaulis** (Hochst. ex Kunze) Boott, *Illustr. Carex* **4**: 155 (1867).

7. C. extensa Good. in *Trans. Linn. Soc. Lond.* **2**: 175 (1794).
Densely caespitose, rhizomes short. Stems 25-70 cm, rigid, obtusely trigonous, solid. Basal sheaths dark brown, becoming slightly fibrous. Leavbes about as long as stems, 1-2 mm wide, canaliculate

or the margins inrolled, glaucous or greyish-green. Male spike 1, 6-10 × 2-3 mm. Females spikes 2-3, 6-12 × 4-5 mm, erect, overlapping or the lowest remote; lowest bract greatly exceeding inflorescence, patent. Female glumes reddish brown. Utricles 3 mm, ascending, greenish brown and speckled with reddish dots, not shining, prominently veined, gradually narrowed into a smooth, emarginate beak. Stigmas 3.
Fl. VII - VIII. *Very rare on cliffs, rocks and walls at the mouth of the São Vicente valley and at Fajã da Areia on the north coast of Madeira.* M

8. C. viridula Michx., *Fl. bor.-amer.* **2**: 170 (1803).
C. demissa Hornem.; *C. oederi* Retz sensu auct. mad.
Somewhat caespitose, with short, creeping rhizomes. Stems 6-35 cm, ascending, obtusely trigonous. Basal sheaths pale yellowish-brown. Leaves as long as stems, 1.5-3(-4) mm wide, those of fertile tillers *c.* 2 mm, flat. Male spike 1(-2), 9-14 × 2 mm, cylindrical, subsessile to shortly pedunculate. Female spikes 2-3, 5-12 × 5-7 mm, ovoid, erect, the lower shortly pedunculate and overlapping, the lowest often remote; lowest bract leaf-like, greatly exceeding inflorescence, erect to deflexed. Female glumes yellowish-brown. Utricles 3-3.7 mm, ellipsoid, yellowish green, prominently veined, patent, somewhat abruptly narrowed into a slightly curved, 2-fid beak. Stigmas 3.
Plate 55
Fl. VI - VIII. *Rare, on damp rocks and along watercourses, mainly in central and eastern Madeira from Rabaçal to Santo da Serra.* M

Madeiran plants belong to (■) subsp. **cedercreutzii** (Fagerstr.) B. Schmid in *Watsonia* **14**: 317 (1983) (*C. tumidicarpa* subsp. *cedercreutzii* Fagerstr.), which otherwise only occurs in the Azores and Canaries and to which the above description and synonymy apply. It differs from the other subspecies principally in the curved stems, flat leaves which are narrow in the fertile tillers, comparatively short male spikes and the comparatively long, slightly curved beak of the utricles.

9. C. pilulifera L., *Sp. pl.* **2**: 976 (1753).
Caespitose, rhizomes short. Stems often curved, trigonous, smooth or weakly (rarely strongly) scabrid above. Basal sheaths yellowish to reddish brown, fibrous. Leaves shorter than stems, flat or slightly canaliculate, pale, usually greyish-green. Male spike solitary. Female spikes 2-4, globose to ovoid, sessile, overlapping or the lowest remote; lowest bract leaf-like, usually patent and exceeding the spike. Female glumes acute to acuminate, dark, usually reddish brown, with narrow scarious margins. Utricles obovoid, greyish-green, sparsely to densely puberulent, ± veinless, abruptly contracted into a short, emarginate, usually conical beak. Stigmas 3.
Fl. ? *Very rare. Recorded only twice, from Funchal and Achadas da Cruz.* M

Plants from the Azores differ from the typical European form in having the lowest bract usually exceeding the inflorescence, the male spike not more than 8 mm, female spikes *c.* 3 mm wide, biconvex utricles and 2 stigmas, and have been separated as subsp. *azorica* (Gay) Franco & Rocha Afonso in *Bot. J. Linn. Soc.* **76**: 366 (1978) (var. *azorica* (Gay) H. Christ.). Madeiran plants may belong to subsp. **pilulifera** but this requires confirmation.

Subgen. **PRIMOCAREX**
Monoecious (in Madeira) with male flowers at top of spike and female flowers below. Spike solitary. Stigmas 2.

10. C. peregrina Link, *Hort. berol.* **1**: 334 (1827).
Densely caespitose, rhizomes absent. Stems 15-50 cm, erect, slender, smooth. Basal sheaths dull brown. Leaves shorter than or equalling the stem, 1-2 mm wide. Spike lax, 2-3.5 mm with 6-11 female flowers. Lower male glumes acute; female glumes caducous, 3.5 mm, obtuse, pale or dark brown, with a broad scarious margin. Utricles 5.5 mm, oblong-ellipsoid, greenish brown,

± abruptly narrowed into a long beak, patent or deflexed. Stigmas 2.
Fl. IV - VII. *Common in valleys in northern and central Madeira.* **M**

CXXI. ZINGIBERACEAE
M.J. Cannon

Perennial herbs with a fleshy, often aromatic, rhizomatous rootstock and aerial stems. Leaves in 2 ranks, with ligulate basal sheaths. Flowers zygomorphic, in dense racemes. Corolla tubular, with 3 lobes. Fertile stamens 1, with 3 fused staminodes forming a petaloid lip, sometimes with 2 additional free lateral staminodes. Ovary inferior, 1- to 3-locular, ovules numerous.

1. Hedychium J. Köenig

Stems elongate, leafy. Leaves in 2 ranks, oblong or lanceolate, sessile or short-stalked. Bracts of the inflorescence each subtending 1-2 flowers. Calyx tubular, 3-dentate. Corolla with a long, slender tube, the lobes equal, linear, spreading. Lateral staminodes linear or oblong, the petaloid lip large and usually bifid. Style long, filiform; stigma sub-globose. Seeds numerous, each with a small, lacerated aril.

1. H. gardneranum Sheppard ex Ker Gawl. in *Bot. Reg.* **9**: t. 774 (1824).
Up to 2 m, leaves oblong, whitish beneath. Inflorescence up to 30 cm, bracts 3-5 cm, rolled around the groups of 1-2 flowers. Flowers bright yellow or white. Corolla-tube a little longer than the bract, lobes reflexed, greenish. Lateral staminodes 2.5-3 cm, the petaloid lip up to 2.5 cm, 2(3-)-lobed, narrowed into a short claw at the base. Filaments 5-6 cm, bright red, anthers up to 1.5 cm, yellow. Capsule valves orange-red within, seeds brownish crimson.
Fl. VI - XII. *Native to the Himalayas and extensively naturalized in the Azores; frequently planted for ornament on Madeira and naturalized in a few places in open areas within the laurisilva. An aggresive, invasive weed, it is capable of spreading rapidly and dominating large areas, especially where the tree canopy is removed.* **M**

CXXII. CANNACEAE
M.J. Cannon

Perennial herbs with swollen, tuberous, underground rhizomes. Leaves large, pinnately-veined. Flowers zygomorphic, each subtended by a bract. Ovary inferior, of 3 fused carpels; ovules numerous, in 2 rows. Seeds with straight embryos and very hard endosperm.

1. Canna L.

Erect perennials with leafy aerial stems. Leaves alternate with sheathing bases. Inflorescence racemose or paniculate. Sepals 3, free, greenish; petals 3, united at the base. Staminal column of 3-5, mainly petaloid staminodes and 1 fertile, petaloid stamen. Ovary 3-locular, style petaloid. Fruit a warty capsule with the persistent sepals erect at the apex. Seeds very small.

1. C. indica L., *Sp. pl.* **1**: 1 (1753). [*Conteira*]
Slender plant up to 120 cm. Leaves glabrous, 20-40 × 20 cm, ovate to lanceolate, sessile. Inflorescence a raceme or loosely branched panicle. Flowers paired, bracts orbicular. Sepals 1-1.5 cm, exceeding the corolla tube. Corolla-lobes 3-3.5 cm, spathulate, much longer than the tube, bright red or orange. Staminodes up to 5 cm, linear to narrowly spathulate or oblanceolate, yellow or orange, the limb curved. Capsule 2.5-3.5 cm, warty.
Fl. III - XI. *Native to tropical America, widely planted in gardens in Madeira, often escaping and becoming naturalized on waste ground.* **M**

CXXIII. ORCHIDACEAE
N.J. Turland

Perennial herbs with rhizomes or tubers. Stems ascending or erect. Leaves alternate, simple, entire to weakly crenulate at margins, sheathing at base. Inflorescence a terminal spike, or a raceme with flowers very shortly pedicellate. Flowers zygomorphic, hermaphrodite. Perianth-segments 6, in 2 whorls, all entire and ± similar except for the larger and often lobed median inner segment (*labellum*). Anthers and stigmas borne on a column; pollen grains coherent in masses (*pollinia*). Ovary inferior. Fruit a capsule dehiscing by longitudinal slits; seeds very numerous, minute, usually requiring association with fungal mycorrhiza to grow successfully.

For simplicity, all inflorescences are referred to as spikes irrespective of whether or not the flowers are sessile. Several tropical and temperate orchids from various countries are grown for ornament in Madeiran gardens, or by plant nurseries to produce material for the cut flower trade. These are mostly epiphytic species often with large and spectacularly coloured flowers; the native plants are all terrestrial.

Literature: G. Frey, Beitrag zur Orchideenflora Madeiras. 1. Teil in *Jber. naturw. Ver. Elberfeld* **29**: 59-61 (1976). G. Frey & C.H. Pickering, Contribution to the knowledge of the orchids of Madeira and the Azores in *Bocagiana* **38**: 1-6 (1975). W.T. Stearn, The nomenclature of the Madeiran orchis (*Orchis maderensis* or *Dactylorhiza foliosa*) and other Orchidaceae of Madeira, in *Bolm Soc. broteriana* II, **49**: 89-96 (1975). H. Sundermann, Beitrag zur Orchideenflora Madeiras. 2. Teil in *Jber. naturw. Ver. Elberfeld* **29**: 62-63 (1976).

1. Leaves 2, cordate at base . . . . **2. Gennaria**
- Leaves 4 or more, not cordate at base 2

2. Spike ± secund; flowers small, with labellum not more than 5 mm 3
- Spike not secund; flowers large and showy, with labellum 8-17 mm 4

3. Leaves abruptly narrowed into a petiole; inflorescence pubescent; labellum entire, concave, without a spur . . . . **1. Goodyera**
- Leaves gradually narrowed into a petiole; inflorescence glabrous; labellum 3-lobed, flat, with a very short spur . . . . **3. Neotinea**

4. Labellum strongly 3-lobed; median lobe extended beyond lateral lobes, emarginate or 2-lobulate at apex . . . . **5. Orchis**
   Labellum shallowly 3-lobed; median lobe not extended beyond lateral lobes, entire at apex . . . . **4. Dactylorhiza**

1. Goodyera R. Br.

Plant with creeping, branching rhizomes. Leaves several, abruptly narrowed into a petiole. Inflorescence pubescent; spike elongate, narrowly cylindric, neither secund nor recurved at apex; flowers numerous, small, crowded. Perianth campanulate, white; segments ± equal; labellum entire, concave, without a spur.

1. G. macrophylla Lowe in *Trans. Camb. phil. Soc.* **4**: 13 (1831).
Plant to 50 cm. Stems ascending, not all terminating in an inflorescence, stout, with circular leaf-scars and thick emergent roots towards base. Leaves 7-20 × 3-5.5 cm, ovate to lanceolate or narrowly elliptic, acute at apex, entire or weakly crenulate at margins, unspotted, withering during flowering. Inflorescence with a few small, bract-like, sheathing leaves below the spike. Flowers 40 to 60 or more, *c*. 12.5 mm. Perianth-segments lanceolate; outer segments pubescent; upper 2 inner segments arching to form a wide hood; labellum fleshy, glabrous. Pollinia 2,

linear-club-shaped. Ovary pubescent. **Plate 57**
Fl. VIII - X. *A very rare endemic of humid laurisilva in a few ravines in central and northern parts of Madeira, from 300-800 m.* ● **M**

2. Gennaria Parl.

Plant with tubers. Leaves 2, cordate at base. Inflorescence glabrous; spike cylindric, secund, often somewhat recurved at apex; flowers numerous, small, crowded. Perianth campanulate, yellowish green; inner segments scarcely larger than outer segments; labellum 3-lobed, ± flat, with a short spur.

1. G. diphylla (Link) Parl., *Fl. ital.* **3**: 405 (1860).
Plant 8-40 cm. Stems erect. Leaves 3-10 × 1.5-7 cm, ovate to broadly so, acute to shortly acuminate at apex, ± entire to weakly crenulate at margins, unspotted, the upper leaf usually smaller. Flowers 15 to 35, 6-8 mm. Perianth-segments elliptic-oblong to narrowly so.
Fl. XII - V. *Laurisilva, rocky and grassy places, preferring moist shady conditions; common from (300-)600-1100 m in Madeira and in the higher parts of Porto Santo; also on Deserta Grande.* **MDP**

3. Neotinea Rchb.f.

Plant with tubers. Leaves several, the lower gradually narrowed into a petiole. Inflorescence glabrous; spike cylindric, ± secund; flowers usually numerous, small, crowded. Perianth pinkish or greenish white, often with darker markings; 5 segments directed forwards and held together to form a pointed hood; labellum directed downwards, 3-lobed, flat, with a very short spur.

1. N. maculata (Desf.) Stearn in *Annls Musei Goulandris* **2**: 79 (1974).
N. intacta (Link) Rchb.f.
Plant to 35 cm. Stems erect. Leaves to 18 × 3 cm, the basal and upper small and sheathing, the lower larger, elliptic to strap-shaped, obtuse and mucronate at apex, ± entire at margins, purple-blotched or unmarked. Flowers 7-30, 10-14 mm. Perianth-segments lanceolate.
Fl. III - VI. *A rare plant of laurisilva, rocky banks and levada sides in humid shady ravines in the central mountains of eastern Madeira.* **M**

4. Dactylorhiza Neck. ex Nevski

Plant with tubers. Leaves several, gradually narrowed towards base. Inflorescence glabrous; spike ovoid to cylindric, neither secund nor recurved at apex; flowers usually numerous, large and showy. Perianth strongly zygomorphic, dark reddish purple to pale pink or white; labellum larger than other segments, broadly elliptic and distinctly wider than long, shallowly 3-lobed, flat, with a distinct spur; median lobe not extended beyond lateral lobes, entire at apex.

1. D. foliosa (Verm.) Soó, *Nom. nova Gen. Dactylorhiza*: 7 (1962).
Orchis foliosa Sol. ex Lowe (1831), non Swartz (1800); *O. maderensis* Summerh.
Plant 20-80 cm. Stems erect. Leaves to 35 × 5 cm, the lowermost small and sheathing, the median largest, ± strap-like to narrowly elliptic or narrowly lanceolate, rounded to acute at apex, ± entire at margins, plain green, rarely dark-spotted, the upper becoming smaller, acuminate and eventually bract-like. Flowers 10-50, 18-30 mm. Outer perianth-segments ± spreading to forward-pointing, lanceolate; inner segments (except for labellum) forward-pointing and often held together to form a hood, lanceolate, often darker than outer segments; labellum directed ± downwards, 8-16 × 9-18 mm, usually with markings darker than base colour. **Plate 57**
Fl. V - VII. *A common endemic of woodland, scrub, grassy, mossy and rocky places as well*

as by levadas and streams from 500-1100(-1500) m, mainly in the central mountains of the eastern half of Madeira, preferring moist conditions. ● **M**

5. Orchis L.

Like *Dactylorhiza* but spike often shorter and fewer-flowered, ovoid to shortly cylindric; labellum suborbicular, strongly 3-lobed; median lobe extended beyond lateral lobes and emarginate or 2-lobulate at apex.

1. O. scopulorum Summerh. in *Bolm Soc. broteriana* II, **35**: 55 (1961).
?*O. mascula* sensu auct. mad., non (L.) L. (1755).
Plant 20-65 cm. Leaves to 15 × 3 cm, dark-spotted or unmarked. Flowers 8 to 18, *c*. 20 mm. Perianth pink to purple; labellum 10-17 mm, often paler than other segments, at least in centre.
Plate 57
Fl. V - VI. *A very rare endemic of moist cliffs and rock-ledges; known only from the high peaks of eastern Madeira, between 1100 m and 1850 m.* ● **M**

O. scopulorum is said to be intermediate between *O. mascula* (L.) L., *Fl. suec.* ed. 2: 310 (1755), which is widespread in Europe, and *O. patens* Desf., *Fl. atlant.* **2**: 318 (1799), which occurs in the western half of the Mediterranean region. *O. mascula* has in fact been recorded from Madeira, but almost certainly in error for *O. scopulorum*.

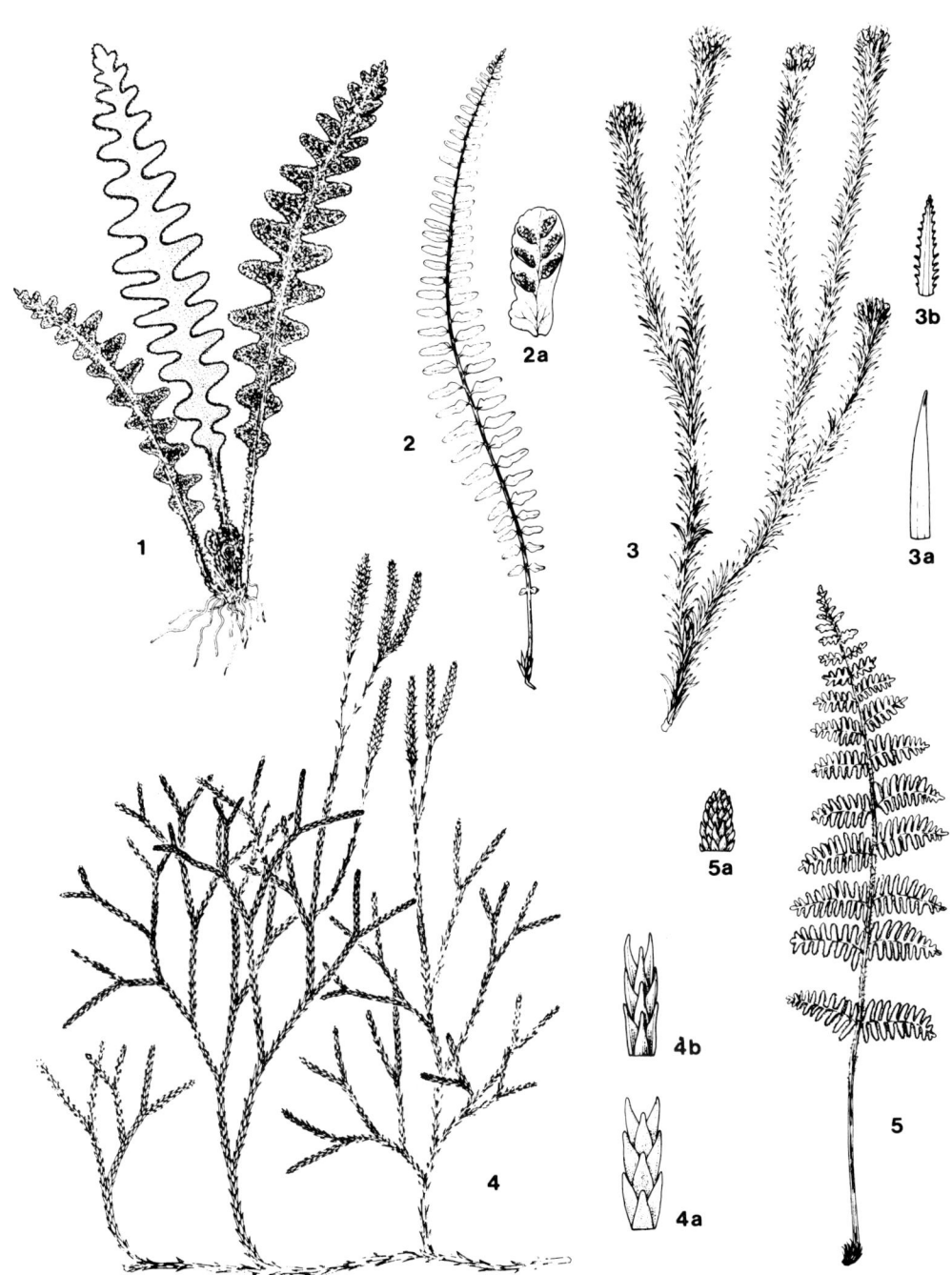

Plate 1 1. *Ceterach lolegnamense*. 2. *Asplenium anceps*. 3. *Huperzia suberecta*, **b**) *H. dentata*. 4. *Diphasiastrum madeirense*. 5. *Notholaena marantae* subsp. *subcordata*. Habits all × ½, details all × 3.

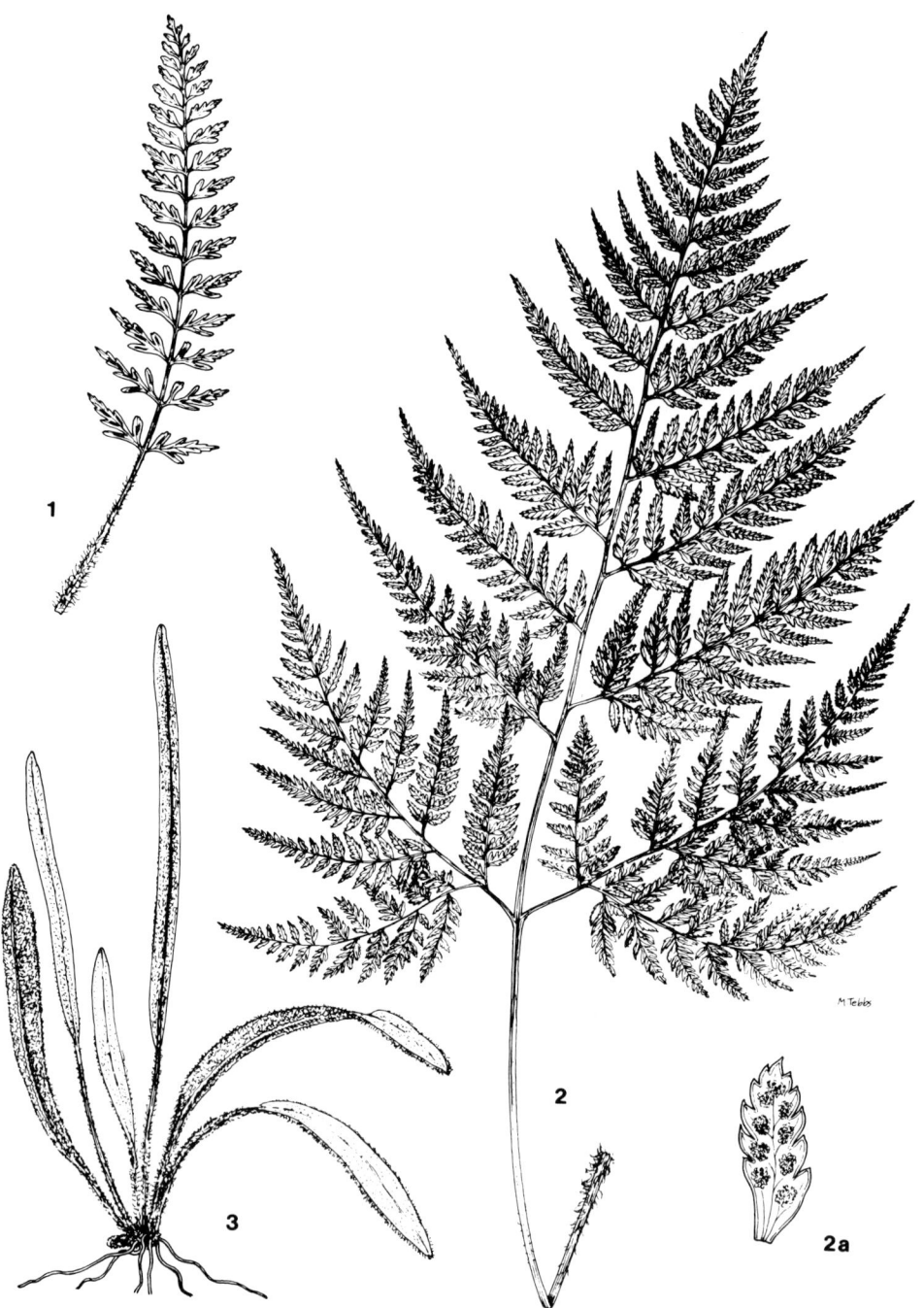

Plate 2 1. *Asplenium aethiopicum* subsp. *braithwaitii*. **2.** *Arachnoides webbianum*. **3.** *Elaphoglossum semicylindricum*. Habits all × ¼, details × 2.

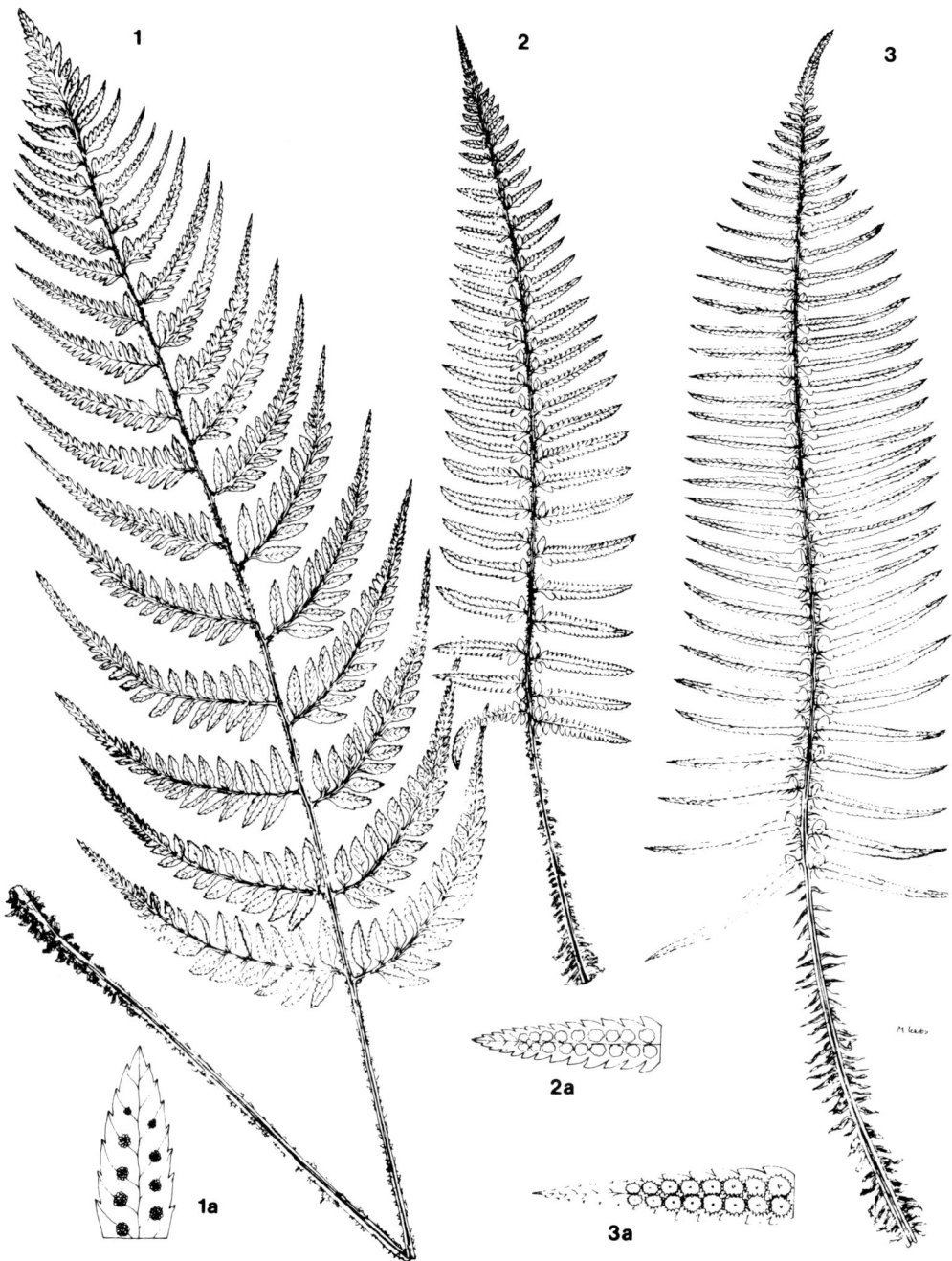

Plate 3 **1**. *Polystichum drepanum*. **2**. *P. falcinellum*. **3**. *P.* × *maderense*. Habits all × ¼, details all × 2.

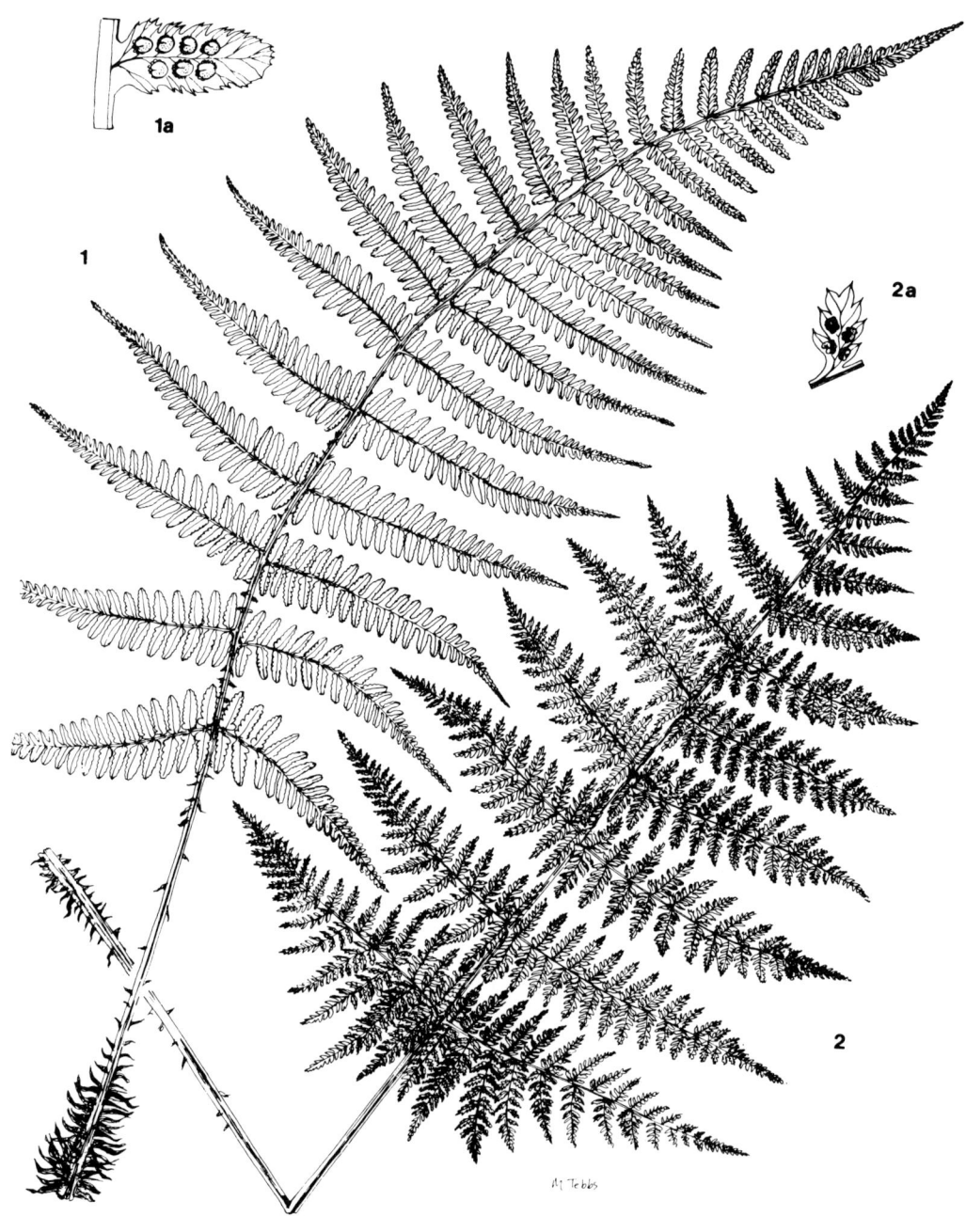

Plate 4 1. *Dryopteris aitoniana*. 2. *D. maderensis*. Habits all × ⅓, details all × 2.

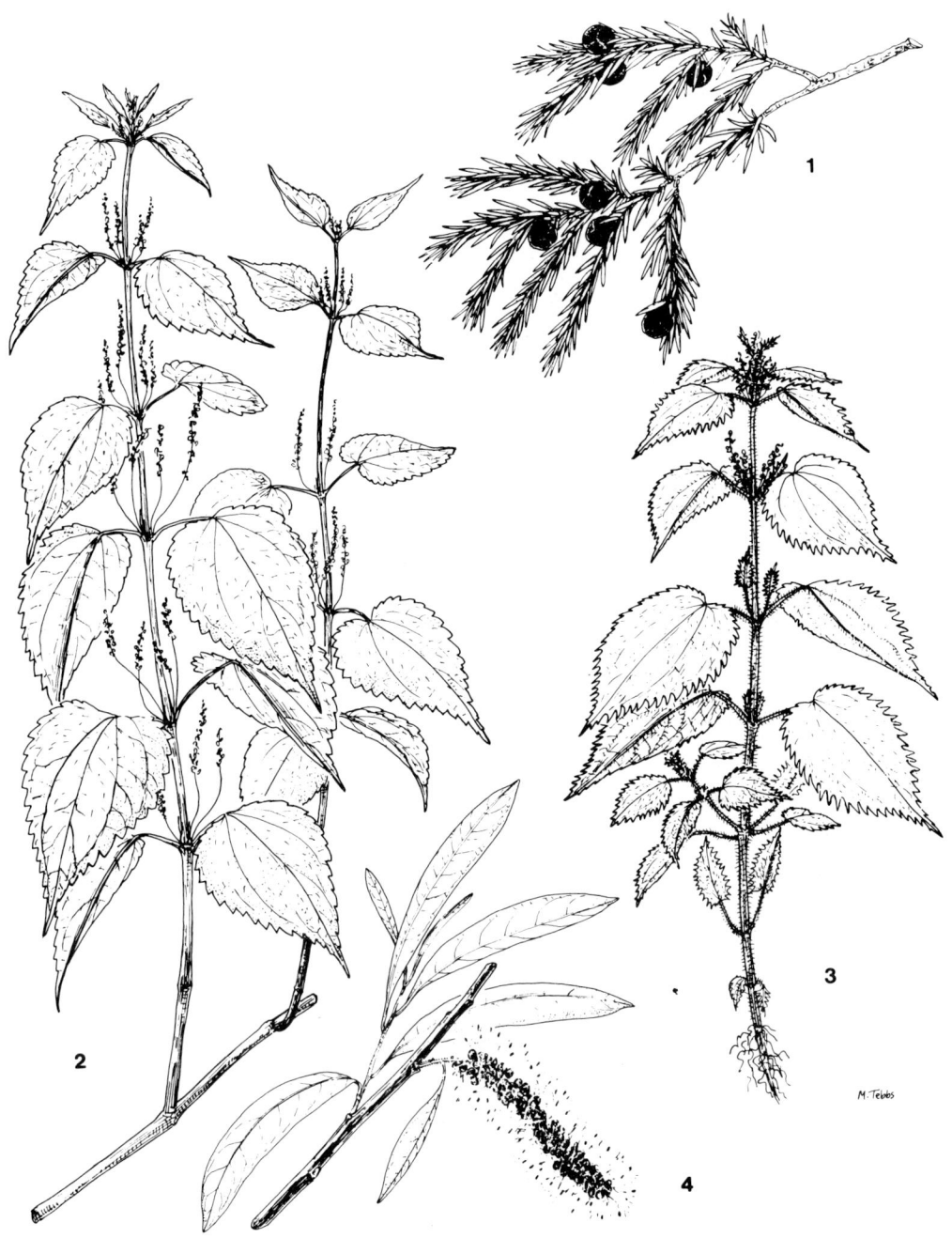

Plate 5 1. *Juniperus cedrus* 2. *Urtica morifolia*. 3. *U. portosanctana*. 4. *Salix canariensis*. All × ½.

Plate 6 **1.** *Rumex maderensis.* **2.** *R. simpliciflorus* subsp. *maderensis.* **3.** *R. bucephalophorus* subsp. *canariensis* var. *canariensis.* **4.** *R. bucephalophorus* subsp. *canariensis* var. *fruticescens.* Habits all × ½, details × 3.

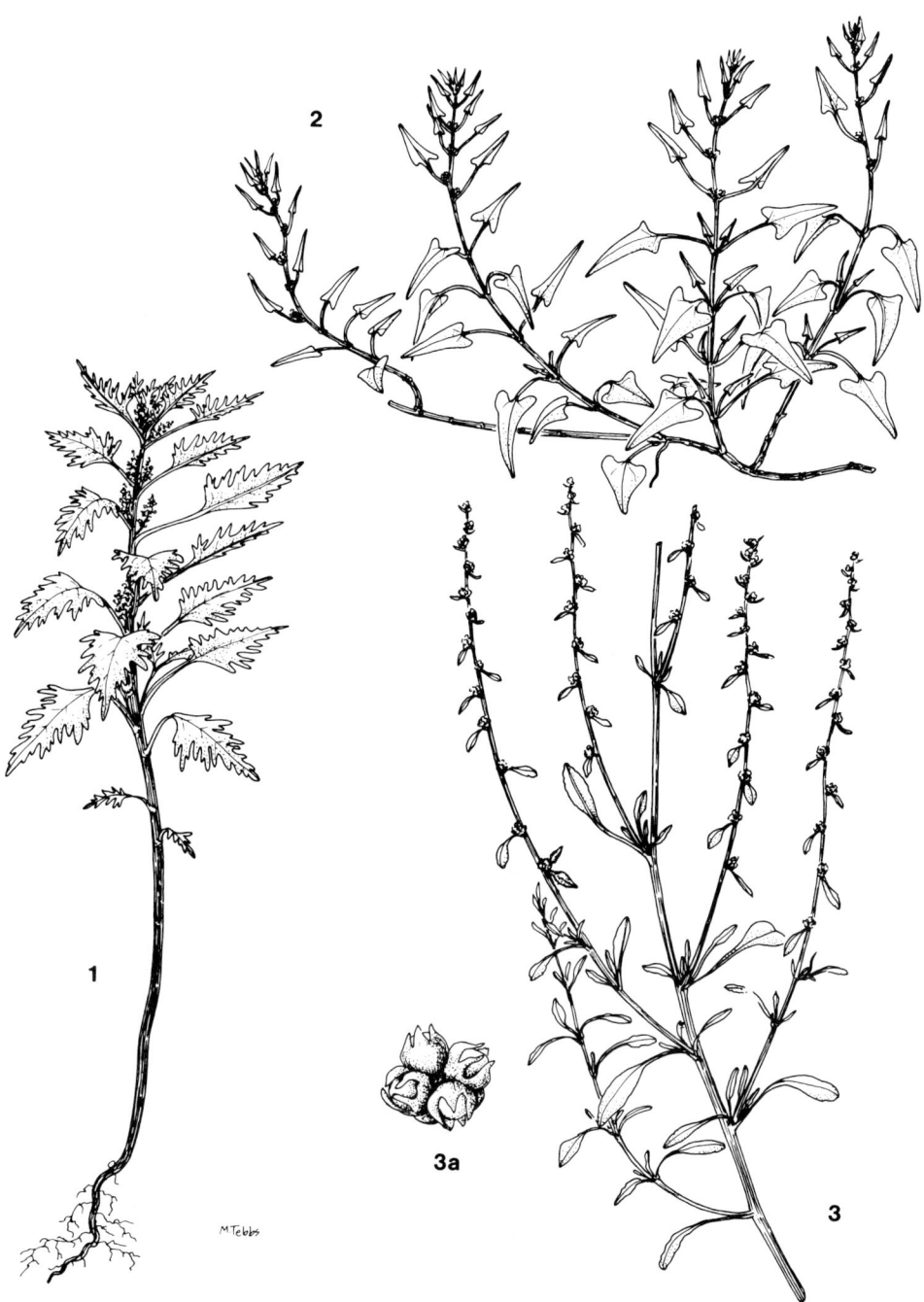

Plate 7 1. *Chenopodium coronopus*. 2. *Beta procumbens*. 3. *B. patula*. Habits all × ½, details × 2

Plate 8 1. *Ranunculus* [*grandiflorus* var. *major*]. **2**. *R.* [*grandiflorus* var. *minor*]. **3**. *Berberis maderensis*. **4**. *Cerastium vagans*. **5**. *Delphinium maderensis*. All × ½.

Plate 9 1. *Apollonias barbujana*. 2. *Laurus azorica*. 3. *Ocotea foetans*. 4. *Persea indica*. All × ½.

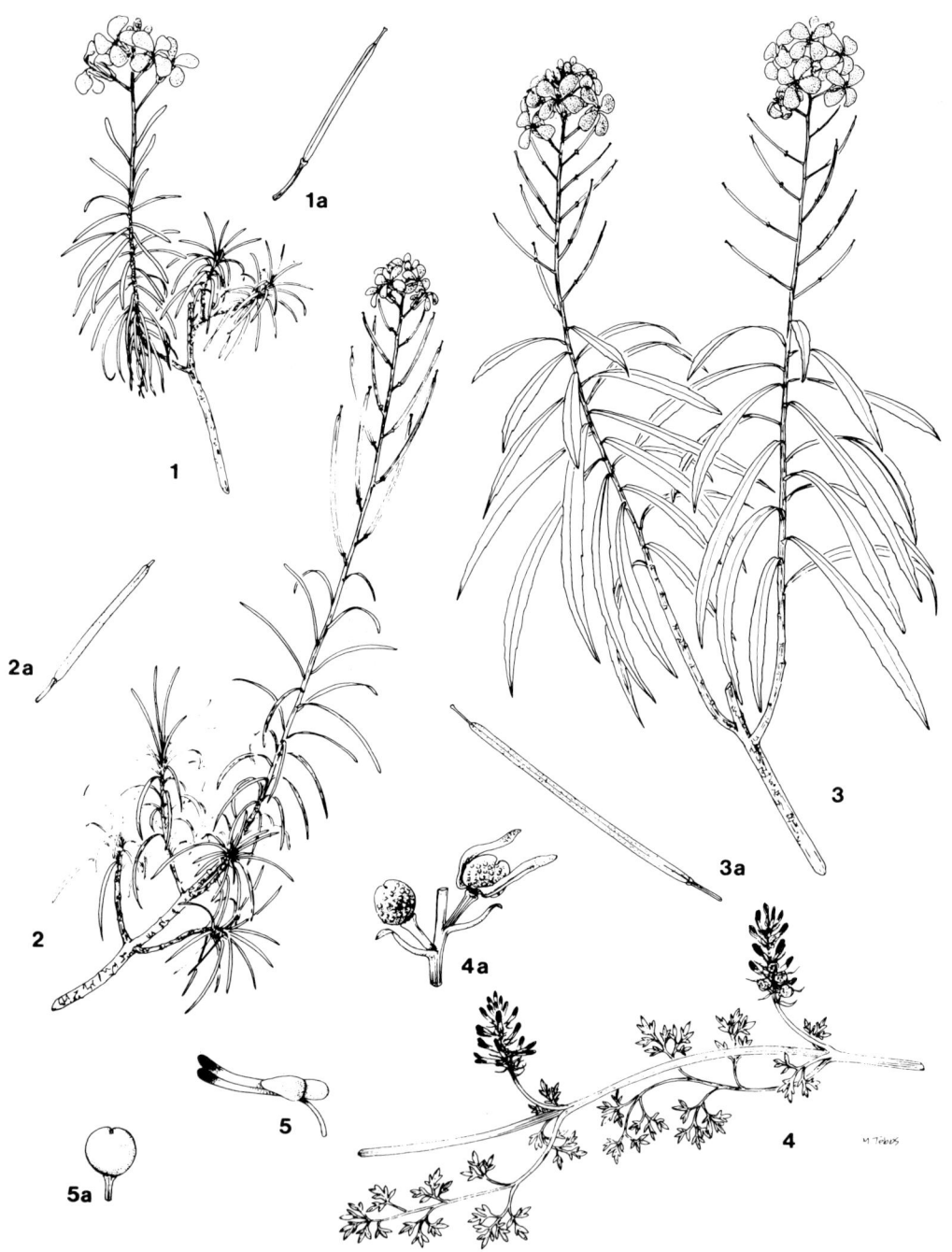

Plate 10 1. *Erysimum arbuscula*. 2. *E. maderense*. 3. *E. bicolo*r. All × ½. 4. *Fumaria montana* × ½, **a)** fruit × 3. 5. *Fumaria muralis* var. *laeta*, flower × 1½ **a)** var. *lowei*, fruit × 3.

Plate 11 1. *Sinapidendron angustifolium*. 2. *S. frutescens* var. *frutescens*. 3. *S. gymnocalyx*. 4. *S. rupestre*. All × ½.

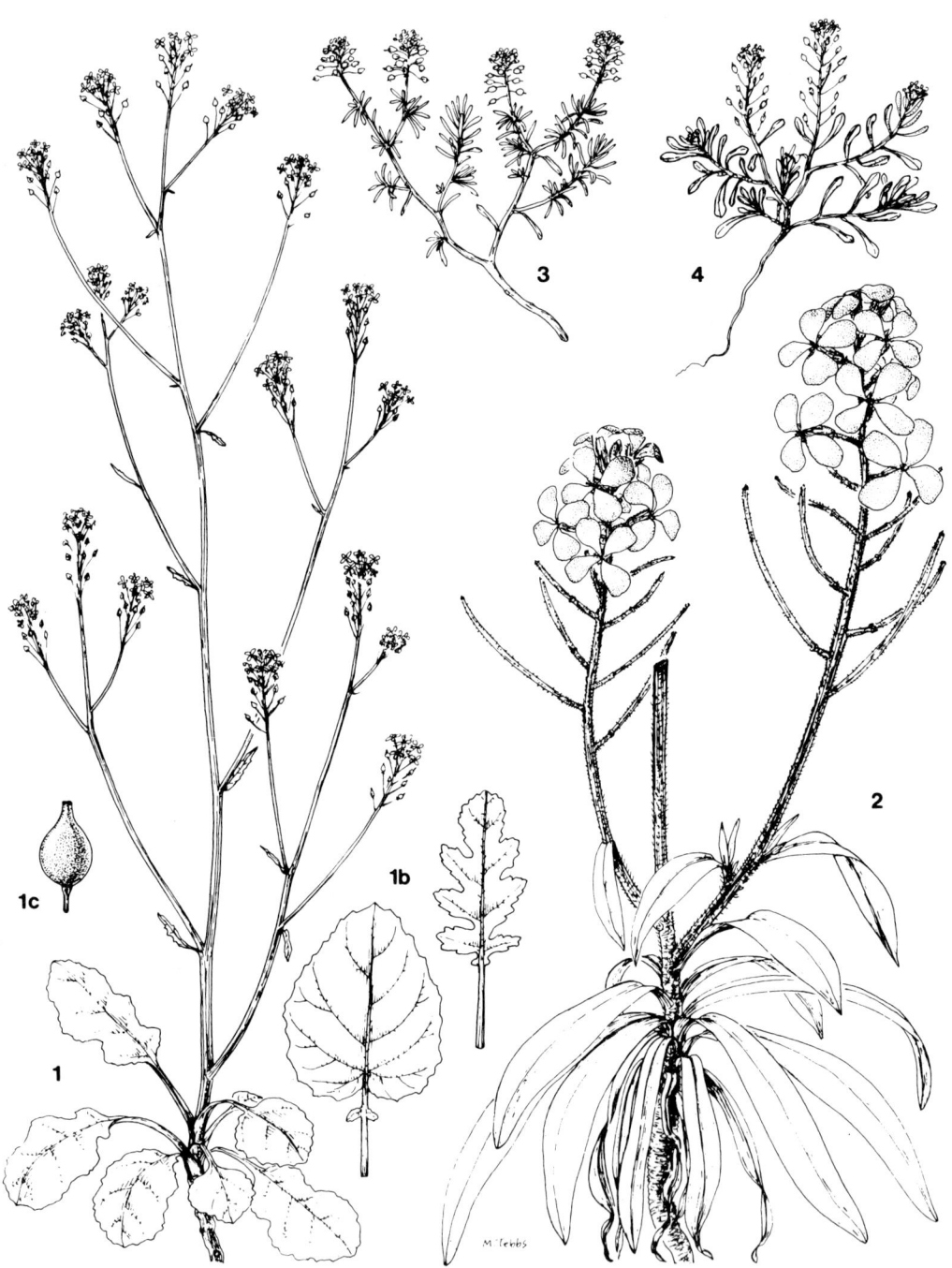

Plate 12 1. *Crambe fruticosa*, **b**) basal leaves, **c**) fruit. 2. *Matthiola maderensis*. 3. *Lobularia canariensis* subsp. *rosula-venti*. 4. *L. canariensis* subsp. *succulenta*. Habits all × ½, 1b × ½, 1c × 5.

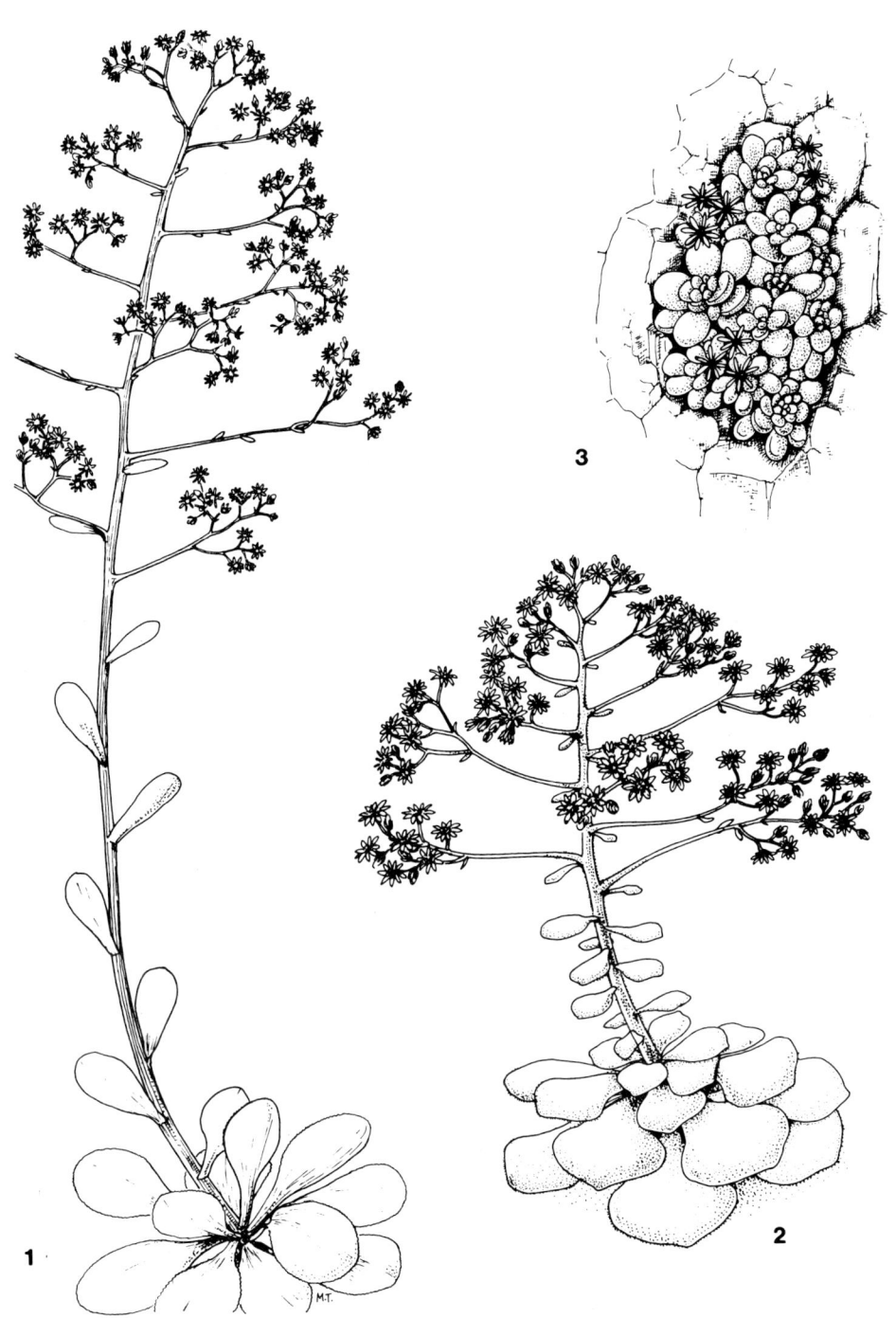

Plate 13 1. *Aeonium glutinosum*. 2. *A. glandulosum*. 3. *Monanthes lowei*. All × ½.

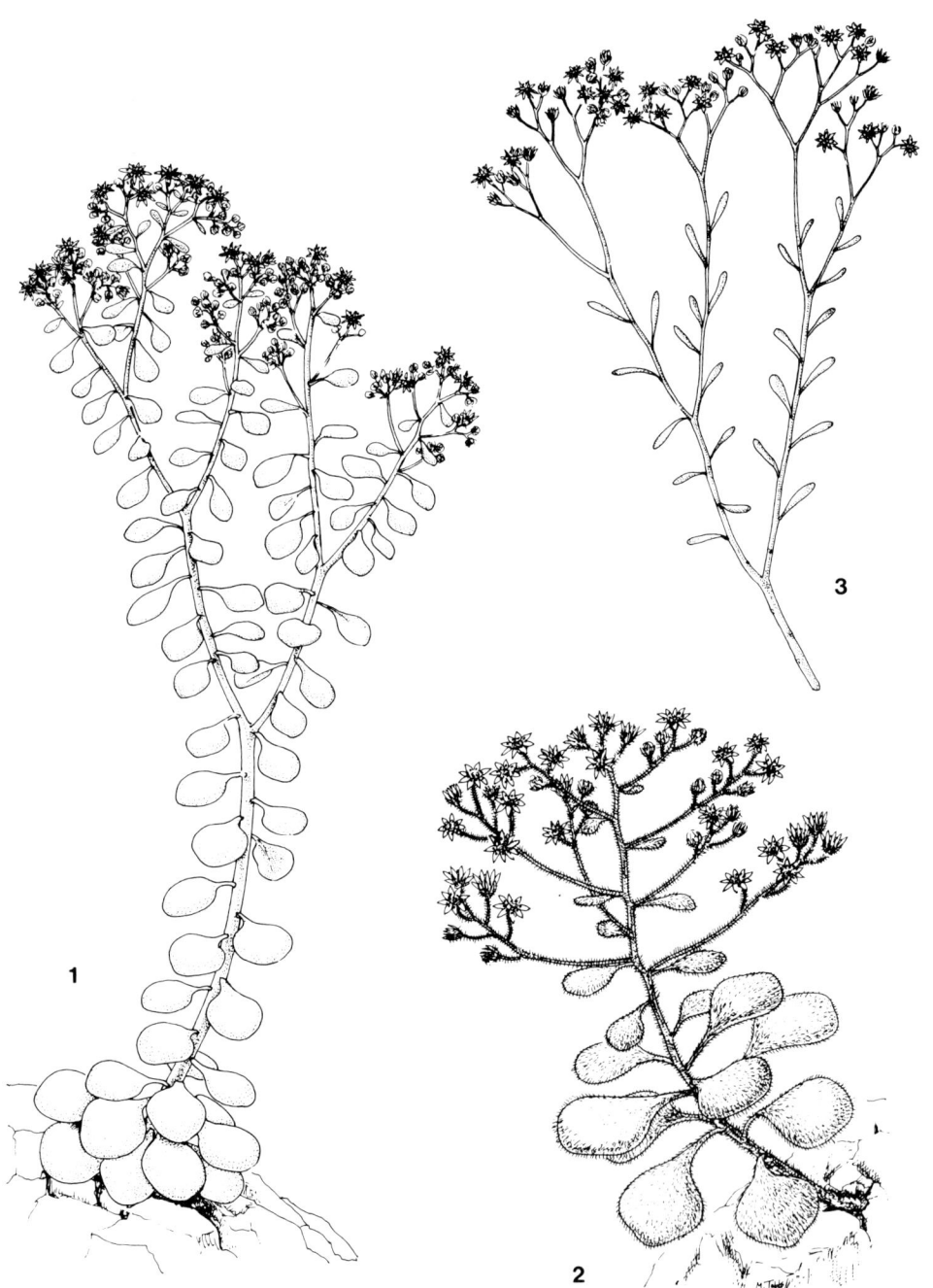

Plate 14 1. *Aichryson divaricatum*. 2. *A. villosum*. 3. *A. dumosum*. All × ½.

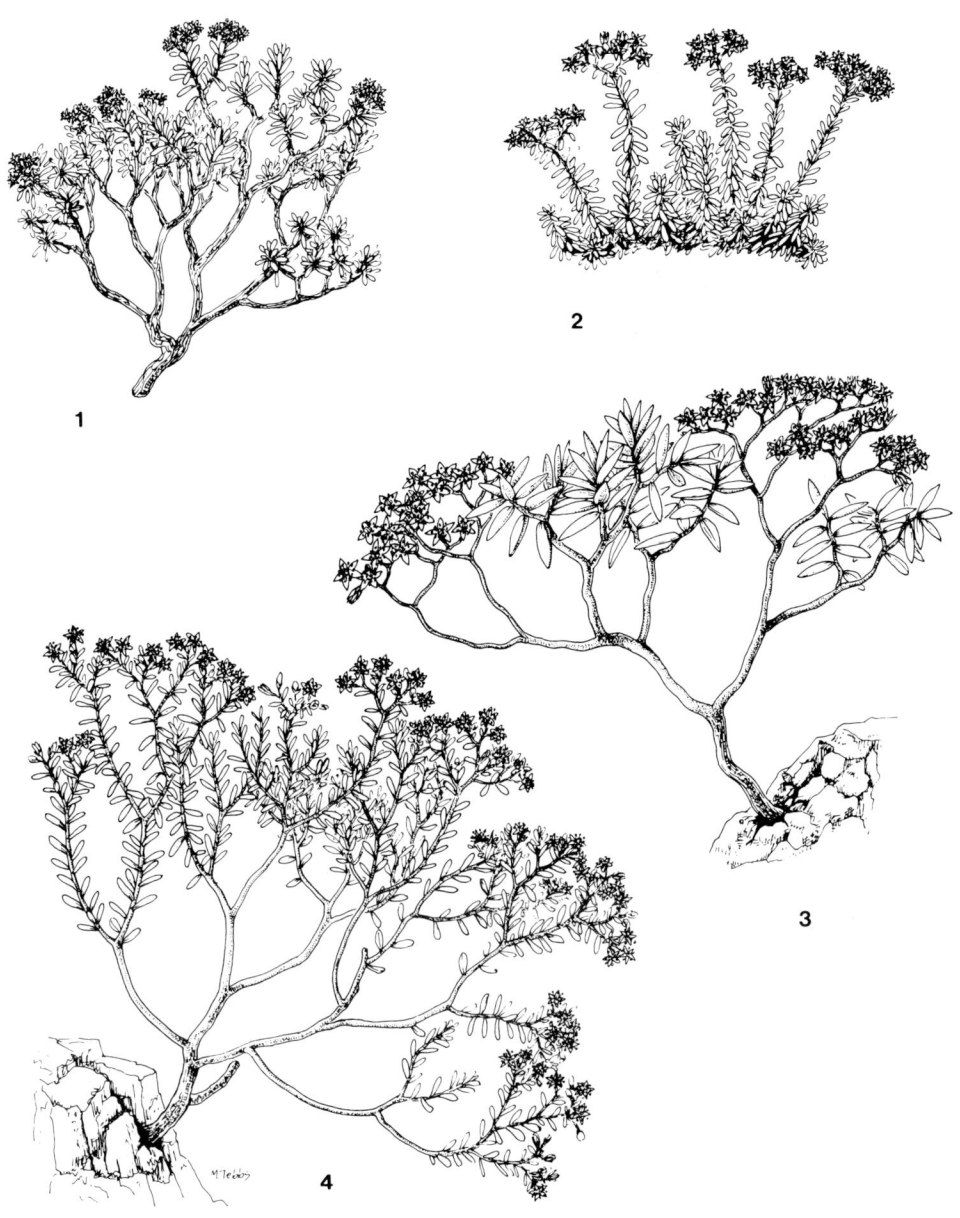

Plate 15 1. *Sedum nudum*. 2. *S. farinosum*. 3. *S. fusiforme*. 4. *S. brissemoretii*. All × ½.

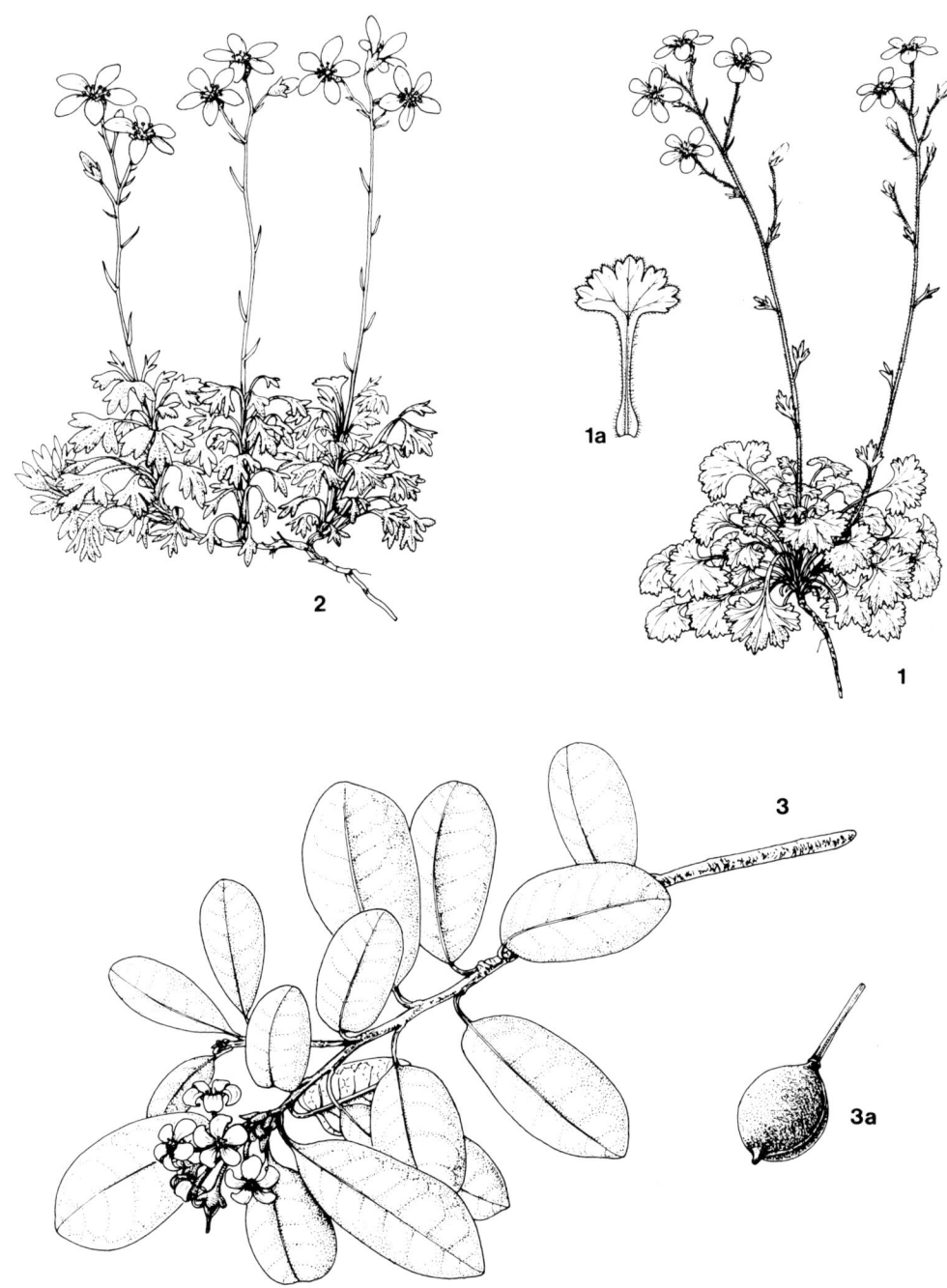

Plate 16 1. *Saxifraga maderensis* subsp. *maderensis*, **a**) subsp. *pickeringii*, leaf. **3.** *S. portosanctana*. **4.** *Pittosporum coriaceum*, **a**) fruit. Habits all × ½, 1a × ½, 3a × 1.

Plate 17 1. *Marcetella maderensis.* **2.** *Sorbus maderensis.* **3.** *Chamaemeles coriacea.* Habits all × ½, 1a, 2a × ½, 3a × 1.

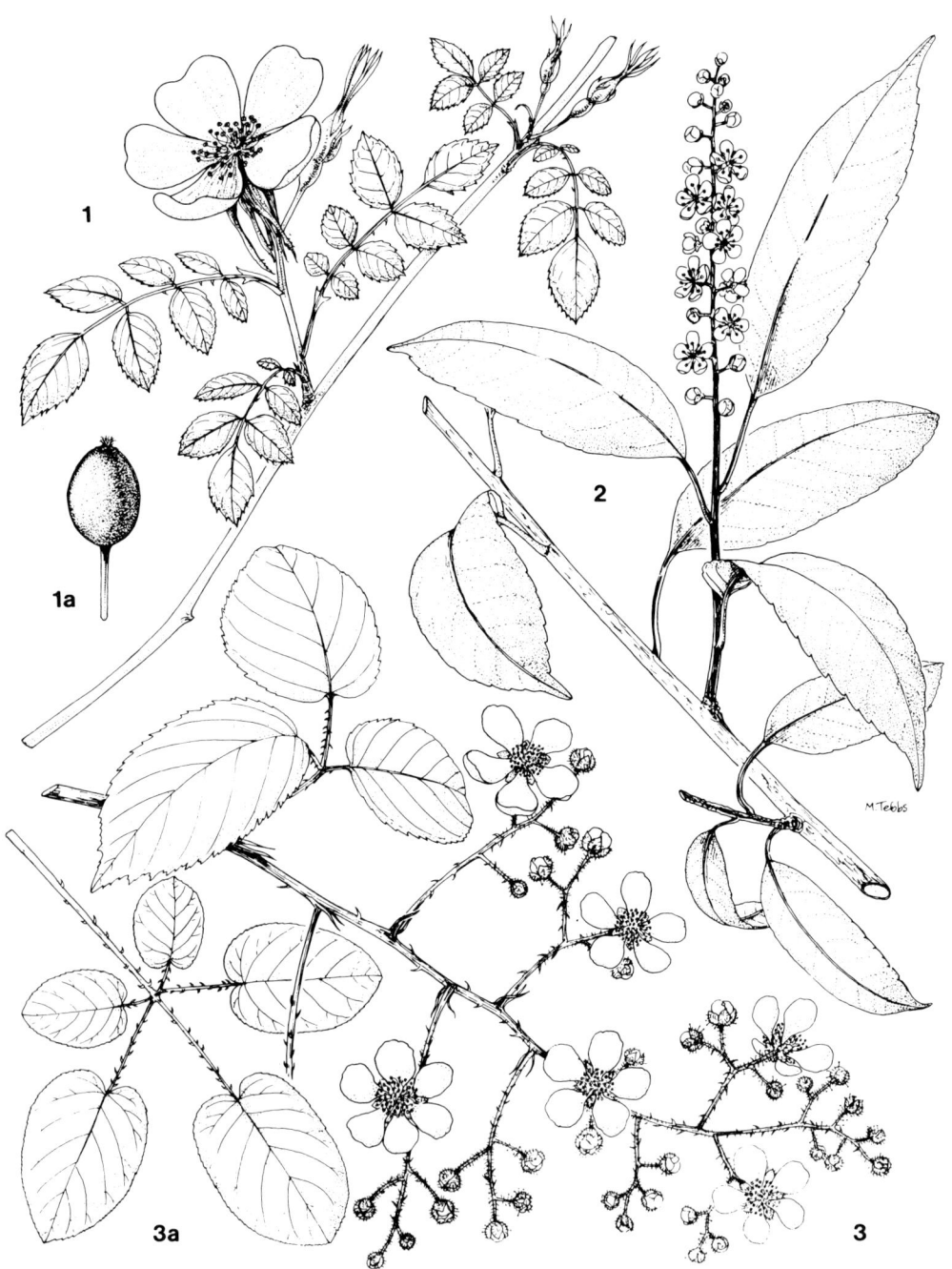

Plate 18 **1**. *Rosa mandonii*. **2**. *Prunus lusitanica* subsp. *hixa*. **3**. *Rubus grandifolius*. All × ½.

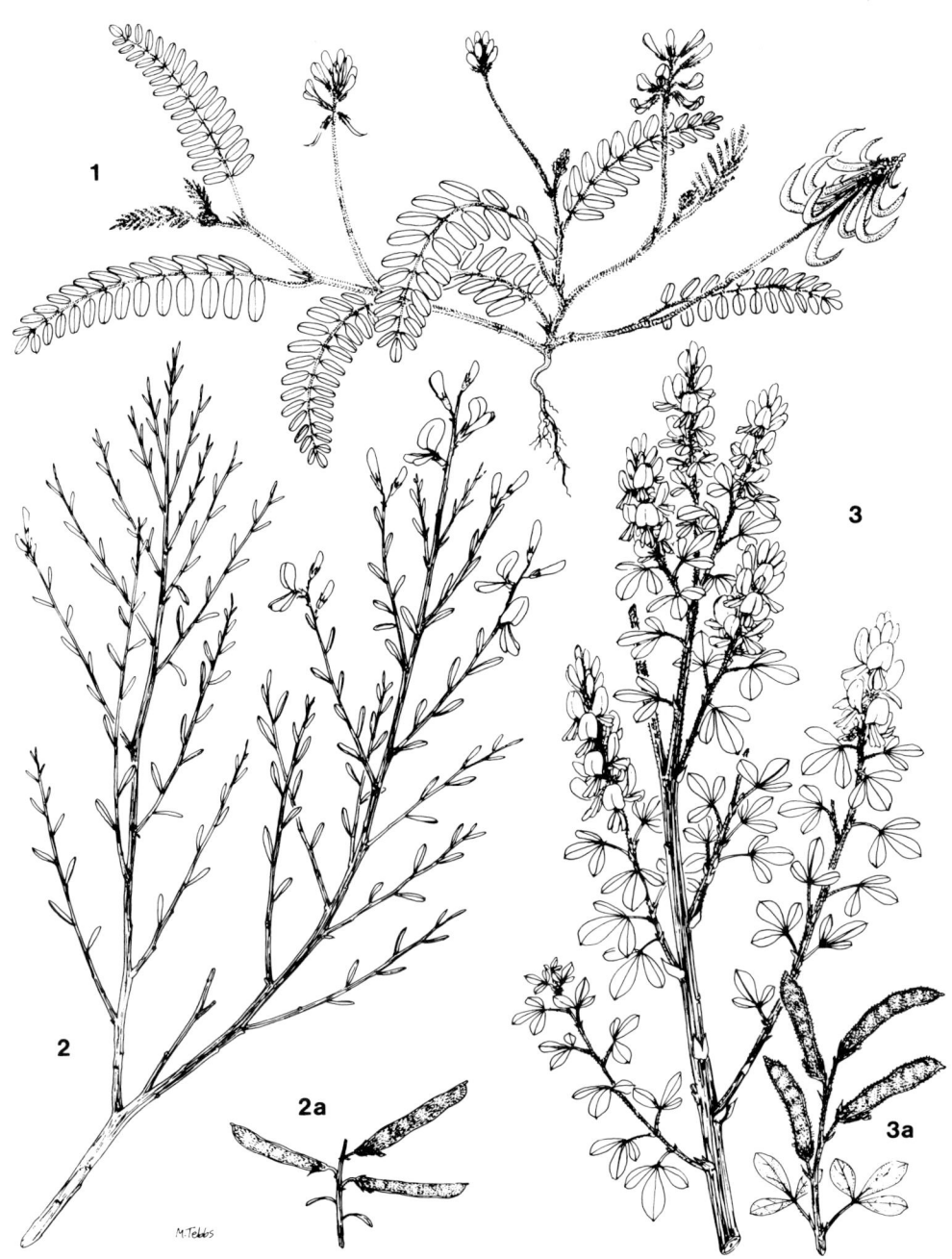

Plate 19 1. *Astragalus solandri*. 2. *Genista tenera*. 3. *Teline maderensis*. All × ½.

Plate 20 1. *Vicia capreolata*. 2. *Anthyllis lemanniana*. All × ½.

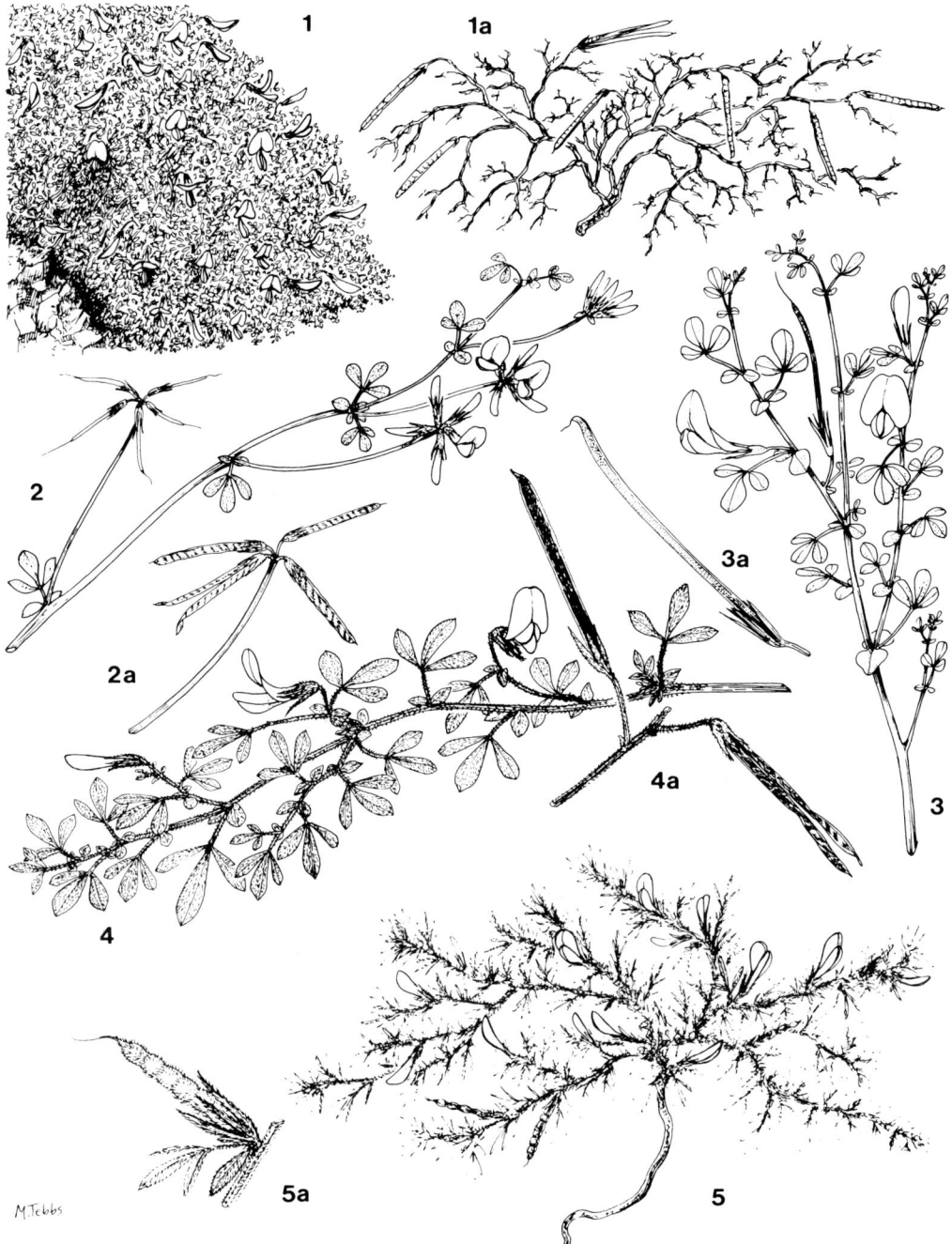

Plate 21 1. *Lotus glaucus*. 2. *L. lancerottensis*. 3. *L. macranthus*. 4. *L. argyrodes*. 5. *L. loweanus*. All × ½.

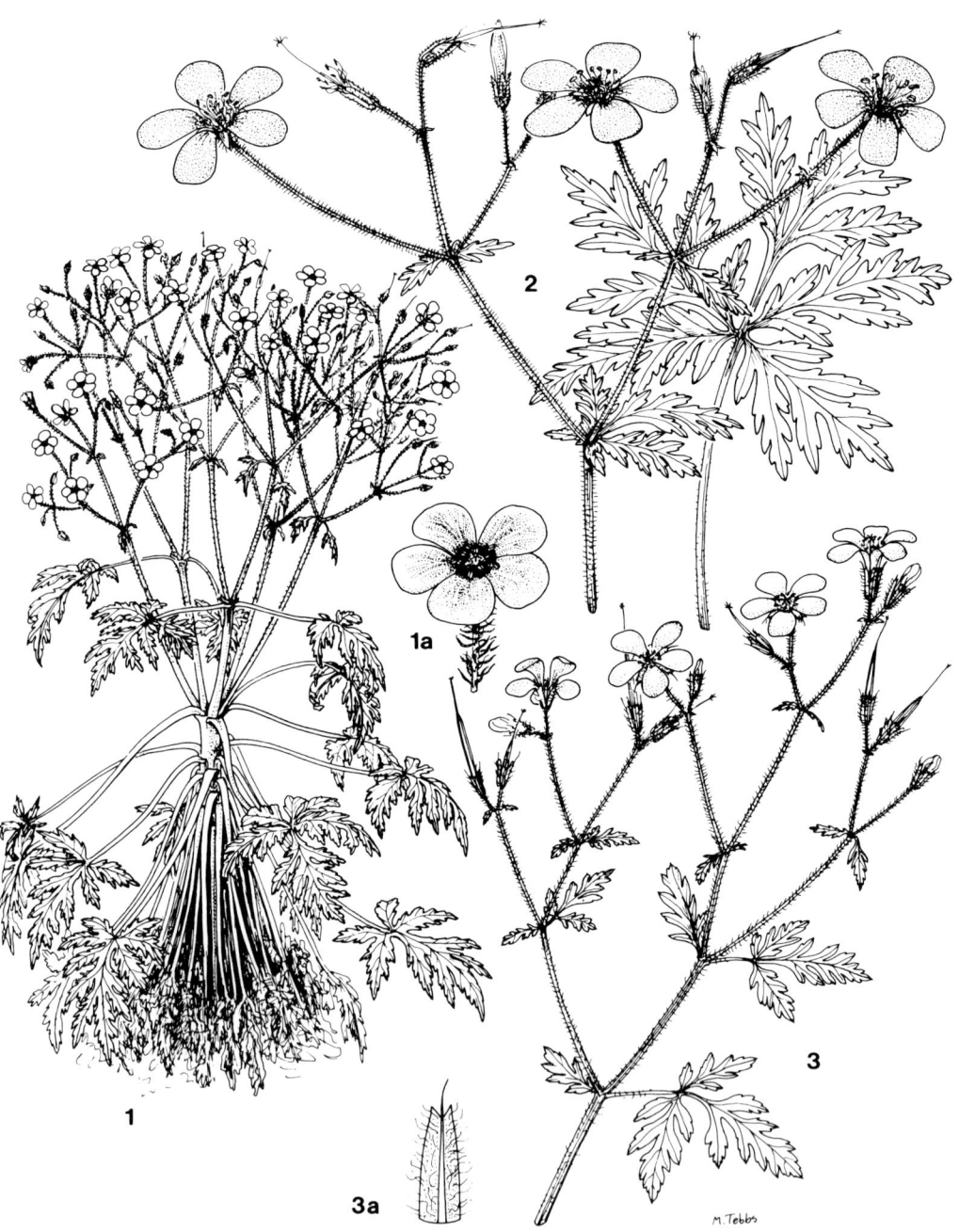

Plate 22 1. *Geranium maderense* × ⅛, **a**) × ½. **2**. *G. palmatum* × ½. **3**. *G. rubescens* × ½, **a**) sepal × 1½.

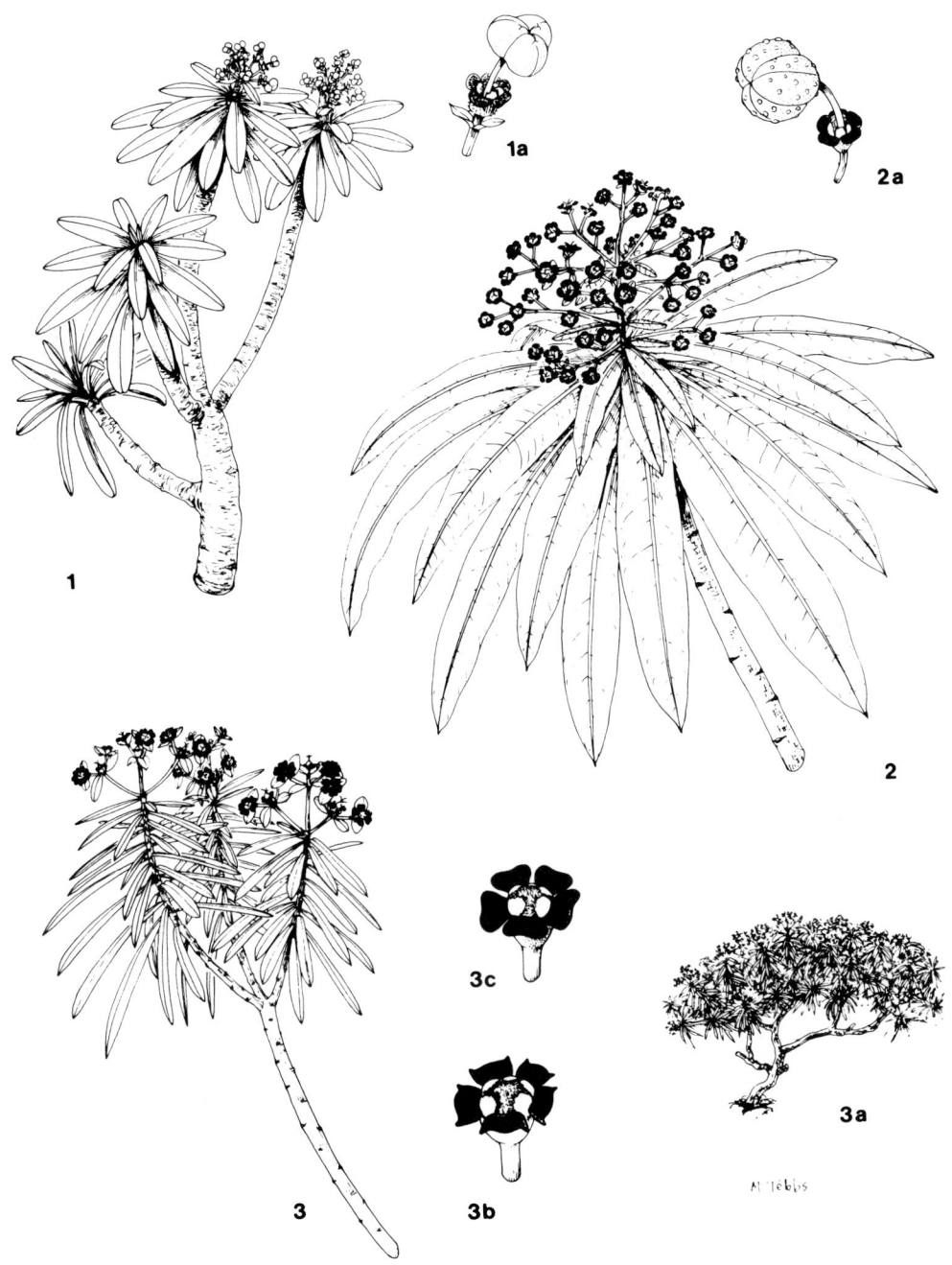

Plate 23 1. *Euphorbia desfoliata* × ½, **a)** × 1. **2**. *E. mellifera* × ½, **a)** × 1. **3**. *E. piscatoria* × ½, **a)** × ¹/₄₀, **b)** × 1, **c)** × 1.

Plate 24 **1**. *Ilex canariensis*. **2**. *I. perado* ssp. *perado*. **3**. *Maytenus umbellata*. All × ½.

Plate 25 1. *Visnea mocanera* × ½, **a)** × 3. **2**. *Rhamnus glandulosa* × ½, **a)** × 1. **3**. *Frangula azorica* × ½, **a)** × 1½.

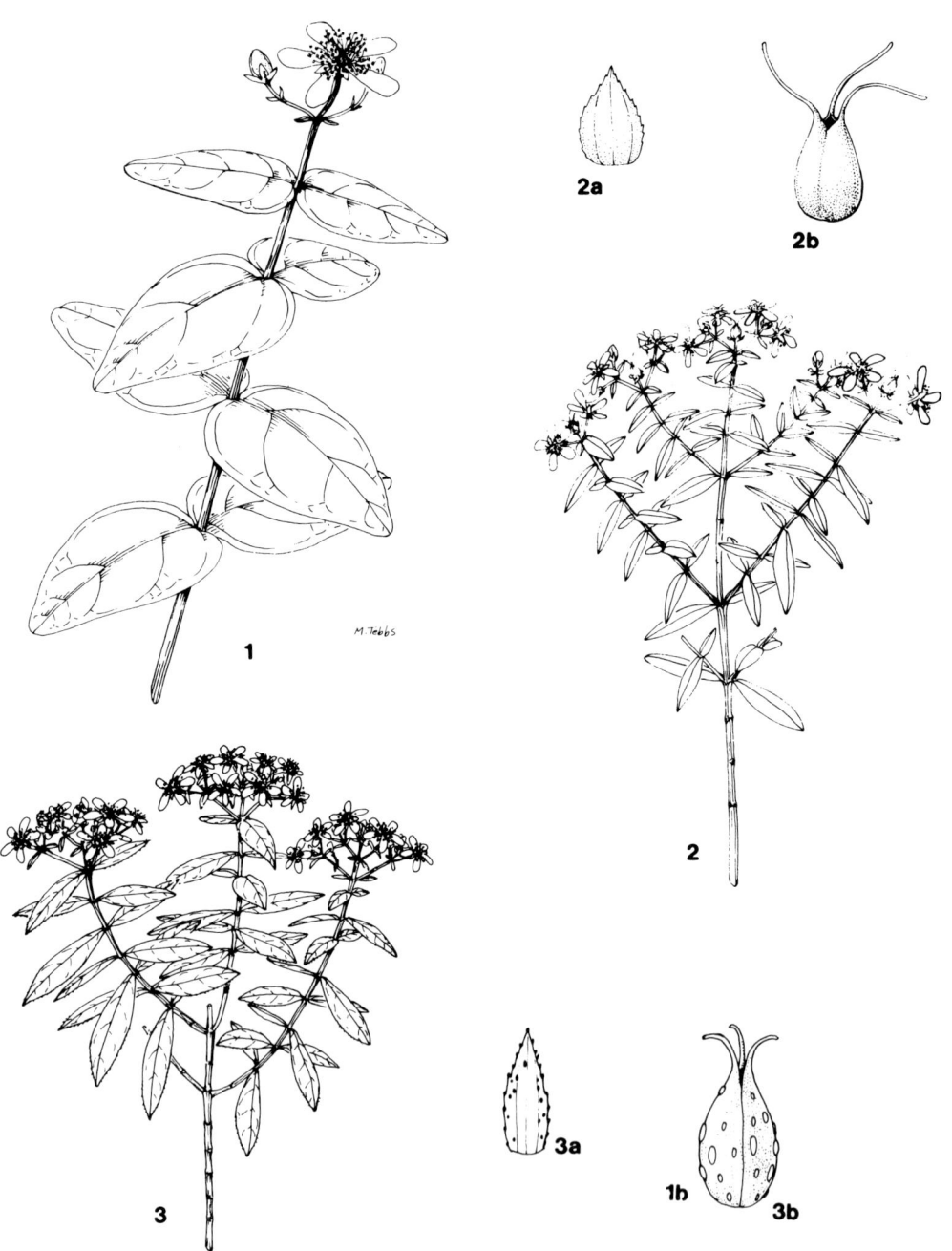

Plate 26 1. *Hypericum grandifolium* × ½. 2. *H. canariense* × ½, **a**) sepal × 3, **b**) fruit × 3. 3. *H. glandulosum* × ⅓, **a**) sepal × 3, **b**) fruit × 3.

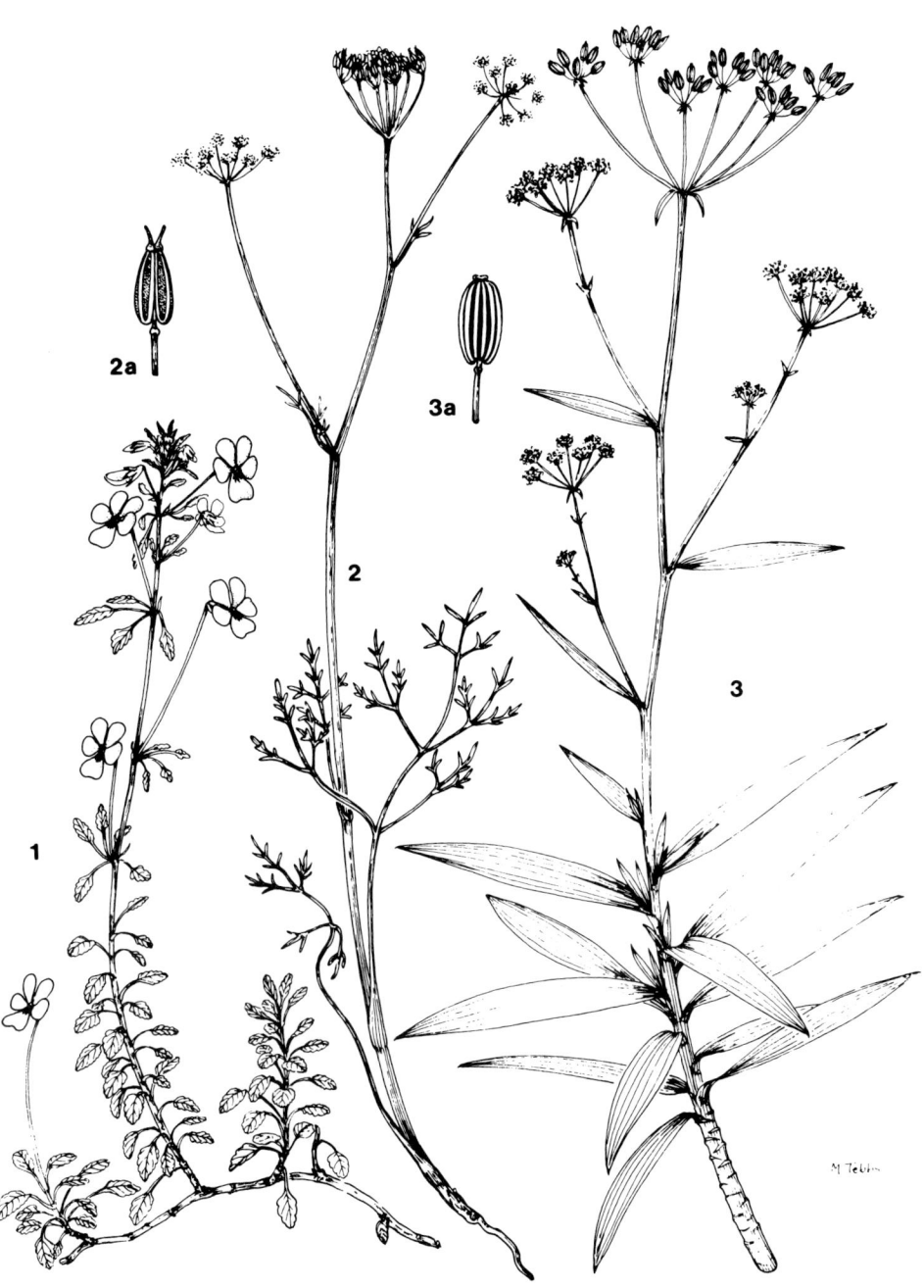

Plate 27 1. *Viola paradoxa* × ⅓. **2.** *Bunium brevifolium* × ½. **3.** *Bupleurum salicifolium* subsp. *salicifolium* var. *salicifolium* × ½. Details all × 2.

Plate 28 **1**. *Peucedanum lowei*. **2**. *Oenanthe divaricata*. **3**. *Capnophyllum peregrinum*. Habits all × ½, details all × 3.

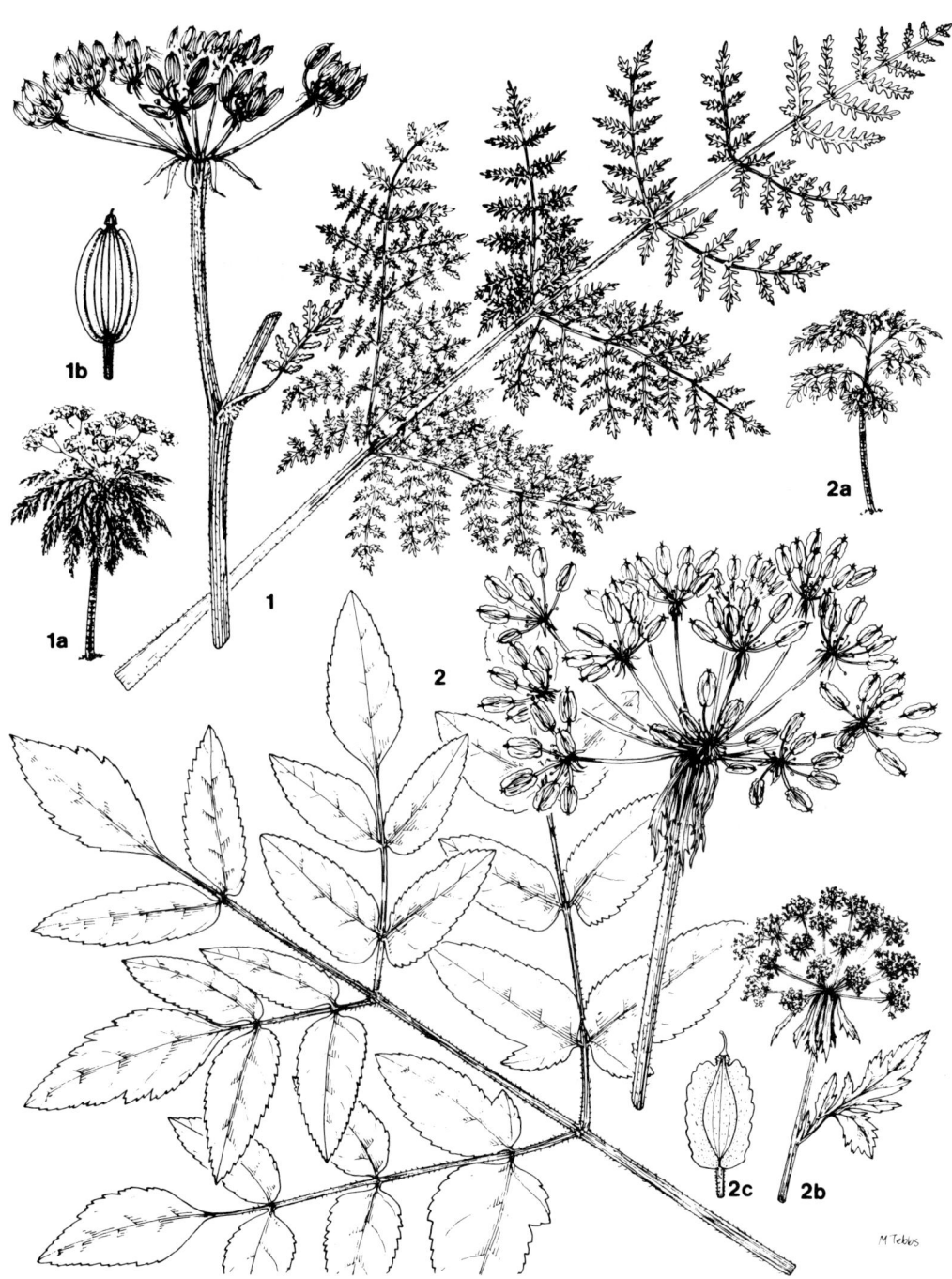

Plate 29 1. *Monizia edulis* leaf × ¼, infructescence × ½, **b)** × 1½. **2**. *Melanoselinum decipiens* leaf × ¼, infructescence × ½, **b)** × ½, **c)** × 1½.

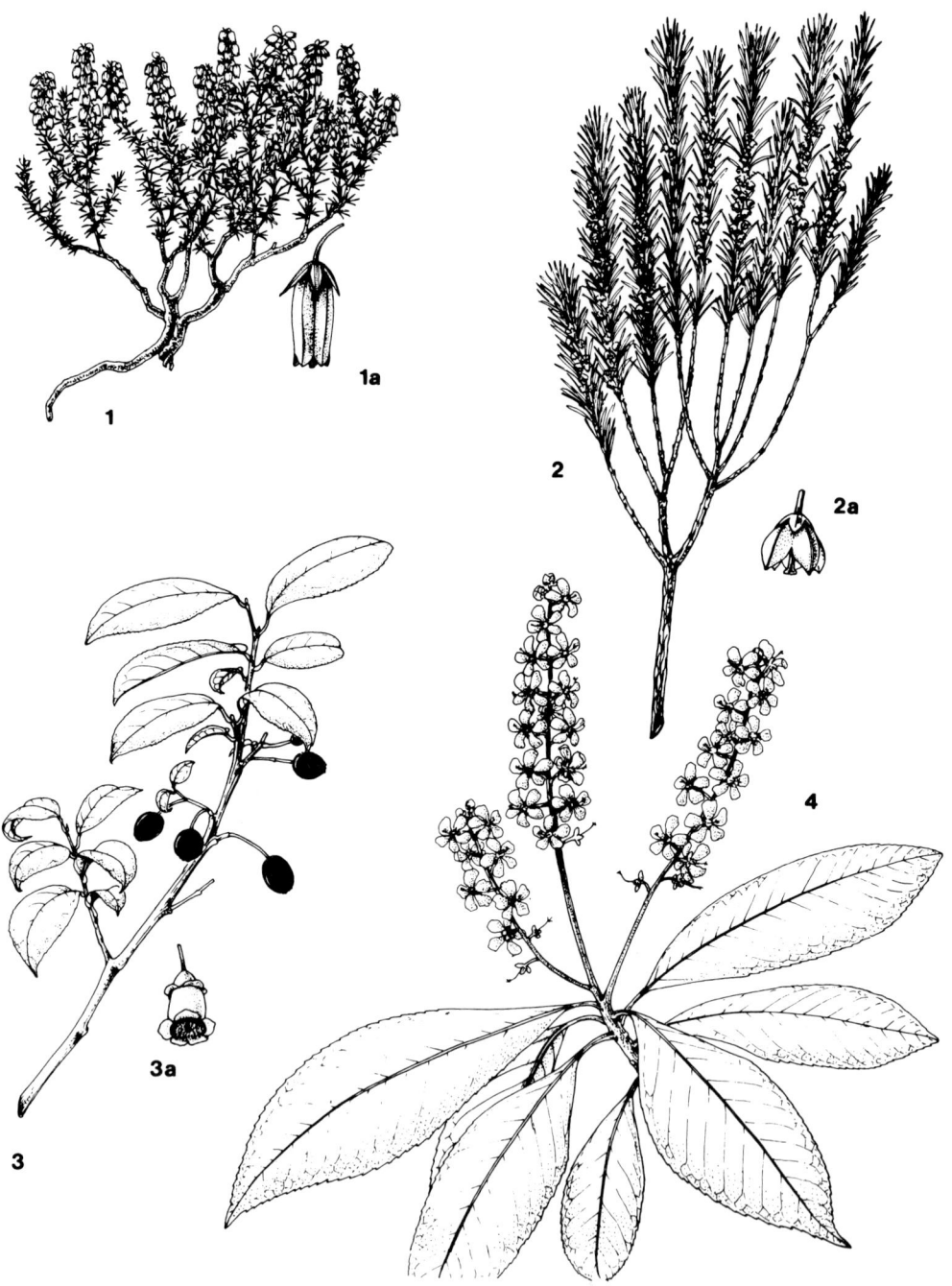

Plate 30 1. *Erica maderensis* × ¼, **a**) × 3. **2**. *E. scoparia* subsp. *maderinicola* × ½, **a**) × 3. **3**. *Vaccinium padifolium* × ½, **a**) × 1. **4**. *Clethra arborea* × ½

Plate 31 1. *Armeria maderensis.* 2. *Limonium ovalifolium* subsp. *pyramidatum.* 3. *L. papillatum* var. *callibotryum.* 4. *Pelletiera wildpretii.* All × ½.

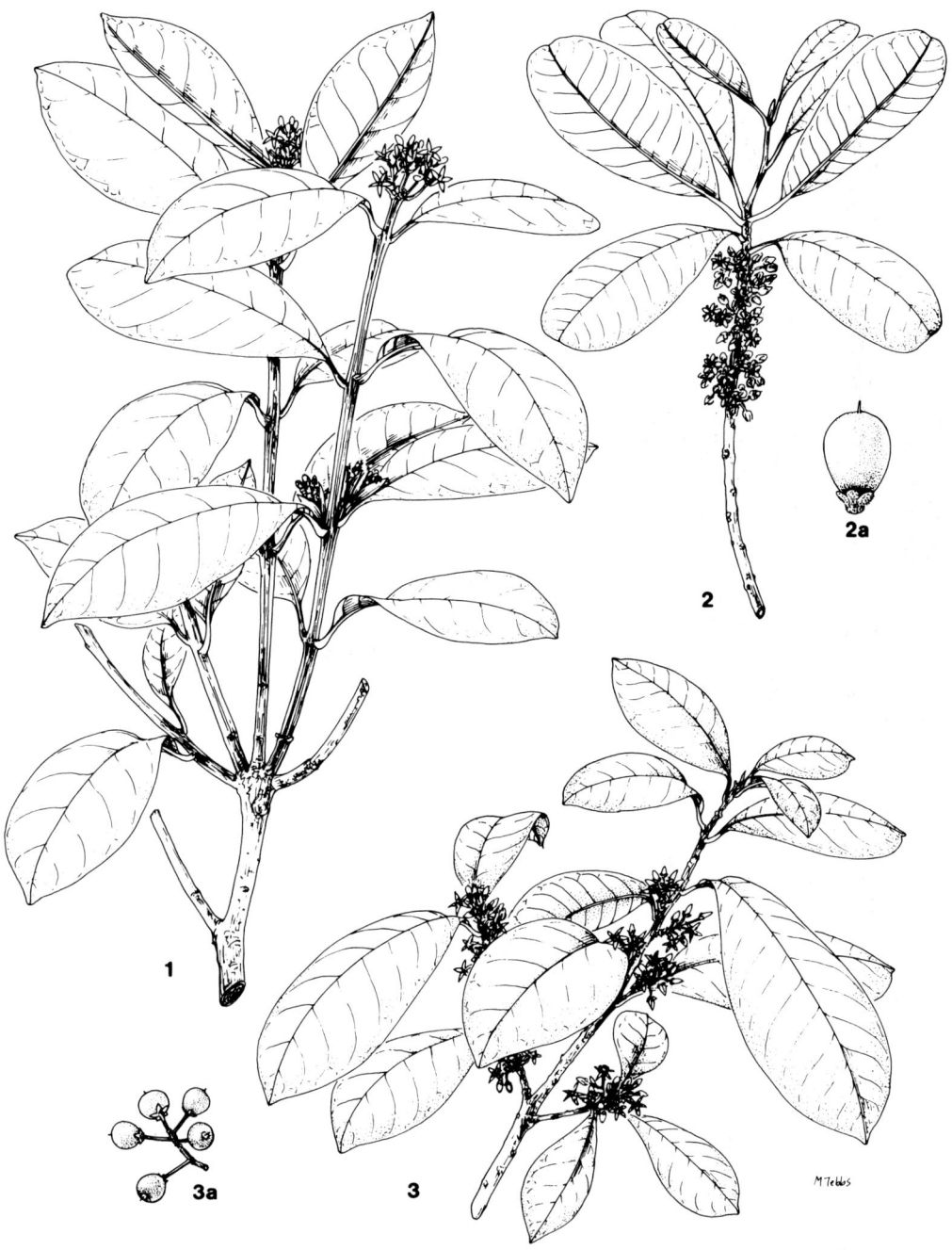

Plate 32 1. *Picconia excelsa.* 2. *Sideroxylon marmulano* var. *marmulano*, **a**) × 1. 3. *Heberdenia excelsa*, **a**) × ½. Habits all × ½.

Plate 33 1. *Jasminum azoricum*. 2. *J. odoratissimum*. 3. *Olea europaea* subsp. *maderensis*. All × ½.

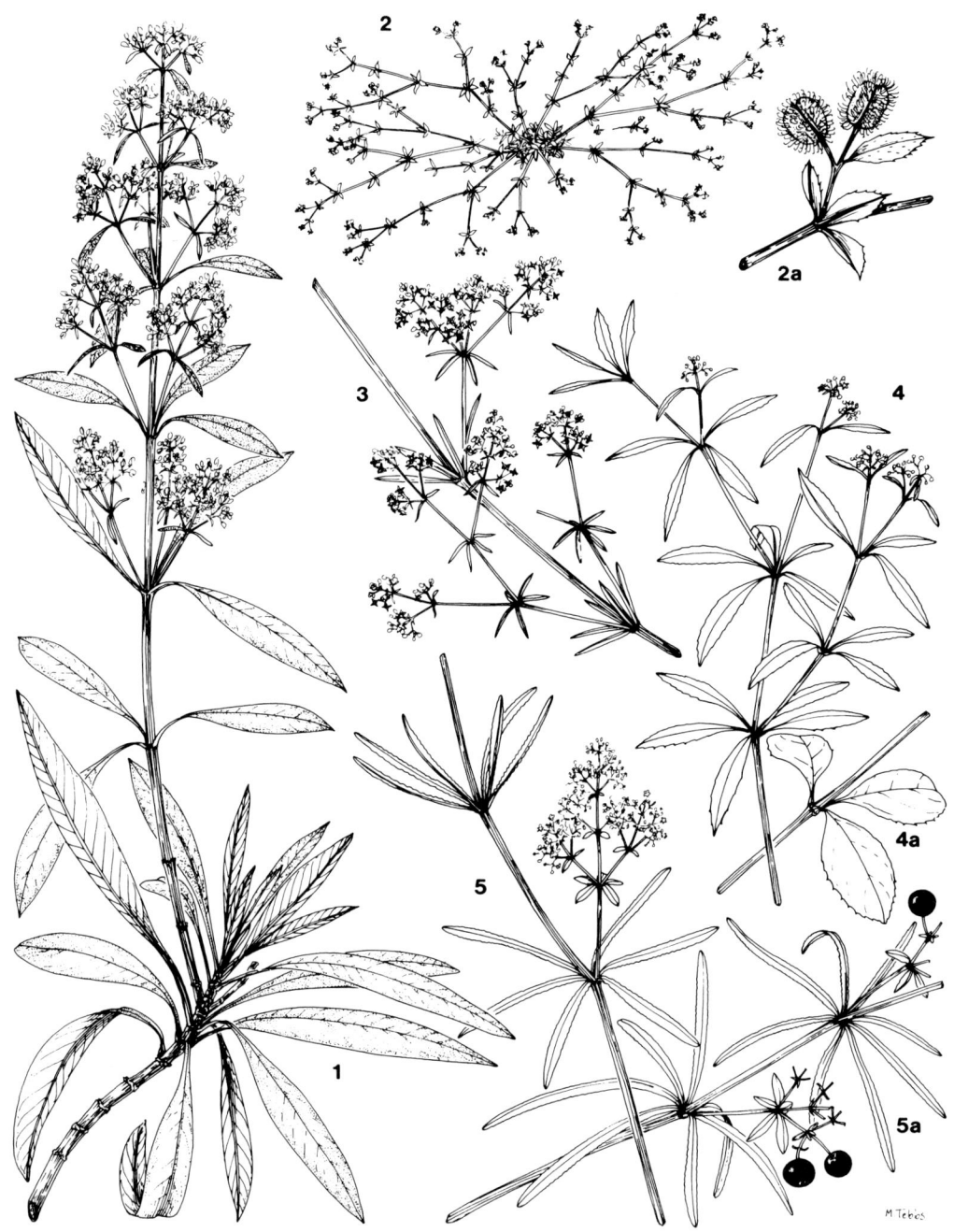

Plate 34 1. *Phyllis nobla.* 2. *Galium geminiflorum,* **a**) × 9. 3. *G. productum.* 4. *Rubia fruticosa* subsp. *fruticosa,* **a**) × ½. 5. *R. agostinhoi,* **a**) × ½. Habits all × ½.

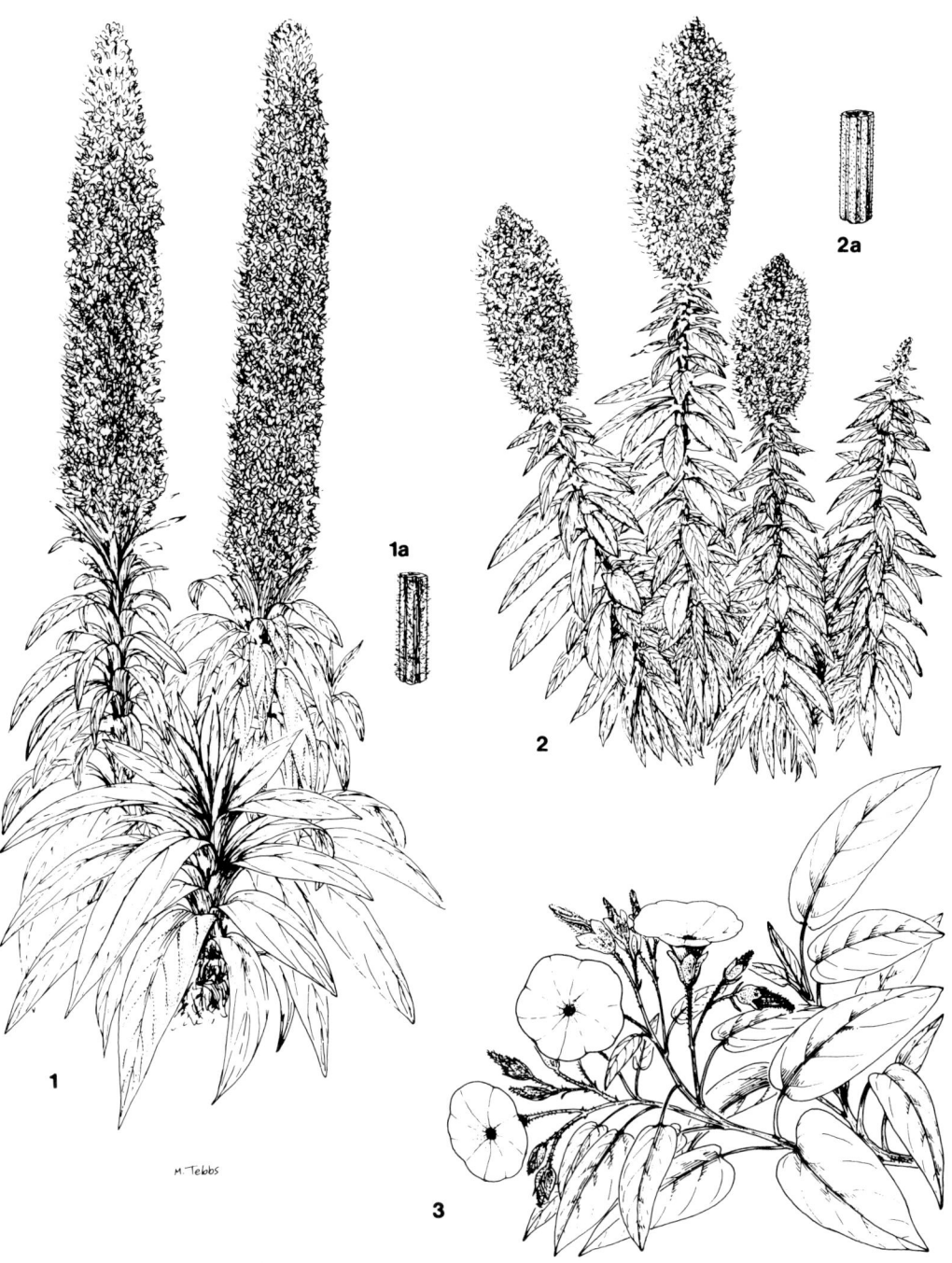

Plate 35 **1**. *Echium candicans* × ¼, **a**) stem × 1. **2**. *E. nervosum* × ¼, **a**) stem × 1. **3**. *Convolvulus massonii* × ½.

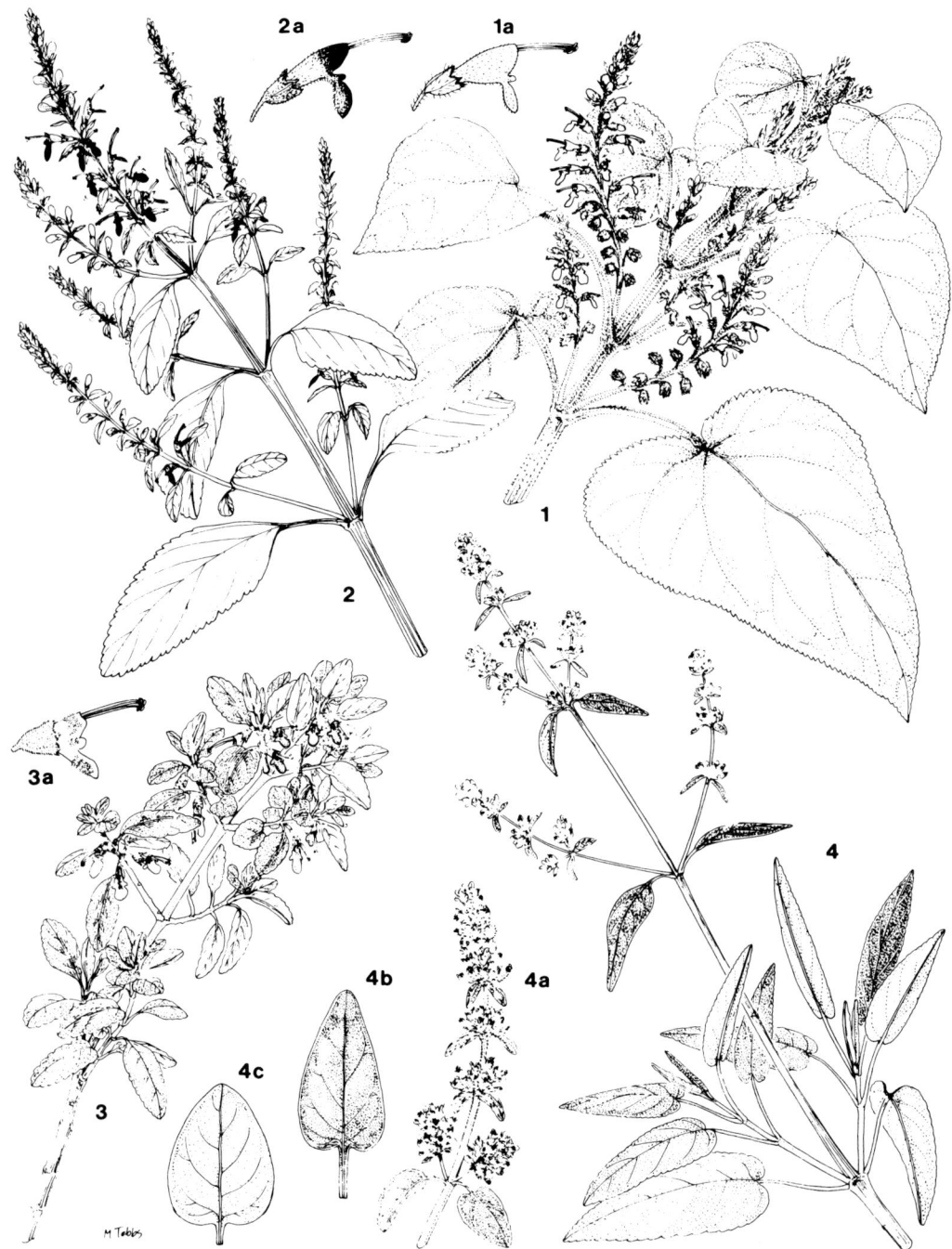

Plate 36 1. *Teucrium abutiloides*, **a**) × 1. **2**. *T. betonicum*, **a**) × 1. **3**. *T. heterophyllum*, **a**) × 1. **4**. *Sideritis candicans* var. *candicans*, **a**) & **b**) var. *multiflora* × ½, **c**) var. *crassifolia* × ½. Habits all × ½.

Plate 37 1. B*ystropogon maderensis*. **2**. *B. punctatus*. **3**. *Satureja varia* subsp. *thymoides*, **a)** var. *thymoides*, **b)** var. *cacuminicolae*. **4**. *Cedronella canariensis*. Habits all × ½, calyx details all × 5.

Plate 38 1. *Lavandula pinnata.* 2. *Normannia trisectum.* 3. *Solanum patens.* All × ½.

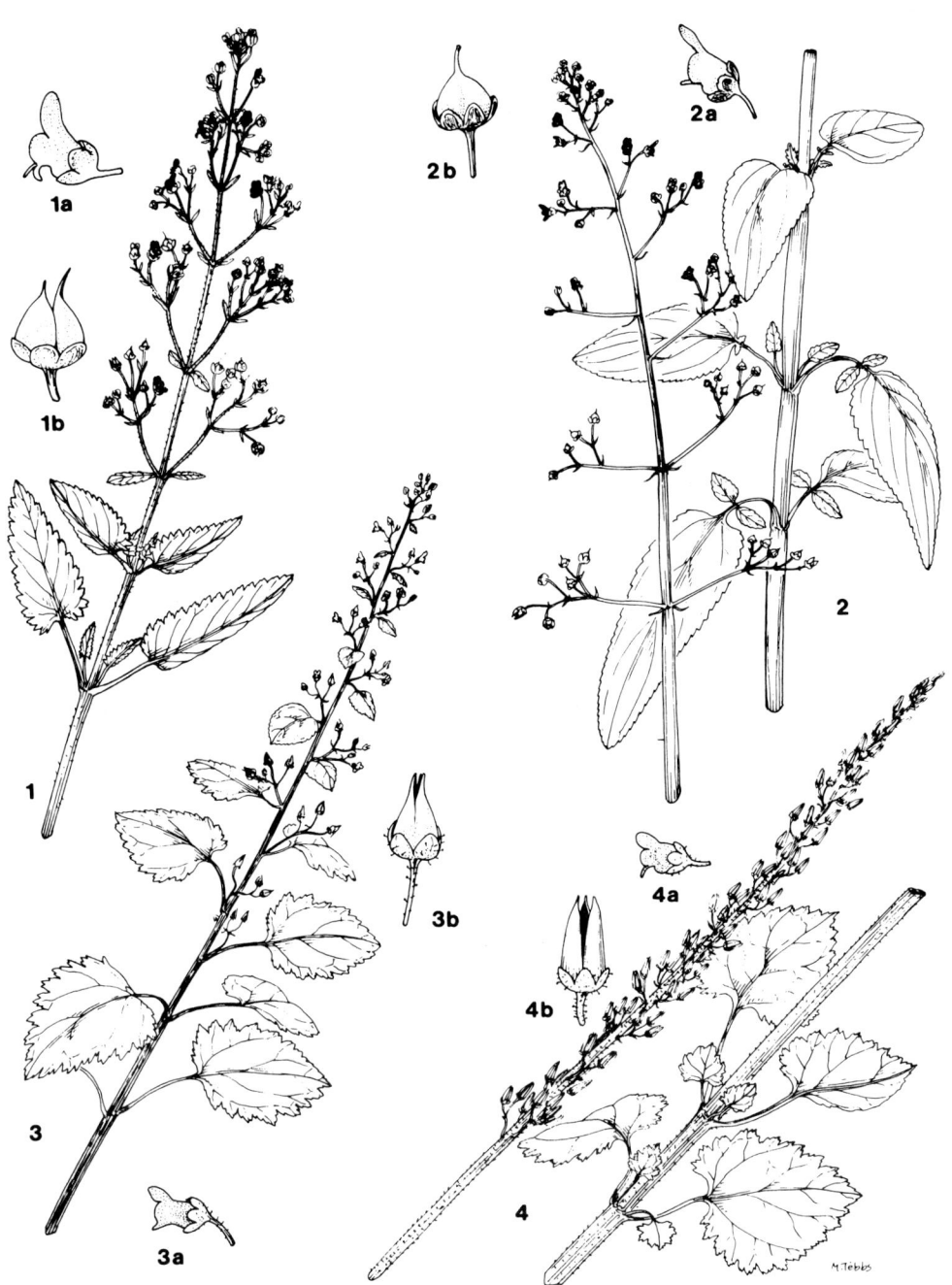

Plate 39 1. *Scrophularia hirta*. **2**. *S. racemosa*. **3**. *S. lowei*. **4**. *S. arguta*. Habits all × ½, flower and capsule details all × 2.

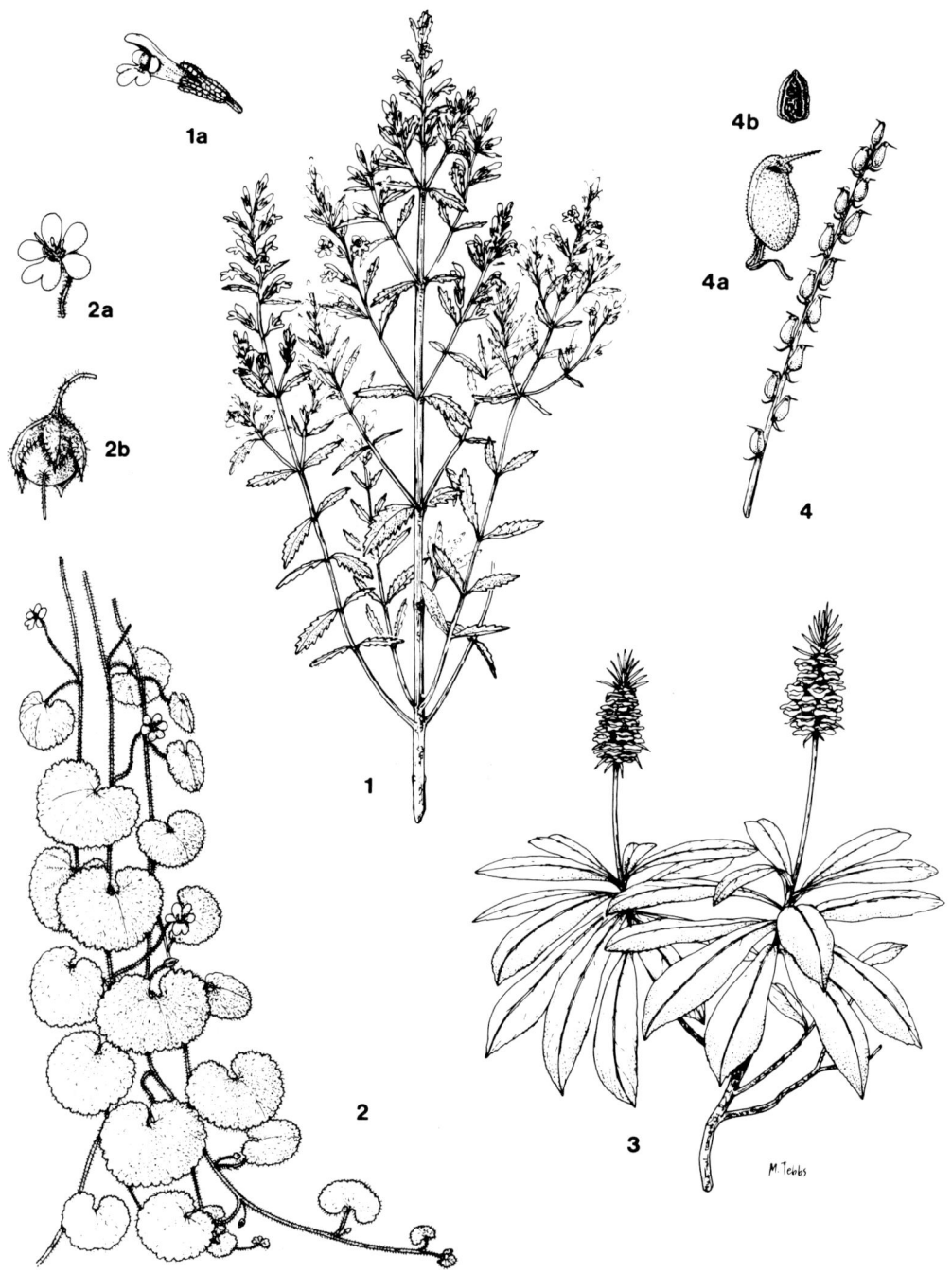

Plate 40 1. *Odontites holliana* × ½, **a**) × 1. **2**. *Sibthorpia peregrina* × ½, **a**) × 1, **b**) × 2. **3**. *Isoplexis sceptrum* × ⅛. **4**. *Misopates salvagensis* × ½, **a**) capsule × 1½, **b**) seed × 10.

Plate 41 1. *Globularia salicina*. 2. *Plantago arborescens* subsp. *maderensis*. 3. *P. leiopetala*. Habits all × ½, detail × 1½.

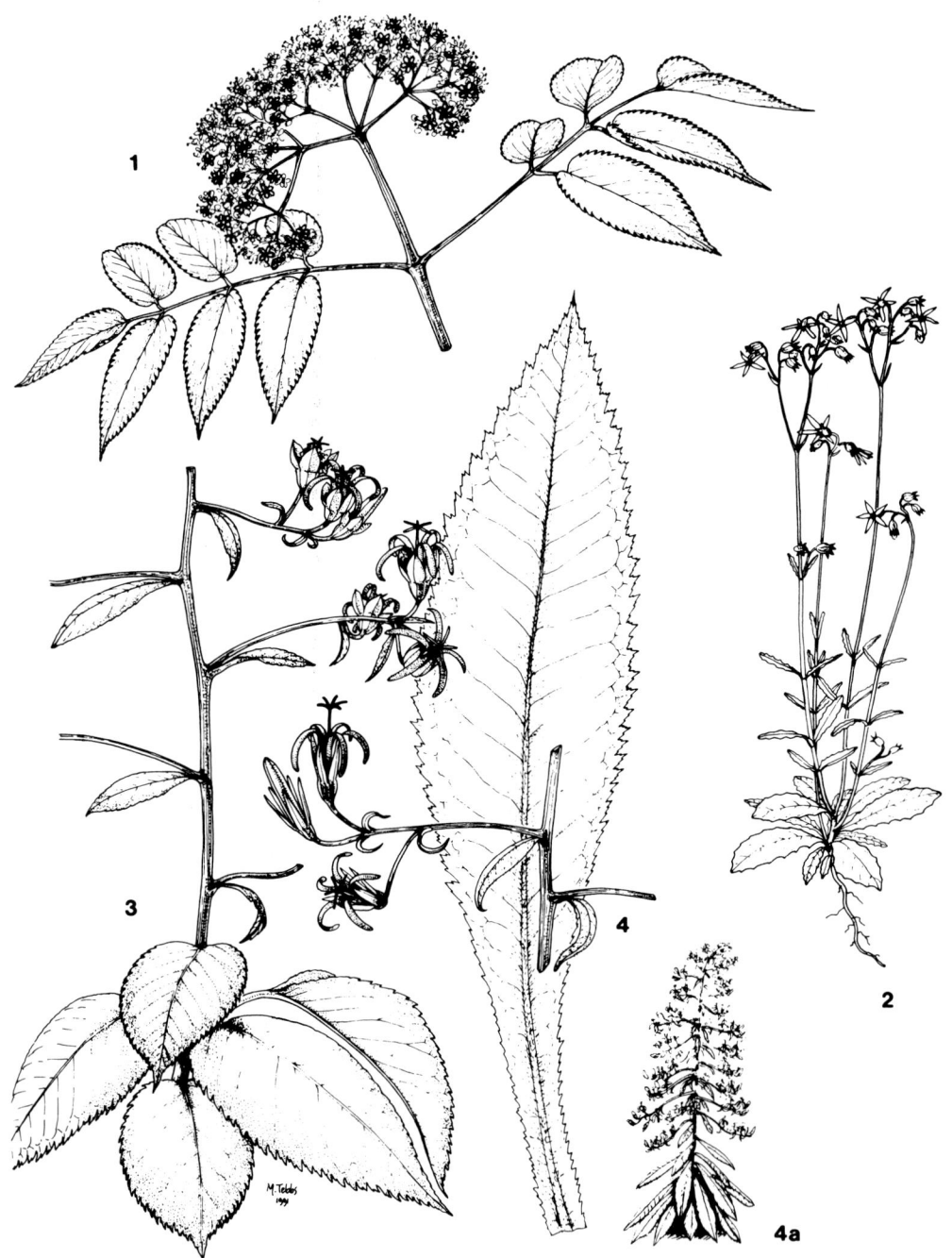

Plate 42 1. *Sambucus lanceolata* × ½. **2**. *Wahlenbergia lobelioides* × ¼. **3** *Musschia aurea* × ½. **4**. *M. wollastoni* × ½.

Plate 43 1. *Helichrysum melaleucum.* **2.** *H. devium.* **3.** *H. monizii*, **a**) capitulum × 2, **b**) bract × 3. **4.** *H. obconicum*, **a**) capitulum × 2, **b**) bract × 3. **5.** *Phagnalon bennettii.* Habits all × ½.

Plate 44 1. *Argyranthemum thalassophilum.* 2. *A. haematomma.* 3. *A. dissectum.* 4. *A. pinnatifidum* subsp. *pinnatifidum,* **b**) subsp. *montanum,* leaf × 1, **c**) subsp. *succulentum,* leaf × ½. Habits all × ½, achene details all × 2½.

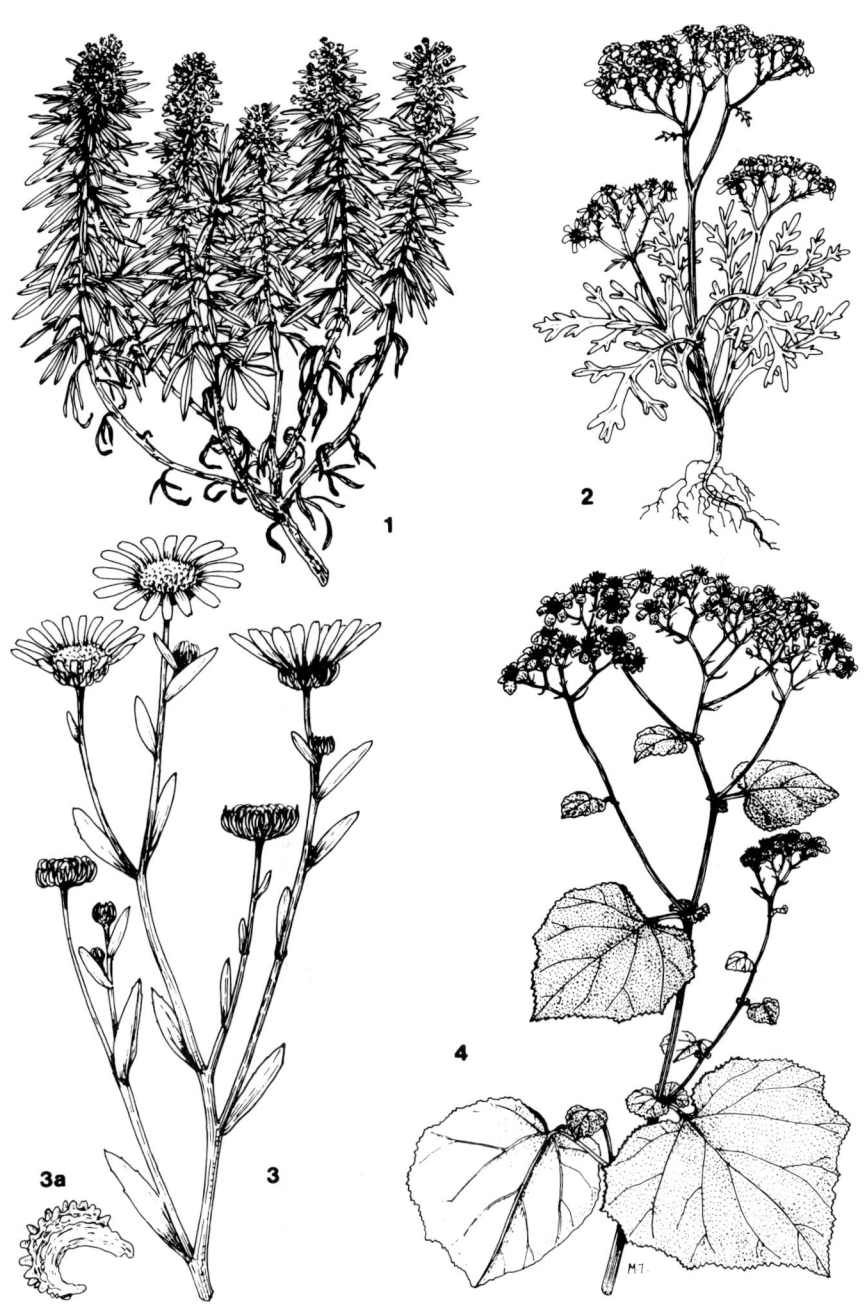

Plate 45 1. *Artemisia argentea* × ¼. **2.** *Senecio incrassatus* × ½. **3.** *Calendula maderensis* × ½, **a)** achene × 1½. **4.** *Pericallis aurita* × ½.

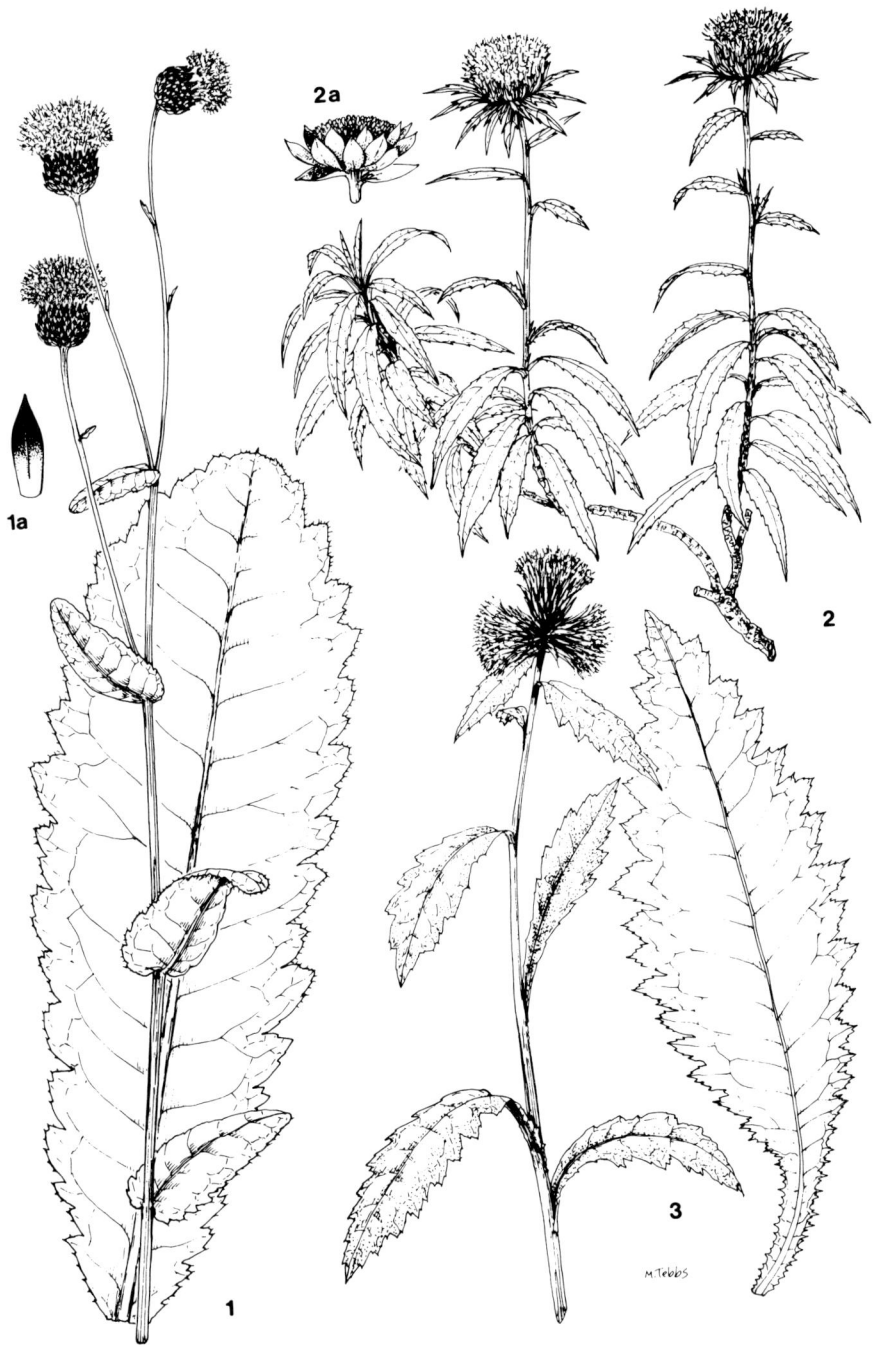

Plate 46 1. *Cirsium latifolium*, **a**) bract × 1½. **2**. *Carlina salicifolia* var. *salicifolia*, **a**) var. *inermis*. **3**. *Carduus squarrosus*. Habits all × ½.

Plate 47 1. *Cheirolophus massonianus*, **a**) involucral bract × 2. **2**. *Tolpis macrorhiza*. **3**. *T. succulenta*, **a - d**) leaf variation × ½. Habits all × ½.

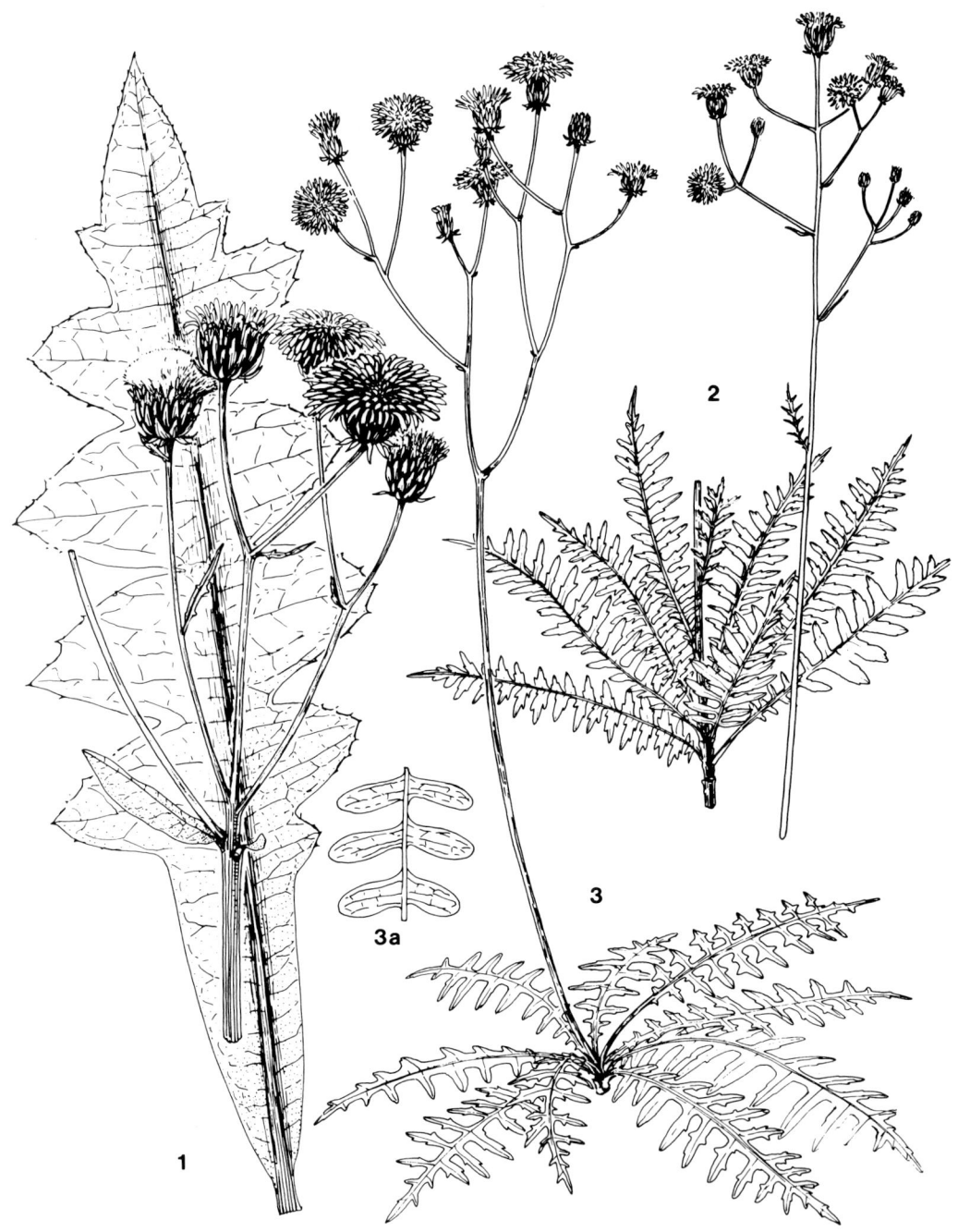

Plate 48 1. *Sonchus fruticosus* × ½. **2**. *S. pinnatus* × ¼. **3**. *S. ustulatus* subsp. *ustulatus* × ¼, **a**) subsp. *maderensis* × ¼.

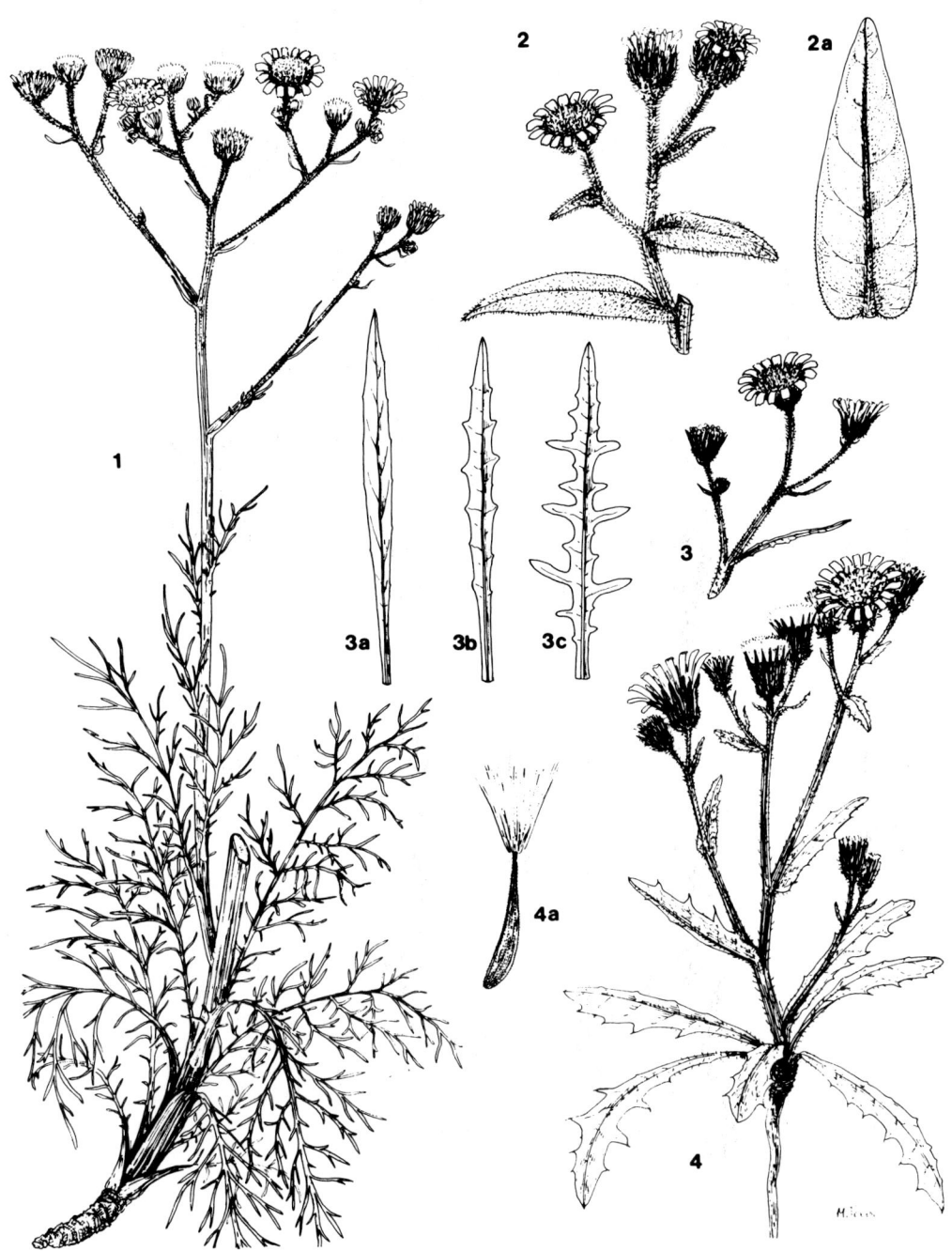

Plate 49 1. *Andryala crithmifolia*. 2. *Andryala glandulosa* subsp. *glandulosa* a) leaf × ½. 3. *A. glandulosa* subsp. *varia*, a - c) leaf variation × ½. 4. *Crepis vesicaria* subsp. *andryaloides*. Habits all × ½.

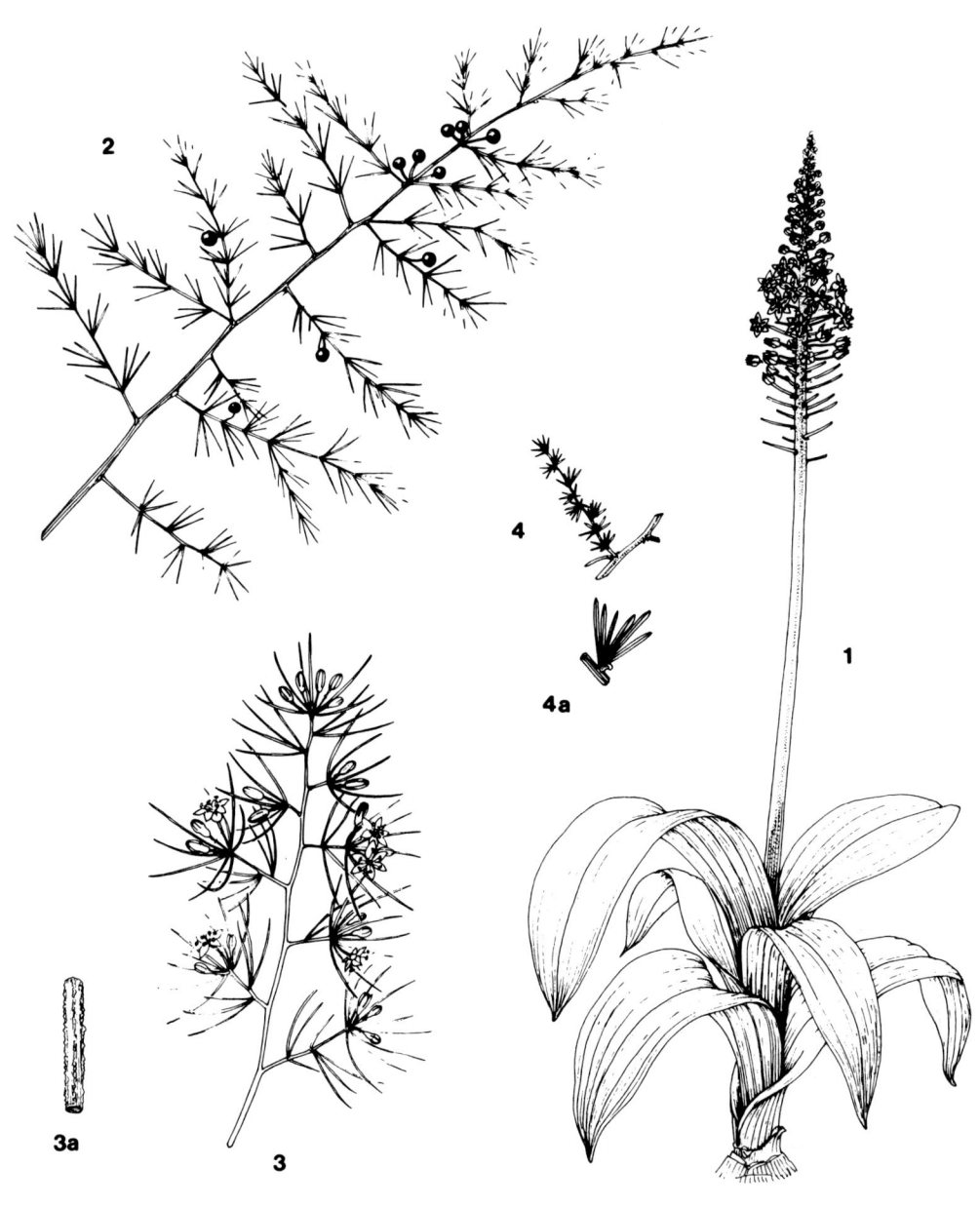

Plate 50 **1**. *Scilla maderensis* subsp. *maderensis*. **2**. *Asparagus scoparius*. **3**. *A. umbellatus* subsp. *lowei,* **a**) branch × 5. **4**. *A. nesiotes*. Habits all × ½.

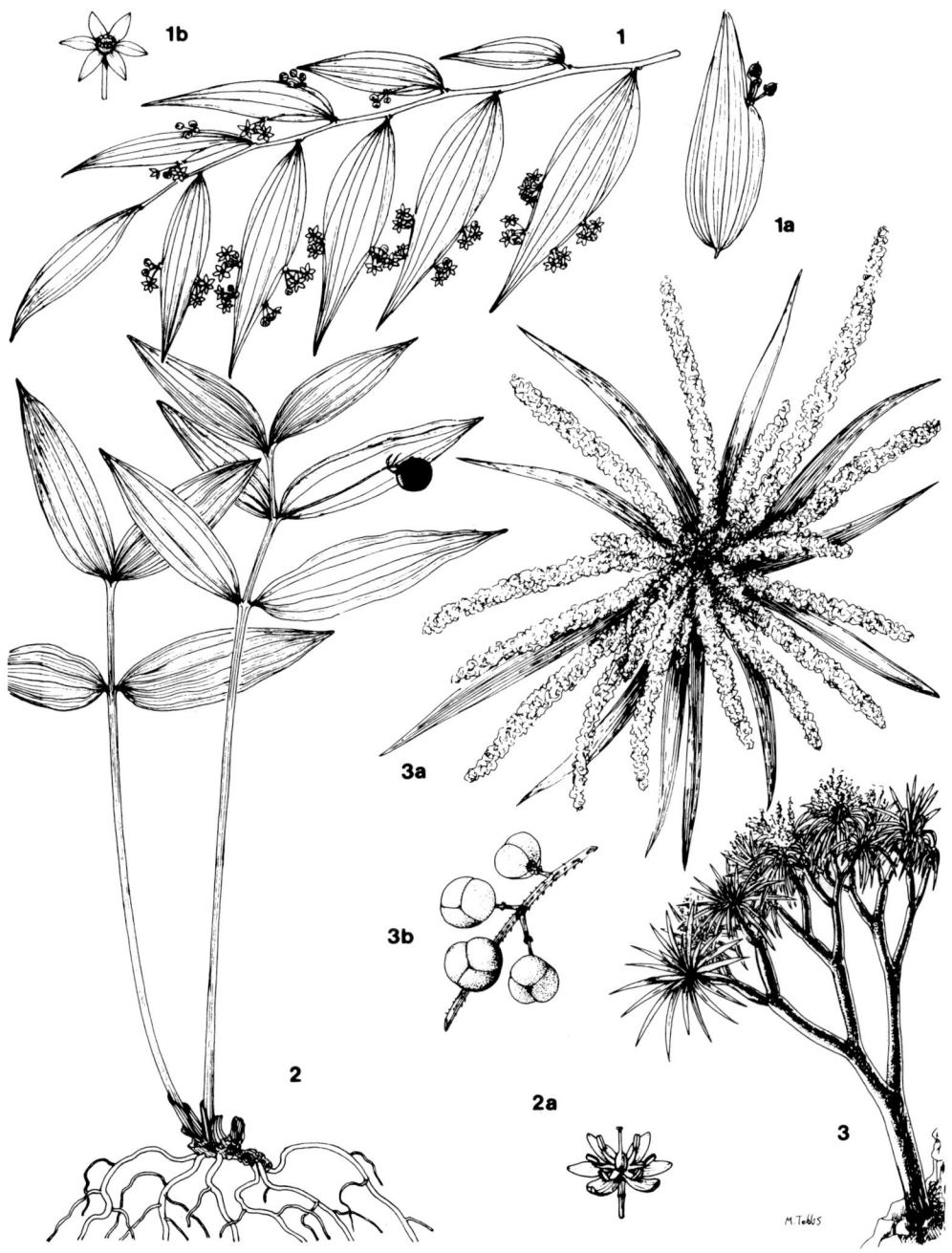

Plate 51 1. *Semele androgyna* × ½, **a**) × ½, **b**) × 2. **2.** *Ruscus streptophyllus* × ½. **3.** *Dracaena draco*.

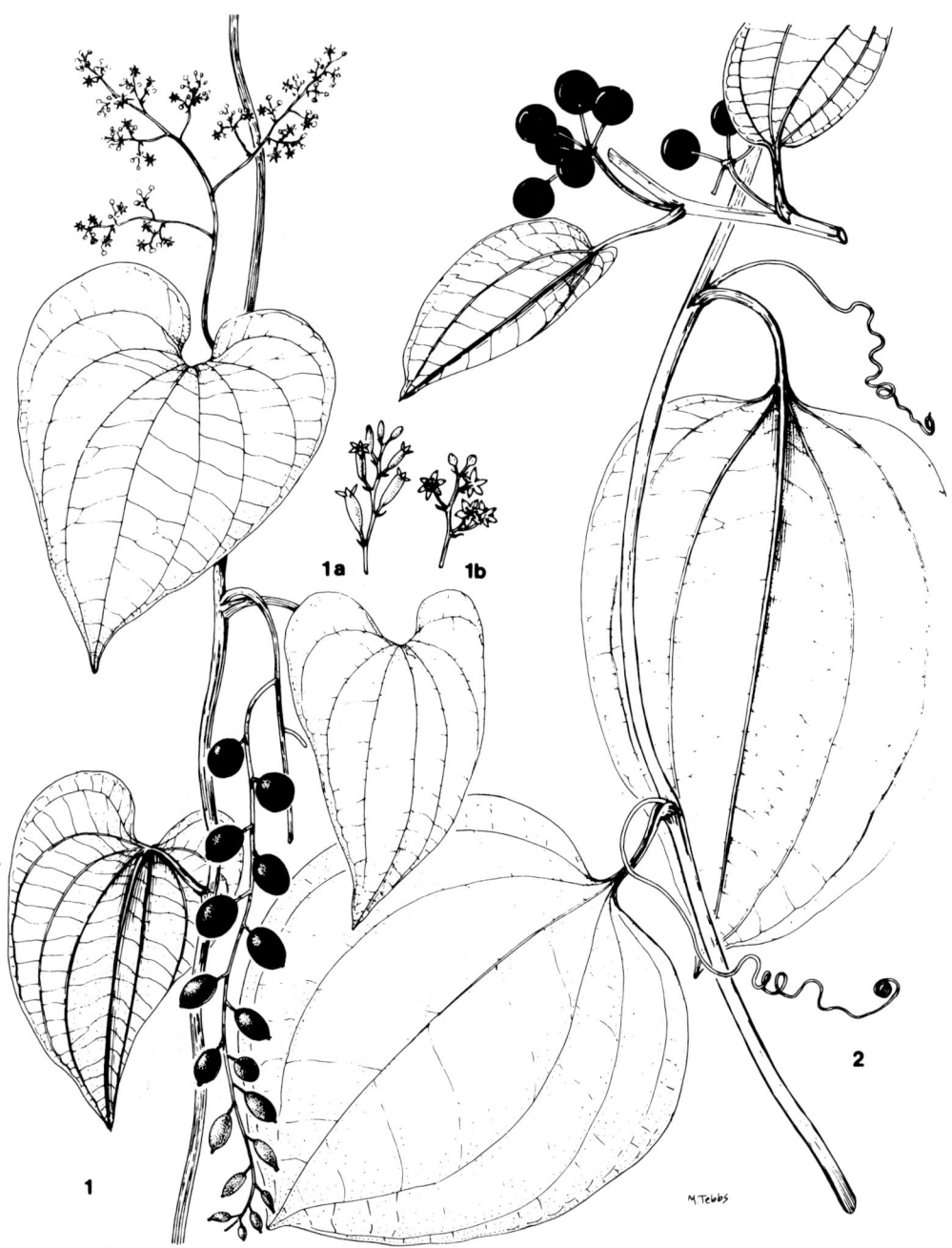

Plate 52 **1**. *Tamus edulis*. **2**. *Smilax canariensis*. Habits × ½, flower details × 2.

Plate 53 1. *Festuca donax*. 2. *F. jubata*. 3. *Luzula elegans*. 4. *L. seubertii*. Habits all × ½, spikelet details × 3.

Plate 54 1. *Parafestuca albida.* 2. *Deschampsia argentea.* 3. *D. maderensis.* 4. *Melica canariensis.* All × ½.

Plate 55 1. *Phalaris maderensis.* 2. *Agrostis obtusissima.* 4. *Carex viridula* subsp. *cerdercreutzii.* 4. *C. lowei.* Habits × ½, utricle details × 2.

Plate 56 **1**. *Arum italicum* subsp. *canariense*. **2**. *Dracunculus canariensis*. Both × ½.

Plate 57 1. *Dactylorhiza foliosa*. 2. *Goodyera macrophylla*. 3. *Orchis scopulorum*. All × ½.

ERRATA

The following are missing from the Index to Scientific Names:

ACANTHACEAE, 317
Acanthus L., 318
    mollis L., 318
Andryala glandulosa subsp. **glandulosa**, 320
Asystasia gangetica (L.) T. Anderson in Thwaites, 318
Bignonia unguis-cati, L. 317
    venusta Ker Gawl., 317
BIGNONIACEAE, 317
Catalpa bignonioides Walter, 317
Cistanche Hoffmanns. ex Link, 318
    phelypaea (L.) Cout., 319
Crepis, 320
Digitalis purpurea, 320
Doxantha unguis-cati (L.) Rehdar, 317
Globularia L., 316
    salicina Lam., 316
GLOBULARIACEAE, 316
Jacaranda mimosifolia D. Don, 317
Kigelia africana (Lam.) Benth., 317
Lytanthus salicinus (Lam.) Wettst., 316
Macfadyena A.DC., 317
    unguis-cati (L.) A.H. Gentry, 317
MYOPORACEAE, 320
Myoporum acuminatum R. Br., 320
    tenuifolium G. Forst., 320
OROBANCHACEAE, 318
Orobanche L., 319
    barbata sensu auct. Mad., 320
    calendulae Pomel, 320
    crenata Forssk., 319
    mauretanica Beck, 320
    minor Sm., 320
    nana (Reut.) Beck, 319
    ramosa L., 319   subsp. **nana** (Reut.) Cout., 319
Pandorea jasminoides (Lindl.) K. Schum., 317
Phelypaea lutea, Desf., 319
PLANTAGINACEAE, 320
Plantago coronopus, 320
Podranea ricasoliana (Tanfani) Sprague, 317
Pyrostegia venusta (Ker Gawl.), 317
Spathodea campanulata P. Beauv., 317
Tecoma capensis (Thunb.) Lindl., 317
    jasminoides Lindl., 317
    stans (L.) Juss. Ex Humb., 317
Tecomaria capensis (Thunb.) Spach, 317
Thunbergia gregorii S. Moore, 318
    alata Bojer ex Sims, 318

INDEX TO SCIENTIFIC NAMES

Abutilon Mill., 216
  grandifolium (Willd.) Sweet, 216
  indicum sensu auct. mad. non (L.) Sweet, 216
  megapotamicum (K. Spreng.) A. St-Hil. & Naudin, 217
  permolle (Willd.) Lowe, 216
  populifolium sensu auct. mad. non (Lam.) Sweet, 216
  sonneratianum (Cav.) Sweet, 216
  striatum J. Dicks. ex Lindl., 217
Acacia Mill., 154
  dealbata Link, 154
  farnesiana (L.) Willd., 154
  julibrissin (Durazz.) Willd., 155
  leucocephala (Lam.) Link, 155
  longifolia (Andrews) Willd., 154
  lophantha Willd., 155
  mearnsii De Wild., 154
  melanoxylon R. Br., 154
  retinodes Schldl., 154
  sophorae (Labill.) R. Br., 154
  verticillata (L Hér.) Willd., 154
Acalypha
  virginica L., 204
Acer L., 211
  campestre L., 211
  pseudoplatanus L., 211
ACERACEAE, 211
Achillea L., 352
  ageratum L., 352
  millefolium L., 352
Achyranthes L., 77
  aspera sensu auct. mad. non L., 77
  aspera var. *sicula* L., 77
  sicula (L.) All., 77
Acrostichum marantae L., 33
  paleaceum Hook. & Grev., 52
  septentrionale L., 46
  squamosum Sw. nom. illegit., 52
  subcordatum Cav., 33
  velleum Aiton, 34
Adenocarpus DC., 158
  complicatus (L.) Gay in Durieu, 158
    subsp. **complicatus**, 159
  divaricatus Sweet, 158
ADIANTACEAE, 32
Adiantum L., 35
  capillus-veneris L., 36

  cuneatum Langsd. & Fisch., 36
  cuneipinnulum N.C. Nair & S.R. Ghosh, 36
  hispidulum Sw., 36
  raddianum C. Presl, 36
  reniforme L., 35
    var. *pusillum* Bolle, 36
Adonis L., 96
  microcarpa DC., 96
Aeonium Webb & Berthel., 132
  arboreum (L.) Webb & Berthel., 133
  glandulosum (Aiton) Webb & Berthel., 133
  glandulosum × glutinosum, 133
  glutinosum (Aiton) Webb & Berthel., 133
Agapanthus L Hér., 387
  orientalis F.M. Leight., 387
  praecox Willd., 387
    subsp. **orientalis** (F.M. Leight.) F.M. Leight., 387
AGAVACEAE, 391
Agave L., 392
  americana L., 392
  attenuata Salm-Dyck, 392
Ageratina Spach, 337
  adenophora (Spreng.) R. King & H. Rob., 337
  riparia (Regel) R. King & H. Rob., 338
Ageratum L., 338
  conyzoides L., 338
    subsp. **conyzoides**, 338
  houstonianum Mill., 338
  mexicanum Sims, 338
Agrimonia L., 145
  eupatoria L., 145
    subsp. **eupatoria**, 145
    subsp. **grandis** (Andrz. ex Asch. & Graebn.) Bornm., 145
Agropyron junceiforme (Á. & D. Löve) Á. & D. Löve, 438
  repens (L.) P. Beauv., 439
Agrostis L., 432
  africana Poir., 444
  alba auct. non L., 433
  castellana Boiss. & Reut., 432
    var. *communis* Menezes, 433
    var. *mixta* Hack., 433

var. *mutica* Hack., 433
hispanica Boiss. & Reut., 432
obtusissima Hack., 433
olivetorum Godr., 432
pourrettii Willd., 433
reuteri auct. macar. non Boiss., 433
salmantica (Lag.) Kunth, 433
spicata Thunb. non Vahl, 444
stolonifera L., 433
Aichryson Webb & Berthel., 134
divaricatum (Aiton) Praeger, 134
var. *pubescens* Lowe, 134
dumosum (Lowe) Praeger, 134
villosum (Aiton) Webb & Berthel., 134
Aira L., 429
caryophyllea L., 429
subsp. *caryophyllea*, 430
subsp. *multiculmis* (Dumort.) Bonnier & Layens, 430
praecox L., 429
AIZOACEAE, 79
Aizoon L., 79
canariense L., 80
hispanicum L., 80
Ajuga L., 281
iva (L.) Schreb., 281
var. **pseudiva** (DC.) Benth., 281
Albizia distachya (Vent.) Macbr., 155
julibrissin Durazz., 155
lophantha (Willd.) Benth., 155
Alcea L., 219
rosea L., 219
Alchemilla arvensis (L.) Scop., 145
Alisma L., 383
lanceolatum With., 383
plantago-aquatica L., 383
lanceolatum With., 383
ALISMATACEAE, 382
Allamanda cathartica L., 258
Allantodia umbrosa R. Br., 48
Allium L., 387
ampeloprasum L., 387
cepa L., 387
fragrans Vent., 388
gracile Aiton, 388
neapolitanum Cirillo, 388
paniculatum L., 388
roseum L., 388
sativum L., 387
triquetrum L., 388
vineale L., 388
Aloe L., 386

barbadense Mill., 386
perfoliata Buch. var. *vera* L., 386
vera (L.) Burm.f., 386
Aloysia triphylla (L'Hér.) Britton, 277
Alstroemeria pulchella L.f., 392
Alternanthera Forssk., 77
achyrantha (L.) Sw., 77
caracasana Kunth, 77
Althea rosea (L.) Cav., 219
AMARANTHACEAE, 74
Amaranthus L., 74
blitum L., 76
caudatus L., 75
cruentus L., 75
deflexus L., 76
graecizans L., 76
subsp. *silvestris* (Vill.) Brenan, 76
hybridus L., 75
var. *acicularis* Thell., 75
hypochondriacus L., 75
lividus L., 76
subsp. *polygonoides* (Moq.) Probst., 76
muricatus Gillies ex Moq., 76
palmeri S. Watson, 76
paniculatus L., 75
var. *ambiguus* Menezes, 75
var. *purpurascens* Moq., 75
patulus Bertol., 75
powellii S. Watson., 75
retroflexus L., 76
spinosus L., 76
viridis L., 77
AMARYLLIDACEAE, 392
Amaryllis L., 393
belladonna L., 393
Ambrosia L., 350
artemisifolia L., 350
elatior L., 350
Ammi L., 244
majus L., 244
procerum Lowe, 244
visnaga (L.) Lam., 244
Amygdalus communis L., 141
persica L., 141
ANACARDIACEAE, 210
Anacyclus radiatus Loisel. subsp. **radiatus**, 352
Anagallis L., 252
arvensis L., 252
subsp. *foemina* (Mill.) Schinz & Thell., 252

caerulea L., 252
foemina Mill., 252
latifolia L., 252
phoenicea Scop., 252
Anchusa L., 274
 azurea Mill., 274
 italica Retz., 274
Andryala L., 381
 cheiranthifolia L Hér., 381
  var. *congesta* Lowe, 382
  var. *sparsiflora* Lowe, 382
  var. *sparsiflora* subvar. *coronopifolia* Lowe, 382
 crithmifolia Aiton, 382
 glandulosa Lam., 381
  subsp. **glandulosa**, 382
  subsp. **varia** (Lowe ex DC.) R. Fern., 382
   var. *varia* forma *coronopifolia* (Lowe) R. Fern., 382
 varia Lowe ex DC., 381
  subsp. *congesta* (Lowe) Menezes, 382
  subsp. *sparsiflora* forma *coronoonopifolia* (Lowe) Menezes, 382
  subsp. *sparsiflora* (Lowe) Menezes, 382
  var. *crithmifolia* DC., 382
Anethum graveolens L., 246
ANGIOSPERMAE, 56
Anogramma Link, 34
 leptophylla (L.) Link, 35
Anomatheca Ker Gawl., 398
 cruenta Lindl., 398
 laxa (Thunb.) Goldblatt, 398
Anredera Juss., 83
 cordifolia (Ten.) Steenis, 83
Anthemis L., 353
 aurea Webb ex Nyman, 353
 cotula L., 353
 mixta L., 352
 nobilis L., 353
  var. *aurea* (L.) Cout., 353
Anthericum comosum Thunb, 385
Antholyza aethiopica L., 397
Anthoxanthum L., 430
 odoratum L., 430
  var. *villosum* Loisel. ex
Anthriscus Pers., 240
 caucalis M. Bieb., 240
 sylvestris (L.) Hoffm., 240
  var. **caucalis**, 240
Anthyllis L., 189
 lemanniana Lowe, 189
 vulneraria L., 189
  subsp. *iberica* (W. Becker) Jalas, 189
Antirrhinum L., 308
 majus L., 308
 orontium L., 309
Aomphocarpus curassavica L., 260
 fruticosa L., 260
Aphanes L., 145
 arvensis L., 145
 arvensis sensu auct. mad. non L., 145
 microcarpa (Boiss. & Reut.) Rothm., 145
APIACEAE, 238
Apium L., 243
 graveolens L., 243
  var. **dulce** (Mill.) DC., 243
  var. **rapaceum** (Mill.) DC., 243
 leptophyllum (Pers.) F. Muell. ex Benth., 243
 nodiflorum (L.) Lag., 243
 tenuifolium (Moench) Thell., 243
APOCYNACEAE, 258
Apollonias Nees, 101
 barbujana (Cav.) Bornm., 101
  subsp. **barbujana**, 102
  subsp. **ceballosi** (Svent.) G. Kunkel, 102
 canariensis (Willd.) Nees, 101
Aptenia N.E. Br., 81
 cordifolia (L.f.) Schwantes, 81
AQUIFOLIACEAE, 212
Aquilegia L., 99
 vulgaris L., 99
Arabidopsis (DC.) Heynh., 112
 thaliana (L.) Heynh., 112
  var. *apetala* O.E.Schulz, 112
  var. **thaliana**, 112
Arabis L., 116
 albida Steven ex M. Bieb., 117
 alpina L., 117
  subsp. **caucasica** (Willd.) Briq., 117
 caucasica Willd. ex Schltdl., 117
ARACEAE, 453
Arachniodes Blume, 50
 webbianum (A. Braun) Schelpe, 50
  subsp. *foliosum* (C. Chr.) Gibby et al., 50
  subsp. **webbianum**, 50

ARALIACEAE, 237
Araujia Brot., 260
  sericofera Brot., 260
Arctium L., 363
  minus Bernh., 363
Arctotis L., 362
  stoechadifolia auct. mad., non Bergius, 363
  venusta Norl., 362
Ardisia bahamensis (Gaertn.) DC., 251
  excelsa Aiton, 251
Arenaria L., 84
  leptoclados (Rchb.) Guss., 85
  serpyllifolia L., 84
    subsp. **leptoclados** (Rchb.) Nyman, 85
Argemone L., 105
  mexicana L., 105
Argyranthemum Webb ex Sch. Bip., 354
  dissectum (Lowe) Lowe, 355
  haemotomma (Lowe) Lowe, 355
  pinnatifidum (L.f.) Lowe, 355
    subsp. **pinnatifidum**, 356
    subsp. **montanum** Rustan, 35656
    subsp. **succulentum** (Lowe) Humphries, 356
    var. *flaccida* Lowe, 356
    var. *succulentum* Lowe, 356
  thalassophilum (Svent.) Humphries, 355
Arisarum Mill., 455
  vulgare Targ.-Tozz., 455
Aristida L., 442
  adscensionis L., 442
  coerulescens Desf., 442
Aristolochia L., 61
  longa auct. non L., 61
  paucinervis Pomel, 61
ARISTOLOCHIACEAE, 61
Armeria Willd., 253
  maderensis Lowe, 253
Arrhenatherum P. Beauv., 425
  album (Vahl) Clayton, 426
  elatius (L.) P. Beauv., 425
    subsp. **bulbosum** (Willd.) Schübl. & M. Martens, 426
    subsp. *elatius* (L.) P. Beauv. ex Presl, 426
    subsp. *erianthum* (Boiss. & Reut.) Trab., 426
Artemisia L., 351
  arborescens L., 351

  argentea L Hér., 351
Arthrochortus loliaceus Lowe, 417
Arthrolobium ebracteatum (Brot.) Lowe, 190
Arum L., 454
  canariense Webb & Berthel., 455
  italicum Mill., 454
    subsp. **canariense** (Webb & Berthel.) P.C. Boyce, 455
    var. *canariense* (Webb & Berthel.) Engler, 455
Arundo L., 441
  australis Cav., 442
  dioeca Spreng. non Lour., 441
  donax L., 441
  selloana Schult., 441
Asarina erubescens (D. Don) Pennell, 303
  scandens (Cav.) Pennell, 303
ASCLEPIADACEAE, 259
Asclepias L., 260
Aspalthium bituminosum (L.) Fourr., 161
Asparagus L., 389
  acutifolius auct. non L., 389
  asparagoides (L.) Druce, 389
  densiflorus (Kunth) Jessop, 389
  lowei Kunth, 390
  medeoloides (L.f.) Thunb., 389
  nesiotes Svent., 389
    subsp. **nesiotes**, 389
    subsp. *purpuriensis* Marrero & Ramos, 389
  scaber Lowenon Brignoli, 390
  scoparius Lowe, 389
  setaceus (Kunth) Jessop, 389
  sprengeri Regel, 389
  umbellatus Link, 390
    subsp. *lowei* (Kunth) Valdés, 390
    subsp. *umbellatus*, 390
    var. *scaber* (Lowe) Baker, 390
Aspera muralis (L.) Lowe, 264
Asphodelus L., 385
  fistulosus L., 385
  madeirensis E. Simon, 385
Aspidium adultum Wikstr., 42
  drepanum Sw., 49
  falcinellum Sw., 49
  frondosum Lowe, 50
  webbianum A. Braun, 50
ASPLENIACEAE, 42
Asplenium L., 42
  acutum Willd., 43

adiantum-nigrum L., 43
　subsp. *onopteris* (L.) Heufl., 43
aethiopicum (Burm.f.) Bech., 44
　subsp. *aethiopicum*, 44
　subsp. **braithwaitii** Ormonde, 44
anceps Lowe ex Hook. & Grev., 44
aquilinum (L.) Bernh., 40
bifurcatum Opiz, 46
billotii F.W. Schultz, 43
canariense sensu auct. mad., non Willd., 44
filare subsp. *canariense* (Willd.) Ormonde, 44
furcatum sensu auct. mad., non Thunb., 44
joncheerei D.E. Mey., 44
lanceolatum Huds., 43
maderense Penny nom. nud., 44
marinum L., 45
monanthes L., 45
obovatum Viv., 43
　var. *billotii* (F.W. Schultz) Bech., 43
　subsp. **lanceolatum** P. Silva, 44
onopteris L., 43
petersenii Kunze, 48
praemorsum sensu auct. mad. non Sw., 44
productum Lowe, 43
rotundatum Kaulf. nom. nud., 43
scolopendrium L., 47
septentrionale (L.) Hoffm., 46
　subsp. **septentrionale**, 46
ticinense D.E. Mey., 43
trichomanes L., 45
　subsp. **maderense** Gibby & Lovis, 45
　subsp. **quadrivalens** D.E. Mey. emend. Lovis, 45
trichomanes sensu auct. mad. non L., 45
Aster L., 339
　squamatus (Spreng.) Hieron., 339
ASTERACEAE, 333
Asteriscus aquaticus (L.) Less., 347
Asterolinon stellatum (L.) Duby, 251
Astragalus L., 160
　boeticus L., 160
　hamosus L., 160
　hamosus sensu auct. mad. non L., 160
　pelecinus (L.) Barneby, 160
　solandri Lowe, 160
Astydamia DC., 246

　latifolia (L.f.) Baill., 246
Athyrium Roth, 47
　filix-femina (L.) Roth, 47
　　var. *subincisum* Menezes, 47
　umbrosum sensu C. Presl non Aiton, 48
Atriplex L., 72
　glauca L., 72
　　var. **ifniensis** (Caball.) Maire, 72
　halimus L., 73
　　var. **halimus**, 73
　hastata auct. mad., 73
　hastata sensu Aellen non L., 73
　prostrata Boucher ex DC., 73
　triangularis Willd., 73
Avena L., 426
　barbata Pott ex Link, 426
　brevis Roth, 426
　fatua L., 427
　lusitanica (Morais) Baum, 426
　marginata Lowe, 425
　occidentalis Durieu, 427
　sativa, 427
　sterilis L., 426
　strigosa Schreb., 426
　sulcata Gay ex Boiss., 425
Avenula marginata (Lowe) Holub, 425

Balantium culcita (L Hér.) Kaulf., 40
Ballota L., 286
　nigra L., 286
　　subsp. **foetida** (Vis.) Hayek, 286
　　var. *foetida* Vis., 286
　　subsp. **uncinata** (Fiori & Bég.) Patzak, 286
BALSAMINACEAE, 212
Barbarea R. Br., 115
　praecox (Sm.) R. Br., 116
　verna (Mill.) Asch., 116
Barkhausia comata Lowe, 380
　divaricata var. *pumila* Lowe, 381
　　var. *robusta* Lowe, 380
　dubia Lowe, 380
　hieracioides Lowe, 380
　laciniata Lowe, 380
Bartsia L., 315
　trixago L., 315
　versicolor (Willd.) Pers., 315
BASELLACEAE, 83
Bassia All., 73
　tomentosa (Lowe) Maire & Weiller, 73
Batatas edulis Choisy, 271
　edulis var. *cordifolia* Lowe, 271

var. *digitata* Lowe, 271
Bellardia trixago (L.) All., 315
Bellis L., 339
  perennis L., 339
Bencomia caudata (Aiton) Webb & Berthel., 146
  caudata sensu auct. mad., non Webb & Berthel., 146
  maderensis Bornm., 146
BERBERIDACEAE, 100
Berberis L., 100
  maderensis Lowe, 100
Beta L., 69
  patellaris Moq., 70
  patula Aiton, 70
  procumbens C. Sm. ex Hornem., 70
  vulgaris L., 69
    subsp. **cicla** (L.) W.D.J. Koch, 70
    var. *cicla* L., 70
    var. *crassa* Alef., 70
    var. *cruenta* Alef., 70
    var. *esculenta* Salisb., 70
    subsp. **maritima** (L.) Arcang., 69
    var. *portosanctana* Gand. ex Menezes, 70
    subsp. **vulgaris**, 70
Betonica officinalis L., 280
Bidens L., 348
  biternata (Lour.) Merr. & Sherff, 349
  pilosa L., 348
    var. **minor** (Blume) Sherff, 349
    var. **pilosa**, 349
    var. **radiata** Sch. Bip. in Webb & Berthel., 349
Biserrula pelecinus L., 160
  var. *glabra* Lowe, 160
  var. *pubescens* Lowe, 160
Bituminaria Fabr., 161
  bituminosa (L.) C.H. Stirt., 161
BLECHNACEAE, 53
Blechnum L., 53
  boreale Sw., 53
  radicans L., 53
  spicant (L.) Roth, 53
Bolboschoenus Asch. ex Palla, 457
  maritimus (L.) Palla, 457
BORAGINACEAE, 271
Borago L., 273
  officinalis L., 273
Bothriochloa Kuntze, 453
  ischaemum (L.) Keng, 453
Botrychium Sw., 32
  lunaria (L.) Sw., 32
Bougainvillea spectabilis Willd., 78
Boussingaultia baselloides auct. mad., non Kunth, 83
  cordifolia Ten., 83
Brachiaria (Trin.) Griseb., 447
  mutica (Forssk.) Stapf, 447
Brachypodium P. Beauv., 437
  distachyum (L.) P. Beauv., 438
  phoenicoides (L.) Roem. & Schult., 43
  sylvaticum (Huds.) P. Beauv., 438
Brassica L., 123
  campestris L., 123
  napus subsp. **rapifera** Metzger, 123
  nigra (L.) W.D.J. Koch, 123
  oleracea L., 123
  rapa L., 123
  sinapistrum subsp. *vulgaris* Cout., 125
BRASSICACEAE, 109
Briza L., 420
  maxima L., 420
  minor L., 420
Bromus L., 435
  catharticus Vahl, 436
  diandrus Roth, 437
  hordeaceus L., 436
    subsp. **hordeaceus**, 436
    subsp. **molliformis** (Lloyd) Maire & Weiller, 436
  lanceolatus Roth, 437
    var. *lanuginosus* (Poir.) Dinsm., 437
  madritensis L., 437
    var. *ciliatus* Guss., 437
  mollis L., 436
  rigidus Roth, 437
  rigens var. *gussonii* (Parl.) Cout., 437
    subsp. *maximus* (Desf.) Menezes, 43
  rubens L., 437
  sterilus L., 437
  unioloides Kunth, 436
  willdenowii Kunth, 436
Brugmansia Pers., 302
  candida Pers., 302
  sanguinea (Ruiz & Pav.) D. Don, 302
  suaveolens (Humb. & Bonpl. ex Willd.) Berchr. & J. Presl, 302
Brunsvigia rosea (Lam.) Hannibal, 393
Bryophyllum delagonensis (Eckl. & Zeyh.) Schinz, 131
  fedtschenkoi (Raym.-Hamet & Perrier) Lauz.-March., 132
  pinnatum (Lam.) Oken, 131

tubiflorum Harv., 131
Bunium L., 244
  brevifolioum Lowe, 244
  bulbocastanum L., 245
Bupleurum L., 242
  lancifolium Hornem, 242
  protractum Hoffmanns. & Link, 242
  salicifolium R. Br. ex Buch, 242
    subsp. *aciphyllum* (Webb ex Parl.) Sunding & G. Kunkel, 243
    subsp. **salicifolium** var. **salicifolium**, 242
      var. *robustum* (Burch.) Cauwet & Sunding, 242
  subovatum Link ex Spreng., 242
BUXACEAE, 214
Buxus sempervirens L., 214
  cv. **'Arborescens'**, 214
Bystropogon L Hér., 290
  ×*indiscretus* Menezes, 291
  maderensis Webb & Berthel., 290
    var. *genuinus* Menezes, 290
    var. *valdehirsutus* Menezes, 290
  piperitus Lowe, 290
  punctatus L Hér., 290
    var. *disjectus* Menezes, 29
    var. *pallidus* Menezes, 290
  ×**schmitzii** Menezes, 291

CACTACEAE, 229
Caesalpinia decapetala (Roth) Alston, 153
  sepiaria Roxb., 153
Cakile Mill., 126
  aegyptiaca Willd., 126
  maritima Scop., 126
    subsp. **maritima**, 126
Calamintha sylvatica subsp. *ascendens* (Jord.) P. W. Ball, 288
Calceolaria L., 311
  chelidonioides Kunth, 312
  pinnata L, 312
  tripartita Ruiz & Pav., 311
Calendula L., 360
  aegyptiaca Desf., 361
  arvensis L., 361
  maderensis DC., 361
  officinalis L., 361
  suffruticosa Vahl, 361
Callianassa sceptrum (L.f.) Webb, 312
CALLITRICHACEAE, 279
Callitriche L., 279
  stagnalis Scop., 279

Calluna Salisb., 249
  vulgaris (L.) Hull, 250
Calystegia R. Br., 267
  sepium (L.) R. Br., 268
    subsp. *americana* (Sims) Brummitt, 268
    subsp. *roseata* Brummitt, 268
    subsp. **sepium**, 268
  soldanella (L.) Roem. & Schult., 268
Camellia L., 220
  japonica L., 220
Campanula L., 329
  aurea L. f., 331
  erinus L., 330
  lusitanica Loefl., 330
CAMPANULACEAE, 329
Canna L., 466
  indica L., 466
CANNACEAE, 466
Capnophyllum Gaertn., 246
  peregrinum (L.) Lange, 246
CAPRIFOLIACEAE, 324
Capsella Medik., 119
  bursa-pastoris (L.) Medik., 119
  rubella Reut., 119
Capsicum frutescens L., 298
Cardamine L., 116
  hirsuta L., 116
Cardiospermum L., 211
  grandiflorum Sw., 212
    forma *hirsutum* (Willd.) Radlk., 212
Cardiospermum halicacabum L., 212
Carduncellus Adans., 368
  caeruleus (L.) C. Presl, 369
Carduus L., 363
  pycnocephalus L., 364
  squarrosus (DC.) Lowe, 364
  tenuiflorus Curtis, 363
Carex L., 462
  Subgen. CAREX, 464
  Subgen. PRIMOCAREX, 465
  Subgen. VIGNEA, 463
  demissa Hornem., 465
  divisa Huds., 463
    var. *moniziana* (Lowe) Menezes, 463
  divulsa Stokes, 463
    subsp. **divulsa**, 463
    subsp. **leersii** (Kneuck.) W. Koch, 463
  echinata auct. mad. non Murray, 463
  elata Lowe non All., 464
  extensa Good., 464

lowei Bech., 464
    forma *simplex* Bornm., 464
    malato-belizii Raymond, 464
    moniziana Lowe ex Boott, 463
    muricata L., 463
        subsp. **lamprocarpa** Čelak., 463
        var. *tenuior* Kük., 463
    oederi Retz sensu auct. mad., 465
    pendula Huds., 464
        var. γ Boott, 464
        var. *myosuroides* (Lowe) Boott, 464
    peregrina Link, 465
    pilulifera L., 465
        subsp. *azorica* (Gay) Franco & Rocha Afonso, 465
        var. *azorica* (Gay) H. Christ., 465
        subsp. **pilulifera**, 465
    punctata Gaudin, 464
        var. **laevicaulis** (Hochst. ex Kunze) Boott, 464
    spicata Huds., 463
    tumidicarpa subsp. *cedercreutzii* Fagerstr., 465
    viridula Michx., 465
        subsp. **cedercreutzii** (Fagerstr.) B. Schmid, 465
Carlina L., 362
    salicifolia (L.f.) Cav., 362
        var. **inermis** Lowe, 363
        var. **salicifolia**, 362
Carpobrotus N.E. Br., 81
    edulis (L.) N.E. Br., 81
Carthamus L., 368
    lanatus L., 368
    tinctorius L., 368
CARYOPHYLLACEAE, 83
Cassia bicapsularis L. Mill., 152
    bicapsularis sensu auct. mad. pro parte, non L., 153
    didymobotrya Fres., 153
    floribunda Cav., 153
    laevigata Willd., 153
    multijuga Rich., 153
    pendula Humb. & Bonpl. ex Willd., 153
    tomentosa L.f., 153
Catapodium Link, 423
    marinum (L.) C.E. Hubb., 423
    rigidum (L.) C.E. Hubb., 423
    tenellum (L.) Trab., 418
Catha dryandri Lowe, 214
Catharanthus roseus (L.) G. Don, 258

Cedronella Moench, 286
    canariensis (L.) Webb & Berthel., 286
    triphylla Moench, 286
CELASTRACEAE, 213
    Celsia cretica L., 305
Celtis australis L., 57
Cenchrus L., 451
    ciliaris L., 452
Centaurea L., 367
    calcitrapa L., 367
    diluta Aiton, 368
    massoniana Lowe, 366
    melitensis L., 367
        var. *conferta* Webb & Berthel., 367
        var. *vulgaris* Webb & Berthel., 367
    salmantica L., 366
    sonchifolia L., 367
    sphaerocephala L., 367
Centaurium Hill, 257
    maritimum (L.) Fritsch, 258
    tenuiflorum (Hoffmans. & Link) Fritsch, 258
        subsp. **acutiflorum** (Schott) Zeltner, 258
        subsp. **tenuiflorum**, 258
        subsp. *viridense* (Bolle) A. Hansen & Sunding, 258
Centranthus DC., 327
    calcitrapae (L.) Dufr., 327
        subsp. **calcitrapae**, 327
    ruber (L.) DC., 327
        subsp. **ruber**, 327
Cerastium L., 85
    diffusum Pers., 87
    fontanum Baumg., 86
        subsp. *triviale* (Spenn.) Jalas, 86
        subsp. **vulgare** (Hartm.) Greuter & Burdet, 86
    glomeratum Thuill., 86
    tetrandrum Curtis, 87
    triviale Link, 86
    vagans Lowe, 86
        [var.] β *calva* Lowe, 86
        var. *ciliatum* Tutin & E.F. Warb., 86
        [var.] α *fulva* Lowe, 86
        [var.] β *subnuda* Lowe, 86
        var. **vagans**, 86
Cerasus avium (L.) Moench, 141
    lusitanica (L.) Loisel., 141
    vulgaris Mill., 141
Ceratonia siliqua L., 153

Cercis siliquastrum L., 153
Cereus triangularis (L.) Mill., 229
Ceterach DC., 46
   aureum (Cav.) Buch, 47
     var. *aureum*, 47
     var. *madeirense* Ormonde nom. nud., 46
     var. *parvifolium* Benl & G. Kunkel, 47
   aureum sensu auct. mad. non, (Cav.) Buch, 46
   lolegnamense Gibby & Lovis, 46
   officinarum sensu auct. mad., non Willd., 46
Chaetonychia cymosa (L.) Sweet, 89
Chamaecytisus proliferus (L.f.) Link, 156
Chamaemeles Lindl., 149
   coriacea Lindl., 149
Chamaemelum Mill., 352
   mixtum All., 352
   nobile (L.) All., 353
     var. **discoideum** (Boiss. ex Willk.) A. Fern., 353
     var. **nobile**, 353
Chasmanthe N.E. Br., 397
   aethiopica N.E. Br., 397
Cheilanthes Sw., 32
   acrostica (Balb.) Tod., 33
   catanensis (Cosent.) H.P. Fuchs, 34
   corsica Reichst. & Vida, 33
   duriensis Mendonça & Vasc., 33
   fragrans (L.) Sw. subsp. *maderensis* (Lowe) Benl, 32
   guanchica Bolle, 33
   maderensis Lowe, 32
   marantae (L.) Domin subsp. *subcorda* (Cav.) Benl & Poelt var. *cupripaleacea*, 33
   pteridioides (Reichard) C. Chr., 32
   pulchella Bory ex Willd., 33
   sventenii Benl, 33
   tinaei Tod., 33
   vellea (Aiton) F. Mueller, 34
Cheiranthus arbuscula Lowe, 114
   mutabilis L Hér. non Brouss. ex Spreng., 114
   scoparius Brouss. ex Willd., 114
   scoparius sensu Menezes non Brouss. ex Willd., 114
   tenuifolius L Hér., 113
Cheirolophus Cass., 366
   massonianus Lowe, 366

Chelidonium L., 105
   majus L., 106
Chenolea lanata (Masson) Moq., 73
Chenoleoides tomentosa (Lowe) Botsch., 73
CHENOPODIACEAE, 68
Chenopodium L., 70
   album L., 71
   ambrosioides L., 71
     subsp. **ambrosioides**, 71
   coronopus Moq., 72
   giganteum D.Don, 71
   murale L., 72
     var. **spissidentatum** Murray, 72
   opulifolium Schrad. ex W.D.J. Koch &Ziz., 71
   vulvaria L., 72
Chloris Sw., 444
   gayana Kunth, 445
   virgata Sw., 445
Chlorophytum Ker Gawl., 385
   capense O. Kunze, 385
   comosum (Thunb.) Jacques, 385
   elatum R. Br., 385
Choisya ternata Humb. Bonpl. & Kunth, 210
Christella H. Lév., 42
   dentata (Forssk.) Brownsey & Jermy, 42
Chrysanthemum L., 353
   barretti Costa, 355
   carinatum Schousb., 354
   coronarium L, 354
   dissectum Lowe, 355
   haemotomma Lowe, 355
   lacustre Brot., 357
   leucanthemum L., 356
   mandonianum Coss., 356
   parthenium (L.) Bernh., 351
   pinnatifidum L.f., 355
   segetum L., 354
Cicer arietinum L., 162
Cichorium L., 369
   endivia L., 369
     subsp. **divaricatum** (Schousb.) P.D. Sell., 370
     subsp. *endivia*, 370
Ciclospermum leptophyllum (Pers.) Sprague, 243
Cirsium Mill., 364
   latifolium Lowe, 364
   syriacum (L.) Gaertn., 364

CISTACEAE, 226
Cistus L., 227
   ladanifer L., 227
   monspeliensis L., 227
   psilosepalus Sweet, 227
   salvifolius L., 227
Citrullus lanatus (Thunb.) Matsum. & Nakai, 229
Citrus L., 209
   aurantifolia (Christm.) Swingle, 209
   aurantium L., 209
   limetta Risso, 209
   limon (L.) Burm.f., 209
   medica L., 209
   reticulata Blanco, 210
   sinensis (L.) Osbeck, 210
Clerodendron fallax Lindl., 278
   speciosissimum Van Geert ex Morr., 278
Clethra L., 248
   arborea Aiton, 248
CLETHRACEAE, 248
Clinopodium L., 287
   ascendens (Jord.) Samp., 288
   vulgare L., 288
      subsp. *arundanum* (Boiss.) Nyman, 288
      subsp. *villosa* (De Noé) Bothmer, 288
Coffea arabica L., 261
Coix lacryma-jobi L., 407
Coleostephus Cass., 357
   myconis (L.) Rchb.f, 357
Coleus blumei Benth., 280
Colocasia Schott, 454
   antiquorum Schott, 454
   esculenta (L.) Schott, 454
Commelina L., 405
   agraria Kunth, 405
   benghalensis L., 405
   diffusa Burm.f., 405
COMMELINACEAE, 405
COMPOSITAE, 333
   Subfam. ASTEROIDEAE, 337
   Subfam. CICHORIOIDEAE, 369
   Tribe ANTHEMIDEAE, 351
   Tribe ARCTOTIDEAE, 362
   Tribe ASTEREAE, 339
   Tribe CALENDULAE, 3660
   Tribe CYNARAE, 362
   Tribe EUPATORIEAE, 337
   Tribe HELENIEAE, 351
   Tribe HELIANTHEAE, 347
   Tribe INULEAE, 341
   Tribe SENECIONEAE, 358
Conium L., 242
Consolida (DC.) Gray, 100
   ajacis (L.) Schur, 100
   ambigua sensu P.W. Ball & Heywood, 100
CONVOLVULACEAE, 266
Convolvulus L., 268
   althaeoides L., 269
      var. *virescens* Lowe, 269
   arvensis L., 269
      var. *arvensis*, 269
      var. *linearifolius* Choisy, 269
   canariensis, 269
      var. *massonii* (F. Dietr.) Sa'ad, 269
   massonii F. Dietr., 269
      var. *uniflorus* Menezes, 269
   siculus L., 270
      subsp. **siculus**, 270
   tricolor L., 269
Conyza Less., 340
   ambigua DC., 340
   bonariensis (L.) Cronquist, 340
   canadensis (L.) Cronquist, 340
   crispa Pourr., 340
   floribunda Humb. Bonpl. & Kunth, 340
   sumatrensis (Retz.) E. Walker, 340
Coriandrum L., 241
   sativum L., 241
Coronilla L., 190
   glauca L., 190
   valentina L., 190
      subsp. *glauca* (L.) Batt., 190
Coronopus Zinn, 122
   didymus (L.) Sm., 122
   procumbens Gilib., 122
   squamatus (Forssk.) Asch., 122
Corrigiola L., 88
   litoralis L., 88
Cortaderia Stapf, 441
   selloana (Schult.) Asch. & P. Graebn., 441
Cosentinia Tod., 34
   vellea (Aiton) Tod., 34
Cosmos bipinnatus Cav., 349
Cotula L., 357
   australis (Sieber ex Spreng.) Hook.f., 358
   coronopifolia L., 358
   leptalea DC., 358

Cotyledon umbilicus sensu Menezes non L., 132
Crambe L., 127
  fruticosa L.f., 127
    var. **brevifolia** Lowe, 127
    var. **pinnatifida** Lowe, 127
Crassula L., 130
  multicava Lem., 131
  tetragona L., 131
  tillaea Lest.-Garl., 130
CRASSULACEAE, 130
Crataegus L., 149
  monogyna Jacq., 149
    subsp. **brevispina** (Kunze) Franco, 149
Crepis L., 379
  andryaloides Lowe, 380
  capillaris (L.) Wallr., 381
    var. **capillaris**, 381
  comata Banks & Sol. ex Lowe, 380
  crinita Sol. ex Lowe, 370
  divaricata (Lowe) F.W. Schultz, 380
    var. *pumila* (Lowe) Lowe, 381
    var. *robusta* (Lowe) Lowe, 380
  dubia (Lowe) F.W. Schultz, 380
  hieracioides (Lowe) F.W. Schultz, 380
    var. *laevigata* Lowe, 380
    var. *nigricans* Lowe, 380
  laciniata (Lowe) F.W. Schultz, 380
    var. *integrifolia* Lowe, 380
    var. *pinnatifida* Lowe, 380
  noronhaea Babc., 381
  pectinata Lowe, 370
  tenuifolia Banks & Sol., 370
  vesicaria L., 380
    subsp. **andryaloides** (Lowe) Babc., 380
    subsp. **haenseleri** (Boiss. ex DC.) P.D. Sell, 380
Crinum bulbispermum (Burm.) Milne-Redh. & Schweick., 392
Crithmum L., 245
  maritimum L., 245
Crocosmia Planch., 398
  × **crocosmiflora** (G. Nicholson) N.E. Br., 398
CRUCIFERAE, 109
CUCURBITACEAE, 229
Culcita C. Presl, 40
  macrocarpa C. Presl, 40
Cullen Medik., 161
  americanum (L.) Rydb., 161

CUPRESSACEAE, 54
Cupressus L., 54
  lusitanica, 55
  macrocarpa Hartw., 55
Cuscuta L., 266
  approximata Bab., 267
    subsp. *approximata*, 267
    subsp. **episonchum** (Webb & Berthel.) Feinbrun, 267
  calycina Webb & Berthel., 267
  epithymum (L.) L., 266
    subsp. **epithymum**, 267
    subsp. *kotschyi* (Des Moul.) Arcang., 267
  planiflora Ten., 267
Cyclosorus dentatus (Forssk.) Ching, 42
Cydonia oblonga Mill., 148
  subsp. *maliformis* (Mill.) Thell., 148
  vulgaris Pers., 148
    var. *oblonga* (Mill.) DC., 148
Cymbalaria Hill, 310
  muralis P. Gaertn. B. Mey. & Scherb., 310
    subsp. **muralis**, 310
Cynara L., 365
  cardunculus L., 365
    var. **ferocissima** Lowe, 365
Cynara horrida Aiton, 365
Cynodon Rich., 445
  dactylon (L.) Pers., 445
Cynoglossum L., 273
  creticum Mill., 273
Cynosurus L., 419
  cristatus L., 419
  echinatus L., 419
  effusus Link in Schrad., 419
  elegans auct. non Desf., 419
    var. *brizoides* (Lowe) Menezes, 419
CYPERACEAE, 456
Cyperus L., 459
  alternifolius sensu auct. mad. non L., 46
  brevifolius (Rottb.) Hassk., 462
  difformis L., 461
  eragrostis Lam., 460
  esculentus L., 460
  flavescens L., 461
  fuscus L., 460
  involucratus Rottb., 460
  laevigatus L., 461
  longus L., 459
    subsp. **badius** (Desf.) Murb., 459

subsp. *genuinus* Cout., 459
rotundus L., 460
vegetus Willd., 460
Cyrtomium C. Presl, 50
    falcatum (L.f.) C. Presl, 50
Cystopteris Bernh., 48
    diaphana (Bory) Blasdell, 48
    diaphana × *fragilis*, 48
    fragilis (L.) Bernh., 48
        subsp. *diaphana* (Bory) Litard., 48
    fragilis sensu auct. mad. non (L.) Bernh. in Schrad., 48
    viridula (Desv.) Desv., 48
Cytisus L., 155
    balansae subsp. *europaeus* (G. López & Jarvis) Muñoz Garmendia, 156
    lusitanicus Quer. ex Willk., 156
    multiflorus (L'Hér.) Sweet, 156
    pendulinus L. f., 155
    proliferus L.f., 156
    purgans auct., 156
    scoparius (L.) Link, 156
        subsp. **scoparius**, 156
    striatus (Hill) Rothm., 155

Dactylis L., 422
    glomerata L., 422
        subsp. *hispanica* (Roth) Nyman, 422
        subsp. *smithii* (Link) Stebbins & Zohary, 422
    smithii Link, 422
        subsp. *hylodes* Parker, 422
        subsp. *marina* (Borrill) Parker, 422
Dactyloctenium Willd., 444
    aegyptium (L.) Willd., 444
    australe Steud., 444
Dactylorhiza Neck. ex Nevski, 468
    foliosa (Verm.) Soó, 468
Danthonia DC., 440
    decumbens (L.) DC., 440
Datura L., 301
    candida (Pers.) Saff., 302
    innoxia Mill., 301
    metel L., 302
    sanguinea Ruiz & Pav., 302
    stramonium L., 301
        var. **tatula** (L.) Torr., 301
    suaveolens Humb. & Bonpl. ex Willd., 302
Daucus L., 247
    carota L., 247
        subsp. **carota**, 247
        subsp. **hispidus** (Arcang.) Heywood, 247
    neglectus Lowe, 247
Davallia Sm., 52
    canariensis (L.) Sm., 52
DAVALLIACEAE, 52
Delphinium L., 99
    ajacis L., 100
    ambiguum auct. non L., 100
    maderense C. Blanché, 99
    peregrinum auct. mad. non L., 99
DENNSTAEDTIACEAE, 40
Deparia petersenii (Kunze) M. Kato, 48
Deschampsia P. Beauv., 428
    argentea (Lowe) Lowe, 428
        var. *gomesiana* Menezes, 428
        var. *prorepens* Hack. & Bornm., 428
    foliosa auct. mad. non Hack., 428
    maderensis (Hack. & Bornm.) Buschm., 428
Dianthus prolifer, 95
Diascia barberae Hook.f., 303
Dichanthium Willemet, 452
    annulatum (Forssk.) Stapf, 453
Dichondra J.R. & G. Forst., 267
    micrantha Urb., 267
    repens sensu auct. mad. non J.R. & G. Forst., 267
Dicksonia L Hér., 40
    antarctica Labill., 40
    culcita L Hér., 40
DICKSONIACEAE, 40
Dietes iridioides (L.) Sweet, 394
    vegeta (L.) N.E. Br., 394
Digitalis L., 312
    purpurea L., 312
Digitaria Haller, 450
    ciliaris (Retz.) Koeler, 450
    sanguinalis (L.) Scop., 450
DIOSCOREACEAE, 393
Diosma ericoides L., 210
Diphasiastrum Holub, 29
    madeirense (J.H. Wilce) Holub, 29
    tristachyum (Pursh) Holub, 29
Diplazium Sw., 48
    allorgei Tardieu, 48
    caudatum (Cav.) Jermy, 48
Diplotaxis DC., 122
    catholica (L.) DC., 122
    siifolia Kunze, 123
Dipogon lignosus (L.) Verdc., 155
DIPSACACEAE, 328

Dipsacus L., 328
   ferox Loisel., 328
Dittrichia Greuter, 346
   viscosa (L.) Greuter, 346
      subsp.**viscosa**, 347
Dolichos lablab L., 161
   lignosus L., 155
Doodia R. Br., 53
   caudata (Cav.) R. Br., 53
Dorotheanthus gramineus (Haw.) Schwantes, 81
Draba L., 117
   muralis L., 117
Dracaena Vand. ex L., 391
   draco (L.) L., 391
Dracunculus Mill., 455
   canariensis Kunth, 455
   vulgaris sensu auct. mad., non Schott, 455
Drosanthemum floribundum (Haw.) Schwantes, 81
Drusa DC., 239
   glandulosa (Poir.) H. Wolff ex Engl., 239
   oppositifolia DC., 239
DRYOPTERIDACEAE, 48
Dryopteris Adans., 50
   aemula (Aiton) Kuntze, 51
   affinis (Lowe) Fraser-Jenk., 51
      subsp. *affinis*, 51
   africana (Desv.) C. Chr., 41
   aitoniana Pic. Serm., 51
   austriaca sensu auct. mad., non (Jacq.) Woyn., 52
   borreri sensu auct. mad., non (Newman) Newman ex Oberholzer & Tavel, 51
   dilatata sensu auct. mad., non (Hoffm.) A. Gray, 52
   elongata (Aiton) Sim, non (Wall. ex Hook.) Kuntze, 51
   furadensis Bennert et al., 52
   intermedia (Muhl. ex Willd.) Gray subsp. *maderensis* (Alston) Fraser-Jenk., 52
   macaronesica Romariz, 51
   maderensis Alston, 52
   pseudomas sensu auct. mad., non Holub & Pouzar, 51
   spinulosa sensu auct. mad., non (O.F. Müll.) Kuntze, 52
Duchesnea Sm., 144

   indica (Andrews) Focke, 144
Duranta erecta L., 278
   plumieri Jacq., 278

Ebingeria elegans (Lowe) Chrtek & Křísa, 400
Echinochloa P. Beauv., 446
   colona (L.) Link, 447
   crus-galli (L.) P. Beauv., 447
Echium L., 276
   candicans L.f., 277
      var. *noronhae* Menezes, 277
   fastuosum sensu auct. mad. non Salisb., Dryand. nec Jacq., 277
   lycopsis auct. non L., 276
   nervosum Dryand., 277
      var. *laxiflorum* Menezes, 277
   plantagineum L., 276
   simplex DC., 277
Eclipta L., 347
   alba (L.) Hassk., 348
   erecta L., 348
   prostrata (L.) L., 348
Elaphoglossum Schott ex Sm., 52
   hirtum sensu auct., non (Sw.) C. Ch., 52
   paleaceum (Hook. & Grev.) Sledge, 52
   semicylindricum (Bowdich) Benl, 52
   squamosum (Sw.) J. Sm. comb. illegit., 52
Elatinoides elatine (L.) Wettst., 310
   lanigera (Desf.) Cout., 311
   spuria (L.) Wettst., 311
Eleocharis R. Br., 458
   palustris (L.) Roem. & Schult., 459
Eleusine Gaertn., 443
   indica (L.) Gaertn., 443
   tristachya (Lam.) Lam., 443
Eleutheranthera ruderalis (Sw.) Sch. Bip., 348
Elymus L., 438
   farctus (Viv.) Runemark ex Melderis, 438
Elytrigia junceiformis . & D. Löve, 438
   repens (L.) Gould, 439
   repens (L.) Nevski, 439
Emex Campd., 65
   spinosa (L.) Campd., 65
Emilia coccinea (Sims) G.Don, 360
   javanica (Burm.f.) C.B. Rob., 360
   sagittata (Vahl) DC., 360

Ephedra L., 56
  fragilis Desf., 56
    var. *dissoluta* Stapf, 56
EPHEDRACEAE, 56
Epilobium L., 236
  lanceolatum sensu auct. mad. non Sebast. & Mauri, 237
    var. *maderense* (Hausskn.) Leveil., 237
  maderense Hausskn., 237
  obscurum Schreb., 237
  parviflorum Schreb., 236
    var. *menezesi* Leveil., 236
    var. *subglabrum* Koch, 236
  tetragonum L., 236
    subsp. **lamyi** (F.W. Schultz) Nyman, 237
    subsp. **tetragonum**, 237
Epimedium pinnatum Fisch., 100
EQUISETACEAE, 30
Equisetum L., 30
  maximum sensu auct. non Lam., 30
  ramosissimum Desf., 31
  telmateia Ehrh., 30
Eragrostis Wolf, 442
  barrelieri Daveau, 443
  cilianensis (All.) Vign. ex Janch., 442
  curvula (Schrad.) Nees, 442
  minor Host, 443
Erica L., 248
  arborea L., 249
  cinerea auct. mad. non. L., 249
    var. *maderensis* Benth., 249
  maderensis (Benth.) Bornm., 249
  scoparia L., 249
    subsp. **maderinicola** D.C. McClint., 249
    subsp. *platycodon* (Webb & Berthel.) A. Hansen & Kunkel, 249
  vagans L., 249
ERICACEAE, 248
Erigeron L., 339
  canadensis L., 340
  karvinskianus DC., 339
  mucronatus DC., 339
Eriobotrya japonica (Thunb.) Lindl., 148
Eriocephalus africanus L., 357
Erodium L Hér., 198
  botrys (Cav.) Bertol., 199
  chium (L.) Willd., 199
    subsp. *chium*, 199
  cicutarium (L.) L Hér., 199
  malacoides (L.) L Hér., 199
  moschatum (L.) L Hér., 200
Eruca Mill., 125
  sativa Mill., 125
  vesicaria (L.) Cav., 125
    subsp. **sativa** (Mill.) Thell., 126
Erucastrum incanum (L.) W.D.J. Koch, 126
Ervum capreolatum (Lowe) Lowe, 165
  ervilia L., 167
  gracile (Loisel.) DC., 166
  hirsutum L., 165
  lens L., 167
  monanthos L., 167
  parviflorum (Loisel.) Bertol., 166
  pubescens DC., 166
    var. *glabrescens* Lowe, 166
    var. *subpilosa* Lowe, 166
  tenuissimum M. Bieb., 166
  tetraspermum L., 166
Erysimum L., 113
  arbuscula (Lowe) Snogerup, 114
  bicolor (Hornem.) DC., 114
  bicolor sensu auct. mad. non (Hornem.) DC., 114
  maderense Polatschek, 113
  maderense sensu A. Hansen & Sunding, non Polatschek, 114
  scoparium (Brouss. ex Willd.) Wettst., 114
  scoparium sensu auct. mad., non (Brouss. ex Willd.) Wettst., 114
  scoparium sensu A. Hansen, non (Brouss. ex Willd.) Wettst., 114
Erythraea maritima (L.) Pers., 258
  pulchella (Sweet) Fries, 258
  ramosissima (Vill.) Pers., 258
  tenuiflora Hoffmans. & Link, 258
Erythrina crista-galli L., 155
Eschscholzia Cham., 106
  californica Cham., 106
Eucalyptus L'Hér., 233
  citriodora Hook., 233
  ficifolia F. Muell., 233
  globulus Labill., 233
  robusta Sm., 233
Eugenia brasiliana (L.) Aubl., 232
  jambos L., 232
  malaccensis L., 232
  uniflora L., 232
Eupatorium adenophorum Spreng., 337
  riparium Regel, 338

Euphorbia L., 205
 anachoreta Svent., 207
 desfoliata (Menezes) Monod, 207
 exigua L., 208
 helioscopia L., 208
 heterophylla Desf., 209
 juncea Aiton, 209
 lathyris L., 208
 mellifera Aiton, 207
 nutans Lag., 206
 obtusifolia var. *desfoliata* Menezes, 207
 paralias L., 209
 peplis L., 206
 peplus L., 208
  var. *genuina* Cout., 208
  var. *peploides* (Gouan) Vis., 208
 piscatoria Aiton, 207
 platyphyllos L., 207
 preslii Guss., 206
 prostrata Aiton, 206
 pterococca Brot., 208
 pulcherrima Willd. ex Klotzsch, 205
 segetalis L., 209
  var. **pinea** (L.) Lange, 209
  var. **segetalis**, 209
 terracina L., 209
EUPHORBIACEAE, 203
Exobasidium laurii Geyl, 103

Faba vulgaris Moench, 164
Fabaceae, 149
Fagonia L., 201
 cretica L., 201
Fallopia Adans., 64
 baldshuanica (Regel) Holub, 64
 convolvulus (L.) Á. Löve, 64
Ferraria crispa Burm., 394
 undulata L., 394
Festuca L., 414
 agustinii subsp. *mandonii* (St.-Yves) A. Hansen, 415
 albida Lowe, 422
 arundinacea Schreb., 414
 barbata L., 441
 brevipila Tracey, 416
 decumbens L., 440
 donax Lowe, 415
 duriuscula L., 415
 elatior L., 414
 indigesta Boiss., 415
 jubata Lowe, 415
 lemanii Bastard, 415
 longifolia Thuill., 416
 ovina L., 415
  subsp. *duriuscula* (L.) Kozlovski, 415
 pratensis, 416
 rubra L., 415
 unioloides Willd., 436
Festulolium loliaceum (Huds.) P. Fourn., 416
Ficus L., 58
 carica L., 58
 elastica Roxb., 58
Filago L., 341
 gallica L., 341
 lutescens Jord., 342
  subsp. **atlantica** Wagenitz, 342
 micropodioides auct. mad., 342
 minima (Sm.) Pers., 341
 montana auct. mad., 341
 pyramidata L., 342
  var. **pyramidata**, 342
Foeniculum Mill., 245
 officinale All., 246
 vulgare Mill., 246
  var. **azoricum** (Mill.) Thell., 246
Fragaria L., 144
 indica Andrews, 144
 vesca L., 144
Frangula Mill., 215
 azorica Grubov, 215
 azorica Tutin nom. superfl., 215
Frankenia L., 228
 hirsuta var. *intermedia* auct. mad., non DC. Boiss., 228
 laevis L., 228
 [var.] α *hebecaulon* Lowe, 228
 pulverulenta L., 228
FRANKENIACEAE, 228
Fraxinus excelsior L., 256
Freesia Eckl. ex Klatt, 396
 refracta (Jacq.) Eckl. ex Klatt, 396
Fuchsia L., 234
 arborescens Sims, 234
 boliviana Carr., 234
 coccinea sensu auct. mad. non Aiton, 234
 magellanica Lam., 234
Fumaria L., 106
 bastardii Boreau, 108
 capreolata L., 107
 laeta Lowe, 108
 montana J.A. Schmidt, 108

muralis Sonder ex W.D.J. Koch, 107
   var. **boraei** (Jord.) Pugsley, 10
   var. **laeta** Lowe, 108
   var. **lowei**Pugsley, 108
   subsp. **muralis**, 108
   var. *vulgaris* Lowe, 108
parviflora Lam., 108
praetermissa Pugsley, 108
sepium Boiss. Reut., 107
   subsp. **sepium**, 107

Gaillardia pulchella Foug., 351
Galactites Moench., 365
   tomentosa Moench, 365
      var. **crinita** Lowe, 365
Galinsoga Ruiz & Pav., 349
   ciliata (Raffin.) S.F. Blake, 350
   parviflora Cav., 349
   quadriradiata Ruiz & Pav., 350
Galium L., 262
   aparine L., 263
   ellipticum Willd. ex Hornem., 263
      var. *lucidum* Lowe, 263
      var. *villosum* (Webb & Berthel.) Lowe, 263
   geminiflorum Lowe, 264
   minutulum Jord., 264
   murale (L.) All., 264
   parisiense L., 264
      var. *leiocarpum* Tausch, 264
   productum Lowe, 263
      var. *latifolium* Menezes, 263
   saccharatum All., 263
   scabrum L., 263
   tricorne Stokes pro parte, 263
   tricornutum Dandy, 263
   valantia Weber, 263
   verrucosum Huds., 263
Gamochaeta Wedd., 342
   calviceps (Fern.) Cabrera, 342
   filaginea (DC.) Cabrera, 343
   pensylvanica (Willd.) Cabrera, 343
Gastridium P. Beauv., 434
   lendigerum (L.) Desv., 434
   phleoides (Nees & Meyen) C.E. Hubb., 43
   ventricosum (Gouan) Schinz & Thell., 43
Gaudinia P. Beauv., 427
   fragilis (L.) P. Beauv., 427
Genista L., 157
   maderensis (Webb & Berthel.) Lowe, 156
   paivae Lowe, 156
   tenera (Jacq. ex Murray) Kuntze, 157
   virgata (Aiton) DC. non Lam., 157
GENTIANCEAE, 257 GERANIACEAE, 193
Geranium L., 194
   anemonifolium L Hér., 196
   dissectum L., 195
   lucidum L., 195
   maderense Yeo, 197
   molle L., 195
   palmatum Cav., 196
   purpureum Vill., 195
   robertianum L., 196
      var. *purpureum* (Vill.) DC., 195
   rotundifolium L., 195
   rubescens Yeo, 196
Geropogon glaber L., 374
Gladiolus L., 396
   hortulanum Bailey, 397
   italicus Mill., 397
   segetum Ker Gawl., 397
Glaucium Mill., 105
   corniculatum (L.) Rudolph, 105
Glechoma hederacea L., 280
Gleditsia triacanthos L., 153
   var. **armata** Lowe, 153
   var. *inermis* DC., 153
Glyceria R. Br., 424
   declinata Bréb., 424
   fluitans var. *pumila* Fr. ex Andersson, non Wimm & Grab., 424
      var. *spicata* (Biv. ex Guss.) Trabut, 424
   loliacea (Huds.) Godr., 416
   max (L.) Merr., 155
   spicata Biv. ex Guss., 424
Gamochaeta Wedd., 342
   calviceps Fern., 342
   luteo-album L., 343
   pensylvanicum Willd., 343
   purpureum auct. mad. non (Willd.) Cabrera, 343
   spathulatum Lam., 343
Gnidia L., 220
   carinata Thunb., 220
   polystachya Bergius, 220
Gomphocarpus R. Br., 260
Gomphrena celosioides Mart., 74
Goodyera R. Br., 467
   diphylla (Link) Parl., 468

macrophylla Lowe, 467
GRAMINEAE, 406
Grammitis aurea (Cav.) Sw., 46
 leptophylla (L.) Sw., 35
Guizotia abyssinica (L.f.) Cass., 347
GUTTIFERAE, 221
Gymnogramma ceterach sensu auct. mad. non (L.) Spreng., 46
 leptophylla (L.) Desv., 35
 lowei Hook. & Grev., 41
GYMNOSPERMAE, 54
Gymnostyles stolonifera (Brot.) Tutin, 357
Gynerium argenteum Nees, 441
Gypsophila elegans M. Bieb, 94

Hainardia Greuter, 424
 cylindrica (Willd.) Greuter, 424
Hakea sericea Schrad., 61
Heberdenia Banks ex A.DC., 250
 excelsa (Aiton) Banks ex DC., 251
Hedera L., 237
 canariensis Willd., 238
 helix L., 237
  subsp. **canariensis** (Willd.) Cout., 238
  maderensis K. Koch ex A. Rutherf., 238
  subsp. *maderensis*, 238
Hedychium J. Köeni, 466
 gardneranum Sheppard ex Ker Gawl., 466
Hedypnois Mill., 371
 cretica (L.) Dum. Cours., 371
 var. *rhagadioloides* (L.) Cout., 371
 rhagadioloides (L.) F.W. Schmidt, 371
Helianthus L., 348
 annuus L., 348
Helichrysum Mill., 344
 devium J.Y. Johnson, 344
 foetidum (L.) Cass., 345
 var. **citreum** Less., 345
 var. *foetidum*, 345
 melaleucum Rchb. ex Holl, 344
 melanophthalmum (Lowe) Lowe, 344
 var. *rosea* Lowe, 344
 monizii Lowe, 345
 obconicum DC., 345
 petiolare Hilliard and B.L. Burtt, 345
 petiolatum auct. mad. non (L.) DC., 345
Helictotrichon Schult., 425
 marginatum (Lowe) Röser, 425

sulcatum (Gay ex Boiss.) Henrard, 425
Heliotropium L., 272
 amplexicaule Vahl, 273
 arborescens L., 273
 erosum Lehm., 272
 europaeum L., 272
 peruvianum L., 273
 ramosissimum (Lehm.) DC., 272
Helminthotheca Vaill. ex Zinn, 373
 echioides (L.) Holub, 373
Helmintia echioides (L.) Gaertn., 373
Helosciadium leptophyllum DC., 243
 nodiflorum (L.) W.D.J. Koch, 243
Hemerocallis lilioasphodelus L., 386
Hemionitis pozoi Lag., 41
Herniaria L., 89
 hirsuta L., 89
 subsp. **cinerea** (DC.) Cout., 90
Hibiscus L., 216
 rosa-sinensis L., 216
 syriacus L., 216
Hippocrepis L., 191
 multisiliquosa L., 191
Hirschfeldia Moench, 126
 incana (L.) Lagr.-Foss., 126
Holcus L., 429
 lanatus L., 429
 mollis L., 429
Hordeum L., 439
 distichon L., 439
 glaucum Steud., 439
 hexastichon L., 439
 leporinum Link, 440
 marinum Huds., 440
 subsp. **gussoneanum** (Parl.) Thell., 440
 subsp. **marinum**, 440
 murinum auct. mad. pro parte, 439-440
 murinum L., 439
 subsp. **glaucum** (Steud.) Tzvelev, 439
 subsp. **leporinum** (Link) Arcang., 440
 secalinum Schreb., 439
 vulgare L., 439
Huperzia Bernh., 29
 dentata (Herter) Holub, 29
 selago sensu auct. mad., non (L.) Bernh. ex Schrank & Mart., 29
 subsp. **dentata** (Herter) Valentine, 29
 subsp. **suberecta** (Lowe) Franco &

Vasc. comb. illegit., 29
suberecta (Lowe) Tardieu, 29
Hydrangea L., 138
  macrophylla (Thunb.) Ser., 138
HYDRANGEACEAE, 138
HYDROPHYLLACEAE, 271
Hylocereus triangularis (L.) Britton & Rose, 229
HYMENOPHYLLACEAE, 37
Hymenophyllum Sm., 37
  maderense Gibby & Lovis, 38
  peltatum sensu auct. non Desv., 38
  tunbrigense (L.) Sm., 38
  unilaterale sensu auct. non Bory, 38
  wilsonii Hook., 38
Hyoscyamus L., 296
  albus L., 297
  major Mill., 297
Hyparrhenia E. Fourn., 453
  hirta (L.) Stapf, 453
HYPERICACEAE, 221
Hypericum L., 221
  acutum Moench, 222
  canariense L., 223
  floribundum Aiton, 223
  glandulosum Aiton, 223
  grandifolium Choisy, 221
  hircinum L. × *androsaemum* L., 222
  humifusum L., 222
  inodorum Mill., 222
  linarifolium Vahl, 222
  perfoliatum L., 222
  perforatum L., 222
  undulatum Schousboe ex Willd., 222
Hypochoeris L., 372
  glabra L., 372
    var. *glabra*, 372
    var. *loiseleuriana* Godr., 372
  radicata L., 372
    var. *rostrata* Moris, 372

Ifloga Cass., 343
  spicata (Forssk.) Sch. Bip. in Webb & Berthel., 343
    susbp. **spicata**, 343
Ilex L., 212
  aestivalis Buch., 213
  aquifolium L., 213
  azevinho Sol. ex Lowe, 213
  canariensis Poir., 213
    subsp. *azevinho* (Sol. ex Lowe) Kunkel, 213
    maderensis Lam., 213
  perado Aiton, 213
    subsp. **perado**, 213
    var. *maderensis* (Lam.) Loes. subvar. *spinulosa-serrata* Loes, 213
Illecebrum L., 90
  verticillatum L., 90
Impatiens L., 212
  balsamina L., 212
Imperata Cirillo, 452
  cylindrica (L.) Reusch., 452
Imperatoria lowei Coss., 247
  ostruthium Lowe, 247
Inula viscosa (L.) Aiton, 346
Ipomoea L., 270
  acuminata (Vahl) Roem. & Schult., 270
  batatas (L.) Poir., 271
  coccinea L., 270
  indica (Burm.) Merr., 270
  ochracea (Lindl.) G. Don, 271
  purpurea (L.) Roth, 271
  quamoclit L., 270
Iresine herbstii Hook., 74
IRIDACEAE, 394
Iris L., 395
  foetidissima L., 395
  japonica Thunb., 395
  pseudacorus L., 396
  xiphium L, 395
Isatis L., 113
  praecox sensu auct. mad. non Kit. ex Tratt., 113
  tinctoria L., 113
Ismelia carinata Sch. Bip, 354
Isolepis R. Br., 458
  cernua (Vahl) Roem. & Schult., 458
  setacea (L.) R. Br., 458
Isoplexis (Lindl.) Benth., 312
  sceptrum (L.f.) Loudon, 312
Ixia maculata L., 394

Jambosa malaccensis (L.) DC., 232
  vulgaris DC., 232
Jasione L., 330
  montana L., 330
Jasminum L., 255
  azoricum L., 255
  grandiflorum L., 255
  odoratissimum L., 256
JUNCACEAE, 399
Juncellus C.B. Clarke, 461
  laevigatus (L.) C.B. Clarke, 461

subsp. **laevigatus**, 461
Juncus L., 401
  acutus L., 402
    subsp. **acutus**, 402
    subsp. **leopoldii** (Parl.) Snogerup, 402
    var. *multibraceatus* (Tineo) Cout., 402
  articulatus L., 404
  bufonius group, 403
  bufonius L., 404
  bulbosus L., 404
  capitatus Weigel, 404
  conglomeratus L., 402
  effusus L., 402
  foliosus Desf., 403
  glaucus Sibth., 402
  hybridus Brot., 404
  inflexus L., 402
  lamprocarpus Ehrh. ex Hoffm., 404
  sorrentinii Parl., 403
  supinus Moench, 404
  tenuis Willd., 403
Juniperus L., 55
  cedrus Webb & Berthel., 55
  oxycedrus subsp. *maderensis* Menezes, 55
  phoenicea L., 55

Kalanchoe Adans., 131
  delagonensis Eckl. & Zeyh., 131
  fedtschenkoi Raym.-Hamet & Perrier, 132
  pinnata (Lam.) Pers., 131
  tubiflora Raym.-Hamet, 131
Kentrophyllum lanatum (L.) DC. & Duby, 368
Kerneria pilosa var. *discoidea* Lowe, 349
  var. *radiata* (Schultz Bip.) Lowe, 349
Kickxia Dumort., 310
  elatine (L.) Dumort., 310
    subsp. *crinita* (Mabille) Greuter, 311
    subsp. **elatine**, 311
  lanigera (Desf.) Hand.-Mazz., 311
  spuria (L.) Dumort., 311
    subsp. **integrifolia** (Brot.) R. Fern., 311
    subsp. *spuria*, 311
Koniga maritima (L.) R. Br., 119
Krubera leptophylla DC., 246
  peregrina (L.) Hoffm., 246
Kyllinga Rottb., 461

brevifolia Rottb., 462

LABIATAE, 279
Lablab purpureus (L.) Sweet, 161
  vulgaris Savi, 161
Lactuca L., 377
  patersonii Menezes, 377
  sativa L., 377
  scariola sensu Lowe non L., 377
  serriola L., 377
  virosa L., 377
Lagurus L., 434
  ovatus L., 434
Lamarckia Moench, 420
  aurea (L.) Moench, 420
LAMIACEAE, 279
Lamiastrum Heist. ex Fabr., 285
  galeobdolon (L.) Ehrend. & Polatschek, 285
Lamium L., 284
  amplexicaule L., 285
    var. **amplexicaule**, 285
    var. *clandestina* Rchb., 285
  hybridum Vill., 285
  purpureum L., 284
    var. **hybridum** (Vill.) Vill., 285
    var. **purpureum**, 284
Lantana L., 278
  camara L., 278
Lapeirousia cruenta (Lindl.) Bak., 398
  laxa (Thunb.) N.E. Br., 398
Lappa minor (Schk.) DC., 363
Lapsana L., 379
  communis L., 379
    subsp. **communis**, 379
Lastrea africana (Desv.) Copel., 41
  dentata (Forssk.) Romariz, 42
  oreopteris (Ehrh.) Bory, 41
Lathyrus L., 167
  Sect. APHACA (Mill.) Dumort., 169
  Sect. LATHYRUS, 168
  Sect. CLYMENUM (Mill.) DC. ex Ser., 170
  Sect. LINEARICARPUS Kupicha., 170
  angulatus L., 170
  annuus L., 168
  aphaca L., 169
  articulatus L., 170
    subvar. *atropurpurea* Lowe, 170
    var. *latifolius* Rouy, 170
    subvar. *rosea* Lowe, 170
  cicera L., 169

subvar. *caerulea* Lowe, 169
subvar. *purpurea* Lowe, 169
clymenum L., 170
  subvar. *albiflora* Lowe, 170
  subvar. *atropurpurea* Lowe, 170
  subvar. *roseopurpurea* Lowe, 17
ochrus (L.) DC., 170
odoratus L., 169
sativus L., 169
sphaericus Retz., 170
sylvestris L., 169
tingitanus L., 169
Launaea Cass., 374
  arborescens (Batt.) Murb., 374
LAURACEAE, 100
Laurus L., 102
  azorica (Seub.) Franco, 102
    var. *longifolia* (Kuntze) G. Kunkel, 102
    var. *lutea* (Menezes) A. Hansen, 102
  canariensis Webb & Berthel. non Willd., 102
Lavandula L., 293
  angustifolia Mill., 294
  dentata L., 294
  pedunculata subsp. *ambigua* Menezes, 293
    subsp.*maderensis* (Benth.) Menezes, 293
  pinnata L. f., 294
  rotundifolia Benth. var. *rotundifolia*, 294
  spica L., 294
  stoechas L., 293
    subsp. **maderensis** (Benth.) Rozeira, 293
  viridis L Hér., 293
Lavatera L., 218
  arborea L., 219
  cretica L., 219
Legousia Durande, 330
  falcata (Ten.) Fritsch, 330
  hybrida (L.) Delarbre, 330
LEGUMINOSAE, 149
  Subfam. CAESALPINIOIDEAE, 152
  Subfam. LOTOIDEAE, 155
  Subfam. MIMOSOIDEAE, 154
Lemna L., 456
  gibba L., 456
  minor L., 456
LEMNACEAE, 455
Lens Mill., 167
  culinaris Medik., 167
Leontodon L., 373
  nudicaulis sensu auct. mad. non (L.) Schinz & R. Keller, 373
  rigens (Dryand. in Aiton) Paiva & Ormonde, 373
  saxatilis subsp. *rothii* (Ball) Maire, 373
  taraxacoides (Vill.) Merat, 373
    subsp.**longirostris** Finch & P.D. Sell, 373
Lepidium L., 120
  bonariense L., 121
  ruderale L., 121
  sativum L., 121
  virginicum L., 121
Leptogramma pilosiusculum (Wikstr.) Alston, 41
  totta (Willd.) J. Sm., 41
Leptospermum J.R. & G. Forst., 233
  scoparium J.R. & G. Forst., 233
Leucaena Benth., 154
  leucocephala (Lam.) de Wit, 155
Leucanthemum Mill., 356
  lacustre (Brot.) Samp., 357
  vulgare Lam., 356
Ligusticum japonicum Thunb., 256
  lucidum, 256
  lucidum W.T. Aiton, 256
LILIACEAE, 384
Lilium candidum L., 384
Limonium Mill., 253
  ovalifolium (Poir.) Kuntze, 254
    subsp. **pyramidatum** (Lowe) O.E. Erikss., A. Hansen & Sunding, 254
  papillatum (Webb & Berthel.) Kuntze, 254
    var. **callibotryum** Svent., 254
  pectinatum (Aiton) Kuntze, 254
  pectinatum sensu auct. mad., non (Aiton) Kunze, 254
  sinuatum (L.) Mill., 253
LINACEAE, 201
Linaria maroccana Hook.f., 311
  spuria (L.) Mill., 311
Linum L., 202
  angustifolium Huds., 202
  bienne Mill., 202
  gallicum L., 202
  strictum L., 203
    var. *cymosum* Gren. & Godr., 203
  trigynum L., 202
  usitatissimum L., 202

Lippia citriodora (Lam.) Kunth, 277
Lobelia L., 332
   erinus L., 332
   laxiflora Kunth, 332
   urens L., 332
Lobularia Desv., 118
   canariensis (DC.) L. Borgen, 118
      subsp. **rosula-venti** (Svent.) L. Borgen, 118
      subsp. **succulenta** L. Borgen, 118
   libyca (Viv.) Meisn., 118
   maritima (L.) Desv., 119
      var. *canariensis* (DC.) Cout., 118
      var. *rosula-venti* Svent., 118
Logfia Cass., 341
   gallica (L.) Coss. & Germ., 341
   minima (Sm.) Dumort., 341
Lolium L., 416
   canariense Steud., 417
   lowei Menezes, 417
   multiflorum Lam., 416
   parabolicae Sennen & Samp., 417
   perenne L., 416
   rigidum Gaudin, 417
      subsp. *lepturoides* (Boiss.) Sennen & Mauricio, 417
      subsp. *rigidum*, 417
      subsp. *rigidum* var. *rigidum*, 417
      var. *rottbollioides* Heldr. ex Boiss., 417
   subulatum Vis., 417
   temulentum L., 417
      var. *arvense* (With.) Junge, 417
Lomaria semicylindrica Bowdich, 52
   spicant (L.) Desv., 53
LOMARIOPSIDACEAE, 52
Lonicera L., 325
   etrusca Santi, 325
      var. **glabra** Lowe, 325
   japonica Thunb., 325
Lophochloa pumila (Desf.) Bor, 427
Lophospermum erubescens D. Don, 303
Lotus L., 184
   Sect. ERYTHROLOTUS Brand., 186
   Sect. LOTEA (Medik.) Willk., 187
   Sect. LOTUS, 186
   Sect. PEDROSIA (Lowe) Brand., 187
   angustissimus L., 186
   argenteus (Lowe) Masferrer non Salisb., 188
   argyrodes R.P. Murray, 188
   argyrodes, 188

   azoricus P.W. Ball, 188
   conimbricensis Brot., 186
   glaucus Aiton, 187
      var. *angustifolius* R.P. Murray, 18
   hispidus sensu auct. mad. non DC., 186
   lancerottensis Webb & Berthel., 187
   loweanus Webb & Berthel., 188
   macranthus Lowe, 188
   macranthus, 188
   neglectus (Lowe) Masf., 187
   ornithopodioides L., 187
   parviflorus Desf., 186
      var. *robustus* Lowe, 186
      var. *tenuis* Lowe, 186
   pedunculatus Cav., 186
   salvagensis R.P. Murray, 187
   suaveolens Pers., 186
   uliginosus Schkuhr, 186
      var. *glabriusculus* Bab., 186
      subsp. *pisifolius* (Lowe) Menezes, 186
      var. *pisifolius* (Lowe) Lowe, 186
Lunaria L., 117
   annua L., 117
Lunathyrium petersenii (Kunze) H. Ohba, 48
Lupinus L., 159
   albus L., 159
      subsp. **albus**, 159
      subsp. *termis* (Forssk.))Cout., 159
   angustifolius L., 159
   luteus L., 159
   termis Forssk., 159
Luzula DC., 399
   campestris (L.) DC., 400
      subsp. *multiflora* (Retz.) Buchenau, 400
   congesta (Thuill.) Lej., 401
   elegans Lowe, 400
   elegans Guthnick non Lowe, 401
   multiflora (Retz.) Lej., 400
      subsp. **congesta** (Thuill.) Hyl., 401
      subsp. **multiflora**, 401
   purpurea (Masson ex Buchenau) Link, 400
   purpureo-splendens Seub., 401
   seubertii Lowe, 400
Lycium L., 296
   europaeum L., 296
Lycopersicon Mill., 301
   esculentum Mill., 301
LYCOPODIACEAE, 28

Lycopodiella Holub, 28
   cernua (L.) Pic. Serm., 28
   veigae (Vasc.) A. Hansen & Sunding, 28
Lycopodium cernuum L., 28
   complanatum sensu auct., mad. non L., 29
   dentatum Herter, 29
   denticulatum L., 30
   madeirense J.H. Wilce, 29
   selago sensu auct. mad. non L., 29
      subsp. *suberecta* sensu Romariz, 29
   suberectum Lowe, 29
LYTHRACEAE, 230
Lythrum L., 230
   flexuosum sensu auct. mad. non Lag., 230
   graefferi Ten., 230
   hyssopifolia L., 230
      var. *acutifolium* DC., 230
         forma *typicum* Cout., 230
   junceum Banks & Sol., 230

Majorana hortensis Moench, 290
Malcolmia maritima (L.) R. Br., 115
Malus domestica Borkh., 148
Malva L., 218
   mauritiana L., 218
   nicaeensis All., 218
   parviflora L., 218
   sylvestris L., 218
MALVACEAE, 216
Malvastrum A. Gray, 217
   coromandelianum (L.) Garcke, 217
Mantisalca Cass., 366
   salmantica (L.) Briq. & Cavill., 366
Marcetella Svent., 146
   maderensis (Bornm.) Svent., 146
Marrubium L., 283
   vulgare L., 283
Matthiola R. Br., 114
   maderensis Lowe, 115
      [var.] γ *albiflora* Lowe, 115
      [var.] β *mitis* Lowe, 115
      [var.] α *muricata* Lowe, 115
   parviflora (Schousb.) R. Br., 115
Maurandya scandens (Cav.) Pers., 303
   semperflorens Ortega, 303
Maytenus Molina, 214
   umbellata (R. Br.) Mabb., 214
Medeola asparagoides L., 389
Medicago L., 173
   ciliaris (L.) All., 177
   helix Willd. var. *calcarata* Lowe, 175
      var. *inermis* sensu Lowe, 175
   hispida Gaertn., 176
      subsp. *lappacea* var. *longispina* sensu Menezes, 176
      subsp. *pentacycla* var. *nigra* (Willd.) Cout., 176
   italica (Mill.) Fiori, 175
      subsp. **tornata** (L.) Emberger & Maire, 175
   laciniata (L.) Mill., 176
   lappacea Desr., 176
      var. *brachycantha* Lowe, 176
      var. *macracantha* (Webb & Berthel.) Lowe, 176
   littoralis Rhode ex Loisel., 175
      var. *breviseta* DC., 175
      var. *inermis* Moris, 175
   lupulina L., 174
   minima (L.) L., 176
      var. *longispina* Benth., 176
      var. **minima**, 176
      var. *mollissima* (Roth) Cout., 176
      subsp. *pulchella* (Lowe) Menezes, 176
      var. **pulchella** (Lowe) Lowe, 176
   obscura subsp. *helix* (Willd.) Batt., 175
      subsp. *helix* var. *aculeata* sensu Menezes, 175
      subsp. *helix* var. *inermis* sensu Menezes, 175
   orbicularis (L.) Bartal., 175
   polymorpha L., 176
   pulchella Lowe, 176
   sativa L., 174
   tribuloides Desr., 175
      var. α. Lowe, 175
      var. β. Lowe, 175
      var. γ. Lowe, 175
      var. *muricata* Menezes, 175
   truncatula Gaertn., 175
Melandrium noctiflorum (L.) Fr., 93
Melanoselinum Hoffm., 245
   decipiens (Schrad. & J.C. Wendl.) Hoffm., 245
   edule (Lowe) Baill., 247
Melica L., 424
   canariensis W. Hempel, 425
   ciliata L., 425
      subsp. *ciliata*, 425
      subsp. **magnolii** (Gren. & Godr.)

Husn., 425
Melilotus Mill., 172
  albus Medik., 173
  elegans Ser., 173
    subsp. *lippoldianus* (Lowe) Menezes, 173
  indicus (L.) All., 173
  lippoldianus Lowe, 173
  parviflorus Desf., 173
  segetalis (Brot.) Ser., 173
  sulcatus Desf., 173
Melinis P. Beauv., 450
  minutiflora P. Beauv., 450
Melissa L., 287
  officinalis L., 287
    subsp. **officinalis**, 287
Mentha L., 291
  aquatica L., 291
    var. *glabrata* Benth., 292
    var. *hirsuta* (Huds.) Willd., 291
    var. *intricata* Menezes, 292
  aromatica sensu Menezes, 293
  longifolia (L.) Huds., 292
  piperita L., 292
    nm. **citrata** (Ehrh.) Boivin, 292
    nm. **piperita**, 292
  pulegium L., 291
    var. *gibraltarica* (Willd.) Batt. & Trab., 291
    var. *tomentella* (Hoffmans. & Link) Cout., 291
    var. *vulgaris* Mill., 291
  rotundifolia auct. non (L.) Huds., 292
    var. *aromatica* Menezes, 293
    var. *maderensis* Menezes, 292
  spicata L., 292
  suaveolens Ehrh., 292
  sylvestris L., 292
  villosa Huds., 293
  viridis L., 292
    var. *hirsuta* Menezes, 292
Mercurialis L., 204
  annua L., 204
    var. *ambigua* (L.f.) Duby, 204
    var. *annua*, 204
Mesembryanthemaceae, 79
Mesembryanthemum L., 80
  cordifolium L.f., 81
  crystallinum L., 80
  edule L., 81
  nodiflorum L., 80
Mespilus germanica L., 149

Microlepia platyphylla (D. Don) Sm., 41
Microlonchus salmantica (L.) DC., 366
Micromeria varia Benth., 288
  subsp. *thymoides* var. *cacuminicolae* P. Pérez, 289
  subsp. *thymoides* (Sol. ex Lowe) P. Pérez var. *thymoides*, 289
Micropyrum (Gaudin) Link, 417
  tenellum (L.) Link, 418
Mimulus L., 304
  moschatus Douglas ex Lindl., 304
Mirabilis L., 78
  divaricata Lowe, 78
  jalapa L., 78
Misopates Raf., 308
  calycinum (Vent.) Rothm., 309
  orontium (L.) Raf., 309
    subsp. **orontium**, 309
  salvagense D.A. Sutton, 309
Modiola Moench, 219
  caroliniana (L.) G.Don, 219
Monanthes Haw., 136
  brachycaulon sensu auct. mad. non (Webb in Webb & Berthel.) Lowe, 136
  lowei (A. Paiva) P. Pérez & Acebes, 136
Monerma cylindrica (Willd.) Coss. & Durieu, 424
Monizia Lowe, 247
MONOCOTYLEDONES, 382
Montanoa bipinnatifida K. Koch, 348
MORACEAE, 58
Morus alba L., 58
  nigra L., 58
Muehlenbeckia Meisn., 68
  complexa (A. Cunn.) Meisn., 68
  platyclados Meisn., 68
  sagittifolia (Ortega) Meisn., 68
Musschia Dumort., 331
  aurea (L.f.) Dumort., 331
    var. *angustifolia* (Ker. Gawl.) DC., 331
  wollastonii Lowe, 331
Myosotis L., 274
  arvensis (L.) Hill, 275
  azorica H.C. Watson, 274
  caespitosa Schultz, 276
  discolor Pers., 275
    subsp. *canariensis* (Pit.) Grau, 275
  intermedia Link, 275
  maritima Hochst., 274
  ramosissima Rochel, 275

repens D. Don, 275
scorpioides L., 276
secunda Al. Murray, 275
stolonifera (DC.) J. Gay ex Leresche & Levier, 276
sylvatica Hoffm., 275
versicolor Sm., 275
Myrica L., 57
faya Aiton, 58
MYRICACEAE, 57
MYRSINACEAE, 250
Myrsiphyllum asparagoides (L.) Willd., 389
MYRTACEAE, 231
Myrtus L., 232
communis L., 232
subsp. **communis**, 232
var. *latifolia* sensu Lowe, 232
var. *lusitanica* L., 232
var. *parvifolia* sensu Lowe, 232
subsp. **tarentina** (L.) Nyman, 232

Narcissus L., 393
jonquilla L., 393
medioluteus Mill., 393
odorus L., 393
pseudonarcissus L., 393
Nardurus lachenalii var. *festucoides* (Bertol.) Cout., 418
Nassella Desv., 414
trichotoma (Nees) Hack. ex Arechav., 414
Nasturtium R. Br., 116
officinale R. Br., 116
var. *genuinum* Gren. & Godr., 116
var. *siifolium* (Rchb.) Steud., 116
Nauplius (Cass.) Cass., 347
aquaticus (L.) Cass., 347
Neotinea Rchb.f., 468
intacta (Link) Rchb.f., 468
maculata (Desf.) Stearn, 468
Nephrodium aemulum (Aiton) Baker, 51
affine Lowe, 51
elongatum (Aiton) Hook. & Grev., 51
foenisecii Lowe, 51
molle (Sw.) R. Br., 42
montanum (J.A. Vogler) Baker, 41
oreopteris (Ehrh.) Desv., 41
Nephrolepis cordifolia (L.) C. Presl, 53
Nerine sarniensis (L.) Herb., 392
Nerium odorum Sol., 258
oleander L., 258

Nicandra Adans., 296
physalodes (L.) Gaertn., 296
Nicotiana L., 302
glauca Graham, 302
tabacum L., 303
wigandioides K. Koch & Fintelm, 303
Nigella L., 96
damascena L., 96
Normania Lowe, 300
triphylla (Lowe) Lowe, 300
Notelaea excelsa (Aiton) Webb & Berthel., 257
Notholaena R. Br., 33
lanuginosa (Desf.) Desv. ex Poir., 34
marantae (L.) Desv., 33
subsp. **subcordata** (Cav.) G. Kunkel, 34
vellea (Aiton) Desv., 34
Nothoscordum Kunth, 388
fragrans (Vent.) Kunth, 388
gracile (Aiton) Stearn, 388
inodorum (Aiton) Nicholson, 388
Notobasis Cass., 364
syriaca (L.) Cass., 364
Notodanthonia tenuior (Steud.) S.T. Blake, 440
Nucularia perrini Batt., 74
NYCTAGINACEAE, 78
Nycterium triphyllum Lowe, 300

Ocimum basilicum L., 280
micranthum Willd., 280
Ocotea Aubl., 102
foetens (Aiton) Baill., 102
Odontites Ludw., 315
holliana (Lowe) Benth., 315
Oenanthe L., 245
divaricata (R. Br.) Mabb., 245
pteridifolia Lowe, 245
Oenothera L., 235
biennis L., 236
grandiflora L Hér. ex Aiton., 236
longiflora L., 235
subsp. **longiflora**, 235
odorata sensu auct. mad. non Jacq., 235
stricta Ledeb. ex Link, 235
suaveolens Pers., 236
tetraptera Cav., 235
Olea L., 256
europaea L., 256
var. *buxifolia* Aiton, 256

var. **maderensis** Lowe, 257
OLEACEAE, 255
ONAGRACEAE, 233
Ononis L., 171
   dentata Sol. ex Lowe, 171
   diffusa Ten., 172
   micrantha Lowe, 172
   reclinata sensu auct. mad. non L., 171
     var. *simplex* Lowe, 171
     var. *tridentata* Lowe, 171
   serrata Forssk., 172
   spinosa L., 171
     subsp. **maritima** (Dumort.) P. Fourn., 171
OPHIOGLOSSACEAE, 31
Ophioglossum L., 31
   azoricum C. Presl, 31
   lusitanicum L., 31
   pennatum Lam., 32
   polyphyllum A. Braun, 31
   polyphyllum sensu auct. mad., non A. Braun, 31
   reticulatum L., 31
Oplismenus P. Beauv., 445
   hirtellus (L.) P. Beauv., 445
Opuntia Mill., 229
   ficus-barbarica A. Berger, 230
   ficus-indica auct. eur., 230
   tuna (L.) Mill., 230
ORCHIDACEAE, 467
Orchis L., 469
   foliosa Sol. ex Lowe non Swartz, 468
   maderensis Summerh., 468
   mascula sensu auct. mad. non (L.) L., 469
   patens Desf., 469
   scopulorum Summerh., 469
Oreopteris Holub, 41
   limbosperma (Bellardi ex All.) Holub, 41
Origanum L., 289
   majorana L., 290
   virens var. *genuinum* Cout., 289
   vulgare L., 289
     subsp. **virens** (Hoffmanns. & Link) Ietsw., 290
Ormenis aureus Coss. & Durieu, 353
   mixta (L.) Dum., 352
Ornithogalum L., 386
   arabicum L., 386
Ornithopus L., 189
   compressus L., 190

   exstipulatus Thore, 190
   perpusillus L., 190
   pinnatus (Mill.) Druce, 190
   sativus Brot., 190
Oryza sativa L., 407
Oryzopsis Mich., 414
   miliacea (L.) Asch. & Schweinf., 414
Osmunda L., 32
   lunaria L., 32
   regalis L., 32
   spicant L., 53
OSMUNDACEAE, 32
OXALIDACEAE, 192
Oxalis L., 192
   cernua Thunb., 193
   corniculata L., 192
   corymbosa DC., 193
   debilis Kunth, 193
   exilis A. Cunn., 192
   intermedia A. Rich., 193
   latifolia Kunth, 193
   pes-caprae L., 193
   purpurea L., 193
   variabilis Jacq., 193
   venusta Lowe, 193

Palhinaea veigae Vasc., 28
   cernua (L.) Franco & Vasc., 28
Panicum L., 445
   barbinode Trin., 447
   capillare L., 446
   crus-galli var. *hostii* (M. Bieb.) K. Richt., 447
   maximum Jacq., 446
   miliaceum L., 446
   repens L., 446
Papaver L., 103
   dubium L., 104
     subsp. **dubium**, 104
   pinnatifidum Moris, 104
   rhoeas L., 104
     var. **rhoeas**, 104
     subsp. *strigosum* (Boenn.) Menezes, 104
     var. *strigosum* Boenn., 104
   somniferum L., 104
     subsp. *nigrum* (Garsault) Thell., 105
     subsp. **setigerum** (DC.) Arcang., 105
     subsp. **somniferum**, 105
PAPAVERACEAE, 103
   Subfam. FUMARIOIDEAE, 106

Subfam. PAPAVEROIDEAE, 103
Paraceterach marantae (L.) R.M. Tryon, 33
Parafestuca Alexeev, 422
   albida (Lowe) Alexeev, 422
Parapholis C.E. Hubb., 423
   filiformis (Roth) C.E. Hubb., 423
   incurva (L.) C.E. Hubb., 423
Parentucellia Viv., 316
   viscosa (L.) Caruel, 316
Parietaria L., 60
   debilis G. Forst., 60
      var. **gracilis** (Lowe) Wedd., 61
      var. **micrantha** (Ledeb.) Wedd., 61
   diffusa Mert. & W.D.J. Koch, 60
   gracilis Lowe, 60
   judaica L., 60
   lusitanica auct. azor. non L., 60
   maderensis Reich., 60
   micrantha Ledeb., 60
   officinalis auct. non L., 60
      var. *diffusa* (Mert. & W.J.D. Koch) Wedd., 60
      subsp. *judaica* (L.) Beg., 60
   ramiflora Moench, 60
Paronychia Mill., 89
   cymosa (L.) DC., 89
   echinata sensu auct. mad. non Lam., 89
Paspalidium Stapf, 449
   geminatum (Forssk.) Stapf, 449
Paspalum L., 447
   dilatatum Poir., 448
   distichum L., 448
   paspalodes (Mich.) Scribn., 448
   vaginatum Sw., 448
Passiflora L., 225
   alba Link & Otto, 226
   antioquiensis H. Karst., 225
   antioquiensis × *mollissima*, 226
   caerulea L., 225
   edulis Sims, 225
   exoniensis hort. ex L.H. Bailey, 226
   ligularis Juss., 225
   lowei Heer, 225
   manicata (Juss.) Pers., 225
   mollissima (Kunth) L.H. Bailey, 226
   quadrangularis L., 225
   subpeltata Ortega, 226
   vanvolxemii (Hook.) Triana & Planch., 225
PASSIFLORACEAE, 225
Patellifolia patellaris (Moq.) Scott, Ford-Lloyd & Williams, 70
   procumbens (C. Sm. ex Hornem.) Scott, Ford-Lloyd & Williams, 70
Pedrosia argentea Lowe, 188
   florida Lowe, 187
      var. *aurantiaca* Lowe, 187
      var. *sulphurea* Lowe, 187
   glauca (Aiton) Lowe, 187
      var. *dubia* Lowe, 187
      var. *intricata* Lowe, 187
   loweana (Webb & Berthel.) Lowe, 188
   macrantha (Lowe) Lowe, 188
   neglecta Lowe, 187
      var. *cinerea* Lowe, 187
      var. *virescens* Lowe, 187
   paivae Lowe, 187
Pelargonium L Hér., 197
   alchemilloides (L.) L Hér., 198
   capitatum (L.) L Hér., 198
   cucullatum (L.) L Hér., 198
   glutinosum (Jacq.) L Hér., 198
   graveolens L Hér., 198
   inquinans (L.) L Hér., 197
   odoratissimum (L.) L Hér., 197
   peltatum (L.) L Hér., 198
   vitifolium (L.) L Hér., 197
   zonale (L.) L Hér., 198
Pellaea viridis (Forssk.) Prantl, 34
Pelletiera A. St.-Hil., 251
   wildpretii Valdés, 251
Pennisetum Rich., 451
   cenchroides Rich., 452
   clandestinum Hochst. & Chiov., 451
   purpureum Schum., 451
   villosum R. Br. ex Fresen, 451
Pereskia aculeata Mill., 229
Pericallis D. Don, 360
   aurita (L Hér.) B. Nord., 360
Periploca L., 259
   laevigata Aiton, 259
Persea Mill., 101
   americana Mill., 101
   gratissima Gaertn., 101
   indica (L.) Spreng., 101
Petrorhagia (Ser. ex DC.) Link, 95
   nanteuilii (Burnat) P.W. Ball & Heywood, 95
Petroselinum Hill, 243
   crispum (Mill.) A.W. Hill, 244
   sativum Hoffmanns, 244
Peucedanum L., 246
   lowei (Coss.) Menezes, 247

Phagnalon Cass., 345
  bennettii Lowe, 346
  hanseni Quaiser & Lack, 346
  rupestre auct. mad. non (L.) DC., 346
  saxatile (L.) Cass., 346
Phalaris L., 430
  altissima Menezes, 431
  aquatica L., 431
  arundinacea L., 431
    var. *picta* L., 431
  brachystachys Link, 432
  canariensis L., 432
  canariensis sensu Brot. non L., 432
  coerulescens Desf., 431
  maderensis (Menezes) Menezes, 431
  minor Retz., 431
  nodosa Menezes non Murray, 431
  paradoxa L., 431
  tuberosa L., 431
Pharbitis learii (Paxton) Lindl., 270
  purpurea (L.) Voigt, 271
Phaseolus coccineus L., 155
  multiflorus Lam., 155
  vulgaris L., 155
Phillyrea lowei DC., 257
  angustifolia L., 257
Phlebodium aureum (L.) Sm., 40
Phoebe indica (L.) Pax, 101
Phragmites Adans., 441
  australis (Cav.) Trin. ex Steud., 442
  communis var. *congesta* (Lowe) Menezes, 442
  congesta Lowe, 442
Phyllanthus L., 203
  niruri L., 204
  tenellus Roxb., 204
Phyllis L., 261
  nobla L., 261
Phyllitis Hill, 47
  scolopendrium (L.) Newman, 47
Phylloxera, 215
Physalis L., 297
  peruviana L., 297
Phytolacca L., 78
  americana L., 78
  dioica L., 79
PHYTOLACCACEAE, 78
Picconia DC., 257
  excelsa (Aiton) DC., 257
Picris echioides L., 373
Pilea microphylla (L.) Liebm., 58
PINACEAE, 54

Pinardia coronaria (L.) Less., 354
Pinus L., 54
  pinaster Aiton, 54
  radiata D. Don, 54
Piptatherum miliaceum (L.) Coss., 414
Pircunia dioica (L.) Moq., 79
Pisum arvense L., 155
  sativum L., 155
    var. *saccharatum* Ser., 155
PITTOSPORACEAE, 138
Pittosporum Banks ex Gaertn., 138
  coriaceum Dryander ex Aiton, 139
  tobira (Thunb.) W.T. Aiton, 139
  undulatum Vent., 139
Pityrogramma calomelanos (L.) Link, 35
  chrysophylla (Sw.) Link, 35
Plantago L., 321
  afra L., 323
    var. *obtusata* (Svent.) A. Hansen & Sunding, 323
  arborescens Poir., 323
    subsp. *arborescens*, 323
    subsp. **maderensis** (Decne.) A. Hansen & G. Kunkel, 323
  aschersonii Bolle, 321
  coronopus L., 321
    var. *latifolia* DC., 321
    var. *pseudo-macrorrhiza* sensu Menezes, 321
    var. *vulgaris* Gren. & Godr., 321
  lagopus L., 322
  lanceolata L., 322
    var. *capitata* Presl, 322
    var. *contigua* Menezes, 322
    var. *eriophora* (Hoffmanns. & Link) Cout., 322
    var. *timbali* (Jord.) Gaut., 322
  leiopetala Lowe, 322
  loeflingii L., 322
  maderensis Decne., 323
  major L., 323
    subsp. **major**, 323
  malato-belizii Lawalrée, 322
  myosurus Lam., 323
    subsp. **myosurus**, 323
  ovata Forssk., 322
PLATANACEAE, 139
Platanus acerifolia (Aiton) Willd., 139
  hispanica Mill. ex Münch, 139
  hybrida Brot., 139
  occidentalis sensu auct. mad. non L., 139

Plectranthus fruticosus L Hér., 280
Plinthanthesis tenuior Steud., 440
PLUMBAGINACEAE, 252
Plumbago auriculata Lam., 253
   capensis Thunb., 253
Plumeria rubra L., 258
Poa L., 420
   annua L., 421
   bulbosa L., 421
      var. **vivipara** Koeler, 421
   compressa L., 421
   eragrostis L., 443
   pratensis L., 421
   trivialis L., 421
POACEAE, 406
Polycarpon Loefl. ex L., 90
   diphyllum Cav., 90
   tetraphyllum (L.) L., 90
      subsp. **diphyllum** (Cav.) O. Bolòs & Font Quer, 90
      subsp. **tetraphyllum**, 90
Polygala myrtifolia L., 210
POLYGALACEAE, 210
POLYGONACEAE, 62
Polygonum L., 62
   arenastrum Boreau, 63
   aviculare L., 63
      var. *commune* Menezes, 63
      var. *confertum* Menezes, 63
   capitatum Buch.-Ham. ex D. Don, 63
   convolvulus L., 64
   hydropiper L., 64
   lapathifolium L., 64
   maritimum L., 63
   patulum M.Bieb., 63
   persicaria L., 63
      var. *biforme* Wahlenb., 64
      var. *elatum* Gren. & Godr., 64
      var. *genuinum* Gren. & Godr., 64
      var. *persicaria*, 64
   salicifolium Brouss. ex Willd., 64
   serrulatum Lag., 64
POLYPODIACEAE, 39
Polypodium L., 39
   aemulum Aiton, 51
   aureum L., 40
   austriacum Jacq., 40
   dentatum Forssk., 42
   diaphanum Bory, 48
   drepanum (Sw.) Lowe, 49
   elongatum Aiton, 51
   falcatum L.f., 50

   filix-femina L., 47
   interjectum Shivas, 39
   leptophyllum L., 35
   limbospermum Bellardi ex All., 41
   setiferum Forssk., 49
   vulgare L., 39
Polypogon Desf., 434
   fugax Nees ex Steud., 435
   maritimus Willd., 435
   monspeliensis (L.) Desf., 435
   semiverticillatus (Forssk.) Hyl., 435
   viridis (Gouan) Breistr., 435
Polystichum Roth, 48
   aculeatum (L.) Roth, 49
   aculeatum sensu auct. mad., non (L.) Roth, 49
   angulare (Kit. ex Willd.) C. Presl, 49
   drepanum (Sw.) C. Presl, 49
   falcatum (L.f.) Diels, 50
   falcinellum (Sw.) C. Presl, 49
   frondosum (Lowe) J. Sm., 50
   maderense J.Y. Johnson, 49
   setiferum (Forssk.) Woyn., 49
   webbianum (A. Braun) C. Chr., 50
Populus L., 56
Portulaca L., 82
   oleracea L., 82
      subsp. **oleracea**
         var. *sylvestris* DC., 82
      sativa Haw., 83
PORTULACACEAE, 82
Potamogeton L., 383
   cyprifolius Lowe, 384
   fluitans Roth, 383
   gramineus L., 383
   leschenaultii Cham. & Schldl., 383
   machicanus Lowe, 383
   natans L., 384
   nodosus Poir., 383
   panormitanus Biv., 383
   polygonifolius Pourr., 384
   pusillus L., 383
POTAMOGETONACEAE, 383
Potentilla L., 144
   anglica Laichard., 144
   procumbens Sibth., 144
   reptans L., 145
Poterium verrucosum Ehrenb. ex Decne., 146
Prasium L., 283
   majus L., 283
      var. *intermedium* Menezes, 283

medium Lowe, 283
PRIMULACEAE, 251
PROTEACEAE, 61
Prunella L., 287
    vulgaris L., 287
Prunus L., 140
    armeniaca L., 141
    avium (L.) L., 141
    cerasus L., 141
    domestica L., 141
    dulcis (Mill.) D.A. Webb, 141
    laurocerasus L., 141
    lusitanica L., 141
        subsp. **hixa** (Brouss. ex Willd.) Franco, 141
    persica (L.) Batsch, 141
Pseudognaphalium Kirp., 343
    luteo-album (L.) Hilliard & B.L. Burtt, 343
Pseudosasa Nakai, 413
    japonica (Siebold & Zucc. ex Steud.) Makino, 413
Psidium L., 231
    cattleianum Sabine, 232
    guajava L., 231
    guineense Sw., 232
    littorale Raddi, 232
        var. *globosum* Heer., 232
    pomiferum L., 231
    pyriferum L., 231
Psoralea americana L., 161
    var. *polystachya* (Poir.) Cout., 161
    bituminosa L., 161
PTERIDACEAE, 36
Pteridium Gled. ex Scop., 40
    aquilinum (L.) Kuhn, 40
PTERIDOPHYTA, 25
Pteris L., 36
    aquilinum L., 40
    arguta Aiton, 37
    incompleta Cav., 37
    longifolia sensu auct. mad. non L., 37
    multifida Poir., 37
    palustris Poir., 37
    serrulata Forssk. non L.f., 37
    serrulata L.f. non Forssk., 37
    tremula R. Br., 37
    vittata L., 37
Pycreus P. Beauv., 461
    flavescens (L.) Rchb., 461
Pyrethrum grandiflorum sensu Holl et auct mad., non Willd., 355

parthenium (L.) Sm., 351
Pyrus aucuparia var. *maderensis* Lowe, 148
    communis L., 148
    maderensis (Dode) Menezes, 148
    pyraster Burgsd., 148

Quamoclit coccinea (L.) Moench, 270
    pennata (Desr.) Bojer, 270

Radiola Hill, 203
    linoides Roth, 203
    millegrana Sm., 203
RANUNCULACEAE, 95
Ranunculus L., 96
    acris L., 97
    arvensis L., 98
    bulbosus L., 97
        subsp. **adscendens** (Brot.) Neves, 97
        subsp. **aleae** (Willk.) Rouy & Fouc., 97
    cortusifolius Willd., 98
        [var. *major*], 98
        [var. *minor*], 98
    flammula L., 99
    grandifolius, 98
        var. *major* Lowe, 98
        var. *minor* Lowe, 99
    muricatus L., 97
    parviflorus L., 98
    repens L., 97
    sardous subsp. *trilobus* (Desf.) Rouy & Fouc., 98
    trilobus Desf., 98
Raphanus L., 127
    raphanistrum L., 128
        subsp. **raphanistrum**, 128
    sativus L., 128
Rapistrum Crantz, 126
    rugosum (L.) All., 127
        subsp. **orientale** (L.) Arcang., 127
        subsp. **rugosum**, 127
Reinwardtia indica Dumort., 201
Reseda L., 128
    lutea L., 129
    luteola L., 129
        var. *australis* (Webb) Walp., 129
        var. *crispata* (Link) Müll. Arg., 129
        var. *gussonii* (Boiss. & Reut.) Müll. Arg., 129
    media Lag., 129
RESEDACEAE, 128

Rhagadiolus stellatus (L.) Gaertn., 371
RHAMNACEAE, 214
Rhamnus L., 214
  glandulosa Aiton, 214
  latifolia L Hér., 215
Rhododendron L., 250
  mucronatum G. Don, 250
  ponticum L., 250
Rhus L., 210
  coriaria L., 211
Richardia aethiopica (L.) Spreng., 454
  africana Kunth, 454
Ricinus L., 204
  communis L., 204
Rivina L., 79
  brasiliensis Nocca, 79
  humilis L., 79
Robinia hispida L., 155
  pseudoacacia L., 155
Romulea Maratti, 398
  columnae Sebast. & Mauri, 399
    subsp. **columnae** 399
    subsp. **grandiscapa** (Webb) G. Kunkel, 399
      var. *grandiscapa* (Webb) Pit., 399
  grandiscapa (Webb) Gay, 399
Rosa L., 147
  bracteata J.C. Wendl., 148
  canina L., 147
    var. *glabra* sensu Lowe, 147
    var. *mandonii* (Déségl.) Menezes, 147
    var. *pubescens* Menezes, 147
  laevigata Michx., 148
  mandonii Déségl., 147
  multiflora Thunb., 148
  rubiginosa L., 147
  stylosa Desv., 147
  wilsoni sensu Lowe, 147
ROSACEAE, 139
Rosmarinus officinalis L., 280
Rostraria Trin., 427
  pumila (Desf.) Tzvelev, 427
Rubia L., 265
  agostinhoi Dans. & P. Silva, 265
  angustifolia sensu auct. mad. non L., 265
  fruticosa Aiton, 265
    subsp. **fruticosa**, 265
    subsp. *melanocarpa* (Bornm.) Bramwell, 266
    subsp. *periclymenon* (Schenck) Sunding, 266
      var. *pendula* Pit., 266
  gratiosa Menezes, 265
  peregrina var. *angustifolia* sensu Webb & Berthel.., 265
    subsp. **agostinhoi** (Dans. & P. Silva) Valdés Berm., & G. Lopez, 265
RUBIACEAE, 261
Rubus L., 142
  bollei group, 143
  bollei Focke, 143
  caesius L., 144
  canariensis Focke, 143
  concolor Lowe, 143
  discolor Weihe & Nees, 142
  grandifolius Lowe, 143
    var. *dissimulatus* Menezes, 143
  idaeus L., 142
  inermis sensu Hansen & Sunding, non Pourr., 142
  pinnatus Willd., 142
  × **suspiciosus** Menezes, 143
  ulmifolius Schott, 142
    subsp. *rusticanus* var. *communis* Menezes, 142
    subsp. *rusticanus* var. *dalmatinus* (Tratt. ex Focke) Menezes, 142
    subsp. *rusticanus* var. *neglectus* Menezes, 142
    subsp. *rusticanus* var. *nutritus* Menezes, 142
  ulmifolius × *R. bollei*, 143
  ulmifolius × *R. vahlii*, 143
  vahlii Frid., 143
Rumex L., 65
  acetosella L., 66
    subsp. **pyrenaicus** (Pourret ex Lapeyr.) Akeroyd, 66
  bucephalophorus L., 68
    subsp. **canariensis** (Stein) Rech.f., 68
      var. **canariensis**, 68
      var. **fruticescens** (Born.) Press, 68
    subsp. *fruticescens* Bornm., 68
  conglomeratus Murray, 67
  crispus L., 67
  maderensis Lowe, 66
    var. *glaucus* Lowe, 66
    var. *virescens* Lowe, 66
  obtusifolius L., 67
    subsp. **obtusifolius**, 67

pulcher L., 67
　subsp. *divaricata* (L.) Murb., 67
　subsp. *pulcher*, 67
simpliciflorus Murb., 66
　subsp. **maderensis** (Murb.) Samuelson, 66
　var. *maderensis* Murb., 66
　var. *rubellianus* Menezes, 66
　vesicarius L., 66
　　var. *rhodophysa* Ball, 66
Ruppia L., 384
　maritima L., 384
　rostellata W.D.J. Koch, 384
RUPPIACEAE, 384
Ruscus L., 390
　androgynus L., 390
　hypophyllum auct. mad., 390
　　var. *lancifolius* Lowe, 390
　　var. *latifolius* Lowe, 390
　streptophyllus P. F. Yeo, 390
Ruta L., 210
　bracteosa DC., 210
　chalepensis L., 210
RUTACEAE, 209
Rytidosperma Steud., 440
　tenuis (Steud.) A. Hansen & Sunding, 440

Saccharum officinarum L., 407
　var. *litteratum* Hassk., 40
　var. *subobscurum* Menezes
　var. *violaceum* Pers., 407
Sagina L., 87
　apetala Ard., 87
　　var. *glabra* Bab., 87
　　[var.] α *glandulosa* Lowe, 87
　procumbens L., 87
　　var. *spinosa* (S. Gibson) Bab., 87
SALICACEAEAE, 56
Salix L., 56
　canariensis C. Sm. ex Link, 57
　fragilis L. × **alba** L. Schrank, 57
　pedicellata subsp. *canariensis* (Buch) A.K. Skvortsov, 57
　rubens Schrank.), 57
Salpichroa Miers, 297
　origanifolia (Lam.) Baill., 297
Salsola L., 74
　kali L., 74
　lanata Masson, 73
Salvia L., 294
　coccinea Juss. ex Murray, 295
　collina Lowe, 294
　farinacia Benth., 294
　fruticosa Mill., 294
　leucantha Cav., 295
　officinalis L., 294
　pseudo-coccinea Jacq., 295
　sessei Benth., 294
　splendens Sellow ex Roem., 294
　triloba L. f., 294
　verbenaca L., 294
　　subsp. *clandestina* (L.) Briq., 294
　　var. *dubia* (Lowe) Menezes, 294
Sambucus L., 324
　ebulus L., 324
　lanceolata R. Br., 324
　maderensis Lowe, 324
　nigra L., 324
　　var. *lanceolata* (R. Br.) Lowe, 324
Samolus L., 252
　valerandi L., 252
Sanguisorba L., 146
　maderensis (Bornm.) Nordborg, 146
　minor Scop., 146
　　subsp. *magnolii* (Spach) Briq., 146
　　subsp. **verrucosa** (Ehrenb. ex Decne.) Holmboe, 146
SAPINDACEAE, 211
Saponaria officinalis L., 94
SAPOTACEAE, 254
Sarothamnus scoparius (L.) Wimmer ex W.D.J. Koch, 156
Satureja L., 288
　calamintha subsp. *sylvatica* Briq., 288
　　subsp. *sylvatica* var. *calaminthoides* (Rchb.) Briq., 288
　clinopodium (L.) Caruel, 288
　varia (Benth.) Webb & Berthel. ex Briq., 288
　　subsp. **thymoides** (Sol. ex Lowe) A. Hansen & Sunding, 288
　　　var. **cacuminicolae** (P. Pérez) A. Hansen & Sunding, 289
　　　var. **thymoides**, 289
Saxifraga L., 137
　maderensis D. Don, 137
　　var. **maderensis**, 137
　　var. **pickeringii** (C. Simon) D.A. Webb & Press, 137
　pickeringii C. Simon, 137
　portosanctana Boiss., 138
　stolonifera Meerb., 138
SAXIFRAGACEAE, 137

Scabiosa L., 328
   atropurpurea L., 329
   maritima L., 329
   succisa L., 328
Scandix L., 240
   pecten-veneris L., 240
     subsp. **pecten-veneris**, 240
Schinus molle L., 210
Schismus P. Beauv., 441
   barbatus (L.) Thell., 441
   marginatus P. Beauv., 441
Schizogyne Cass., 347
   obtusifolia var. *sericea* DC., 347
   sericea (L. f.) Schultz Bip., 347
Schoenoplectus (Rchb.) Palla, 457
   triqueter (L.) Palla, 458
Schufia arborescens (Sims) Spach, 234
Scilla L., 386
   hyacinthoides Aiton, non L., 386
   hyacinthoides L., 387
   latifolia Willd., 387
   madeirensis Menezes, 386
     var. *melliodora* Svent., 386
     var. *melliodora* forma *pallida* Svent., 386
Scirpus cernuus Vahl, 458
   maritimus L., 457
     var. *genuinus* Gren & Godr., 457
     var. *monostachyus* sensu auct. mad., non Webb & Berthel., 458
     var. *monostachyus* Webb & Berthel., 457
   pungens var. *sarmentoi* Menezes, 458
   savii Sebast. & Mauri, 458
   setaceus L., 458
   triqueter L., 458
Scleranthus L., 88
   annuus L., 88
     subsp. **annuus**, 88
     subsp. **polycarpos** (L.) Thell. in Schinz & R. Keller, 88
   polycarpos L., 88
Scleropoa rigida (L.) Griseb., 423
Scolopendrium officinarum Sw., 47
   septentrionale (L.) Roth, 46
   vulgare Sm., 47
Scolymus L., 369
   maculatus L., 369
Scorpiurus L., 191
   muricatus L., 191
     var. *sulcatus* (L.) Fiori, 191
   sulcatus L., 191
   vermiculatus L., 191
Scrophularia L., 306
   arguta Sol. ex Aiton, 307
   auriculata L., 307
   confusa Menezes, 306
   hirta × **scorodonia**, 307
   hirta Lowe, 306
     subsp. *ambigua* Menezes, 306
   laevigata Vahl, 307
   langeana Bolle, 307
   laxiflora Lange, 307
   longifolia Benth., 306
   lowei Dalgaard, 308
   moniziana Menezes, 307
   pallescens Lowe ex Menezes, 306
   racemosa × **scorodonia**, 307
   racemosa Lowe, 307
   scorodonia L., 306
   smithii Hornem., 307
   spuria Menezes, 307
SCROPHULARIACEAE, 303
Secale cereale L., 407
Sechium edule (Jacq.) Sw., 229
Sedum L., 135
   brissemoretii Raym.-Hamet, 136
   farinosum Lowe, 136
   forsterianum Sm., 135
   fusiforme Lowe, 136
   lancerottense R.P. Murray, 136
   nudum Aiton, 135
   praealtum A.DC., 135
Selaginella P. Beauv., 30
   denticulata (L.) Spring, 30
   kraussiana (Kunze) A. Braun, 30
SELAGINELLACEAE, 30
Semele Kunth, 390
   androgyna (L.) Kunth, 390
   maderensis G.C. da Costa, 390
   menezesi G.C. da Costa, 390
   pterygophora G.C. da Costa, 390
   tristonis G.C. da Costa, 390
Sempervivum arboreum L., 133
   divaricatum Aiton, 134
   dumosum Lowe, 134
   glandulosum Aiton, 133
   glutinosum Aiton, 133
   villosum Aiton, 134
Senebiera DC., 122
   coronopus (L.) Poir., 122
   didyma (L.) Pers., 122
   pinnatifida DC., 122
Senecio L., 358

auritus (L Hér.) Lowe, 360
incrassatus Lowe, 359
lividus L., 359
maderensis DC., 360
mikanoides Otto ex Walp., 358
petasitis (Simms) DC., 359
populifolia auct mad., 360
sylvaticus L., 359
vulgaris L., 360
Senna Mill., 152
  bicapsularis (L.) Roxb., 152
    var. **bicapsularis**, 153
  didymobotrya (Fresen.) Irwin & Barneby, 153
  floribunda (Cav.) Irwin & Barneby, 153
  multiglandulosa (Jacq.) Irwin & Barneby, 153
  multijuga (Rich.) Irwin & Barneby, 153
  pendula (Humb. & Bonpl. ex Willd.) Irwin & Barneby, 153
    var. **glabrata** (Vogel) Irwin & Barneby, 153
  septemtrionalis (Viv.) Irwin & Barneby, 153
Sesamoides Ortega, 129
  canescens var. *suffruticosa* (Lange) Abdallah & de Wit., 130
  clusii (Spreng.) Greuter & Burdet, 129
  pygmaea (Scheele) Kuntze, 130
Setaria P. Beauv., 448
  adhaerens (Forssk.) Chiov., 449
  geniculata auct. non (Willd.) P. Beauv., 449
  glauca auct. non (L.) P. Beauv., 449
  gracilis Kunth, 449
  megaphylla (Steud.) T. Durand & Schinz, 448
  palmifolia (König) Stapf, 448
  parviflora (Poir.) Kerguélen, 449
  pumila (Poir.) Roem. & Schult., 449
  verticillata (L.) P. Beauv., 449
  viridis (L.) P. Beauv., 449
Sherardia L., 261
  arvensis L., 262
Sibthorpia L., 315
  peregrina L., 280, 315
Sida L., 217
  carpinifolia auct. mad. non L.f., 217
  rhombifolia L., 217
    var. **canariensis** Lowe, 217

    var. **maderensis** Lowe, 217
Sideritis L., 283
  candicans Aiton, 283
    var. **candicans**, 284
    var. **crassifolia** Lowe, 284
    var. **multiflora** (Bornm.) Mend.-Heur, 284
  massoniana var. *crassifolia* Lowe, 284
    var. *longifolia* Lowe, 284
Sideroxylon L., 254
  marmulano Banks ex Lowe, 254
    var. *edulis* Chev., 255
    var. **marginata** (Pierre) Chev., 255
    var. **marmulano**, 255
Sieglingia decumbens (L.) Bernhardt, 440
Silene L., 92
  armeria L., 93
  behen L., 93
  gallica L., 94
    var. *anglica* (L.) Mert. & Koch, 94
    var. *quinquevulnera* (L.) Mert. & Koch, 94
  ignobilis Lowe, 93
  inaperta L., 93
  inflata Sm., 93
  maritima With., 93
  noctiflora L., 93
  nocturna L., 94
  uniflora Roth, 93
  venosa Asch., 93
  vulgaris (Moench) Garcke, 93
    subsp. *maritima* (With.) Á. & D. Löve, 93
Silybum Adans., 365
  marianum (L.) Gaertn., 366
Sinapidendron Lowe, 123
  angustifolium (DC.) Lowe, 124
  frutescens (Sol. in Aiton) Lowe, 124
    [var.] α *diffusa* Lowe, 124
    var. **frutescens**, 124
    subsp. *succulentum* (Lowe) Rustan, 124
    var. **succulentum** Lowe, 124
  gymnocalyx (Lowe) Rustan, 125
  rupestre Lowe, 124
    [var.] α *chaetocalyx* Lowe, 124
    [var.] β *gymnocalyx* Lowe, 125
  salicifolium Lowe, 124
  sempervivifolium Menezes, 124
Sinapis L., 125
  alba L., 125
  arvensis L., 125

Sisymbrium L., 111
  columnae Jacq., 111
  erysimoides Desf., 112
  irio L., 111
  officinale (L.) Scop., 112
    var. **leiocarpum** DC., 112
  orientale L., 111
  thalianum (L.) J. Gay, 112
Smilax L., 390
  aspera L., 391
    var. *altissima* Morris & de Not., 391
  canariensis Brouss. ex Willd., 391
  mauritanica Poir., 391
  pendulina Lowe, 391
  pseudochina Buch, 391
Smyrnium olusatrum L., 242
SOLANACEAE, 295
Solanum L., 298
  alatum Moench, 299
  auriculatum Aiton, 300
  chenopodioides Lam., 299
  erianthum D. Don, 300
  linnaeanum Hepper & Jaeger, 300
  luteum Mill., 299
  lycopersicum L., 301
  mauritianum Scop., 300
  miniatum Bernh. ex Willd., 299
  nigrum L., 298
    [var.] α *glabrum* Lowe, 298
    [var.] β *hebecaulon* Lowe, 299
    subsp. **nigrum**, 298
    subsp. **schultesii** (Opiz) Wessely, 299
  patens Lowe, 299
  pseudocapsicum L., 299
  sodomeum L., 300
  sublobatum Willd. ex Roemer & Schultes, 299
  trisectum Dunal, 300
  tuberosum L., 300
  villosum Mill., 299
    [var.] β *laevigata* Lowe, 299
    subsp. **miniatum** (Bernh. ex Willd.) Edmonds, 299
    [var.] α *velutina* Lowe, 299
    subsp. **villosum**, 299
Soleirolia Gaudich., 61
  soleirolii (Req.) Dandy, 61
Soliva Juss., 357
  stolonifera (Brot.) R. Br. ex G. Don, 357
Sonchus L., 374

  Subgen. DENDROSONCHUS Webb ex Sch. Bip., 376
  Subgen. SONCHUS, 375
  asper (L.) Hill, 375
    subsp. **asper**, 376
    subsp. **glaucescens** (Jord.) Ball, 376
    var. *integrifolius* Lowe, 376
    var. *vulgaris* Coss. & Germ., 376
  fruticosus L.f., 376
  oleraceus L., 375
    var. *integrifolius* Wallr., 375
    var. *lacerus* Wallr., 375
    var. *laciniatus* Lowe, 375
    var. *triangularis* Wallr., 375
    var. *rotundifolius* Hoff. & Link, 375
  pinnatus Aiton, 376
    var. angustilobus *Lowe,* 376
    var. *latilobus* Lowe, 376
  tenerrimus L., 376
  ustulatus Lowe, 376
    var. *angustifolia* Lowe, 377
    var. *imbricata* Lowe, 377
    var. *latifolia* Lowe, 377
    subsp. **maderensis** Aldridge, 377
    subsp. **ustulatus**, 377
Sorbus L., 148
  maderensis Dode, 148
Sorghum Moench, 452
  bicolor (L.) Moench, 452
  halepense (L.) Pers., 452
  vulgare Pers., 452
Sparaxis Ker Gawl., 396
  grandiflora (D. Delaroche) Ker Gawl., 396
  tricolor (Schneev.) Ker Gawl., 396
Spartium junceum L., 157
Specularia falcata (Ten.) A. DC., 330
  hybrida (L.) DC., 330
Spergula L., 91
  arvensis L., 91
  fallax (Lowe) E.H.L. Krause, 91
  vulgaris Boenn., 91
Spergularia (Pers.) J. & C. Presl, 91
  bocconii (Scheele) Graebn., 92
  fallax Lowe, 91
  marina (L.) Griseb., 92
  marina, 92
  rubra sensu auct. mad., non (L.) J. & C. Presl, 92
  salina J. & C. Presl, 92
SPERMATOPHYTA, 54
Sphenopus Trin., 423

divaricatus (Gouan) Rchb., 423
Sporobolus L., 444
   africanus (Poir.) Robyns & Tournay, 444
Sporobolus indicus auct. non (L.) R. Br., 444
Stachys L., 285
   arvensis (L.) L., 285
   ocymastrum (L.) Briq., 286
   officinalis (L.) Trevir., 280
   sylvatica L., 286
Statice ovalifolia Poir., 254
   var. *pyramidata* (Lowe) Menezes, 254
   pectinata Aiton, 254
   pectinata sensu auct. mad. non Aiton, 254
   pyramidata Lowe, 254
Stegnogramma Blume, 41
   pozoi (Lag.) K. Iwats., 41
Stellaria L., 85
   alsine Grimm, 85
   media (L.) Vill., 85
   uliginosa Murray, 85
Stenophragma thalianum (L.) elak., 112
Stenotaphrum Trin., 450
   americanum Schrank, 450
   secundatum (Walter) Kuntze, 450
Stipa L., 413
   capensis Thunb., 413
   eminens Nees non Cav., 413
   neesiana Trin. & Rupr., 413
   tortilis Desf., 413
Suaeda Forsk. ex Scop., 73
   fruticosa auct., 73
   laxifolia var. *crassifolia* Lowe, 73
      var. *tenuifolia* Lowe, 73
   tomentosa Lowe, 73
   vera J.F. Gmel., 73
Succisa Haller, 328
   praemorsa Asch., 328
   pratensis Moench, 328
Syzygium Gaertn., 232
   jambos (L.) Alston, 232
   malaccense (L.) Merr. & Perry, 232

Tacsonia exoniensis Hort., 226
   manicata Juss., 225
   mollissima Kunth, 226
   vanvolxemii Hook., 225
Tagetes minuta L., 351
Talinum paniculatum (Jacq.) Gaertn., 83
TAMARICACEAE, 227

Tamarix L., 228
   gallica L., 228
Tamus L., 393
   edulis Lowe, 394
Tanacetum L., 351
   parthenium (L.) Sch. Bip., 351
Taraxacum F.H. Wigg., 378
   Sect. OBOVATUM Soest, 378
   Sect. SPECTABILIA Dahlst. emend. A.J. Richards, 379
   Sect. VULGARIA Dahlst., 378
   adamii Claire, 379
   cacuminatum G.E. Haglund, 378
   duplidentifrons Dahlst., 378
   hamatum Raunk., 378
   lainzii Soest, 379
   lidianum Soest, 378
   maderense Sahlin & Soest, 378
   obovatum DC., 379
   officinale agg., 379
   praestans H. Lindb., 379
   raunkieri Wiinst., 378
TAXACEAE, 55
Taxus L., 55
   baccata L., 56
Tectaria caudata Cav., 48
Teesdalia R. Br., 119
   coronopifolia (J.P. Bergeret) Thell., 120
   lepidium DC., 120
   nudicaulis (L.) R. Br., 120
Teline Medik., 156
   maderensis Webb & Berthel., 156
      var. *paivae* (Lowe) Arco, 156
   monspessulana (L.) K. Koch, 157
   paivae (Lowe) Gibbs & Dingwall, 156
Tetragonia L., 82
   expansa Murray, 82
   tetragonoides (Pall.) Kuntze, 82
Tetragoniaceae, 79
Teucrium L., 281
   abutiloides L Hér., 282
   betonicum L Hér., 282
   heterophyllum L Hér., 282
   scorodonia L., 282
THEACEAE, 220
THELYPTERIDACEAE, 41
Thelypteris dentata (Forssk.) E.P. St. John, 42
   limbosperma (Bellardi ex All.) H.P. Fuchs, 41
   oreopteris (Ehrh.) Sloss., 41
   pozoi (Lag.) C.V. Morton, 41

Thlaspi L., 120
   arvense L., 120
Thrincia hispida var. *chaetocephala* Lowe, 373
      var. *gymnocephala* Lowe, 373
THYMELAEACEAE, 220
Thymus L., 289
   caespititius Brot., 289
   micans Lowe, 289
   vulgaris L., 289
Tigridia pavonia (L.f.) DC., 394
Tillaea muscosa L., 130
Tinantia Scheidw., 405
   erecta (Jacq.) Schltdl., 406
   fugax Jacq., 406
Tolpis Adans., 370
   barbata (L.) Gaertn., 370
      subsp. **barbata**, 370
   crinita (Sol. ex Lowe) Lowe, 370
   fruticosa Schrank, 370
   macrorhiza (Lowe ex Hook.) DC., 371
   succulenta (Dryand. in Aiton) Lowe, 370
      var. *ligulata* Lowe, 370
      var. *linearifolia* Lowe, 370
      var. *multifida* Lowe, 370
      var. *oblongifolia* Lowe, 370
   umbellata Bertol., 370,
Torilis Adans., 240
   arvensis (Huds.) Link, 241
      subsp. **arvensis**, 241
      subsp. *elongata* (Hoffmanns. & Link) Cannon, 241
      subsp. **neglecta** (Schult.) Thell., 241
      subsp. **purpurea** (Ten.) Hayek, 241
   leptophylla (L.) Rchb.f., 241
   nodosa (L.) Gaertn., 241
Trachelium L., 331
   caeruleum L., 331
      subsp. **caeruleum**, 331
Trachynia distachya (L.) Link, 438
Tradescantia L., 406
   fluminensis Vell., 406
   virginiana L., 406
   zebrina Hort. ex Bosse Vollst., 406
Tragopogon L., 374
   hybridus L., 374
Trichomanes L., 39
   aethiopicum Burm.f., 44
   brevisetum R. Br., 39
   canariensis L., 52
   radicans sensu auct. non Sw., 39
   speciosum Willd., 39
   tunbrigense L., 38
Trifolium L., 177
   Sect. LOTOIDEA Crantz., 178
   Sect. TRICOCEPHALUM W.D.J. Koch., 184
   Sect. TRIFOLIUM, 181
   Sect. VESICARIA Crantz., 180
   Sect.CHRONOSEMIUM Ser., 181
   agrarium L., 181
   angustifolium L., 183
   arvense L., 183
   bocconei Savi, 182
      var. **bocconei**, 182
   campestre Schreb., 181
   cernuum Brot., 179
   cherleri L., 183
   dubium Sibth., 181
   fragiferum L., 180
      var. **pulchellum** Lange, 180
   glomeratum L., 179
   incarnatum L., 182
   isthmocarpum Brot., 179
   lappaceum L., 183
      var. **lappaceum**, 183
   ligusticum Balb. ex Loisel., 182
   maritimum Hudson, 184
   minus Sm., 181
   ornithopodioides L., 179
   pratense L., 181
   procumbens L., 181
   repens L., 179
      var. **repens**, 179
   resupinatum L., 180
   scabrum L., 182
   squamosum L., 184
   squarrosum L., 184
   stellatum L., 181
   striatum L., 182
      subsp. *genuinum* (Lan.) Cout., 182
   subterraneum L., 184
      subsp. **brachycalycinum** var. **flagelliforme** Guss., 184
      subsp. **subterraneum** var. **subterraneum**, 184
   suffocatum L., 180
   tomentosum L., 180
Trigonella ornithopodioides (L.) DC., 179
Triodia decumbens (L.) P. Beauv., 440
Triplachne Link, 433
   nitens (Guss.) Link, 433
Triticum L., 407

aestivum L., 407
compactum Host, 407
turgidum L., 407
Tritonia crocosmiflora (Lemoine) G. Nicholson, 398
TROPAEOLACEAE, 200
Tropaeolum L., 200
majus L., 200
Tunica prolifera sensu auct. mad., 95

Ulex L., 158
europaeus L., 158
subsp. **europaeus**, 158
minor Roth, 158
ULMACEAE, 57
UMBELLIFERAE, 238
Umbilicus DC., 132
horizontalis (Guss.) DC., 132
pendulinus DC., 132 [var.] Lowe, 132
rupestris (Salisb.) Dandy, 132
Urospermum Scop., 371
picroides (L.) Scop. ex F.W. Schmidt, 371
var. *asperum* DC., 371
Urtica L., 59
azorica Seub., 59
dubia Forssk., 59
elevata Banks ex Lowe, 60
membranacea Poir., 59
morifolia Poir., 60
var. *elevata* (Banks ex Lowe) Menezes, 60
var. *genuina* Menezes, 60
portosanctana Press, 59
subincisa Benth. var. *floribunda* Wedd., 59
urens L., 59
URTICACEAE, 58

Vaccaria hispanica (Mill.) Rauschert, 95
pyramidata Medik., 95
Vaccinium L., 250
maderense Link, 250
padifolium Sm., 250
Vachellia farnesiana (L.) Wight & Arn., 154
VALERIANACEAE, 325
Valerianella Mill., 325
bracteata Lowe, 326-327
carinata Loisel., 326
dentata (L.) Pollich, 326
var. *dasycarpa* Rchb., 326
var. *leiocarpa* (DC.) W.D.J. Koch, 326
locusta (L.) Laterr., 326
microcarpa Loisel., 326
var. *puberula* sensu auct. mad., non (Bertol. ex Guss.) Gaut., 326
morisonii (Spreng.) DC, 326
var. *lasiocarpa* (W.D.J. Koch) Lowe, 326
var. *leiocarpa* DC., 326
olitoria (L.) Pollich, 326
puberula sensu auct. mad. non (Bertol. ex Guss.) DC., 326
Verbascum L., 304
blattarioides Lam., 305
creticum (L.) Cav., 305
densiflorum Bertol., 305
floccosum Waldst. & Kit., 305
haemorrhoidale Aiton, 305
pulverulentum Vill., 305
sinuatum L., 305
thapsiforme Schrad., 305
thapsus L. subsp. **thapsus**, 305
virgatum Stokes in With., 305
Verbena L., 278
bonariensis L., 278
bonariensis sensu auct. mad. pro parte, non L., 278
litoralis sensu Menezes non Kunth, 278
officinalis L., 279
rigida Spreng., 278
venosa Gillies & Hook., 278
VERBENACEAE, 277
Veronica L., 312
agrestis L., 315
anagallis auct., 313
anagallis-aquatica L., 313
var. *elata* Hoffmanns., 313
arvensis L., 314
hederifolia L., 314
officinalis L., 313
peregrina L., 314
persica Poir., 314
polita Fr., 314
serpyllifolia L., 313
transiens Rouy, 313
Vicia L., 162
Subgen. VICIA, 163
Subgen. VICILLA (Schur) Rouy., 164
Sect. FABA (Mill.) Ledeb., 164
Sect. HYPECHUSA (Alef.) Asch. & Graebn., 163

Sect. VICIA, 163
Sect. CRACCA Dumort., 164
Sect. ERVILIA (Link) W.D.J. Koch, 167
Sect. ERVOIDES (Godr.) Kupicha, 166
Sect. ERVUM (L.) Taub., 166
albicans Lowe, 165
articulata Hornem., 167
atlantica J.G. Costa non Pomel, 165
atropurpurea Desf., 165
benghalensis L., 165
capreolata Lowe, 165
conspicua Lowe, 163
   var. *dumetorum* Lowe, 163
   var. *lactea* Lowe, 163
   var. *laeta* Lowe, 163
cordata Wulfen ex Hoppe, 163
costae A. Hansen, 165
disperma DC., 166
ervilia (L.) Willd., 167
faba L., 164
   var. *major* Cout., 164
ferreirensis Goyder
gracilis Loisel. non Banks & Sol., 166
hirsuta (L.) Gray, 165
laxiflora Brot., 166
lutea L., 164
   subsp. *genuina* Cout., 164
   subsp. **lutea**, 164
   subsp. *muricata* (Ser.) Guinea, 164
   var. *pallidiflora* DC., 164
   var. *purpurascens* Lowe, 164
   subsp. **vestita** (Boiss.) Rouy, 164
monanthos (L.) Desf., 167
narbonensis L., 164
   var. **serratifolia** (Jacq.) Ser, 164
parviflora Cav., 166
pectinata Lowe, 163
peregrina L., 164
peregrina sensu auct. mad. non L., 164
portosanctana Gand., 164
portosanctana Menezes, 165
pubescens (DC.) Link, 166
sativa L., 163
   var. *bobartii* (E. Forst.) Koch, 163
   subsp. **cordata** (Wulfen ex Hoppe) Batt., 163
   subsp. **devia** J.G. Costa, 163
   var. *maculata* sensu Menezes non (C. Presl) Burnat, 163
   subsp. **nigra** (L.) Ehrh., 163
   subsp. *sativa* sensu A. Hansen & Sunding, 163
   var. *segetalis* (Thuill.) Burnat, 163
tenuissima Schinz & Thell., 166
tetrasperma (L.) Schreb., 166
villosa Roth, 165
   subsp. **varia** (Host) Corb., 165
Vinca L., 259
   major L., 259
   rosea L., 258
Viola L., 223
   arvensis Murray, 224
   odorata L., 224
      subsp. *maderensis* (Lowe) G. Kunkel, 224
   paradoxa Lowe, 224
   riviniana Rchb., 224
   sylvatica Fr. ex Hartm., 224
   sylvestris var. *riviniana* (Rchb.) W.D.J. Koch, 224
   tricolor sensu auct. mad. non L., 224
VIOLACEAE, 223
Visnea L.f., 220
   mocanera L.f., 220
VITACEAE, 215
Vitis L., 215
   aestivalis Michx., 215
   berlanderi Planch., 215
   labrusca L., 215
   riparia Michx., 215
   vinifera L., 215
Vulpia C.C. Gmel., 418
   bromoides (L.) Gray, 418
   geniculata (L.) Link, 418
   ligustica All., 418
   muralis (Kunth) Nees, 418
   myuros (L.) C.C. Gmel., 419
   sciuroides (Roth) C.C. Gmel., 418

Wahlenbergia Schrad. ex Roth, 332
   lobelioides (L. f.) Schrad., 332
      subsp. **lobeliodes**, 332
      subsp. *nutabunda* (Guss.) Murb., 332
      subsp. *riparia* (A. DC.) Thulin, 332
Watsonia Mill., 397
   ardernei Sanders, 397
   borbonica (Pourr.) Goldblatt subsp. **ardernei** (Sanders) Goldblatt, 397
   meriana (L.) Mill., 397
Wigandia caracasana Kunth, 271
Wisteria sinensis (Sims) Sweet, 155
WOODSIACEAE, 47

Woodwardia Sm., 53
caudata Cav., 53
radicans (L.) Sm., 53

Xanthium L., 350
strumarium L., 350
subsp. **italicum** (Monetti) D. Löve, 350
subsp. **strumarium**, 350
Xerotium gallicum (L.) Bluff & Fingerh., 341

Zantedeschia Spreng., 454
aethiopica (L.) Spreng., 454
Zea mays L., 407
Zebrina pendula Schenizl., 406
ZINGIBERACEAE, 466
ZYGOPHYLLACEAE, 200
Zygophyllum L., 201
fontanesii Webb & Berthel., 201

INDEX TO PORTUGUESE NAMES

Abacate 101
Abacateira 101
Abrotôna 282
Abundancia 337
Acacia 154, 155
Acacia branca 154
Açafrão 368
Açafroa 368
Acanto 318
Acelga 70
Açucena 384
Agrião 116
Agrião da rocha 117
Agulha 199
Agulheta 199
Aipo da serra 245
Aipo do gado 245
Aipo preto 245
Alamanda 258
Alecrim 280
Alecrim da serra 289
Alecrim de fora 227
Alegra campo 390
Alegra campo de folho miuda 389
Alfabaca 346
Alfaca 377
Alface da terra 326
Alfarrôba 153
Alfarrobeira 153
Alfavaca 346, 382
Alfavaca de cobra 60
Alfazema 294
Alfinête 199
Alfinetes de Senhora 351
Alho 387
Alho americano 388
Alho bravo 388
Alho porro 387
Alindres 207
Almeirante 369, 380
Almeirão 369, 380
Almirante 369
Aloendro 258
Ameixieira 141
Ameixieira de espinho 100
Amendoeira 141
Amoreira 58
Amoricos 145
Amor perfeito 224
Apténia 81

Araça amarelo 232
Araça roxo 232
Armores de burro 348
Arôma amarelo 154
Arôma branco 155
Aroz 407
Arroz da rocha 135 136
Arrozinho 81
Arruda 210
Artemija 351
Árvore das salsichas 317
Arvore de Judas 153
Ârvore de seda 260
Arvore do incenso 139
Avenca 36
Avenca das fontes 36
Avoadeira, 340
Azeda 66, 191
Azedinha 66
Azevim 213
Azevinho 213
Azinhaga do forte 323

Babosa 386
Bacaira 187
Bálsamo 81
Balsamo da rocha 290
Balsamo de cheiro 273
Barbesano 290
Barbusano 101
Barbuzano 101, 254
Bardana 363
Barrilha 73, 80
Batateira 271
Bebereira 58
Beija mão 367
Beladona 393
Beleno 297
Bella donas 393
Bellas noites 302
Berradura 299
Beterraba 70
Betouro 249
Bigalhó 455
Bignónia 317
Bignónia unha-de-gato 317
Boca de peixe 309
Bocca de peixa 308, 309
Bofe de Burro 115
Boffe de burro 382

Bois de pissenlit 317
Bolsa de pastor 192
Bonina 78
Borragem 273
Bredos 74
Broculos 123
Bufareira 301
Buganvilha 78
Buxo 214
Buxo da rocha 149, 214

Cabelleira 188
Cabelleira de coquinho 188
Cabreira 191, 261
Cabriera 332
Cabrinhas 52
Cacho roxo 155
Cafeeiro 261
Caiota 229
Camarões 317
Camélia 220
Cana de asugar 407
Cardeal vermelho 216
Cardeal violeta 216
Cardo 364, 365, 366, 368
Cardo da gente 365
Cardo de Santa Maria 366
Carqueja 158
Carqueja brava 74
Carqueja mansa 126
Carrapateira 204
Cavalinho 30
Cebolhino 388
Cebolinho de burro 385
Cebolo 387
Cedro 55, 228
Cedro da Madeira 55
Cedronha 106
Celga 70
Celidonia 106
Cenoura da rocha 247
Ceredonha 106
Cerejeira 141
Cevadilha 258
Cha bravo 217
Chagas 200
Chama da floresta 317
Cheiros 289
Chícaros 169
Chicharinha 93
Chícharo 170
Chícharo branco 169

Chícharos 169
Chorão 81
Chorão baguinho de arroz 81
Cidra 209
Cidreira 209
Cigerão 165, 166
Ciumes 100
Codeço 158
Codeso 158
Coenha 67
Cóleos 280
Conteira 466
Coquinho 460
Corriola da praia 268
Corriola brava 269
Corriola da praia 268
Corriola de balões 212
Corriola mansa 269
Couve 123
Couve da rocha 124, 125, 127
Couve flôr 123
Crássula 131
Craveiro 95
Cravo de burro 115
Cravo de seara 374
Crista de galo 75
Cuidados 361

Damasqueiro 141
Dedaleira 312
Deiradinha 46
Diabelha 321
Doiradinha 359
Douradhina 98
Dragoeiro 391

Endros 293
Engos 324
Ensaião 133, 134, 135
Ensaião da festa 133
Ensaião de pasta 133
Ensayão 133
Erva arroz 135, 136
Erva branca 282, 283
Erva de coelho 360
Erva de Santa Maria 298, 299
Erva gigante 318
Erva mel 94
Erva moira 298
Ervilha 155, 167
Ervilhaca 163
Esparto 389

Espatódea 317
Espinafres 82
Espinheiro 296
Eucalypto 233
Eufórbia marítima 209

Faia 57
Farrobo 133
Fausse bignone 317
Fava 164
Faveira 164
Fedegoso 71, 72, 161
Fedorênte 125
Fedorento 125
Fedorento manso 112
Feijoa 155
Feijoeiro 155
Feiteirinha macellão 352
Feiteiro 40
Feto abrum 40
Feto de calvalto 48
Feto de escoumas 45
Feto de palma 37
Feto de tres bicos 46
Feto do botão 53
Feto frisado 39
Feto lanegero 34
Feto manso 47
Feto maritimo 45
Feto pente 53
Feto redondo 35
Figueira 58
Figueira do inferno 207
Figueirinha 209
Flôr da paixão 225
Flor de coelho 382
Flor de manteiga 258
Focinho de burro 308, 309
Folhadeiro 248
Folhado 248
Formigueira 71
Freixo 256
Fustete 100

Gaitas 317
Gaitinhas 317
Gerânio folha-de-anémona 196
Giesta 156
Giesta de piorno 157
Gingeira brava 215
Ginjeira 141
Ginjeira brava 141

Girasol 348
Globulária 316
Goiaba 231
Goiabeira 231
Goivo 115
Goivo da rocha 115
Goivo da serra 114
Goivos 114
Gorda 91
Grão de bico 162

Hera 237
Hera terrestre 315
Herva cidreira 287
Herva de São João 222
Herva ferrea 287
Herva menina 405
Herva redonda 315
Herva pecegueira 63
Herva pombinha 107
Herva terrestre 315
Hipericão 223
Hortelã 292
Hortelã da serra 290
Hortelã de burro 286
Hortelã de cabra 286
Hortelã de leite 292
Hortelan pimenta 292
Hortensias 138
Hysopo 288

Inça muito 337
Inhame 454
Inhame de galatixa 132
Inhame de lagartixa 132
Inhame do Enxerto 454
Invejosa 276
Isabella 215
Isca 346
Isoplexis 312

Jacarandá 317
Jamboeiro 232
Japonesa 220
Jarro 454
Jarvão 279, 294
Jasmineiro 255
Jasmineiro amarello 256
Jasmineiro branco 255
Junça 459
Junco 402, 403
Junquilho 461

Labaça 65, 67
Lagartixa 132, 135
Laranja 210
Laranja azeda 209
Laranja tangerina 210
Laranjeira 210
Laranjeira azeda 209
Laranjeira do México 210
Leitua 371
Leituga 373, 376
Leituga de burro 371
Lentilha 167
Lentilha d'água 279
Letubra mansa 380
Lilaz 155
Lima 209
Lima de chêiro 209
Limão 209
Limão de gallinha 209
Limão doce 209
Limoeira 209
Lingua cervina 47
Lingua de cobra 31
Lingua de vaca 52, 373, 376
Linheio 266
Linho 202
Linho bravo 202
Lirio 129
Lodão bastardo 57
Loendro 258
Loireiro 102
Lombrigueira 71
Losna 351
Loureiro 102
Louro 102
Louro cerejo 141
Louro inglês 141
Luvas de Nossa Senhora 99

Maçacota 72, 73
Maçela 352, 353
Maçela de botão 353
Macieira 148
Madeira 288
Madresilva 325
Malfurada 221, 222, 223, 316
Malmequer 354, 355
Malpica 348
Malva 197, 198, 218
Mangericão 280
Mangerona 290
Manhãs de páscoa 205

Maracujá 225
Maracujá amarelo 225
Maracujá banana 226
Maracujá inglez 225
Maracujá roxo 225
Maranta 33
Maravilhas 212
Margaça 352, 353
Marmeleiro 148
Marmulano 254
Marroios 167, 382
Marroiso 167
Marruiço 167
Martyrio 225
Marugem 85
Massaroco 277
Mastruço 121
Meimendro 297
Meiomento 297
Mentastro 291, 286, 292
Mimos 234
Mioporo 320
Mocano 139, 220
Molarhina 107
Morangueiro 144
Morrão 345
Morugem 85
Mostarda 123, 125
Murrão 345
Murrião 252, 345
Murruca 363
Murta 232
Murta da India 232
Músquia 331

Nabo 123
Nespera de Japão 148
Nespereira 149
Nêveda 288
Norça 394
Novelos 138
Nozelha 244, 247
Nozelhina 244

Olaia 153
Oliveira 256
Oregãos 289, 290
Orelha de boi 93
Orelha de cabra 322
Orelha de rato 311
Orga 91

INDEX - PORTUGUESE

Pajeita 354
Pajito 354
Palha de amarrar vinha 464
Pampilho 354, 357
Pampilhos 355
Pão branco 257
Papoila 104
Papoila branca 105
Papoila vermelha 104
Pássaras 197
Pastel 113
Pastinha 133
Patinha de agua 456
Peceguaia 230
Pecegueiro 141
Pecegueiro Inglez 277
Pelicão 222
Pencas 365
Pepinella 229
Perado 213
Pereira 148
Pereira abacate 101
Pereiro 148
Perpetua 344, 345
Pimenta encarnada 298
Pimenteira 298
Pimenteira bastarda 210
Pimenteira brava 299
Pinheirinho 30
Pinheiro 54
Piorno 156, 157
Pitangueira 232
Piteira 392
Pitosporo 139
Planta da seda 260
Planta dos dentes 258
Platano 139
Poejos 291
Poinsétia 205
Prados 322
Propéta 344
Pulgueira 64

Quebra panella 290
Queiranto 114
Quigélia 317

Rainúnculo 98
Raspa-lingua 263
Rasteira 228
Rasteyro 228
Resteira 228

Ricasoliana 317
Rinchão 127
Rinchões 123
Riparia 215
Rosa brava 147
Rosa de toucar 148
Rosa mosquêta 148
Roseira 147
Rosmaninho 293, 294
Ruivinho 265

Saboia 326
Saboneteira 90
Sabugueiro 324
Saião 133, 135
Salgueiro 73
Salva 294
Samouco 57
Sandalos 291, 292
Sanguinho 214
Saramago 125, 128
Sarmento 208
Saudades 329
Seda 260
Segurelha 289
Seisim 261
Seisinho 261
Seixeiro 57
Seixo 57
Sejamos amigos 354
Selvageira 283
Sementeira 57
Semilhas 300
Sempre noiva 63
Serralha 375
Serralha da rocha 376
Setas 348
Silvado 142
Silvado da serra 143
Sinapidendro de folha estreita 124
Solda 144
Sumagre 211

Tabaco 303
Tabaiba 207
Tabaibeira 230
Tabaqueira 300, 302, 303
Tabaqueira azul 302
Tamargueira 228
Tanchagem 322, 323
Tangerão manso 364
Tangerineira 210

Teijeira 363, 312
Teixo 56
Tigarro 369
Til 102
Tinjeira 363
Tintureira 215
Tomate 297, 301
Tomate inglez 297
Tomateiro 301
Tomateiro inglez 297
Tremoço 159
Tremoço amarello 159
Trevina 187
Trevo 172, 173, 177
Trevo branco 172
Trevo de namorado 173
Trevo de pé de passaro 184
Trevo de seara 173
Trevo massaroco 183
Trevo preto 176
Trombetas 302
Trombeteira 302
Trovisco 206, 207, 208, 209
Tudesco 158
Tumbérgia 318

Urgebão 279
Urtiga 59

Urtiga morta 204
Urze de cheiro 210
Urze durazia 249
Urze mollar 249
Uva da serra 250
Uva de galatixa 135
Uva de rato 135

Vacoa 361
Vaqueira 361
Verbasco 305
Vermelhão 276
Vidêira 215
Vigândia 271
Vimeiro 57
Vinha 215
Vinha americana 215
Vinha silvado 215
Vinhático 101
Violeta 224
Violeta amarela da Madeira 224
Violeta da Madeira 224
Viuvas 99

Zambujeiro 256
Zimbreiro 55
Zimbro 55